WILLIAM F. MAAG LIBRARY
YOUNGSTOWN STATE UNIVERSITY

ANNUAL REVIEW OF PHYSICAL CHEMISTRY

EDITORIAL COMMITTEE (1977)

R. B. BERNSTEIN
V. BLOOMFIELD
J. M. DEUTCH
W. H. FLYGARE
R. NOYES
B. S. RABINOVITCH
J. M. SCHURR
H. L. STRAUSS

Responsible for the organization of Volume 28
(Editorial Committee, 1975)

R. B. BERNSTEIN
C. J. CHRISTENSEN
J. M. DEUTCH
H. EYRING
W. H. FLYGARE
H. S. JOHNSTON
H. M. McCONNELL
R. NOYES

Assistant Editor	M. CANTLON
Indexing Coordinator	M. A. GLASS
Subject Indexers	F. M. HALL
	V. E. HALL

ANNUAL REVIEW OF PHYSICAL CHEMISTRY

B. S. RABINOVITCH, *Editor*
University of Washington

J. M. SCHURR, *Associate Editor*
University of Washington

H. L. STRAUSS, *Associate Editor*
University of California, Berkeley

VOLUME 28

1977

ANNUAL REVIEWS INC.
Palo Alto, California, USA

COPYRIGHT © 1977 BY ANNUAL REVIEWS INC., PALO ALTO, CALIFORNIA. ALL RIGHTS RESERVED. No part of this book may be reproduced in any form or by any means without permission in writing from the publisher.

International Standard Book Number: 0-8243-1028-4

Library of Congress Catalog Card Number: A-51-1658

Annual Reviews Inc. and the editors of its publications assume no responsibility for the statements expressed by the contributors to this Review.

REPRINTS

The conspicuous number aligned in the margin with the title of each article in this volume is a key for use in ordering reprints. Available reprints are priced at the uniform rate of $1 each post paid. The minimum acceptable reprint order is 10 reprints and/or $10.00, prepaid. A quantity discount is available.

PRINTED AND BOUND IN THE UNITED STATES OF AMERICA

PREFACE

Time continues to work its changes. Harold Johnston was Associate Editor of this Review for twenty-one years, 1955 to 1976. He has contributed notably to its success through his invaluable service and counsel. The Editorial Committee and staff are most grateful to him and wish him success in his further scientific work.

This volume continues to record the broad and pervasive domain of physical chemistry. We thank the contributing authors for their signal contributions. We are also glad to acknowledge the assistance of Assistant Editor Martha Cantlon and former Assistant Editor Rosalie West. We thank Victor and Frances Hall who prepared the subject index and all staff who have assisted in the preparation of this volume.

THE EDITORIAL COMMITTEE

CONTENTS

MEN, MINES, AND MOLECULES, *Henry Eyring*	1
THEORETICAL STUDIES OF NEGATIVE MOLECULAR IONS, *Jack Simons*	15
NMR RELAXATION IN THERMOTROPIC LIQUID CRYSTALS, *Charles G. Wade*	47
INITIATION OF GASEOUS DETONATION, *John H. S. Lee*	75
ELECTRON TUNNELING IN CHEMISTRY AND BIOLOGY, *W. F. Libby*	105
THERMAL REARRANGEMENTS, *Jerome A. Berson*	111
LASER SEPARATION OF ISOTOPES, *V. S. Letokhov*	133
PHOTOELECTRON SPECTROSCOPY: STUDY OF VALENCE BANDS IN SOLIDS, *Jennifer C. Green*	161
RHEOLOGY AND KINETIC THEORY OF POLYMERIC LIQUIDS, *R. Byron Bird*	185
PICOSECOND SPECTROSCOPY, *Kenneth B. Eisenthal*	207
HYDRODYNAMICS IN BIOPHYSICAL CHEMISTRY, *Victor A. Bloomfield*	233
VIBRATIONAL STATE ANALYSIS OF ELECTRONIC-TO-VIBRATIONAL ENERGY TRANSFER PROCESSES, *Stephen Lemont and George W. Flynn*	261
TIME DOMAIN REFLECTOMETRY, *Robert H. Cole*	283
STATISTICAL MECHANICS OF MOLECULAR MOTION IN DENSE FLUIDS, *James T. Hynes*	301
PHOTODISSOCIATION DYNAMICS OF POLYATOMIC MOLECULES, *William M. Gelbart*	323
LASER-INDUCED FLUORESCENCE, *James L. Kinsey*	349
LIQUIDS OF LINEAR MOLECULES: COMPUTER SIMULATION AND THEORY, *W. B. Streett and K. E. Gubbins*	373
DECORATED LATTICE-GAS MODELS OF CRITICAL PHENOMENA IN FLUIDS AND FLUID MIXTURES, *John C. Wheeler*	411
ION THERMOCHEMISTRY AND SOLVATION FROM GAS PHASE ION EQUILIBRIA, *P. Kebarle*	445
SURFACE SCATTERING, *Sylvia T. Ceyer and Gabor A. Somorjai*	477
RESONANCE EFFECTS IN VIBRATIONAL SCATTERING FROM COMPLEX MOLECULES, *Thomas G. Spiro and Paul Stein*	501
INDEXES	
AUTHOR INDEX	523
SUBJECT INDEX	544
CUMULATIVE INDEX OF CONTRIBUTING AUTHORS, VOLUMES 24–28	565
CUMULATIVE INDEX OF CHAPTER TITLES, VOLUMES 24–28	567

ANNUAL REVIEWS INC. is a nonprofit corporation established to promote the advancement of the sciences. Beginning in 1932 with the *Annual Review of Biochemistry*, the Company has pursued as its principal function the publication of high quality, reasonably priced Annual Review volumes. The volumes are organized by Editors and Editorial Committees who invite qualified authors to contribute critical articles reviewing significant developments within each major discipline.

Annual Reviews Inc. is administered by a Board of Directors whose members serve without compensation.

BOARD OF DIRECTORS
1977

Dr. J. Murray Luck
Founder Emeritus, Annual Reviews Inc.
Department of Chemistry
Stanford University

Dr. Joshua Lederberg
President, Annual Reviews Inc.
Department of Genetics
Stanford University Medical School

Dr. James E. Howell
Vice President, Annual Reviews Inc.
Graduate School of Business
Stanford University

Dr. William O. Baker
President
Bell Telephone Laboratories

Dr. Sidney D. Drell
Deputy Director
Stanford Linear Accelerator Center

Dr. Eugene Garfield
President
Institute for Scientific Information

Dr. William D. McElroy
Chancellor
University of California, San Diego

Dr. William F. Miller
Vice President and Provost
Stanford University

Dr. John Pappenheimer
Department of Physiology
Harvard Medical School

Dr. Colin S. Pittendrigh
Director
Hopkins Marine Station

Dr. Esmond E. Snell
Department of Microbiology
University of Texas at Austin

Dr. Harriet Zuckerman
Department of Sociology
Columbia University

Annual Reviews are published in the following sciences: Anthropology, Astronomy and Astrophysics, Biochemistry, Biophysics and Bioengineering, Earth and Planetary Sciences, Ecology and Systematics, Energy, Entomology, Fluid Mechanics, Genetics, Materials Science, Medicine, Microbiology, Nuclear Science, Pharmacology and Toxicology, Physical Chemistry, Physiology, Phytopathology, Plant Physiology, Psychology, and Sociology. The *Annual Review of Neuroscience* will begin publication in 1978. In addition, two special volumes have been published by Annual Reviews Inc.: *History of Entomology* (1973) and *The Excitement and Fascination of Science* (1965).

MEN, MINES, AND MOLECULES ✤ 2636

Henry Eyring

MEN, MINES, AND MOLECULES ✤ 2636

Henry Eyring
Department of Chemistry, The University of Utah, Salt Lake City, Utah 84112

My ancestors were drawn together from northern Europe by the new Mormon religion. My mother's people, the Romneys and Cottams, migrated from around Preston, England, arriving in Nauvoo, Illinois in 1839. My grandfather, Henry Eyring, came from Coburg in Germany, and Grandmother Eyring came from German Switzerland. My Eyring grandparents met while crossing the plains and arrived in Salt Lake City in the same pioneer company in the fall of 1860. My mother's people had reached Salt Lake City 10 years earlier. During the next three decades, colonization of the intermountain area spread from Salt Lake City to Alberta on the North and to Chihuahua and Sonora on the South, wherever the water could be turned out onto the parched land. As a result of these migrations, all my grandparents ended up in the late 1880s in Colonia, Juarez, in northern Mexico, about 100 miles straight south of Columbus, New Mexico.

This part of Chihuahua lies in the foothills of the Sierra Madre Mountains at 5000–6000 feet of elevation in an excellent region for raising cattle. At that time Don Luis Terrasas owned about a third of the state of Chihuahua and had one of the largest herds of cattle in the world. My father owned the Tenaja ranch of 10,000 fenced acres where he raised high-grade cattle and had about 4000 acres of pasture and farmland nearer town where he kept 50–100 head of horses. The 10,000-acre ranch seven miles from town pastured about 600 head of shorthorn Durham cattle. The bull calves were sold to surrounding ranchers to upgrade the herds of Spanish longhorn cattle.

I was born into this well-to-do family in 1901 in Colonia Juarez. I was the third child and the first son. I have no memory of learning to ride, since I was riding as soon as my legs were long enough to straddle a horse. My earliest memories are of my father coming home from the ranch, putting me on his horse after the horse had been unsaddled, and leading him to the river to drink. Mother told a story of such an occasion when she accompanied us to the river. After drinking, the horse shook himself the way a horse wet with sweat often does, and I tumbled off into the river. After being fished out of the river, my first remark (as reported by my mother) was, "Put me back on the horse." This was at about the age of three.

At four I almost died of typhoid fever. Our prosperous little town had electric lights and telephones, but the clear water from the river was piped directly into our homes with no previous purification. After the first cases of typhoid appeared, the

1

practice was to boil the drinking water, but this was rather a case of locking the barn after the horse was stolen. One of my vivid memories of the spring of 1905 is of the magnolias in our front yard in full bloom while I was recovering from typhoid.

At about this time my father gave me two small goats, but they soon became a nuisance by getting into the neighbor's lot. My father's way of getting rid of them while leaving me contented was interesting. One day he came riding into the yard leading a pretty, little sorrel horse and asked me if I would be willing to trade my two goats for the sorrel horse. I was delighted with the proposition. My goats disappeared, and I started riding my sorrel horse, which was named Chivo, the Spanish word for goat. However, Chivo turned out to be too much horse for a small boy to handle, so Father traded me a dun horse, Grullo, for Chivo, and finally a pretty, little, black horse for Grullo. The black horse was called the Spanish equivalent of "black baby" and was the horse that I rode the most until we left Mexico in 1912. We left Mexico because of the unrest accompanying the Mexican revolution, which started in 1910. About 4800 colonists from our area left Mexico at this time. This migration to El Paso, Texas, occurred within about a week in mid-July 1912. Since everyone expected to return within a short time, everything was left behind except the requirements for a few weeks' stay. However, we, like most of the colonists, never returned as a result of the continued unrest. Our large, refugee family was thus suddenly transformed from affluence to humble circumstances. My first job in Calisher's department store paid $2 for a six-day week. At that time five cents would buy a loaf of bread or a quart of milk. After a year in El Paso waiting for the Mexican situation to improve and a second year spent in Safford and Thatcher, Arizona, we moved to nearby Pima in southeastern Arizona, where Father had purchased a farm by making a small down payment. Part of the farm was under cultivation. Brush was cleared from the rest by hitching two teams, one to each end of a steel rail, and dragging it across the land after the mesquite had been dug up with ax, pick, and shovel. This involved a lot of hard work, but we were all healthy and anxious to get on our feet again.

I had finished the fifth grade in Mexico, missed a year's schooling in El Paso, and graduated from the eighth grade in Pima in 1914, having skipped the first and seventh grades.

My high school had an important bearing on my subsequent career. I went to a church academy in Thatcher, Arizona, six miles from Pima. My predilection for mathematics and the sciences showed up early. My science teacher, Alma Sessions, who had been a star basketball player at the University of Arizona, and was much admired, gave me career advice. He said I should study either electrical or mining engineering at the University of Arizona. I chose mining as the less hazardous. A $500 state fellowship, won in the competition in Graham County, launched my mining career at the University of Arizona.

My father's advice upon departure for the university made a strong impression on me. He said the constraints our religion placed on one are "to be dedicated to the truth, wherever one finds it, and to live in such a way as to make one comfortable in the company of good people." This advice has had an appeal for me that has lasted through the years.

Arriving at the university in 1919, the same time as the veterans from the First World War were returning to school, made this an interesting time. The first day at the university, I saw President Rufus von Kleinschmidt driving his carriage, drawn by two well-matched black horses, through the campus grounds. This was still the customary way to travel. Hazing was also still in full swing. One resisted it, but the battle was often lost. The war veterans in the freshman class added a disciplined resistance to hazing which I found refreshing. I found it distasteful that, as a freshman, I was expected to wear a green "beanie." The sophomores would throw freshmen not wearing their beanies into the pool. However, it was against the rules to carry hazing into the dormitories. Since I thought I could outrun my tormentors, I left the dormitory without my beanie, believing they would have to quit the pursuit when I reached the door, or else they would be in trouble themselves. It was a miscalculation. I beat them to the dormitory and then to my room where I slammed the door, but they kept coming. Since my door was now locked, they tried to climb in through the transom, but I blocked their entry by working on their hands with a broom handle. By this time, they were annoyed and threatened to break down my door if I did not come out. I was pretty sure I would be paying for the door if they broke it down, so I came out. There were plenty of sophomores, and I was soon face down in the air with someone holding on to each arm and leg and another fellow enthusiastically swinging a large wooden paddle where it would do the most good. This did not engender love for authority in me, but it did engender respect for it when it is backed up by sufficient "lynch law."

The four years I spent studying mining at the university were very pleasant. I made my own way by assisting in classes and waiting on tables. I was able to send a little money home to help with payments on the farm. I enjoyed my studies and made high grades. I particularly liked mathematics and wrote a senior thesis with Professor Cressy of the mathematics department on the theory of the aerial tramway.

The summer after my junior year I worked as a miner in the Inspiration Copper Company in Miami, Arizona. The work was interesting and paid about twice as much as farming. Since I was a prospective mining engineer, they changed me rapidly from one job to another to give me added experience. After I had been underground for only a few weeks, I was given the job of timberman, with a man assigned as my helper who had been mining for 15 years.

The caving system used in the mine was economical, but extremely hazardous. One half of a square mile of rock would be blown up at one time and drained by gravity down raises (tunnels) into square sets where a man would be stationed to regulate the rate of flow to a "grizzly" 20 feet below. At the grizzly, a second man would break up the larger boulders with a sledgehammer, or with dynamite, until the fragments passed between the six steel rails of the grizzly, spaced about 10 inches apart. The ore then continued its downward course another 20 feet, where it was drained into cars of a train driven by compressed air. The train carried the ores to a tipple which turned the cars upside down, dumping the ore into a bin. The ore was drained out of the bin into elevators that carried it to the surface, where trains carried it to the mill. Fifteen thousand tons of ore and gangue would be taken from the mine in three eight-hour shifts. There were many separate operations paralleling the one just described. On one memorable occasion a shift boss assigned my helper and me to repair one of the square sets, which had been all but destroyed by the tumbling boulders that had passed through it. We climbed up the raise to inspect the square set, and my veteran helper declared that he had no intention of being trapped in this particular hell hole and, if it was to be repaired, it was up to me. He would wait down at the grizzly, and I could call when and if I needed help. I had the usual ax and crowbar with which I could gingerly pry out the loose rocks until I had cleared enough space to put in new timber before it all caved in on me. I went about it as carefully as I could, but it was not long before a rock somewhat bigger than my head fell from the ceiling and hit my boot. I soon had a boot partly filled with blood, and I was taken out of the square set and brought up on top. When I returned to the mine some days later, I was assigned to another part of the mine.

I was once on another shift where there were three separate, fatal accidents. There were two other boys from Pima working in the mine that summer, and one of these had his arm crushed and later amputated. This accident occurred as he was driving his train onto the tipple. The customary indemnity paid to the family of a deceased miner was $6000. This rapid method of mining born of wartime and economic needs has been replaced by the much safer opencut mining of today.

Although as a mining engineer I would not personally have to take these risks much longer, I would still have to send others to take them. Therefore, I reluctantly decided to graduate in mining but planned to change over to metallurgy after receiving my degree. The next summer, 1923, my younger brother, Edward, and I worked together as miners at Sacramento Hill in Bisbee, Arizona, for the Phelps Dodge Corporation. This work was interesting and a much safer type of operation than the caving system since it involved driving drifts into undisturbed terrain. This summer ended pleasantly, terminating my active mining career. In the fall I began my thesis on *The Separation of Heavy Sulfide Ores by Selective Flotation*. I had a Bureau of Mines fellowship and worked under the direction of Thomas Chapman, professor of metallurgy, at the University of Arizona.

After completing my master's degree, I spent the summer of 1924 working in the United Verde smelter at Clarksdale, Arizona. Again I had a favored position as a prospective metallurgist and was rapidly shifted among different phases of the smelting operation. After being there a few weeks I was assigned to take samples from the blast furnaces. The sulfur dioxide smoke was especially strong, and I was holding a handkerchief soaked in baking soda over my face when the smelter superintendent came by, slapped me on the shoulder, and said, "Eyring, I plan to put you in charge of the blast furnaces in a few weeks." The problems were intriguing, but the sulfur smoke made it easy for me to return to the University of Arizona as a chemistry instructor.

In 1925, toward the end of the teaching year, Professor Theophyl Buehrer recommended me to Berkeley, where he had taken his Ph.D. Dr. Lathrop E. Roberts recommended me to the University of Chicago, where he had worked under Harkins. I accepted the invitation to Berkeley, where I went in August 1925 and finished my Ph.D. in June 1927. This was a stimulating experience. The emphasis was on research and, since I was no stranger to work, everything went well. I started on a problem involving the lowering of freezing points with Professor Merle Randall, but because Dr. Vanselow needed the equipment longer than expected to finish his Ph.D., I changed to working on an exciting problem with Professor George E. Gibson. A long vacuum tube filled with hydrogen at low pressure was bombarded with an 11 MV high-frequency discharge from a Tesla coil. The protons were expected to pass through a thin aluminum window at the bottom of the tube and strike a beryllium target, which we hoped would emit interesting radiation. Unfortunately, one got lots of unspecified radiation and a frequently punctured, evacuated tube. Nevertheless, I had gained useful experience from this study. We next turned to the study of the amount of ionization, the stopping power, and the straggling of alpha particles from polonium in various gases. We found that stopping power depends very little on how the atoms are bound together, but the total number of ions formed depends very much on the bonds being broken. This is readily understandable since the primary ionization induced by the alpha particles involves a large energy transfer to the electron being ionized, with the result that the fraction of energy lost by the alpha particle as a result of a difference in the molecular bonding of the atoms is negligible. Conversely, in the ionization by the fast secondary electrons only about half of the energy transferred from the secondary electrons goes into ionization, so that differences in the bonding energy are an appreciable part of the total energy expended in ionization.

In the fall of 1927, as a brand new Berkeley Ph.D., I became an instructor at the University of Wisconsin. During my first year there, I continued my research on ionization, stopping power, and straggling of alpha particles in different gases. The second year I took a full-time experimental research position with Professor Farrington Daniels, studying the decomposition of N_2O_5 in a wide variety of solvents. This was the beginning of my active interest in reaction kinetics, which has continued unabated.

In 1929 I was granted a national research fellowship to work with Professor Bodenstein in the University of Berlin. However, before my expected date of departure, I received word that Professor Bodenstein was going to be at Princeton at the dedication of the new Frick chemical laboratory. Professor Frumkin, who was visiting the University of Wisconsin at the time, suggested that, in view of Bodenstein's absence, I should work with Professor Michael Polanyi at the Kaiser Wilhelm Institute in Berlin. Acting on this advice, my wife, Mildred Bennion, and I sailed to Europe by way of Bergen and Oslo, Norway, through southern Sweden, to Copenhagen and Berlin. Arriving in Berlin, we found Professor Polanyi had likewise gone to the Frick dedication. Just at this time Bonhöffer and Harteck were front page news with their study of the rate of conversion of *para*- to *ortho*-hydrogen. Fritz Haber was directing his laboratory effectively. Workers in Haber's laboratory included Fritz London, Eugene Wigner, the Farkas brothers, and Hubert Alyea, along with many others. We had not been in Berlin long before I was visited by Professor Robbins from the Paris branch office, who was responsible for the National Research fellows. He was greatly disturbed to find that neither Professor Bodenstein nor Polanyi was in Berlin to greet me. Professor Polanyi, however, returned soon thereafter.

My first research was the study of light-emitting reactions. A sodium vapor jet meeting a jet of chlorine precipitated NaCl with the emission of a bright light. Spectroscopic examination of the light told the story behind the mechanism. Although we were intrigued by this subject, Professor Polanyi and I turned our attention to the construction of a potential surface for the reaction $H_2(para) + H = H_2(ortho) + H$. We made use of Fritz London's approximate equation for the potential energy as a function of the distance between the atoms. This involved using the Heitler-London-Sugiura exchange and coulombic integrals for the energy of attraction between atomic pairs. The results were disappointing. We then changed to a spectroscopic estimation of the attraction between pairs of atoms using Morse curves. The theoretical calculations gave us the needed guidance in apportioning the bonding energy between the exchange and coulombic integrals. This way we got an exciting, if only approximate, potential surface and with it gained entrance into a whole new world of chemistry, experiencing all the enthusiasm such a vista inspired. We perceived immediately the role of zero-point energy in reaction kinetics, and our method of using Morse curves made it possible to extend our calculations to all kinds of reactions. I continued this work with enthusiasm whenever opportunity permitted.

I received a disturbing letter from my father early during my stay in Berlin. He informed me that, contrary to the opinion of lawyers whom he had consulted when we returned to the United States, the State Department had ruled that my younger brother, Joseph, was a Mexican citizen. This meant that I, too, was a Mexican citizen. On that basis I held a passport and fellowship to which I was not entitled, and I had voted for Hoover. I spent a very restless night, to say the least. Early the next morning, I consulted the American consul, who ruled that I had acted in good faith and should continue my stay in Berlin. He recommended that when I returned to the United States, I should turn in my passport, at which time I could proceed with naturalization. This I did. Five years later, I took out my naturalization papers with a judge in Trenton, New Jersey, who had just recently naturalized Professor Einstein. I was then on the faculty at Princeton, and the judge was so interested in telling me about his earlier experience with Professor Einstein that I passed the tests without difficulty.

My year with Professor Polanyi was fruitful and altogether delightful. He was a very gracious and gifted human being. Toward the end of our stay in Berlin, Professor Wendel Latimer of the University of California at Berkeley visited me and, upon hearing of our involvement in applications of quantum mechanics to chemistry, suggested to Professor G. N. Lewis that I be invited back to Berkeley for a year as lecturer, to take over some of the duties that Professor Hildebrand's impending absence would leave open. As a result of Latimer's proposal, Mildred and I found ourselves back in Berkeley at the beginning of the fall quarter of 1930.

This was an exciting year. My duties at Berkeley were not heavy, so I was able to develop further the quantum mechanical attack on reaction kinetics. Our oldest son, Edward, was born in Oakland on January 7, 1931. By spring I had a paper to present at the Indianapolis meeting of the American Chemical Society, using potential surfaces to explain why iodine was the only halogen-hydrogen reaction to involve four atoms in the activated complex, while the other reactions all went by three-atom complexes. The paper engendered much excitement, and Professor Hugh Taylor invited me on the spot to go to Princeton and present two lectures on quantum mechanical calculations of reaction rates. This visit led to an invitation to go to Princeton, which was the beginning of an exciting fifteen years to be spent in what turned out to be an ideal scientific environment. Hugh Taylor was an inspiring departmental chairman, incisive in chemical discussions and always generous in his encouragement of others.

It will not be possible to speak of more than a few of my 560 scientific papers and 9 books, but a few highlights may be interesting. Twenty years as editor of the *Annual Review of Physical Chemistry* were useful in keeping me conversant with the field. Also valuable was my coeditorship, with Douglas Henderson and Wilhelm Jost, of the 14 volumes of *Physical Chemistry: An Advanced Treatise*. Douglas Henderson and I are continuing to edit volumes entitled *Theoretical Chemistry: Advances and Perspectives*.

My second paper at Princeton was inspired by Professor Charles P. Smyth's suggestion that it would be useful to calculate the effective resultant dipole for a molecule having various dipoles lying along bonds that rotate with respect to each other. This is also essentially the same problem as calculating the distance between ends of a flexible chain, since lengths also lie along bonds. The procedure adopted was to choose coordinate systems such that the origins were at successive atoms along a chain, and the x axis coincided with the bonds of interest. One could then transform all coordinates to an initial set of coordinates and add vectorially the transformed dipoles, or lengths, in this final coordinate system. The rotations around axes could be appropriately averaged according to the potential energies of rotation. The procedure has since been extensively used by others for high polymers, both in calculating the mean lengths of a chain and in evaluating partition functions in thermodynamic calculations.

The task of developing methods of calculating potential energy surfaces for more than the four electrons treated by Fritz London was straightforward but tedious. In the April 1933 issue of the *Journal of Chemical Physics*, George Kimball and I presented a quick way of getting the secular equation for any number of electrons. This, together with the use of the Morse equation, enabled us to construct approximate potential surfaces for any molecular system of interest. It is interesting that, in the same issue, Pauling published a parallel procedure for getting the secular equation for any number of electrons. At this time George Kimball and I tried using difference equations to send a wave packet over the $H_2 + D = HD + H$ surface. This effort was premature. Joseph Hirschfelder carried out the first classical trajectory calculation in his doctoral thesis.

In 1932 Bethe's group theory treatment of orbitals in fields of different symmetry caught my attention. Arthur Frost, John Turkevich, and I incorporated group theory into our bond eigenfunction solution of the methane problem. From then on, group theory became standard procedure with us and naturally found its way, as a chapter, into the book, *Quantum Chemistry*, by Eyring, Walter, and Kimball. Interestingly enough, a contract for the book was signed in 1933. Publication, however, must always await completion of the manuscript, and the book did not come out until 1944. (Not unseemly haste!) This book has had wide use, there having been over 20 printings of the unrevised first edition and a considerable number of translations into other languages.

Our second son, Henry Bennion, was born in Princeton on May 31, 1933. At about this time I was invited to participate in a symposium on molecular quantum mechanics held by the Physical Society, over which Niels Bohr presided. The three other participants were Mulliken, Pauling, and Slater. The audience was most attentive. I was immersed in all kinds of approximate quantum mechanical calculations that could be carried out at that time, without computers. Papers on the relative rates of isotopic reactions followed naturally from my first paper with Polanyi and from experiments with Professor Hugh Taylor. Very early in our experiments, Taylor and I prepared large amounts of heavy water by electrolysis. We found that 92% heavy water proved fatal to tadpoles of the green frog *Rana clamitans*, the aquarium fish *Lebistes reticulatus*, flatworms (*Planaria maculata*), and the protozoan, *Paramecium caudatum*. This work was reported by Taylor, Swingle, Eyring, and Frost in October 1933. During this period, my associates and I were investigating potential energy surfaces for a wide variety of reactions. In November 1934 I submitted my paper on "The Activated Complex in Chemical Reactions" to the *Journal of Chemical Physics*. I showed that rates could be calculated using quantum mechanics for the potential surface, the theory of small vibrations to calculate the normal modes, and statistical mechanics to calculate the concentration and rate of crossing the potential energy barrier. This procedure provided the detailed picture of the way reactions proceed that still dominates the field.

The activated complex has a fleeting existence of only about 10^{-13} sec and is situated at the point of no return or of almost no return. It is much like any other molecule except that it has an internal translational degree of freedom and is flying apart. This concept describes any elementary reaction involving the crossing of a potential barrier. If the activated state is really a point of no return, there is no perturbation of the forward rate by the backward rate, so that the rate at equilibrium applies unchanged to the rate away from equilibrium. For thin barriers, as in the inversion of ammonia, leakage through the barrier is faster than the rate of surmounting the barrier and so must be taken into account for cases involving light atoms and thin barriers. Of the nine books I have coauthored, all but *Quantum Chemistry* and *The Theory of Optical Activity* with Dennis Caldwell involve reaction rates. They deal with such diverse subjects as rates in plastic deformation, liquids, gases, solids, biology, physics, and engineering.

In 1937, E. U. Condon, W. Altar, and I, working at Princeton, published our paper on one-electron optical activity. This was precipitated by the general belief, current then, that a one-electron transition, even in a dissymmetric field, would not contribute significantly to optical activity. Werner Kuhn had shown earlier that Drude's calculation of the optical activity of an electron moving in a spiral neglected a term that reduced the outcome for his model to zero. The result was that Kuhn, Max Born, and other workers in the field adopted the coupled oscillator model as the sole source of optical activity. This seemed unrealistic to us in view of the success of treating spectroscopy as approximately due to one-electron transitions. Adding the simplest perturbing potential that would give optical activity, $Axyz$, to the potential $k_1 x^2 + k_2 y^2 + k_3 z^2$, a system which could be readily solved, we found one-electron transitions were indeed optically active for reasonable values of A if the k's were unequal. This treatment gave rise to what has later been popularized as the octant rule. Interestingly enough, theory showed that this one-electron optical activity also persists in the classical limit. These considerations were extended in my work with Walter Kauzmann, John Walter, Daniel Miles, Dennis Caldwell, and many others. This interest has evolved into a major concern of ours with absorption spectra, circular dichroism, magnetic circular dichroism, and, not surprisingly, nuclear magnetic resonance. These tools are of course invaluable in establishing molecular structure.

In the autumn of 1937 I went with my wife, Mildred, and two oldest sons for a four-month visit to Manchester, England, where Professor Polanyi headed the chemistry department. This visit was prompted by invitations to address the Faraday Society and the Chemical Society, and included lectures at various English universities. The trip included memorable envelopments in the famous English fog, but was altogether delightful.

Returning to Princeton in December, I continued my active involvement in research and teaching. Our third son, Harden Romney, was born in Princeton on August 20, 1939.

In early 1942, I was talking to Professor Newton Harvey, head of the biology department at Princeton, about problems of shock brought on by broken bones when Professor Frank Johnson came into the room. Harvey immediately began to discuss the problem that Johnson, Brown, and Marsland had encountered. They had been studying bioluminescence of bacteria as affected by temperature and pressure. Bioluminescence was a field in which Harvey was preeminent. Bioluminescence is absent at ice temperature and becomes maximal about halfway to blood temperature. It then drops to a very low value at blood temperature due to inactivation of the enzyme at higher temperatures. The problem that interested Johnson could be stated as follows: Why, when the bacteria in a suitable solution are subjected to 200 atm pressure, is there a marked decrease in luminescence in the low-temperature range with a rise of luminescence in the high-temperature range? Since, as Braun and Le Chatelier pointed out long ago, increased pressure shifts an equilibrated system toward lower volumes, and since the activated complex is in equilibrium with reactants, the luminescent response must follow the same laws governing other equilibria. Hence, the reactants are less voluminous than the activated complex in the low-temperature range, and the reverse is true in the high-temperature range.

Soon after this discussion John Magee and I published a paper that quantitatively explained this behavior. This discussion started a collaborative investigation with Frank Johnson into biological reactions that has continued and eventuated in our writing two books dealing with reactions important in biology and medicine. Study of the pressure, temperature, and narcotic effects on bioluminescence has been a powerful tool in understanding many physiological problems.

Prior to the Second World War, I was able to show that gases that penetrated gas masks more readily than those used in the First World War were not to be expected. This conclusion derived from the study of various freshly prepared and aged smokes with the new electron microscope available in Camden, New Jersey. Fairly frequent conversations with Dr. Irving Langmuir at this time revealed that his was a brilliant, determined mind uncompromising on principle.

During the war I worked with the Navy as a consultant on high explosives. Dr. Stephen Brunauer was in charge of high-explosive research for the Navy and suggested that we consult with Professor Einstein, who also lived in Princeton. After a very pleasant morning in discussion with Professor Einstein at the Institute for Advanced Study, we walked together at noon through what had been a rose garden but was now planted with a field crop. I plucked a sprig and asked Professor Einstein what it was. He did not know. We walked a little farther and encountered the gardener sitting on his wheelbarrow. His reply to the same query was, "It is soybeans." Even for a first-rate mind, what gains attention is not just propinquity but interest. Professor Einstein's mind was too busy with more important things. Einstein's manner was never ostentatious but, indeed, on the very infrequent occasions when I talked to him, always kindly.

I also talked frequently with Professor Hugh Taylor about the nickel barriers to be used for separation of uranium isotopes, and I occasionally discussed the properties of uranium with Professor Wigner. I must confess that I did not expect the atom bomb to materialize soon enough to influence the war's outcome.

During the war the study of detonations took up a lot of my time and eventuated in a 112-page report entitled *The Stability of Detonations*, written with Richard Powell, George Duffey, and Ransom Parlin. The article was published in the August 1946 issue of *Chemical Reviews*. The curved front of the detonation was related to the diameter of the cylindrical explosive and the thickness of the reaction zone behind the wave front. The curved front theory, which established these relationships, is still widely used. The fact that the individual solid particles in an explosive burn from the outside, in layer after layer, accounts for the slow burning rate. The consequent outward burning of layer after layer of the bubbles in liquid explosives gives this same type of delayed reaction. This is a case where reaction rate is slowed down by delay in heat conduction and is to be expected, generally, at very high temperatures.

In 1944 I was asked by Professor Taylor to head up research for the Textile Research Institute, which was coming to Princeton. The position was vacant because Dr. Milton Harris, who had made distinguished contributions in this field, had accepted a position in industry. The position carried with it the responsibility for appointing 15 fellows who would work for their Ph.D. at Princeton University, and I would have Professor Eugene Pacsu, Professor John Whitwell, and Dr. Robert Rundell as my associates. This activity required 40% of my time; the remaining 60% was devoted to my regular duties at Princeton. This was a rewarding experience and drew on my engineering as well as my chemical experience.

The Institute was housed on the shore of Carnegie Lake, with extensive grounds whose graveled roads were never meant to carry the increased traffic. As a result, we would receive an occasional call for help from a stalled motorist, which we would answer with enthusiasm and dispatch, even though the consequences were often detrimental to our shoes and trousers. I enjoyed sharing in these rescues as much as my fellow workers did.

The study of spinning, weaving, dyeing, and the measurement of the physical properties of fibers and fabrics involved us in new problems of physical chemistry. My active interest in deformation kinetics dates from this time. Professor A. S. Krausz, of the University of Ottawa, and I coauthored the book, *Deformation Kinetics*, which was published in 1975. George Halsey made notable contributions to the study of the physical properties of fibers, as did Howard White and many others. I carried on parallel work with Arthur Tobolsky and others in the chemistry department. I became a member of the National Academy of Sciences in 1945.

In the spring of 1946, Ray Olpin, newly installed president of the University of Utah, visited us in Princeton to offer me the job of dean of the School of Mines or of the graduate school. My wife said that I should be the one to decide what we should do. Since at Princeton I had nine graduate students working with me, had been a professor since 1938, and had what was considered a high salary at that time, the choice was difficult. At Utah I would have to start the doctoral program and build up my own research program. As a result I decided not to go and wrote a letter to President Olpin to that effect. The next day Mildred asked what decision I had made. I told her, and she was crushed. We had lived away from her family and her Salt Lake mountains for 19 years. Although Princeton had been most pleasant, she wanted to go home. She said nothing at the time I told her of my decision, but prepared a nice letter that she asked me to read when I got to the university. When I realized her need to return home, I naturally agreed.

I immediately told Professor Taylor of the change in plan. His response was, "I told President Dodds the storm wasn't over. We have more money than the University of Utah. What do you want?"

"Nothing," I replied. "We're going."

He asked, "Do you want the Jones Professorship?"

"No. My wife wants to go, and we are going."

"One can't do business with a crazy man," he exclaimed. "Do you mind if I talk with your wife?"

"Help yourself," I said.

He and Mildred were well acquainted, and they had a pleasant conversation, but nothing was changed. So in August 1946 we started our new adventure at Utah, where I was dean of the graduate school and professor of chemistry.

Although the University of Utah had not granted Ph.D.'s before my arrival as dean of the graduate school in 1946, the transition went smoothly since the university already had a strong master's program. I found that two general administrative policies are possible: One involves strict administration from the top, the other encourages individual initiative as long as it is successful. When appropriate procedures are well understood, most decisions can be made at the departmental level, and routine decisions can be made by capable secretaries in the dean's office. Following such a policy of decentralization left me free to devote the required time to administration without seriously curtailing my research during the 20 years I served as graduate dean.

One of my first graduate students at Utah was Tracy Hall, who, after studying chromium complexes at Utah, went on to solve the problem of making diamonds at General Electric in Schenectady. Bruno Zwolinski, who had commenced his work on the transmission coefficient of reaction kinetics at Princeton, finished up at Utah and got his degree from Princeton. Thirty-five of my collaborators completed their doctorates at Princeton and over a hundred at Utah.

Most of the avenues of research I began at Princeton, as well as new ones, were carried out at Utah. In 1935 it was clear to me that the law of rectilinear diameters was understandable in terms of fluidized vacancies in liquids that mimic the behavior of molecules in the vapor both in concentration and behavior. That this is reasonable follows from the fact that to form a molecular-sized vacancy without vaporizing a molecule requires the same number of broken bonds as volatilizing a molecule without leaving a vacancy. Accordingly, the heat required to form a vacancy is the same as the heat required to vaporize a molecule. Further, molecules falling domino-like into a vacancy move the vacancy much as a gas molecule moves. Thus, the fluidized vacancy converts three molecular vibrations into translations. A molecule behaves like a gas for the fraction of the time, $V - V_s/V$, in which there is no neighbor to prevent its free fall. The remaining fraction of the time, V_s/V, it behaves like a solid. Here, V_s and V are the molal volume of the solid at the melting point and the volume of the liquid, respectively. These considerations led to the development, with Professors T. Ree and N. Hirai, of the significant structure theory of liquids, which has evolved into a quantitative treatment of the thermodynamic and transport properties of many types of liquids. Since a liquid is intermediate between solid and vapor, it seems natural to treat it as a mixture of solid-like clusters intermingled with vapor. A quantitative description of a liquid in terms of solid and vapor properties actually rests on first principles insofar as solids and liquids can be so described. The result is a single partition function for solid, liquid, and vapor.

My wife, Mildred, died of cancer June 25, 1969 after a lingering illness that she endured gracefully with a serene faith in happier times to come. During the five years and four major operations spent fighting cancer, we sought help in every quarter where there seemed any prospect of curbing the disease. The most that could be accomplished was to slow, somewhat, its relentless course.

In 1970 Dr. Betsey Jones Stover came to me with some of the results of radiating beagles with radium and plutonium, which caused them to die of bone cancer. She pointed out that the survival curve reminded her of the curve for Fermi-Dirac statistics. The equation that fitted her data for the fraction of a population surviving, S, plotted against the age, t, is

$$S = [1 + \exp k(t - \tau)]^{-1}. \qquad 1.$$

The death rate, $-dS/dt$, is given by

$$-\frac{dS}{dt} = kS(1 - S) = \frac{d(1 - S)}{dt}. \qquad 2.$$

The fraction not surviving is, of course, $(1 - S)$. From Equation 1 we see τ is the age at which half of the population still survives. From Equation 2 we see also that $-k/4$ is the slope of Equation 1 at age $t = \tau$, where $S = \frac{1}{2}$. Chemists will recognize that the equations for S are those for an autocatalytic reaction. If S is the fraction of healthy cells and $(1 - S)$ the fraction of sick ones, one would expect Equation 2 to represent the rate of spread of the disorder as it likewise should represent the growth of any ecological population.

Dr. Stover and I also pointed out that under certain conditions Equation 2 would also represent the rate of mutation. This would be true if the rate of mutation were proportional to the product $S (1 - S)$, where S is the probability one gene is undamaged and $(1 - S)$ is the probability that a neighboring gene is damaged. By use of absolute reaction rate theory, a meaning was given to the parameters k and τ. The last chapter in *The Theory of Rate Processes in Biology and Medicine*, which I coauthored with Frank Johnson and Betsey Stover, develops these considerations further.

In 1971 I married Winifred Brennan and added her daughters Eleanor, Patricia, Joan, and Bernice to our family. This has been a rewarding experience. At the time of writing, January 1977, I still have my regular professorship at the University of Utah and am enjoying excellent health. My research and teaching are going ahead at an undiminished rate with almost 20 collaborators.

As I look back over my efforts, I would characterize my contributions as being largely in the realm of model building. To test a model it is usually advantageous to cast it in mathematical form so that quantitative predictions can be used to compare calculations with experimental findings. Ideally, agreement should be quantitative and complete. Unfortunately, this never happens in the real world. Even Newtonian mechanics must be amended in the realms of relativity and quantum mechanics, and Maxwell's electromagnetic theory fails to predict stationary electronic orbits. The usual statement that the entropy always increases is not mended very successfully by Boltzmann's proposal of known theories of fluctuations. The observed cosmological departures from equilibrium boggle the imagination when considered in terms of fluctuation theory. A better statement of the second law of thermodynamics would seem to be that living things never exist in an environment where there are large decreases in the total entropy. This statement seems to include all we really know about entropy. We still need to find the gigantic Maxwellian demon that winds up worlds and consequently exists.

In model building it is convenient to start out with the following hypotheses: (*a*) There is always a model that will explain any related set of bonafide experiments. (*b*) Models should start out simple and definite enough that predictions can be made. (*c*) A model is of limited value except as it correlates a substantial body of observable material. (*d*) Models that suggest important new experiments, even if the theory must be modified, can be useful.

To be a Newton or a Maxwell it is very convenient to be stimulated by a Kepler or a Faraday, but if one were gifted enough he could still be a Gibbs with very little outside interaction.

Self-analysis is always hazardous but can be amusing. I perceive myself as rather uninhibited, with a certain mathematical facility and more interest in the broad aspects of a problem than the delicate nuances. I am more interested in discovering what is over the next rise than in assiduously cultivating the beautiful garden close at hand. In any event, the study of chemistry is still both exciting and rewarding to me.

THEORETICAL STUDIES OF NEGATIVE MOLECULAR IONS

✤ 2637

Jack Simons[1]

Department of Chemistry, The University of Utah, Salt Lake City, Utah 84112

INTRODUCTION

In 1968 Berry (1) reviewed the experimental and theoretical progress that had been made toward understanding the stabilities and bonding characteristics of small, isolated (gas phase), negative ions. In this review Berry commented:

> For the theorist, electron affinities and other properties of negative ions pose greater difficulties than do properties of neutrals or positives, insofar as electron correlation plays a relatively larger part in determining the properties of a negative ion than it does in other species. In fact, electron affinities are frequently about the same size as the differences between correlation energies in atoms and in the corresponding negative ions.

As an example of the magnitude of electron correlation effects, one need only consider the results of our calculations on the vertical ($R = 1.718$ au) electron detachment energy ($X^1\Sigma^+OH^- \rightarrow X^2\Pi_i OH$) of OH^-. Using an atomic orbital basis consisting of 20 Slater-type orbitals (STOs), we obtained a Koopmans' theorem approximation to the detachment energy equal to 3.06 eV. The energy difference between two separate SCF calculations (ΔSCF), one on OH^- and one on OH, carried out within the same basis, was equal to -0.2 eV. The difference between Koopmans' theorem and ΔSCF represents the effects of allowing the orbitals to relax upon removal of the π electron. Our best computed energy difference (2), which contains effects of electron correlation through third order, was 1.76 eV, which is in good agreement with both Branscomb's early experiments (3) and Lineberger's more recent laser detachment results (4). The difference between 1.76 eV and the ΔSCF value of -0.2 eV represents the effects of electron correlation. These effects are indeed as large as the entire electron affinity of OH; moreover, this result is not atypical.

---

[1] Alfred P. Sloan Foundation Fellow, Camille and Henry Dreyfus Fellow.

Because the treatment of both orbital relaxation and electron correlation effects in a sufficiently rigorous manner is an absolute necessity in any reliable scheme for computing properties of anions, theoretical progress toward understanding negative ions has been made rather slowly. Quite simply put, it is difficult to include correlation effects to a high enough order to guarantee precision of ± 0.2 eV in computed ion-neutral energy differences. In Berry's review article, his assessment of the state of quantum chemical research on anions involved briefly mentioning the works of Pekeris (5) on H^-; Weiss (6) on Li^-, Na^-, and K^-; Clementi and co-workers (7) on several atomic ions; Sinanoglu (8) on C^-, O^-, and F^-; Taylor & Harris (9) on H_2^-; Wahl & Gilbert (10) on halogen diatomics; and Cade (11) on OH^-, CH^-, SiH^-, SH^-, and PH^-. Therefore, in 1968 it would have been fair to say that negative molecular ions could not yet be conveniently studied by existing quantum chemical methods. On the other hand, the development of modern laser technology was making new tools available to the experimentalist to use in carrying out high-precision photodetachment and photoelectron spectroscopy studies of gas-phase anions. Thus, even in 1968 a great deal of experimental progress had begun. These experimental developments made the parallel development of theoretical methods and models aimed at better understanding negative ions a necessary and quite natural step in the scientific progress in this area.

In 1973 Simons & Smith (12) published an article in which they attempted to use equation-of-motion (EOM) techniques to express the vertical electron affinity (EA) or detachment energy (DE) of a closed-shell species in a manner that treated orbital relaxation and electron correlation through third order in perturbation theory (the difference between the coulombic interaction and the Hartree-Fock interaction is the perturbation). This developmental paper was followed by other formal papers by Simons, Jørgensen, Jordan, and Chen (13–15) in which small deficiencies in the original theory were corrected and connection made with the recent Green's function developments of Cederbaum (16), Pickup & Goscinski (17), Purvis & Öhrn (18), and Freed (19). The result of these papers was a method that permits the direct calculations of EAs and DEs of closed-shell species, which are accurate through third order.

In succeeding publications (2, 20–26), the third-order EOM theory was applied to studies of the stability and bonding characteristics of several molecular anions (and cations). Calculations of the electron affinities of OH, BeH, NH_2, CN, and BH provided theoretical support (to within ± 0.2 eV) for existing experimental measurements. Studies of the EAs of BO and Li_2 resulted in theoretical predictions for species where good experimental data is not available. Calculations of the EAs of LiF, LiCl, LiH, NaH, and BeO have led to predictions of both the existence and the stabilities (with respect to electron loss) of the anions of these species. Jordan (92) has also examined the dimer anion $(LiH)_2^-$; Simons & Jordan (93) have recently found Be_2^- to be stable even though Be_2 is unbound. Very recently, the ion $LiCl^-$ was observed by Lineberger (27), thereby verifying the theoretical prediction of Jordan et al (77). Later in this review we treat the precise nature of some of the calculations mentioned above, together with the principal conclusions of these works.

In the time since Berry's 1968 review article was completed as well as for a few years previously, several ab initio calculations, in addition to those mentioned above, were performed on molecular anions that are of chemical interest. These studies include the following works: Clementi (28) (N_3^-), Lipscomb (29) (PO^-), Krauss (30) (BH_4^-, O_2^-), Kaufman (31) (O_3^-), Csizmadia (32) (CH_3^-, NH_2^-), Pfeiffer (33) (NO_2^-), Popkie (34) (C_2^-), McLean (35) (OCN^-, SCN^-), Wahl (36) (Cl_2^-, F_2^-), Fink (37) (OH^-, NH_2^-, CH_3^-, BH_4^-), Geller (38) (BH_2^-), Thulstrup (39) (NO^-), Peyerimhoff (40) (BeH_3^-), Heaton (41) (NH_2^-), Schaefer (42) (NO_2^-), Thulstrup (84) (C_2^-), O'Hare (85) (NF^-, NS^-, PF^-, SF^-).

The above list is by no means a complete tabulation of all work done on negative molecular ions; it is simply meant to indicate the kinds of systems that have been studied as well as the approximate number of calculations that have been performed to date. Although it is true that a reasonably large number of ab initio calculations have been carried out for diatomic and small polyatomic species, very few of the studies listed above include any electron correlation effects. Most of these calculations have been done at the SCF level within small- to moderate-size bases. Therefore, the EAs that have been obtained in this manner are probably not reliable. On the other hand, the equilibrium geometries and charge densities obtained in the above SCF-level works may not be any less accurate than the results of analogous calculations on neutral species; electron correlation effects are not as dominant in determining charge densities and geometries as they are in determining EAs. Nevertheless, it is my feeling that most of the reliable work on negative ions has been, and will continue to be, characterized by a careful treatment of electron correlation and charge relaxation. For this reason, the remainder of this review will be restricted to discussion of results of studies which treat correlation in an ab initio manner.

With this brief survey of the developments made since 1968 as a background, let us now turn to a more detailed discussion of the most recently utilized methods as well as of the results that have been obtained with these methods. In the following section we review the foundation of the direct-calculation approach of References 12–19. The third section of this review contains a survey of results on OH^-, NH_2^-, BeH^-, MgH^-, CN^-, BO^-, LiF^-, $LiCl^-$, LiH^-, NaH^-, BeO^-, and NO_2^-, in which the effects of orbital relaxation and correlation have been included. Specific attention is paid to stabilities (EA or DE), geometries (R_e, θ_e), vibrational frequencies, dissociation energies, and charge densities. In the final section we review the conclusions that have been reached thus far, and we suggest areas which seem to show special promise for future development.

THEORETICAL METHODS

The electron propagator, or the one-electron Green's function (15), has been used for some time (16, 18, 20–26, 43–46) in the study of electron spectroscopy. The advantages of using the electron propagator arise because the transition energies and the transition strengths are obtained directly as poles and residues of the propagator, respectively. Several alternative procedures for decoupling the equation of motion (EOM) for the electron propagator have been developed. In this work we use the superoperator formalism of Goscinski & Lukman (47) as the framework for our development of an electron propagator that is consistent through third order. In an alternative derivation using the equation-of-motion formalism of Rowe (48), Simons & Smith (12) attempted to obtain an equation of motion that was consistent through third order. Purvis & Öhrn (49) pointed out some deficiencies in the theory of Simons & Smith; these deficiencies are mentioned again in this section. We show further how the electron propagator can be obtained consistent through third order. The consistency is made more apparent by demonstrating that all second- and third-order self-energy diagrams of Cederbaum (16) are included in our formalism.

The definition of the spectral electron propagator (44) can be written within the superoperator formalism as

$$\mathbf{G}(E) = [\mathbf{a}|(E\hat{I} - \hat{H})^{-1}|\mathbf{a}], \qquad 1.$$

where $\hat{I}$ and $\hat{H}$ are the superoperator identity and Hamiltonian respectively, and the **a** are a set of annihilation operators $\mathbf{a} = \{a_i\}$ that are arranged in a superrow vector. The superoperator scalar product is defined in the conventional fashion (47). The superoperator resolvent $(E\hat{I} - \hat{H})^{-1}$ can be approximated via the inner projection technique, and the propagator then takes the form

$$\mathbf{G}(E) = (\mathbf{a}|\mathbf{h})(\mathbf{h}|E\hat{I} - \hat{H}|\mathbf{h})^{-1}(\mathbf{h}|\mathbf{a}), \qquad 2.$$

where **h** is a projection manifold that, if chosen to be complete and orthonormal, makes Equations 1 and 2 identical. The operator space

$$\{\mathbf{h}_1, \mathbf{h}_3, \mathbf{h}_5 \ldots\} = \{a_i, a_i^+ a_k a_l, a_i^+ a_j^- a_k a_l a_r \ldots\}, k > l, i > j, k > l > r \ldots \qquad 3.$$

spans the manifold **h**. We now discuss appropriate selections of **h** that, in conjunction with our choice of the reference state, ensure that the electron propagator is calculated correctly through third order in the electronic interaction.

It is well known (17, 45) that the projection manifold $\mathbf{h}_1$, $\mathbf{h}_3$, in connection with the Hartree-Fock (HF) ground state, is able to give the electron propagator correct through second order in the electronic interaction. Our experience (2, 20–26) tells us that second-order calculation of EAs is not sufficiently precise to be useful, except in well-understood special cases that are discussed below. We demonstrate how, using a correlated ground state and the same projection manifold, we are able to get the electron propagator correct through third order in the electronic interaction.

The effect of including $\mathbf{h}_5$ in the projection manifold, where the HF ground state is used as reference state, has been discussed by Tyner et al (50), and from their analysis it is clear that $\mathbf{h}_5$ introduces terms that are at least fourth order in the electronic interaction, independent of the choice of reference state. We therefore concentrate on using $\mathbf{h}_1$, $\mathbf{h}_3$ as our projection manifold in our search for a theory that is consistent through third order.

As the reference state in our analysis we use a correlated wave function given by:

$$|0\rangle = N^{-1/2} \left\{ 1 + \sum_{\substack{p\delta}} (K_\delta^p a_p^+ a_\delta) + \sum_{\substack{m>n \\ \alpha>\beta}} (K_{\alpha\beta}^{mn} a_m^+ a_n^+ a_\beta a_\alpha) \right.$$

$$\left. + \sum_{\substack{m>n>p \\ \alpha>\beta>\delta}} (K_{\alpha\beta\delta}^{mnp} a_m^+ a_n^+ a_p^+ a_\alpha a_\beta a_\delta) + \ldots \right\} |HF\rangle, \qquad 4.$$

where the $\mathbf{a}^+$ are a set of HF creation operators, and where indices m, n, p, q (α, β, δ, γ) refer to unoccupied (occupied) spin orbitals in the HF ground state, and i, j, k, l, r are unspecified spin orbitals. We take the correlation coefficients from Rayleigh-Schrödinger perturbation theory:

$$K_\delta^p = \sum_{\substack{m>n \\ \alpha>\beta}} \{\langle p\alpha||mn\rangle \delta_{\delta\beta} - \langle p\beta||mn\rangle \delta_{\alpha\delta} + \langle \beta\alpha||\delta m\rangle \delta_{np} - \langle \beta\alpha||\delta n\rangle \delta_{pm}\}$$

$$\times \frac{\langle mn||\alpha\beta\rangle}{(\varepsilon_\delta - \varepsilon_p)(\varepsilon_\alpha + \varepsilon_\beta - \varepsilon_n - \varepsilon_m)} + \begin{array}{c} \text{higher order terms} \\ \text{in the electronic} \\ \text{interaction} \end{array} = K_\delta^p(2, 3, \ldots), \qquad 5.$$

$$K_{\alpha\beta}^{mn} = \frac{\langle mn||\alpha\beta\rangle}{\varepsilon_\alpha + \varepsilon_\beta - \varepsilon_m - \varepsilon_n} + \begin{array}{c} \text{higher order terms in} \\ \text{the electronic interaction} \end{array} = K_{\alpha\beta}^{mn}(1, 2, \ldots), \qquad 6.$$

$$K_{\alpha\beta\delta}^{mnp} = K_{\alpha\beta\delta}^{mnp}(2, 3, \ldots), \qquad 7.$$

where the first number in the bracket indicates the lowest order in the electronic interaction. The ε_i indicate HF orbital energies, and the two-electron integral $\langle mn|\alpha\beta\rangle$ refers to the charge densities $m\alpha$ and $n\beta$; we have

$$\langle mn||\alpha\beta\rangle = \langle mn|\alpha\beta\rangle - \langle mn|\beta\alpha\rangle. \qquad 8.$$

In our analysis we consider the projection manifold $\{\mathbf{h}_1, \mathbf{h}_3\}$, where the $\mathbf{h}_3$ space has been redefined for convenience as

$$\mathbf{h}_3 = \{a_i^+ a_k a_l + \langle a_i^+ a_l\rangle a_k - \langle a_i^+ a_k\rangle a_l\}. \qquad 9.$$

The choice of the subspace $\mathbf{h}_3$ ensures that this space is orthogonal to $\mathbf{h}_1$:

$$(\mathbf{h}_1|\mathbf{h}_3) = \mathbf{0}, \qquad 10.$$

even for a correlated reference state. We also have the following orthogonality relations:

$$(\mathbf{h}_1|\mathbf{h}_1) = \mathbf{1}, \quad (\mathbf{h}_3|\mathbf{h}_3) = \mathbf{S}(0, 2, 3, \ldots). \qquad 11.$$

Using Equations 10 and 11, Equation 2 can be partitioned into the form

$$\mathbf{G}^{-1}(E) = (\mathbf{h}_1|E\hat{I} - \hat{H}|\mathbf{h}_1) - (\mathbf{h}_1|\hat{H}|\mathbf{h}_3)(\mathbf{h}_3|E\hat{I} - \hat{H}|\mathbf{h}_3)^{-1}(\mathbf{h}_3|\hat{H}|\mathbf{h}_1)$$
$$= E\mathbf{1} - \mathbf{A} - \mathbf{BD}^{-1}\mathbf{C}, \qquad 12.$$

where the matrices **A**, **B**, **C**, and **D** are defined as:

$$\begin{aligned}\mathbf{A} &= (\mathbf{h}_1|\hat{H}|\mathbf{h}_1), \\ \mathbf{B} &= (\mathbf{h}_1|\hat{H}|\mathbf{h}_3), \\ \mathbf{C} &= (\mathbf{h}_3|\hat{H}|\mathbf{h}_1), \\ \mathbf{D} &= (\mathbf{h}_3|E\hat{I} - \hat{H}|\mathbf{h}_3),\end{aligned} \qquad 13.$$

We will now make an order-by-order analysis of Equation 12, in which we retain only those terms that are zeroth, first, second, or third order in the electronic interaction. Since the **B** and **C** matrices are at least of first order (12), we need to consider only that part of the **D** matrix that is zeroth and first order. This constrains the indices in the projection manifold $\mathbf{h}_3$ to be of the form $a_m^+ a_\alpha a_\beta \alpha > \beta$ or $a_\alpha^+ a_m a_n m > n$, since operators such as $a_m^+ a_n a_\alpha$ lead to matrix elements in the **D** matrix that are at least of second order. The resulting subspace $\mathbf{h}_3$ is thus identical to that used in calculating the electron propagator correctly through second order. To calculate the the electron propagator through third order in the electron repulsions, we thus need to obtain the **B** and **C** matrices through second order, the **D** matrix through first order, and the **A** matrix through third order.

Previous attempts (12) to obtain the electron propagator correctly through third order have used as the reference state:

$$|0\rangle = N_0^{-1/2}\left(1 + \sum_{\substack{m>n \\ \alpha>\beta}} K_{\alpha\beta}^{mn} a_m^+ a_n^+ a_\alpha a_\beta\right)|HF\rangle, \qquad 14.$$

where the $K_{\alpha\beta}^{mn}$ are determined from first order Rayleigh-Schrödinger perturbation theory. In calculating the **B**, (**C**), and **D** matrix elements correctly through second and first order, respectively, no changes are obtained from considering the higher correlated ground state in Equation 4. The matrix elements of **B**, **C**, and **D** are given by:

$$B_{i,\alpha m\beta} = -\langle im|\alpha\beta\rangle - 1/2 \sum_{p,q} \langle im|pq\rangle K_{\alpha\beta}^{pq} + \sum_{\gamma,p} [\langle i\gamma|p\alpha\rangle K_{\beta\gamma}^{mp} - \langle i\gamma|p\beta\rangle K_{\alpha\gamma}^{mp}], \qquad 15.$$

$$B_{i,n\alpha m} = \langle i\alpha|mn\rangle + 1/2 \sum_{\gamma\delta} \langle i\alpha|\delta\gamma\rangle K_{\delta\gamma}^{mn} + \sum_{\gamma p} [\langle ip|\gamma n\rangle K_{\alpha\gamma}^{mp} - \langle ip|\gamma m\rangle K_{\alpha\gamma}^{np}], \qquad 16.$$

$$\mathbf{C}^+ = \mathbf{B} \text{ (through second order).} \qquad 17.$$

$$D_{n\alpha m,\delta p\gamma} = 0, \qquad 18.$$

$$\begin{aligned}D_{n\alpha m,q\beta p} = {}& \delta_{nq}\delta_{\alpha\beta}\delta_{mp}(\varepsilon_m + \varepsilon_n - \varepsilon_\alpha) - \delta_{qn}\langle m\beta||p\alpha\rangle - \delta_{pm}\langle n\beta||q\alpha\rangle \\ & + \delta_{qm}\langle n\beta||p\alpha\rangle + \delta_{\alpha\beta}\langle mn||pq\rangle + \delta_{pn}\langle m\beta||q\alpha\rangle,\end{aligned} \qquad 19.$$

$$\begin{aligned}D_{\delta p\gamma,\alpha q\beta} = {}& -\{\delta_{\delta\alpha}\delta_{pq}\delta_{\gamma\beta}(\varepsilon_p - \varepsilon_\delta - \varepsilon_\gamma) - \delta_{\gamma\beta}\langle\delta q||\alpha p\rangle - \delta_{\delta\alpha}\langle\gamma q||\beta p\rangle \\ & + \delta_{\gamma\alpha}\langle\delta q||\beta p\rangle - \delta_{pq}\langle\delta\gamma||\beta\alpha\rangle + \delta_{\delta\beta}\langle\gamma q||\alpha p\rangle\}.\end{aligned} \qquad 20.$$

In the **A** matrix we need to include all terms up to third order. The **A** matrix elements obtained by using the state defined in Equation 14 as a reference state need to be modified by third-order terms that result from interaction between the singly excited states and the HF ground state. The triply excited states that also result from a second-order Rayleigh-Schrödinger perturbation calculation do not introduce third-order terms. We thus have to add to the **A** matrix elements given in Reference 12 the terms δA_{ij} defined by:

$$\delta A_{ij} = \sum_{\substack{\delta\beta \\ pmn}} \frac{\langle jp||i\delta\rangle\langle\delta\beta||mn\rangle\langle mn||p\beta\rangle}{(\varepsilon_\delta - \varepsilon_p)(\varepsilon_\delta + \varepsilon_\beta - \varepsilon_m - \varepsilon_n)}, \quad (A3)$$

$$+ \sum_{\substack{\delta\beta \\ pmn}} \frac{\langle j\delta||ip\rangle\langle p\beta||mn\rangle\langle mn||\delta\beta\rangle}{(\varepsilon_\delta - \varepsilon_p)(\varepsilon_\delta + \varepsilon_\beta - \varepsilon_m - \varepsilon_n)}, \quad (A4)$$

$$+ \sum_{\substack{\delta\alpha\beta \\ pn}} \frac{\langle jp||i\delta\rangle\langle\delta n||\beta\alpha\rangle\langle\alpha\beta||pn\rangle}{(\varepsilon_\delta - \varepsilon_p)(\varepsilon_\alpha + \varepsilon_\beta - \varepsilon_p - \varepsilon_n)}, \quad (A5)$$

$$+ \sum_{\substack{\delta\alpha\beta \\ pn}} \frac{\langle j\delta||ip\rangle\langle\beta\alpha||\delta n\rangle\langle pn||\alpha\beta\rangle}{(\varepsilon_\delta - \varepsilon_p)(\varepsilon_\alpha + \varepsilon_\beta - \varepsilon_p - \varepsilon_n)}. \quad (A6)$$

The **A** matrix of Reference 12 is:

$$A_{ij} = \delta_{ij}\varepsilon_i + \sum_{k,l} \langle ik|jl\rangle F_{kl}, \qquad 21.$$

$$E_{kl} = \sum_{\alpha>\beta,p} K_{\alpha\beta}^{pk} K_{\alpha\beta}^{pl} - \sum_{p<q,\alpha} K_{\alpha l}^{pq} K_{\alpha k}^{pq}. \qquad 22.$$

We have thereby calculated the electron propagator consistently through third order.

A comparison with a diagrammatic perturbation expansion of the self-energy (86) makes it further evident that we have really included all terms through third order in our analysis of the electron propagator. In Figure 1 we have displayed the terms A3–A6 as diagrams, using the rules of Brandow (51), which combine the Goldstone

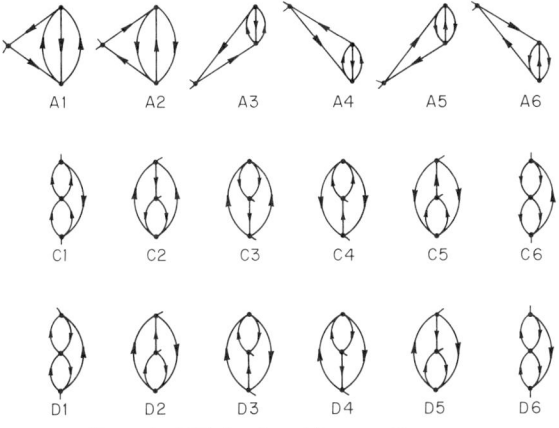

Figure 1 Third-order self-energy diagrams.

diagrams with the antisymmetrized vertices of Hugenholtz. The diagrams labeled A3–A6 are identical to the third order diagrams given by Cederbaum (16), in which dots refer to antisymmetrized vertices. The diagrams A3–A6 were shown by Purvis & Öhrn (49) to be the only missing third-order diagrams that evolve from a theory where the reference state is given by Equation 14. The analysis of Purvis & Öhrn considered $(\mathbf{h}_3|E\hat{I} - \hat{H}|\mathbf{h}_3)$ as two matrices; $\mathbf{u}$, which contains E and orbital energies (zeroth-order terms), and $\mathbf{M}$, which contains the two-electron integrals (first-order terms). Making use of the identity $(\mathbf{u} + \mathbf{M})^{-1} = \mathbf{u}^{-1} - \mathbf{u}^{-1}\mathbf{M}(\mathbf{u} + \mathbf{M})^{-1}$ to expand the inverse, Purvis & Öhrn identified the diagrams C1–C6 and D1–D6 of Cederbaum (16) as originating from the $\mathbf{BD}^{-1}\mathbf{C}$ term, while the diagrams A1–A2 were found to arise from the $\mathbf{A}$ matrix previously given in Reference 12. We have thus accounted for all third-order diagrams that arise from an expansion of the self energy. The electron propagator calculation thus needs to use the second-order correlated ground state given in Equation 4 as a reference state and the subspace $\{\mathbf{h}_1, \mathbf{h}_3\}$ given by Equation 3 as the projection manifold in order to be correct through third order.

We have shown how the electron propagator can be obtained correctly through third order. Our development stresses the fact that a complete treatment of the inverse of the $\mathbf{D}$ matrix is needed to guarantee that all desired terms are included. Computational applications have only been carried out so far by using a diagonal approximation to the $\mathbf{D}$ matrix. This situation is unsatisfactory and should be improved. A unitary transformation that brings the $\mathbf{D}$ matrix closer to diagonal form is related to the theory of linear response as discussed by Jørgensen & Purvis (52). By use of this kind of procedure, we would expect to get an approximation to the propagator that would be nearly complete through third order. The energy-shifted denominators that result from evaluating the $\mathbf{D}$ matrix correspond to the result of summing certain classes of diagrams to infinite order, which implies that the electron propagator treatment has the computational advantage of expressing these summations in closed form. In a diagrammatic summation of self-energy diagrams, one has to account explicitly for each energy-shifted denominator through each order. We note finally that the side-shifted diagrams given in Figure 1 do not appear in the original third-order theory of Simons & Smith (12) for calculating ionization potentials and electron affinities. These diagrams result from using a more highly correlated wave function than the one considered by Simons & Smith as the reference state.

With this derivation of the needed third-order equation accomplished, let us now turn to a more detailed description of selected results that have been obtained by making use of the propagator approach. Recall that we are limiting the discussion to studies that have treated electron correlation effects in an ab initio manner. The anions that we have chosen to discuss in some detail can be divided into several classes. NH_2^-, OH^-, SH^-, BeH^-, and MgH^- are hydrides whose parents have a half-filled orbital to which the "extra" electron is added. CN^-, BO^-, and NO_2^- are ions whose parents also have half-filled orbitals. All of the above ions are closed-shell species. Li_2^- and Be_2^- are open-shell ions that are formed by adding an electron either to an antibonding (Li_2^-) or bonding (Be_2^-) molecular orbital of the closed-shell parent. LiF^-, $LiCl^-$, LiH^-, NaH^-, and BeO^- are each formed by adding an electron to an essentially nonbonding orbital of the very polar neutral parent. Correlation and relaxation are not very important in these ions because the "extra" electron resides in an orbital that is localized on the "back" end of the electropositive atom where it encounters little dynamic interaction with the other electrons.

SURVEY OF RESULTS

OH^-

In carrying out the calculations on OH^- described here (2), we employed an atomic orbital basis consisting of Slater-type functions whose orbital exponents were taken from the bases of Cade (53) for OH^-, and of Cade & Huo (54) for OH. Information describing our basis as well as the essential results of the SCF calculation on the parent $X^1\Sigma^+OH^-$ for this basis are given in Table 1. Note that the basis used in this work is not very large.

As shown in Table 2, the vertical detachment energies computed using the basis of Table 1 are within 0.1 eV of the experimental result quoted by Lineberger et al (4). An important observation that should be made here is that the basis given in Table 1 is capable of yielding a very accurate detachment energy. Our results show that the theory of molecular electron affinities and ionization potentials in Reference 12 is capable of yielding the vertical electron detachment energy of $X^1\Sigma^+OH^-$ to within

Table 1 20-function Hartree-Fock wave function for OH^- [a]

σ atomic orbitals	1σ	2σ	3σ	π atomic orbitals	1π
$O1s$ (7.0168)	0.9721	−0.1645	0.1213	$O2p$ (2.0624)	0.5857
$O2s$ (2.8646)	0.1268	0.8711	−0.0336	$O2p'$ (3.7529)	0.1949
$O2p$ (2.1172)	−0.0349	0.0490	0.3761	$O2p''$ (0.7128)	0.3246
$O1s'$ (12.3850)	0.0961	0.0081	−0.0111	$O3d$ (1.2500)	0.0133
$O2s'$ (1.5729)	0.0141	0.3714	−0.3687	$H2p$ (0.9250)	0.0958
$O2p'$ (1.0227)	0.0067	−0.0065	0.2047	—	—
$O2p''$ (3.7596)	−0.0053	0.0074	0.1777	—	—
$H1s$ (1.1986)	−0.0027	0.1507	0.4723	—	—
$H2s$ (2.3003)	−0.1816	−0.3336	−0.1403	—	—
$H1s'$ (2.4385)	−0.0014	0.0494	0.0024	—	—

[a] $R = 1.781$ au, $E = -75.3801$ au, $\varepsilon_{1\sigma} = -20.22091$, $\varepsilon_{2\sigma} = -0.94178$, $\varepsilon_{3\sigma} = -0.27867$, $\varepsilon_{1\pi} = -0.12616$.

Table 2 Summary of detachment energies for OH^-

Detachment energy (eV)	Source
1.773	EOM (2)
1.825 ± 0.002	Hotop, Patterson, Lineberger (4)
1.83	Branscomb (3)

0.1 eV. It has also been demonstrated that a highly accurate description of the core orbital of OH⁻ is not essential to an accurate calculation of the $^2\Pi_i$ valence electron detachment energy. Finally, an investigation of the roles of orbital relaxation and correlation-energy change in determining the ion-molecule energy difference has led to the conclusion (2) that both of these effects must be treated properly in any study of negative molecular ions unless one knows that the "extra" electron is essentially uncorrelated (perhaps by spatial localization) with the other electrons. As discussed in a later section on LiF⁻, LiCl⁻, LiH⁻, NaH⁻, and BeO⁻, such is the case for the family of anions formed by adding an electron to a highly polar, closed-shell molecule.

BeH⁻

An initial basis set for the closed-shell ($^1\Sigma^+$) BeH⁻, consisting of 20 Slater-type orbitals (STOs), was adapted from the optimized BeH basis set reported by Cade & Huo (54). To accommodate the extra electron correlation, $2p_\pi$ functions and diffuse s and $2p_\sigma$ functions were added to the "sigma only" BeH basis set to replace functions contributing nominally to the description of the occupied BeH molecular orbitals.

The orbital exponents of the four BeH⁻ STOs in the original basis set having the largest expansion coefficients in the 3σ highest-occupied molecular orbital (HOMO) were optimized at the initially calculated BeH⁻ equilibrium internuclear distance of 2.660 au. The initial and optimized BeH⁻ basis sets and expansion coefficients for occupied molecular orbitals are listed in Table 3. Basis functions that were also used in the Cade & Huo BeH basis set have been marked with an asterisk. From this table we observe that optimization of the BeH⁻ basis set caused dramatic increases in the importance of the diffuse $2s$ Be and $1s$ H basis functions describing the 3σ HOMO.

Table 3 Original and optimized 20 STO basis sets for BeH⁻ [a]

BeH basis	Orbital	ζ (original)	ζ (optimized)	$C_{1\sigma}^{(orig)}$	$C_{1\sigma}^{(op)}$	$C_{2\sigma}^{(orig)}$	$C_{2\sigma}^{(op)}$	$C_{3\sigma}^{(orig)}$	$C_{3\sigma}^{(op)}$
*	1sBe	2.9448	—	0.84377	0.85361	−0.16987	−0.15081	−0.09822	−0.08297
*	1s'Be	5.7480	—	0.23092	0.22616	−0.00186	−0.01022	−0.01053	−0.01590
—	2sBe	0.4000	0.4250	0.01243	−0.03928	0.01391	−0.09120	0.63441	1.08018
*	2s'Be	0.8925	1.1500	0.07154	0.10249	0.29861	0.39163	0.63860	0.56833
—	2s"Be	1.7238	—	−0.11599	−0.16573	0.12175	−0.02368	0.06845	−0.05475
—	$2p_\sigma$Be	0.4000	—	0.01257	−0.01303	0.03299	−0.03817	−0.13953	−0.01752
*	$2p'_\sigma$Be	0.8080	—	0.03336	0.01491	0.03954	−0.01068	−0.11016	−0.28577
*	$2p''_\sigma$Be	1.0460	—	−0.04211	−0.00897	0.12792	0.13322	−0.08124	0.05031
*	$2p'''_\sigma$Be	1.5000	—	0.01356	0.00198	0.09628	0.08734	−0.10446	−0.12637
—	$2p_\pi$Be	0.8080	—	0	0	0	0	0	0
—	$2p'_\pi$Be	1.0460	—	0	0	0	0	0	0
—	$2p''_\pi$Be	1.5000	—	0	0	0	0	0	0
—	1sH	0.4000	0.3000	−0.05489	0.03915	0.05648	0.15939	−0.47443	−0.73778
—	1s'H	1.0000	1.0500	0.01772	−0.02409	0.71242	0.64325	−0.19299	−0.19021
*	2sH	2.5000	—	−0.00019	0.01496	0.01453	−0.01514	0.00898	0.00635
—	$2p_\pi$H	1.4500	—	0	0	0	0	0	0

[a] Original basis set; $R_e = 2.660$ au, $E = -15.12104$ au, $\varepsilon_{1\sigma} = -4.51210$ au, $\varepsilon_{2\sigma} = -0.28926$ au, $\varepsilon_{3\sigma} = -0.02032$ au; $IP = 0.02751$ au. Optimized basis set; $R_e = 2.670$ au, $E = -15.12308$ au, $\varepsilon_{1\sigma} = -4.50693$ au, $\varepsilon_{2\sigma} = -0.27730$ au, $\varepsilon_{3\sigma} = -0.01877$ au, $IP = 0.02913$ au.

Table 4 Energy vs internuclear separation data in au for BeH and BeH⁻, original and optimized basis sets

R(au)	$E_{BeH}^{(orig)}$	$E_{BeH}^{(op)}$	$E_{BeH^-}^{(orig)}$	$E_{BeH^-}^{(op)}$	$IP_{BeH^-}^{(orig)}$	$IP_{BeH^-}^{(op)}$	$-\varepsilon_{3\sigma}^{(orig)}$	$-\varepsilon_{3\sigma}^{(op)}$
2.380	−15.11607	−15.11785	−15.09207	−15.09201	0.02400	0.02584	—	0.01559
2.420	−15.11750	−15.11935	−15.09307	−15.09311	0.02443	0.02624	0.01781	0.01595
2.460	−15.11868	−15.12056	−15.09381	−15.09392	0.02487	0.02664	0.01818	0.01633
2.500	−15.11959	−15.12150	−15.09422	−15.09441	0.02537	0.02709	0.01857	0.01673
2.538	−15.12023	−15.12217	−15.09439	−15.09465	0.02584	0.02752	0.01895	0.01715
2.580	−15.12070	−15.12269	−15.09429	−15.09467	0.02641	0.02802	0.01940	0.01763
2.620	−15.12096	−15.12297	−15.09401	−15.09446	0.02695	0.02851	0.01985	0.01812
2.660	−15.12104	−15.12308	−15.09353	−15.09409	0.02751	0.02899	0.02032	0.01864
2.700	−15.12097	−15.12303	−15.09284	−15.09348	0.02813	0.02955	0.02081	0.01918
2.740	−15.12076	−15.12284	−15.09201	−15.09275	0.02875	0.03009	0.02133	0.01975
2.780	−15.12042	−15.12251	−15.09101	−15.09184	0.02941	0.03067	0.02188	0.02035

In Table 4 we present our calculated BeH and BeH⁻ energies for both the original and optimized basis sets. SCF calculations for $E_{BeH^-}(R)$ were executed on the University of Utah Univac 1108 computer by use of a modified version of the Harris DIATOM program. Execution time for each run was approximately 4.50 min. Vertical ionization energies of BeH⁻ and the $IP_{BeH^-}(R)$ were calculated by our third-order equations-of-motion program. Execution time for each ionization energy calculation on the Univac 1108 was approximately 25 sec. BeH energies were calculated by adding the vertical ionization energy of BeH⁻ to the BeH⁻ energy,

$$E_{BeH}(R) = E_{BeH^-}(R) + IP_{BeH^-}(R). \qquad 23.$$

Introduction of angular correlation into the BeH⁻ basis set, while necessary for a good description of the electron correlation effects associated with negative-ion ionization energy calculations, resulted in total electronic energy values for both BeH and BeH⁻ that were not as good as those obtained from calculations employing a sigma-only HF basis set. To see this we compare the BeH minimum energy of −15.15312 au, determined by the HF calculations of Cade & Huo (54), with our optimized minimum SCF-EOM BeH energy of −15.09446 au.

In Figure 2 the calculated BeH and BeH⁻ potential curves are presented for the original and optimized basis sets. The energy scale shown in this figure is relative energy in au where the zeros of the curves have all been adjusted so that their shapes may be compared. It is evident from these figures that BeH⁻ has a shallower potential and a larger equilibrium internuclear distance than BeH.

The differences in energies ΔE_{BeH^-}, ΔIP_{BeH^-}, and ΔE_{BeH} between the two basis sets are plotted as a function of R in Figure 3. Since orbital exponents were optimized at the initially calculated equilibrium internuclear distance of BeH⁻ to give the best BeH⁻ energy, it is not surprising that E_{BeH} was more sensitive to this basis set optimization than E_{BeH^-}. For both BeH and BeH⁻ the qualitative effect of BeH⁻ basis set optimization was to increase slightly the slope of the potential curves for $R \lesssim R_e$ and to decrease the slope for $R \gtrsim R_e$, with R_e^{BeH} and $R_e^{BeH^-}$ becoming slightly larger. We can visualize this by considering the effect of subtracting the ΔE curves in Figure 3 from the original basis BeH and BeH⁻ curves in Figure 2.

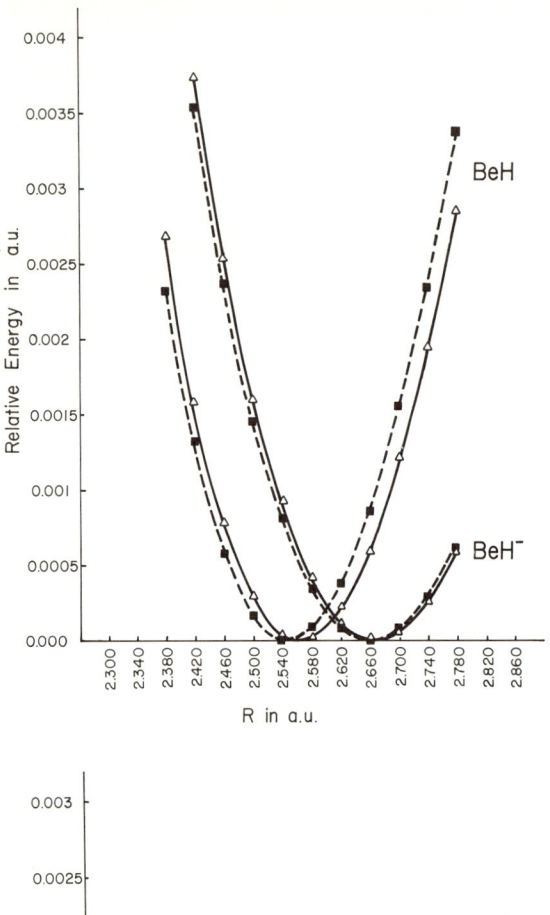
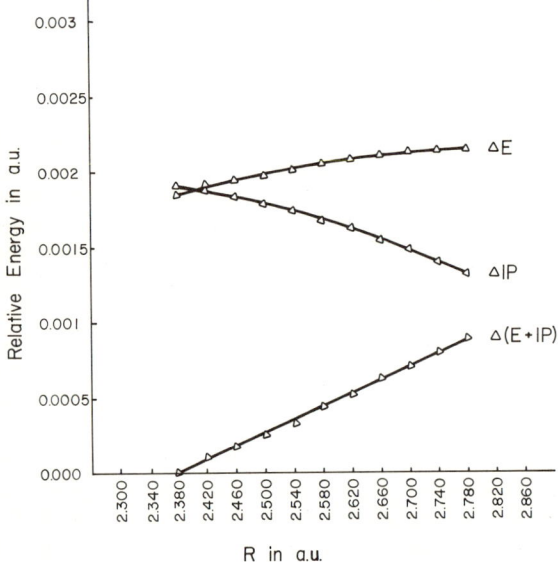

Figure 2 (*Top*) The original (dashed) and optimized (solid) basis calculation of the BeH$^-$ and BeH energies.

Figure 3 (*Bottom*) The difference between the original and optimized basis energies of BeH$^-$ (ΔE), IP_{BeH^-} (ΔIP), and BeH [$\Delta(E + IP)$] as functions of R.

Table 5 Calculated and experimental spectroscopic parameters for BeH and BeH⁻

Parameter	Units	BeH$^{(exp)}$	BeH$^{(HF)}$	BeH$^{(orig)}$	BeH$^{(op)}$	BeH$^{-(orig)}$	BeH$^{-(op)}$	BeH$^{-(exp)}$
R_e	au	2.538^a	2.528^c	2.540	2.560	2.660	2.670	—
k_e	dyne cm^{-1} × 10^{-1}	2.246^{a,h}	2.461^c	2.345	2.118	1.903	1.753	—
v_e	cm^{-1}	2058.6^a	2154.6^{c,h}	2103.2^h	1998.6^h	1894.9^h	1818.5^h	—
D_e	eV	(2.33)^{a,b}	2.18^c	—	—	2.15^e 2.31^f	2.20^{d,e} 2.35^{d,f}	—
EA_{BeH} (R_e^{BeH})	eV	—	—	0.7031	0.7560^d	—	—	—
IP_{BeH^-} ($R_e^{BeH^-}$)	eV	—	—	—	—	0.7485	0.7926^d	—
IP_{BeH^-} (thermo)	eV	—	—	—	—	0.7253	0.7727^d	0.74^g

^{a} Herzberg (55).
^{b} Gaydon (56), reports 2.3 eV with an uncertainty of ±0.3 eV.
^{c} Cade & Huo (54).
^{d} Results obtained from SCF and EOM calculations at R values not reported in Table 2 of Reference 21.
^{e} Calculated from Cade & Huo HF D_e^{BeH} of 2.18 eV.
^{f} Calculated from Herzberg's D_e^{BeH} of 2.33 eV.
^{g} Feldmann (57), photodetachment energy value corrected for zero-point vibrational energy difference by Equation 19.
^{h} Calculated from k_e or v_e assuming $v_e = 1/2\pi c \sqrt{k_e/\mu}$.

Our calculated spectroscopic parameters for BeH and BeH⁻ are presented in Table 5. Hartree-Fock values and experimental values have been included in this table for comparison with our theoretical results.

Dissociation energies for BeH⁻ were calculated according to the procedure depicted in Figure 4, from which the following can be written:

$$D_e^{BeH^-} = D_e^{BeH} + \Delta D_e, \qquad 24.$$

$$\Delta D_e = IP_{BeH^-}(R_e^{BeH}) + E_{BeH^-}(R_e^{BeH}) - E_{BeH^-}(R_e^{BeH^-}) - EA_H. \qquad 25.$$

Ab initio approximations to $D_e^{BeH^-}$ were obtained using the HF value for D_e^{BeH} of 2.18 eV reported by Cade & Huo. Semi-empirical BeH⁻ dissociation energy results were calculated using Herzberg's BeH dissociation energy (55) of 2.33 eV. Referring to the $D_e^{BeH^-}$ values in Table 5, we can see that the ion-molecule dissociation energy difference ΔD_e is small, and that it changes sign with basis set optimization. Herzberg and, more recently, Gaydon (56) both note the large uncertainty in experimental D_e^{BeH} values. Gaydon reports for D_e^{BeH} 2.3 ± 0.3 eV, an uncertainty that is, of course, much larger than our calculated ΔD_e value. We must therefore assume that a corresponding uncertainty is introduced into our semi-empirical calculations of the negative-ion dissociation energy.

Approximate vibrational force constants and fundamental vibrational frequencies for the BeH and BeH⁻ systems were obtained by fitting a least squares quadratic polynomial to each potential energy curve. Agreement of these calculations with the experimental BeH values was quite good. For the original basis set, $v_e^{(calc)} - v_e^{(exp)}$ was +44.6 cm^{-1}, a +2.17% deviation, and for the optimized basis set the deviation was −60 cm^{-1} or −2.91%. For comparison, the value for v_e^{BeH} reported by Cade & Huo deviated from the experimental value by +96 cm^{-1} or +4.66%. It is reasonable to assume that our calculated fundamental vibrational frequencies for BeH⁻

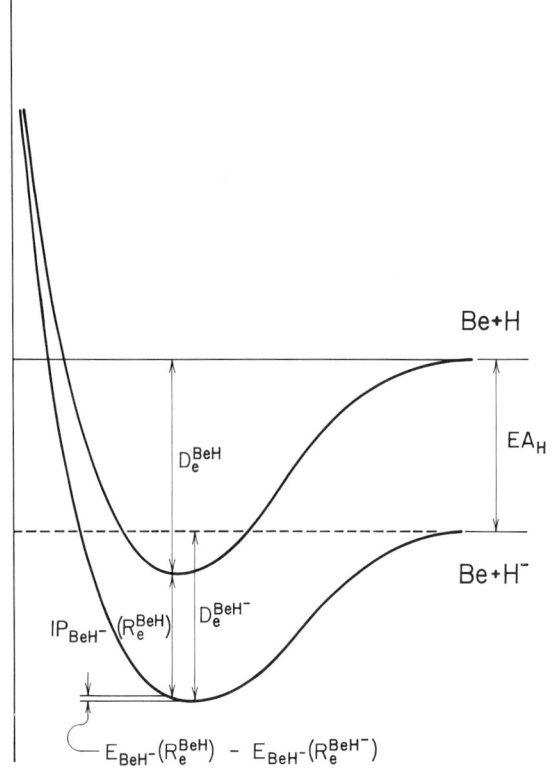

Figure 4 The potential curves of BeH⁻ and BeH as functions of R.

are at least this accurate. These results for BeH⁻ should prove extremely useful to experimentalists in the rapidly developing area of negative-ion spectroscopy. The smaller vibrational force constant and the correspondingly lower vibrational frequency for BeH⁻ as compared to BeH are expected from the relative shapes of their potential energy curves as shown in Figure 2. The fundamental vibrational frequencies for both the negative ion and the neutral molecule decrease with basis set optimization. This trend follows from the widening of the optimized potential wells of the two species described earlier.

The thermodynamic ionization energy of BeH⁻ was calculated using the relationship:

$$IP_{\text{BeH}^-}(\text{thermo}) = E_{\text{BeH}}(R_e^{\text{BeH}}) - E_{\text{BeH}^-}(R_e^{\text{BeH}^-}), \qquad 26.$$

in which $E_{\text{BeH}}(R_e^{\text{BeH}})$ was determined indirectly from our EOM results and the BeH⁻ energies of the SCF calculation. Vertical ionization energies of BeH⁻ and vertical electron affinities of BeH were calculated directly from our EOM theory. The threshold photodetachment energy for BeH⁻, $\Delta E_{\text{BeH}^-}^{h\nu}(v^{\text{BeH}^-} = 0 \to v^{\text{BeH}} = 0)$, recently determined by Feldmann (57), is related to IP_{BeH^-}(thermo) by the following equation:

$$IP_{BeH^-}(\text{thermo}) = \Delta E_{BeH^-}^{hv}(0, 0) - h/2(v_e^{BeH} - v_e^{BeH^-}). \qquad 27.$$

The second term in Equation 27, which gives the difference in zero-point vibrational energies between BeH and BeH$^-$, takes into account the fact that the photodetachment energy is the energy difference between zero-point vibrational levels of the negative ion and the neutral, rather than the difference between their potential minima. Calculation of this zero-point vibrational frequency correction gives -0.0129 eV and -0.0111 eV for the original and optimized basis set data respectively. In reporting Feldmann's value for IP_{BeH^-}(thermo) of 0.74 eV in Table 5, we have thus subtracted 0.01 eV from the BeH$^-$ photodetachment energy value of 0.75 eV. Calculated thermodynamic ionization potentials for BeH$^-$ deviate from the experimental value by $(-0.01$ eV, $-14\%)^{(\text{orig})}$, $(+0.03$ eV, $+4.1\%)^{(\text{op})}$. Vertical ionization energies and electron affinities differ no more than 7% from the experimental thermodynamic value. These results indicate that for reasonable approximations to the thermodynamic ionization energy of BeH$^-$, our vertical EOM ionization energy calculations at R_e^{BeH} and $R_e^{BeH^-}$ are quite good.

Our SCF calculations on the $^1\Sigma^+$ BeH$^-$ molecular ion and our third-order EOM (BeH$^- \to$ BeH) ionization energy calculations are capable of producing ab initio results for R_e^{BeH}, v_e^{BeH}, and IP_{BeH^-}(thermo) in excellent agreement with experiment. These theoretical methods are also capable of yielding new information about BeH$^-$, such as $R_e^{BeH^-}$, $v_e^{BeH^-}$, that should be of great value to current experimental efforts in negative ion spectroscopy. Comparisons of the occupied MOs of BeH, BeH$^-$, and BH make it possible to better understand the influences of electronic and nuclear charge changes in the bonding of these systems.

Our studies of basis set optimization in BeH$^-$ show that good approximations to ion and neutral potential curves, ionization energies, and spectroscopic parameters may be obtained with a carefully chosen nonoptimized basis set. The small differences between $IP_{BeH^-}(R_e^{BeH^-})$, $IP_{BeH^-} - (R_e^{BeH})$, and IP_{BeH^-}(thermo) indicate threshold photodetachment energies and thermodynamic ionization potentials can be predicted to within ± 0.05 eV by single EOM vertical ionization energy calculations near the minima of the ion or neutral potential wells.

The EOM-Koopmans' theorem correlation we have observed in our BeH$^-$ and BH ionization energy calculations suggests that we may be able to account for orbital reorganization and electron correlation effects at many internuclear distances by calculating the EOM correction to Koopmans' theorem ionization energy at a single internuclear distance. We feel that the generality of this effect and its theoretical implications are worthy of much more extensive study.

CN^- and BO^-

The starting point for the construction of the double zeta basis sets used in these calculations was Roetti & Clementi's excellent set (58) of double zeta functions for the component atoms. To better describe the charge distribution in the resultant negative molecular ions, the orbital exponents of the functions with large expansion coefficients in the highest occupied molecular orbital (HOMO) of CN^- and of BO^- were varied to maximize the ionization energies. The greatest changes in each case were produced by modifying the exponents of the $2s$ and $2p\sigma$ functions on the less electronegative atom in the ion.

Since little is known about the geometry of BO^-, its basis set was optimized at $R = 2.278$ bohr, which corresponds to the equilibrium separation of the neutral BO molecule. The total energy before optimization was -99.55 hartree and the corresponding $X^1\Sigma^+ \rightarrow X^2\Sigma^+$ ionization potential was 2.16 eV. The final basis gave a slightly lower SCF energy, -99.554 hartree, and a vertical ionization potential of 2.81 eV.

The CN^- basis set optimizations were performed at $R = 2.2$ bohr, the internuclear separation corresponding to the lowest energy found after doing a few preliminary SCF calculations on CN^- with the starting basis. The initial energy at 2.2 bohr was -92.2634 hartree, and the vertical ionization energy for the starting basis was 3.04 eV. After optimization of $2s$ and $2p\sigma$ functions on carbon and $2s$ functions on nitrogen, the total energy was -92.2645 hartree and the $X^1\Sigma^+ \rightarrow X^2\Sigma^+$ vertical ionization energy was 3.69 eV. The diffuse $2p\pi$ function on nitrogen was varied slightly in an attempt to obtain a reasonable $^1\Sigma \rightarrow {}^2\pi$ ionization potential. However, the $^2\pi$ state of CN is not expected to be well described in our basis since the optimization of the pi functions was not extensive. The final optimized CN^- and BO^- basis sets shown in Table 6 were used to compute ion-molecule energy differences at several internuclear separations; these differences were added to the total SCF energies at corresponding R values of the negative ions to generate SCF-level potential curves for BO and CN.

Examination of the ground-state potential curves for BO, BO^-, CN, and CN^- allows us to determine the adiabatic electron affinities of BO and CN, which we

Table 6 Basis sets (STOs) for CN^- and BO^-

	CN^-						BO^-					
	σ orbitals			π orbitals			σ orbitals			π orbitals		
center	nl	ζ	center	nl	ζ	center	nl	ζ	center	nl	ζ
C	1s	5.1231	C	2p	1.2566	B	1s	4.2493	B	2p	0.9500
C	1s	7.5223	C	2p	2.7304	B	1s	6.5666	B	2p	2.2173
C	2s	0.9750	N	2p	1.3380	B	2s	0.8250	O	2p	1.5200
C	2s	1.9400	N	2p	3.2493	B	2s	1.6500	O	2p	3.6944
C	2p	1.2566				B	2p	0.8500			
C	2p	2.8700				B	2p	2.2173			
N	1s	5.9864				O	1s	6.8377			
N	1s	8.4960				O	1s	9.4663			
N	2s	2.3500				O	2s	2.8200			
N	2s	1.3750				O	2s	1.6754			
N	2p	1.4992				O	2p	1.6586			
N	2p	3.2493				O	2p	3.6944			

can compare with existing experimental data. Our predicted electron affinity for BO, 2.79 ± 0.2 eV, is within the range of experimental estimates (59) that vary from 2.4 eV to 3.1 ± 0.1 eV. Experimental studies of CN have yielded more precise results. Chupka et al (60) have reported an electron affinity of 3.82 ± 0.02 eV that was obtained from photodissociation measurements of HCN. Our calculated electron affinity, 3.70 ± 0.2 eV, is in good agreement with this value.

The results of our EOM calculations and Koopmans' theorem estimates for vertical ionization potentials of CN^- and BO^- are shown in Table 7 for selected internuclear separations. The Koopmans' theorem values deviate considerably from EOM results for the $^1\Sigma^+ \to {}^2\Sigma^+$ ionization of CN^-, less so for the $^1\Sigma^+ \to {}^2\Sigma^+$ BO^- ionization, and are actually very close to EOM results for the $^1\Sigma^+ \to {}^2\pi$ ionization of CN^-.

The large difference between Koopmans' theorem and the EOM result for the $^1\Sigma^+ \to {}^2\Sigma^+$ ionization of CN^- indicates that orbital relaxation and electron correlation effects are important in the process of removing an electron from the 5σ orbital of CN^-. Analogously, relaxation and correlation effects appear to be less important in the ionization of BO^-.

We have used the EOM method to study the stability of BO^- and CN^- and to investigate the nature of the highest occupied molecular orbital in each of these species. Our calculations show a $^1\Sigma^+ \to {}^2\Sigma^+$ ionization energy of 2.88 eV for BO^- at 3.35 bohr, the equilibrium internuclear separation (R_e) of the ion, and an ionization potential of 2.81 eV at the R_e of BO, 2.278 bohr. The resulting adiabatic electron affinity of BO, 2.79 eV, falls within the range of experimental values (59) obtained for this quantity. The EOM ionization potential ($^1\Sigma^+ \to {}^2\Sigma^+$) of CN^- was found to be 3.69 eV at $R = 2.25$ bohr, the equilibrium separation of both the ion and the molecule. This result is very close to the experimental electron affinity of CN determined from photodissociation experiments (60).

In each of these ions, the electron is ionized out of a nonbonding 5σ molecular orbital consisting mainly of diffuse $2s$ and $2p_\sigma$ functions on the less electronegative

Table 7 Selected CN^- and BO^- ionization potentials

| | CN$^-$ | | | | | BO$^-$ | |
| | $^1\Sigma^+ \to {}^2\Sigma^+$ | | $^1\Sigma^+ \to {}^2\Pi$ | | | $^1\Sigma^+ \to {}^2\Sigma^+$ | |
R (au)	IP(eV)	$-\varepsilon_{5\sigma}$(eV)	IP(eV)	$-\varepsilon_{1\pi}$(eV)	R (au)	IP(eV)	$-\varepsilon_{5\sigma}$(eV)
2.75	3.74	4.62	3.60	3.53	2.80	3.10	3.56
2.70	3.74	4.70	3.76	3.70	3.35[a]	2.88	3.23
2.35[c]	3.71	5.14	4.93	5.01	2.278[b]	2.81	3.14
2.30	3.70	5.18	5.11	5.21	2.20	2.72	3.04
2.25[a,b]	3.69	5.21	5.29	5.43			
2.00	3.65	5.23	6.28	6.60			

[a] R_e of the ion.
[b] R_e of $^2\Sigma^+$ state of the neutral.
[c] R_e of $^2\Pi$ state of the neutral.

atom in the ion. For CN⁻, there is also some contribution from the diffuse $2p_\sigma$ nitrogen function. The character of the 5σ orbital in the isoelectronic sequence, N_2, CN⁻, CO, BO⁻, changes in a regular fashion; the electron density in the 5σ orbital becomes more polarized toward the less electronegative atom as the electronegativity difference between the constituent atoms increases.

The $^2\Sigma^+$ and $^2\pi$ states of CN were found to cross at about 2.75 bohr, whereas the Koopmans' theorem ionization energies predict the crossing at 2.35 bohr. The situation in the region from 2.35 to 2.75 bohr is analogous to the observed energy-ordering of N_2 orbitals and N_2^+ states, and is interpreted in terms of the larger correlation-energy correction to Koopmans' theorem for ionization from the 5σ orbital of CN⁻ rather than from the 1π orbital.

Li_2

The EOM method discussed in Reference 12 requires as input the results of an SCF calculation on the closed-shell parent, Li_2. These SCF results were obtained using the Harris DIATOM program, which is run on the University of Utah Univac 1108 computer. The basis set was formed by starting with a basis for the lithium atom reported by Clementi (61). To this set of Slater-type functions we added one full set of diffuse p functions and another set of p_π-type functions. The basis was then optimized to give the maximum value for the electron affinity of Li_2. Based upon our experience with other molecules, we find that this procedure of optimization leads to the best balanced description of the parent, Li_2, and the anion, Li_2^-.

During the optimization the need for a more diffuse $2s$ function became apparent. The first three $2p$ functions on each lithium atom were optimized together as a set. After attempting to optimize the p_σ and s orbitals to yield the maximum value for the electron affinity, we added another set of diffuse p_π functions whose exponents were then optimized. All optimization was performed at 5.3 bohrs, which is near our computed equilibrium bond length of Li_2. The resulting basis set is shown in Table 8. The results of this optimization gave a vertical electron affinity of 0.80 eV for Li_2 at $R = 5.3$ bohr (94).

The spectroscopic parameters of Li_2 and Li_2^- were determined by least squares fitting a parabola to the bottom of the respective potential curves. The results of this calculation are given in Table 9.

The dissociation energy of Li_2^- was calculated from Equation 28:

$$D_0(Li_2^-) = D_0(Li_2) + E.A.(Li_2, thermo) - E.A._{Li}, \qquad 28.$$

where $D_0(Li_2)$ is the chemical dissociation energy of Li_2; $D_0(Li_2^-)$ is the chemical dissociation energy of Li_2^-; E.A.(Li_2, thermo) is the thermodynamic electron affinity; and E.A.$_{Li}$ is the electron affinity of the lithium atom. The dissociation energy of Li_2 was obtained from the experimental determination of Velasco et al (65). The electron affinity for atomic Li has also been determined experimentally (66). The thermodynamic electron affinity appearing in Equation 28 was obtained from the calculated Li_2 and Li_2^- potential curves by making harmonic zero-point energy corrections to the difference in the minimum electronic energies. This procedure results in a

Table 8 The optimized 20 STO basis set for Li_2 and the coefficients for the occupied orbitals at 5.3 bohrs[a]; $E = -14.86325$[b]

Orbital	ζ	$C_{1\sigma}$	$C_{2\sigma}$	$C_{3\sigma}$
1sLi	2.4739	0.63248	0.63664	−0.07663
1sLi'	2.4739	0.63687	−0.63224	−0.07663
1s'Li	4.6925	0.07906	0.07974	−0.02727
1s'Li'	4.6925	0.07968	−0.07919	−0.02727
2sLi	0.3523	0.00004	−0.00752	0.19206
2sLi'	0.3523	−0.00001	0.00752	0.19208
2s'Li	1.0287	−0.00016	−0.00232	0.56164
2s'Li'	1.0287	−0.00017	0.00232	0.56164
2s"Li	1.6350	0.00501	0.00573	−0.26926
2s"Li'	1.6350	0.00505	−0.00570	−0.26926
$2p_\sigma$Li	0.4066	−0.00054	−0.00352	0.18068
$2p_\sigma$Li'	0.4066	0.00057	−0.00352	−0.18068
$2p_\pi$Li	0.4066	0	0	0
$2p_\pi$Li'	0.4066	0	0	0
$2p_\pi$Li	0.6449	0	0	0
$2p_\pi$Li'	0.6449	0	0	0

[a] The asymmetry in the coefficients is due to convergence criteria problems.
[b] This SCF energy is not as good as that reported by Das (62).

Table 9 Molecular properties for Li_2 and Li_2^-

Parameter	Unit	Li_2(calc)	Li_2(exp)	Li_2^-(calc)
R_e	au	5.29	5.051[a]	6.3
K_e	dyne cm^{-1} × 10^{-4}	2.112	—	0.7849
v_e	cm^{-1}	319.7	351.43[a]	194.9
D_e	eV	—	1.026 ± 0.006[b]	1.31
E.A.(vert)	eV	0.8	—	—
IP	eV	—	5.15 eV[c]	1.06
E.A.(thermo)	eV	0.9	4.94 eV[d]	—

[a] Herzberg (63).
[b] Velasco et al (65).
[c] P. J. Foster, R. E. Leckenby, E. J. Robbins. 1969. *J. Phys. B* 2:478. (Isotopic Li_2).
[d] A. M. Emel'yanov, V. A. Peredvigina, L. N. Gorokhov. 1971. *High Temp.* 9:164. (Isotopic Li_2).

predicted thermodynamic electron affinity of 0.9 eV and a calculated $D_0(Li_2^-)$ of 1.31 eV.

The ground state of Li_2^- has been reported by Linnett et al (67) to be $^2\Pi_u$. In the Linnett calculations, the $^2\Pi_u$ state is more stable than the $^2\Sigma_u^+$ state by 0.016 hartree (0.44 eV). We find in this work that the $^2\Sigma_u^+$ state is more stable than the $^2\Pi_u$ by 0.6 eV at 5.3 bohrs, based upon a difference of optimized vertical electron affinities for the two states.

By using the potential curves discussed above, together with the calculated vibrational frequency of Li_2^- and the measured frequency of Li_2, we determined the thermodynamic electron affinity of Li_2 to be 0.9 eV. This is substantially larger than the 0.27 eV reported by Linnett (67) for the $^2\Pi_u$ state of Li_2^-. The calculation of the vertical photodetachment energy of Li_2^- can also be achieved from the above-mentioned potential energy curves. For this quantity, we obtain a value of 1.06 eV.

It is very interesting to make use of the results of the present calculations to compare the bonding and spectroscopic parameters of Li_2^+ and Li_2^-, both of which have a bond order of one half. Calculations on Li_2^+ give a fundamental frequency (68) of 254.7 cm^{-1}, which is not very different from the fundamental frequency of Li_2^- determined in this work (195.9 cm^{-1}). Several values of the dissociation energy of Li_2^+ have been tabulated by Wahl et al (68). One of the more recent values (1.31 eV) (69) is identical to our computed dissociation energy of Li_2^-.

By comparing these spectroscopic parameters for Li_2^+ and Li_2^- to those of the neutral molecule, we see that Li_2 has a stronger bond (larger ω_e, shorter R_e) than the ions; however, both of the ions are more stable (larger D_0) than the neutral Li_2. The positive and negative ions have very similar bond lengths (both ~ 6 bohrs) that are 1 bohr longer than that of the neutral. This is in line with the difference in bond order between the neutral Li_2 and its ions. The fact that the ions have a larger dissociation energy may be related to the long-range ion-atom interaction. A set of minimal-basis valence bond calculations (64) on Li_2, Li_2^+, and Li_2^- provides data that tend to support our findings. In Reference 64 the resulting dissociation energies of Li_2, Li_2^+, and Li_2^- are 0.76 eV, 1.06 eV, and 0.92 eV, respectively.

Previous calculations (2, 20–25) have shown how well the EOM method has succeeded in obtaining electron affinities of molecules in cases where experimentally determined values were known. This indicates that by applying the EOM method valuable predictions can be made. From this work we obtain a prediction that the $^2\Sigma_u^+$ ground state of Li_2^- is stable with respect to dissociation and electron loss. We find that the ground state of Li_2^- should be as stable to dissociation as Li_2^+. The overlap of the $v = 0$ Li_2 and Li_2^- vibrational wave functions is so small that we conclude that the vertical electron affinity and photodetachment measurements should give significantly different values for this system. Furthermore, we have found the Li_2–Li_2^- system to be very interesting because the large change in internuclear separation in going from Li_2 to Li_2^- is accompanied by a decrease in vibrational frequency and an increase in dissociation energy. Such a seemingly anomalous situation also occurs in going from Li_2 to Li_2^+.

LiF^-, $LiCl^-$, LiH^-, NaH^-, BeO^-

In a communication (70) we made theoretical predictions of stable negative ions of LiH and NaH. Later, we described in a full paper (25) the research that led us to the prediction of the existence of these stable anions. We also presented results that showed that the BeO^- and LiF^- anions are stable with respect to dissociation and autodetachment. We reexamined the HF^- anion, and concluded that it is unstable at the equilibrium configuration of HF. In an extension of this work, Jordan (90) has investigated the possible existence of H_2O^-, HCN^-, and $(HF)_2^-$, all of which he finds to be unstable with respect to electron detachment near the equilibrium geometry of the parent.

It has been demonstrated by several researchers (71–76) that an electron in the field of a fixed, finite dipole greater than 1.625 Debye possesses an infinity of bound states. Although LiH, NaH, BeO, and LiF all have dipole moments in excess of this value, stable negative ions of these species have not been detected experimentally of predicted from ab initio calculations prior to our investigations.

A stable negative ion of LiCl has been detected by Carlsten, Peterson & Lineberger (27), and the Hartree-Fock predictions on LiCl$^-$ of Jordan & Luken (77, 80) are in good agreement with the experimental results. In view of the nonbonding nature of the orbital occupied by the extra electron demonstrated by this theoretical study, inclusion of electron correlation would be expected to result in only a small correction to the description of the binding of an electron to LiCl. That this expectation is confirmed by the good agreement with the Hartree-Fock method in describing the binding of an electron to LiCl suggests the possibility of similar results with other highly polar molecules.

The experimental and theoretical studies on LiCl$^-$ indicated that the simple, fixed, finite dipole model provides an inadequate description of the binding of an electron to LiCl. One of the motivations of our work is to provide a more extensive evaluation of the validity of the dipole model.

The most essential aspect of our investigation of the binding of electrons to polar molecules is the choice of basis sets in which diffuse functions are added to the electropositive atom to permit the "extra" electron to "attach" itself to the positive end of the polar parent molecule. Most of the calculations in the present paper employ Slater-type orbitals (STOs); the unnormalized STOs are functions of the type $r^{n-1}e^{-\zeta r}$, where n is the principal quantum number and ζ the orbital exponent. In Table 10 we list the STO basis sets employed in our calculations.

In Table 11 the KT (Koopmans' theorem) and EOM electron affinities are presented for the equilibrium bond length of the parent neutral. Perhaps the most striking observation to be drawn from these data is that orbital relaxation and correlation corrections to the EAs are small, in marked contrast to the situation encountered with covalent molecules such as OH, O_2, and NO, for which orbital relaxation and correlation corrections to the electron affinities are typically an eV or more. The small correlation and relaxation energy changes that occur with the formation of the ions from the neutrals are in accord with a description in which the "extra" electron resides in a region of space that is essentially unoccupied by other electrons. Apparently, the orbital picture provides a good first approximation to the binding of electrons to highly polar molecules.

The second unoccupied σ orbital of LiCl was found to have a negative energy (~ -0.007 eV) when very diffuse $3s$ and $3p_\sigma$ functions were placed on the lithium. A similar result is observed for LiF, and probably would be observed for NaH, BeO, and LiH if sufficiently diffuse functions of the appropriate symmetry were included in the basis sets. However one should not attach much physical significance to such weakly bound states, since they would probably become unbound if corrections to the Born-Oppenheimer approximation were included. It is possible that

Table 10 STO basis set for LiH, LiF, NaH, HF, and BeO[a]

LiH		LiF		NaH		BeO	
$1s_\sigma$Li	(4.6990)	$1s_\sigma$Li	(4.6925)	$1s_\sigma$Na	(11.1543)	$1s_\sigma$Be	(6.4072)
$1s'_\sigma$Li	(2.5212)	$1s'_\sigma$Li	(2.4739)	$2s_\sigma$Na	(2.0006)	$1s'_\sigma$Be	(3.5297)
*$2s_\sigma$Li	(1.2000)	$2s_\sigma$Li	(1.6350)	$2p_\sigma$Na	(4.1786)	$2s_\sigma$Be	(1.1956)
*$2s'_\sigma$Li	(0.7972)	*$2s'_\sigma$Li	(1.0287)	$2p'_\sigma$Na	(2.2798)	*$2s'_\sigma$Be	(0.8557)
*$2s''_\sigma$Li	(0.6000)	*$2s''_\sigma$Li	(0.5352)	$3s_\sigma$Na	(6.2601)	*$2s''_\sigma$Be	(0.4677)
*$2s'''_\sigma$Li	(0.3000)	*$2s'''_\sigma$Li	(0.3223)	$3s'_\sigma$Na	(0.9106)	$2p_\sigma$Be	(2.0717)
$2p_\sigma$Li	(2.7500)	*$2p_\sigma$Li	(0.4066)	$3s''_\sigma$Na	(0.4000)	*$2p'_\sigma$Be	(0.8557)
$2p'_\sigma$Li	(1.2000)	$2p_\pi$Li	(0.4066)	*$3s'''_\sigma$Na	(0.2000)	*$2p''_\sigma$Be	(0.4677)
$2p''_\sigma$Li	(0.7369)	$1s_\sigma$F	(14.1095)	$3p_\sigma$Na	(1.2631)	$2p_\pi$Be	(1.0846)
$2p'''_\sigma$Li	(0.6000)	$1s'_\sigma$F	(7.9437)	$3p'_\sigma$Na	(0.7108)	$1s_\sigma$O	(7.6092)
*$2p''''_\sigma$Li	(0.3000)	$2s_\sigma$F	(3.2563)	*$3p''_\sigma$Na	(0.4000)	$2s_\sigma$O	(3.1394)
$2p_\pi$Li	(0.7369)	$2s'_\sigma$F	(1.9346)	$3p'''_\sigma$Na	(0.1500)	$2s'_\sigma$O	(0.8792)
$2p'_\pi$Li	(0.3500)	$2p_\sigma$F	(4.2784)	$2p_\pi$Na	(4.1742)	$2p_\sigma$O	(3.4198)
$2s_\sigma$H	(1.5657)	$2p'_\sigma$F	(2.3732)	$2p'_\pi$Na	(2.2828)	$2p'_\sigma$O	(1.7405)
$1s'_\sigma$H	(0.8877)	$2p''_\sigma$F	(1.4070)	$3p_\pi$Na	(0.9636)	$2p''_\sigma$O	(1.0626)
*$2s_\sigma$H	(0.4000)	$2p_\pi$F	(2.3291)	$1s_\sigma$H	(0.7808)	$2p_\pi$O	(3.4198)
$2p_\sigma$H	(1.3765)	$2p'_\pi$F	(1.3584)	*$2s_\sigma$H	(0.4000)	$2p'_\pi$O	(1.7405)

[a] LiH: $R_e = 3.015$ au; $E = -7.9866$ au; LiF: $R_e = 2.955$ au; $E = -106.584$ au; NaH: $R_e = 3.566$ au; $E = -161.9422$ au; BeO: $R_e = 2.51$ au; $E = -89.3432$ au.

Table 11 Equations-of-motion electron affinities and Koopmans' theorem estimates (eV)

	LiH(5.88)[a]	LiF(6.33)	LiCl(7.13)	NaH(6.98)	BeO(7.41)
EOM[b]	0.2986	0.4645	[0.54/0.61]	0.3618	1.7692
$-\varepsilon$LUMO[b]	0.1997	0.4252	0.48	0.2897	1.4144
Orbital relaxation and correlation corrections[c]	0.0989	0.0393	0.13	0.0721	0.3548

[a] The quantities in parenthesis are the dipole moments in Debye. The dipole moment of BeO has not been experimentally determined. The value listed is that calculated by Yoshimine (78). The experimental dipole moments for the other molecules are listed. These values are taken from the tables of R. D. Nelson, D. R. Lide, A. A. Maryott, *Selected Values of Electric Dipole Moments for Molecules in the Gas Phase*, U.S. Dept. Commerce, Natl. Bur. Stand.

[b] The LiCl LUMO orbital energy is from Jordan, Herzenberg & Luken (77). An EOM calculation has not been performed on LiCl. The value of 0.54 eV was obtained from a ΔE_{HF} [$\Delta E_{HF} = E_{HF}(LiCl) - E_{HF}(LiCl^-)$] calculation, while the value of 0.61 eV is the experimental value of Carlsten, Peterson & Lineberger (27).

[c] For LiH, LiF, NaH, and BeO the orbital relaxation and correlation corrections are obtained by subtracting the Koopmans' theorem estimate of the electron affinity from the EOM value. For LiCl the experimental electron affinity is used in place of the EOM value.

molecules with substantially larger dipole moments such as CsCl ($\mu = 10.4$ Debye), could have two stable negative ion states near the equilibrium separation of the parent molecule. Certainly if the dipole moment becomes large enough, the first excited anion state will remain stable even when corrections to the Born-Oppenheimer approximation are included.

Figures 5–8 display the ground state potential energy curves of LiH, LiF, NaH, BeO, and their anions. The potential energy curves of the neutral parent molecules

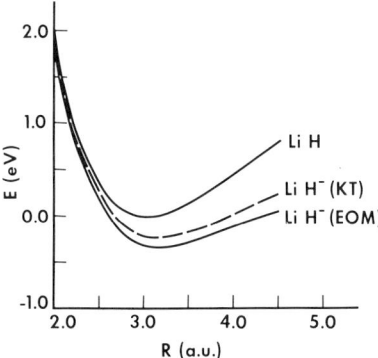

Figure 5 The potential energy curves of LiH and LiH$^-$ (obtained from Koopmans' theorem and EOM) as functions of R.

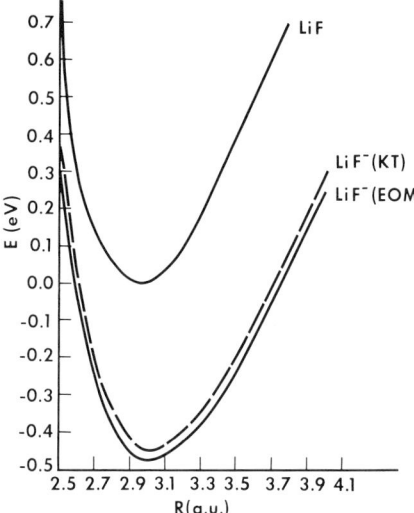

Figure 6 The potential energy curves of LiF and LiF$^-$ (obtained from Koopmans' theorem and EOM) as functions of R.

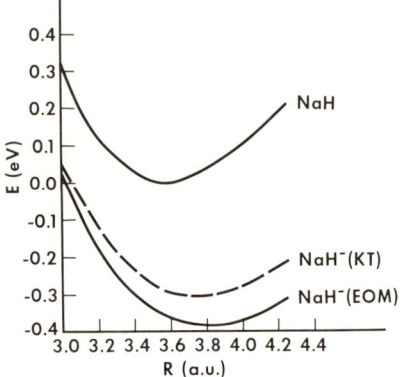

Figure 7 The potential energy curves of NaH and NaH⁻ (obtained from Koopmans' theorem and EOM) as functions of R.

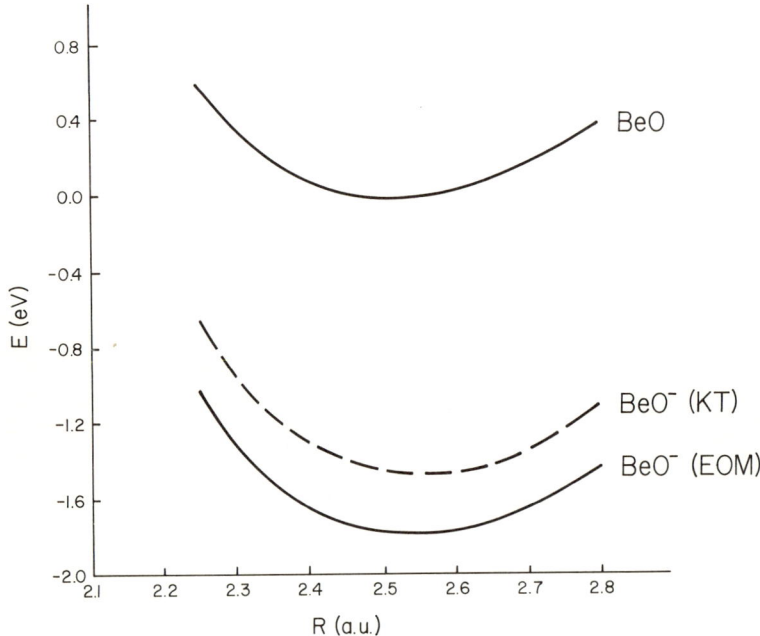

Figure 8 The potential energy curves of BeO and BeO⁻ (obtained from Koopmans' theorem and EOM) as functions of R.

are constructed from experimental data via the Padé approximant procedure (81):

$$U_{PA}(X) = a_0 X^2 / [1 - a_1 X + (a_1^2 - a_2) X^2], \qquad 29.$$

where $X = (R - R_e)/R_e$ and a_0, a_1, a_2 are determined from the spectroscopic constants ω_e, $\omega_e \chi_e$, B_e, and α_e as follows:

$$a_0 = \omega_e^2/4B_e, \qquad 30.$$
$$a_1 = -(\omega_e \alpha_e/6B_e^2) - 1, \qquad 31.$$
$$a_2 = \tfrac{5}{4} a_1^2 - (\tfrac{2}{3})(\omega_e \chi_e / B_e). \qquad 32.$$

The Padé approximant given by Equation 29 provides a particularly good representation of the potential curves of ionic molecules near their equilibrium configurations. For example, for the range of R considered in Figure 5, the potential curve for LiH given by Equation 29 is essentially indistinguishable from the RKR curve and is very close to the potential curve of Docken & Hinze (82) obtained from a configuration interaction calculation. In Table 12 we list the spectroscopic constants ω_e, $\omega_e \chi_e$, B_e, α_e, and R_e for the molecules being considered.

It is appropriate at this point to compare our calculated electron affinities with those of the fixed finite dipole model. In Figure 9 we have plotted the binding energy of the first three states of an electron in the field of a fixed dipole moment as a function of the dipole moment. Here we have specifically considered the case of a dipole arising from two charges, $+q$ and $-q$, where $q = 1$. For LiH, LiF, LiCl,

Table 12 Experimental spectroscopic constants for the neutral ions, and calculated spectroscopic constants for the anions[a]

Parameter	Units	LiH	LiF	LiCl	NaH	BeO
R_e	au	3.015	2.956	3.814	3.566	2.516
ω_e	cm^{-1}	1405.6	910.3	662	1172.2	1487.3
$\omega_e \chi_e$	cm^{-1}	23.20	7.929	4.501	19.72	11.83
B_e	cm^{-1}	7.5131	1.3454	0.625	4.9001	1.6510
α_e	cm^{-1}	0.2132	0.0201	0.0079	0.1353	0.0190

Parameter	Units	LiH$^-$	LiF$^-$	LiCl$^-$	NaH$^-$	BeO$^-$
R_e	au	3.20	3.00	4.01	3.80	2.54
ΔR_e	Å	0.08	0.02	0.10	0.12	0.01
ω_e[b]	cm^{-1}	1250	840	525	960	1360
% change in ω_e	%	11	8	21	18	8
D_e	eV	1.94	3.02	1.84	1.65	4.86

[a] The values of $\omega_e \chi_e$ for LiCl, and the values of α_e for LiF and LiCl are from P. Brumer and M. Karplus. 1973. J. Chem. Phys. 58:3903. The rest of the data for the neutral species are from Reference 23.

[b] The ω_e are obtained by fitting points on the negative ion curve obtained by subtracting the EOM electron affinity from the potential curve of the neutral parent molecule. The uncertainty in ω_e due to our fitting procedure, is probably $\sim \pm 50$ cm^{-1}.

and NaH q is close to one, so the choice of $q = 1$ introduces only a small error. As was pointed out in Reference 77, one should not attempt to correlate the ground state negative ions of these molecules with the (0, 0, 0) ground state of the dipole model, since this state of the dipole model does not have the correct nodal behavior. There is, however, a qualitative correlation between the ground state negative ions where lithium is the electropositive species and the first excited (1, 0, 0) state of the dipole model. Similarly, for species where sodium is the electropositive species there is a qualitative correlation of the negative ions with the second excited state (2, 0, 0) of the dipole model. This correlation is apparent from Figure 9. These correlations would become much more quantitative if core penetration effects were incorporated in the dipole model. For example, near R_e LiCl is well described as Li^+Cl^-, where the Li^+ has a $(1s)^2$ core. In the negative ion the nonbonding LUMO has a node ~ 0.8 au behind the lithium, while the dipole model, for a dipole moment of 7.2 Debye, locates the node 2.7 au behind the positive center. The ab initio calculations allow lithium $2s$ and $2p$ electrons to penetrate the $(1s)^2$ core and "feel" the $+3$ nuclear charge; this important effect is not accounted for in the dipole model. This is a major source of the discrepancy between the dipole model and the more realistic ab initio calculations.

We know that the ground-state negative ion of BeO should also correlate with the (1, 0, 0) state of the dipole model. The electron affinity of BeO has not been included in Figure 9 since BeO is not a monovalent species. To compare the electron

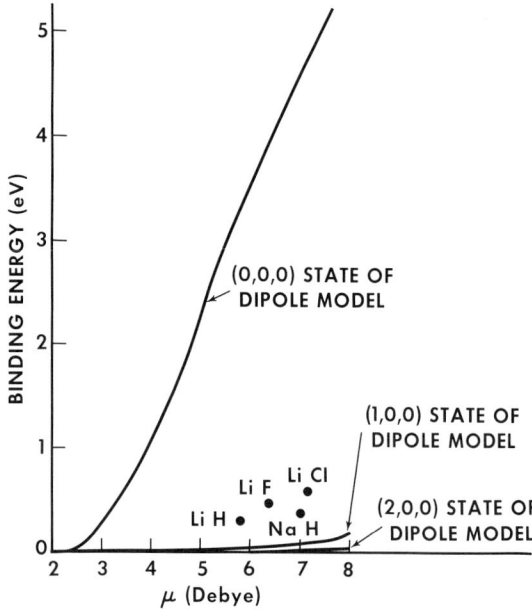

Figure 9 Correlation between dipole moment and electron affinity, and comparison with dipole model.

affinity of BeO with the dipole model predictions, we would have to repeat the calculations for $q \simeq 1.5$.

LiH, LiF, LiCl, and NaH all have dipole moments between 5.8 and 7.2 Debye, and all have computed (vertical) electron affinities between 0.3 and 0.6 eV. It would be of interest to determine the electron affinities of molecules with dipole moments in the 3–6 Debye range. For dipole moments $\lesssim 6$ Debye the electron affinities should be $\lesssim 0.3$ eV. These cases will pose special difficulties. The considerations of this work have assumed the validity of the Born-Oppenheimer approximation. As stressed by Garrett (75), one should allow for the fact that the dipole is nonstationary. Anions that are weakly bound under the assumption of the validity of the Born-Oppenheimer approximation may turn out to be unbound when corrections to the Born-Oppenheimer approximation are included.

In this section we have presented calculations on the ground state anions of LiH, LiF, NaH, and BeO. In each case, the extra electron is found to occupy a predominantly nonbonding orbital. This results in small orbital relaxation and correlation corrections, as well as only a small increase of the internuclear separation and a small (10–20%) decrease in ω_e upon formation of the ion. The electron affinities as predicted by the dipole model are found to be in poor agreement with those of our ab initio calculations. We attribute this disagreement to the neglect of core penetration in the dipole model.

There are several possible extensions and applications of the work presented in this paper. We intend to report on negative ions of various polar polyatomic molecules in future publications. We are also investigating whether the stable negative ions of ionic molecules may influence the mechanisms of certain chemical reactions. In collisions between an ionic molecule and an atom or molecule with a low ionization potential, an electron-jump mechanism may play a role. For example, an excited cesium atom could lose an electron to LiCl. Furthermore, this would tend to favor a "backside" attack, proceeding through a $Cs^+(LiCl)^-$ intermediate as opposed to LiClCs. The possibility of formation of Li from the reaction of Li^- with X_2 is particularly interesting. Although alkali-halogen reactions have long been a favorite in studies of collision dynamics, to date the experimental studies have dealt with neutral or positively charged species.

It is hoped that the calculations and suggested new reaction pathways presented in this paper will stimulate experimental investigations of the anions of highly polar molecules.

Other Recent Results

In addition to the investigations that are described in some detail in the preceding parts of this section, studies of numerous other anions have been carried out by ourselves and others since 1968. We have completed preliminary studies of MgH^-, SH^-, NH_2^-, and NO_2^- that tie in with our work on BeH^-, OH^-, and with the works of Schaefer (42), Lineberger (89), and Brauman (88) on NO_2^-, and of Brauman (87) and Hall (87) on NH_2^-. For these systems, our computed vertical detachment energies, together with the experimental and theoretical results of others are shown in Table 13.

Table 13 Computed and experimental vertical detachment energies (eV)[a]

Anion	Our D E	Experimental D E	Other theoretical D E
MgH^-	0.85	0.8 (57)	—
SH^-	2.20	2.32 (83)	2.25 (11)
NH_2^-	0.42	0.74 (87)	0.30 (41)
NO_2^-	2.68 (79)	2.8 (88)	3.45 (33)
	2.42 (adiabatic)	2.36 (89) (adiabatic)	

[a] Numbers in parentheses designate reference sources.

Quite recently, Jordan (90) has undertaken the study of a number of anions whose neutral parents have very large dipole moments. This is an extension of the work of Jordan and ourselves described in the preceding part of this section. Jordan has predicted the stability and spectroscopic characteristics of the following anions: $LiCN^-$, $LiNC^-$, Li_2O^-, $(LiH)_2^-$ (92), $LiCH_3^-$, $LiOH^-$, and LiN^-. He has also carried out some very nice work, in collaboration with Burrow (91) on metastable organic molecular anions (ethylene, butadiene, substituted benzenes). Jordan & Simons (93) have found the lowest $^2\Sigma_g$ state of Be_2^- to be stable with respect to dissociation and electron detachment. The potential curve of this state has a minimum at $R = 4.52$ au and crosses the repulsive Be_2 ground-state curve near 5.9 au; for longer R values, Be_2^- is unstable with respect to electron loss. The vertical electron detachment energy of Be_2^- was predicted to be 0.38 eV.

CONCLUSION

Since 1968 a great deal of progress has been made toward understanding the stability and bonding characteristics of small anions in the gas phase. Quantum chemical methods that adequately treat electron correlation and orbital relaxation effects have been shown to be useful tools for investigating anions. A third-order treatment of correlation seems to be accurate to ± 0.2 eV; second-order calculations are not useful because they have an accuracy of ± 0.6 eV. Koopmans' theorem or ΔSCF-level electron affinities are generally not of sufficiently high accuracy to be useful, although the shapes of SCF-level potential energy surfaces may not be any worse for anions than for neutral molecules. A good choice of basis set, with sufficient diffuse functions, is an essential ingredient of any reasonable calculation on a negative ion. With a flexible basis and a method for treating correlation and relaxation effects one can, as has been demonstrated in the work reviewed here, reliably carry out good quantum mechanical studies of stable negative molecular ions.

In the opinion of this reviewer, the challenges of the immediate future are as follows:

1. the accurate, physically clear, and computationally tractable treatment of metastable negative molecular ions;

2. extension of the work described in this review to anions involving more complex functional groups and the study of substituent effects;
3. development of accurate models that include the effects of solvents so that hydrated gas-phase ions and anions in solution can be quantitatively investigated.

ACKNOWLEDGMENTS

The author acknowledges the financial support of the U.S. Army Research Office (Grant #DAAG-29-74-G-022) and of the National Science Foundation (Grant #CHE 75-19376).

He also thanks Professor Ken Jordan, Dr. George Purvis, and Professor Yngve Öhrn for many helpful discussions.

Literature Cited

1. Berry, R. S. 1969. *Chem. Rev.* 69:533–42
2. Smith, W. D., Chen, T. T., Simons, J. 1974. *Chem. Phys. Lett.* 27:499–502
3. Branscomb, L. M. 1966. *Phys. Rev.* 148:11–18
4. Hotop, H., Patterson, T. A., Lineberger, W. C. 1974. *J. Chem. Phys.* 60:1806–12
5. Pekeris, C. L. 1958. *Phys. Rev.* 112:1649–58
6. Weiss, A. W. 1968. *Phys. Rev.* 166:70–74
7. Clementi, E., McLean, A. D. 1964. *Phys. Rev. A* 133:419–23; Clementi, E., McLean, A. D., Raimondi, D. L., Yoshimine, M. 1964. *Phys. Rev. A* 133:1274–79; Clementi, E. 1964. *Phys. Rev. A* 135:980–84
8. Sinanoglu, O., Oksuz, I. 1968. *Phys. Rev. Lett.* 21:507–11
9. Taylor, H. S., Harris, F. E. 1963. *J. Chem. Phys.* 39:1012–16
10. Wahl, A. C., Gilbert, T. L. 1965. *Bull. Am. Phys. Soc.* 10:1097; Zemke, W. T., Das, G., Wahl, A. C. 1972. *Chem. Phys. Lett.* 14:310–14
11. Cade, P. E. 1967. *J. Chem. Phys.* 47:2390–2406; 1967. *Proc. R. Soc. London Ser. A* 91:842
12. Simons, J., Smith, W. D. 1973. *J. Chem. Phys.* 58:4899–4907
13. Simons, J. 1974. *Chem. Phys. Lett.* 25:122–28
14. Jordan, K. D., Chen, T. T., Simons, J. 1976. *Chem. Phys.* 14:145–47
15. Jørgensen, P., Simons, J. 1975. *J. Chem. Phys.* 63:5302–4
16. Cederbaum, L. S. 1973. *Theor. Chim. Acta* 31:239–60; Cederbaum, L. S., Holneicher, G., Peyerimhoff, S. 1971. *Chem. Phys. Lett.* 11:421–24; Cederbaum, L. S., Holneicher, G., von Niessin, W. 1973. *Chem. Phys. Lett.* 18:503–8
17. Pickup, B. T., Goscinski, O. 1973. *Mol. Phys.* 26:1013–35
18. Purvis, G., Öhrn, Y. 1974. *J. Chem. Phys.* 60:4063–69; Purvis, G., Öhrn, Y. 1975. *J. Chem. Phys.* 62:2045–49
19. Tsui, F., Freed, K. F. 1975. *Chem. Phys. Lett.* 32:345–50
20. Griffing, K., Simons, J. 1975. *J. Chem. Phys.* 62:535–40
21. Kenney, J., Simons, J. 1975. *J. Chem. Phys.* 62:592–99
22. Griffing, K., Simons, J. 1976. *J. Chem. Phys.* 64:3610–14
23. Andersen, E., Simons, J. 1976. *J. Chem. Phys.* 64:4548–50
24. Griffing, K., Kenney, J., Simons, J., Jordan, K. D. 1975. *J. Chem. Phys.* 63:4073–75
25. Jordan, K. D., Griffing, K. M., Kenney, J., Andersen, E. L., Simons, J. 1976. *J. Chem. Phys.* 64:4730–40
26. Andersen, E., Simons, J. Submitted to *J. Chem. Phys.*
27. Carlsten, J. L., Peterson, J. R., Lineberger, W. C. 1976. *Chem. Phys. Lett.* 37:5–8
28. Clementi, E., McLean, A. D. 1963. *J. Chem. Phys.* 39:323–25
29. Boyd, D. B., Lipscomb, W. N. 1967. *J. Chem. Phys.* 46:910–19
30. Krauss, M. 1964. *J. Res. Natl. Bur. Stand.* 63A:635–39; Krauss, M., Neumann, D., Wahl, A. C., Das, G., Zemke, W. 1973. *Phys. Rev.* A7:69–77
31. Heaton, M. M., Pipano, A., Kaufman, J. J. 1972. *Int. J. Quantum Chem. Symp.* 6:181–86
32. Kari, R. E., Csizmadia, I. G. 1972. *J. Chem. Phys.* 56:4337–44; Robb, M. A., Csizmadia, I. G. 1971. *Int. J. Quantum Chem.* 5:605–35

33. Pfeiffer, G. V. 1967. Unpublished data, Princeton University
34. Popkie, H. E., Henneker, W. H. 1971. *J. Chem. Phys.* 55:617–28
35. McLean, A. D., Yoshimine, M. 1967. *J. Chem. Phys.* 46:3682–83; McLean, A. D., Yoshimine, M., 1967. *Int. J. Quantum Chem.* 15:313–19
36. Gilbert, T. L., Wahl, A. C. 1965. *Bull. Am. Phys. Soc.* 10:1097
37. Fink, W. H., Allen, L. C. Unpublished data
38. Geller, M., Sachs, L., Kaufman, J. J. Unpublished data, RIAS and Jet Propulsion Lab.
39. Thulstrup, P. W., Thulstrup, E. W., Andersen, A., Öhrn, Y., 1974. *J. Chem. Phys.* 60:3975–80
40. Peyerimhoff, S. D., Buenker, R. J., Allen, L. C. 1966. *J. Chem. Phys.* 45:734–49
41. Heaton, M., Cowdry, R. 1975. *J. Chem. Phys.* 62:3002–9
42. Pearson, P. K., Schaefer, H. F. III, Richardson, J. H., Stephenson, L. M., Brauman, J. I. 1974. *J. Am. Chem. Soc.* 96:6778–79
43. Linderberg, J., Öhrn, Y. 1967. *Chem. Phys. Lett.* 1:295–96
44. Linderberg, J., Öhrn, Y. 1972. *Propagators in Quantum Chemistry.* London: Academic.
45. Doll, J. D., Reinhardt, W. P. 1972. *J. Chem. Phys.* 57:1169–84
46. Schneider, B., Taylor, H. S., Yaris, R. 1970. *Phys. Rev. A* 1:855–67; Yarlagadda, B. S., Csanak, G., Taylor, H. S., Schneider, B., Yaris, R. 1973. *Phys. Rev. A* 7:146–54
47. Goscinski, O., Lukman, B. 1970. *Chem. Phys. Lett.* 7:573–76
48. Rowe, D. J. 1968. *Rev. Mod. Phys.* 40:153–66; Rowe, D. J. 1968. *Phys. Rev.* 175:1283–92
49. Purvis, G., Öhrn, Y. 1975. *Chem. Phys. Lett.* 33:396–98
50. Redmon, L. T., Purvis, G., Öhrn, Y. 1975. *J. Chem. Phys.* 63:5011–17
51. Brandow, B. H. 1967. *Rev. Mod. Phys.* 39:771–828
52. Jørgensen, P., Purvis, G. *J. Chem. Phys.* 1977. In press
53. Cade, P. E. 1967. *J. Chem. Phys.* 47:2390–2406
54. Cade, P. E., Huo, W. M. 1967. *J. Chem. Phys.* 47:614–48
55. Herzberg, G. 1961. *Spectra of Diatomic Molecules*, 2nd ed. Princeton: Van Nostrand. p. 508
56. Gaydon, A. G. 1968. *Dissociation Energies and Spectra of Diatomic Molecules*, 3rd ed. London: Chapman & Hall. p. 264
57. Feldmann, D. Private communication
58. Roetti, C., Clementi, E. 1974. *J. Chem. Phys.* 60:4725–29
59. Jensen, D. E., 1970. *J. Chem. Phys.* 52:3305–6; Srivastava, R. D., Uy, O. M., Farber, M., 1971. *Trans. Faraday Soc.* 67:2941–44
60. Berkowitz, J., Chupka, W. A., Walter, T. A. 1969. *J. Chem. Phys.* 50:1497–1500
61. Clementi, E. 1963. *J. Chem. Phys.* 38:996–1000
62. Das, G. 1967. *J. Chem. Phys.* 46:1568–79
63. Herzberg, G. 1961. *Spectra of Diatomic Molecules*, 2nd ed. Princeton: Van Nostrand. p. 546
64. Khrustov, V. F., Stepanov, N. F., Yarovoy, S. S., Abramenkov, A. V., Poshyunaite, N. P., Tsirul, Z. Y. 1971. *Vestn. Mosk. Univ. Khim.* 12:221–23
65. Velasco, R., Ottinger, C., Zare, R. N. 1969. *J. Chem. Phys.* 51:5522–32
66. Patterson, T. A., Hotop, H., Kasdan, A., Norcross, D. W., Lineberger, W. C. 1974. *Phys. Rev. Lett.* 32:189–92
67. Blustin, P. H., Linnett, J. W. 1974. *J. Chem. Soc. Faraday Trans. 2,* 70:826–36
68. Henderson, G. A., Zemke, W. T., Wahl, A.C. 1973. *J. Chem. Phys.* 58:2654–56
69. Bottcher, C., Dalgarno, A. 1975. *Chem. Phys. Lett.* 36:137–44
70. Griffing, K. M., Kenney, J., Simons, J., Jordan, K. D. 1975. *J. Chem. Phys.* 63:4073–75
71. Wallis, R. F., Herman, R., Milnes, H. W. 1960. *J. Mol. Spectrosc.* 4:51–74
72. Crawford, O. H. 1967. *Proc. Phys. Soc.* 91:279–84
73. Coulson, C. A., Walmsley, M. 1967. *Proc. Phys. Soc. London* 91:31–32
74. Crawford, O. H., Koch, B. J. D. 1974. *J. Chem. Phys.* 60:4512–19
75. Garrett, W. R. 1971. *Phys. Rev. A* 3:961–72
76. Turner, J. E., Anderson, V. E., Fox, K. 1968. *Phys. Rev.* 174:81–89
77. Jordan, K. D., Luken, W. 1976. *J. Chem. Phys.* 64:2760–66
78. Yoshimine, M. 1964. *J. Chem. Phys.* 40:2970–76; Yoshimine, M. 1968. *J. Phys. Soc.* 25:1100–19
79. Andersen, E. A. Unpublished data
80. Hehre, W. J., Lathan, W. A., Ditchfield, R., Newton, M. D., Pople, J. A. 1976. *Quantum Chem. Program Exch.* program 236, Indiana Univ.
81. Jordan, K. D., Kinsey, J. L., Silbey, R. 1974. *J. Chem. Phys.* 61:911–17; Jordan, K. D. 1975. *J. Mol. Spectrosc.* 56:329–31
82. Docken, K. K., Hinze, J. 1972. *J. Chem. Phys.* 57:4928–36
83. Steiner, B. 1968. *J. Chem. Phys.* 49:5097–5104

84. Thulstrup, P. W., Thulstrup, E. W. 1974. *Chem. Phys. Lett.* 26:144–48
85. O'Hare, P. A. G. 1971. *J. Chem. Phys.* 54:4124–26; O'Hare, P. A. G., Wahl, A. C. 1970. *J. Chem. Phys.* 53:2834–46; ibid 1971. 54:4563–77
86. Cederbaum, L. 1973. *Theor. Chim. Acta* 31:239–60
87. Smyth, K. C., Brauman, J. I. 1972. *J. Chem. Phys.* 56:4620–25; Celotta, R. J., Bennett, R. A., Hall, J. L. 1974. *J. Chem. Phys.* 60:1740–45
88. Richardson, J. H., Stephenson, L. M., Brauman, J. I. 1974. *Chem. Phys. Lett.* 25:318–20
89. Herbst, E., Patterson, T. A. Lineberger, W. C. 1974. *J. Chem. Phys.* 61:1300–4
90. Jordan, K. D. 1976. *J. Chem. Phys.* 65:1214–15, and private communication
91. Burrow, P. D., Jordan, K. D. 1975. *Chem. Phys. Lett.* 36:594–98; Jordan, K. D., Burrow, P. D., Michejda, J. A. 1976. *J. Am. Chem. Soc.* 98:1295–96
92. Jordan, K. D. 1977. *Chem. Phys. Lett.* In press
93. Jordan, K. D., Simons, J. 1976. *J. Chem. Phys.* 65:1601
94. Dixon, D. A., Gole, J. L., Jordan, K. D. Unpublished

NMR RELAXATION IN THERMOTROPIC LIQUID CRYSTALS

✤ 2638

Charles G. Wade

Laboratory of Chemical Biodynamics, Lawrence Berkeley Laboratory, Berkeley, California 94720[1]

INTRODUCTION

Liquid crystals have symmetry and mechanical properties that are intermediate between those of crystals (order) and liquids (disorder) (1–6). All liquid crystal phases are characterized by short-range and extended long-range orientational order (4–6); this gives them qualities quite distinct from those of other fluids. In a liquid crystal phase (or mesophase), the molecules experience a preferred direction of alignment with respect to a director, $\mathbf{N}(\mathbf{r}, t)$, which may vary spatially and temporally over the sample. A general definition is

$$\mathbf{N}(\mathbf{r}, t) = \mathbf{N}_0 + \delta\mathbf{N}(\mathbf{r}, t), \qquad 1.$$

where $\mathbf{N}_0$ is the equilibrium director. Although the molecules may tumble rapidly and randomly about $\mathbf{N}(\mathbf{r}, t)$, the ordering creates motional anisotropy usually defined in terms of an order parameter, S. In the general case (4, 5), the ordering is described by a tensor order parameter, Q_{ab}, that is real, symmetric, and of zero trace. For the simple models considered here, however, we use the scalar S defined as

$$S = \langle 3\cos^2\beta - 1\rangle/2, \qquad 2.$$

where $\langle \ \rangle$ indicates an ensemble average and β is the angle between the molecular axis and $\mathbf{N}(\mathbf{r}, t)$. For isotropic fluids we have $S = 0$, while for a perfectly ordered system of molecules parallel to the director we have $S = 1$. For most liquid crystals, S ranges from 0.4 to 0.7, implying that while some anisotropy is present, the molecules can exist on the average at large angles ($\sim 40°$) with respect to the director. In general, the $\mathbf{N}_0$ will be isotropically distributed over a bulk sample, but surface effects, the presence of applied fields, or both, can cause a generally preferred macroscopic orientation of $\mathbf{N}_0$.

This review focuses upon thermotropic liquid crystals, which are single-component systems that form mesophases upon heating from the solid phase. The solid to mesophase transition has a heat of fusion typical of those found for large, organic molecules (a few kcal mole^{-1}), and a volume increase ($\sim 10\%$) typical of solid to

---

[1] Present address: Department of Chemistry, University of Texas, Austin, Texas 78712.

liquid transitions. Because the orientation fluctuations scatter light strongly, the liquid crystal phase is turbid. As temperature is increased, thermal disorder increases. The system may pass through more than one mesophase, but at some critical temperature, T_c, (the "clearing point"), the system undergoes a transition to the liquid phase, a phase without long-range order. The heat of transition (~ 200 cal mole^{-1}) and the volume increase ($\sim 0.5\%$) associated with T_c are small; consequently it is reasonable to assume that the short-range correlations between neighboring molecules are not greatly changed at the transition. Indeed, it is now firmly established that short-range order and pretransition fluctuations exist in the liquid phase a few degrees above T_c; thus even the liquid phase of liquid crystal compounds is unusual.

Liquid crystals have undergone several periods of scientific inquiry during the 80 years since their discovery. The ordering of liquid crystals and the changes in that ordering induced by surfaces and applied fields account for most of the resurgence of interest in these compounds over the last decade (2–7). Electrooptic effects, digital displays, thermography, light-scattering experiments, critical phenomena investigations, and spectroscopic studies have focused on one or more of the aspects of ordering and anisotropy of liquid crystals.

NMR relaxation is a versatile probe of molecular dynamics (6–13). The relaxation rate measures the coupling between the nuclear spins and the other degrees of freedom (the "lattice") of the system. Because the coupling usually involves fluctuating electromagnetic fields modulated by molecular motions, studies of the frequency and temperature dependence of relaxation can provide, in principle, significant information on molecular dynamics. Variable-frequency NMR relaxation studies were first made on liquid crystals in the late 1960s. The results, perhaps not surprisingly, were strikingly different from those found in more conventional fluids; the ordering in the mesophases had a pronounced effect on NMR relaxation as well. Very recently, theoretical work on relaxation in mesophases has advanced sufficiently to allow information on molecular dynamics in these systems to be obtained. This review discusses experimental and theoretical studies of nuclear spin relaxation in thermotropic liquid crystals. It is not an exhaustive survey of the now considerable body of literature on the topic. Rather, it is a selection of results chosen to illustrate present knowledge (and shortages thereof) in the field. Relaxation in nematic, smectic, and cholesteric liquid crystals is covered. Literature through 1976 is covered.

This review gives very little attention to relaxation in the "isotropic" liquid phases of compounds forming liquid crystals, although the short-range correlations in this phase make relaxation in that region very interesting. Two recent articles (6, 7) have reviewed NMR relaxation in the liquid phase near T_c.

Lyotropic liquid crystals are formed when a compound is dissolved in a solvent. Examples include soaps in water, micellar systems, and lipid-water model membrane systems (1–3, 9a,b). This review does not cover lyotropic phases, although an enormous number of NMR relaxation studies have been done on these. An introduction to that literature can be found in several of the general references (1–3, 9a,b) and in standard reviews (12).

Several excellent books (4, 9a,b) and review articles (2, 5–7, 13) on general properties of liquid crystals have appeared. In particular, the book by de Gennes (4) provides material basic to an understanding of NMR relaxation and a variety of other physical phenomena in liquid crystals.

Other aspects of NMR in liquid crystals, including relaxation, can be found in review publications (9b, 10, 11a,b).

This review begins with a brief survey of the properties of liquid crystals and the fundamentals of NMR relaxation. Representative theories of relaxation in liquid crystals are then surveyed. Finally, we review relaxation data in liquid crystals; applications and comparisons of the various theories are given where appropriate.

General Properties of Liquid Crystals

Several thousand compounds are known to form mesophases (1–3, 14). This section summarizes a conventional classification of the phases by symmetry and physical properties. The treatment follows de Gennes (4), and further details can be found in his book.

NEMATICS A simple model of the nematic structure is an arrangement of approximately parallel molecules (Figure 1). The centers of gravity have no long-range order; they are separated by distances typical of conventional liquids. Apparent viscosities are low (see later discussion) so nematics flow like liquids. The molecules tend to be parallel to some common axis, such as the director (Equation 1), whose orientation in space is arbitrary unless surface effects or applied fields establish a generally preferred direction over the entire sample. Optically, an oriented nematic is a uniaxial medium with the optic axis along $\mathbf{N}_0$. The states of the molecule parallel and antiparallel to the director are indistinguishable, and there is complete rotational symmetry about the director.

CHOLESTERICS: DISTORTED NEMATICS These phases are formed by chiral compounds, especially esters of cholesterol, or by the addition of a chiral "impurity" to a nematic phase. Locally a cholesteric is very similar to a nematic: parallel ordering along $\mathbf{N}_0$ with no long-range ordering of the centers of gravity. However, $\mathbf{N}_0$ takes on a helical arrangement in space (Figure 1); both the magnitude and the sign of

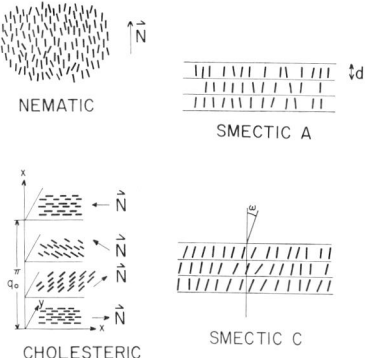

Figure 1 Simple models of liquid crystal structures.

the pitch of the helix are meaningful. Cholesteric mesophases possess a multitude of interesting optical properties (1–5, 13). Cholesteric liquid crystals formed by the esters of cholesterol do not tend to align in magnetic fields, but cholesterics formed by adding chiral molecules to nematics tend to align with the helix perpendicular to the magnetic field. In such systems the field can affect the pitch, and the cholesteric can be "unwound" by the field to form (at a critical field, H_c) a nematic structure.

SMECTICS All smectic liquid crystals are layered structures with a well-defined interlayer distance (Figure 1). Inside each layer the centers of gravity may or may not show long-range order; each layer is a two-dimensional liquid. Various subclasses of smectics exist. In the smectic A, the molecular axes are perpendicular to the layers; the system is optically uniaxial. A smectic C structure has molecular axes tilted at a definite angle with respect to the perpendicular to the layers. Smectic C phases are optically biaxial.

In contrast to smectics A and C, the smectic B phase has the periodicity and rigidity of a solid within each layer. Even more exotic smectics have been described (4).

STATIC DISTORTIONS IN LIQUID CRYSTALS In macroscopic samples, the order parameter varies spatially; surface effects and external fields produce small distortions in alignment. The distances over which the distortions occur are much greater than the molecular dimensions, so the continuum theory can be used to describe the system. Using $\mathbf{N}(\mathbf{r}, t)$ to describe the distorted state, the hydrodynamic properties of mesophases can be formulated. In nematic systems, the fluctuations in $\mathbf{N}(\mathbf{r}, t)$ are produced by "splay" (K_1), "twist" (K_2), and "bend" (K_3) elastic deformations (Figure 2). The elastic constants, K_i, have been measured in some nematics. The fluctuations affect $\mathbf{N}(\mathbf{r}, t)$, as in Equation 1, with $\delta \mathbf{N} = (\delta \mathbf{N}_x, \delta \mathbf{N}_y, 0)$ perpendicular to $\mathbf{N}_0$. The analysis of the effects of the elastic deformations is simplified if the "one-constant approximation" is used: $K \approx K_1 \approx K_2 \approx K_3$. This approximation is quite valid for nematics; it is not valid for smectics. The amplitudes and relaxation times of the deformations are required when an analysis of NMR relaxation is attempted. From a free-energy development, de Gennes and the Orsay liquid crystal group (4) have shown that the mean-square amplitude of the fluctuations can be expressed as

$$\langle |\delta \mathbf{N}_q|^2 \rangle = 2kT/(Kq^2 V), \qquad 3.$$

and the relaxation time, τ_q, for the relaxation of the modes is given by

$$\tau_q^{-1} = Kq^2/\eta. \qquad 4.$$

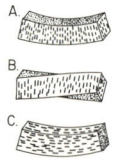

Figure 2 The three distinct curvature strains of a liquid crystal. A, splay; B, bend; C, twist. (After Stephen & Straley, Reference 5.)

In these equations, V is the sample volume, $\mathbf{q}$ is the wave vector, and η is the appropriate viscosity (defined later).

In cholesteric phases the same deformations are allowed, but K_3 is larger than K_2. The analysis of deformations in smectics is much more complicated than in nematics. In smectic C phases, static deformations similar to those in nematics are allowed, but the sizes of the elastic constants may differ from the nematic case. Bend and twist deformations are forbidden in zero order in smectic A phases; consequently K_2 and K_3 increase as the smectic A phase is approached. Almost no experimental data on the sizes of the elastic constants in smectic liquid crystals exist.

FLUCTUATIONS NEAR PHASE TRANSITIONS As liquid–liquid crystal and liquid crystal–liquid crystal phase transitions are approached from the high-temperature side, a number of interesting thermodynamic and critical phenomena occur. Some of the phase transitions may be second order, for example. In addition, just above the nematic–smectic A transition, small domains with a smectic organization appear. These have been termed cybotactic groups. The nematic elastic constants K_2 and K_3 increase approximately in proportion to the size of the cybotactic clusters as the temperature is lowered.

DYNAMICAL PROPERTIES The dynamics of an incompressible nematic liquid crystal involves a second-rank, symmetric tensor with zero trace. Consequently, five independent components (the Leslie components), each with the dimensions of viscosity, are required for a description. Values have been measured for nematics; these are typically 0.05 to 2 poise.

The set of equations that describes the dynamics of cholesterics is identical to the set used for nematics. In spite of the formal similarities, the physical flow and orientational effects are much more complex in the helical structure. In particular, the apparent bulk viscosity of cholesteric mesophases often seems to be several orders of magnitude larger than the five coefficients resulting from the Leslie treatment.

NMR Relaxation

Nuclear spin relaxation reflects the spectral density, $J_h(p\omega)$, of the fluctuating fields created when nuclear interactions are modulated by molecular motions (8). The spectral density measures the intensity of the fluctuations at the frequency ω,

$$J_h(p\omega) = \int_{-\infty}^{\infty} \langle F_h(0)F_h^*(t)\rangle e^{-ip\omega t}\, dt, \qquad 5.$$

where $\langle F_h(0)F_h^*(t)\rangle$ is the correlation function of the Hamiltonian, $\mathscr{H}$, involved in the relaxation process, and the F_h are determined by the form of $\mathscr{H}$. The spin lattice relaxation time, T_1, measures the rate at which the spin system returns to equilibrium with the static applied magnetic field, H_0. The relaxation rate is defined as

$$T_1^{-1} = C_1[J_1(\omega_0) + J_2(2\omega_0)]. \qquad 6.$$

The nuclear Larmor frequency, ω_0, equals $-\gamma H_0$, where γ is the gyromagnetic ratio of the nucleus, and the constant C_1 is determined by the form of $\mathscr{H}$. A calculation of T_1 involves the evaluation of the correlation function of the interaction responsible for relaxation. Because $\mathcal{H}$ involves either nuclear magnetic dipolar or electric quadrupolar interactions, calculation of the correlation functions usually requires averaging spherical harmonics of order two over molecular motions. For some mechanisms, averaging over the radial separation, **r**, may also be required.

In the next section, we present relatively standard relaxation expressions derived from simple models of isotropic systems.

INTRAMOLECULAR DIPOLAR RELAXATION For nuclei of spin $I = \frac{1}{2}$, the principal interaction is dipolar. The relaxation rate for a pair of nuclei separated by a fixed distance can be derived. If the nuclei are located on the surface of a sphere, their dipole-dipole interaction is modulated as the sphere undergoes random, rapid molecular tumbling (Brownian motion). This calculation approximates the intramolecular contribution to T_1 of the protons of a water molecule. In this case, the F_h are the time-dependent, angular portions of the dipolar $\mathcal{H}$. The calculation begins with Equation 6, and produces $C_1 = \frac{9}{8}\gamma^4\hbar^2 r^{-6}$. If the tumbling is a Markoff process described by a single correlation time, τ_c, the following well-known expression obtains:

$$T_{1a}^{-1} = \tfrac{3}{20}\gamma^4\hbar^2 r^{-6}[2\tau_c(1 + \omega_0^2\tau_c^2)^{-1} + 8\tau_c(1 + 4\omega_0^2\tau_c^2)^{-1}]. \qquad 7.$$

Equation 7 predicts that in the rapid-motion limit we have $(\omega_0^2\tau_c^2 \ll 1)$, $T_{1a}^{-1} = 1.5\gamma^4\hbar^2 r^{-6}\tau_c$. Under these conditions, which usually apply to most fluids, T_{1a} is independent of the Larmor frequency and strongly temperature dependent, usually increasing as temperature increases. Both conditions are observed in experimental work in ordinary liquids. A further prediction of Equation 7 is that in the slow-motion limit $(\omega_0^2\tau_c^2 \gg 1)$, T_{1a} will vary as $\omega_0^2\tau_c$. Consequently T_{1a} will increase as the square of the Larmor frequency and, because of its relationship to τ_c, probably decrease as temperature increases. Such behavior is observed in molecular solids.

INTERMOLECULAR DIPOLAR RELAXATION Diffusion also modulates dipolar interactions, and Torrey (16a) has presented a theory for the intermolecular contribution to relaxation through self-diffusion (T_{1SD})

$$T_{1SD}^{-1} = 0.6\gamma^4\hbar^2\pi n(a^3\omega_0)^{-1}I(b, x), \qquad 8.$$

where a is the distance of closest approach of two spins, n is the number of spins per cm^3, $b = \langle r^2 \rangle/12a^2$, and $\langle r^2 \rangle$ is the mean-square flight distance for the diffusion process. The lengthy expression for $I(b, x)$, where $x = (\omega_0 a^2/D)^{1/2}$ and D is the diffusion coefficient, can be found in Torrey's paper. Applications of this theory require values of a, and $\langle r^2 \rangle$. Often these will be obtained from fits of the experimental T_{1SD} data. In the low-frequency limit, Equation 8 reduces to

$$T_{1SD}^{-1} = C - F\omega_0^{1/2}, \qquad 9.$$

where C and F are positive constants. A diffusion correlation time $\tau_{SD} = a^2/6D$ is implicit in the theory. Corrections to the Torrey theory are discussed in (16b), but the developments in that reference have not been applied to intermolecular nuclear spin relaxation in liquid crystals.

QUADRUPOLAR RELAXATION Nuclei with a spin $I \geq 1$ may possess an electric quadrupole moment and hence have another relaxation mechanism in addition to dipolar interactions. The interaction between the nuclear quadrupole moment and the local electric field gradient is modulated by molecular motions; this mechanism nearly always dominates the dipolar contribution to T_1. Consequently, quadrupolar relaxation, T_{1Q}, is almost completely intramolecular. For isotropic systems, the quadrupolar $\mathscr{H}$ for $I = 1$ (deuterium) is formally equivalent to the intramolecular dipolar expression (Equation 7) if $(e^2qQ/2h)^2$ replaces $(\gamma^4 h^2 r^{-6})$. Such a revised expression would provide T_{1Q}^{-1} for the deuterons in D_2O. The quantity e^2qQ/h is the quadrupole coupling constant.

COMBINED MECHANISMS Many mechanisms may contribute to relaxation in fluids and solids. If the motions (interactions) are uncoupled (i.e. $\mathscr{H}$ is separable), the relaxation rates are additive:

$$T_1^{-1} = \sum_i T_{1i}^{-1}. \qquad 10.$$

Both T_{1a} and T_{1SD} contribute to relaxation in water, for example, and $T_1^{-1} = T_{1a}^{-1} + T_{1SD}^{-1}$. Other terms would need to be added to the relaxation rate if other mechanisms were present. If the possible mechanisms are coupled, that is if $\mathscr{H}$ is not separable, then the calculation of T_1 is in general more difficult; separate relaxation rates may not be defined in this case.

The material in this section indicates that by careful analysis of the frequency and temperature dependence of T_1 in a well-defined system, it is in principle possible to identify the mechanisms responsible for relaxation and obtain information on molecular dynamics. We should note that relaxation time is shortest when the spectral density at the Larmor frequency is a maximum. Consequently, the most effective mechanism is that with the strongest Fourier component at ω_0. Possible mechanisms with Fourier frequencies either much faster or much slower than ω_0 will not be as effective.

The key to the interpretation of NMR relaxation data is the evaluation of correlation functions.

Relaxation in Liquid Crystals

HISTORICAL ASPECTS Current interest in NMR relaxation in liquid crystals resulted in large part from observations in the 1960s (17a–c) that the proton T_1 in para-azoxyanisole (PAA) varied as the square root of the frequency,

$$T_1^{-1} = A'\omega_0^{-1/2} + B', \qquad 11.$$

where A' and B' are constants. Furthermore, in many nematic phases, proton relaxation showed very little temperature dependence (17a–c, 18), in marked contrast to ordinary liquids. The theoretical explanation, advanced by Pincus (19) then modified by many others (20–25), especially Doane et al (21a,b, 22), is that modulation of the dipolar interactions by fluctuations of the order director (OF mechanism) could lead to such behavior. The general development has been centered on the premise that the molecular tumblings with respect to the director are fast with respect to the Larmor frequency; they are thus less effective at causing relaxation than the slower fluctuations of the director. The OF mechanism is unique to liquid crystals, and it has remained the center of attention in liquid crystal NMR relaxation theory.

Of course, other relaxation mechanisms exist for liquid crystals. For methoxy-benzylidene-n-butylaniline (MBBA) for example, relaxation over a resonant frequency range of 5 MHz to 270 MHz fits quite well to a relationship similar to Equation 8 (23a–c). Consequently, diffusion (the SD mechanism) is often proposed as a significant mechanism for relaxation in MBBA (23a–c). Controversies exist in the earlier literature over which mechanism dominates in certain liquid crystals. In particular, MBBA relaxation has been interpreted in both diffusion-dominated (23a–c) and OF-dominated (21b, 22). In part, these controversies arose because the range of resonant frequencies then available made a clear separation of the various mechanisms virtually impossible. Very recently, however, with the development of field cycling techniques that provide much wider (4 kHz $< v_0 <$ 270 MHz) frequency ranges for T_1 studies (24b, 26, 27), a relatively well-defined separation of the relaxation mechanisms has emerged. It is now evident that the rapid molecular tumblings, the slower OF mechanism, and an intermolecular (perhaps SD) mechanism all contribute to T_1 in most liquid crystals, with one or more of the mechanisms probably dominating at some given temperature and resonant frequency.

Order Director Fluctuation (OF) Mechanism

The simplest approach to relaxation in liquid crystals is to assume that the rapid molecular tumbling and the slower collective modes are completely uncoupled. We begin with this approach in order to investigate the OF mechanism, which is unique to liquid crystals. We briefly consider its derivation in order to gain insight into its role in the more general treatments of relaxation that follow.

The Fourier components most effective in nuclear relaxation are those with correlation times of the order of the Larmor period, ω_0. This limits the hydrodynamic modes of interest to collective modes with wavelengths of the order of 100 to 1000 Å. In view of the wavelengths involved, the elastic continuum theory of liquid crystals can often (though not always) be used to derive the correlation functions.

We begin by treating T_1 for the intramolecular relaxation of a proton pair in a nematic liquid crystal (21a,b, 22). For simplicity, the internuclear vector, $\mathbf{r}$, is assumed to be parallel to the molecular axis, $\mathbf{M}$ (Figure 3). The molecule reorients with respect to the director, $\mathbf{N}(\mathbf{r}, t)$, which is displaced by $\delta\theta(\mathbf{r}, t)$ from $\mathbf{N}_0$ by collective-mode fluctuations. In the general case, $\mathbf{N}_0$ may make an angle θ_0 with $\mathbf{H}_0$. The transformation in terms of Euler angles of the correlation functions from the unprimed frame $(\mathbf{z}\|\mathbf{H}_0)$ to the primed frame of $\mathbf{N}_0(\mathbf{z}'\|\mathbf{N}_0)$ is given in (21b):

$$\langle F_p(0)F_p^*(t)\rangle = \sum_{q=0}^{2} f_{pq}(\theta_0)\langle F_q'(0)F_q'^*(t)\rangle. \qquad 12.$$

The F_q' describe the internuclear vector in the primed frame of $\mathbf{N}_{0'}$ and f_{pq} is a time-independent transformation matrix element. Because $\mathbf{N}(\mathbf{r}, t)$ thermally fluctuates about $\mathbf{N}_0$ and modulates the interactions, an infinitesimal transformation of the

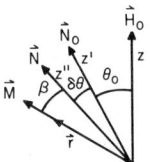

Figure 3 Coordinate system for order-director fluctuation mechanism.

F' to the doubly primed frame, F'', of $\mathbf{N}(\mathbf{r}, t)$ is required (details in Reference 22). The result is

$$\langle F_p(0)F_p^*(t)\rangle = C_2(f_{pq'}F'')\langle \delta\theta(0)\delta\theta^*(t)\rangle, \qquad 13.$$

where C_2 is given in (22).

The F'' describe the internuclear vector in the frame $\mathbf{N}(t)$. For small fluctuations, $\langle \delta\theta(0)\delta\theta^*(t)\rangle \simeq \langle \delta\mathbf{N}(0)\cdot\delta\mathbf{N}(t)\rangle$. Equation 13 can be simplified by applying Khinchin's theorem and by assuming a single correlation time for the system, so that

$$\langle F_p''(0)F_p''(t)\rangle = \langle F_p''(0)\rangle^2 + \langle |F_p''(0)|^2\rangle e^{-t/\tau_c}. \qquad 14.$$

Equation 14 is a relatively standard expression in relaxation theory, but its evaluation emphasizes the unique properties of liquid crystals. For isotropic fluids, the $\langle F_p''(0)\rangle^2 = 0$ for all p. For liquid crystals, however, only the terms for $p = 1, 2$ vanish, while $\langle F_0''(0)\rangle^2 \simeq 4S^2$ within our approximations. Consequently,

$$\langle F_p(0)F_p^*(t)\rangle = f_{p1}S^2\langle \delta\mathbf{N}(0)\cdot\delta\mathbf{N}(t)\rangle. \qquad 15.$$

Evaluation of $\langle \delta\mathbf{N}(0)\cdot\delta\mathbf{N}(t)\rangle$ has been done by de Gennes and the Orsay group. Using the continuum theory and a Fourier superposition of the modes, expressions for the thermal amplitudes (Equation 3) and the relaxation times of the modes (Equation 4) were obtained by use of the one-constant approximation and lumping together all the viscosity components. This yields an expression:

$$J_p(p\omega) = 4^{-1}f_{p1}\langle F_0''(0)\rangle^2 \int_{-\infty}^{\infty}\int_0^{q_c}\langle |\delta\mathbf{N}_q|^2\rangle e^{-|t|/\tau_q}e^{ip\omega t}\,dt V(2\pi)^{-3}4\pi q^2\,dq. \qquad 16.$$

The upper limit, q_c, is to account for the fact that the long-range modes should be cut off near the molecular size. Although a closed-form expression using q_c has been derived (7, 24a,b) it is lengthy. Because it has been demonstrated that the cutoff is unimportant in most cases (7, 22, 24a,b), we shall take the limiting case of an infinite limit on the wave vectors. This yields

$$T_{1OF}^{-1} = AS^2\omega_0^{-1/2}[f_{11} + f_{21}/\sqrt{2}], \qquad 17.$$

where

$$A = 9\gamma^4\hbar^2kT\eta^{1/2}(8\sqrt{2}\pi r^6 K^{3/2})^{-i}. \qquad 18.$$

For the case $\theta_0 = 0$, this becomes the familiar expression

$$T_{1OF}^{-1} = AS^2\omega_0^{-1/2}. \qquad 19.$$

This result, despite the approximations involved, predicts the desired frequency dependence of relaxation.

For the general case, where θ_0 is nonzero and the individual elastic constants and viscosities are included,

$$J_h(p\omega) = C' \sum_{\alpha=1}^{2} K_\alpha^{-1}(2\eta_\alpha^{1/2}/K_3 p\omega)^{1/2}, \qquad 20a.$$

$$C' = f_h(\theta_0)S^2 kT/(2\pi^2), \qquad 20b.$$

$$f_0(\theta_0) = 18(\cos^2\theta_0 - \cos^4\theta_0), \qquad 20c.$$

$$f_1(\theta_0) = (1 - 3\cos^2\theta_0 + 4\cos^4\theta_0)/2, \qquad 20d.$$

$$f_2(\theta_0) = 2(1 - \cos^4\theta_0). \qquad 20e.$$

The expression for T_1 can be found in (7, 24a,b).

Derivations of the OF mechanism for smectic phases have been reported (7, 24a,b). They are quite lengthy; consequently only their qualitative features are described in this article as needed.

Diffusion (SD) Mechanism

The treatment of the intermolecular (diffusional) contribution to relaxation in liquid crystals has involved the adaptation of Torrey's theory (Equation 8) (16a). Until very recently (28a–c), there were no explicit data on frequency dependence of the intermolecular contributions. The data that now exist (28a–c) for PAA indicate that the diffusional contribution might not be a simple one. Diffusion coefficients in liquid crystals are in the range of 10^{-6} to 10^{-8} cm^2 sec^{-1}; consequently, the correlation times for diffusion and the collective modes are comparable. Therefore, there may be considerable coupling of diffusion with nematic order insofar as NMR relaxation is concerned.

Vilfan & Zumer (29) have adapted Torrey's theory to the anisotropic diffusion conditions existing in liquid crystals in a magnetic field (diffusion parallel to $\mathbf{H}_0$ is about twice as fast as diffusion perpendicular to $\mathbf{H}_0$). The predicted dispersion does not differ markedly from Torrey's theory (Equation 8), but the angular (θ_0) dependence of T_1 differs markedly from that predicted by the collective modes mechanism (Equation 18). Some aspects of the diffusional contribution to relaxation are discussed by Freed and co-workers (16b, appendix A in 25). Doane et al (22) have an expression for an intermolecular contribution involving diffusion, molecular tumbling, and order fluctuations. This is discussed below, where we review relaxation in PAA.

Generalized Treatments of Relaxation in Nematics

The anisotropy of liquid crystal phases presents significant challenges in the calculation of correlation functions when incorporation of the complete range of molecular dynamics and frequencies is considered. Exact solutions do not exist, but various approximate approaches have been developed. In this section we consider in some detail three theories applicable to nematics.

The work of Freed (25) is the most recent and most comprehensive treatment of NMR relaxation processes in liquid crystals. This work also presents a theory for electron spin resonance line-widths in nematics (25, 30). His development treats intramolecular processes, with the objective of properly analyzing the interdependence of the two frequency domains discussed earlier: molecular reorientation and the (slower) director fluctuations. Generally these are coupled; the hydrodynamic modes represent fluctuations in the potential of mean torque experienced by an individual molecule in the nematic phase. Consequently, the molecular reorientation is affected by the hydrodynamic modes. Freed's treatment differs from earlier theories in that he attempts to develop statistical interdependence of the various motions. Earlier treatments, some of which are discussed below, have assumed independence of various contributions or have added relaxation mechanisms heuristically. Freed begins with the Smoluchowski equation for motion of a single particle with an orientation potential that slowly fluctuates in time. The entire analysis involves a comprehensive Markov process and results in a sum of spectral density terms. In the derivation it is assumed that the molecular rotational diffusion tensor is isotropic; this could be corrected for anisotropic rotational diffusion (e.g. an ellipsoid), but the changes in the results would probably be minor. Freed also uses the one-constant approximation. The results yield a T_1 with three contributions: (a) a molecular reorientation term (relative to the potential of mean torque), (b) a collective modes term (to zero order the same as Equations 17–19), and (c) a negative cross term between these. Inclusion of a cutoff, $q_c = (\omega_c \eta / K)^{1/2}$, in the integrations over the wave vectors was only required for the cross term. Where the molecular reorientation time, τ_R, obeys $\tau_R \gg \omega_c \geq \tau_q^{-1}$, this theory yields

$$T_1^{-1} = C_F \sum_{K=-2}^{2} |f^{(K)}|^2 [J_{K1}(\omega_0) + 4J_{K2}(2\omega_0)], \qquad 21a.$$

$$|f^{(0)}|^2 = [(1 - 3\cos^2 \beta')]^2/4, \qquad 21b.$$

$$|f^{(\pm 1)}|^2 = (3\sin^2 \beta' \cos^2 \beta')/2, \qquad 21c.$$

$$|f^{(\pm 2)}|^2 = (3\sin^4 \beta')/8 \qquad 21d.$$

and

$$J_{KM}(\omega) = \kappa(K, M)\tau_R(1 + \omega^2 \tau_R^2)^{-1} + 1.5 S^2 \mathscr{A} \delta_{K,0} \delta_{M, \pm 1} [(\pi/2)^{1/2} U(\omega/\omega_c)\omega^{-1/2} \\ - 2\tau_R \omega_c^{1/2} \pi^{-1/2}(1 + \omega^2 \tau_R^2)^{-1}], \qquad 22.$$

where $\mathscr{A} = kT\eta^{1/2}(2\pi^{3/2} K^{3/2})^{-1}$.

For the interaction of a pair of protons, $C_F = \frac{3}{4}\gamma^4 \hbar^2 r^{-6}$ and β' is the angle between the internuclear vector and the molecular axis. For deuteron relaxation, $C_F = \frac{3}{8}(e^2 qQ/\hbar)^2$ and β' is the angle between the carbon-deuteron bond and the molecular axis. In Equation 22 the coefficient $\kappa(K, M)$ can be obtained from tables in (25) as a power series in S for isotropic rotational diffusion in the high frequency limit; corrections for the general case are discussed in (30). The function $U(\omega/\omega_c)$ is approximately 1 when $\omega \ll \omega_c$.

Other cross terms are possible in Freed's theory. Except for the discussions that follow, this theory has not yet been applied extensively to experimental data.

Consequently, the relative importance of the other cross terms or the extent of applicability of the entire theory have yet to be determined.

Ukleja, Pirs & Doane (UPD) (22) have developed a theory beginning with the OF derivation (as outlined above), but including the averaged effects of molecular reorientations as well as a cross term between the OF and the molecular reorientation term. The result gives, for θ_0 equal to zero, three contributions to the relaxation: (a) T_{1OF}^{-1}, (b) a molecular reorientation term, T_{1M}^{-1}, and (c) a cross term between these,

$$T_1^{-1} = T_{1OF}^{-1} + T_{1M}^{-1} + A\tau_c^{1/2}[\tfrac{1}{3}(2S + 1)g(\omega_0\tau_c) + \tfrac{8}{15}(1 - S)g(2\omega_0\tau_c)], \qquad 23.$$

where T_{1OF}^{-1} is given by Equation 17 and

$$T_{1M}^{-1} = 2A(1 - S)(15)^{-1} d\tau_c^{1/2}[(1 + \omega_0^2\tau_c^2)^{-1} + 4(1 + 4\omega_0^2\tau_c^2)^{-1}], \qquad 24.$$

where $d = (2\sqrt{2\pi}/kT)(K^3/\eta)^{1/2}$ and the other terms are as previously defined. The function $g(x)$ becomes $\sqrt{2}$ in the rapid-motion limit.

Freed has pointed out that Equation 23 does not reduce to the expected isotropic relaxation behavior in the limit as S approaches zero; this may indicate a fundamental problem in the derivation of the cross terms.

An additional feature of the UPD treatment is an expression for an intermolecular contribution to T_1. It is the only one of the theories in this section to do this.

Nordio et al (31a–c) have developed a third approach to relaxation by assuming that $T_1^{-1} = T_{1OF}^{-1} + T_{1M}^{-1}$, and deriving an expression for T_{1M}^{-1} by solving a differential equation in the presence of a time-independent orientational pseudopotential, V. Their approach follows applications to liquid crystals of the mean field theory by Maier and Saupe. A diffusion equation identical (except for the form of V) to the time-dependent Schrödinger equation is solved for $V = 3\lambda\langle\cos^2\beta - 1\rangle/2$, where β is the same as Figure 3. For any given S, the potential strength λ is determined so that the diffusion equation can be solved and the correlation function for the rotational diffusion can be obtained. Solutions to the equation are obtained in terms of the axial component ($D_\perp$) of the rotational diffusion tensor, Wigner matrix elements (D_{mn}^2), and the eigenvalues (λ_{mn}^2) of the diffusion equation. We have

$$T_{1M}^{-1} = \tfrac{3}{2}\gamma^4\hbar^2 r^{-6} D_\perp C_4, \qquad 25.$$

where C_4 is a known function of D_{mn}^2, $D_\perp^{-2}$, and ω_0^{-2}. If S is known, all parameters except $D_\perp$ can be calculated. Although $D_\perp$ can be obtained from dielectric relaxation studies, some ambiguity is involved in its application to NMR relaxation, since dielectric relaxation is a collective phenomenon, whereas T_1 is much more indicative of the properties of an individual molecule.

In comparison, Freed's theory is the most comprehensive of these; it allows formally for the presence of molecular reorientation terms. UPD theory includes reorientation effects in a quite different fashion; the effects of the order fluctuations (the A term) affect the molecular reorientation term as well. The UPD result is similar in some respects to Nordio's, but the averaging procedures and the method of including the cross terms differ considerably in the two theories. All three treat intramolecular effects, although a special variation of the UPD theory can give an intermolecular T_1 expression. All three reduce to Equations 17–19 when the OF mechanism dominates. All three predict that the molecular reorientation and cross terms will become frequency-dependent as the slow-motion limit is approached. Consequently, the calculated dispersion of T_1 can be a very complicated function.

In the remainder of this review, we consider experimental data on relaxation in liquid crystals (Figure 4), applying and comparing these and other theories.

Proton Relaxation in Liquid Crystals

INTERMOLECULAR AND INTRAMOLECULAR EFFECTS Because the intramolecular (T_{1a}) and intermolecular (T_{1r}) contributions to T_1 have rather complicated and quite different dispersive behavior, the separation of the two effects from relaxation vs frequency plots is a very formidable problem. In most liquid crystals (Figure 4), the OF mechanism might be expected to apply to the aromatic protons, but perhaps not to the more flexible end chains. The relaxation effects of the end-chain protons and the effects of diffusion on the entire relaxation process are difficult to treat theoretically. Because the static dipolar forces in liquid crystals are not motionally averaged to zero, broad NMR lines are obtained. The end-chain resonances and the aromatic resonances cannot be separately resolved, so the relaxation of the individual moieties in a completely protonated molecule cannot be measured.

It is preferable to identify and separate the intermolecular and intramolecular contributions to relaxation before applying theories. The only molecule for which these effects have been unambiguously separated is PAA (28a–c). A series of experiments on methyl-deuterated (PAA-d6), ring-deuterated (PAA-d8), and perdeuterated (PAA-d14) molecules was conducted (28a–c) to separate these effects. Deuteration has no effect on the transition temperatures and, as far as is known, no effect on the elastic constants or viscosities. An experimentally well-established procedure in NMR holds that for proton relaxation in a molecule studied as a function of dilution in a perdeuterated solvent, $T_1^{-1} = T_{1a}^{-1} + XT_{1r}^{-1}$, where X is the mole fraction of the protonated species. (Minor corrections to this must be made in some cases to account for the proton-deuteron interaction.) Studies of relaxation as a function of X allow T_{1a} and T_{1r} to be determined. Such studies (28b) on PAA at 30 MHz and 120°C found that $T_{1a}/T_{1r} = 0.8$; consequently, the intermolecular

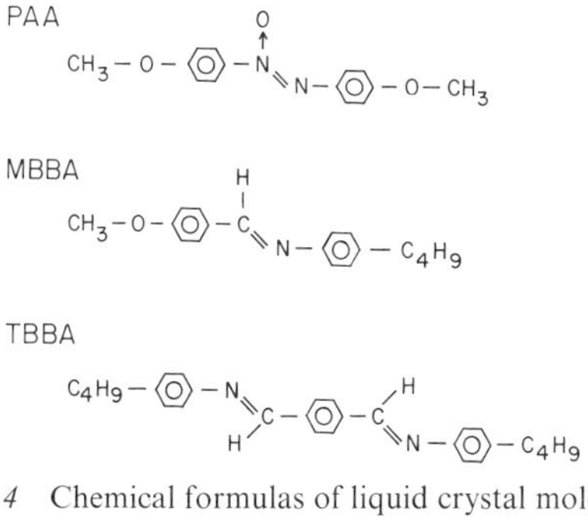

Figure 4 Chemical formulas of liquid crystal molecules.

contribution to T_1 is not negligible. In contrast, a much better molecule to study is PAA-d6, wherein the methyl contribution to proton T_1 is negligible. Furthermore, the proton relaxation is dominated by the pair-wise interaction of the *ortho* ring pairs, so this molecule is an excellent prototype for testing theories that almost invariably treat a proton pair. In PAA-d6, $T_{1a}/T_{1r} = 0.39$ at 30 MHz, so intermolecular effects are less of a problem than in PAA, although still not negligible for many applications. The frequency dependences of T_{1a} and T_{1r} were also obtained over a frequency range of 5 to 100 MHz at 120°C (Figure 5). The results could be fitted to an OF plot (Equation 11). As discussed in the next section, the agreement with OF theoretical predictions is an indication that ring proton T_{1a} is dominated

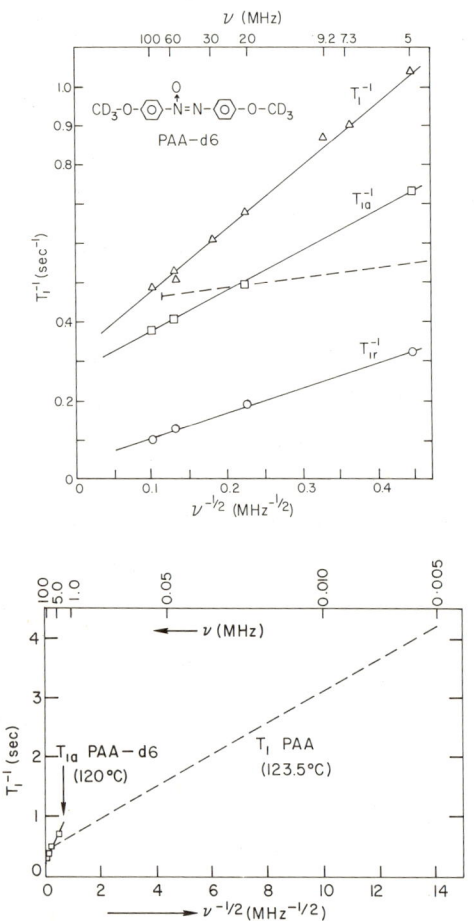

Figure 5 Proton relaxation in PAA and PAA-d6. Dashed line is the best fit of data on PAA quoted by Wolfel et al (26). △, □, ○ are data on PAA-d6 from Fung et al (28a).

by this mechanism. Rather surprisingly, the T_{1r} do not plot linearly in a simple diffusion plot (Equation 9), but do plot linearly in an OF plot (Equation 11). This study indicates that the intermolecular contribution to diffusion may be more involved than a mere extension of isotropic diffusion theories to liquid crystals would indicate.

The intermolecular and intramolecular contributions to T_1 of the methyl protons in the ring-deuterated (PAA-d8) molecule have also been obtained (28c).

PROTON RELAXATION IN PAA AND PAA-d6 An enormous body of work on relaxation in PAA has been published; Reference 26 contains a rather complete listing of these. In this review, we select those studies that appear to represent the present level of understanding of T_1 in this molecule. By far the most extensive range of frequencies studied was reported by Wolfel, Noack & Stohrer (26). Using a field cycling technique, they obtained T_1 data on PAA (Figure 5) over the range of Larmor frequencies from 3.8 kHz to 75 MHz at 123.5°C. Over the entire range, the data can be fitted to $T_1^{-1} = 0.268 v_0^{-1/2} + 4.31$, where v_0 is in MHz. The results are rather astonishing in view of the fact that T_{1r} is presumably important (28a–c), and it may have a different frequency dependence than the OF mechanism. The OF mechanism (Equations 17–19) predicts a slope of 1.3 sec$^{-3/2}$. Extension of this work to different temperatures is forthcoming [see the authors' comment in (27)], and this should provide additional information on relaxation mechanisms.

A number of papers have discussed relaxation in PAA (6, 7, 17a–31c, for example) and, because of the frequency dependence of T_1, there is general agreement that the OF mechanism is dominant. Because the PAA-d6 system is somewhat better defined insofar as theoretical interpretation is concerned, we confine our analysis to that molecule even though the frequency range covered is much narrower than for PAA. The data (28a–c) yield $T_{1a}^{-1} = 1.0 v_0^{-1/2} + 0.27$ and $T_{1r}^{-1} = 0.70 v_0^{-1/2} + 0.02$, where v_0 is in MHz (Figure 5).

Using the single-constant approximation and the rapid-motion limit, all of the theories discussed above predict a slope of $AS^2 = 1.30$ sec$^{-3/2}$, within 30% of the experimental value. The calculation uses $S = 0.5$, $r = 2.45$ Å, $\eta = 0.75$ poise (32), $K = 0.82 \times 10^{-6}$ dyn (33). The theoretical expressions (rapid-motion limit) become

Freed $\quad T_1^{-1} = 1.3 v_0^{-1/2} + 1.31 \times 10^9 \tau_R,$ $\qquad$ 26a.

UPD $\quad T_1^{-1} = 1.3 v_0^{-1/2} + 1.8 \times 10^9 \tau_c + 1.7 \times 10^4 \tau_c^{1/2},$ $\qquad$ 26b.

Nordio $\quad T_1^{-1} = 1.3 v_0^{-1/2} + 2.04 \times 10^8/D_\perp,$ $\qquad$ 26c.

where v_0 is in MHz. Equation 26a uses the isotropic diffusion approximation from (25), which was discussed earlier. The experimental intercept yields a $\tau_R = 2 \times 10^{-10}$ sec from Freed's theory and a $\tau_c = 8 \times 10^{-11}$ from UPD. Nordio's theory is fit with $D_\perp = 7.7 \times 10^8$ sec^{-1} for the perpendicular component of the rotational diffusion tensor, in good agreement with an independent estimate of $D_\perp$ from dielectric relaxation studies.

All three approaches thus fit the data semiquantitatively but with different correlation times. The cross terms and the $\omega_0^2 \tau^2$ in both the Freed and UPD expressions are relatively unimportant for PAA-d6. The correlation times are in order of magnitude agreement with those found by other methods.

Theoretical predictions for T_{1r} are much less satisfactory. UPD theory (Equation 25 in Reference 22) is consistent with the observed frequency dependence (Figure 5), but predicts a slope of 0.03 $\sec^{-3/2}$, whereas the experimental value is 0.70. The intercept of the plot yields (22) a diffusion correlation time considerably shorter than that estimated from diffusion coefficient measurements. Application of the Torrey theory (16a, 29) in both its low-frequency limit (Equation 9) and its complete form (Equation 8) does not provide an explanation since neither form describes the data. Because diffusion in PAA is fast, it may provide a relaxation contribution that is approximately frequency independent; consequently, it might account for the zero-frequency intercept of T_{1r} in Figure 5 (M. Vilfan, private communication).

The temperature dependence of T_1 in PAA has also been investigated (6, 7, 17a–21b). Goren et al (34a) measured relaxation at 47 MHz into the supercooled nematic phase and found that T_1 varied exponentially with reciprocal temperature in a manner similar to that of MBBA and two other liquid crystals. They found that the relaxation rate of the supercooled region continued into the nematic phase and argued that the temperature dependence of relaxation was determined by the viscosity in Equation 18 since the remainder of that expression has very little temperature dependence. In contrast, a similar experiment on relaxation in PAA-d6 (35) at 30 MHz showed that the temperature dependence of the supercooled region did not extend into the nematic phase for the ring protons. The activation energy for relaxation in the supercooled region was ~ 3.8 kcal mole^{-1} in both studies. Martins (34b) has argued that the temperature dependence is due to the end chains, a view supported by Dong et al (36a).

The molecule *p*-methoxybenzylidene aminophenylacetate shows a temperature and frequency dependence (36b,c) similar to that of PAA.

PROTON RELAXATION IN MBBA Graf, Noack & Stohrer (GNS) (27) have provided an extensive study of T_1 over a frequency range 1 kHz $\leq v_0 \leq$ 270 MHz, exceeding the ranges of earlier studies (see references in GNS) by more than three orders of magnitude. The study, done at 18, 27, 35, and 45°C, requires at least three relaxation mechanisms to explain the data adequately. GNS fitted the data assuming the three mechanisms were (*a*) an OF mechanism, (*b*) a diffusion contribution (T_{1SD}^{-1}, Torrey isotropic theory), and (*c*) a molecular reorientation mechanism (T_{1M}^{-1}, anisotropic tumbling of an ellipsoid). The best-fit values for the terms representing these three contributions are surprisingly close to theoretical predictions for mechanisms *a* and *c*.

Earlier work on MBBA, done over considerably narrower frequency ranges (4 to 270 MHz), gave contrasting views of the principal relaxation mechanism. Vilfan et al (23a,b) felt a diffusion mechanism best represented the data. UPD theory (22), on the other hand, explains the data on the basis of Equation 23 with $\omega_0^2 \tau_c^2 > 1$; in this limit the cross terms become important contributors to the dispersion of T_1. (In comparison with PAA, the viscosity coefficients of MBBA are at least an order of magnitude larger, in part because of the temperature difference, so a much longer correlation time for molecular reorientation in MBBA would in fact be expected.) The advantage of the GNS study is the wider range of frequencies used, which provides a better definition of relaxation mechanism. The low-frequency results are rather different from the behavior estimated from the rotating-frame relaxation (37) and dipolar relaxation studies (38) of other investigators. Because the T_1 studies are less subject to experimental problems, and because the GNS results are internally consistent, we concentrate on the GNS results for the low-frequency behavior.

The GNS analysis assumes that

$$T_1^{-1} = \sum_j p_j T_{1j}^{-1}, \qquad 27a.$$

$$T_{1j}^{-1} = [H_0^2 T_{zj}^{-1} + H_L^2 T_{dj}^{-1}]/[H_0^2 + H_L^2], \qquad 27b.$$

$$T_1^{-1} = p_{OF} T_{1OF}^{-1} + p_{SD} T_{1SD}^{-1} + p_M T_{1M}^{-1}, \qquad 27c.$$

where p_i are the fractions of protons concerned with each mechanism, T_{dj} is the dipolar relaxation time, and T_{1M}^{-1} is the relaxation of an ellipsoid. T_{zj}^{-1} is essentially T_1 (Equation 6) for the j mechanism. Work at kHz range necessitated the inclusion of the effects of the local dipolar field, H_L, and T_{dj} in the expression for T_{1OF}. GNS fitted the data using a least-squares regression technique with six adjustable parameters (A, H_L, D, a, τ, and ξ) defined as follows.

Because the OF term is dominant at lower frequencies, GNS used

$$T_{1OF}^{-1} = [H_0^2 T_{zOF}^{-1} + H_L^2 T_{dOF}^{-1}]/[H_0^2 + H_L^2], \qquad 28.$$

where $T_{zOF}^{-1} = A\omega_0^{-1/2}$, T_{dOF} is assumed equal to $T_{zOF}/3$, H_0 is the applied magnetic field, and H_L is typically 1 gauss. When $H_0 \gg H_L$, which holds for all but the lowest frequencies, Equation 28 reduces to the collective modes expression (Equation 19). On the assumption that only the ring protons were affected by the collective modes, $p_{OF} = \frac{8}{21}$.

T_{SD}^{-1} is fitted by varying any two of the parameters $\tau_{SD} = a^2/6D$, with a the distance of closest approach and D the self-diffusion coefficient in the isotropic diffusion theory of Torrey (Equation 8). The weighting factor for the mechanism was assumed to be 1.

Woessner's ellipsoidal model of relaxation (39) was adapted to the T_{1M} term; it was assumed that only motion around the short ellipsoidal axis was effective; the parameter for this fit was the correlation time τ_M. To account for the anisotropy of motion in the liquid crystal, GNS used ξ as an adjustable parameter in the proportionality ($T_{1M}^{-1} \alpha \xi \tau_M$) (valid in the high frequency limit) in which the constant of proportionality was calculable from first principles.

Although the GNS analysis is a fitting procedure, the parameters (especially D, A, and a) obtained from their procedure agree sufficiently well with theoretical predictions that the relative contributions of the various mechanisms to T_1 as functions of frequency and temperature should be given close attention. Less emphasis, however, should probably be given (27) to details of molecular motion (especially diffusion and molecular reorientation) derived from their model because of the approximations used.

The GNS data and the analyzed contributors to relaxation are shown in Figure 6. Their analysis indicates that at low frequencies the OF term dominates, while at high frequencies the SD term dominates. Only at the lowest temperature (supercooled nematic at 18°C) could the dispersion curves be adequately explained using only these two mechanisms. At higher temperatures the molecular reorientation mechanism was also required at frequencies between the OF and SD limits. The

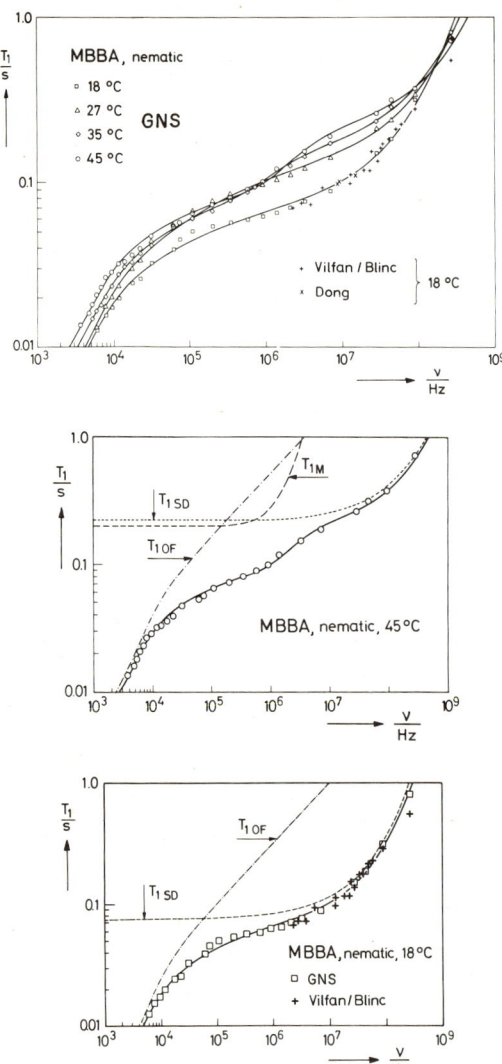

Figure 6 Proton relaxation in MBBA. (*a*) Data of GNS (27), Blinc et al (23a–c), and Dong et al (36a). (*b, c*) Contributions of relaxation mechanisms by GNS (27). [After GNS (27).]

parameters resulting from the fits are shown in Table 1. The reader should consult (27) for details of the various mechanisms.

The calculated value of A at $18°C$ (Equation 18) is $22,400$ sec$^{-3/2}$ using $\eta = 1$ poise (32) and $K = 5.8 \times 10^{-7}$ dyn (33). The best fit value of A from GNS at $18°C$ is $20,840$ sec$^{-3/2}$, within 10% of the calculated value. Figure 7 presents a comparison of the various theoretical approaches to T_1 in MBBA. The original papers of Blinc et al (23b), UPD (22), Dong et al (36a), and Nordio & Segre (31a–c) fit their theories to MBBA relaxation data (23b) from 2 to 270 MHz; the fits obtained for that range were quantitative to within experimental error. Blinc et al used the Torrey theory and fitted the distance of closest approach, the self-diffusion coefficient, $\langle r^2 \rangle / a^2$, and an added frequency-independent term B in Equation 8; the resulting best fit value of D was within 50% of the independently measured value. UPD varied A and τ_c in Equation 23 and obtained (for a resonant frequency in H_z) $A = 2.1 \times 10^4$ sec$^{-3/2}$ and $\tau_c = 4.5 \times 10^{-9}$ sec. Application by UPD of their intermolecular mechanism gave physically unrealistic values for the fitting parameters (22). The paper by Dong et al incorporates molecular reorientation effects by adding, somewhat heuristically, a molecular motion term (similar to Equation 7) to T_1 by assuming that $T_1^{-1} = T_{1OF}^{-1} + B_1\tau_d(1 + \omega_0^2\tau_d^2)^{-1}$. Their fit resulted in a value of A within 10% of the calculated value, of $B_1\tau_d = 6.1$ sec^{-1}, and of $\tau_d = 2.9 \times 10^{-9}$ sec. Nordio fit the MBBA data by using a literature value for $D_\perp$ plus values from his theory for the other terms in Equation 25 to obtain a numerical expression for

Table 1 Parameters obtained by GNS (27) for MBBA

Temperature ($T/°C$)	Order fluctuation term		Self-diffusion term		Rotational term		Average error (δ'')
	(A/sec$^{-3/2}$)	(H_L/Oe)	(τ_{SD}/sec)	(d/Å)	(τ_M/sec)	(ξ)	
18	20840	1	1.21×10^{-9}	2.67	2.43×10^{-7}	3.5×10^{-4}	5
27	19940	1	0.65×10^{-9}	2.64	1.45×10^{-7}	9.5×10^{-4}	4.6
35	15980	1	0.50×10^{-9}	2.61	0.94×10^{-7}	6.8×10^{-3}	2.1
45	12670	1	0.30×10^{-9}	2.50	0.57×10^{-7}	1.6×10^{-2}	4.3

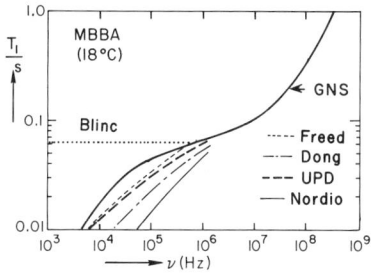
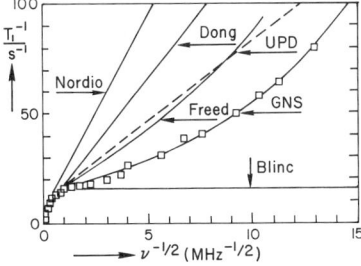

Figure 7 Comparison of theoretical treatments of relaxation in MBBA. Curves labeled "Blinc", "UPD", and "Dong" are reproduced from (27). Curves labeled "Freed" and "Nordio" are merely meant to be representative. They were calculated as explained in the text.

T_{1M}. This expression was frequency dependent above 20 MHz. By subtracting T_{1M}^{-1} from the experimental data, Nordio obtained a straight line that, in his theory, represents the OF contribution: $(T_{1OF}^{-1})_{fit} = 1.4 \times 10^4 v_0^{-1/2} + 1.36 \text{ sec}^{-1}$. These results are extrapolated in Figure 7.

While all of these fits might be improved if applied to the more recent data in (27), it is doubtful whether completely satisfactory agreement would be obtained. There is ample evidence (27) in the GNS fitting attempts, which used terms qualitatively similar to those in most of these other theories, that all three mechanisms must be included; at the very least diffusion and collective modes must be applied. Freed's theory is derived for intramolecular effects, so one anticipates that many of the problems that appear in other efforts to treat the MBBA data will be present with this theory as well. In order to provide a crude comparison with the other theories, however, we apply Equation 21 to the 2 MHz datum at 18°C and obtain a $\tau_R = 1 \times 10^{-8}$ sec. This result is then extrapolated to other frequencies in Figure 7. It should be emphasized that this calculation is very risky, because it assumes isotropic diffusion and that κ can be calculated using the high frequency limit; both assumptions are suspect for MBBA. The same curve correctly predicts T_1 at 10 MHz, but underestimates the experimental T_1 at 270 MHz by about 30%.

In general, none of the theories (with the exception of the fitting of GNS) correctly describes proton relaxation in MBBA over the entire range of experimental conditions.

Dong et al (36a) have measured relaxation for the chain protons of MBBA-$d8$ at 18°C at resonant frequencies from 4.5 to 60 MHz. The behavior shows many similarities to the results for MBBA over the same region. Their analysis argues that the temperature dependence of relaxation in MBBA arises from the relaxation of the end-chain protons, rather than from the effect of viscosity on the OF term. On the other hand, Goren et al (34a) contended that the temperature dependence arose primarily from the viscosity.

PROTON RELAXATION IN SMECTIC LIQUID CRYSTALS Virtually all of the work on relaxation in smectic phases has been done in a series of elegant papers by Blinc and co-workers (24a,b). Formulae have been derived for relaxation by the OF mechanism and for a wide variety of other mechanisms (e.g. smectic C fluctuations in smectic A phases) that may be important in these phases (24a,b). Because smectics are in general more viscous than nematics, molecular motion and diffusion may be slower in smectics, and motional narrowing may not be valid. Consequently, relaxation in these phases is more involved than in nematics.

Figure 8 shows the temperature dependence of T_1 at 60 MHz (24a) in TBBA (Figure 4), a compound with six mesophases. A preliminary report of the relaxation frequency dependence (0.14 to 90 MHz) helps define the processes involved (24b).

The nematic-phase relaxation behavior is unusual relative to most nematics in that T_1 increases as temperature decreases. An explanation can be found in Equation 20a and in the knowledge that as the smectic A is approached, K_2 and K_3 increase rapidly; thus T_1 is expected to increase.

The dispersion of T_1 in the smectic A phase obeys, at higher frequencies (1–90 MHz), the frequency (Equation 11) and angular (θ_0 in Figure 3) predictions

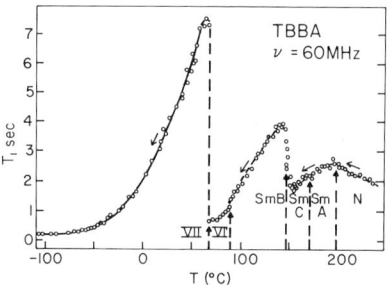

Figure 8 Proton T_1 in TBBA. (After Reference 24a.)

of the OF mechanism; at lower frequencies (below 0.5 MHz), T_1 deviates from $v_0^{-1/2}$. An interpretation postulates two mechanisms: nematic-type director fluctuations in the smectic phase as well as a coupling between the nematic and smectic order (24b). All three elastic constants must be present for this mechanism to be effective, and frequency cutoffs are required to fit the experimental curves adequately. The same mechanism appears to be effective in the smectic C phase.

At the smectic C to B transition, self-diffusion is "frozen out" (as evidenced by changes in the dipolar relaxation time) (24a), and T_1 changes drastically. The structures of the lower-temperature smectic phases are as yet unknown, but it appears they have order within the layers (that is they do not behave as two-dimensional liquids). The relaxation frequency dependence in these phases differs markedly from that in the higher-temperature smectics. At lower frequencies, T_1 is proportional to v_0^2; at higher frequencies (above 20 MHz for smectic B) T_1 is independent of frequency. This behavior is describable as a superposition of two sets of relaxation processes, one in the fast-motion limit and one in the slow-motion limit. These may reflect motions of the chains that are not frozen out. In the higher-temperature phases, these motions would be too fast to affect relaxation. In the smectic B, then, relaxation is postulated to be dominated by a gradual slowing down of the molecular chain segments. At about 84°C, the phase transition to VI (in Figure 8) occurs, with a decrease in T_1; a discontinuous change in molecular motion is inferred.

At 68°C, the VI–VII change freezes out the motions of the chains, although the terminal methyl-group rotation may continue to contribute to T_1 at lower temperatures.

PROTON RELAXATION IN CHOLESTERICS Only a few papers on relaxation in cholesteric phases have been published; these are summarized in (40). Many esters of cholesterol form mesophases (a cholesteric and often a smectic) only upon cooling from the liquid phase. Very little theoretical work on relaxation in these systems has been done, and the experimental work shows a general trend toward nonexponential relaxation in the liquid and cholesteric phases. The problems inherent in the analysis of nonexponential relaxation preclude a definitive analysis of relaxation mechanisms. Relaxation in the smectic phases of all cholesteric esters studied obeys a simple exponential relation and shows little temperature dependence.

Two papers on T_1 in chiral-doped nematics have appeared (41a,b). The cholesteric-nematic "unwinding" transition is a function of H_0, T, and the concentration of the chiral "impurity." Relaxation is affected by the transition in these (41a,b) and in 4,4′-di-*n*-hexyloxyazoxybenzene doped with 4,4′-bis-(2-methyl-1-butoxy)-azoxybenzene (28c). The helix pitch affects rotating-frame relaxation in a chiral-doped nematic for fields below H_c (41a). An adaptation of OF theory (21b) was used to explain the data (41a).

Deuteron Relaxation in Liquid Crystals

Deuteron relaxation (T_{1Q}) is useful because it is almost entirely intramolecular. Only a few applications of T_{1Q} to liquid crystals have been published, but most of these show rather unusual behavior.

Ring deuteron relaxation in PAA-d8 (42) in 4,4-di-*n*-heptyloxyazoxybenzene (HOAB-d8) (42) and in MBBA-d8 (35) show a strong temperature dependence (Figure 9). Arrhenius plots of the data yield an activation energy of ≈ 5.5 kcal mole^{-1} for PAA-d8 in the nematic phase and ≈ 9.3 kcal mole^{-1} for MBBA-d8. In HOAB-d8, oriented sample studies show that T_{1Q} is independent of the angle θ_0.

Compared to proton relaxation, another unusual aspect of these studies is the lack of frequency dependence of T_{1Q} in PAA-d8 (4.5 to 15.5 MHz) and MBBA-d8 (4.5 to 9.2 MHz) (Figure 9). The OF mechanism predicts a frequency dependence for deuterons similar to that for protons; however, β' (Equation 21), the angle between the C–D bond and the molecular axis, is approximately 60° in these systems. This reduces the strength of the collective modes term, so other mechanisms might be more effective. For the ring deuterons, internal rotations and rotations about the long molecular axis can modulate the electric quadrupole interaction and lead to relaxation; these motions are of marginal significance for ring proton relaxation (17c).

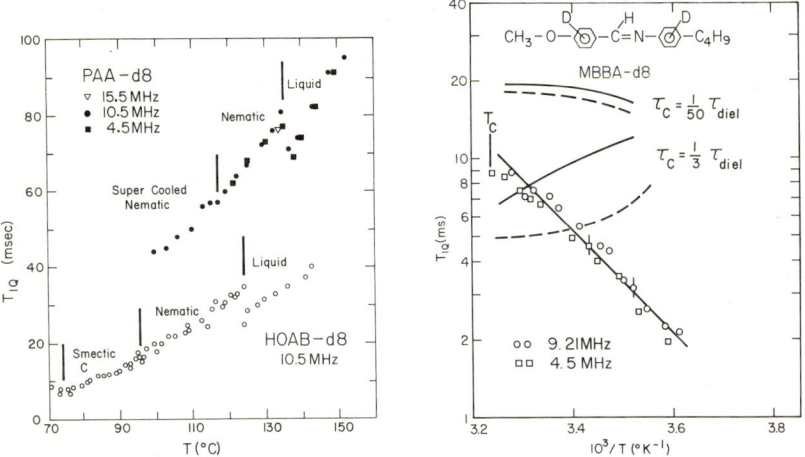

Figure 9 Ring deuteron relaxation in (*a*) PAA-d8 and HOAB-d8 (42); (*b*) MBBA-d8 (35). Solid and dashed lines are theoretical plots at 9.21 and 4.5 MHz, respectively, predicted by UPD theory (22) using dielectric relaxation correlation times, τ_{diel}.

Theoretical limitations become apparent when the UPD (22) theory is applied to Figure 9; Nordio's formulation (31a–c) may present problems as well. UPD predictions depend critically upon the size of the correlation time (35). Qualitatively correct behavior is obtained for MBBA if, in Equation 23 (corrected for the quadrupole case with $\beta' = 66°$), τ_c is 10^{-2} times the dielectric relaxation correlation time. However, the resulting T_{1Q} is about four times too large using this approach, and τ_c is about 10^{-11} sec at 18°C, compared to a value of 4.5×10^{-9} sec from proton studies. Nordio's theory has not been applied to all the data, but Reference 31b quotes agreement with T_{1Q} at 10.5 MHz for PAA-d8 at 120°C if the OF term is neglected and a $D_\perp = 0.16 \times 10^{10}$ sec is used; the proton case for PAA-d6 for this theory gave a $D_\perp$ of 7.7×10^8 sec.

Freed's theory (25) provides a possible explanation of T_{1Q} through the molecular reorientation term. His formulation for T_{1Q} (with $\beta' = 60°$ and the other parameters used above) yields for PAA-d8 $\tau_R \approx 2 \times 10^{-11}$ sec at 120°C and 1.5×10^{-11} sec at 132°C. Within experimental error, both are independent of frequency. The value at the lower temperature is within a factor of ten of the τ_R calculated for the protons of PAA-d6 (above). For MBBA at 18°C, this theory predicts $\tau_R \approx 3 \times 10^{-10}$ sec, which is considerably smaller than the proton estimate for MBBA.

In all three theories, the frequency dependence depends on the size of β', since the OF term vanishes for $\beta' \approx 54°$; however, Freed's formulation is the least sensitive to that angle. Rather interestingly, Freed's theory predicts that T_{1Q} will become frequency dependent at frequencies *lower* than those studied.

Methyl deuteron T_{1Q} in PAA-d6 has been measured at 9 MHz (40) and is approximately independent of temperature in the nematic phase.

Deloche & Cabane (44) studied the deuteron relaxation of the dimer molecules of the liquid crystal *p-n*-hexyloxybenzoic acid (PHBA) deuterated at the acidic hydrogen position. For this molecular system, relaxation of the deuterons is affected by the molecular orientational motions and the rapid exchanges of hydrogen bonding. In the nematic phase, T_{1Q}^{-1} varies with temperature in a manner similar to S^{-1} and diverges as T_c is approached. T_{1Q} is frequency dependent over 2.7 to 13.8 MHz. The theory to explain this behavior (44) proposes a coupling between the dynamics of the hydrogen bonding and the orientational motions of the molecules. For example, in the nematic phase, collisions of approximately parallel dimer molecules would not tend to break the dimers. In an isotropic fluid, however, collisions would have a greater tendency to disrupt the dimer structure. Relaxation in PHBA is ultimately controlled by the OF modes. However, the temperature and frequency dependence of T_{1Q} is produced in the nematic phase by fluctuations of the local concentrations of broken hydrogen bonds. These fluctuations are coupled with the OF modes and controlled by them. Deuteron relaxation in the liquid phase was also studied.

Deuteron relaxation in the nematic phase of 4-4'-d^{17}-*n*-octylbiphenyl at 38°C and 15.35 MHz has been reported (45). Relaxation times could be separately measured for the CD_3 group on carbon eight and for the methylene carbons four through seven. Along the chain, T_{1Q} increases from 20 msec at carbons four and five to 270 msec at carbon eight. The dominant relaxation mechanism appears to come from the internal motions of the chain rather than from the OF mechanism.

Carbon-13 Relaxation in Liquid Crystals

Carbon-13 has a spin of $\frac{1}{2}$ and is relaxed through modulation of the dipolar interactions with bonded protons. The natural abundance of ^{13}C is low ($\sim 1\%$) and its chemical shift is much larger than that of the proton. Consequently, the ring and methyl carbon signals can be resolved in an NMR experiment in liquid crystals. Schwartz et al (46) measured the carbon-13 relaxation times for the ring and methyl carbons at 25 MHz for PAA in the nematic and liquid phases. For the methyl carbons, T_1 was approximately 5.8 sec and independent of temperature over the entire temperature range. The ring carbon T_1 values showed Arrhenius behavior with an activation energy of about 6.5 kcal mole^{-1}; the relaxation behavior was unaffected by the T_c transition. Methyl relaxation occurs via rotation about the methyl axis combined with reorientation of this axis about the major molecular axis. The lack of temperature dependence for the methyl carbon T_1 can be explained by a partial contribution to T_1 of a spin-rotational interaction mechanism. Such is known to be of some significance in rapidly rotating methyl groups, and it has a temperature dependence opposite that of the dipolar mechanism.

Ring carbon relaxation is affected primarily by modulations affecting the C–H bond, which forms an angle of approximately 60° with the major axis. Consequently, rotations of the molecule about the long axis and internal rotations of the aromatic rings can lead to relaxation. The measured activation energy is typical of hindered rotation, so the authors favored the interpretation that internal rotation was the dominant mechanism.

CONCLUSIONS AND FUTURE POSSIBILITIES

Interest in relaxation in liquid crystals has often focused upon a unique mechanism: fluctuations in the order director. It is well-known, however, that other mechanisms, especially molecular reorientations and self-diffusion, are also important in relaxation in these systems.

It appears from the present literature that the OF mechanism may be a significant contributor to the intramolecular T_1 of the ring protons in nematics. The theory for this mechanism is now firmly established. Theories that treat intramolecular relaxation in liquid crystals (combining OF and molecular tumbling mechanisms) (22, 25, 31a–c) provide a semiquantitative description of the frequency dependence of the intramolecular relaxation of the ring protons of PAA-d6, the one case in which the intramolecular contributions have been experimentally separated. In addition, Freed's theory (25) provides, through the molecular reorientation term, an explanation of the lack of frequency dependence and of the strong temperature dependence of the ring deuteron relaxation in PAA and MBBA.

There are indications that present theories predict the proper temperature dependence for T_1 for the OF mechanism, but more experimental work needs to be done to verify this.

Much more work, both theoretical and experimental, needs to be done on the role of diffusion as a relaxation mechanism in all types of liquid crystals. No present general theory adequately includes intermolecular effects. The fragmentary experimental evidence (28a) that exists on this mechanism indicates that it may be more involved than the diffusional contribution to relaxation in isotropic systems.

Wide-range frequency-dependence studies, such as the field cycling work of Noack et al (26a,b, 27) on PAA and MBBA, and of Blinc et al (24b) on TBBA, have made a significant contribution to the field. Similar studies on other molecules would provide valuable information.

The full potential of deuteron, carbon-13, and nitrogen relaxation to provide detailed information on the dynamics of specific atomic positions on a molecule have yet to be exploited, although a few deuteron and carbon-13 studies were described above. A significant and increasing number of NMR line-width studies using these nuclei have been reported in the literature.

Of the various phases, nematics have received the greatest attention, and relaxation in these is reasonably well understood. Detailed studies of smectics are just beginning; cholesterics have received comparatively little attention.

ACKNOWLEDGMENTS

This review was prepared at the Laboratory of Chemical Biodynamics of the Lawrence Berkeley Laboratory; the hospitality and stimulating discussions of Melvin Calvin, Mel Klein, and the entire membership of that laboratory are greatly appreciated. Discussions with C. R. Dybowski, J. W. Doane, and J. H. Freed, and correspondence with P. Nordio, V. Graf, and F. Noack, were helpful. Tony Moore and Mirko Hrovat at the University of Texas gave special assistance in the theoretical aspects of this review, and their efforts are especially appreciated. Thanks are due to Gloria Goldberg for typing and Marlynn Amann for technical illustrations.

This work was supported in part by the National Institutes of Health (Grant HL-12528), the Robert A. Welch Foundation (Grant F-370), and the Division of Biomedical and Environmental Research of the Energy Research and Development Administration.

Literature Cited

1. Gray, G. W. 1962. *Molecular Structure and the Properties of Liquid Crystals.* New York: Academic. 314 pp; Priestly, E. B., Wojtowicz, P. J. 1975. *Introduction to Liquid Crystals.* New York: Plenum. 356 pp.
2. Brown, G. H., Doane, J. W., Neff, J. D. 1971. *A Review of the Structure and Physical Properties of Liquid Crystals.* Cleveland: Chem. Rubber Co. 94 pp.
3. Swedish Institute for Surface Chemistry. *5th International Conference on Liquid Crystals.* 1974. *J. Phys. Paris Colloq. No. 1.* 421 pp.
4. de Gennes, P. G. 1974. *The Physics of Liquid Crystals.* London: Oxford Univ. Press. 333 pp.
5. Stephen, M. J., Straley, J. P. 1974. *Rev. Mod. Phys.* 46:617–704; Chistyakov, I. G. 1967. *Sov. Phys. Usp.* 9:551–73; Chandrasekhar, S. 1976. *Rep Prog. Phys.* 39:613–92
6. Cabane, B. 1972, *Adv. Mol. Relaxation Processes* 3:341–53
7. Blinc, R. 1976. In *NMR Basic Principles and Progress*, ed. M. Pintar, 13:97–111. Berlin: Springer
8. Abragam, A. 1962. *The Principles of Nuclear Magnetism*, Chap. 8 Oxford; Clarendon. 599 pp; Goldman, M. 1970. *Spin Temperature and Nuclear Magnetic Resonance in Solids.* Oxford: Clarendon. 246 pp; Carrington, A., McLachlan, A. D. 1967. *Introduction to Magnetic Resonance.* New York: Harper & Row. 266 pp.

9a. Friberg, S., ed. 1976. *Lyotropic Liquid Crystals and the Structures of Biomembranes*. Washington DC: Am. Chem. Soc. 156 pp.
9b. Khetrepal, C. L., Kunwar, A. C., Tracey, A. S., Diehl, P. 1975. See Ref. 7. Vol. 9. 85 pp.
10. Waugh, J. S., ed. *Adv. Magn. Reson.*; Mooney, E. F., ed. *Annu. Rep. NMR Spectrosc.*
11a. The Chemical Society. *Nuclear Magnetic Resonance, A Specialist Periodical Report*. London: Chem. Soc.
11b. Emsley, J. W., Feeney, J., Sutcliffe, L. H., eds. *Prog. Nucl. Magn. Reson. Spectrosc.*
12. Tiddy, G. J. T. See Ref. 11a, Chap. 7, 4:233–52, 6: In press
13. Brown, G. H. 1973. *J. Opt. Soc. Am.* 63:1505–14; 1969. *Anal. Chem.* 41: 26A–39A; Saupe, A. 1968. *Angew. Chem. Int. Ed. Engl.* 7:97–112
14. Kast, W. 1960. *Landolt-Bornstein Tables*, Vol. 2, Pt. 2a, pp. 266–335. Berlin: Springer. 974 pp.
15. Leslie, F. M. 1966. *Q. J. Mech. Appl. Math.* 19:357–70
16a. Torrey, H. 1953. *Phys. Rev.* 92:962–69; Harmon, J., Muller, B. H. 1969. *Phys. Rev.* 182:400–10
16b. Hwang, L-P., Freed, J. H. 1975. *J. Chem. Phys.* 63:4017–25
17a. Doane, J. W., Visintainer, J. J. 1969. *Phys. Rev. Lett.* 23:1421–23
17b. Weger, M., Cabane, B. 1969. *J. Phys. Paris Colloq.* C4, 30:72–73
17c. Blinc, R., Hogenboom, D. L., O'Reilly, D. E., Peterson, E. M. 1969. *Phys. Rev. Lett.* 23:969–72
18. Dybowski, C. R., Smith, B. A., Wade, C. G. 1971. *J. Phys. Chem.* 75:3834–36
19. Pincus, P. 1969. *Solid State Commun.* 7:415–17
20. Lubensky, T. C. 1971. *Phys. Rev. Sect. A* 2:2497–2514; Sung, C. C. 1971. *Chem. Phys. Lett.* 10:35–38
21a. Doane, J. W., Johnson, D. L. 1970. *Chem. Phys. Lett.* 6:291–95
21b. Doane, J. W., Tarr, C. E., Nickerson, M. A. 1974. *Phys. Rev. Lett.* 33:620–24. There are several errors in this paper; see (22, 23a) for corrections.
22. Ukleja, P., Pirs, J., Doane, J. W. 1976. *Phys. Rev. Sect. A* 14:414–23
23a. Blinc, R. Vilfan, M., Rutar, V. 1975. *Solid State Commun.* 17:171–74
23b. Vilfan, M., Blinc, R., Doane, J. W. 1972. *Solid State Commun.* 11:1073–75
23c. Zupancic, I., Zagar, V., Rozmarin, M., Levstik, I., Kogovsek, F., Blinc, R. 1976. *Solid State Commun.* 18:1591–93
24a. Blinc, R., Luzar, M., Vilfan, M., Burgar, M. 1975. *J. Chem. Phys.* 63:3445–51
24b. Blinc, R., Luzar, M., Mali, M., Osredkar, R., Seliger, J., Vilfan, M. 1976. *J. Phys. Paris Colloq.* C3, 37:C3–73
25. Freed, J. H. 1977, *J. Chem. Phys.* 66:4183–99
26. Wolfel, W., Noack, F., Stohrer, M. 1975. *Z. Naturforsch. Teil A* 30:437–41
27. Graf, V., Noack, F., Stohrer, M. 1977. *Z. Naturforsch. Teil A* 32:61–72. The equation for A in this paper is a factor of two larger than that used in Refs. 22 and 23. The difference is perhaps due to different limits in the integration over q.
28a. Fung, B. M., Wade, C. G., Orwoll, R. D. 1976. *J. Chem. Phys.* 64:148–49
28b. Samulski, E. T., Dybowski, C. R., Wade, C. G. 1972. *Phys. Rev. Lett.* 29:340–42, 29:1050–51; 1973. *Mol. Cryst. Liq. Cryst.* 22:309–15
28c. Dewey, D. B. 1975. *Proton Relaxation Studies of Liquid Crystals*. MA thesis. Univ. Tex., Austin. 45 pp.
29. Vilfan, M., Zumer, S. 1976. *Intermolecular spin-lattice relaxation due to translational diffusion in liquid crystals—nematic phase*. Presented at 19th Congr. Ampere, Heidelberg
30. Polnaszek, C. F., Freed, J. H. 1975. *J. Chem. Phys.* 79:2283–2306
31a. Nordio, P., Segre, U. 1976. *Mol. Cryst. Liq. Cryst.* In press
31b. Nordio, P., Segre, U. 1976. *Gazz. Chim. Ital.* 106:431–37
31c. Agostini, G., Nordio, P. L., Rigatti, G., Segre, U. 1975. *Atti Accad. Naz. Lincei Mem. Cl. Sci. Fis. Mat. Nat. Sez. 2a.* 13:1–20
32. Meiboom, S., Hewitt, R., 1974. *Phys. Rev. Lett.* 30:261–63. The values of γ_1 in this reference are used for T_1 calculations.
33. Gruler, H. 1973. *Z. Naturforsch. Teil A* 28:474–83
34a. Goren, S. D., Korn, C., Marks, S. B. 1975. *Phys. Rev. Lett.* 34:1212–13
34b. Martins, A. F. 1972. *Phys. Rev. Lett.* 28:289–92
35. Visintainer, J. J., Dong, R. Y., Bock, E., Tomchuk, E., Dewey, D. B., Kuo, A-L., Wade, C. G. 1977. *J. Chem. Phys.* 66:3343–46
36a. Dong, R. Y., Tomchuk, E., Visintainer, J. J., Bock, E. 1976. *Can. J. Phys.* 54:1600–5
36b. Dong, R. Y., Forbes, W. F., Pintar, M. M. 1971. *J. Chem. Phys.* 55:145–51
36c. Easwaran, K. R. K. 1973. *J. Magn. Reson.* 9:190–97

37. Visintainer, J., Doane, J. W., Fishel, D. 1971. *Mol. Cryst. Liq. Cryst.* 13:69–84; Dong, R., Forbes, W., Pintar, M. 1972. *Mol. Cryst. Liq. Cryst.* 16:213–22
38. Dong, R., Wiszniewska, M., Tomchuk, E., Bock, E. 1974. *Mol. Cryst. Liq. Cryst.* 27:259–67
39. Woessner, D. 1962. *J. Chem. Phys.* 37:647–54
40. Matthews, R. C., Wade, C. G. 1975. *J. Magn. Reson.* 19:166–72
41a. Tarr, C. E., Feld, M. E. 1975. *Mol. Cryst. Liq. Cryst.* 30:143–48
41b. Dong, R. Y., Pintar, M., Forbes, W. 1971. *J. Chem. Phys.* 55:2449–51
42. Orwoll, R. D., Wade, C. G., Fung, B. M. 1975. *J. Chem. Phys.* 63:986–90
43. Wiszniewska, M., Dong, R. Y., Tomchuk, E., Bock, E. 1974. *Can. J. Phys.* 52:2294–95
44. Deloche, B., Cabane, B. 1972. *Mol. Cryst. Liq. Cryst.* 19:25–61
45. Emsley, J., Lindon, J. C., Luckhurst, G. 1977. *Mol. Phys.* 32:1187–90
46. Schwartz, M., Fagerness, P. E., Wang, C. H., Grant, D. 1974. *J. Chem. Phys.* 60:5066–68

INITIATION OF GASEOUS DETONATION

✥ 2639

John H. S. Lee

Department of Mechanical Engineering, McGill University, Montreal, Canada

INTRODUCTION

Most fuel-oxygen gas mixtures can be detonated. If a detonating mixture is diluted with an inert gas such as nitrogen, there then exists a particular oxygen to nitrogen ratio below which the mixture can no longer be detonated. If this oxygen to nitrogen ratio is less than about 0.25 (composition of air), the fuel will also detonate when mixed with air. If a mixture detonates, the detonation states can be predicted quite adequately for most practical purposes by the classical Chapman-Jouguet theory. However, no exact quantitative theory currently exists whereby one can predict, a priori, whether a given fuel-air mixture can detonate, and if so, what the detonability limits are. Neither can one predict whether a flame can accelerate to a detonation in this mixture, or whether the detonation can be initiated directly via a powerful explosive charge. The practical significance of the above questions is self-evident, and considerable effort has been directed in recent years toward answering these questions in connection with vapor cloud explosions, accidental or otherwise. Our qualitative understanding of the detonation phenomena is almost complete. The aim of this article is to report primarily on the initiation aspect and to demonstrate the central role it plays in the overall detonation phenomena. The article is not meant to be an extensive review of the current literature. It expresses my personal view of how the various pieces of the problem are now fitting together.

Generally speaking, there are two modes of initiation: a slow mode where the detonation is formed via an accelerating flame and a fast mode where the detonation is formed "instantaneously" when a sufficiently powerful igniter is used. The slow mode is usually referred to as the transition from deflagration to detonation. Turbulence and interactions between pressure waves and flame are the principle flame-acceleration mechanisms that generate the critical states for the onset of detonation. In general, the ignition source plays no role in the transition processes. On the other hand, the ignition source plays the dominant role in the fast mode of initiation. The blast wave generated by the igniter energy produces the necessary critical states for the onset of detonation. The fast mode is referred to as *direct initiation*, since the detonation is formed directly without a predetonation deflagration regime. It is also referred to as *blast initiation* in some recent literature to emphasize the role that the blast wave plays in the initiation processes. It would be appropriate to call the slow mode of transition from deflagration to detonation *self-initiation* because the detonation is caused solely by the energy release from the combustion of the mixture itself in the predetonation regime. The parameters that characterize these two modes of initiation are the transition distance for self-initiation and the igniter energy for direct initiation. The basic initiation mechanisms associated with these two modes are understood quite well on a qualitative basis. Quantitative models are also being formulated to predict the initiation parameters with some success (e.g. the energy required for direct initiation). The subsequent sections of this article are devoted to the exposition of the qualitative and quantitative aspects of these two modes of initiation as well as their roles in the overall detonation phenomena.

SELF-INITIATION

For most fuel-oxygen mixtures confined in a closed-end tube, a flame ignited by a weak spark or a glow wire will rapidly accelerate to a detonation after a flame travel of the order of a meter. Except for a few very sensitive fuels like acetylene, ethylene, hydrogen, etc, most hydrocarbon-air mixtures generally require a flame travel several orders of magnitude larger, thus exceeding the length of most laboratory-scale detonation tubes. The flame travel or the transition distance is dependent on a very large array of parameters (e.g. tube dimension and geometry, wall roughness, the nature, strength, and location of the igniter, boundary conditions at the ends of the tube, flame speed, heat of combustion, and sound speed of the mixture, etc). Qualitatively, the influence of these parameters on the transition distance is quite well understood. They are generally considered to contribute toward one or both of the two main flame-acceleration mechanisms of turbulence and shock-flame interaction.

To illustrate the general features of the self-initiation processes, consider the typical self-luminous streak record of flame acceleration to detonation in a tube, shown in Figure 1. Ignition is effected by the hot gas jet formed by the reflection of a detonation wave *1* on an orifice plate *2*. The reflected shock *3* is clearly visible as it propagates back into the combustion products. Upon ignition, the initially laminar flame *4* accelerates and becomes turbulent *5*. Further amplification of the turbulent flame results in the generation of shock waves and eventual transition to a detonation *7*. At the point of transition *6* a retonation wave *9* is seen to originate and propagate back into the burned gases. Transverse vibrations *8*, associated with the onset of detonation, are shown by the horizontal luminous bands. Transition of the initially laminar flame to the turbulent flame stems from two mechanisms. One is due to the onset of turbulence in the flow of unburned gases ahead of the flame when the Reynolds number becomes sufficiently high. The other mechanism is due to the interaction of pressure waves with the flame. The latter is beautifully

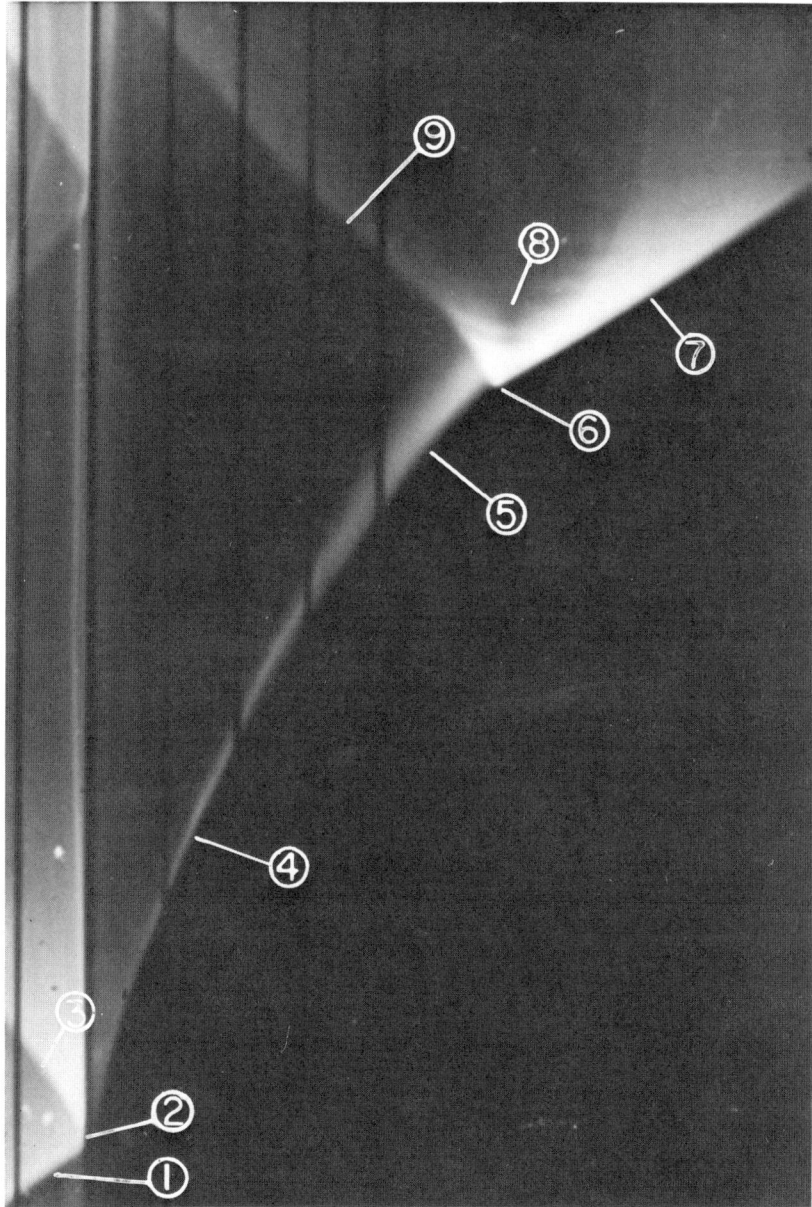

Figure 1 Self-luminous streak record of the transition from deflagration to detonation in $C_2H_2 + O_2$ at $p_0 = 100$ torr in a closed-end one-inch diameter tube.

illustrated by the sequence of schlieren photographs taken by Markstein (1), shown in Figure 2. The interaction of the shock with the flame is characterized by the folding of the flame to a tulip shape due to the Taylor interface instability mechanism, namely, the acceleration of the cold, dense, unburned gases behind the shock toward the hot, light, combustion products behind the flame front. The subsequent formation of a highly wrinkled flame front is clearly illustrated in these photographs. Associated with the increase in the flame area due to folding or wrinkling is an increase in the rate of energy release. This gives rise to a higher displacement rate of unburned gases and generates stronger pressure waves. The positive feedback

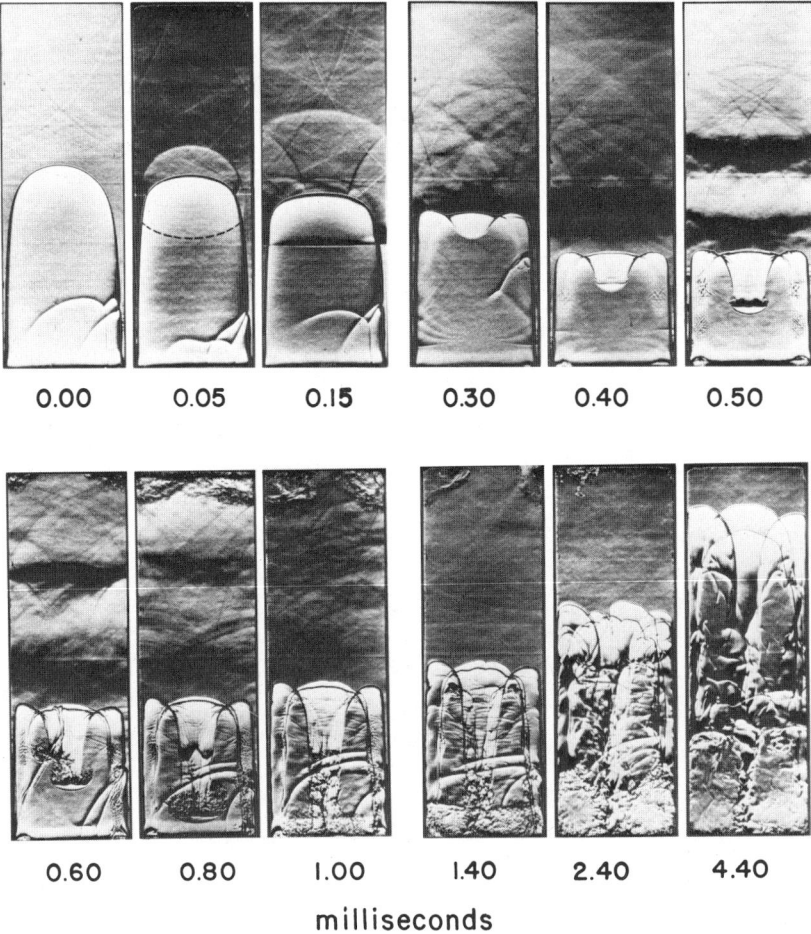

Figure 2 Interaction between stoichiometric butane-air flame and shock wave of pressure ratio 1.3. Ignition 23.6 msec before origin of time scale (1). (Courtesy of G. Markstein.)

mechanism between the burning rate of the flame and the gas-dynamic flow structure it generates eventually leads to the formation of the critical state for the onset of detonation. With turbulence playing the dominant role in this predetonation regime of the flame motion, it is expected that the transition distance is a difficult-to-reproduce experimental parameter that is generally apparatus dependent. Attempts to correlate the transition distance with the experimental parameters have met with limited success (2, 3). It is doubtful that correlations of sufficient generality can be obtained at present. Even in turbulent flames, universal correlations of experimental flame speeds with turbulence parameters have yet to be achieved. However, the qualitative aspects of the predetonation regime can be said to be fairly well understood. The contributions of Oppenheim and co-workers to the transition problem are particularly noteworthy (4–9).

The termination of the predetonation regime is marked by the "sudden" formation of the detonation. The onset of detonation is not a continuous process from flame to detonation. Just prior to the onset of detonation, the universal picture is one of a highly turbulent flame brush propagating behind a train of intense shock waves. The genesis of the detonation always occurs at some *localized* region in the unburned gases between the shock and the turbulent flame brush. This localized "explosion in the explosion," as Oppenheim called it, results from the formation of a "hot spot" due to finite gas-dynamic fluctuations in the hot, unburned gases. The exact physical nature of the processes prior and subsequent to the explosion in the explosion, leading to the eventual establishment of a fully developed detonation, is universal to all modes of initiation. It is therefore important to examine this in some detail.

Perhaps the most illuminating photographic records of the onset of detonation are those taken by Urtiew & Oppenheim (9) using the stroboscopic-laser schlieren technique. A typical sequence of such schlieren photographs of the onset of detonation is shown in Figure 3. The first frame shows the universal feature of a highly turbulent flame brush behind a shock front just prior to the onset of detonation. The localized explosion originating in the flame brush is quite evident in the second frame. A hemispherical blast wave from this localized explosion then propagates from the upper wall downwards. The coalescence of this blast wave with the leading shock front then forms the detonation. The reflection of the blast wave from the bottom wall sets up the transverse vibrations shown by the luminous bands in Figure 1. The part of the blast that propagates back into the burned gases is the retonation wave. In Figure 3 the localized explosion is the volume explosion of a "bubble" of unburned gases in the turbulent flame brush when its temperature rises to the autoignition limit. This mechanism was first suggested by Brinkley & Lewis (10).

Urtiew & Oppenheim also demonstrated that the localized explosion can originate either at the shock front, the boundary layer, or the contact surface formed when two shock waves merged ahead of the flame brush. The latter case has been analyzed in some detail (11), and the localized explosion has been shown to be a chemical kinetic branched-chain explosion.

Finite amplitude gas-dynamic fluctuations that trigger off the localized explosion in the hot, unburned gases can have numerous causes (e.g. local perturbations by the confining wall, wave interactions, volume explosions in the bubbles of trapped gases in the flame brush, etc). However, the average condition of the unburned gases between the time of shock and flame brush must be close to the autoignition limit of the mixture before transition can occur. Thus the essence of self-initiation is the generation of the critical thermodynamic states by the turbulent flame brush so that the localized explosion can be triggered by gas-dynamic fluctuations, which are always present in the vicinity of an intense turbulent flame (e.g. noise emission).

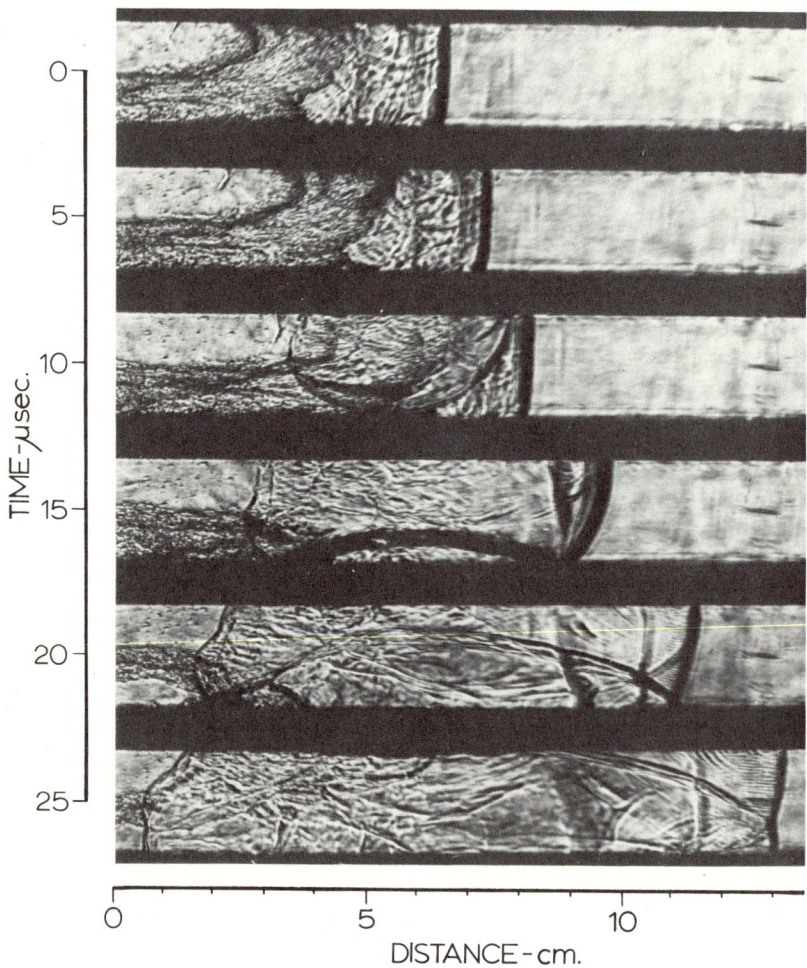

Figure 3 Stroboscopic schlieren photographs of the onset of detonation in $2H_2 + O_2$ at $p_0 = 554$ torr with the localized explosion originating at the flame brush (9). (Courtesy of A. K. Oppenheim.)

Criterion for Self-Initiation

Since the critical state for the onset of detonation is fairly universal and corresponds to the autoignition limit of the mixture itself, it should be possible to deduce the minimum energy-release rate in the turbulent flame required to generate this critical condition. Thus, a criterion may be formulated whereby at least qualitative estimates can be made concerning the possibility of self-initiation in a certain fuel-air mixture under a given set of conditions, and the relative ease with which self-initiation can occur in different mixtures can be ascertained. Such a criterion for transition has been proposed by Lee et al (12). The model assumes that the critical states are achieved via the adiabatic compression of a single shock wave generated by the turbulent flame. The energy release rate in the turbulent flame is linked directly to the rate of work done by the piston that maintains the required shock strength. Thus, for the energy release rate per unit area of the shock we may write:

$$dE/dt = \rho_u Q u_t = P_p U_p, \qquad 1.$$

where P_p and U_p denote the pressure and velocity at the piston surface, while ρ_u, Q, and u_t represent the density of the unburned gas in front of the turbulent flame, the chemical energy per unit mass, and the turbulent flame velocity, respectively. For a planar shock, the pressure and particle velocities at the piston surface are identical to their values at the front, which are given by the Rankine-Hugoniot relationships,

$$P_p = P_s = P_0[2\gamma M_s^2 - (\gamma - 1)]/(\gamma + 1),$$
$$U_p = U_s = 2\dot{R}_s(1 - 1/M_s^2)/(\gamma + 1),$$

where $M_s = \dot{R}_s/C_0$ is the shock Mach number, $\dot{R}_s$ is the shock velocity, and C_0 is the sound speed ahead of the shock. Equation 1 can be written, with the help of these relationships, as

$$\rho_u Q u_t = \frac{2P_0 C_0}{(\gamma + 1)^2} M_s [2\gamma M_s^2 - (\gamma - 1)](1 - 1/M_s^2). \qquad 2.$$

For most hydrocarbon-oxygen mixtures, the autoignition limits are of the order of $1100°K$. Thus the minimum shock strength required is $M_s \sim 4$ so that terms like $1/M_s^2$ and $(\gamma - 1)$ can be ignored in Equation 2. For hydrocarbon-air mixtures, the autoignition limit is even higher ($T \simeq 1400°K$ to $1500°K$). Hence Equation 2 reduces to

$$\rho_u Q u_t \simeq 4\gamma P_0 C_0 M_s^3/(\gamma + 1)^2. \qquad 3.$$

The density ρ_u ahead of the flame brush corresponds to the density behind the shock, which is given by the Rankine-Hugoniot equations as

$$\rho_u = \rho_s = \rho_0(\gamma + 1)/[(\gamma - 1) + 1/M_s^2] \simeq \rho_0(\gamma + 1)/(\gamma - 1),$$

where ρ_0 is the initial density of the mixture ahead of the shock. Substituting the above into Equation 3 leads to

$$u_t^* = 4(\gamma - 1)M_s^3/[(\gamma + 1)^3 Q^*], \qquad 4.$$

where $u_t^* = u_t/C_0$ and $Q^* = Q/C_0^2$. For hydrocarbon-air mixtures $Q \simeq 600 \text{ kcal kg}^{-1}$, while for the corresponding fuel-oxygen mixtures the value of Q is almost doubled. At a sound speed of about 330 m sec^{-1}, the representative values for Q^* would be about 20 and 40 for hydrocarbon-air and hydrocarbon-oxygen mixtures, respectively. Taking a value of $\gamma \simeq 1.4$ and the autoignition shock strength $M_s \simeq 5$ and 4 for fuel-air and fuel-oxygen mixtures respectively, Equation 4 gives the value for the dimensionless turbulent flame speed $u_t^* \simeq 0.7$ for hydrocarbon-air mixtures and $u_t^* \simeq 0.2$ for hydrocarbon-oxygen mixtures. At a sound speed of 330 m sec^{-1}, the turbulent flame speed required for self-initiation of fuel-air mixtures is about 230 m sec^{-1}, while for fuel-oxygen mixtures it is about three times less, ~ 65 m sec^{-1}. The laminar flame speed for most hydrocarbon-air mixtures is of the order of 0.5 m sec^{-1} (acetylene and hydrogen are about three times higher). Thus an amplification factor $u_t/u_l \simeq 460$ is required to produce the critical states for the onset of detonations. On the other hand the laminar flame speed is about 10 m sec^{-1} for the fuel-oxygen mixtures. The amplification factor required for fuel-oxygen mixtures is $u_t/u_l \simeq 6$. Thus a mere sixfold increase in the flame velocity is sufficient for the onset of detonation. This explains the great ease with which self-initiation can occur in the fuel-oxygen mixtures, as compared to the fuel-air mixtures. For mixtures of acetylene and hydrogen with air, the laminar flame speed is of the order of 1.5 m sec^{-1}. The amplification (u_t/u_l) required is therefore about 150, which is much lower than for most hydrocarbon-air mixtures. This may account for the fact that transitions in mixtures with high laminar flame speeds can be observed in most laboratory-scale experiments.

At this point, it is appropriate to clarify what is meant by a turbulent flame speed. In general, any departure of the flame from a smooth planar surface is called a turbulent flame. Thus, for a flame propagating in a constant-area tube, the folded flame due, say, to interaction with a pressure wave (Figure 2) may be called turbulent even though the folded structure is still smooth. Of course, if based on the rate of energy release per unit area of tube, the folded laminar flame will have a higher burning velocity. If the flame were to pass through a perforated plate, its surface area might be significantly folded. Hence, the burning velocity based on the heat release per unit area of the tube downstream of the orifice plate can be very high and will decrease as the "folds" burn themselves out. Thus, depending on the folded area of the flame, its burning velocity can be arbitrarily high even though the folded structure is still smooth, and the local burning velocity normal to the surface may still be about the same as the laminar burning velocity with curvature taken into consideration. The above discussions indicate the importance of flame folding, due to interaction of the flame with obstructions in its path and with reflected pressure waves, on the self-initiation process.

For very fine-scale turbulence, the flame sheet itself may have a rough structure. The local, normal burning velocity of the folded flame may now be significantly higher than the laminar flame velocity. Thus we may ascribe the turbulent burning velocity u_t based on the energy release rate per unit area of the tube to both large-scale flame folding and small-scale turbulent effect. Such a distinction is not generally made in turbulent flame theory where the large-scale wrinkled structure is considered turbulent. However, it appears appropriate to separate these two effects, as far as the transition phenomenon is concerned, to emphasize the importance of flame folding due to the interaction of the flame with obstructions and pressure waves.

It would be of interest to estimate the large-scale flame folding required for self-initiation under extremely fine-scale turbulence. From the recent studies of Abdel-Gayed & Bradley (13) on H_2-air flames, a value of $u_t/u_l \simeq 20$ is obtained in the limit of very large turbulent intensities ($u_l/u' \to 0$) over a wide range of turbulent Reynolds number. Taking their value as indicative of the limiting turbulent flame speed due to fine-scale turbulence, we see that for an amplification $u_t/u_l \simeq 150$, the large-scale flame folding required is about seven times. If we assume the characteristic tulip-shaped flame to be a cone, the cone height would be about 3.5 times the base diameter for a ratio of cone surface to base area of seven. This can readily be achieved via shock-flame interactions. On the other hand, the slow-burning hydrocarbon-air mixtures for which $u_t/u_l \simeq 460$ would require an effective area increase via flame folding of about 23 times, even with the same fine-scale turbulent velocity of 20 times the laminar value. This would be difficult to achieve naturally, but could be induced artificially if the flame were to propagate through a perforated plate or suitable wire grids. The use of hemispherical grids to induce transition in enriched ethylene-air mixtures has been reported by Wagner and co-workers (14). The effectiveness of obstructions in promoting self-initiation was first demonstrated by Shchelkin (15). He showed that a short wire coil placed inside the flame tube could lead to a significant decrease in the transition distance.

Since flame folding is a transient process, the shock wave generated as a result of the increased rate of energy release is also transient. For the onset of detonation, it is evident that the critical states must be maintained for a certain minimum duration for the autoignition to occur. If the duration of the shock is too short, the expansion waves associated with the shock decay will again cool the gas to below its autoignition limit. The minimum duration for which the shock must be maintained should be at least of the order of magnitude of the induction time corresponding to the autoignition temperature. For fuel-oxygen mixtures, the induction time is of the order of microseconds. However, for fuel-air mixtures the induction time is one to two orders of magnitude larger (i.e. tens of microseconds to milliseconds). Thus, for certain fuel-air mixtures, self-initiation may not be possible at all. Even though the required turbulent flame speed can be generated artificially, the folded structure cannot be maintained long enough for the onset of detonation to occur.

DIRECT INITIATION

Direct or blast initiation is the fast mode in which the detonation is formed in the immediate vicinity of the powerful igniter. The igniter must be capable of not only generating a strong shock wave, but of maintaining the shock above a certain minimum strength for some required duration. For a given igniter, the energy of the igniter characterizes the phenomenon. Below a certain threshold value of the ignition energy, it is found that the blast wave generated by the igniter will progressively decouple from the reaction front. The blast wave decays to a sound wave, and the subsequent propagation of the reaction front is identical to an ordinary flame. This has been referred to as the subcritical regime (16), and a sequence of schlieren photographs illustrating the phenomenon is shown in Figure 4a.

If the ignition energy exceeds the critical threshold value, the blast and reaction front are always coupled in the form of a multiheaded detonation wave that starts at the source and expands at about the Chapman-Jouguet detonation velocity. This is shown in Figure 4b and is referred to as the supercritical regime (16).

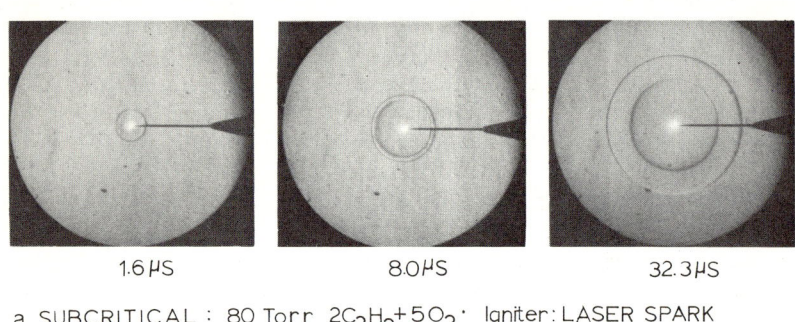

a. SUBCRITICAL : 80 Torr $2C_2H_2 + 5O_2$; Igniter: LASER SPARK

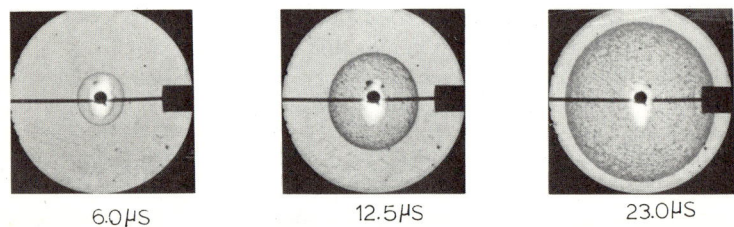

b. SUPERCRITICAL: 120 Torr $H_2 + Cl_2$; Igniter: ELECTRICAL SPARK

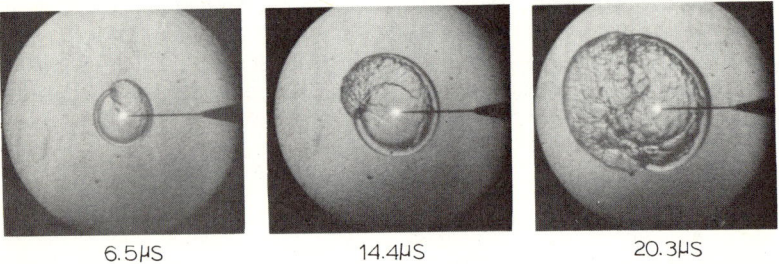

c. CRITICAL: 100 Torr $2C_2H_2 + 5O_2$; Igniter: LASER SPARK

Figure 4 Spark schlieren photographs of a spherical detonation illustrating the subcritical, supercritical, and critical regimes of initiation.

When the ignition energy is at the critical threshold value, the phenomenon is more interesting. For very early times, the blast and reaction front are coupled. As the blast expands, the decoupling occurs and the reaction front recedes from the shock. However, the decoupling process soon terminates when the chemical energy released by combustion begins to contribute significantly to the blast motion. The blast no longer decays, and the shock wave and the reaction front then propagate as a coupled complex at a constant velocity. This is called the quasi-steady period of the blast motion, and during this period the blast strength corresponds approximately to the autoignition limit of the mixture (i.e. $M_s \simeq 4$). The duration of the quasi-steady regime corresponds approximately to the induction time at the autoignition temperature. The termination of the quasi-steady regime is marked by the sudden appearance of a localized explosion in the shocked layer of gas bounded by the shock and the reaction front. The blast from the localized explosion immediately forms a multiheaded detonation "bubble" that then grows to engulf the entire layer of the shock-heated mixture, resulting in the formation of an asymmetrical detonation. These processes in the critical regime are illustrated in Figure 4c. The formation of the detonation from the localized explosion is referred to as "detonation reestablishment" because the shock and reaction front are initially coupled as in a detonation. Then decoupling occurs, and eventually a coupled, multiheaded detonation is regenerated by the action of the localized explosion. However, it is evident that reestablishment is identical to the onset of detonation in self-initiation. In direct initiation, the conditions for the onset of detonation are formed by the reacting blast-wave generated by the igniter. For self-initiation or the transition mode discussed previously, these same critical conditions are derived from the acceleration of the flame itself.

The localized explosion is formed when a hot spot resulting from gas-dynamic fluctuations is induced in the shocked layer of unburned mixture near its autoignition temperature. Figure 5 illustrates the formation of the localized explosion

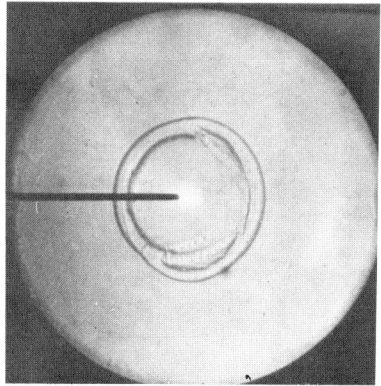 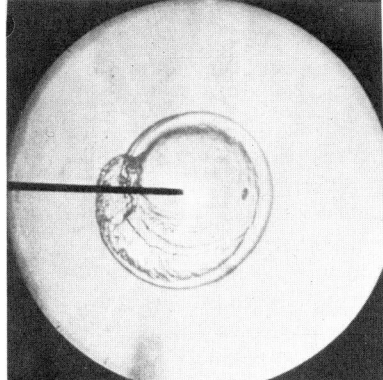

Figure 5 Spark schlieren photographs of the onset of detonation at the end of the quasi-steady period of the critical regime. The mixture is stoichiometric C_2H_2–O_2 at $p_0 = 100$ torr, and ignition is by a laser spark.

when the gas-dynamic perturbation is in the form of a shock wave generated by a "mild" volume explosion at the lower portion of the spherical blast. The shock wave propagating tangentially to the spherical blast induces the localized explosion, which generates the "detonation bubble" shown in the left frame of Figure 5. This detonation bubble then spirals around the blast sphere and causes the entire blast to become a detonation.

The three propagation regimes, namely, subcritical, critical, and supercritical, that characterize the direct initiation phenomena are based on the total energy of the igniter. The igniter energy can in general be readily measured experimentally, and for each mixture at given initial conditions there corresponds a critical energy for direct initiation. The earlier studies of direct initiation consist mainly of the experimental measurement of the critical energy for various fuel-oxygen mixtures (17). For a given igniter and geometry (e.g. spherical), the general behavior of the critical energy vs mixture composition at a given pressure is in the form of a U-shaped curve with the critical energy approaching infinity at the lean and rich limits. For a fixed composition, the critical energy increases as the pressure decreases toward the low-pressure detonation limit. However, quantitative comparisons between the critical energies obtained for the same mixture at the same initial conditions using different types of igniters have not been possible. It was found that even for the same type of igniter (i.e. electrical spark) the critical energy for the same mixture can differ by three orders of magnitude when the energy-time characteristics are varied (18). My colleagues and I have carried out a series of investigations to determine the dependence of the critical energy on the energy-time characteristics of the source (19, 20). The study is based primarily on the electric spark, since its energy-time characteristics can readily be controlled over a wide range through the R-L-C parameters of the discharge circuit. The spark energy can be obtained from a simultaneous measurement of the discharge current and the voltage drop across the spark. Cylindrical detonations are studied that are compatible with the spark geometry (i.e. a line source).

To establish what part of the spark energy actually contributes toward the initiation process, the "crowbar" or the clamped discharge technique is used to shunt the spark current through a bypass. In this way the total spark energy can be controlled by firing the crowbar switch at the desired time, thus terminating the main discharge. The results indicate that only the energy released up to the first quarter-cycle of the ringing damped discharge is relevant to initiation. The first quarter-cycle corresponds to the peak power of the spark. Thus, it may be concluded that only the energy released before the igniter attains maximum power is important in the initiation process.

The discharge frequency of the electrical spark has also been varied to determine the dependence of this meaningful spark energy on the time it takes to reach the peak power itself. These results show that the critical energy decreases with the duration of the energy release (i.e. time to reach quarter-cycle). A minimum value of energy is reached, below which any further decrease in the duration of the discharge does not change this minimum value. Thus, it appears that a limiting value of the critical energy is obtained as the power of the source approaches infinity (i.e. ideal line source). This minimum limiting value would then be independent of the source characteristics and represents the true critical energy for direct initiation of the explosive mixture. For the case of $C_2H_2-O_2$ and H_2-O_2 mixtures, the variations of the critical energy with the duration of the discharge are shown in Figure 6. The values for the H_2-O_2 mixture were obtained recently by Ramamurthi (21).

The question of spark geometry was investigated by Matsui & Lee (22). The linear electrical spark with different electrode spacings and configurations was used to initiate spherical detonations. The results are shown in Figure 7. For large electrode spacing the critical energy per unit length of the spark approaches a constant value of about 0.1 J cm^{-1} for a 100 torr oxyacetylene mixture. This indicates that although the detonation is eventually spherical, it is first formed as a cylindrical wave when the electrode spacing is large. There exists a range of electrode spacings where the critical energy per unit length increases linearly with decreasing spark length, thus giving a constant value of about 0.3 J for the total spark energy.

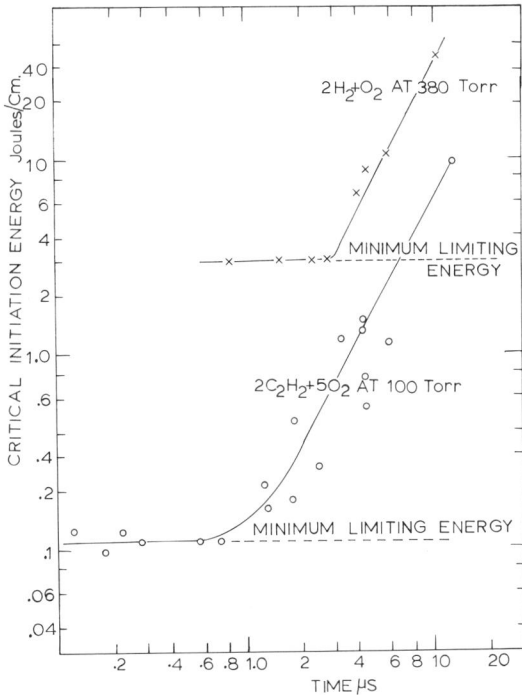

Figure 6 The variation of the critical energy for direct initiation with the time of its release for cylindrical detonations in stoichiometric $C_2H_2-O_2$ at $p_0 = 100$ torr (20) and in $2H_2 + O_2$ at $p_0 = 380$ torr (21).

This regime corresponds to the spherical geometry for which the energy should be independent of the spark length. For very small electrode spacing, the losses to the electrodes become important, and the critical energy increases sharply to compensate for the losses. For flanged electrodes, the initiation corresponds to the cylindrical geometry throughout. For electrode spacings less than 1.0 cm, the critical energy rises sharply. This indicates that confinement by the walls becomes important when the flange spacing is less than about five times the characteristic transverse wave spacing of the mixture. Further studies of the influence of electrode geometry on the initiation energy were carried out by Ramamurthi (21). He concluded that the geometry of the initiation can be established by comparing the explosion length associated with the source energy [i.e. $R_0 = (E_0/p_0)^{1/(j+1)}$, where $j = 0, 1, 2$ for planar, cylindrical, and spherical geometries respectively, E_0 is the source energy per unit area, per unit length, or just the energy itself for the three geometries, and p_0 is the initial gas pressure] with the characteristic length L of the igniter. For $R_0 > L$, the initiation is spherical since the onset of detonation occurs too far from the source for the igniter geometry to influence the initiation processes. For $R_0 < L$, the geometry of the igniter will dictate whether the initiation corresponds to a planar or a cylindrical geometry. For example, if L denotes the spark length, $R_0 < L$ corresponds to a cylindrical initiation since detonation is formed at a distance that is small in comparison to the length of the linear electrical spark.

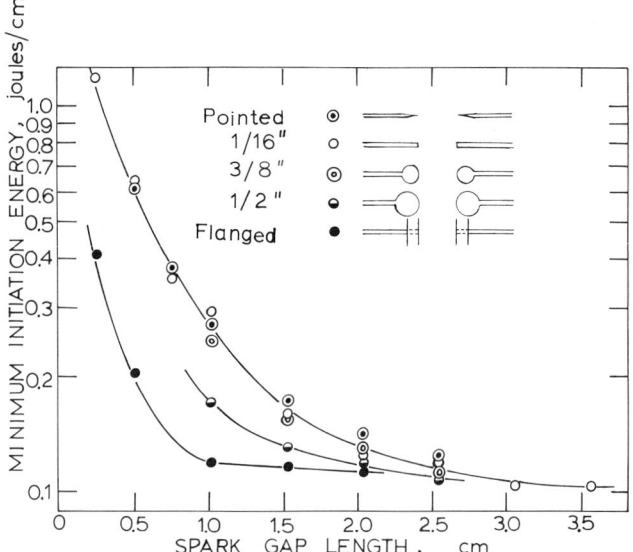

Figure 7 The dependence of the critical energy for direct initiation of spherical detonations in stoichiometric $C_2H_2-O_2$ at $p_0 = 100$ torr on the spark length for different electrode configurations (22).

The equivalence of the explosion length R_0 in the different geometries of initiation is also borne out by the experiments of Ramamurthi (21). In other words, if the critical initiation energy per unit area for a given mixture for the case of planar initiation is known, the corresponding critical energy for spherical initiation can be found since the explosion length is the same for all geometries (i.e. $E_{spherical}/E_{cylindrical} = E_{cylindrical}/E_{planar} = R_0$). The experimental data that support the above conclusion are shown in Table 1. The invariance of the explosion length with geometry is quite useful in practice. It enables the critical energy for direct initiation of spherical unconfined detonations to be estimated from the corresponding critical energy for the planar case. Thus, expensive, large-scale, unconfined experiments can be replaced by simpler, confined experiments in tubes where the detonation can be observed over a longer travel to guarantee that a self-sustained wave is indeed formed. Of course the tube diameter must be chosen so that wall effects on the initiation energy are negligible. In general, the tube diameter is of the order of about 10 transverse wave spacings of the detonation wave.

A comparison between the critical energies for acetylene-oxygen mixtures obtained using various igniters is shown in Figure 8. The critical energy versus composition curve demonstrates a characteristic U-shape. The minimum limiting value of the spark energies of (23) is generally an order of magnitude less than the exploding wire energies reported in (26) on the basis of the total $CV^2/2$ energy stored in the capacitor. For the same type of igniter (i.e. exploding wire) the critical energies obtained by Lee et al (26) agree with those reported by Litchfield et al (17). With the use of a piston model (23) to reduce the critical tube diameter data in Matsui's study of spherical detonation initiation by a linear detonation (24), the resulting equivalent energies may be seen to correspond closely to the minimum spark energy. The dependence of the critical energy on composition is qualitatively the same for most detonating gases. The sharply increasing trends in the initiation energy for fuel-lean and fuel-rich compositions, namely, the vertical arms of the U-shaped curve, are in fact used to determine experimentally the composition limits of detonability of explosive gas mixtures.

Table 1 Value of R_0 in the different geometries of initiation

Gas mixture	R_0 with spherical initiation (cm)	R_0 with cylindrical initiation (cm)	R_0 with planar initiation (cm)
100 torr $2C_2H_2 + 5O_2$	2.8	3.0	—
200 torr $2C_2H_2 + 5O_2$	1.23	1.3	—
300 torr $2C_2H_2 + 5O_2$	0.83	0.82	—
40 torr $2C_2H_2 + 5O_2$	—	7.2	7.5
80 torr $2C_2H_2 + 5O_2 + 6Ar$	7.0	7.4	—
380 torr $2H_2 + O_2$	6.4[a]	7.4	—
800 torr $2H_2 + O_2$	7.2	6.9	7.5

[a] Deduced on the basis of detonation diffraction experiments.

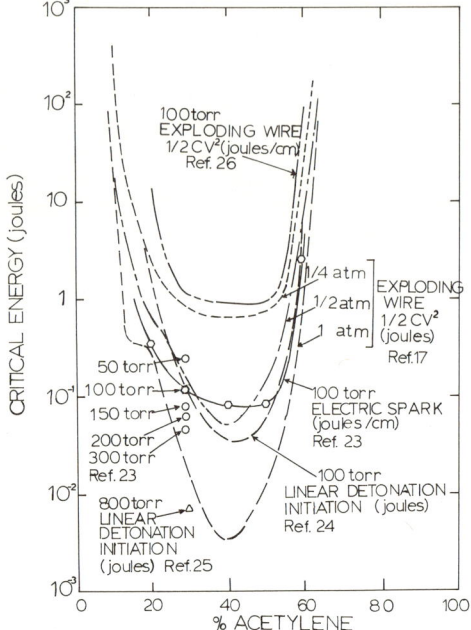

Figure 8 The dependence of the critical energy for direct initiation on mixture composition for C_2H_2–O_2 (23).

Theoretical Models for Direct Initiation

The fast mode of detonation initiation is more amenable to quantitative analysis in that the turbulence is not responsible for the creation of the critical conditions for the onset of a detonation. The critical conditions are a direct result of the coupling between the chemical energy release and the transient gas-dynamic flow structure generated by the igniter. However, due to the nonlinearity of the gas-dynamic equations, the problem is not straightforward. Nevertheless, the solution is much simpler than the self-initiation problem where turbulence and shock flame interaction play the dominant role.

A phenomenological criterion linking the critical energy to the chemical properties of the explosive gas mixture was given as early as 1956 by Zel'dovich (25). His argument was based on the fact that the blast wave generated by the igniter decays rapidly. Thus, in order for the chemical energy release in the wake of the blast to actually sustain the shock wave, a sufficient time must be available for the fluid particles to release their chemical energy after being processed by the shock wave. The time taken for a blast wave to decay to a certain strength is a function of the blast energy. Choosing the induction time as the characteristic time for the chemical reactions, Zel'dovich postulated that for direct initiation, the time for the blast to decay to the Chapman-Jouguet-Mach number (M_{CJ}) must be at least equal to the induction time for the mixture. From the above criterion he demonstrated that the critical energy for direct initiation of spherical detonations must be proportional to the cube of the induction time (i.e. $E_0 \sim \tau^3$).

Lee et al (26) attempted to use the Zel'dovich criterion for the quantitative determination of critical energies. They used the ideal strong blast solution of Taylor (27) to give the blast trajectory and shock strength. For the spherical geometry, these are given by

$$R_s = At^{2/5},$$
$$\dot{R}_s = 2At^{-3/5}/5, \qquad 5.$$

where

$$A = [25E_0/(16\pi\rho_0 I)]^{1/5},$$

ρ_0 is the density, and I is a numerical constant equal to 0.423 for $\gamma = 1.4$. Replacing $\dot{R}_s$ by $\dot{R}_{CJ}$ where $\dot{R}_{CJ}$ is the C-J velocity and t by the induction time τ, Equation 5 immediately yields

$$E_0 = 125\pi\rho_0 I C_0^5 M_{CJ}^5 \tau^3/2, \qquad 6.$$

where $M_{CJ} = \dot{R}_{CJ}/C_0$; C_0 is the sound speed in the undisturbed mixture.

Generalizing the above formula to other geometries we obtain

$$E_0 = k_j \rho_0 I (M_{CJ} C_0)^{j+3} [(j+3)/2]^{j+1} \tau^{j+1}. \qquad 7.$$

The numerical constant I is dependent on the specific heat ratio γ of the mixture. For $\gamma = 1.4$, $I = 0.423$, 0.626, and 1.212 for $j = 2, 1$, and 0, respectively. For a given mixture whose M_{CJ} and C_0 are known, the critical energy for direct initiation can be obtained from Equation 7 once the induction time τ is specified.

Since the blast wave is a transient wave, when its strength has decayed to M_{CJ} the induction zone thickness actually corresponds to the induction period for the gas particles that crossed the shock wave much earlier. In addition, the fluid particles upon crossing the shock are subjected to an adiabatic expansion. Hence, we need an exact solution of the flow structure behind a reacting blast wave in order to compute the actual induction time τ. If τ is determined on the basis of steady flow properties behind a shock of strength M_{CJ}, the calculated values of the critical energies are about three orders of magnitude less than those determined from experiments. However, if the experimentally observed reaction zone thickness (28) is used in Equation 7, the prediction of the correct order of magnitude of initiation energies is shown to be possible. The experimentally observed reaction zone thickness is generally referred to as the hydrodynamic thickness and represents an average distance between the shock front and an equivalent Chapman-Jouguet surface, namely, a one-dimensional length scale characterizing a three-dimensional detonation (29).

Edwards et al (30) and Sichel (31) have recently proposed certain relationships for obtaining the critical energy for direct initiation. In the expression given by Edwards et al, experimental values of the hydrodynamic thickness of the detonation wave must be used to evaluate the critical energy. In Sichel's proposal, an experimental value for the critical energy of a reference composition of the mixture is required to determine the critical energy corresponding to other compositions. Qualitatively, both Edwards' and Sichel's results are identical to Equation 7.

A more general model with greater fundamental significance has been developed by Lee & Ramamurthi (32). The idea behind the model is developed from an analogy with the flame ignition problem. In flame ignition, the energy source generates a volume of hot gases called a flame kernel. Successful ignition requires the flame kernel to be of a certain minimum size so that the rate at which heat is generated by the chemical reactions within the kernel offsets the rate of heat loss from this volume. In the case of detonation initiation, the energy source must be capable of producing a strong blast wave. By the time the blast wave has decayed to some critical value M_s^* (e.g. the autoignition limit), it is postulated that the blast radius must not be below a certain minimum size R_s^*. If the size of the so-called "detonation kernel" R_s^* is too small, the chemical energy released inside the kernel is negligible as compared to the ignition energy. This implies that the blast motion is still dominated by the igniter energy, and the shock wave will continue to decay in strength for $R_s > R_s^*$. Since the mixture subsequently processed by the blast wave can no longer spontaneously autoignite for $M_s < M_s^*$, the chemical energy release also terminates, and hence a deflagration results.

If, however, the detonation kernel size R_s^* is sufficiently large that the chemical energy release is significant compared to the ignition energy itself, subsequent motion of the blast wave for $R_s > R_s^*$ will be strongly influenced by the chemical energy release. The shock then couples to the chemical processes to produce a quasi-stationary regime that is necessary for the onset of detonation. Using the concept of the critical size of the detonation kernel, a simple theory can be developed for the prediction of the critical energy for direct initiation.

The starting point of the analysis is the conservation of the total energy enclosed by the blast wave at any instant, i.e.

$$E_0 + \int_0^{R_s(t)-d} Q\rho k_j r^j \, dr = \int_0^{R_s(t)} (u^2/2 + e)\rho k_j r^j \, dr - \int_0^{R_s(t)} e_0\rho_0 k_j r^j \, dr, \qquad 8.$$

where E_0 is the energy deposited by the igniter, d is the induction distance, Q is the chemical energy per unit mass of the mixture, e_0 is the initial internal energy of the gas mixture, $k_j = 1$, 2π, and 4π for $j = 0$, 1, and 2, corresponding to the planar, cylindrical, and spherical geometries respectively, and p, ρ, and u have their usual meaning. In Equation 8 the right-hand side denotes the increase in the kinetic and internal energies of the gas particles enclosed by the shock. This energy is contributed by the source energy E_0 and the energy from the chemical reaction, which together form the left-hand side of Equation 8.

Making a straightforward transformation to blast wave coordinates (33), Equation 9 can be written as

$$M_s^2 = \frac{1}{I}\left[\frac{E_0}{\rho_0 C_0^2 k_j R_s(t)^{j+1}} + \frac{Q}{C_0^2}\int_0^{1-d/R_s(t)} \psi\xi^j \, d\xi + \frac{e_0}{C_0^2}\int_0^1 \psi\xi^j \, d\xi\right], \qquad 9.$$

where I is the integral defined by

$$I = \int_0^1 [f/(\gamma - 1) + \psi\phi^2/2]\xi^j \, d\xi, \qquad f = P/\rho_0 \dot{R}_s^2, \qquad \psi = \rho/\rho_0,$$

$$\phi = u/\dot{R}_s, \qquad \text{and} \qquad \xi = r/R_s.$$

No assumptions have been made thus far, and Equation 9 is an exact statement of the conservation of the total energy. From Equation 9 we can obtain the appropriate limits of energy-dominated and chemical-energy-dominated blast wave motion. Thus, for instance, for small $R_s(t)$, the source energy term on the right-hand side of Equation 9 dominates the chemical energy and initial internal energy terms so that we recover the strong blast limit for which $M_s \sim R_s^{-(j+1)/2}$ as given by the classical self-similar solution of Taylor (27). For large shock radius $R_s(t)$, i.e. $R_s(t) \to \infty$ and $d/R_s(t) \to 0$, the chemical energy dominates over the other terms, and we get a shock wave sustained by chemical heat release, i.e. $M_s \to M_{CJ}$. The criterion $d/R_s(t) \to 0$ implies that the shock and chemical reactions are coupled. If initiation fails, the reaction front propagates as a flame, and $d/R_s(t)$ tends to unity since the flame speed is of the order of a few cm sec^{-1}, in contrast to the shock, whose minimum velocity is the speed of sound, about 350 m sec^{-1}.

A further examination of Equation 9 reveals that the strength of the blast wave at any instant is the consequence of two competing terms; the first term on the right-hand side of Equation 9 is a decreasing function of the shock radius, while the second term, due to the chemical energy release, has an increasing trend. Thus, analogous to the flame kernel used in the flame ignition theory, we define a minimum size R_s^* for the detonation kernel by the condition that the two competing terms be of the same order. Formally we equate the two terms and write:

$$\frac{E_0}{\rho_0 C_0^2 k_j R_s^{*j+1}} = \frac{Q}{C_0^2} \int_0^{1-d/R_s^*} \psi \xi^j \, d\xi, \qquad \qquad 10.$$

when $R_s(t) \to R_s^*$ and $M_s \to M_s^*$. Since the two terms are equal we now write Equation 9 as

$$M_s^{*2} = \frac{1}{I}\left[\frac{2Q}{C_0^2} \int_0^{1-d/R_s^*} \psi \xi^j \, d\xi + \frac{e_0}{(j+1)C_0^2}\right]. \qquad 11.$$

The integral I in the above equation is not sensitive to the details of the gasdynamic flow structure behind the shock front and tends to have a constant value. This constant value can be obtained from the condition that $M_s \to M_{CJ}$ when $R_s(t) \to \infty$, namely,

$$I = (Q + e_0)/[(j+1)C_0^2 M_{CJ}^2]. \qquad 12.$$

The mass integral in Equation 11 can also be readily evaluated using the experimental fact that the onset of a detonation is preceded by a quasi-stationary regime in which the shock and the reaction zone move as a coupled complex at constant velocity ($M_s \simeq M_s^*$). The portion of the shocked mass of gas that releases chemical energy corresponds to the mass enclosed by the reaction front, i.e.

$$\int_0^{R_s^*-d} \rho k_j r^j \, dr = k_j \rho_0 [R_s^* - \dot{R}_s^* \tau(M_s^*)]^{j+1}/(j+1). \qquad 13.$$

In blast wave coordinates the above equation reduces to:

$$\int_0^{1-d/R_s^*} \psi \xi^j \, d\xi = [1 - \dot{R}_s^* \tau(M_s^*)/R_s^*]^{j+1}/(j+1). \qquad 14.$$

If we replace the integral term in Equation 11 by the above expression and solve for the detonation kernel size R_s^*, we have

$$R_s^* = \frac{M_s^* C_0 \tau(M_s^*)}{1 - \left[\dfrac{j+1}{2} \dfrac{C_0^2}{Q}\left(M_s^{*2} I - \dfrac{e_0}{(j+1)C_0^2}\right)\right]} \qquad 15.$$

Once the critical conditions corresponding to the autoignition condition are specified [i.e. M_s^*, $\tau(M_s^*)$], the above equation gives the detonation kernel size R_s^*.

With the minimum kernel size R_s^* known, the initiation energy can be found from Equation 9 to be

$$E_0 = \frac{k_j \rho_0 C_0^2 R_s^{*j+1}}{2} \{M_s^{*2} I - 1/[(j+1)\gamma(\gamma-1)]\}. \qquad 16.$$

In general, the initial energy e_0 of the gas is small compared to the chemical energy Q. Equation 16 may be written in the following simplified form if e_0 is neglected:

$$E_0 = \frac{\rho_0 k_j I}{2} \frac{(C_0 M_s^*)^{j+3} \tau(M_s^*)^{j+1}}{\left\{1 - \left[\dfrac{1}{2}\left(\dfrac{M_s^*}{M_{CJ}}\right)^2\right]^{1/(j+1)}\right\}^{j+1}}. \qquad 17.$$

In the above expression, Equation 12 has been used to eliminate the chemical energy term. From Equation 17 we note that the dependence of initiation energy on induction time is identical to the result based on Zel'dovich's criterion (i.e. Equation 7).

Using Equation 16 and the shock-tube induction time data obtained from Strehlow & Cohen (34) and White (35), a comparison between the experimental results for the critical energy for direct initiation of cylindrical and spherical detonations in stoichiometric oxyhydrogen and oxyacetylene mixtures is shown in Figure 9. The experimental data are based on the true minimum spark energy taken at the peak power of the ignition source. The agreement between theory and experiment is extremely good within the uncertainty in the induction kinetic data.

Since $E_s \sim \tau^3$ for the unconfined detonation, the critical energy for fuel-air mixtures will be about 3 to 6 orders of magnitude greater than for the fuel-oxygen mixtures. This is because the induction times for the fuel-air mixtures are one to two orders of magnitude greater while the other detonation parameters are not radically affected. This observation is readily borne out by experiments. Thus, for example, to form a spherical detonation in an explosive methane-air mixture at atmospheric pressure requires an explosive charge slightly in excess of 1 kg of high explosive (36) ($E_0 > 4 \times 10^6$ J). Using Equation 16 and the induction time data of Lifshitz et al (37) we obtain a value of E_0 of 8×10^6 J. Similarly, for the propane-air mixture, Equation 16 gives the critical energy for direct initiation to be 2.5×10^4 J with the induction time data of Burcat et al (38). Recent experiments at my laboratory gave a value of $E_0 \simeq 6 \times 10^3$ J for propane-air detonations confined in tubes. For both the methane-oxygen and propane-oxygen mixtures, the initiation energy is of the order of joules.

The theories discussed so far do not give details of the reacting blast wave motion as it approaches the self-sustained C-J detonation. A phenomenological theory developed by Lee & Bach (18) and reviewed by Lee (33), and Fry & Nicholls (39) demonstrates the existence of steady sub–C-J detonations and the development of instabilities during the approach to a steady detonation. The numerical calculations of Kyong (40) and more recently of Levin & Markov (41) also show the development of severe flow oscillations before a self-sustained detonation is formed in the critical

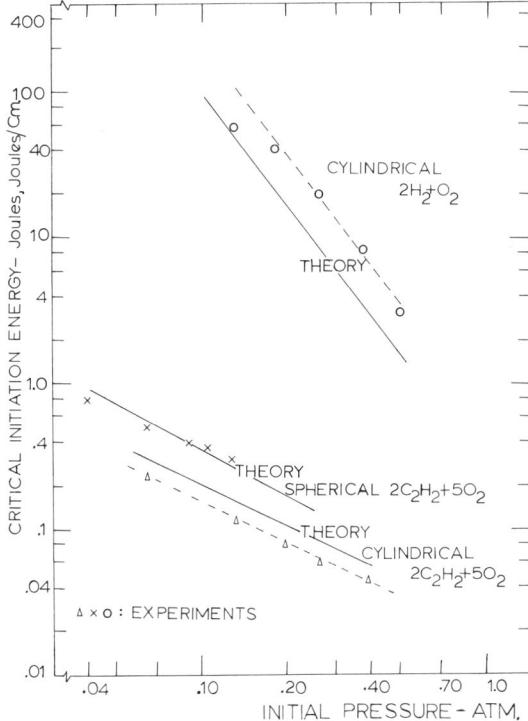

Figure 9 Comparison of the experimental results for the critical energies for direct initiation in $2C_2H_2$–$5O_2$ and $2H_2 + O_2$ with the theoretical values from the Lee and Ramamurthi theory.

regime of initiation. The numerical computations are rather involved and time consuming even with the modern high-speed computers. Except under special circumstances where details of the predetonation blast motion are required, such computations are difficult to justify. Computations based on a one-dimensional model cannot describe the localized phenomenon of the onset of the detonation nor the subsequent motion of the multiheaded detonation wave itself. Thus the numerical solution is valid only over a limited regime.

Attempts have also been made to describe the predetonation blast propagation analytically, based on a perturbation theory. These theories have limited success in that only certain qualitative features of the reacting blast motion can be described. I have already reviewed these studies (33) and will not elaborate on them here.

DETONABILITY LIMITS

In practice, it is of great importance to know if a gas can detonate and, if so, what the composition limits of detonability are. Thus far no quantitative theory exists for predicting the limits, and experimental determination of the limits is not a simple task. However, an intimate relationship exists between the initiation and propagation mechanisms of a detonation wave since, by definition, initiation implies the formation of a self-sustained detonation in the mixture. If the mixture is outside the detonability limits, initiation will not be possible. In fact, the experimental determination of detonability limits for unconfined fuel-air detonations is based on the characteristic U-shaped curve of initiation energy vs the composition of the mixture. For confined detonations in tubes, where the wave can be observed over a long distance, the detonability limits in principle could be specified by noting the eventual decay of the wave far from the source. However, for fuel-air mixtures, the initiation of confined detonations still requires very high ignition energies [e.g. ~ 70 g TNT for CH_4-air in a 30 cm diameter tube (42)]. Thus, within the distance of most laboratory-scale detonation tubes, of the order of 10 meters, the influence of the ignition energy on the wave motion may not be negligible. Hence, the characteristic U-shaped curve would eventually still have to be used to set approximate detonability limits in a given-size apparatus. This establishes that initiation determines the limits experimentally. However, the relationship goes deeper than that, because the propagation mechanism of a multiheaded detonation wave is one of continuous reinitiation.

In the past two decades, the excellent experimental studies on the detailed structure of cellular detonation waves have demonstrated conclusively that the detonation front consists of a number of reacting blast "wavelets" whose collisions with each other form a series of waves transverse to the direction of propagation. The reacting blast wavelets themselves are in a process of decay and are energized cyclically upon the collision of the transverse waves. The trajectories of the triple points (i.e. the point of interaction of the incident, Mach, and transverse waves) along the walls of a tube can be recorded readily by smoked foils or open-shutter photography of detonation propagation in thin channels. Figure 10 shows the reconstruction of the wave processes at different times during a cycle of interaction, along with the variation in the velocity of the wavelet along its centerline. We note that the shock wave and the reaction zone are coupled initially, and the shock velocity corresponds to about twice the C-J value. The blast wavelet decays with progressive decoupling of the reaction zone from the shock front. Just prior to the end of the cycle, where the transverse waves intersect, the velocity of the wavelet is about 0.6 times the C-J velocity, which is about the limiting condition for auto-ignition. A localized explosion occurs at the end of the cycle, and the process repeats itself.

The life history of the blast wavelet in a self-sustained cellular detonation front is remarkably similar to that of the reacting blast wave generated by the igniter during the predetonation phase in the critical regime (Figure 4c). In fact, it appears that the initiation energy is supplied just to start the first cycle. Thereafter, the propagation is effected through a series of periodic reinitiations due to the energy derived from the localized chemical explosions. In other words, the propagation mechanism of a detonation wave is one of continuous reinitiation. The complex interactions of transverse waves and the lead shock within the cell are responsible for producing the critical state for localized explosion to occur in order to start the next cycle. In making a formal postulate on this reinitiation concept, we note that the length of a detonation cell (i.e. the distance traveled by the blast wavelet

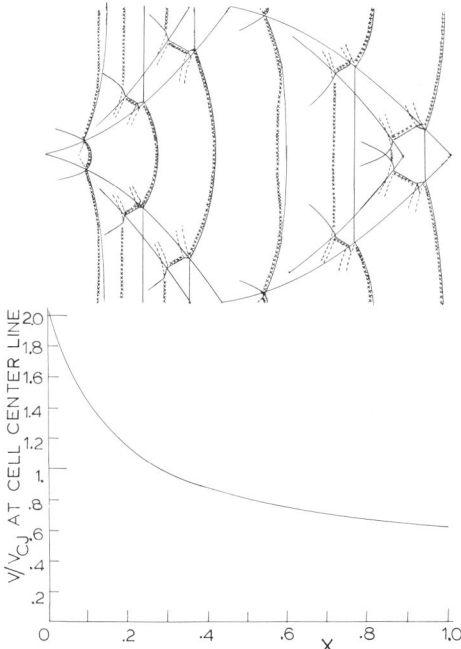

Figure 10 Reconstruction of the reacting blast "wavelet" at various positions (from Reference 43).

in one cycle) should correspond to the detonation kernel size R_s^*, given by Equation 15. For H_2–O_2–Ar and $2C_2H_2 + 5O_2$ mixtures, whose transverse wave-spacing data (hence the length of the detonation cell) have been measured by Strehlow (44) and Voitsekhovskii et al (45) respectively, a comparison can be made between the theoretically calculated cell spacings (from R_s^*) and experiments. This is shown in Figure 11. The agreement is remarkably good and indicates the fundamental significance of the detonation kernel concept. It lends support to the important postulate that the propagation mechanism is in fact identical to the initiation mechanism. Just how nature achieves the correct energy release to reinitiate the blast wavelet for each cycle of motion remains a mystery.

From the above discussion, we see that if the vertical arms of the U-shaped curve are used to set the limits (i.e. $E_0 \to \infty$), this implies also the condition on the detonation kernel size, $R_s^* \to \infty$, as the detonability limits are approached. With the one-to-one correspondence between the kernel size and the length of a detonation cell in a multiheaded detonation front, we see that as the limits are approached, the size of the detonation cell (or, equivalently, the transverse wave spacing) also approaches infinity. Equation 15 indicates that for $R_s^* \to \infty$, either the induction time $\tau \to \infty$ or the chemical energy release becomes so small that an adequate

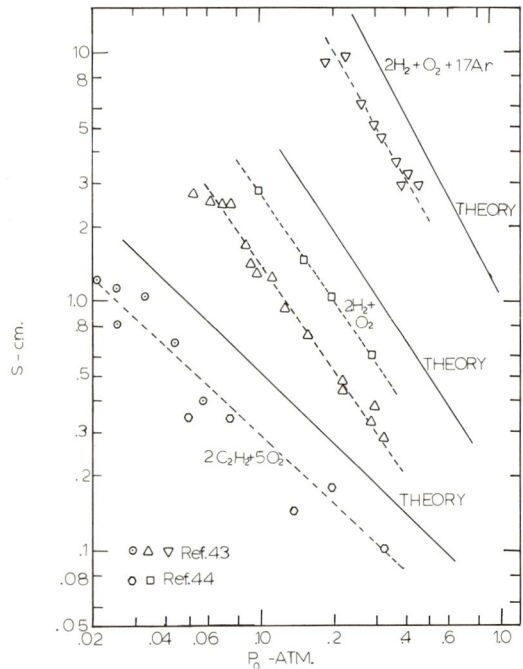

Figure 11 Comparison of the transverse wave spacing data for H_2–O_2–Ar and $2C_2H_2 + 5O_2$ mixtures with the theoretical values predicted by the Lee and Ramamurthi theory.

kernel can never be formed to sustain the shock motion. In general, the autoignition limit (i.e. $\tau \to \infty$) is the key parameter in determining the detonation limits via Equation 16, since the chemical energy of a detonable mixture differs insignificantly from that of flammable mixtures. The autoignition limit is strongly dependent on the losses in the system. Thus losses play a role in establishing the limits just as they do in influencing the initiation energy.

From the above discussions on initiation and limits, one could explain qualitatively the narrow limits generally found for unconfined spherical waves as compared to planar waves (46). Experiments have established that the average cell size is a constant for a self-sustained detonation. Thus, for a planar wave propagating in a tube, the total number of cells of the detonation front is, on the average, a constant. However, for diverging waves, the surface area of the front increases with radius (e.g. area $\sim R_s^2$ for spherical waves). Thus, to keep the average cell dimension the same, the total number of cells have to multiply continuously as the wave expands. This requires the formation of more than one localized explosion at the end of the cycle of a decaying blast wavelet. In this way the cell may divide to form more new cells. If multiplication does not occur in a diverging wave, the cell size gets progressively larger, and the increase in the time for the blast wavelet to decay means that the thermodynamic states at the end of the cycle, when the transverse waves finally collide, may drop below the autoignition limit required for localized explosion to occur. Hence, reinitiation is not possible, and the wave fails. The open-shutter photographs in Figure 12 clearly demonstrate the above discussions. In Figure 12a a continuous multiplication of cells occurs; the average cell dimension thus remains constant as the wave expands. In Figure 12b no such multiplication occurs, and the cells get progressively larger; as a result, the detonation wave eventually fails.

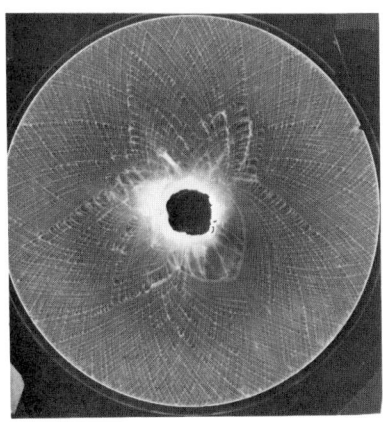 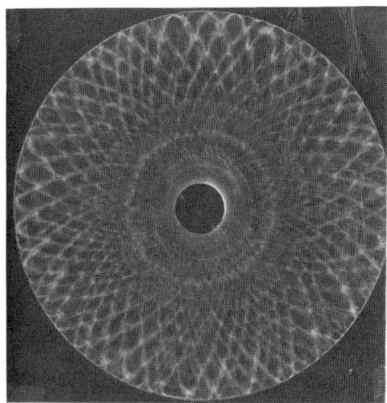

A. SELF-SUSTENANCE   B. FAILURE

Figure 12 Open-shutter photographs of the propagation of a cylindrical detonation in a thin cylindrical chamber 1 mm thick and 25 cm in diameter in equimolar C_2H_2–O_2 mixtures. (a) $p_0 = 80$ torr; (b) $p_0 = 30$ torr.

To achieve the formation of more than one localized explosion, a higher temperature is required. From ignition experiments in reflected shock waves, we note that for conditions well within the explosion limit, the reaction front appears homogeneously as a plane. For conditions close to the limit, reactions originate from a few, localized, discrete centers (Figure 13). If one considers a homogeneous front to be the limit when the number of discrete reaction centers approach infinity, i.e. they are so large as not to be individually discernible, the number of reaction centers depends on how close the conditions are to the explosion limit. Thus, for unconfined detonations where a multiplication of the number of transverse waves is necessary, it is evident that at the end of the cycle of the blast wavelet, the condition must be such that more than one reaction center can be formed. In other words, on the average, the thermodynamic states at the end of the cycle for spherical waves should be higher than for the corresponding planar waves.

Thus, if the transverse wave spacing for planar and spherical waves is about the same as that indicated by experiments, the blast energy for reinitiation must be higher and perhaps more difficult to achieve in the spherical case. From this qualitative argument we can account for the fact that unconfined detonations should have

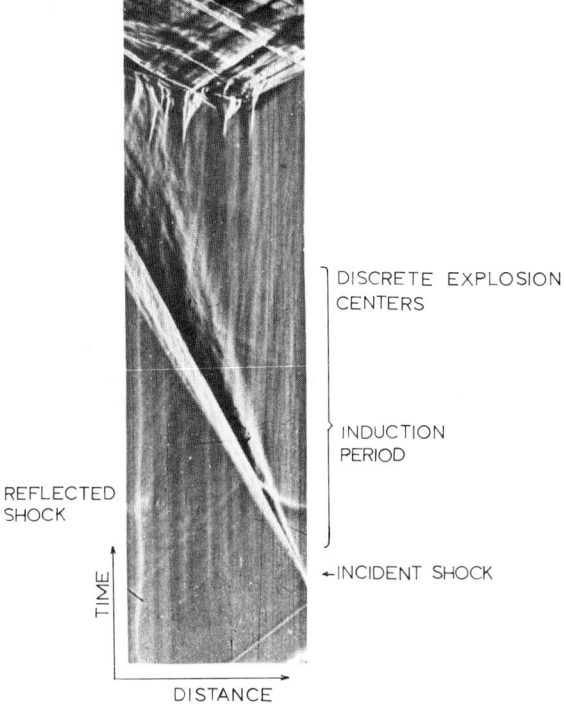

Figure 13 Streak schlieren record of the ignition of H_2–O_2 mixtures behind a reflected shock, which illustrates the formation of discrete explosion centers. (Courtesy of R. I. Soloukhin.)

narrower limits. Quantitative analysis must wait until the exact nature of formation of the localized explosion is discovered. For a given homogeneous reacting mixture, the formation of discrete explosion centers is associated with the instability of the reacting medium to gas-dynamic fluctuations. The number of discrete centers is related to the wave number of the perturbations, which get amplified under the given chemical and thermodynamic states. The studies of acoustic wave propagation in a reactive medium by Toong and co-workers (47) are most appropriate in furthering our basic understanding of the instability mechanisms leading to the formation of explosion centers in the reactive medium itself. Once formed, the wave dynamics associated with the subsequent growth of the explosion centers have been studied numerically by Oppenheim and co-workers (48). A quantitative theory for propagation and detonability limits is, however, still lacking.

CONCLUSIONS

The qualitative aspects of the initiation of detonations in gaseous mixtures are currently fairly well understood. The essence of all modes of initiation is the generation of the critical states for the onset of detonation. These critical states correspond to those at the autoignition limit of the mixture. In the transition or self-initiation mode, the critical states are generated by an accelerating flame. The principal mechanisms of flame acceleration are turbulence and the interaction of the flame with pressure waves and obstacles in its path. Our current understanding and ability to describe turbulent flames are inadequate to permit formulation of quantitative models for the self-initiation of detonation. However, on the basis of the critical conditions necessary for the onset of detonation itself, an analysis has been developed for estimating the turbulent flame speed required for self-initiation. The analysis is of great practical significance in assessing the potential hazards of a given fuel.

For the fast mode of direct or blast initiation, the required critical states are the result of the coupling between the blast generated by the ignition source and the chemical reactions that occur in the flow behind it. This mode of initiation is fairly well understood, and adequate quantitative models exist whereby the critical energy for direct initiation can be predicted from the properties of the mixture. Detailed quantitative description of the predetonation periods of the propagation of the reacting blast wave itself would require extensive numerical computations.

The propagation mechanisms (i.e. the particular manner in which a detonation propagates) are directly linked to the initiation mechanisms by the postulation that the life cycle of a blast wavelet in a cellular detonation corresponds to that of the reacting blast in the predetonation regime. Agreement between the size of the detonation kernel in the Lee and Ramamurthi theory of direct initiation and the experimental data for cell length of multiheaded detonations supports the postulate. Thus, propagation of a self-sustained detonation wave is said to be effected via a process of continuous reinitiation of the blast wavelets toward the end of their respective cycles. Detonability limits (outside which detonation cannot propagate) are obtained from the limits within which it is possible to initiate a detonation in a given mixture (i.e. with finite initiation energy). This is compatible with the use of the characteristic U-shaped critical energy versus mixture composition curve to establish experimentally the detonability limits. While direct initiation utilizes an external energy source to produce the critical conditions for the onset of detonation, the reinitiations of the cyclic blast wavelets derive their energies from the mixture itself. The underlying principle by which nature achieves the exact partition of the appropriate energy for the reinitiation of the blast wavelets through a control of the cell sizes is still a mystery. The good prediction of the cell size obtained by equating it to the detonation kernel size does not tell us why this has to be so. However, by acknowledging that the propagation mechanisms are similar to the initiation mechanisms, we are closer to the eventual solution of the problem.

The critical states corresponding to the autoignition limit of the mixture can be seen to occupy the central role in detonative combustion: namely, in self-initiation, direct initiation, and in the propagation of the wave itself. In the reinitiation of the blast wavelets, the critical states are achieved via the collision of the transverse waves. The onset of detonation in the mixture at the critical state is associated with the instability mechanisms in reactive media. The consequence of the instability is the formation of discrete localized explosion centers, which then leads to the development of the detonation wave. The essence of the problem involves structure formation or heterogeneities in a homogeneous reacting system. Such a problem is universal, and the outstanding problems of current research in all fields of physics, chemistry, and biology are of this type (e.g. phase transitions, morphogenesis, etc). Gaseous detonation waves are but one example of such universal phenomena.

ACKNOWLEDGMENTS

The immense contributions of K. Ramamurthi to this article are gratefully acknowledged. He rewrote various paragraphs on blast initiation to include the recent results from his Ph.D. dissertation, pointed out a number of inaccuracies and raised a number of points for discussion, all of which resulted in a clearer, sharper, and more up-to-date presentation. He is also responsible for all the necessary details in producing the article from my manuscript. Much of the work on which this article is based are the joint efforts of my collaborators (G. Bach, C. Guirao, and R. Knystautas) over the past decade. Financial support for detonation research is from the National Research Council of Canada and the US Air Force Office of Scientific Research (B. Wolfson, technical monitor). Special thanks go to A. K. Oppenheim, G. Markstein, R. I. Soloukhin, and R. A. Strehlow for providing photographic records of their results. It has not been possible to cite all the important works that have contributed directly or indirectly to my views on this article's subject. However, I have learned much from the works and the numerous discussions I had with the following individuals and their colleagues over the past 15 years: Professors N. Manson and H. Guenoche (France), W. Jost and H. G. Wagner (Germany), R. I. Soloukhin and V. Korobeinikov (USSR), D. H. Edwards (Wales), T. Fujiwara (Japan), A. K. Oppenheim, R. Strehlow, T. Y. Toong, A. Nichols, M. Sichel, E. Dabora, H. Barthel, and Drs. R. Duff, P. Urtiew and G. Schott (USA), B. Ahlborn and I. I. Glass (Canada).

Literature Cited

1. Markstein, G. 1956. *6th Symp. Int. Combust. Proc.*, p. 387
2. Bollinger, L. E., Fong, M. C., Edse, R. 1961. *ARS J.* 31:588
3. Pawel, D., Van Tiggelen, P. J., Vasatko, H., Wagner, H. G. 1970. *Combust. Flame* 15:173–77
4. Oppenheim, A. K., Stern, R. A. 1959. *7th Symp. Int. Combust. Proc.*, pp. 837–50
5. Laderman, A. J., Oppenheim, A. K. 1961. *Phys. Fluids* 4:778
6. Laderman, A. J., Urtiew, P. A., Oppenheim, A. K. 1962. *Combust. Flame* 6:325–35
7. Laderman, A. J., Oppenheim, A. K. 1962. *Proc. R. Soc. London Ser. A* 268:153–80
8. Laderman, A. J., Urtiew, P. A., Oppenheim, A. K. 1962. *9th Symp. Int. Combust. Proc.*, pp. 265–74
9. Urtiew, P. A., Oppenheim, A. K. 1966. *Proc. R. Soc. London Ser. A* 295:13–28
10. Brinkley, S., Lewis, B. 1959. *7th Symp. Int. Combust. Proc.*, pp. 807–11
11. Urtiew, P. A., Oppenheim, A. K. 1967. *11th Symp. Int. Combust. Proc.*, pp. 665–70
12. Lee, J. H., Pangritz, D., Wagner, H. G. 1976. *General Considerations for the Initiation of Gaseous Detonations. Rep. Max Planck Inst. Strömungforschung,* Göttingen. 45 pp.
13. Abdel-Gayed, R. G., Bradley, D. 1976. *16th Symp. Int. Combust. Proc.* In press
14. Dorge, K. J., Pangritz, D., Wagner, H. G. 1976. *Acta Astron.* In press
15. Shchelkin, K. I. 1940, *J. Exp. Theor. Phys. USSR* 10:823; also in Zel'dovich, Ya. B., Kompaneets, A. S. 1960. *Theory of Detonations*, p. 201. New York: Academic
16. Bach, G., Knystautas, R., Lee, J. H. 1968. *12th Symp. Int. Combust. Proc.*, pp. 853–67
17. Litchfield, E. L., Hay, M. H., Forshey, D. R. 1962. *9th Symp. Int. Combust. Proc.*, pp. 282–87
18. Bach, G., Knystautas, R., Lee, J. H. 1970. *13th Symp. Int. Combust. Proc.*, pp. 1097–1110
19. Lee, J. H., Knystautas, R., Guirao, C. 1974. *15th Symp. Int. Combust. Proc.*, pp. 53–68
20. Knystautas, R., Lee, J. H. 1976. *Combust. Flame* 27:221–28
21. Ramamurthi, K. 1976. *On the blast initiation of gaseous detonations.* Ph.D thesis. McGill Univ., Montreal
22. Matsui, H., Lee, J. H. 1976. *Combust. Flame* 27:217–20
23. Lee, J. H., Matsui, H. 1977. *Combust. Flame*. 28:61–66
24. Matsui, H. 1973. *Decaying of Gaseous Detonation by Expansion, Res. Rep. Res. Inst. Ind. Saf. RIIS-RR-22-2.* (In Japanese.) 11 pp.
25. Zel'dovich, I., Kogarko, S. M., Semonov, N. N. 1956. *Sov. Phys. Tech. Phys.* 1(8):1689–1713
26. Lee, J. H. S., Lee, B. H. K., Knystautas, R. 1966. *Phys. Fluids* 9:221–22
27. Taylor, G. I. 1950. *Proc. R. Soc. London Ser. A* 201:159–64
28. Kelly, J. Private communication; see also Lee, J. H. 1965. *The Propagation of Shocks and Blast Waves in a Detonating Gas. Rep. 65-1,* McGill Univ., Montreal
29. Soloukhin, R. I. 1969. *12th Symp. Int. Combust. Proc.*, p. 799
30. Edwards, D. H., Hooper, G., Morgan, J. M. 1975. *The experimental investigation of the direct initiation of spherical detonations.* Presented at Int. Colloq. Gasdynamics Explos. React. Syst., 8th, Bourges, France
31. Sichel, M. 1977. *Acta Astron.* In press
32. Lee, J. H. S., Ramamurthi, K. 1976. *Combust. Flame.* 27:331–40
33. Lee, J. H. S. 1972. *Acta Astron.* 17:455
34. Strehlow, R. A., Cohen, A. 1962. *Phys. Fluids* 5:97
35. White, D. R. 1971. *11th Symp. Int. Combust. Proc.*, p. 147
36. Vanta, E., Foster, J., Parsons, G. H. 1973. *Detonability of Some Natural Gas–Air Mixtures. Air Force Armament Lab. Tech. Rep. AFATL-TR-74-80*
37. Lifshitz, A., Scheller, K., Burcat, A., Skinner, G. B. 1971. *Combust. Flame* 16:311
38. Burcat, A., Lifshitz, A., Scheller, K., Skinner, G. B. 1971. *13th Symp. Int. Combust. Proc.*, p. 745
39. Fry, R. S., Nicholls, J. A. 1975. *15th Symp. Int. Combust. Proc.*, p. 43
40. Kyong, W. H. 1972. *A Theoretical Study of Spherical Gaseous Detonation Waves.* PhD thesis. McGill Univ., Montreal. 105 pp.
41. Levin, V. A., Markov, V. V. 1976. *Fluid Dyn. USSR* 9:754
42. Kogarko, S. M. 1959. *Sov. Phys. Tech. Phys.* 3:1904–16
43. Crooker, A. J. 1969. *Phenomenological Investigation of Low Mode Marginal Planar Detonations. Univ. Illinois, Tech. Rep. FAE 69-2*

44. Strehlow, R. A., Engel, C. D. 1969. *AIAA J.* 7:492
45. Voitsekhovskii, B. V., Mitrofanov, V. V., Topchian, M. E. 1966. *The Structure of a Detonation Front in Gases. Rep. FTD MT 64-527 (AD 633-821)*, Wright Patterson AFB, Ohio
46. Manson, N., Ferrie, F. 1953. *4th Symp. Int. Combust. Proc.*, p. 486
47. Toong, T. Y., Arbeau, P., Garris, C. A., Patureau, J. P. 1974. *15th Symp. Int. Combust. Proc.*, p. 87
48. Cohen, L. M., Oppenheim, A. K. 1975. *Combust. Flame* 25:207

ELECTRON TUNNELING IN CHEMISTRY AND BIOLOGY

✣ 2640

W. F. Libby

Department of Chemistry, University of California, Los Angeles, California 90024

INTRODUCTION

Electron tunneling is well recognized in physics. It was first proposed by Nordheim (1) in 1927 and was quickly applied to electron movements by many early theorists. For example, Nordheim (1) applied it to thermal emission of electrons. Oppenheimer (2) applied it to the problem of field emission of electrons in 1928. Nordheim (3), Frenkel & Joffe (4), and Wilson (5) applied it to rectifying barriers in 1932. Zener (6) used it to explain avalanche breakdown in semiconductor junctions in 1934. Giaever (7) first observed tunneling currents from superconducting metals separated through oxide barriers. Josephson (8) predicted, and it has been confirmed experimentally, that electrons tunneling through junctions in superconducting rings exhibit certain quantum-mechanical oscillatory phenomena. Mead and co-workers (9) have studied electron tunneling through barriers nearly 100 Å thick and confirmed the quantum-mechanical relationships.

Practical appliances have resulted such as tunnel diodes [Esaki (10) 1958], and numerous devices (11) for accurately measuring small voltages or magnetic fields, or detecting infrared radiation that use the Josephson effect. Still being developed are computer memories (12) that use the Josephson effect, and a method of spectroscopy of minute amounts of organic or biological materials embedded in a tunnel barrier (13).

In chemistry, the tunneling of electrons is also well recognized. Gurney (14) proposed in 1931 that electrolysis involved electron tunneling through the electrode surface and showed how the necessity to match energy levels could produce the phenomenon of overvoltage. His theory was extended by Horiuti & Polanyi (15) in 1935. In 1939 Mott (16) suggested that the oxidation of aluminum is controlled by electrons tunneling through the oxide layer from the metal to adsorbed O_2. In 1940 Libby (17a) suggested electron tunneling to explain the rapid exchange between MnO_4^- and MnO_4^{2-} in solution. In 1952 (17b), Libby used tunneling, along with a requirement of solvent rearrangement, to satisfy the Franck-Condon principle and explain such facts as that ferri- and ferrocyanide ions exchange electrons much more rapidly than ferric and ferrous ions. R. A. Marcus (18), in a series of papers, greatly extended the theory of electron transfer both at electrodes and in solution. Gaseous atoms and ions were noticed to exchange electrons over distances almost twice as large as those required for momentum transfer in collisions (19, 20). Eley et al (21) suggested tunneling between molecules as an explanation for the electrical conductivity properties of organic crystals. Weissman (22) in 1958 and Voevodskii et al (23) in 1959 used EPR to measure the exchange of electrons between two aromatic groups held apart on the same molecule by chains of CH_2 groups of various lengths. In 1965, Ruff (24) proposed tunneling to explain data (25, 26) that pointed to the exchange of electrons between ions in water or frozen in ice an average of 100 Å apart. Kowalsky (27) used nuclear magnetic resonance methods to measure the rate of exchange between ferri- and ferrocytochrome c in 0.02 M solution at 100 sec^{-1}.

Early theoretical work on electron tunneling in chemistry has been done by Weiss (28), R. J. Marcus et al (29), Hush (30), Conway (31), Saloman & Conway (32), Gerischer (33), McConnell (34), Gouterman (35), Bockris et al (36), and Glaeser & Berry (37). More recently, Marcus' theory has been extended with greater quantum-mechanical detail and sophistication by Levich (38, 39) and Dogonadze (40) and co-workers. An apparently rather comprehensive theory has been given by Schmidt (41).

Recently, Brockelhurst (42) and Miller (43) have studied the recombination by tunneling of free and solvated electrons trapped in a frozen aqueous solution after bombardment with high-energy electron beams.

Electron tunneling is also known in biology. The first evidence for the phenomenon is perhaps the 1960 discovery of Chance & Nishimura (44) that the bacterium *Chromatium* oxidizes its cytochrome in a light-induced reaction at liquid nitrogen temperatures.

Almost simultaneously, Arnold & Clayton (45) observed light-induced electronic movement in dried chromatophores at 4°K. In 1966 DeVault & Chance (46) measured the rates of cytochrome oxidation with a pulsed laser and fast spectrophotometer; they found the reaction to be temperature-dependent above 100°K (activation energy = 3.3 kcal $mole^{-1}$) and strictly temperature-independent (at 2.3 msec half-time) below 100°K—down to 4°K. This was taken as proof of electron tunneling, since the reaction involves only an electron transfer from cytochrome to bacteriochlorophyll. Furthermore, the reaction is too slow to be barrierless, and there is no activation energy to mount the barrier. Tunneling does not have to be temperature-independent, but the fortunate discovery of this temperature-independent case greatly facilitated the diagnosis of tunneling. The work just described was done with whole cells of *Chromatium*. *Chromatium* chromatophores and subchromatophore preparations made by Thornber (47) showed the same temperature profile.

These observations were closely followed by a similar study by McElroy, Mauzerall & Feher (48), who observed a temperature-independence from 77°K to 4°K of the back reaction of the primary process in bacterial photosynthesis, namely, the transfer of an electron from primary acceptor back to oxidized reaction-center bacteriochlorophyll. Kihara (49) found numerous other photosynthetic bacteria that show low-temperature (77°K) light-induced cytochrome oxidation. We can presume that these will be found to be temperature-independent upon further measurement at lower temperatures. In 1963, Chance & Bonner (50, 51) observed cytochrome oxidation in green leaves at 77°K. Knaff & Arnon (52) and Floyd, Chance & DeVault (53) showed that it was cytochrome b_{559} that is oxidized. This is a low-temperature oxidation of a b-type cytochrome, in contrast to those mentioned above, which are c-type.

An excellent presentation of the physics and physical chemistry of electron tunneling is given by Reynolds & Lumry (54). Other reviews have been given by Libby (55) and by Zamaraev et al (56).

RECENT DEVELOPMENTS

The number of applications of electron tunneling is growing rapidly, and I fear that I may have omitted important papers.

Electron tunneling has been studied in redox reactions between metal complexes at 4.2 and 77°K (57). Also, substantial work has been done on electron tunneling in reactions of organic-anion radicals at 77°K (58).

Electron tunneling in the reactions

$$\overset{\downarrow \qquad \quad \downarrow e-}{O^- + K_4Fe(CN)_6^{4-}}, \quad \overset{\downarrow \qquad \downarrow e-}{SO_4^- + FeSO_4}$$

at 4.2 and 77°K (59) has been studied.

Kinetic equations for tunnel electron-transfer reactions in solids for various types of spatial distribution of reacting species have been given (60).

The reaction $e_{tr}^- + O^- \rightarrow O^{2-}$ in a 10 M NaOH-water matrix at 4.2 and 120°K has been investigated (61), as has the reaction $Nh(S^1) + CCl_4 \rightarrow Nh^+ + CCl_3 + Cl^-$, where $Nh(S^1)$ is naphthalene in the first singlet excited state (62).

Thus we have abundant evidence of tunneling in solid glasses at 4.2 and 77°K, and it seems to be reasonable to expect it to occur in the solid and liquid states at higher temperatures. Solutions of hydrocarbons in liquid argon and xenon show yields for the radiolysis by Co^{60} gamma radiation in keeping with complete neutralization of the argon and xenon ions by electron transfer from the solute hydrocarbon at concentrations as low as 0.001 M, where tunneling would appear to be required (63, 64).

The important work of the Pennsylvania Group (44, 46, 48–50) on biological systems strongly indicates that electron tunneling is important in biophysics. This work suggests that the hydrophobic nature of the interiors of many globular proteins with internal electron donor-receptor systems, as in cytochrome c, may be required by the electronic Franck-Condon principle, which requires that the aqueous phase be excluded from the vicinity of the donor (or receptor) iron-containing heme group.

Both phenothiazine and N,N,N',N¹-tetramethyl benzidine (TMB) have been shown to photoionize with 347.1 nm light when the dyes are incorporated into negatively charged micelles (sodium lauryl sulfate). Positively charged micelles (dodecyl trimethyl ammonium chloride) showed no effect (65–68).

In cyclohexane and methanol solutions, photoionization of TMB occurred only 17% as frequently as triplet excited state formation. The yield in the anionic micellar solutions (NaLS) rose 36-fold, and formed hydrated electrons and TMB^+ ions stable in the micelles for as long as two days. The authors explain these observations as the result of the tunneling of photoelectrons from the micellar into the aqueous phase (65–67). The gas ionization potential of TMB is thought to be between 6.1 and 6.8 eV (69), whereas 347.1 nm light has only 3.6 eV per photon. The authors report a low yield of photoelectrons in methanol solution, the yield varying with the square of the light intensity as it would for a double-photon excitation. The yield of hydrated electrons in aqueous 0.1 M NaLS with 5.10^{-5} M TMB appears to be nearly 85%, relative to quanta absorbed.

The exposure of these anionic micellar systems to sunlight appears to give good yields of TMB^+ ions in the micelles and hydrated electrons in the aqueous phase that, of course, produce atomic and molecular hydrogen. Thus we are faced with tantalizing possibilities of using sunlight to generate hydrogen.

J. R. Miller (70) has studied electron transfer between aromatic molecules in glassy ethanol at 77°K. His data appear to fit well with simple quantum-mechanical calculations. He worked with diphenyl anion and triphenyl ethylene and used pulse radiolysis with 13 MeV electrons in the Argonne Linac and 0.15 M diphenyl and 0.01 M ϕ_2^- with prime absorption at 407 nm, which decreases with time and concentration of ϕ_3Et in accordance with the simplest of tunneling principles. At 514 nm the principal absorption of ϕ_3Et^- occurs, and it proceeds to increase at the calculated rate.

He concludes: "The simplified model works reasonably well, but why it works is not known. An appropriate theory must treat not only the electronic and nuclear states of the electron donor and acceptor molecules but also those of the intervening solvent."

I agree completely with this assessment of our theoretical needs, particularly in view of the successes of tunneling spectroscopy. It is probably this need to satisfy an unexpressed intuition that has retarded the adoption of electron tunneling in chemistry and biology as completely as it has been taken into physics. We need a means to calculate barrier heights and widths and penetration probabilities.

CONCLUSIONS

Although the theory is incomplete, there remains no doubt that electron tunneling is an extremely important principle for both chemistry and biophysics. For many years it has been accepted as such in solid state and nuclear physics. Every effort to develop an appropriate barrier theory should be made.

As Gurney pointed out nearly fifty years ago (14), electrochemistry may well hinge upon electron tunneling at the electrode interfaces. In the rates of redox reactions, it is almost certainly a factor and is sometimes a rate-determinant. It appears to play a commanding role in biophysics.

Literature Cited

1. Nordheim, L. 1927. *Z. Phys.* 46:833
2. Oppenheimer, J. R. 1928. *Phys. Rev.* 31:66
3. Nordheim, L. 1932. *Z. Phys.* 75:434
4. Frenkel, J., Joffe, A. 1932. *Phys. Z. Sowjetunion* 1:60
5. Wilson, A. H. 1932. *Proc. R. Soc. London Ser. A* 136:487
6. Zener, C. 1934. *Proc. R. Soc. London Ser. A* 145:523
7. Giaever, I. 1960. *Phys. Rev. Lett.* 5:147; 1974. *Science* 183:1253–58
8. Josephson, B. D. 1962. *Phys. Lett.* 1:251; 1965. *Adv. Phys.* 14:419
9. Mead, C. A. 1969. In *Tunneling Phenomena in Solids*, Chap. 9, pp. 127–34. New York: Plenum; Kurtin, S. L., McGill, T. C., Mead, C. A. 1971. *Phys. Rev. B* 3:3368–79
10. Esaki, L. 1958. *Phys. Rev.* 109:603; 1974. *Science* 183:1149–55
11. Clarke, J. 1974. *Science* 184:1235–42
12. Fulton, T. A., Dynes, R. C., Anderson, P. W. 1973. *Proc. IEEE* 61:28–35
13. Hansma, P. K., Coleman, R. V. 1974. *Science* 184:1369–71
14. Gurney, R. W. 1931. *Proc. R. Soc. London Ser. A* 134:137–54
15. Horiuti, J., Polanyi, M. 1935. *Acta Electrochem. USSR* 2:505
16. Mott, N. F. 1939. *Trans. Faraday Soc.* 35:1175; 1940. *Trans. Faraday Soc.* 36:473
17a. Libby, W. F. 1940. *J. Am. Chem. Soc.* 62:1930
17b. Libby, W. F. 1952. *J. Phys. Chem.* 56:863–68
18. Marcus, R. A. 1956. *J. Chem. Phys.* 24:966; 1964. *Ann. Rev. Phys. Chem.* 15:155–96
19. Hornbeck, J. A. 1952. *J. Phys. Chem.* 56:829–31
20. Holstein, T. 1952. *J. Phys. Chem.* 56:832–36
21. Eley, D. D., Parfitt, G. D., Perry, M. H., Taysum, D. H. 1953. *Trans. Faraday Soc.* 49:79; Eley, D. D., Parfitt, G. D. 1955. *Trans. Faraday Soc.* 51:1529; Eley, D. D., Willis, M. R. 1961. In *Symposium on Electrical Conductivity in Organic Solids*, ed. H. Kallmann, M. Silver, pp. 257–76. New York: Wiley
22. Weissman, S. I. 1958. *J. Am. Chem. Soc.* 80:6462
23. Voevodskii, V. V., Solodovnikov, S. P., Chibrikin, V. M. 1959. *Dokl. Akad. Nauk SSSR* 129:1082
24. Ruff, I. 1965. *J. Phys. Chem.* 69:3183
25. Silverman, J., Dodson, R. W. 1952. *J. Phys. Chem.* 56:846
26. Horne, R. A. 1963. *J. Inorg. Nucl. Chem.* 25:1139
27. Kowalsky, A. 1965. *Biochemistry* 4:2382
28. Weiss, J. 1954. *Proc. Roy. Soc. London Ser. A* 222:128
29. Marcus, R. J., Zwolinski, B. J., Eyring, H. 1954. *J. Phys. Chem.* 58:432
30. Hush, N. S. 1958. *J. Chem. Phys.* 28:962; 1961. *Trans. Faraday Soc.* 57:557
31. Conway, B. E. 1958. *Can. J. Chem.* 37:178
32. Saloman, M., Conway, B. E. 1965. *Discuss. Faraday Soc.* 39:223
33. Gerischer, H. 1960. *Z. Phys. Chem.* 26:223; 1966. *J. Electrochem. Soc.* 113:1174–82
34. McConnell, H. M. 1961. *J. Chem. Phys.* 35:508
35. Gouterman, M. 1962. *J. Chem. Phys.* 36:2846–53
36. Bockris, J. O., Matthews, D. F. 1965. *J. Electroanal. Chem.* 9:325; 1966. *Proc. R. Soc. London Ser. A* 292:479; Bockris, J. O. M., Sen, R. K., Conway, B. E. 1972. *Nature Phys. Sci.* 240:143–44
37. Glaeser, R. M., Berry, R. S. 1966. *J. Chem. Phys.* 44:3797–3810
38. Levich, V. G. 1965. *Adv. Electrochem. Electrochem. Eng.* 4:00–000; 1970. In *Physical Chemistry, an Advanced Treatise*, ed. H. Eyring, D. Henderson, W. Jost, Vol. 9B, Chap. 12, pp. 985–1074. New York:Academic
39. Levich, V. G., Dogonadze, R. 1959. *Dokl. Akad. Nauk SSSR* 124:125
40. Dogonadze, R. 1971. In *Reactions of Molecules at Electrodes*, ed. N. S. Hush, New York:Wiley
41. Schmidt, P. P. 1972. *J. Chem. Phys.* 57:3749–62
42. Brockelhurst, B. 1973. *Chem. Phys.* 2:6–18; 1974. *Chem. Phys. Lett.* 28:357–60
43. Miller, J. R. 1973. *Chem. Phys. Lett.* 22:180–82
44. Chance, B., Nishimura, M. 1960. *Proc. Natl. Acad. Sci. USA* 46:19
45. Arnold, W., Clayton, R. K. 1960. *Proc. Natl. Acad. Sci. USA* 46:769–76
46. DeVault, D., Chance, B. 1966. *Biophys. J.* 6:826–47; DeVault, D., Parkes, J. H., Chance, B. 1967. *Nature* 215:642–44
47. Thornber, J. P. 1970. *Biochemistry* 9:2688
48. McElroy, J. D., Mauzerall, D. C., Feher, G. 1974. *Biochim. Biophys. Acta* 333:261–78

49. Kihara, T., Chance, B. 1969. *Biochim. Biophys. Acta* 189:116–24; Dutton, P. L., Kihara, T., Chance, B. 1970. *Arch. Biochem. Biophys.* 139:236–40
50. Chance, B., Bonner, W. D. Jr. 1963. In *Photosynthesis Mechanisms in Green Plants*, NAS-NRC Publ.1145, p. 66
51. Chance, B., Kihara, T., DeVault, D., Hildreth, W., Nishimura, M., Hiyama. T. 1969. In *Progress in Photosynthesis Research*, ed. H. Metzner, p. 1321. Tubingen: Int. Union Biol. Sci.
52. Knaff, D. B., Arnon, D. I. 1969. *Proc. Natl. Acad. Sci. USA* 63:956
53. Floyd, R. A., Chance, B., DeVault, D. 1971. *Biochim. Biophys. Acta.* 226:103–12
54. Reynolds, W. L., Lumry, R. W. 1966. *Mechanism of Electron Transfer.* New York: Ronald. 175 pp.
55. Libby, W. F. 1971. *The Chemical Impact of the Franck-Condon Principle and of Electron Tunneling, A Tribute to Edward U. Condon*, ed. W. E. Brittinand, H. Odabasi, pp. 205–18. Boulder: Colorado Assoc. Univ. Press
56. Zamaraev, K. I., Khairutdinov, R. F. 1974. *Chem. Phys.* 4:181
57. Khairutdinov, R. F., Zamaraev, K. I. 1975. In *Dokl. Akad. Nauk SSSR* Vol. 224
58. Khairutdinov, R. F., Zamaraev, K. I. 1975. *Izv. Akad. Nauk SSSR Ser. Khim.*
59. Khairutdinov, R. F., Zamaraev, K. I. 1975. *Fiz. Tverd. Tela* 17:929
60. Parmon, V. N., Khairutdinov, R. F., Zamaraev, K. I. 1974. *Fiz. Tverd. Tela* 16:2572
61. Zamaraev, K. I., Khairutdinov, R. F. Mikhailov, A. I., Goldanskii, V. I. 1971. *Dokl. Akad. Nauk SSSR* 199:640
62. Khairutdinov, R. F., Sadovskii, N. A., Parmon, V. N., Kuzmin, M. G., Zamaraev, K. I. 1975. *Dokl. Akad. Nauk SSSR* 220:888
63. Davis, D. R., Libby, W. F., Kevan, L. 1965. *J. Am. Chem. Soc.* 87:2766
64. Sheridan, M. E., Greer, E., Libby, W. F. 1972. *J. Am. Chem. Soc.* 94:2614
65. Alkaitis, S. A., Beck, G., Grätzel, M. 1975. *J. Am. Chem. Soc.* 97:5723
66. Alkaitis, S. A., Grätzel, M., Henglein, A. 1975. *Ber. Bunsenges. Phys. Chem.* 79:541
67. Alkaitis, S. A., Grätzel, M. 1977. *J. Am. Chem. Soc.* In press
68. Alkaitis, S. A., Grätzel, M., Henglein, A. 1977. *J. Am. Chem. Soc.* In press
69. Fulton, A., Lyons, L. E. 1969. *Aust. J. Chem. Soc. Jpn.* 42:3030
70. Miller, J. R. 1975. *Science* 189:221

THERMAL REARRANGEMENTS ✦ 2641

Jerome A. Berson
Department of Chemistry, Yale University, New Haven, Connecticut 06520

INTRODUCTION

When chemists finally acquire the long-sought power to predict the rate and course of any given reaction, much of the intellectual motivation for research on reaction mechanisms will fade. For the present, the risk of such self-disemployment seems small. As activity in the field continues, there is fairly general agreement that the study of thermal unimolecular reactions may well offer the best chance to isolate and identify the specifically *molecular* (as opposed to environmental) features underlying reactivity.

Several aspects of thermal rearrangements have been discussed extensively in recent reviews. An incomplete list of those that have appeared since 1968 includes two monographs on the theory of unimolecular reactions (1, 2), chapters on the kinetics and descriptive chemistry of thermal reorganizations (3, 4), and several discussions of biradicals and their role in these and related processes (5–13).

The recent great interest in orbital symmetry methods for the qualitative theoretical analysis of concerted reactions has generated many reviews, some of which are of special significance in the context of this article because of their coverage of sigmatropic rearrangements and pericyclic reactions (14–20). With these reviews as introduction, and with the coverage provided by annual publications (e.g. 21, 22), the interested reader should find access to the literature of this large and growing field.

This article has a narrower scope. It is mainly concerned with recent experimental and theoretical attempts to define the reaction pathway in thermal reorganizations of cyclopropanes and cyclobutanes. It provides information on these reactions supplementary to that in Bergman's stimulating 1973 review (11), as well as coverage of several topics not reviewed elsewhere. The treatment is selective rather than exhaustive, a result of limitations of both editorial space and authorial competence. I apologize to the many active investigators in this area whose work I have been unable to include.

THERMAL REARRANGEMENTS OF CYCLOPROPANES

One hopes to ease the task of theory by minimizing the number of particles. For this reason, the structural simplicity of the reactants has made the pyrolytic structural

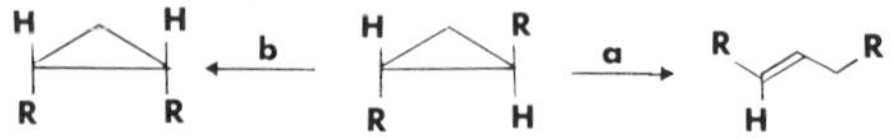

Figure 1 Structural rearrangement (*a*) and stereomutation (*b*) of a disubstituted cyclopropane.

isomerization (*a*), or stereomutation (*b*), of cyclopropane and its derivatives (Figure 1) the most intensively studied of all unimolecular reactions.

Although gas-phase reactions of these small molecules at low pressures offer valuable comparisons of the abilities of the rival Slater and RRKM (Rice-Rampsperger-Kassel-Marcus) theories to account for the shape and position of the Lindemann fall-off curve of the rate constant (23, 24), the results of the tests are not very sensitive to the model chosen for the structure of the transition state (23, 25–28). Hence, they are of limited applicability to the question of reaction mechanism.

Theoretical descriptions of the mechanisms have usually originated from one of two approaches. The older of these, advocated primarily by Benson and his coworkers (5–9, 29), invokes a biradical as a metastable intermediate and estimates the energy and entropy of the rate-determining transition state by increments added to or subtracted from the corresponding thermodynamic quantities of the biradical. The second approach calculates the transition-state energy by quantum mechanical methods.

Quantum Mechanical Calculations: General Aspects

One should not be surprised if the absolute value of the activation energy obtained from the quantum mechanical calculation is sometimes in poor agreement with experiment. The calculation is necessarily approximate, although modern techniques seem to be making good progress in the description of systems with one weak bond (biradicals) (30). In fact, the experimental activation energy for the stereomutation of cyclopropane, 64.2 (104) or 65.1 (105) kcal mole^{-1}, is matched surprisingly well by generalized valence-bond (31) and CNDO (38) calculations, which give values of 60.5 and 63 kcal mole^{-1}, respectively. Extended Hückel (55) and ab initio calculations (35, 37) give values somewhat lower, 44 and 52 kcal mole^{-1}, respectively.

These apparent triumphs, however, do not provide the basis for unrestrained rejoicing. It is simultaneously more demanding and more significant to ask that theory be able to discriminate among mechanisms by predicting *differences* among the transition-state energies of competing pathways. The hope that this goal may be achieved is based upon the assumption that many of the large energy terms will be common to the competing pathways and hence will cancel out in an examination of differences.

Thermochemical-Kinetic Calculations: General Aspects

A basic assumption of the thermochemical-kinetic approach is that the energy of the transition state can be calculated from that of a nearby, hypothetical reactive intermediate. In the cyclopropane rearrangements, the intermediate is assumed to be a trimethylene biradical, Figure 2, **1** (5–9, 29). The heat of formation, ΔH_f°,

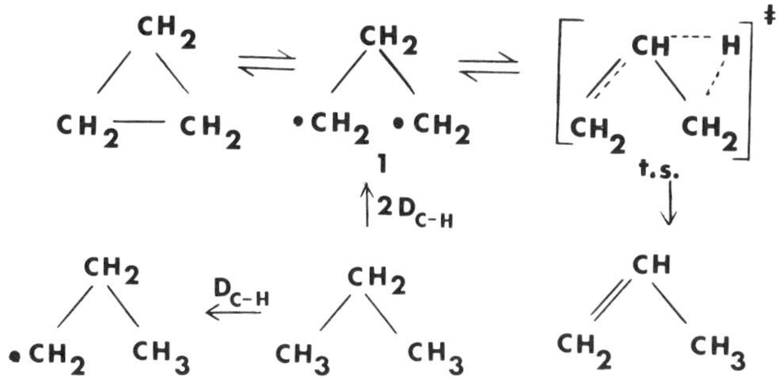

Figure 2 Fixed points in the biradical mechanism for the pyrolysis of cyclopropane (see Figure 3).

of **1** can be calculated by any of a number of thermochemical cycles, the results of which would be the same as equating it to the sum of $\Delta H_f°$ of propane plus twice the C–H bond-dissociation energy in propane minus the bond-dissociation energy of H_2 (Figure 3).

The use of $2D_{C-H}$ in Figure 2 implies the assumption (29) that in **1** no significant interaction between the terminal CH_2 groups survives the bond-rupture step. This method of calculating biradical $\Delta H_f°$ values had been advocated earlier by Seubold (39, 40), who justified it on the grounds that, with appropriate conformational corrections, it accounted for the strain energies of saturated alicyclic hydrocarbons.

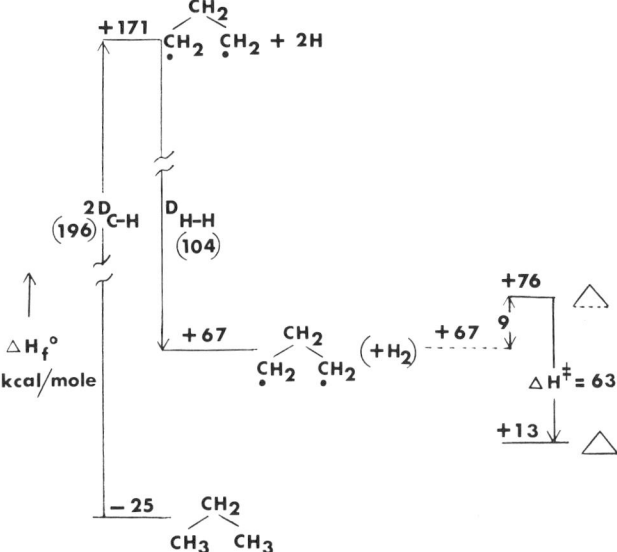

Figure 3 Thermochemical analysis of the pyrolysis of cyclopropane.

Some Specific Correlations and Predictions of the "Biradical Mechanism" (Thermochemical-Kinetic Approach)

According to O'Neal & Benson (6–8), the estimates of the Arrhenius kinetic parameters A and E_a by thermochemical additivity methods are probably reliable to $10^{0.3}$ sec^{-1} and 1.5 kcal mole^{-1}, respectively. Suppose that one wished to calculate by these methods the product proportions from the *relative* rates of two competing reactions that, in actuality, happen to have the same rate constants under the particular experimental conditions, e.g. 600°K. The root-mean-square uncertainties in the transition-state estimates would then permit the predicted rate ratio to lie anywhere between 1:16 and 16:1, corresponding to a predominance of one or the other product to the extent of 94%. Clearly, the procedure is not useful for estimates of such product ratios.

Rather, the experimental validation of the biradical mechanism for cyclopropane is supposed to rest upon the correspondence between measured and thermochemically calculated Arrhenius parameters for a series of thermal reactions of substituted cyclopropanes. This predictive power, however, may be at least partly illusory, because the calculated parameters are not very sensitive to mechanism.

For example, the analysis of the stereomutation and structural isomerization of *cis*-1,2-dideuteriocyclopropane, based on thermochemical-kinetic arguments (6), is shown in Figure 4. The reaction coordinates for the two processes, respectively are assumed to be the contraction of the C-C-C angle of the biradical and the hydrogen migration. The thermochemical estimation of the Arrhenius parameters for the biradical mechanism then reduces to calculation of the heat and entropy of formation (ΔH_f° and ΔS_f°) of the proposed biradical intermediate and, by increments from them, calculation of the ΔH_f° and ΔS_f° of the transition states (‡ in Figure 4).

In this treatment (6), ΔH_f° values for the transition states are not actually calculated numerically, but are expressed as the sum of the biradical ΔH_f° and a

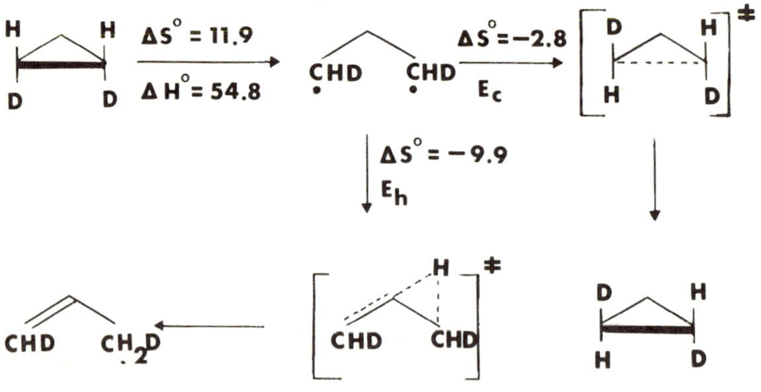

Figure 4 Entropy and enthalpy changes in the biradical mechanism for pyrolysis of cyclopropane-1,2-d_2.

disposable parameter E_c or E_h. The latter increments are evaluated as 9.3 and 10.8 kcal mole^{-1} from the difference between the experimental ΔH_f° of the transition states (+76.3 and +77.8 kcal mole^{-1}, Figure 3) and the calculated biradical ΔH_f° (+67 kcal mole^{-1}). These values of 9.3 and 10.8 kcal mole^{-1}, derived from data on unsubstituted cyclopropane, are assumed (6) to serve also as a "standard" value of the activation energies for cyclization and hydrogen shift in substituted trimethylene biradicals.

The estimation of values for ΔS_f° of the biradical and transition states involves the assignment of vibrational frequencies for each, as well as an internal rotational barrier height for the biradical. These assumed values are then used to extrapolate to the reaction temperature the ΔS_f° values calculated from group-additivity tables.

A-FACTOR ESTIMATES The predictive power of the biradical mechanism for the cyclopropane pyrolyses thus can be tested with respect to the Arrhenius preexponential term A, which is given by $A = (ekT/2h)\exp(\Delta S^\ddagger/R)$, where $S^\ddagger = \Delta S_f^\circ$ (transition state) $- \Delta S_f^\circ$ (ground state). In general, quite good agreement is found (6–8) between the calculated and experimental A-factors by this approach.

As has been pointed out (41), however, the "success of such calculations does not in itself provide any real evidence that the assumed biradical mechanism is correct." In particular, any method of calculating the entropy of the transition state would suffice. The biradical intermediate could then be deleted, and the same A-factor estimate would be obtained. The biradical hypothesis provides, therefore, a stepwise method for calculating ΔS_f° of the transition state, but concordance with experiment does not require that the reaction occur by a stepwise pathway.

The consequences of a specific alternative hypothesis on the A-factor estimate illustrates the insensitivity of the calculation to mechanistic details. We first examine some properties of the hypothetical biradical intermediate.

The barrier to ring closure is taken as 9.3 kcal mole^{-1}. Although internal rotational barriers in biradicals are unknown, it has been suggested (6, 8, 42) that they should be assumed to be about two thirds of that in the corresponding alkane, which would mean that the barrier in trimethylene is about 2 kcal mole^{-1}. There seems to be no experimental basis for this hypothesis. In fact, the sizes of the internal rotational barriers for several monoradicals are now known from β-proton-coupling constants and from line-broadening effects observed in direct ESR spectroscopy (43, 44; reviewed in 45), and the barriers are much smaller than in the corresponding alkanes.

Two pertinent examples are the 1-propyl radical and the isobutyl radical, which have barriers for rotation of the CH_2 group of 0.4–0.5 and 0.3 kcal mole^{-1} (43, 44) as compared to 2.8 and 3.6 kcal mole^{-1} for rotation of the CH_3 groups in the alkanes. Even a radical as highly substituted as 2,3-dimethyl-2-butyl still shows a barrier of only 1.2 kcal mole^{-1} (44).

Using an A-factor of $10^{12.9}$ sec^{-1} and a barrier V of 2.3 kcal mole^{-1}, corresponding to an activation energy $E_a = V + RT = 3.8$ kcal mole^{-1} at 750°K, O'Neal & Benson (6) calculate a rate constant for internal rotation, k_{rot}, in trimethylene biradical. Similarly, with $A = 10^{13.6}$ sec^{-1} and $E_a = 9.3$ kcal mole^{-1}, they calculate k_{cycl}, the ring-closure rate constant. On this basis, the ratio k_{rot}/k_{cycl} would be about 8. Using the more plausible rotational barrier of about 0.4 kcal mole^{-1}, we calculate $k_{rot}/k_{cycl} \sim 28$.

In other words, internal rotation in trimethylene should be much faster than ring closure, even when the higher (and less likely) rotational barrier is used. Note that this is a direct consequence of two basic assumptions of the model. The first is that $\Delta H_f°$ of the biradical can be calculated from additive C–H bond-rupture increments in propane, and the second, (which is really a consequence of the first) is that there is no interaction between the radical centers.

Now suppose that we postulate another mechanism in which there is no local minimum for the biradical. Internal rotation and ring closure would become competitive, yet the structure and hence the $\Delta S_f°$ of the transition state for stereomutation would look very much the same as before. Therefore, the A-factor estimated for this mechanism would also fit the observed value. Clearly, a complete spectrum of mechanisms, from a stereorandom intermediate to a stereospecific process, could be entirely compatible with the observed A-factor.

E_a ESTIMATES   There are, however, more serious deficiencies of the biradical model for the thermal rearrangements of cyclopropanes. One of these emerges in the case of 1,1,2,2-tetramethylcyclopropane (Figure 5). The experimental activation parameters for the structural isomerization to 2,4-dimethylpent-2-ene, (5-4) are reported as $A = 10^{15.8}$ sec^{-1}, $E_a = 64.4$ kcal mole^{-1} (46). These values are very close to those of cyclopropane itself, which seems surprising at first, since the tetrasubstitution on one bond might have been expected to weaken it by ~ 8–10 kcal mole^{-1} (5, 8, 9).

In accord with this, transition-state thermodynamic parameters calculated from the additivity tables (6, 47) lead to a predicted value for A of $10^{14.4}$ sec^{-1} and for E_a of 56.0. Although blame for the large discrepancy has been placed (6) on experimental inaccuracy arising from measurements made at temperatures "only over a 30° range," the rate measurements in fact span temperatures between 435 and 484°C, a range considered adequate (6) in other cases. An experimental reinvestigation of the structural isomerization of 1,1,2,2-tetramethylcyclopropane (48) reveals a minor side-reaction, but the new data, although permitting a better fit of the preexponential factor, still give an activation energy, $E_a = 62.3$ kcal mole^{-1}, much higher than

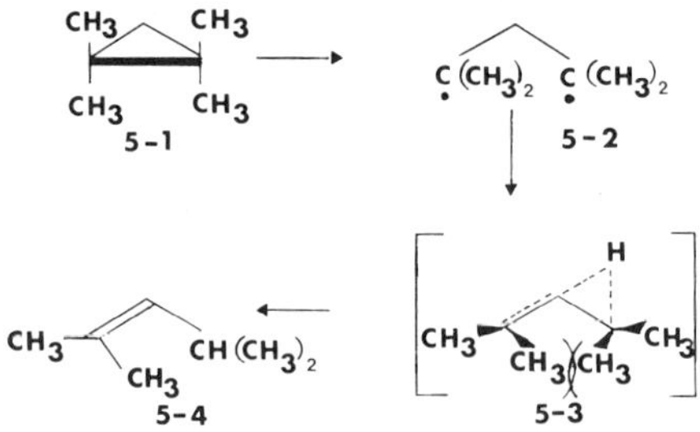

Figure 5 Hypothetical biradical mechanism for the structural isomerization of 1,1,2,2-tetramethylcyclopropane.

that calculated, 56.0 kcal mole^{-1}. The discrepancy is rationalized by postulation of a 3–6 kcal mole^{-1} steric repulsion in the transition state (48).

It seems likely that a steric effect should be present (see 5–3), but we know of no good way to evaluate its magnitude a priori. The conclusion (48) that the transition-state estimate should be given as $E_a = 59 \pm 3$ kcal mole^{-1} means that the predicted rate constant at 730°K would be uncertain by a factor of 60, so that the model provides a rather blunt predictive tool for such cases.

Attempts to interpret the stereomutation of *trans*-1,1,2,2-tetramethylcyclopropane-d_6 (Figure 6) in terms of the biradical mechanism encounter even more severe difficulties. Optically active **6-5** undergoes two thermal stereomutation processes (49): loss of optical activity (rate constant k_α), and conversion to the equilibrium mixture (50:50) with the *cis* isomer **6-6** (rate constant k_i). The Arrhenius parameters of the k_i process are $A = 10^{15.0}$ sec^{-1} and $E_a = 54.5$ kcal mole^{-1}. At 350°C, the ratio k_i/k_α equals 1.74.

On the basis of the biradical mechanism (Figure 6), the experimental ratio, k_i/k_α, may be expressed in terms of the two mechanistic rate constants k_{cycl} and k_{rot}, which characterize the ring closure and interconversion by internal rotation of each of the hypothetical biradical intermediates **6-8**, **6-9**, and **6-10**. Note that the molecular symmetry ensures that, aside from a negligibly small configurational isotope effect, the same rate constants apply to all three intermediates. The competition between ring closure and internal rotation then controls the observed rate-constant ratio according to the equation $k_{cycl}/k_{rot} = 4(1-n)/(n-2)$, where $n = k_i/k_\alpha$. In the framework of the biradical mechanism, cyclization of the intermediate is much faster than internal rotation, because the value $k_i/k_\alpha = 1.74$ means that $k_{cycl}/k_{rot} \cong 11$ (49).

Whereas ring closure is considered to be rate-determining in the biradical mechanism for stereomutation of cyclopropane-1,2-d_2 (see Figure 4) and several substituted cyclopropanes (6–8), the experimental results in the tetramethylcyclopropane system are incompatible with this. It is suggested (48) that a high barrier in the tetramethyl biradical slows down the internal rotation rate and makes that process

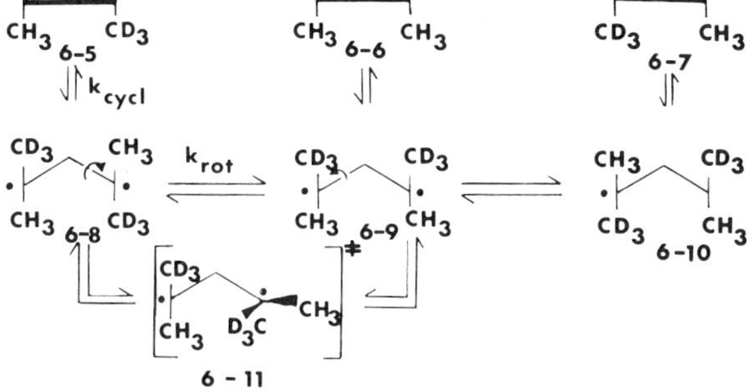

Figure 6 Hypothetical biradical mechanism for the stereomutation of optically active *trans*-1,2-dimethyl-1,2-bis-trideuteriomethylcyclopropane.

rate-determining. If the "standard" ring-closure barrier of $V_{cycl} = 9.3$ kcal mole^{-1} is imagined to persist here, the internal rotation barrier V_{rot} may be derived by fitting the experimentally deduced ratio of cyclization to rotation rates according to the equation:

$$\log (k_{cycl}/k_{rot}) = \log 11 = \log (1/2) + (\Delta\Delta S^{\ddagger}/2.3R) - (V_{cycl} - V_{rot})/2.3RT,$$

where $\Delta\Delta S^{\ddagger}$ is the difference in activation entropies for rotation and cyclization, calculated by the usual additivity methods.

On this basis, V_{rot} for the biradical is 9.8 kcal mole^{-1} (48). Actually, this should be considered a minimum value, since it would be reasonable to increase V_{cycl} by two "*cis*-methyl" interactions (2 × 1 kcal mole^{-1}), which are postulated (6, 8) in similar instances.

Thus, a rotational barrier of 10–12 kcal mole^{-1} for **6-8** must be invoked to preserve the biradical interpretation. It is argued (48) that since barriers of 9.3 and 11.0 kcal mole^{-1} are reported (50) for the corresponding hydrocarbon, 2,4-dimethylpentane, a high barrier in the biradical **6-8** is "physically reasonable." However, the data already described on internal rotational barriers in monoradicals (43, 44) offer little support for this view.

Some Specific Predictions and Correlations Derived from Quantum Mechanics

This subject has been reviewed elsewhere (11, 51), but it is useful to examine a few aspects of it for the purposes of the present discussion. The efflorescence of theoretical attention to the chemistry of cyclopropane and trimethylene originated in experimental observations on the thermal decomposition of pyrazolines. Although the full mechanistic interpretation of such reactions remains a puzzling challenge even to subtly designed current investigation (52), early experiments (53, 54) suffice to establish a key observation, the stereochemical "crossover effect," outlined in Figure 7, in which a *cis*-3,5-disubstituted pyrazoline gives mainly a *trans*-1,2-disubstituted cyclopropane, and vice versa. The stereospecificity is not high, since the "crossover" pathway is preferred only by a factor of about 3, but the result still demands under-

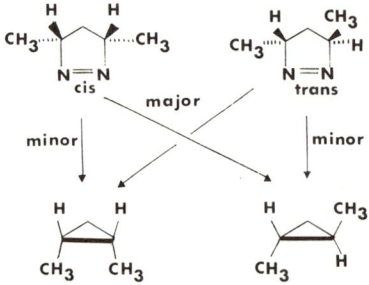

Figure 7 Stereochemical cross-over in the pyrolysis of *cis* and *trans*-3,5-dimethyl-1-pyrazoline.

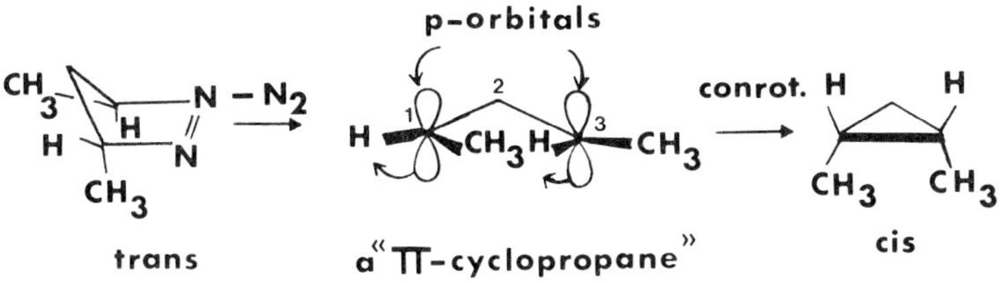

Figure 8 Hypothetical π-cyclopropane from the pyrazoline thermolysis.

standing. Part of a rationalization is provided by the insights (53) that an intermediate might be a trimethylene in which the two-terminal CH_2 groups occupy the plane of the three carbons (a "π-cyclopropane") (Figure 8), and that if, for some reason, it were constrained to cyclize in a conrotatory manner, the result would be a "crossover" of stereochemistry. A quantum mechanical basis for conrotation emerges from Hoffmann's extended Hückel calculations on unsubstituted trimethylene (55).

The calculations show that in the "π-cyclopropane," there is appreciable hyperconjugative interaction between one of the filled bonding orbitals of the central CH_2 group (at C_2) and the *p*-orbitals. This interaction has two main consequences. First, it splits the degeneracy of the in-phase and out-of-phase π-combinations (Figure 9) of the *p*-orbitals by selective destabilization of the in-phase level. For C_1-C_2-C_3 angles $>100°$, this leaves the out-of-phase π-combination as the highest occupied molecular orbital (MO) and therefore, on orbital symmetry grounds, favors conrotation. Second, for C_1-C_2-C_3 angles near 125°, the hyperconjugative interaction stabilizes the "π-cyclopropane" conformation (0,0, with coplanar terminal methylenes) relative to other conceivable rotamers of trimethylene, e.g. 0,90 or 90,90 (Figure 9).

The calculations (55) suggest that whereas no barrier should oppose the synchronous conrotatory closure of the 0,0 species to cyclopropane, the rotation of one CH_2 group out of the plane (process 0,0 → 0,90) should require a substantial activation energy. This should ensure the preservation of the stereochemical relationship between the labels in a substituted case.

Although neither theory nor experiment requires that the pyrazoline-derived intermediate be involved in the cyclopropane pyrolysis, it is not unreasonable to

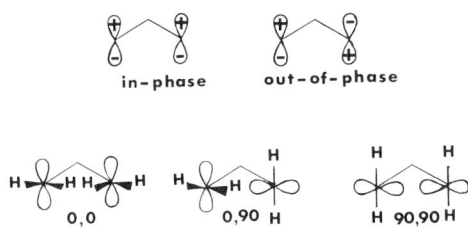

Figure 9 In-phase and out-of-phase combinations of the nonbonding π-cyclopropane orbitals, and three extreme configurations of trimethylene.

guess (55) that the same 0,0 species lies on the pathway for the pyrolysis of cyclopropanes. Later and more extensive theoretical examinations of the energy surface for cyclopropane (31–38, 56) tend to confirm qualitatively the 0,0 species as the lowest energy trimethylene, although the preference is not as large as the 6–10 kcal mole^{-1} shown by extended Hückel calculations (55).

Microscopic reversibility requires that if trimethylene cyclizes by synchronous conrotation, it must also be formed by the same path. The quantum mechanically developed picture of stereomutation in cyclopropane thus predicts strongly coupled motions of the terminal CH_2 groups of the reacting bond, in contrast to the stereorandom behavior predicted by the bond-additivity scheme.

The electron pair in the filled C–H bonding level of the π-cyclopropane is analogous to a lone pair in a cyclopropyl anion (Figure 10, **10-1**) or its isoelectronic analogs, aziridine, **10-2**, and oxirane, **10-3**. Strong experimental evidence for synchronous conrotation in substituted examples of both of the latter systems is available (58–62), but in the azomethine and carbonyl ylids, **10-5** and **10-6**, the conjugative interaction between the nonbonding heteroatom orbitals and the terminal p-orbitals is stronger than the hyperconjugative interaction involving the C–H orbital in the π-cyclopropane (Figure 9) (55, see also 56). The crucial question, therefore, is whether this hyperconjugation really is strong enough to stabilize π-cyclopropanes and couple the rotations of the terminal methylene groups.

Theoretical explorations of the energy surface by far more elaborate methods agree with the original extended Hückel conclusion that the conrotatory ring closure of the π-cyclopropane should occur without activation energy. The most ambitious of the calculations are those of the Salem (35, 37) and Goddard (31) groups.

Salem uses a minimal basis set of Slater orbitals in a self-consistent field (SCF) molecular orbital (MO) method to calculate the full 21-dimensional energy surface for cyclopropane stereomutation. In the region of the trimethylene biradical, the calculation employs a 3-by-3 configuration interaction among the ground, singly excited, and doubly excited electronic configurations. The energies are further improved by inclusion in the main program of an open-shell program that modifies the Nesbet-Hartree-Fock formalism (37).

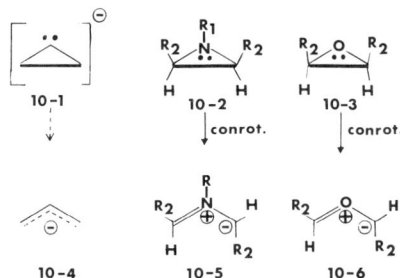

Figure 10 Thermal ring-openings of cylopropyl anion, aziridine, and oxirane.

The Goddard calculation (31) uses an ab initio generalized valence-bond method (see 34 for review) and, although a complete surface is not available, thorough attention is given to the energy changes upon rotation and angular deformation in the trimethylene.

According to the results of these calculations (31, 35, 37), it is not quite accurate to refer to the 0,0 biradical as a "π-cyclopropane," because it is likely that the geometry of the terminal methylenes would relax to a more stable pyramidal configuration. Nevertheless, the calculations confirm that rotation of one of the methylenes to a "0,90" configuration (also pyramidalized) is uphill from the pyramidalized "0,0" configuration; the energy difference is 0.6 kcal mole^{-1} in the Salem work.

Another serious conflict with the thermochemical additivity model arises from the failure of all of the quantum mechanical calculations to reveal any local energy minimum near the transition state.

Experimental Tests of the Stereorandom Biradical and π-Cyclopropane Mechanisms of Stereomutation

Most of the tests involve substituted cyclopropanes, but since the quantum mechanical studies refer to the parent compound, the application of the predictions to the substituted cases requires some assumptions, not all of which are invariably stated explicitly. The predictions of the thermochemical-kinetic model, of course, grow directly out of the data on substituted cases.

The pyrolysis of *anti*-vinylcyclopropane-2,3-*cis*-d_2 (Figure 11, **11-1**) provides the only example of one kind of experimental approach that could be exploited fruitfully

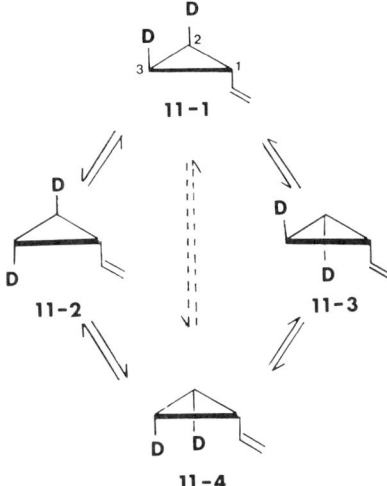

Figure 11 Stereomutation in the thermolysis of 1-vinylcyclopropane-2,3-d_2. Experimentally, reactant **11-1** gives a product mixture in which the ratio (**11-2** + **11-3**)/**11-4** is 2:1 and is independent of time.

in additional cases (63). At 325°C and 100–150 torr, the original stereospecific labeling pattern of **11-1** becomes scrambled because of the formation of **11-2**, **11-3**, and **11-4** (analysis by NMR). The ratio of the *trans* (**11-2** + **11-3**) and *syn-cis* (**11-4**) isomers produced is 2:1, at all times between 0.6 to 10 half-lives for achievement of the equilibrium distribution of the label.

To explain this behavior by synchronous double rotations would require that the reactions **11-1** → **11-2** and **11-1** → **11-3**, which involve C_1–C_3 and C_1–C_2 rotation, respectively, must not only have equal rates by symmetry, but the rates must also be fortuitously equal to that of the **11-1** → **11-4** reaction, which involves C_2–C_3 rotation. This seems unlikely on the grounds that the C_1–C_2 and C_1–C_3 bonds should be weakened by allylic resonance in the transition state, whereas the C_2–C_3 bond should not. If the weakening in question is even a substantial fraction of the approximately 10–12 kcal mole^{-1} frequently observed in other potential allylic systems (64 and references cited therein), the contribution of reaction at the "wrong" C_2–C_3 bond should be negligible.

A biradical intermediate with internal rotation much faster than cyclization, however, accounts well for the observations (63). Incidentally, in this case no assumption is necessary regarding which bond breaks. As long as stereoequilibrium in the intermediate is achieved before ring closure, the product ratio will be (**11-2** + **11-3**)/(**11-4**) = 2:1, as observed.

A different kinetic technique forms the basis of several other searches for the synchronous mechanism. Coupled (con- or dis-) rotation of the two substituted carbons in a chiral 1,2-disubstituted cyclopropane leads only to racemization. *Trans-cis* isomerization can occur by this mechanism through coupled rotations of either of the substituted carbons with the unsubstituted one (C_3) (Figure 12), or, by an alternative mechanism, through single rotations of C_1 or C_2. These events cannot be analyzed in the general case, because the system is "underdetermined," that is, there are ten mechanistic rate constants but only six kinetic observables. The difficulty is often bypassed with the plausible but unproven assumption that reaction occurs much more readily at the disubstituted ring bond, which should be

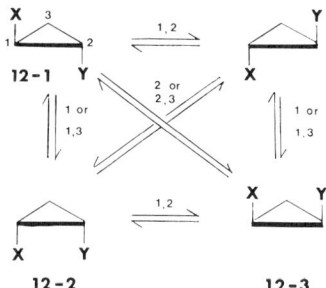

Figure 12 Single and double rotation mechanisms in the stereomutation of a disubstituted cyclopropane. The numbers alongside the arrows indicate which carbons are rotating in that pathway. For example, **12-1** → **12-2** can occur by single rotation of C_1 or by double rotation of C_1 and C_3.

the weakest, because its cleavage would give the most stable biradical. Although this crucial assumption is logically circular and lacks independent experimental support, it forms the basis for most of the kinetic analyses.

Pyrolyses of optically active *trans*-1,2-diphenylcyclopropane (65), *cis*- and *trans*-1-methyl-2-ethylcyclopropane (66), *cis*- and *trans*-1-cyano-2-isopropenylcyclopropane (67), *cis*- and *trans*-1,2-diphenyl-1-carbomethoxycyclopropane (68), and *trans*-1,2-dimethyl-1,2-bis-trideuteriomethyl-cyclopropane (69) each occur with a *trans* ⇌ *cis* interconversion rate that is at least competitive with, and in some cases larger than, the rate of racemization. The results can be interpreted by a mechanism in which C_1 and C_2 of a biradical intermediate experience single, noncoupled rotations at rates dependent on the nature of the substituent. A general trend is observed (11, 67) in which the more massively substituted carbons rotate slower, in accord with the prediction (6) of a ponderal retardation of internal rotations.

Sharply different behavior emerges from the pyrolysis of optically active *trans*-1-phenylcyclopropane-2-*d* (**12-1**, $X = $ Ph, $Y = $ D) (70, 71). The molecular symmetry enables a partial kinetic analysis free of the "most-substituted bond" assumption. In particular, it is possible to extract from the data the rate constant (k_1) for a single, unaccompanied rotation of the phenyl-bearing carbon (C_1), and the rate constants (k_{12} and k_{13}) for the synchronous double rotations of C_1 with C_2 and C_1 with C_3, respectively. The constants k_{12} and k_{13} differ only by a small secondary isotope effect. The values ($\times 10^5$ sec at 309.5°C) are $k_{12} = 0.96$, $k_{13} = 1.06$, and $k_1 \leqslant 0.098$. The contribution of single rotation to the stereomutation at C_1 thus amounts to less than 4% $[100k_1/(k_1 + k_{12} + k_{13})]$. The major pathway is a reaction in which every rotation of CHPh is accompanied by a synchronous rotation of $CH_2(k_{13})$ or $CHD(k_{12})$. Although the distinction between con- and disrotations remains an experimental challenge, the coupling of the motions of the ring carbons is just as predicted by the π-cyclopropane mechanism.[1]

The stereomutation at the deuterated carbon C_2 cannot yet be completely analyzed. However, the data available permit the conclusion (70, 71) that k_{12}, the rate constant that characterizes the double rotation of C_2 in synchrony with C_1, is 80% of the total C_2 stereomutation. The remainder of the C_2 stereomutation is made up of single (k_2) and double (k_{23}) rotations in a ratio that is not determinable from these experiments.[2] It is already clear, however, that the major reaction at C_2, just as at C_1, is a synchronous double rotation.

A thermochemical-kinetic analysis would predict an activation energy for a biradical mechanism of the phenylcyclopropane stereomutation about 10–12 kcal mole^{-1} lower than the value of 65 kcal mole^{-1} found for cyclopropane itself, because of the expected radical-stabilizing effect of the phenyl group (6–9). Although the reaction has not been studied over a range of temperatures, the assumption of a

---

[1] The possibility that the overall double rotation of 1-phenylcyclopropane-2-*d* might be achieved by reversible 1,3-sigmatropic rearrangements passing through bicyclo [4.3.0] nonatriene intermediates must be considered as unlikely, since this would require an unfavorable (89, 121, 122) antarafacial shift.

[2] In principle, the complete analysis of the C_2 stereomutation could be carried out by a study of the pyrolysis of the 1-phenylcyclopropane-2, 3-d_2 isomers.

preexponential factor of 10^{15} sec^{-1}, a rough average value for other cyclopropanes (1–9, 11), and the observed rate at 309.5°K (70, 71) lead to $E_a \sim 54$ kcal mole^{-1} for phenylcyclopropane, which is just about the expected value.

The hypothetical "uncoupled" biradical of the thermochemical-kinetic analysis would predict fast single rotation of the deuterated carbon (C_2) and slow single rotation of the phenyl-substituted carbon (C_1). The strongly coupled motions observed, instead, demonstrate how misleading a "correct" prediction of the activation parameters can be. Here, the transition state has exactly the energy predicted by the biradical mechanism, but the reaction occurs by a completely different pathway.

The pyrolysis of optically active *trans*-cyclopropane-1,2-d_2 (**12-1**, $X = Y = D$) gives a similar result (71, 72). In this case, the observable events are loss of optical activity (rate constant k_α) and *trans-cis* isomerization (rate constant k_i). Except for the small secondary isotope effect, all the ring bonds are equivalent and must break at the same rate. The symmetry now permits a complete analysis of the kinetic data, since the mechanistic extremes each predict different values of the ratio k_i/k_α. It is a justifiable assumption that the rate of cleavage of each bond is retarded by about 10% per deuterium, although the conclusions are not significantly changed if there is assumed to be no isotope effect. With $k_H/k_D = 1.10$, the single rotation pathway, the stereorandom intermediate of the thermochemical additivity scheme, the synchronous double rotation, and a previously unconsidered triple rotation respectively predict k_i/k_α values of 2.00, 1.53, 1.05, and 0.00. The observed value of 1.07 ± 0.04 can be interpreted in only two ways. Either the reaction occurs by a mixture of triple rotation with random intermediate and/or single rotation mechanisms, or, more plausibly, it is essentially pure double rotation (71, 72).

The experimental error in the *trans*-cyclopropane-d_2 study would permit perhaps 10% of a single rotation pathway to escape undetected. In the case of 1-phenyl-cyclopropane-2-d, single rotation of C_1 and C_2 might constitute as much as 4% and 20%, respectively, of the total turnover of these centers. Thus, the single rotation transition states might lie only 3.2, 3.7, and 1.6 kcal mole^{-1} above the corresponding double-rotation transition states.[3] This is still somewhat larger than the gap predicted by the most sophisticated theories (34, 35). More recent theoretical studies (74, 75) suggest that the competition between single and double rotation may not be controlled simply by the transition-state energy differences as $k_s/k_d = \exp(\Delta\Delta G^\ddagger/RT)$, but that dynamical effects may enhance the preference for double rotation. In any case, it would not be surprising if substituents could modify the preference, and in fact, recent CNDO calculations (57) rationalize the apparent differences in behavior between some of the earlier examples (63, 65–69) and the recent ones (70–72) in terms of substituent-induced perturbations of the separation and ordering of the single- and double-rotation transition states.

There remains the question of whether some of the results in the early studies (63, 65–69) may be clouded by contributions from cleavage of the "wrong" (less-

---

[3] It is not clear whether the π-cyclopropane is an intermediate or the transition state in the stereomutation. If it is the transition state, it cannot also be the transition state for the structural isomerization to propylene, since a true activated complex cannot correspond to the intersection of three or more valleys of the potential surface (84).

substituted) bond. Thus, the recent suggestion (76) that the stereomutation of the spirodiene **13-1** may occur with double-rotation rate constants in the ratio $k_{13}:k_{12}:k_{23} \approx 40:30:30$ implies a startlingly large "wrong" bond-reaction component. Because the kinetic analysis of the **13-1** system is not yet complete (76), this set of relative rates is only one of several possible interpretations of the data, but should it be confirmed, the biradical hypothesis would be shown to have erred by many orders of magnitude in predicting the relative "strengths" of the C_1–C_2, C_1–C_3, and C_2–C_3 bonds.

The Question of a Local Minimum

Application (7, 77a) of the thermochemical-additivity approach to the stereomutation (77b) of 2-bicyclo[2.1.0]pentane (**13-2** ⇌ **13-3**, Figure 13) suggests a metastable biradical intermediate **13-4**, $R = CH_3$, which is protected from reclosure by an activation energy of about 9 kcal mole^{-1}. This barrier to cyclization is about the same size as that proposed for trimethylene itself (6, 8, 9, 29). The reader will recall that quantum mechanical calculations do not confirm the existence of such a minimum on the trimethylene potential surface (31, 35, 37). Although similar calculations have not been performed for the hypothetical biradical **13-4**, $R = H$, a relevant experimental result has recently become available (78).

Irradiation of the diazene **13-6** at 1°K generates the triplet ground state of the biradical **13-4**, $R = H$, which can be recognized by its electron spin resonance (ESR) spectrum. Warming to a few degrees Kelvin causes disappearance of the ESR signal as the biradical cyclizes to the hydrocarbon **13-5**. Analysis of the temperature dependence of the kinetics of the signal decay suggests that at least some of the cyclization occurs by tunneling through a barrier of less than 1–2 kcal mole^{-1} for closure of the singlet biradical (78). Here again, the thermochemical approach has seriously overestimated the stability of a 1,3-biradical.

It has been suggested (79) that inaccuracies in the thermochemical estimates of $\Delta H_f°$ of trimethylene may result in part from the use (6, 8, 9) of incorrect values for D_{C-H} in propane (see Figures 2 and 3). This seems unlikely to be the cause of the discrepancies, since recent redeterminations of D_{C-H} for hydrocarbons tend to confirm the previous values (80–83a–d).

The thermally generated intermediates in small ring pyrolyses presumably are singlet species, but the thermochemical additivity approach neglects any spin correlation effects. It has been argued (97) that since the hypothetical C–H homolysis

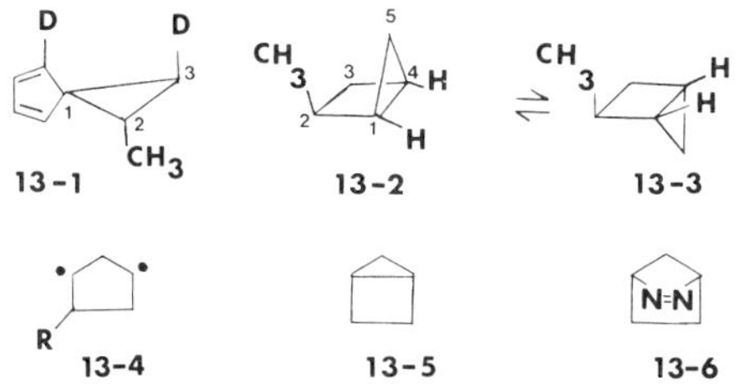

Figure 13

reaction by which the monoradical is converted to the biradical has a 3:1 chance of forming the triplet, the additivity assumption leads to calculated $\Delta H_f°$ values for the singlet that are too low by a weighted average amount, which is three fourths of the singlet-triplet splitting.

The correct method of taking spin into account in the framework of the thermochemical approach is not obvious. The most reliable quantum mechanical calculations (37) place the triplet of trimethylene in a local minimum 4.4 kcal mole^{-1} below the transition state (SRTS) for the hypothetical single-rotation mechanism of stereomutation. The experimental facts (71, 72) place the double-rotation transition state (DRTS) 3.2 kcal mole^{-1} below SRTS, which would mean that the triplet is only 1.2 kcal mole^{-1} below DRTS. Since the thermochemical biradical is 9.3 kcal mole^{-1} below DRTS (Figure 3), the triplet is calculated to lie 9.3 − 1.2 = 8.1 kcal mole^{-1} *above* the thermochemical biradical. To place the energy of the thermochemical biradical close to that of the triplet would require that the singlet-triplet splitting for trimethylene be much larger than that calculated theoretically (37) and much larger than that observed experimentally (78) (<1 kcal mole^{-1}) in the only available model biradical **13-4**.

More probably, the difficulty lies in the additivity assumption itself. That is, the C_3–H bond of the propyl radical is stronger by several kcal mole^{-1} than the C_1–H bond of propane. A clear theoretical explanation of this effect would be welcome.

Whatever the cause, if such overestimates of the stabilities of 1,3-biradicals prove to be general, there will be serious consequences for the interpretation of kinetic data. Thus, heretofore, some observed activation energies would have placed the transition state $\Delta H_f°$ well above that of the hypothetical biradical and hence would have permitted a biradical mechanism. An upward revision of the biradical $\Delta H_f°$ would narrow the energy gap and, in some cases, reverse the ordering, so that the transition-state energy might actually fall below that of the biradical. Thus, there may be many more concerted reactions involving cyclopropanes than have been previously suspected. Obviously, it would be helpful to have more direct experimental determinations of ring-closure barriers, not only for 1,3-biradicals (see 78) but also for 1,4- and higher biradicals. In the meantime, the reliability of the thermochemical criterion of mechanism will be questionable, not so much because of the experimental limits of accuracy of the Arrhenius parameters, but rather because of uncertainty about the validity of the calculated $\Delta H_f°$ values of the hypothetical biradical intermediates.[4]

---

[4] Recent contributions in the cyclopropane area include a study of a triply labeled system and a numerical integration procedure for solution of the kinetic scheme that takes into account the constraints of microscopic reversibility (85). Also, the overall 1,3-sigmatropic rearrangements of 1-styryl-2,2-diphenylcyclopropanes have been interpreted in terms of a conrotatory opening to a π-cyclopropane followed by a disrotatory closure to rearranged product (86). I am indebted to Professors A. H. Andrist and R. Huisgen for providing preprints of these publications. Other recent work on the vinylcyclopropane rearrangement is reported elsewhere (87–89).

THERMAL FORMATION, REARRANGEMENT, AND DECOMPOSITION OF CYCLOBUTANES

Space limitations preclude all but a cursory mention of the rich store of experimental and theoretical information on this subject (90–103, and references cited therein). The suggestion, based on thermochemical-kinetic grounds (93), that the thermal decomposition of cyclobutanes to two ethylenes occurs by a stepwise mechanism is supported by the experimental observations that such reactions are not stereospecific (99–101) and are accompanied by stereomutation in the cyclobutane (99) analogous to those observed (63, 65–72, 104–107) in cyclopropanes. The $\Delta H_f°$ of the hypothetical tetramethylene biradical intermediate (Figure 14, **14-2**) calculated by the usual additivity assumptions places it in a potential well about 4 kcal mole^{-1} below the transition state for decomposition. Quantum mechanical calculations at the extended Hückel level (94) show only a broad flat region near the transition state, but a later SCF calculation at the STO-3G level with 15-dimensional configuration interaction does show two true minima, for the *gauche* and *trans* (extended) conformations of **14-2** (95). The *gauche* form is protected by barriers to dissociation and cyclization of 3.6 and ≥ 2.0 kcal mole^{-1}, respectively.

The same kind of intermediate can be imagined in the 2 + 2 cycloaddition of two olefins **14-3**, in the thermolysis of tetrahydropyridazines **14-1**, and in the Norrish Type II photocleavage of ketones **14-4** → **14-5** (109, 110). It was postulated at one time that because of the energy liberated by release of N_2, photochemical *or thermal* decompositions of azo compounds such as **14-1** should give rise to vibrationally

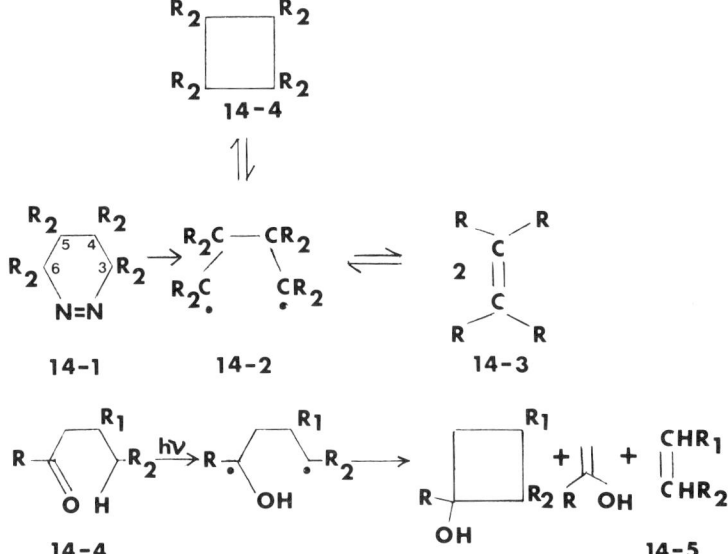

Figure 14 Alternative pathways involving 1,4-biradicals.

excited tetramethylene biradicals, which should cyclize more rapidly than the thermally equilibrated biradicals generated by the alternative pathways of Figure 14 (108, reviewed in 11). The highly stereospecific formation of the isomeric 1,2-dimethyl-1,2-diethylcyclobutanes from the corresponding 3,6-dimethyl-3,6-diethyltetrahydropyridazines (111) was offered in support of this idea. The stereospecificity, however, could also have been the result of slow rotations of the heavily encumbered terminal carbons in the 1,4-biradical intermediate (6, 11, 112–114), and recent studies (115) of less substituted examples in the 3,6 and 3,4-dimethyltetrahydropyridazine series now show the marked decrease in the stereospecificity of ring closure that would be consistent with the expected diminution of such a ponderal effect. Stereoretention thus is not an intrinsic property of 1,4-biradicals but depends upon substitution.

Kinetic analysis (115) reveals that the competition between ring closure and internal rotation in the tetramethylene biradicals generated from the dimethyltetrahydropyridazines is essentially identical with that in the species formed during thermolyses (99) of 1,2-dimethylcyclobutane. This work deprives the hypothetical, vibrationally excited biradical (108) of an experimental basis and, moreover, provides a rare example of concordance between the behavior of supposed common intermediates from two biradical precursors.[5,6]

In another system, such a concordance is predicted by the thermochemical-kinetic assignment (9, 116) of stepwise mechanisms via the common biradical **15-1a** (Figure 15) for the epimerization of 4-vinylcyclohexene **15-4a** (117), for the formation of **15-4a** by the Diels-Alder dimerization of butadiene **15-2a**, and for the sigmatropic rearrangement of *trans*-1,2-divinylcyclobutane **15-3a** (118). It is a reasonable extrapolation to apply this prediction also to the closely related case of piperylene **15-2b**, dimerization of which gives **15-4b**. The latter compound is also formed from *trans*-1,2-dipropenylcyclobutane by sigmatropic rearrangement (121, 122).

The experimental facts are at variance with the predictions. Stereochemical labeling studies show that the Diels-Alder reactions both occur with high retention of configuration at the dienophilic site (the 3,4-bond), 90% in the case of **15-2a** (119) and >98% in the case of **15-2b** (120, 121), but the sigmatropic rearrangements both occur with weak predominance of *inversion* at the same site (121, 122) in what is proposed (116) as the "common" intermediate **15-1a** or **b**. The rearrangement clearly has an orbital symmetry-controlled concerted suprafacial-inversion component, despite the fact that the transition-state energy is appropriate for a reaction with a biradical intermediate (121, 122). Similar incompatibility between the thermochemical-kinetic prediction and the behavior revealed by stereochemical experiment has been noted in other vinylcyclobutane pyrolyses (123–125).

---

[5] The data permit the further reasonable conclusion that the decomposition of 3,4-dimethyltetrahydropyridazine to N_2, ethylene, and 2-butenes has a detectable concerted component (115).

[6] For a similar result in the 2,3-dimethylcyclohexandiyl system, see W. R. Roth, unpublished, as cited in (11).

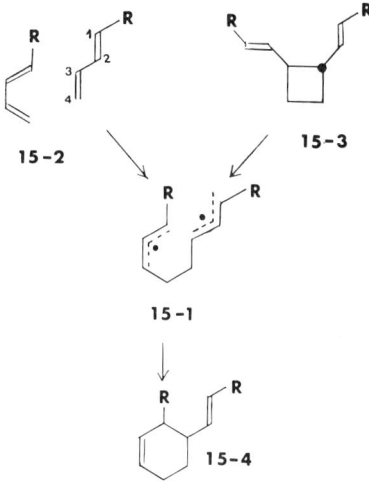

Figure 15 Hypothetical biradical common intermediate (**15-1**) in the Diels-Alder dimerization of simple dienes (**15-2**) and in the thermal 1,3-sigmatropic rearrangement of dialkenylcyclobutanes (**15-3**). In the text, compounds **a** have $R = H$; **b**, $R = CH_3$.

ACKNOWLEDGMENTS

Research from the author's laboratory reviewed here was supported by the National Science Foundation (GP-33909X and GP-11017X), the Hoffmann-LaRoche Foundation, and the Petroleum Research Fund.

Literature Cited

1. Robinson, P. J., Holbrook, K. A. 1972. *Unimolecular Reactions.* New York: Wiley-Interscience. 371 pp.
2. Forst, W. 1973. *Theory of Unimolecular Reactions.* New York: Academic. 445 pp.
3. Laidler, K. J., Loucks, L. F. 1972. *Comp. Chem. Kinet.* 5:1–148
4. Willcott, M. R., Cargill, R. L., Sears, A. B. 1972. *Prog. Phys. Org. Chem.* 9:25–98
5. Frey, H. M., Walsh, R. 1969. *Chem. Rev.* 69:103–24
6. O'Neal, H. E., Benson, S. W. 1968. *J. Phys. Chem.* 72:1866–87
7. O'Neal, H. E., Benson, S. W. 1970. *Int. J. Chem. Kinet.* 2:423–56
8. Benson, S. W., O'Neal, H. E. 1970. *Kinetic Data on Gas Phase Unimolecular Reactions*, NSRDS-NBS 21. Washington DC: Nat. Bur. Stand., US Dep. Commer. 628 pp.
9. Benson, S. W. 1971. *Thermochemical Kinetics.* New York: Wiley. 223 pp. 1st ed.; 1976. 320 pp. 2nd ed.
10. Jones, G. II. 1974. *J. Chem. Educ.* 51:175–81
11. Bergman, R. G. 1973. In *Free Radicals*, ed. J. Kochi, 1:191–237. New York: Wiley. 713 pp.
12. Bartlett, P. D. 1968. *Science* 159:833–38
13. Bartlett, P. D. 1970. *Q. Rev. Chem. Soc. London* 24:473–97
14. Woodward, R. B., Hoffmann, R. 1970. *The Conservation of Orbital Symmetry.* New York: Academic. 178 pp.
15. Gilchrist, T. L., Storr, R. C. 1972. *Organic Reactions and Orbital Symmetry.* Cambridge: Cambridge Univ. Press. 271 pp.
16. Trong Anh, Nguyen. 1970. *Les Regles de Woodward-Hoffmann.* Paris: Ediscience. 196 pp.

17. Lehr, R. E., Marchand, A. P. 1972. *Orbital Symmetry: A Problem-Solving Approach*. New York: Academic. 190 pp.
18. Gill, G. B., Willis, M. R. 1974. *Pericyclic Reactions*. London: Chapman & Hall. 240 pp.
19. Dolbier, W. R. Jr. 1971. In *Mechanisms of Molecular Migrations*, ed. B. S. Thyagarajan, 3:1–66. New York: Wiley. 464 pp.
20. Gajewski, J. J. 1971. In *Mechanisms of Molecular Migrations*, ed. B. S. Thyagarajan, 4:1–54. New York: Wiley. 326 pp; Brown, J. M. 1973. In *MTP Review of Science*, ed. W. Parker, pp. 159–204. College Park, Md: Univ. Maryland Press; Spangler, C. W. 1976. *Chem. Rev.* 76:187–217
21. Capon, B., Rees, C. W. 1965. *Organic Reaction Mechanisms*. London: Interscience. 352 pp.
22. The Chemical Society. 1967–1975. *Annu. Rep. Prog. Chem. Sect. B*
23. Robinson, P. J., Holbrook, K. A. See Ref. 1, pp. 238–67
24. Forst, W. See Ref. 2, pp. 380–86
25. Wieder, G. M., Marcus, R. A. 1962. *J. Chem. Phys.* 37:1835–52
26. Schlag, E. W., Rabinovitch, B. S. 1960. *J. Am. Chem. Soc.* 82:5996–6000
27. Lin, M. C., Laidler, K. J. 1968. *Trans. Faraday Soc.* 64:927–44
28. Simons, J. W., Rabinovitch, B. S. 1964. *J. Phys. Chem.* 68:1322–35
29. Benson, S. W. 1961. *J. Chem. Phys.* 34:521–26
30. Salem, L., Rowland, C. 1972. *Angew. Chem. Int. Ed. Engl.* 2:92–111
31. Hay, P. J., Hunt, W. J., Goddard, W. A. III. 1972. *J. Am. Chem. Soc.* 94:638–40
32. Dewar, M. J. S., Kirschner, S., Kollmar, H. W., Wade, L. E. 1974. *J. Am. Chem. Soc.* 96:5242–44
33. Siu, A. K. Q., St. John, W. M. III, Hayes, E. F. 1970. *J. Am. Chem. Soc.* 92:7249–52
34. Goddard, W. A. III, Harding, L. 1977. Unpublished data
35. Horsley, J. A., Jean, Y., Moser, C., Salem, L., Stevens, R. M., Wright, J. S. 1972. *J. Am. Chem. Soc.* 94:279–82
36. Salem, L. 1971. *Acc. Chem. Res.* 4:322–28
37. Jean, Y., Salem, L., Wright, J. S., Horsley, J. A., Moser, C., Stevens, R. M. 1971. *Proc. 23rd Int. Congr. Pure Appl. Chem. Spec. Lect.* 1:197–217
38. Kollmar, H. 1973. *J. Am. Chem. Soc.* 95:966–67
39. Seubold, F. H. Jr. 1953. *J. Chem. Phys.* 21:–1616–17
40. Seubold, F. H. Jr. 1954. *J. Chem. Phys.* 22:945–46
41. Robinson, P. J., Holbrook, K. A. See Ref. 1, p. 187
42. O'Neal, H. E., Benson, S. W. 1973. In *Free Radicals*, ed. J. Kochi, 2:275–359. New York: Wiley. 906 pp.
43. Fessenden, R. W. 1964. *J. Chim. Phys. Phys. Chim. Biol.* 61:1570–75
44. Krusic, P. J., Meakin, P., Jesson, J. P. 1971. *J. Phys. Chem.* 75:3438–53
45. Fischer, H. See Ref. 42, pp. 435–91
46. Frey, H. M., Marshall, D. C. 1962. *J. Chem. Soc.*, pp. 3052–55
47. See Ref. 48, footnote on p. 6
48. Blumstein, C., Henfling, D., Sharts, C. M., O'Neal, H. E. 1970. *Int. J. Chem. Kinet.* 2:1–10
49. Berson, J. A., Balquist, J. M. 1968. *J. Am. Chem. Soc.* 90:7343–44
50. Zirnitis, U., Sushchinskii, M. M. 1964. *Opt. Spectrosc. USSR* 16:489–90, cited in Ref. 48
51. Berson, J. A., Pedersen, L. D., Carpenter, B. K. 1976. *J. Am. Chem. Soc.* 98:122–43
52. Clarke, T. C., Wendling, L. A., Bergman, R. G. 1975. *J. Am. Chem. Soc.* 97:5638–40
53. Crawford, R. J., Mishra, A. 1965. *J. Am. Chem. Soc.* 87:3768–69; 1966. 88:3963–69
54. McGreer, D. E., Chiu, N. W. K., Vinje, M. G., Wong, K. C. K. 1965. *Can. J. Chem.* 43:1407–16
55. Hoffmann, R. 1968. *J. Am. Chem. Soc.* 90:1475–85
56. Hayes, E. F., Siu, A. K. Q. 1971. *J. Am. Chem. Soc.* 93:2090–91
57. Gavezzotti, A., Simonetta, M. 1975. *Tetrahedron Lett.* pp. 4155–58
58. Huisgen, R., Scheer, W., Huber, H. 1967. *J. Am. Chem. Soc.* 89:1753–55
59. Huisgen, R. 1971. *Proc. 23rd. Int. Congr. Pure Appl. Chem. Spec. Lect.* 1:175–95
60. Dahmen, A., Hamburger, H., Huisgen, R., Markowski, L. 1971. *Chem. Commun.* pp. 1192–94
61. Pommelet, J. C., Manisse, N., Chuche, J. 1972. *Tetrahedron* 28:3929–41
62. McDonald, H. J. H., Crawford, R. J. 1972. *Can J. Chem.* 50:428–33
63. Willcott, M. R. III, Cargle, V. H. 1969. *J. Am. Chem. Soc.* 91:4310–11
64. Doering, W. v. E., Beasley, G. H. 1973. *Tetrahedron* 29:2231–43

65. Crawford, R. J., Lynch, T. R. 1968. *Can. J. Chem.* 46:1457–58
66. Carter, W., Bergman, R. G. 1968. *J. Am. Chem. Soc.* 90:7344–46; Bergman, R. G., Carter, W. 1969. *J. Am. Chem. Soc.* 91:7411–25
67. Doering, W. v. E., Sachdev, K. 1974. *J. Am. Chem. Soc.* 96:1168–87
68. Chmurny, A., Cram, D. J. 1973. *J. Am. Chem. Soc.* 95:4237–44
69. Berson, J. A., Balquist, J. M. 1968. *J. Am. Chem. Soc.* 90:7343–44
70. Berson, J. A., Pedersen, L. D., Carpenter, B. K. 1975. *J. Am. Chem. Soc.* 97:240–41
71. Berson, J. A., Pedersen, L. D., Carpenter, B. K. 1976. *J. Am. Chem. Soc.* 98:122–43
72. Berson, J. A., Pedersen, L. D. 1975. *J. Am. Chem. Soc.* 97:238–40
73. Deleted in proof
74. Jean, Y., Chapuisat, X. 1974. *J. Am. Chem. Soc.* 96:6911–20
75. Chapuisat, X., Jean, Y. 1975. *J. Am. Chem. Soc.* 97:6325–37
76. Gilbert, K. E., Baldwin, J. E. 1976. *J. Am. Chem. Soc.* 98:1594–95
77a. Benson, S. W. See Ref. 9, 2nd ed., p. 128
77b. Chesick, J. P. 1962. *J. Am. Chem. Soc.* 84:3250–54
78. Buchwalter, S., Closs, G. L. 1975. *J. Am. Chem. Soc.* 97:3857–58
79. Freeman, G. R. 1966. *Can. J. Chem.* 44:245–47
80. Larson, C. W., Tardy, D. C., Rabinovitch, B. S. 1968. *J. Chem. Phys.* 49:299–312
81. King, K. D., Golden, D. M., Benson, S. W. 1970. *Trans. Faraday Soc.* 66:2794–97
82. McKean, D., Duncan, J. L., Batt, L. 1973. *Spectrochim. Acta Part A* 29:1037–41
83a. Skinner, H. A. 1967. *Annu. Rep. Prog. Chem. Sect. A* 64:3–22; ibid 1968. 65:49–62
83b. Wiberg, K. B. 1971. In *Determination of Organic Structures by Physical Methods*, ed. F. C. Nachod, J. J. Zuckerman, 3:207–45. New York: Academic. 472 pp.
83c. Egger, K. W., Cocks, A. T. 1973. *Helv. Chim. Acta* 56:1516–52
83d. McKean, D. C., Biedermann, S., Buerger, H. 1974. *Spectrochim. Acta Part A* 30:845–57
84. Murrell, J. N., Laidler, K. J. 1968. *Trans. Faraday Soc.* 64:371–77
85. Zabramski, J. M., Wilburn, B. E., Knox, K., Andrist, A. H. 1976. Papers presented at Am. Chem. Soc. Natl. Meet., 172nd, San Francisco. Abstr. ORGN 126
86. Huisgen, R. 1976. Paper presented at Centen. Am. Chem. Soc. Meet., New York. Abstr. ORGN 16
87. Mazzocchi, P., Tamburin, H. J. 1975. *J. Am. Chem. Soc.* 97:555–61
88. Doering, W. v. E., Sachdev, K. 1975. *J. Am. Chem. Soc.* 97:5512–20
89. Baldwin, J. E., Gilbert, K. E. 1976. Papers presented at Am. Chem. Soc. Natl. Meet., 172nd, San Francisco. Abstr: ORGN 124; *J. Am. Chem. Soc.* 98:8283–84
90. Roberts, J. D., Sharts, C. M. 1962. *Org. React.* 12:1–56
91. Huisgen, R., Steiner, G. 1973. *J. Am. Chem. Soc.* 95:5054–55
92. Huisgen, R., Schug, R., Steiner, G. 1974. *Angew. Chem. Int. Ed. Engl.* 13:80–81
93. Benson, S. W., Nangia, P. S. 1963. *J. Chem. Phys.* 38:18–25
94. Hoffmann, R., Swaminathan, S., Odell, B. G., Gleiter, R. 1970. *J. Am. Chem. Soc.* 92:7091–97
95. Segal, G. A. 1974. *J. Am. Chem. Soc.* 96:7892–98
96. Stephenson, L. M., Gibson, T. A. 1972. *J. Am. Chem. Soc.* 94:4599–4602
97. Dewar, M. J. S., Kirschner, S. 1974. *J. Am. Chem. Soc.* 96:5246–47
98. Wright, J. S., Salem, L. 1972. *J. Am. Chem. Soc.* 94:322–29
99. Gerberich, H. R., Walters, W. D. 1961. *J. Am. Chem. Soc.* 83:3935–39, 4884–88
100. Baldwin, J. E., Ford, P. W. 1969. *J. Am. Chem. Soc.* 91:7192
101. Cocks, A. T., Frey, H. M., Stevens, I. D. R. 1969. *J. Chem. Soc. D*, pp. 458–59
102. Halevi, E. A. 1975. *Helv. Chim. Acta* 58:2136–51
103. Jones, G. II. 1974. *J. Chem. Educ.* 51:175–81
104. Rabinovitch, B. S., Schlag, E. W., Wiberg, K. B. 1958. *J. Chem. Phys.* 28:504–5
105. Schlag, E. W., Rabinovitch, B. S. 1960. *J. Am. Chem. Soc.* 82:5996–6000
106. Setser, D. W., Rabinovitch, B. S. 1964. *J. Am. Chem. Soc.* 86:564–69
107. Flowers, M. C., Frey, H. M. 1960. *Proc. R. Soc. London Ser. A* 257:122–31
108. Stephenson, L. M., Brauman, J. I. 1971. *J. Am. Chem. Soc.* 93:1988–91
109. Yang, N. C., Elliott, S. P. 1969. *J. Am. Chem. Soc.* 91:7550–51; Wagner, P. J.,

Liu, K.-C. 1974. *J. Am. Chem. Soc.* 96:5952–53
110. Stephenson, L. M., Cavigli, P. R., Parlett, J. L. 1971. *J. Am. Chem. Soc.* 93:1984–88
111. Bartlett, P. D., Porter, N. 1968. *J. Am. Chem. Soc.* 90:5317–18
112. Berson, J. A., Tompkins, D. C., Jones, G. II. 1970. *J. Am. Chem. Soc.* 90:5799–5800
113. Jones, G. II, Fantina, M. E. 1973. *J. Chem. Soc. D*, pp. 375–76
114. Casey, C. P., Boggs., R. A. 1972. *J. Am. Chem. Soc.* 94:6457–63
115. Dervan, P. B., Uyehara, T. 1976. *J. Am. Chem. Soc.* 98:1262–64
116. Benson, S. W. 1967. *J. Chem. Phys.* 46:4920–26
117. Doering, W. v. E., Franck-Neumann, M., Hasselmann, D., Kaye, R. L. 1972. *J. Am. Chem. Soc.* 94:3833–44
118. Hammond, G. S., DeBoer, C. D. 1964. *J. Am. Chem. Soc.* 86:899–902
119. Stephenson, L. M., Gemmer, R. V., Current, S. 1975. *J. Am. Chem. Soc.* 97:5909–10
120. Berson, J. A., Malherbe, R. 1975. *J. Am. Chem. Soc.* 97:5910–12
121. Berson, J. A., Dervan, P. B., Malherbe, R., Jenkins, J. 1976. *J. Am. Chem. Soc.* 98:5937–68
122. Berson, J. A., Dervan, P. B. 1973. *J. Am. Chem. Soc.* 95:267–70
123. Berson, J. A., Nelson, G. L. 1967. *J. Am. Chem. Soc.* 89:5503–4; 1970. 92:1096–97
124. Berson, J. A. 1968. *Acc. Chem. Res.* 1:152–60; 1972. 5:406–14
125. Berson, J. A., Holder, R. 1973. *J. Am. Chem. Soc.* 95:2037–38

LASER SEPARATION OF ISOTOPES ✣ 2642

V. S. Letokhov

Institute of Spectroscopy, Academy of Sciences USSR, Moscow,
142092, Podol'skii rayon, Akademgorodok, USSR

INTRODUCTION

The creation of laser-coherent light sources that have tunable radiation opened the prospect of selective excitation of practically any quantum state of atoms and molecules with an excitation energy in the range 0.1–10 eV. At present it is possible to obtain coherent radiation in the range of wavelengths from 2000 Å to 20 mcm with sufficient intensity to excite the significant part of atoms and molecules into a chosen quantum state. It is this quantitative increase in the art of quantum electronics, beginning in 1969–1970, that provided the means for the systematic studies of the effect of selective laser radiation on matter. Laser isotope separation is undoubtedly one of the most important current problems. An attractive field of study for scientists and engineers was provided by the uniquely important role that materials whose isotopic composition differs from that naturally occurring play in nuclear technology and energetics, since all existing methods of isotope separation had considerable disadvantages and new separation methods using lasers seemed possible. These studies apply the latest data on molecular and atomic structures and their interaction with the coherent laser light with the latest achievement in the development of tunable lasers. Researchers sought to develop new isotope separation methods that would be cheaper, more productive, more flexible, and less power-consuming, than the existing ones. Many hundreds of investigators in dozens of laboratories around the world are working in this vital field. Hundreds of works on this very problem have been published subsequent to the pioneer works of 1969–1972. Most of these works have been reviewed in (1–3). A more detailed study was later published in References 4–7. Since this article necessarily takes a survey approach, I consider only the most important ideas, methods, and experimental results, without trying to examine secondary problems and doubtful results.

CONCEPT, REQUIREMENTS, AND CLASSIFICATION OF METHODS

Concept and Requirements

The general concept of laser isotope separation is illustrated in Figure 1. When two types of atoms and molecules of different isotopic composition (A and B) have even

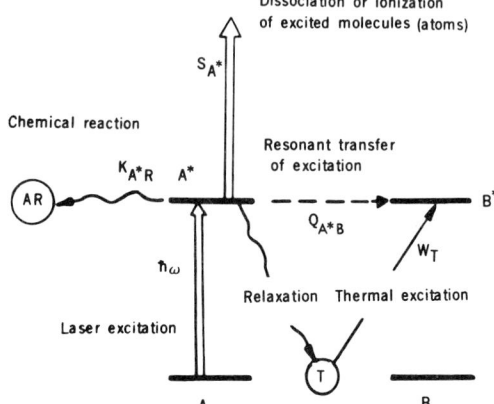

Figure 1 General concept of the selective laser excitation of A particles and classification of processes causing the loss or conservation of selectivity.

one spectral line that does not overlap with others, it is possible to excite selectively with laser light an atom or molecule ("particle" A) of a needed isotopic composition. The excitation of the A particles changes their chemical and physical properties and, hence, may be used to separate substances by any method based on differences in the characteristics of excited and unexcited particles. Selective laser excitation of A particles is followed by the chemical reaction of A^* to give products designated AR, or by absorption of a second photon to ionize or dissociate A^*. Nearly resonant transfer of energy to B, yielding B^*, and loss of energy to the thermal reservoir with subsequent thermal population of B^* are diagrammed in Figure 1. The excitation energy is meant here, of course, to be much higher than the thermal energy of the particles kT, which is responsible for nonselective thermal population of the lower energy states of all particles in the mixture. This is the general concept of laser separations at the atomic and molecular level.

Successful application of a specific concept of photoseparation requires that the following four conditions be met:

1. There must be at least one absorption line, $\omega_A^{(i)}$, of the A particles being separated that does not overlap significantly with any absorption lines of the other particles in the mixture.
2. Monochromatic radiation at the chosen frequency of selective absorption, $\omega_A^{(i)}$, must be available with the characteristics of power, duration, divergence, and monochromaticity necessary for the separation method in use.
3. A primary photophysical or photochemical process must be found that transforms the excited particles into species easily separated from the mixture.
4. The selectivity obtained for the A particle must be maintained against all competing photochemical and photophysical processes throughout the entire separation.

Laser Separation Methods

The selectivity of laser excitation may be preserved by using the changes in the properties of atoms and molecules caused by photoexcitation. Laser separation schemes may fulfill Requirement 3, based on the following properties (Figure 2):

1. The chemical reactivity of atoms or molecules is often increased by excitation.
2. The ionization energy of an excited atom or molecule is smaller than that of unexcited ones.
3. The dissociation energy of an excited molecule is smaller than that of an unexcited one.
4. Predissociation occurs when an excited molecule passes spontaneously into a dissociative state.
5. The excitation of a molecule may result in isomerization. The isomer, because of its different internal structure, has different chemical properties.
6. The recoil of an atom or molecule occurs when it absorbs a photon with momentum $\hbar\omega/c$. This gives a very small but observable particle photodeflection.
7. Excited atoms and molecules may have a higher polarizability, a different symmetry of wave function, etc. This may cause changes in cross sections for scattering by other particles, in motion in external fields, etc.

Many possible photochemical and photophysical methods of laser separation follow from this classification.

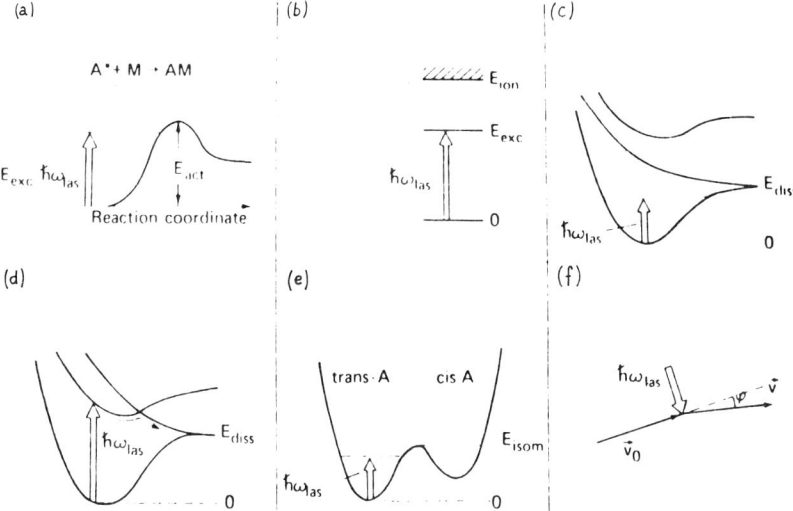

Figure 2 The atomic and molecular properties varying during laser excitation: (*a*) enhancement of reactivity; (*b*) reduction in ionization energy; (*c*) reduction in dissociation energy; (*d*) predissociation; (*e*) isomerization; (*f*) change in mechanical trajectory.

Changes in the reactivity of atoms and molecules due to photon absorption are well known and have long been used in photochemistry, including isotopically selective photochemistry. The first successful experiment was conducted by Kuhn & Martin (8), who exposed phosgene ($CO^{35}Cl_2$) molecules to a strong spectral line of an aluminum spark at $\lambda = 2816.2$ Å. With the advent of laser sources of intense monochromatic radiation, it became possible to selectively excite many atoms and molecules without depending upon accidental coincidences between strong lines of spontaneous radiation and absorption lines of atoms and molecules. The first attempt to effect photochemical isotope separation with a laser was made by Tiffany, Moos & Schawlow in 1966 (9). High-power infrared (IR) lasers permitted excitation of molecular vibrational levels and made possible vibrational photochemistry. The first attempt at laser isotope separation by vibrational photochemistry was carried out by Mayer et al in 1970 (10). Both of these experiments relied on the photochemically induced, increased chemical reactivity of excited molecules. Both failed because the overall chemistry did not preserve the initial excitation selectivity (Requirement 4).

Selective photophysical methods based on dissociation and ionization have met with greater initial success. The first proposal for laser isotope separation appears in a 1965 patent by Robieux & Auclair (11) for two-step photoionization of UF_6 molecules. While this method remains impractical, selective two-step atomic ionization was suggested by Letokhov (12) and demonstrated experimentally (13, 14). Selective two-step molecular dissociation was also suggested by Letokhov (15) and demonstrated experimentally (14, 16). Selective photopredissociation is another simple unimolecular process for isotope separation (17, 18). In 1974 isotopically selective molecular dissociation by high-power IR laser radiation was discovered (19, 20). All of these methods have yielded good isotope separations and are being developed for commercial application.

Separation Selectivity

For quantitative comparison of the separation methods let us define the coefficient of separation selectivity of A and B atoms or molecules by the relation

$$K(A/B) = \frac{[N_{AR}]_f}{[N_{BR}]_f} : \frac{[N_A]_o}{[N_B]_o} = R_f(A/B) : R_o(A/B), \qquad 1.$$

where $[N_A]_o$ and $[N_B]_o$ are the initial concentrations of A and B particles in the mixture, $[N_{AR}]_f$ and $[N_{BR}]_f$ are the final concentrations of new resultant molecules in the mixture that contains the A and B atoms or molecules being separated. The parameter $K(A/B)$ is defined as the final isotopic ratio $R_f(A/B)$ divided by the initial isotopic ratio $R_o(A/B)$. With no separation effect present, we have the selectivity coefficient $K(A/B) = 1$. The values $K(A/B) \gg 1$ correspond to a high selectivity of separation. For example, in natural uranium, which contains one part ^{235}U in 140 of natural uranium, the isotope ratio $R_o(235/238) = N_o(235)/N_o(238)$, is equal to 0.0072. The mole fraction is $\chi_o = N_o(235)/[N_o(235) + N_o(238)] = 0.0071$. For reactor fuel the mole fraction needs to be about 3% or $N_f(235)/[N_f(235) + N_f(238)] = 0.031$. So, for uranium, a $K(235/238) = 4.25$ yields a reactor-grade product. Now this factor can be achieved through a series of cascaded stages such as occur in the gaseous diffusion processes. For the laser separation process this value of the separation selectivity may be achieved through a single-stage process.

However, in every step of the laser separation process from excitation through final physical removal of the desired component, there are processes that degrade the initial excitation selectivity. Figure 1 outlines the processes that result in either conservation or loss of the selectivity attained in exciting the particles of type A. The following processes cause the loss of selectivity:

1. Thermal nonselective excitation. Excited molecules (all this applies in equal measure to atoms) may relax to the lowest state before A^* is acted upon by the primary separation process. For vibrational excitation the relaxation is radiationless, and hence causes the gas mixture to heat to some temperature $T > T_0$. As a result, the vibrational levels of both A and B molecules at energy $E^i_{A,B}$ are thermally populated with the Boltzmann probability $\exp(-E^i_{A,B}/kT)$. Thermal excitation is completely nonselective, and the molecules of types A and B are therefore both processed at the same rate under radiation. The energy of laser excitation must be significantly higher than the thermal energy kT.

2. Resonant excitation transfer. This process occurs when an excited A molecule collides with an unexcited B molecule. The cross section of the resonance excitation transfer for molecules with nearly resonant levels is often gas-kinetic, and for atoms it may be hundreds of times larger. Thus it is essential for the primary separation process to occur on a time scale short compared to that for A–B collisions.

3. Scrambling processes. Selectivity may be reduced by secondary processes that act after the primary separation step of chemical reaction, dissociation, ionization, or particle deflection has occurred. Free-radical chain reactions initiated by the products of $A^* + R$ or the dissociation fragments of A^* may attack B. In the ionization approach the charge exchange between ion A^+ and neutral atom B ($A^+ + B \to A + B^+$) occurs with large cross sections.

General Comparison of Different Approaches

To carry out a highly selective separation process with atoms or molecules of one type, A, there are several principal possibilities.

PHOTOCHEMICAL REACTION To produce high selectivity of separation, a chemical reaction between excited A^* molecules and suitable acceptors R must be found that occurs at a rate, K_{A^*R} (total time-averaged production rate for A^*R), substantially exceeding the rate of selectivity loss by resonance transfer of excitation, Q_{A^*B}, and by thermal excitation, W_T:

$$K_{A^*R} \gg Q_{A^*B}, W_T. \qquad 2.$$

It is also necessary that the reaction rate of the reagent R with unexcited A and B molecules be much less than that of its reaction with excited A^* molecules:

$$K_{A^*R} \gg K_{AR} \simeq K_{BR}. \qquad 3.$$

In principle, the fulfillment of Conditions 2 and 3 is quite possible even though in each specific case R must be specially selected. To ensure a high degree of selectivity with A and B molecules, it is important to fulfill Conditions 2 and 3 with a fairly large margin. It is worthwhile to emphasize that such a process is not fully controlled by laser radiation because it depends to a critical extent on the relation between rates of chemical reaction, deexcitation, and excitation transfer in elementary collisions. The efficiency and selectivity of photochemical separations is limited by our ability to find reaction systems that satisfy Conditions 2 and 3 well and do not give scrambling through subsequent undesired reactions.

TWO-STEP PHOTOIONIZATION AND PHOTODISSOCIATION A second laser is used to induce the transition of A^* particles to another state for which the selectivity loss rate is much smaller. This can be done by photoionization of selectively excited atoms or photodissociation of selectively excited molecules. The second laser radiation does not select between A and B, only between excited and unexcited particles. To preserve the initial selectivity the rate of laser-induced transition of A^*, S_{A^*}, must exceed the rate of resonance transfer of excitation and of thermal excitation,

$$S_{A^*} \gg Q_{A^*B}, W_T, \qquad\qquad 4.$$

as well as the rate of photoionization and photodissociation of unexcited A and B particles,

$$S_{A^*} \gg S_A \simeq S_B. \qquad\qquad 5.$$

Unlike the approach of excited particle chemical binding, this method is universal for the production of high separation selectivity. A high selectivity can always be achieved here by using laser power high enough to ensure that photoionization or photodissociation of the excited particles occurs rapidly compared to collision processes. This all but final stage of separation is fully controlled by laser radiation and is a very important feature of these photophysical methods. When the rate of ionization or dissociation is high enough, the selectivity of photophysical methods is much higher than that of photochemical ones.

$$K_{\text{photophys}}(A/B) \gg K_{\text{photochem}}(A/B). \qquad\qquad 6.$$

The scrambling processes that compete in the final stage of separation are often analogous to those that destroy selectivity in photochemical processes but are usually less effective. The great advantage of two-step excitation is that the ionized or dissociated particle has properties that differ much more than B itself from the properties of unexcited A and B.

PHOTOPREDISSOCIATION AND PHOTOISOMERIZATION One-photon unimolecular processes of molecular photopredissociation and photoisomerization (Figure 2, d and e) seem to combine the advantages of one-step (single-laser) photochemical processes and two-step (or multistep) collision-free photophysical processes. The excitation and primary separation steps are combined because the dissociation or rearrangement of the selectively excited molecule occurs spontaneously without collisions. Since the spontaneous rate may be slow and is not controlled by laser radiation, it is often more effective to induce the process by collisions. In practice, these methods are intermediate between the photochemical and two-step photophysical ones.

PARTICLE DEFLECTION   The sixth and seventh effects (Figure 2, f) use small changes in particle properties during excitation, which may be seen in experiments with atomic and molecular beams. Isotopically selective excitation of particles in a beam leads to a small change in trajectory of chosen isotopic particles and can be used for isotope separation (21). However, even weak, grazing collisions are sufficient to alter trajectories and destroy selectivity. These methods thus call for low-density atomic and molecular beams in a very high vacuum. The mass throughput is much less than for the first three approaches. It should be emphasized that these shortcomings are not due to the use of atomic or molecular beams but to the fact that selective excitation brings about only very small changes in particle properties. For instance, with the methods of selective photoionization or photodissociation, the accelerated ions or dissociation fragments possess drastically altered trajectories. In these cases relatively high-density atomic or molecular beams may be used efficiently under much less stringent requirements for the vacuum. In the present review I do not consider the method of photodeflection because it is uncompetitive in comparison to the others.

Isotopically Irreversible Methods

An entirely different concept of separation may be employed in processes where the selectivity does not exist as the result of selective excitation, but is due to an inherent isotopic selectivity in the kinetics of the process. Kinetic isotope effects in chemical reactions are a well-known example. For instance, isotope separation is possible in molecular mixtures with high vibrational temperatures and low translational temperatures. Nonselective vibrational heating may be accomplished by laser or by electric discharge excitation (22–25). In these systems the isotope selected is defined by kinetic rate constants. The selection of isotope cannot be reversed. This approach is not universal when compared to those discussed above. It can be used for a limited number of isotopes in simple molecules; therefore it is not discussed in detail in this article.

ISOTOPE EFFECT AND SELECTIVE EXCITATION

The first step of any laser isotope separation scheme is absorption of a laser photon by the desired isotopic atom or molecule at a wavelength for which the undesired isotopic species is relatively transparent. Thus, a transition with an isotope shift greater that its absorption linewidth is needed. The absorption spectrum should not be so dense that the shift only results in coincidence with another absorption feature of the unwanted isotopic species.

Generally speaking, these requirements relate to the absorption spectrum of atoms and molecules under action of a laser field, but not to the absorption in a weak field typical of spectroscopy. An evident isotope excitation selectivity is certainly important, even in a weak field. However, in cases where the isotope effect does not exist in the linear absorption, the isotope selectivity can sometimes be obtained in the spectrum of nonlinear absorption, multiphoton absorption in particular.

Atomic Spectra

The isotope shift in atomic spectra is caused by changes in the nuclear mass with a varying number of neutrons, in the nuclear volume and hence in the charge distribution, and in the nuclear spin, i.e. in the hyperfine structure (26, 27). The variation of isotope shift for various atoms is shown in Figure 3. The isotope shift for most elements, heavy elements especially, is usually several times more than the Doppler width

$$\Delta v_D = v_o\, 7.16 \times 10^{-7}(T/A)^{1/2},\qquad\qquad 7.$$

where T is the vapor temperature and A is the weight (in amu). Since the atomic isotope separation is commonly obtained at comparatively low gas pressures (less than several torr), the Doppler effect is the main mechanism for the spectral line broadening. For several alkali metal and alkaline earth elements the isotope shift is less than the Doppler width. In such cases selective excitation becomes impossible without taking additional measures to increase the selectivity. There are at least two methods of excitation selectivity increase based on elimination of Doppler broadening.

First, one can use atomic beams in which the Doppler width is determined by the angular divergence of the beam (Figure 4a). This method is particularly convenient in isotope separation through selective photoionization when, for many other considerations, the use of an atomic beam appears most reasonable. Experiments on laser separation of Ba (21) and Ca (29) isotopes with very small isotope shifts used the atomic beam method.

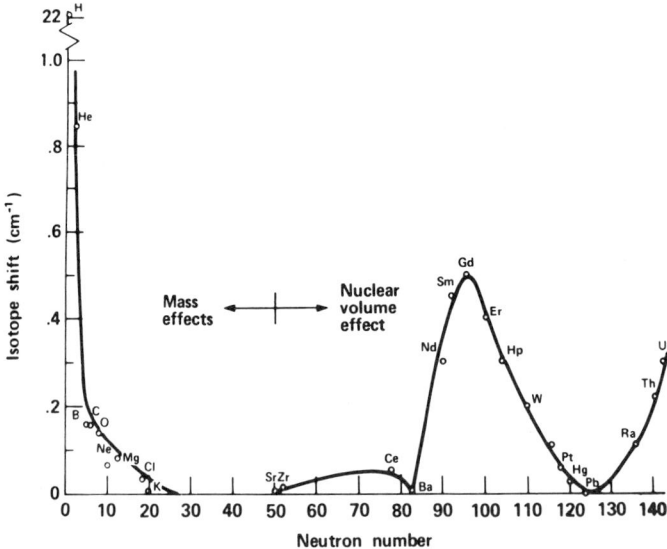

Figure 3 Atomic isotope shifts vs neutron number (28).

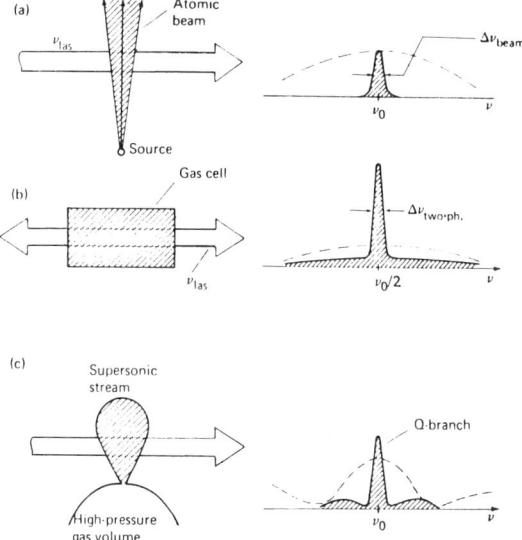

Figure 4 Methods of enhancement of the excitation selectivity for overlapped absorption zones: (*a*) the atomic or molecular beam method; (*b*) the two-quantum excitation method in a standing light wave; (*c*) the method of molecular gas-dynamic cooling during adiabatic expansion. (*Left*) geometry of the method; (*right*) changes of the absorption spectral line.

Second, the methods of two-quantum excitation in the field of two counter-running light waves with equal frequencies may be applied (30) (Figure 4*b*). A narrow absorption peak at the center of a Doppler-broadened absorption line permits, in principle, selective excitation of the atoms for which the isotope shift is hundreds of times smaller than the Doppler width. This method necessitates higher intensities in comparison to the method of single-photon excitation, and requires consideration of the possible shift of the absorption line in a strong field. Such a method of enhancing excitation selectivity has been considered theoretically (31–33) and demonstrated experimentally for Doppler-free spectroscopy [see review (34)], but it has not yet been applied in isotope separation experiments.

Molecular Spectra

The isotope shift in molecular spectra shows up most vividly in the variation of molecular vibration frequency with the change in mass of one of the nuclei. In vibrational molecular spectra the isotope shift is much larger than the Doppler width, but, owing to a rich rotational structure of heavy complex molecules, a strong overlapping appears in rotational-vibrational bands of various isotopes. In addition, due to intramolecular interactions the degeneracy of many lines is removed so that it is hardly possible to find isolated nonoverlapped rotational-vibrational lines. In this case we may use the isotope shift of the whole vibrational band. The half-width of rotational structure is determined by the approximate expression:

$$\Delta v_{rot} \simeq 4.6(B_0 T_{rot})^{1/2}, \qquad 8.$$

where B_0 is the rotational constant, T_{rot} is the rotational temperature (in cm^{-1}). Only in light molecules (^{i}BCl$_3$, ^{i}SF$_6$) does the isotope shift for the i-atom exceed the rotational bandwidth. The situation is somewhat better for the molecules with a well-pronounced Q-branch in their vibrational band. The bandwidth of Q-branch is usually much smaller than the rotational bandwidth, Δv_{rot}, and, therefore, even for some heavy molecules (for instance, ^{i}OsO$_4$) it is possible to ensure isotopically selective excitation of the vibrations.

However, for many polyatomic molecules even the Q-branch width is much larger than the isotopic shift. In this case the rotational broadening of the Q-branch may be reduced by cooling the gas. To prevent the liquefaction of the gas, it is advisable to make use of dynamic cooling during adiabatic expansion of molecular gas (35, 36) (Figure 4c). The low translational temperature narrows the Doppler width and the low rotational temperature brings all of the molecules into the lowest molecular rotational states. These effects produce a great simplification of the spectrum. The first successful spectral measurements on dynamic cooling gas have been done for SF$_6$ and UF$_6$ (37). It appears that this method may make possible highly isotopically selective excitation of the v_3 band of ^{235}UF$_6$.

The multiple-photon excitation of vibrational levels opens a new possibility for isotopically selective excitation without isotopic shift in linear IR absorption (38). Figure 5a shows linear absorption spectra of molecules CH$_3^{14}$NO$_2$ and CH$_3^{15}$NO$_2$ over the range 900–1100 cm^{-1} (v_7 and v_{13} band) taken with a resolution of about 0.5 cm^{-1}. There is no isotope shift for v_{13} band within the experimental errors. In the intense IR field (above 10^7 W/cm^2) the molecules' absorption spectra change differently in form, which is equivalent to an isotope shift appearance of about 5 cm^{-1} (Figure 5b). This effect has been used for nitrogen isotope separation in an isotopic mixture of nitromethane molecules.

The isotope shift in molecular electronic spectra appears only in the cases where there are narrow electronic-vibrational-rotational lines corresponding to transitions between discrete energy levels. The isotope effect has been studied only for diatomic

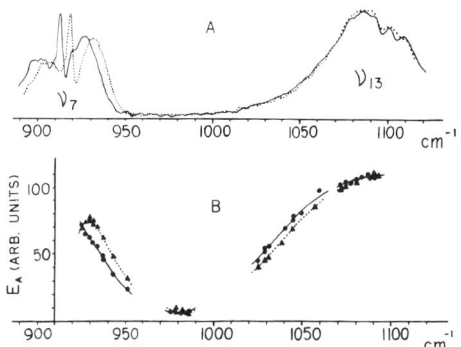

Figure 5 Isotope effects for the v_7 and v_{13} bands of CH$_3$NO$_2$ (solid curves, CH$_3^{15}$NO$_2$; dashed curves, CH$_3^{14}$NO$_2$): (*a*) linear absorption spectrum at a pressure of 20 torr; (*b*) multiple-photon absorption spectrum at a power density of 10^9 W cm^{-2} and a pressure of 2 torr (38).

and some simple polyatomic molecules, but for most molecules there are no experimental data available (39–41). The H_2CO molecule, whose spectrum contains isotope shifts for H, C, and O atoms, is an important exception.

Our discussions thus far have considered a gaseous medium only. Certainly it is possible to use a low-temperature molecular solution in a matrix. In this case one can obtain considerably simplified electronic-vibrational and vibrational spectra, which provide high excitation selectivity. The first experiment on laser isotope separation in a low-temperature (lower than 4°K), condensed medium was carried out on the electron excitation of the s-tetrazene molecule (42). However, this method is hardly expected to be highly productive or industrially applicable.

PHOTOPHYSICAL METHODS

Among the basic photophysical methods of practical interest under active development in many research centers are the following: 1. selective multistep photoionization of atoms; 2. selective two-step photodissociation of molecules by IR and UV radiation; 3. photopredissociation of molecules; and 4. multiple photon molecular dissociation in an intense IR field. Each of these methods has been demonstrated by a laboratory experiment, has its advantages and disadvantages, and is either at the stage of providing evidence for possible application in pilot installation (methods 2 and 3), or is at the stage of direct realization in pilot installation (methods 1 and 4).

Selective Multistep Photoionization

Selective photoionization of atoms is the most universal photophysical method for selective separation of substances, particularly isotopes, at the atomic level. A common feature of all schemes for selective ionization is the sequence of the two processes: 1. isotopically selective excitation and 2. ionization of the excited atoms. Figure 6 illustrates some schemes of selective atomic ionization of special interest

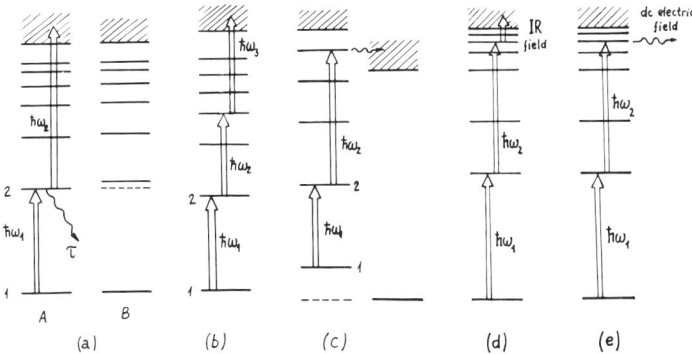

Figure 6 Schemes of selective step-wise photoionization of atoms by laser radiation: (*a*) two-step photoionization; (*b*) three-step photoionization; (*c*) two-step photoionization through an autoionization state; (*d*) two-step selective excitation of a Rydberg state and its photoionization by IR laser radiation; (*e*) two-step selective excitation of a Rydberg state and its ionization by dc electric field.

for laser isotope separation. The two-step photoionization scheme is the simplest (12, 13). The three-step scheme may be of use, say, for atoms with a high ionization potential. The photoionization cross section can be increased by tuning the frequency of last-step radiation to that of the transition to an autoionization [spontaneous (43) or electric-field-induced (44)] state (Figure 6c). Finally, high-lying Rydberg states can be ionized by IR radiation (45) or a pulsed electric field (44) (Figure 6d,e).

All these schemes have already been tested for isotopically selective ionization. Those interested in information concerning the advantages and disadvantages of these methods should consult (46) or brief sections of (4–6) for review. Here we give only the general requirements, which should be satisfied by a scheme of selective ionization for isotope separation on a practical scale:

1. All atoms in an unexcited beam should be in the ground state, and there should be no ions. If atoms of a selected isotope are distributed over several levels or sublevels, multifrequency radiation is required to excite atoms from every sublevel in order to completely remove the selected isotope from the mixture. Any thermal ions existing in the atomic vapor must be removed before laser excitation.
2. The laser radiation should perform selective photoionization for each atom of a selected isotope. This requirement determines the power of the exciting and ionizing radiation and depends on the cross sections for the excitation and ionization processes.
3. The laser radiation intensity should practically be used in full to excite and ionize atoms of a selected isotope. Certain requirements arise in this connection concerning the geometry of the atomic beam (flow), the laser beams, and the density of the atoms.
4. There should be no transfer of excitation or charge between the isotopes being separated. This condition limits the density allowed.

Most attention has been paid to uranium isotope separation. Results have been published from research programs at Avco Everett Research Laboratory and the Lawrence Livermore Laboratory. In 1974, the results of the first Livermore experiments were presented at the 8th International Conference on Quantum Electronics (47). In these experiments, CW dye lasers were used for excitation, and ^{235}U and the UV radiation of mercury lamps were used for ionization. The selectivity separation factor was about 100. In 1975 the Livermore Laboratory presented the results of its experiments on two-step ionization of uranium atoms by xenon and krypton ion lasers (48). The ion yield rate of ^{235}U$^+$ in their experiments was 2×10^{-3} g hr^{-1}, which is 10^7 times higher than the rate obtained in the early experiments.

An industrial research and development program is being carried out jointly by Avco Everett Research Laboratory and Exxon Nuclear Co. Some results of this work, obtained, according to the authors, in 1971 in the early stage of investigation of the method, were reported in 1975 (49). In an experiment in which the exciting, pulsed dye laser was scanned over a range broad enough to cover both the ^{235}U and the ^{238}U transitions, and the pulsed N$_2$ laser was used to photoionize the excited ^{i}U, the selectivity separation factor $K(235/238)$ was about 140.

Besides the uranium isotope separation, which demands the rather difficult conditions indicated above, laboratory experiments are being carried out on selective ionization of other isotopes, potentially useful in much smaller quantities, such as K, Ca, Rb, rare-earth, and of course, transuranium elements. An effective ionization of excited atoms at an average, moderate intensity of the ionizing radiation presents a serious problem in these cases. The ionization schemes given in Figure 6a–c cannot be practically applied here, so greater attention is paid to the ionization scheme of highly excited atoms shown in Figure 6d,e (see 46).

From the viewpoint of the comparative ease of ionization, the highly excited states of atoms are of great interest for isotope separation. The first successful experiment on increasing ionization cross sections by electric-field-induced auto-ionization of the Rydberg state was carried out in studies (50) using Na vapor. A comprehensive study of the autoionization of highly excited Na atoms is reported in (51, 52). A quantum state with the principal quantum number n will fall within the continuum if electric field strength (in amu) is:

$$\mathscr{E}_{cr} \geq (16n^4)^{-1}. \qquad 9.$$

The results of critical ionization field measurements for different principal quantum numbers, obtained for sodium (51, 53) and uranium (54), are compared to the theoretical dependence (Equation 9) shown in Figure 7. All these experiments demonstrate the efficiency of using the electric field to ionize highly excited states.

Selective Two-Step $(IR + UV)$ Photodissociation

The process of selective two-step molecular photodissociation is possible if the excitation of a molecule shifts the band of continuous photoabsorption and results in the photodissociation of a molecule. Then, selecting the laser frequency ω_2 in the region of the shift where the ratio of the absorption coefficients of the excited

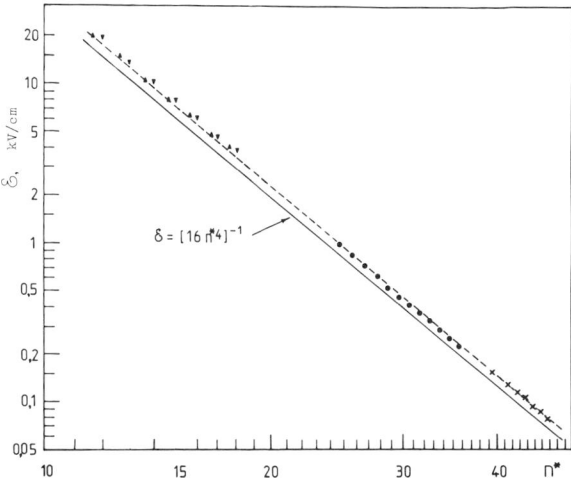

Figure 7 Log-log plot of critical ionization electric field vs effective principal quantum number n^* for $n = 12, \ldots 50$ (▼, and ▲ show the results for S- and D-states of Na; ●, the results for Na; ×, the results for U) (53).

and unexcited molecules has its maximum, we can accomplish photodissociation of molecules excited selectively by laser radiation of frequency ω_1 via an excited, unstable electronic state.

The intermediate state may be a stable, excited vibrational or electronic state. Each scheme has its advantages and shortcomings. When a vibrational state is excited, the shift of the electronic absorption band is sometimes fairly small, and the low vibrational levels are populated appreciably by nonselective thermal excitation. Therefore, it is sometimes difficult to achieve preferential photodissociation of the laser-radiation-excited molecules. On the other hand, the isotope shift is manifested clearly in the vibrational spectrum. The photoabsorption band may exhibit a very large shift for a electronically excited intermediate state, but the electronic spectra of many molecules do not have lines with a sharp structure in which the isotope shift would appear clearly. Moreover, when the isotope structure does appear, the method of two-step photodissociation competes with the methods of one-step selective photopredissociation and selective electronic photochemistry. Therefore, in practice, the most interesting method is the photodissociation via an intermediate vibrational state by IR and UV radiation (Figure 8a).

The process of two-step photodissociation of molecules is more complex than the process of two-step photoionization of atoms because of the following effects, which influence the selectivity and rate of the process (14): 1. the thermal nonselective excitation of vibrational levels; 2. the broadening of the edge of the electronic photoabsorption band of the molecule; and 3. the bottleneck effect due to the rotational structure of the vibrational levels. The first two effects decrease the dissociation selectivity, and the third limits the absorption rate of IR radiation by the molecule (55) and, consequently, the rate of the two-step photodissociation process. These effects were discussed in detail in reviews (4, 6) and in the original papers. At this juncture, we can only indicate that the influence of the first effects may be

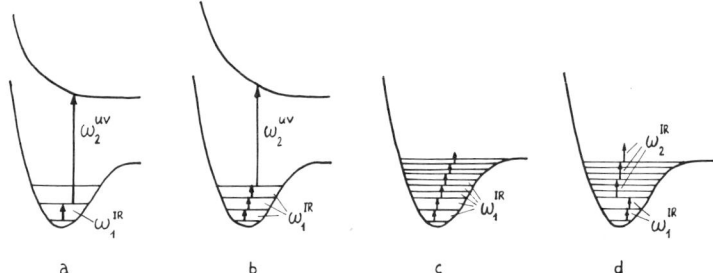

Figure 8 Schemes of selective multistep photodissociation of molecules by laser radiation through an excited electronic state (a, b) and ground electronic state (c, d): (a) two-step IR + UV photodissociation; (b) multistep selective excitation of high vibrational levels and their photodissociation by UV radiation; (c) multiple-photon selective excitation and dissociation by single frequency intense IR field; (d) multiple-photon selective excitation of vibrational levels by a resonant IR field and their multiphoton dissociation by a nonresonant, intense IR field.

essentially reduced by selective excitation of high-lying vibrational levels (Figure 8b). This may be performed by one of several methods: 1. direct excitation of high levels with laser radiation at the overtone transition frequency as was done, for example, in work on HCl molecules (56); 2. subsequent step-wise excitation of high levels with multifrequency IR radiation; 3. resonance excitation of high levels of polyatomic molecules due to multiple absorption of IR photons from a pulsed IR field; this was demonstrated for the SF_6 (20, 57) and OsO_4 (58) molecules using a CO_2 laser.

The first experiments on the separation of isotopes by the two-step selective photodissociation method were described in (16, 59). These experiments were carried out on $^{14}NH_3$ and $^{15}NH_3$ molecules because, first, they could be excited selectively by CO_2 laser radiation and, second, IR and UV absorption spectra and photochemical decomposition were thoroughly investigated. The enrichment coefficient, $K(^{15}N/^{14}N)$ for the final product N_2 varied from 2.5 to 6. These results were confirmed by Japanese investigators (60). Experiments in which the boron isotopes ^{10}B and ^{11}B were separated by the two-step selective photodissociation of BCl_3 molecules were carried out using similar apparatus (a CO_2 laser and a conventional UV source emitting in the 2000 Å region) (61). Only a 10% enrichment of a mixture with the light boron isotope, comparable to the typical value of the kinetic isotope effect, was achieved in these experiments.

The universal character of two-step photodissociation of a molecule by joint action of IR and UV radiation is limited to molecules having an unstable excited electronic state with energy <4–5 eV, which can be excited in the accessible UV range ($\lambda_2 > 2000$ Å). From this point of view the multiphoton dissociation of polyatomic molecules by an intense IR field due to vibrational transitions within the ground electronic state is more universal. However, as two-step IR-UV dissociation is applicable in principle to a number of molecules of practical interest with heavy isotopic atoms, this method is being actively developed in many laboratories (37).

One-Step Selective Photopredissociation

Isotope separation by photopredissociation requires a molecular excited state that exhibits a resolvable isotope shift, decays primarily by dissociation, and whose dissociation products are simply removed from the starting material. This method is not as general as the two-step photoprocesses. Moreover, for most molecules, sufficient spectroscopic and photochemical data is not available to show whether these requirements are satisfied.

Photopredissociation has been studied most extensively in formaldehyde, primarily by C. B. Moore and co-workers. Near the origin of the first excited singlet state, H_2CO dissociates with high quantum yield to H_2 and CO, i.e. absorption of a single photon leads to chemically stable dissociation products. Separation of hydrogen from deuterium has been demonstrated using 1:1 mixtures of H_2CO and D_2CO (17, 62–64). Enrichments, limited by the excitation selectivity of the laser source, were as high as 9:1 (63). An experiment on hydrogen isotope separation in the natural mixture of H_2CO and HDCO has been conducted (65), in which the enrichment coefficient $K(D/H)$ is about 14 under irradiation by the CW He-Cd laser line at 325.03 nm. The 80-fold enrichment of CO in the isotope ^{12}C has been obtained (66) by means of photopredissociation of a mixture $H_2^{12}CO:H_2^{13}CO$ = 1:10. Spectroscopic and photochemical research is under way (67) that should lead to a full understanding of the photoprocesses in formaldehyde and the development of practical systems for the separation of ^{13}C, ^{18}O, and ^{17}O.

A number of experiments have been done with other molecules. Leone & Moore (68) excited Br_2 to the predissociated $^3\Pi_{0+u}$ state, and avoided the problems of many of the potential scrambling processes by observing the IR chemiluminescence from HBr formed in vibrationally excited states in the reaction of the Br fragment with HI. The enrichment coefficient $K(^{81}Br/^{79}Br)$ was about 5. By selective predissociation of *ortho*-I_2 molecules with a 514.5 nm argon ion laser light, workers (69, 70) were able to convert *ortho*- to *para*-I_2 with an enrichment factor of about 2–4. This process can also be used to separate iodine isotopes. Hochstrasser & King (42) and Karl & Innes (71) have independently demonstrated high enrichments of carbon and nitrogen isotopes in low-temperature condensed-phase (42) and gas-phase (71) irradiation of *s*-tetrazene having the naturally occurring isotopic composition. Evidence is given for the dissociation reaction *s*-tetrazene → N_2 + 2HCN.

Photopredissociation promises to become a practical method of isotope separation. Work on H_2CO and probably *s*-tetrazene may lead to economically viable methods of enriching ^{13}C, ^{14}C, ^{17}O, and ^{18}O. While the method is not as generally applicable as two-step excitation methods, in some situations it may be the most practical.

Despite the considerable simplicity of the photopredissociation approach, it has not yet entered the stage of pilot plants for any isotopes. This may be connected with the energetics of narrow-band tunable lasers of UV range. The creation of excimer lasers with high efficiency can contribute greatly to rapid, practical implementation of this method.

Multiple–Photon Dissociation of Polyatomics

All the laser methods of isotope separation discussed above are based on the excitation of electronic states of atoms and molecules by visible or UV laser radiation. The isotope separation method discussed below is quite different because it only uses intense IR laser radiation for direct excitation of very high vibrational levels in electronic ground states (Figure 8c, d). The method is based on the isotopically–selective dissociation of polyatomic molecules [BCl_3 (19), SF_6 (20), OsO_4 (58)] by intense CO_2-laser pulses. The effect was discovered in our laboratory in 1974. Although for some people our discovery was perhaps unexpected, for us it was in fact a logical result of our work on the isotopically selective dissociation of molecules by laser radiation. The discovery of this effect was preceded by several studies (72–77) of the interaction of powerful IR radiation pulses with molecular gases. A comprehensive discussion of these early works may be found in (7, 78).

The essence of the effect consists in the following. When the CO_2 laser, with an intensity of about 10^7 to 10^9 W/cm^2, is tuned to a molecular vibrational band whose isotope shift is comparable to, or larger than, the width of the Q-branch of the vibrational band, irreversible dissociation of the irradiated isotopic molecules occurs. This is reflected by changes in the isotopic composition (enrichment) of both the undissociated and the dissociated molecules. An enrichment of over 3000 of the residual SF_6 gas with the isotope ^{34}S was obtained in the first experiments (20).

The chemical composition of some pure gases (BCl_3, OsO_4, etc) remains constant even under rather intense radiation in which visible molecular luminescence can be observed and, thus, their dissociation continues steadily. Investigations have shown that this is caused by the reverse reaction, i.e. recombination of the dissociation products forming the initial molecule. When an acceptor is introduced that reacts with the dissociation products before they recombine, irreversible, isotopically selective dissociation of the initial molecules occurs when high-power IR radiation acts on the mixture [$BCl_3 + O_2$ (19, 79), $OsO_4 + C_2H_4$ (58)]. In this more general sense, dissociation under high-power IR laser radiation is typical of all, rather than some, polyatomic molecules.

At present, investigators have obtained much information on isotope separation by dissociation caused by the multiple IR photon absorption method in many polyatomic molecules [$^iBCl_3$ (19, 79, 80); $^iSF_6$ (20, 81, 82); $^iOsO_4$ (58); $^iC^jCl_4$ (83, 84); $^iSiF_4$ $^iC_2F_2Cl_2$ (80); $CH_3{}^iNO_2$ (38); $^iMoF_6$ (85); $^iH_2Cl_2C_2$ (86); ^iH_2CO (87)]. The dissociation process of SF_6 was studied the most carefully (81, 88). These studies have enabled us to understand the nature and basic characteristics of both selective dissociation and isotope separation by this process. Multiple IR photon laser photochemistry has been described comprehensively in a special review (7). We can only discuss briefly the most important features of this approach in the present review.

Of primary interest is the dependence of the dissociation selectivity S on the IR laser frequency at which dissociation occurs and its correlation with the low-power IR absorption band shape of SF_6 (the dissociation selectivity $S = W_a/W_b$ is the ratio of dissociation rates of two isotopically different molecules, a and b). Such a frequency dependence of the enrichment coefficient of SOF_2 formed as a result of dissociation SF_6 is given in Figure 9a (88). This figure also shows the dissociation rate $W(^{32}SF_6)$ dependence of $^{32}SF_6$ on laser frequency. It is obvious that when the dissociation rates W_i for both isotopic molecules become equal, no enrichment is to be expected. The dashed line in Figure 9a is for the assumed frequency dependence of the $^{34}SF_6$ dissociation rate. The comparison with Figure 9a shows that near the intersection point (at 931 cm^{-1}) of the dissociation rate curves, the enrichment coefficient equals unity, which means that there is no enrichment. A shift up or down from this frequency gives enrichment of one or another isotope in the reaction product (SOF_2). The highest dissociation selectivity measured at 0.05 torr is $S = 14$. Actually S is much higher, since the mass analysis of SOF_2 has been carried out with low resolution, and the measured value therefore depends on the natural $^{34}S/^{18}O$ ratio. Within the framework of this paper it is difficult to describe theoretically the phenomenon of polyatomic molecule dissociation in the intense IR field. This problem has been analyzed in many works (7, 78, 89, 90).

At this point it is necessary to emphasize the following conclusion concerning the dissociation mechanism, which is important for isotope separation. A *polyatomic molecule* can absorb a large amount of energy (3–5 IR photons) in a comparatively moderate intensity IR field (about 10^5–10^6 W cm^{-2}) due to the "soft" compensation

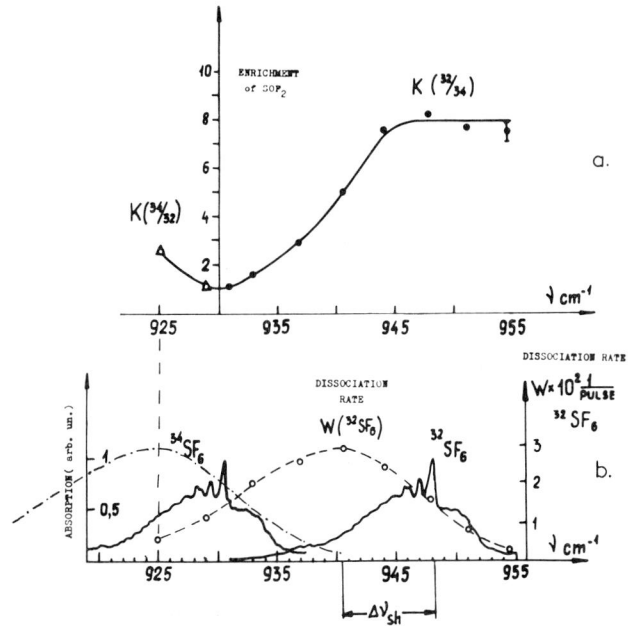

Figure 9 Frequency characteristics for multiphoton selective dissociation of SF_6 molecules by IR laser radiation. The measurements were made in a focused beam, the average radiation power density was 31 MW cm^{-2}, and the SF_6 pressure in the cell was 0.2 torr. (*a*) Dependence of the enrichment coefficients $K(32/34)$ (●) and $K(34/32)$ (△) on laser frequency. The enrichment coefficients were measured in the dissociation products with the fragmentation ion SOF_2^+. (*b*) Dependence of $^{32}SF_6$ dissociation rate on laser radiation frequency (dashed curve) and the linear absorption spectrum of $^{32}SF_6$ and $^{34}SF_6$ molecules (solid curves) (88).

of vibrational anharmonicity on lower transitions. The intensity threshold (or, more correctly, energy density threshold for laser pulses with a duration of $<10^{-6}$ sec) of dissociation seems to be related to saturation of the vibrational high-lying transitions in the vibrational quasi-continuum, which is an inherent feature of polyatomic molecules. Thus it would be quite natural to separate the functions of selective excitation and dissociation between two fields with different frequencies (Figure 8*d*). In this case we get an enhancement in selectivity, since a nonresonant strong field does not cause power broadening due to the dynamic Stark effect of lower transitions; however, dissociation of selectively excited molecules requires a strong field with rather coarse tunability.

The first successful experiments on the dissociation of SF_6 molecules by the field of two IR frequencies ω_1 and ω_2 have been recorded (57, 91). The low-intensity field at ω_1 selectively excites the molecules and transfers the excited molecules to above the lowest boundary of the vibrational quasi-continuum. The frequency ω_1 is in resonance with the molecular vibration v_3 SF_6. The field at ω_2 is tuned 130 cm^{-1}

away from the absorption band of v_3 and serves to excite the molecules in the quasi-continuum up to the dissociation limit. The frequency ω_2 and the intensities of both frequencies are selected so that the molecules of SF_6 in the reaction vessel dissociate only with both fields present. Figure 10 presents some measurements of the frequency dependence of the dissociation rate in a two-frequency field when ω_1 is scanned. For comparison, the dissociation rate by a single-frequency intense field is also shown. It is seen from Figure 10 that in a two-frequency field there is no dissociation at all if ω_1 is tuned to the transition frequencies of SF_6 in the R-branch; molecules can dissociate only if the Q- and P-transitions are selectively excited. The resonance characteristic of $W(\omega_1)$ is narrowed here, leading to an increase of dissociation selectivity.

To date this method seems to be the most promising for the isotope separation of heavy elements by IR radiation. Recently the enrichment of Os isotopes by the dissociation of OsO_4 molecules of natural isotopic composition was achieved (92). The enrichment coefficient obtained was $K = 1.6$, while no enrichment occurred at the molecular dissociation by a single-frequency intense IR field.

In order to attain a low dissociation threshold, it is very important to roughly tune the ω_2 laser frequency to the maximum of the multiple IR absorption in the vibrational quasi-continuum (92). Quite an interesting peculiarity of multiple photon dissociation is the above-indicated possibility (Figure 5) of isotope separation influencing the vibration band, which has no isotope shift in the IR absorption spectrum for a weak field (38).

Powerful IR radiation is easy and inexpensive to obtain from molecular lasers, and multiphoton dissociation by an IR field is quite simple, hence this method is the most appropriate for development in pilot plants.

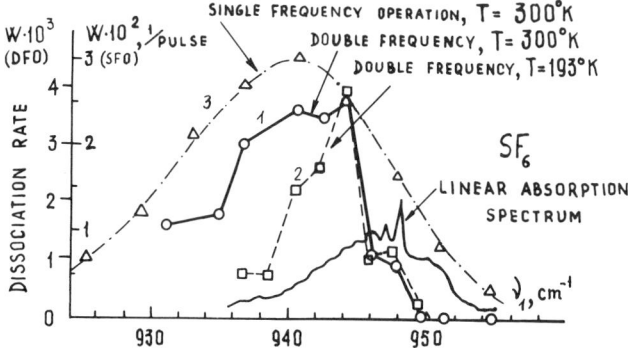

Figure 10 The dependence of the dissociation (scheme in *Figure 9d*) rate W in SF_6 on the frequency v_1, which is in resonance with the IR absorption band; the off-resonant frequency $v_2 = 1084 \text{ cm}^{-1}$. Intensities of both IR pulses: $I_1 = 4 \text{ MW cm}^{-2}$, $I_2 = 60 \text{ MW cm}^{-2}$ (averaged over the irradiation volume; focused beam). Curves 1 and 2 correspond to two different temperatures of SF_6: 300° and 190°K. The linear absorption of SF_6 and the dissociation rate dependence on frequency in the single-frequency case at $T = 300°K$ (curve 3) is also shown (91).

PHOTOCHEMICAL METHODS

To date, the possibilities of photochemical isotope separation have been successfully demonstrated, using the excitation of atomic and molecular electronic states and of molecular vibrational levels. However, in spite of the optimism of early research, the actual advances made appear to be much less impressive than for the photophysical methods discussed above.

Electronic Photochemistry

The chemical reactions of electronically excited atoms and molecules have been an active subject of research for many years—especially before the creation of the laser. The photochemical isotope separation of Hg excited by the 253.7 nm resonance line has been successfully demonstrated with a variety of reagents (8, 93, 94). For example, Pertel & Gunning (93) were able to enrich ^{202}Hg from 30% of its natural abundance to 85% in a mixture of Hg, H_2O, and butadiene. The kinetics of the photochemical reactions of Hg are sufficiently complex that even the extensive work of Gunning and his collaborators, reviewed in (95), does not yield a complete mechanism. Isotopic enrichment in diatomics has been carried out by Harteck and co-workers (96, 97), who used atomic resonance lamps for excitation: an ^{i}NO molecule with a Br lamp and an ^{i}CO molecule with an iodine lamp. The enrichment factor is about 4–6.

Obviously, laser sources for photochemical isotope separation have many advantages over incoherent sources. The broad tunability and high ultimate resolution of lasers gives a comparatively free choice of absorption lines in the visible, ultraviolet, and probably far-ultraviolet ranges, while permitting the highest possible selectivity.

Several successful schemes of laser photochemical enrichment have been reported, mostly by R. Zare and co-workers. The most interesting results have been obtained in experiments on photochemical separation of ^{35}Cl and ^{37}Cl through selective excitation of $I^{37}Cl$ molecules by CW dye laser radiation (98). The laser radiation only excited the $I^{37}Cl$ molecules to states below the predissociation limit. The excited molecules were subjected to two reactions. In one case, the $I^{37}Cl$ molecules reacted with *trans*-ClHC=CHCl, forming *cis*-ClHC=CHCl and causing 10% enrichment with ^{37}Cl. In the other case, they reacted with 1,2-dibromoethylene, forming *trans*-ClIC=CHCl and causing 50% enrichment with ^{37}Cl. Much higher enrichment in this scheme at a pressure of mixture 7.5 torr was reported by Zare in 1976 (99). This result is most promising in isotopically selective electronic photochemistry.

Another successful experiment for a diatomic halogen has been reported (70, 100). The *ortho*-I_2 molecules in the mixtures with 2-hexene were excited with the 514.5 nm line of the CW Ar laser. The excited *ortho*-I_2 molecules react with 2-hexene and *para*-I_2 remains unreacted. The selective photochemical reaction of *ortho*-I_2 molecules studied (70, 100) is a repetition, at a new level, of the experiment in prelaser selective photochemistry (101), and it can be applied directly to iodine isotope separation.

Enrichment with Cl has also been achieved by selective excitation of Cl_2CS in mixtures with diethoxyethylene (102). Mass-spectroscopic analysis of the remaining Cl_2CS after irradiation with either argon or dye laser light showed the concentration of ^{35}Cl altered from its natural abundance of 75% to 64% or 80%, depending on the isotopic species initially excited.

The great prospects of electronic photochemistry for laser isotope separation have still not been as sufficiently demonstrated as might be expected for such an old and classical approach. There are some possible explanations. First, as noted in the introduction to this review, the problem of selectivity loss in secondary photochemical processes is more difficult for photochemical methods than for photophysical ones. For example, even in such an ideal case as the reaction of selectively excited metastable mercury atoms, the enrichment coefficient due to secondary processes does not exceed 14, in spite of numerous experiments (95). Second, the method of electronic photochemistry, judging by the publications, shows no prospect for uranium isotope separation, which is the focus of interest for most investigators. Third, the energetics of visible and UV lasers is far from being economically effective, and the powerful lasers of this range are still hardly accessible.

Vibrational Photochemistry

The rate of a chemical reaction may be substantially enhanced by vibrational excitation of the reactant molecules. Gibert (103) suggested this as a method for isotope separation in 1963. However, the first successful experiment by this approach was not done until more than ten years later (104).

A wide variety of vibrational excitation processes may be used (Figure 11). The IR active fundamental vibrations of a molecule may be excited by absorption of a single photon (Figure 11*a*). Excitation of combination and overtone bands gives

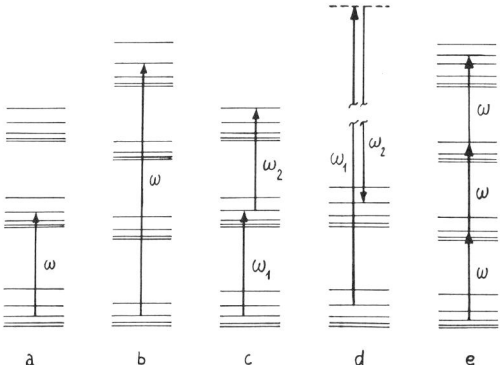

Figure 11 Schemes of selective excitation of vibrational levels of molecules by laser radiation: (*a*) single-photon absorption on a fundamental band; (*b*) single-photon absorption on a second overtone band; (*c*) two-step excitation by two-frequency IR field; (*d*) Raman excitation of an IR-inactive absorption vibration by two-frequency visible laser field; (*e*) multiple-photon excitation of highly excited levels by a single-frequency, intense IR field.

two or more quanta of vibrational excitation on absorption of a single higher-energy photon (Figure 11b). Higher vibrational levels may also be reached by step-wise excitation through one or more intermediate levels by multifrequency radiation (Figure 11c). Raman excitation may also be used (Figure 11d). It is the only method for excitation of vibrations with a zero-transition dipole (e.g. homonuclear diatomics). The selective excitation of high vibrational levels by multiple IR photon absorption at comparatively moderate intensities (10^6–10^7 W/cm^2) give us one more—and probably the only effective—method for direct excitation of vibrational levels with energies of several eV (Figure 11e). All these excitation schemes were tested in experiments on laser isotope separation. Positive results were obtained by the latter three excitation schemes (Figure 11c–e). However, the actual experiments are small in number, so we cannot state the fundamental disadvantages of any of them. The success or failure of an experiment is evidently more connected with the experimental conditions and the right scavenger than with the particular excitation scheme used. But, of course, the schemes that excite the high vibrational levels are more favorable, because they provide potentially a freer choice of chemical reactions and a lesser influence of heat mechanism on the separation selectivity.

Arnoldi et al (104) have reported the enrichment of ^{35}Cl by enhancement of the reaction of Br atoms with HCl. They used an HCl pulsed chemical laser to excite HCl sequentially from $v = 0$ to $v = 1$ to $v = 2$ (Figure 11c). Selective excitation of HCl led to an acceleration of the reaction $Br + H^{35}Cl$ ($v = 2$) $\rightarrow$ $HBr + ^{35}Cl$ by a factor of eleven orders of magnitude over the rate of reaction of Br with HCl ($v = 0$). Isotopic enrichment was demonstrated by time-resolved mass spectroscopy of BrCl formed in the secondary process $Cl + Br \rightarrow BrCl + Br$. For equal pressures of Br and HCl a separation factor of 2 was observed.

Raman excitation of N_2 in air at 77°K was reported by Basov et al (24, 105) to produce NO enriched 100 times in ^{15}NO. This might result from $N_2^* + O_2 \rightarrow 2NO$. Basov et al interpreted their results in terms of a mechanism (22, 23) that requires considerable vibrational energy transfer among N_2 molecules in a Treanor pumping process (106). The same separation method and a similar interpretation was used by Basov et al in the experiments on isotope separation in an electric discharge, where the enrichment of NO by ^{15}N isotope reached 10. But the results of this work contradict those of (25), so their interpretation in (24, 105) cannot be considered final.

Pumping of BCl_3 molecules in a mixture of $BCl_3 + H_2S$ and $BCl_3 + D_2S$ by a focused, pulsed CO_2 laser, which probably excites high vibrational levels, caused separation of boron isotopes (107). Starting with natural boron isotopic composition, irradiation of the mixture with either the $P(16)$ or the $R(20)$ line of the 10 nm CO_2 laser gave enrichment of the residual $BCl_3 K(10/11)$ on the order of 1.7 and $K(10/11) \simeq 0.7$, respectively. It is clear now that this experiment is an intermediate one between single-photon IR photochemistry in a weak field and multiphoton IR photochemistry in an intense field.

The attractive feature of vibrational photochemistry for isotope separation is the prospect of using low-energy IR photons from an efficient molecular laser to get a good yield of the product. However, from the point of view of efficient usage of IR radiation, the method of IR molecular photochemistry will be able to compete with the method of multiphoton excitation and dissociation of polyatomic molecules in an intense IR field. The multiple-photon approach has a number of disadvantages in comparison with vibrational photochemistry (reaction area restricted by the intense field area and application to polyatomic molecules only), but is much less restrictive in many other aspects. Future experiments will undoubtedly indicate the application preferable for each approach.

CONCLUSION

The present review attempts to consider briefly the main methods of laser isotope separation already demonstrated in research laboratories. These methods should be developed further not only in research but also in industrial laboratories. Quite recently, some methods were proposed for isotope separation on the basis of heterogeneous chemical reactions (108), condensation of vibrationally excited molecules (109), selective dissociation of polyatomic molecules in the low-temperature solid matrix by intense IR pulses (110), selective unimolecular photoisomerization by electronic (111), and multiple photon vibrational excitation (112), among others. It would be premature to consider all these methods in the present review before experimental verification of their predictions.

It would be rather difficult to compare in detail the different methods of laser isotope separation without considering specific isotopes or classes of isotopes. Comparison of the methods under development with those already existing is especially complicated. Laser methods are clearly advantageous for the separation of isotopes that cannot be satisfactorily separated by existing methods, i.e. less abundant isotopes and the transuranium elements. Essential progress in the technology of separation of many stable isotopes in comparatively moderate quantity seems likely to occur in the near future. In the case of isotopes demanded on a large scale for nuclear energy (uranium, deuterium), the situation is still somewhat vague. In many cases of comparing various separation methods for specific isotopes we must keep in mind the level of existing laser technology. In fact, by the time industry implements some of the methods, progress in laser technology may change the situation drastically. Accurate prediction of the most practicable and profitable method of future large-scale laser separation of isotopes cannot be made with certainty given the present art of laser technology.

To conclude this review, I would like to emphasize that the selective photophysical and photochemical methods for laser isotope separation will find fundamental and practical applications well outside the isotope separation problem. Essentially, we are dealing with a new approach to materials technology at the atomic and molecular level when one can use laser radiation to manipulate atoms and molecules of particular kinds directly, i.e. one can collect macroscopic amounts of matter atom by atom and molecule by molecule. In a recent report I briefly discussed examples of such broader future applications (113): the possibility of laser production of very pure substances, laser separation of nuclear isomers, laser-selective biochemical reactions, selective detection of single atoms and molecules, and spatial localization of molecular bonds.

In conclusion I would like to express gratitude to my co-workers: Dr. R. V. Ambartzumian from the Institute of Spectroscopy, USSR Academy of Sciences, and Prof. C. B. Moore from the University of California, Berkeley, who collaborated with me in writing reviews (4, 6, 7). These special reviews were used during the preparation of the present review.

Literature Cited

1. Moore, C. B. 1973. *Acc. Chem. Res.* 6:323-28
2. Letokhov, V. S. 1973. *Science* 180:451-58
3. Gross, R. W. F. 1974. *Opt. Eng.* 13:506-15
4. Letokhov, V. S., Moore, C. B. 1976. *Kvantovaya Elektron. Moscow* 3:248-87, 3:485-516 (in Russian); *Sov. J. Quantum Electron.* 6:129-50, 6:259-76
5. Aldridge, J. P. III, Birely, J. H., Cantrell, C. D. III, Cartwright, D. C. 1976. In *Laser Photochemistry, Tunable Lasers, and Other Topics, Vol. 4, Physics of Quantum Electronics*, ed. S. F. Jacobs, M. Sargent III, M. O. Scully, C. T. Walker, pp. 57-144. Reading, Pa: Addison-Wesley. 470 pp.
6. Letokhov, V. S., Moore, C. B. 1977. In *Chemical and Biochemical Applications of Lasers*, ed. C. B. Moore. 3:1-165. New York: Academic
7. Ambartzumian, R. V., Letokhov, V. S. 1977. See Ref. 6, 3:166-316
8. Kuhn, W., Martin, H. 1933. *Z. Phys. Chem. Abt. B* 21:93-137
9. Tiffany, W. B., Moos, H. W., Schawlow, A. L. 1967. *Science* 157:40-43
10. Mayer, S. W., Kwok, M. A., Gross, R. W. F., Spencer, D. J. 1970. *Appl. Phys. Lett.* 17:516-19
11. Robieux, J., Auclair, J. M. 1965. *French Patent No. 1, 391, 738*
12. Letokhov, V. S. 1970. *Soviet Patent No. 65743* (publ. 1977)
13. Ambartzumian, R. V., Kalinin, V. P., Letokhov, V. S. 1971. *Pis'ma Zh. Eksp. Teor. Fiz.* 13:305-8 (in Russian); *Sov. Phys. JETP Lett.* 13:217-19
14. Letokhov, V. S., Ambartzumian, R. V. 1971. *IEEE J. Quantum Electron.* QE-7:305-6; Ambartzumian, R. V., Letokhov, V. S. 1972. *Appl. Opt.* 11:354-58
15. Letokhov, V. S. 1970. *Soviet Patent No. 65744* (publ. 1977)
16. Ambartzumian, R. V., Letokhov, V. S., Makarov, G. N., Puretskii, A. A. 1972. *Pis'ma Zh. Eksp. Teor. Fiz.* 15:709-11 (in Russian); (*Sov. Phys. JETP Lett.* 15:501-3; 1973. 17:91-94, 63-65
17. Yeung, E. S., Moore, C. B. 1972. *Appl. Phys. Lett.* 21:109-10
18. Letokhov, V. S. 1972. *Chem. Phys. Lett.* 15:221-22
19. Ambartzumian, R. V., Letokhov, V. S., Ryabov, E. A., Chekalin, N. V. 1974. *Pis'ma Zh. Eksp. Teor. Fiz.* 20:597-600 (in Russian); *Sov. Phys. JETP Lett.* 20:273
20. Ambartzumian, R. V., Gorokhov, Yu. A., Letokhov, V. S., Makarov, G. N. 1975. *Pis'ma Zh. Eksp. Teor. Fiz.* 21:375-78 (in Russian); *Sov. Phys. JETP Lett.* 21:171-73
21. Bernhardt, A., Duerre, D., Simpson, J., Wood, L. 1974. *Appl. Phys. Lett.* 25:617-19
22. Dubost, H., Abouaf-Marguin, L., Legay, F. 1972. *Phys. Rev. Lett.* 29:145-48
23. Belenov, E. M., Markin, E. P., Oraevsky, A. N., Romanenko, V. I. 1973. *Pis'ma Zh. Eksp. Teor. Fiz.* 18:196-98 (in Russian); *Sov. Phys. JETP Lett.* 18:116-17
24. Basov, N. G., Belenov, E. M., Gavrilina, L. K., Isakov, V. A., Markin, E. P., Oraevskii, A. N., Romanenko, V. I., Ferapontov, N. B. 1974. *Pis'ma Zh. Eksp. Teor. Fiz.* 20:607-8 (in Russian); *Sov. Phys. JETP Lett.* 20:277-79
25. Manuccia, Y. J., Clark, M. D. 1976. *Appl. Phys. Lett.* 28:372-74
26. Kuhn, H. G. 1969. *Atomic Spectra*, Chap. 6. New York: Academic
27. Sobel'man, I. I. 1972. *An Introduction to the Theory of Atomic Spectra*, Chap. 6. Oxford: Pergamon
28. Stern, R. C., Snavely, B. B. 1976. *Ann. NY Acad. Sci.* 267:71-80
29. Brinkmann, U., Hartig, W., Telle, H., Walther, H. 1974. *Appl. Phys.* 5:109-15
30. Vasilenko, L. S., Chebotayev, V. P.,

Shishaev, A. V. 1970. *Pis'ma. Zh. Eksp. Teor. Fiz.* 12:161–64 (in Russian); *Sov. Phys. JETP Lett.* 12:113–15
31. Kelley, P. L., Kildal, H., Schlossberg, H. R. 1974. *Chem. Phys. Lett.* 27:62–65
32. Chebotayev, V. P., Golger, A. L., Letokhov, V. S. 1975. *Chem. Phys.* 7:316–20
33. Shimoda, K. 1976. *Appl. Phys.* 9:239–47
34. Bloembergen, N., Levenson, M. D. 1976. In *High Resolution Laser Spectroscopy*, ed. K. Shimoda, p. 315. Berlin-Heidelberg: Springer. 378 pp.
35. Hagens, O., Henkes, W. 1960. *Z. Naturforsch. Teil A* 15:851–65
36. Kataev, D. I., Mal'tzev, A. A. 1973. *Zh. Eksp. Teor. Fiz.* 64:1527–33 (in Russian); *Sov. Phys. JETP* 37:772–76
37. Rockwood, S. D. 1976. In *Tunable Lasers and Applications*, ed. A. Mooradian, T. Jaeger, P. Stokseth, p. 140. Berlin-Heidelberg: Springer. 404 pp.
38. Chekalin, N. V., Doljikov, V. S., Kolomiisky, Yu. R., Letokhov, V. S., Lokhman, V. N., Ryabov, E. A. 1976. *Phys. Lett. A* 59:243–44
39. Herzberg, G. 1950. *Molecular Spectra and Molecular Structure. Vol. 1, Spectra of Diatomic Molecules.* New York: Van Nostrand. 2nd ed. 658 pp.
40. Herzberg, G. 1945. *Molecular Spectra and Molecular Structure. Vol. 2, Infrared and Raman Spectra.* New York: Van Nostrand. 632 pp. 1st ed.
41. Herzberg, G. 1966. *Molecular Spectra and Molecular Structure. Vol. 3, Electronic Spectra and Electronic Structure of Polyatomic Molecules.* New York: Van Nostrand. 1st ed.
42. Hochstrasser, R. M., King, D. S. 1975. *J. Am. Chem. Soc.* 97:4760–62
43. Levy, R., Janes, G. S. 1973. *US Patent No. 3,722,519*
44. Ivanov, L. N., Letokhov, V. S. 1975. *Kvantovaya Elektron. Moscow* 2:585–90; (in Russian); *Sov. J. Quantum Electron.* 4:1391–95
45. Nebenzahl, I., Levin, M. 1973. *FRG Patent No. 2,312,194*
46. Letokhov, V. S., Mishin, V. I., Puretzkii, A. A. 1977. *Prog. Quantum Electron.* 6: In press
47. Tuccio, S. A., Dubrin, J. W., Peterson, O. G., Snavely, B. B. 1974. *IEEE J. Quantum Electron.* QE-10:790
48. Tuccio, S. A., Foley, R. J., Dubrin, J. W., Krikorian, O. 1975. *IEEE J. Quantum Electron.* QE-11:101D
49. Janes, G. S., Itzkan, I., Pike, C. T., Levy, R. H., Levin, L. 1975. *IEEE J. Quantum Electron.* QE-11:101D
50. Ambartzumian, R. V., Bekov, G. I., Letokhov, V. S., Mishin, V. I. 1975. *Pis'ma Zh. Eksp. Teor. Fiz.* 21:595–98 (in Russian); *Sov. Phys. JETP Lett.* 279–81
51. Ducas, T. W., Littman, M. C., Freeman, R. R., Kleppner, D. 1975. *Phys. Rev. Lett.* 35:366–70
52. Littman, M. C., Zimmerman, M. L., Kleppner, D. 1976. *Phys. Rev. Lett.* 37:486–89
53. Bekov, G. I., Letokhov, V. S., Mishin, V. I. 1977. *Zh. Eksp. Teor. Fiz.* (in Russian). 77:N1 In press
54. Paisner, J. A., Carlson, L. R., Worden, E. F., Johnson, S. A., May, C. A., Solarz, R. W. 1976. Preprint UCRL-78034, Lawrence Livermore Lab.
55. Letokhov, V. S., Makarov, A. A. 1972. *Zh. Eksp. Teor. Fiz.* 63:2064–76 (in Russian); *Sov. Phys. JETP.* 36:1091–1101
56. Ambartzumian, R. V., Apatin, V. M., Letokhov, V. S. 1972. *Pis'ma Zh. Eksp. Teor. Fiz.* 15:336–39 (in Russian); *Sov. Phys. JETP Lett.* 15:237–39
57. Ambartzumian, R. V., Gorokhov, Yu. A., Letokhov, V. S., Makarov, G. N., Puretzkii, A. A., Furzikov, N. P. 1976. *Pis'ma Zh. Eksp. Teor. Fiz.* 23:217–20 (in Russian)
58. Ambartzumian, R. V., Gorokhov, Yu. A., Letokhov, V. S., Makarov, G. N. 1975. *Pis'ma Zh. Eksp. Teor. Fiz.* 22:96–100 (in Russian); *Sov. Phys. JETP Lett.* 22:43–44
59. Ambartzumian, R. V., Letokhov, V. S., Makarov, G. N., Puretzkii, A. A. 1974. In *Laser Spectroscopy* ed. R. G. Brewer, A. Mooradian, p. 611. New York: Plenum. 671 pp.
60. Noguchi, N., Izawa, Y. 1974. *Progress Report X*, p. 63. Osaka, Japan: Osaka Univ.
61. Rockwood, S., Rabideau, S. W. 1974. *IEEE J. Quantum Electron.* QE-10:789
62. Yeung, E. S., Moore, C. B. 1973. *J. Chem. Phys.* 58:3988–98
63. Ambartzumian, R. V., Apatin, V. M., Letokhov, V. S., Mishin, V. I. 1975. *Kvantovaya Elektron. Moscow* 2:337–42 (in Russian); *Sov. J. Quantum Electron.* 5:191–95
64. Bazhin, N. M., Skubnevskaya, G. I., Sorokin, N. I., Molin, Yu. N. 1974. *Pis'ma Zh. Eksp. Teor. Fiz.* 20:41–44 (in Russian); *Sov. Phys. JETP Lett.* 20:18–20

65. Marling, J. B. 1975. *Chem. Phys. Lett.* 34:84–87
66. Clark, J. H., Haas, Y., Houston, P. L., Moore, C. B. 1975. *Chem. Phys. Lett.* 35:82–85
67. Baranovski, A. P., Cabello, A., Clark, J. H., Haas, Y., Houston, P. L., Kung, A. H., Moore, C. B., Reilly, J., Weisshaar, J. C., Zughul, M. B. 1976. In *Tunable Lasers and Applications*, ed. A. Mooradian, T. Jaeger, P. Stokseth, p. 108. Berlin-Heidelberg: Springer. 404 pp.
68. Leone, S. R., Moore, C. B. 1974. *Phys. Rev. Lett.* 33:269–72
69. Bazhutin, S. A., Letokhov, V. S., Makarov, A. A., Semchishen, V. A. 1973. *Pis'ma Zh. Eksp. Teor. Fiz.* 18:515–19 (in Russian) *Sov. Phys. JETP Lett.* 18:303–5
70. Balikin, V. I., Letokhov, V. S., Mishin, V. I., Semchishen, V. A. 1976. *Chem. Phys.* 17:111–21
71. Karl, R. R. Jr., Innes, K. K. 1975. *Chem. Phys. Lett.* 36:275–79
72. Isenor, N. R., Richardson, M. C. 1971. *Appl. Phys. Lett.* 18:224–27
73. Letokhov, V. S., Ryabov, E. A., Tumanov, O. A. 1972. *Opt. Commun.* 5:168–70
74. Letokhov, V. S., Ryabov, E. A., Tumanov, O. A. 1972. *Zh. Eksp. Teor. Fiz.* 63:2025–32 (in Russian); *Sov. Phys. JETP* 36:1069–73
75. Lyman, J. L., Jensen, R. J. 1972. *Chem. Phys. Lett.* 13:431–34
76. Isenor, N. R., Merchant V., Hallsworth, R. S., Richardson, M. C. 1973. *Can. J. Phys.* 51:1281–90
77. Ambartzumian, R. V., Chekalin, N. V., Doljikov, V. S., Letokhov, V. S., Ryabov, E. A. 1974. *Chem. Phys. Lett.* 25:515–18
78. Ambartzumian, R. V., Letokhov, V. S. 1977. *Acc. Chem. Res.* 10:61–70
79. Ambartzumian, R. V., Gorokhov, Yu. A., Letokhov, V. S., Makarov, G. N., Ryabov, E. A., Chekalin, N. V. 1975. *Kvantovaya Elektron. Moscow* 2:2197–2201 (in Russian)
80. Lyman, J. L., Rockwood, S. D. 1976. *J. Appl. Phys.* 47:595–602
81. Ambartzumian, R. V., Gorokhov, Yu. A., Letokhov, V. S., Makarov, G. N. 1975. *Zh. Eksp. Teor. Fiz.* 69:1956–70 (in Russian); *Sov. Phys. JETP* 63:993–1000
82. Lyman, J. L., Jensen, R. J., Rink, J., Robinson, C. P., Rockwood, S. D. 1975. *Appl. Phys. Lett.* 27:87–89
83. Ambartzumian, R. V. Gorokhov, Yu. A., Letokhov, V. S., Makarov, G. N., Puretzkii, A. A. 1975. *Pis'ma Zh. Eksp. Teor. Fiz.* 22:374–77 (in Russian); *Sov. Phys. JETP Lett.* 22:177–78
84. Ambartzumian, R. V., Gorokhov, Yu. A., Letokhov, V. S., Makarov, G. N., Puretzkii, A. A. 1976. *Phys. Lett. A* 56:183–85
85. Lyman, J. L., Rockwood, S. D. 1976. *Invited Rep. 9th Int. Conf. Quantum Electron., Amsterdam, June 14–18*
86. Yogev, A., Benmair, R. M. J. 1975. *J. Am. Chem. Soc.* 97:4430–31
87. Koren, G., Oppenheim, U. P., Tal, D., Okon, M., Weil, R. 1976. *Appl. Phys. Lett.* 29:40–42
88. Ambartzumian, R. V., Gorokhov, Yu. A., Letokhov, V. S., Makarov, G. N., Puretzkii, A. A. 1976. *Zh. Eksp. Teor. Fiz.* 71:440–53 (in Russian)
89. Ambartzumian, R. V., Gorokhov, Yu. A., Letokhov, V. S., Makarov, G. N., Puretzkii, A. A. 1976. *Pis'ma Zh. Eksp. Teor. Fiz.* 23:26–30 (in Russian); *Sov. Phys. JETP Lett.* 23:22–25
90. Bloembergen, N., Cantrell, C. D., Larsen, D. M. 1976. In *Tunable Lasers and Applications*, ed. A. Mooradian, T. Jaeger, P. Stokseth, p. 162. Berlin-Heidelberg: Springer. 404 pp.
91. Ambartzumian, R. V., Gorokhov, Yu. A., Furzikov, N. P., Letokhov, V. S., Makarov, G. N., Puretzky, A. A. 1976. *Opt. Commun.* 18:517–21
92. Ambartzumian, R. V., Gorokhov, Yu. A., Makarov, G. N., Puretzky, A. A., Furzikov, N. P. 1977. *Kvantovaya Elektron. Moscow* (in Russian). 4:N7 In press
93. Pertel, R., Gunning, H. E. 1959. *Can. J. Chem.* 37:35–42
94. Billings, B. H., Hitchcock, W. J., Zelikoff, M. 1953. *J. Chem. Phys.* 21:1762–66
95. Gunning, H. E., Strausz, O. P. 1963. *Adv. Photochem.* 1:209
96. Liuti, G., Dondes, S., Harteck, P. 1966. *J. Chem. Phys.* 44:4052–53
97. Dunn, O., Harteck, P., Dondes, S. 1973. *J. Phys. Chem.* 77:878–83
98. Liu, D. D., Datta, S., Zare, R. N. 1975. *J. Am. Chem. Soc.* 97:2557–59
99. Zare, R. N. 1976. *Invited Rep. at 9th Int. Conf. Quantum Electron. Amsterdam, June 14–18*
100. Letokhov, V. S., Semchishen, V. A. 1975. *Dokl. Akad. Nauk SSSR* 222:1071–74 (in Russian), *Sov. Phys. Dokl.* 20:423–25; *Spectrosc. Lett.* 8:263–74
101. Badger, R. M., Urmston, J. W. 1930. *Proc. Natl. Acad. Sci. USA* 16:808–10
102. Lamotte, M., Dewey, H. J., Keller,

R. A., Ritter, J. J. 1975. *Chem. Phys. Lett.* 30:165–69
103. Gibert, R. 1963. *J. Chim. Phys. Physicochim. Biol.* 60:205–15
104. Arnoldi, D., Kaufmann, K., Wolfrum, J. 1975. *Phys. Rev. Lett.* 34:1597–1600
105. Basov, N. G., Belenov, E. M., Isakov, V. A., Markin, E. P., Oraevskii, A. N., Romanenko, V. I., Ferapontov, N. B. 1975. *Kvantovaya Elektron. Moscow* 2:938–45 (in Russian); *Sov. J. Quantum Electron.* 5:510–16
106. Treanor, C. E., Rich, J. W., Rehm, R. G. 1968. *J. Chem. Phys.* 48:1798–1801
107. Freund, S. M., Ritter, J. J. 1975. *Chem. Phys. Lett.* 32:255–60
108. Gochelashvili, K. S., Karlov, N. V., Orlov, A. N., Petrov, R. P., Petrov, Yu. N., Prokhorov, A. M. 1975. *Pis'ma Zh. Eksp. Teor. Fiz.* 21:640–43 (in Russian); *Sov. Phys. JETP Lett.* 21:302–4

109. Basov, N. G., Belenov, E. M., Isakov, V. A., Leonov, Yu. S., Markin, E. P., Oraevskii, A. N., Romanenko, V. I. 1975. *Pis'ma Zh. Eksp. Teor. Fiz.* 22:221–24 (in Russian); *Sov. Phys. JETP Lett.* 22:102–4
110. Ambartzumian, R. V., Gorokhov, Yu. A., Makarov, G. N., Puretzkii, A. A., Furzikov, N. P. 1976. *Pis'ma Zh. Eksp. Teor. Fiz.* 24:287–89 (in Russian)
111. Brauman, J. I., O'Leary, T. J., Schawlow, A. L. 1974. *Opt. Commun.* 12:223–24
112. Ambartzumian, R. V., Chekalin, N. V., Doljikov, V. S., Letokhov, V. S., Lokhman, V. N. 1976. *Opt. Commun.* 18:400–2; *J. Photochem.* 6:55–67
113. Letokhov, V. S. 1976. In *Tunable Lasers and Applications*, ed. A. Mooradian, T. Jaeger, P. Stokseth, p. 122. Berlin-Heidelberg: Springer. 404 pp.

PHOTOELECTRON SPECTROSCOPY: STUDY OF VALENCE BANDS IN SOLIDS

✫ 2643

Jennifer C. Green

Inorganic Chemistry Laboratory, South Parks Road, Oxford OX1 3QR, United Kingdom

INTRODUCTION

In photoelectric experiments, monochromatic radiation of frequency ω is used to ionize a material, and the energy distribution of the emitted electrons is measured. Use of the Einstein equation,

$\hbar\omega = IE + KE$,

converts the measured kinetic energy (KE) distribution of the emitted electrons into an ionization energy (IE) spectrum for the material. To maximize the information from this type of experiment, the photon energy may be varied, and, in addition, the angular distribution of the electron emission may be studied. The aim of most workers in this field is to obtain an experimentally based description of the electronic structure of the material they are studying.

The studies may be classified in various ways: by the type of material studied, which is normally either a solid or a gas; by the category of electron ionized, either core or valence; and by the photon energy, which is either in the X-ray or ultraviolet region. Two previous reviews in this series (1, 2) have dealt primarily with valence electrons in gas-phase molecules studied by ultraviolet radiation. Other major literature sources are the books by Turner et al (3), Siegbahn et al (4, 5), Eland (6), and Carlson (7), and the series of reviews by Orchard et al (8–10). This review concerns itself with the study of valence bands in solids by photoelectron spectroscopy, a field that is often referred to as photoemission.

THE THREE-STEP MODEL

Whereas in a gas-phase experiment an electron is ejected into a simple continuum orbital, in the solid state the photoemission process is more complex. It is envisaged as a three-step process (11): 1. excitation from an initial state i to a final state f,

2. transport of the electron to the surface, 3. escape from the surface. The final state is a vacant band whose energy distribution and shape is determined by the crystal and, in the energy region immediately above the band gap or Fermi level, is likely to be heavily structured. The first part of the process is analogous to an optical excitation, and it is normal to assume direct transitions; that is, excitation may only occur between states with the same reduced **k** vector. If $\varepsilon_i(\mathbf{k})$ and $\varepsilon_f(\mathbf{k})$ are the energies of the initial and final states, and $\hbar\omega$ is the photon energy,

$$\Omega_{fi}(\mathbf{k}) = \varepsilon_f(\mathbf{k}) - \varepsilon_i(\mathbf{k}) - \hbar\omega = 0,$$

defines the range of allowed excitations. For these to result in photoemission, ε_f must lie above the Fermi level E_F. In a photoelectric experiment the number of photoelectrons with a particular energy is measured. If consideration is restricted to step 1, this is represented by the energy distribution of the joint density of states, commonly abbreviated as EDJDOS, and given by the function

$$\mathscr{D}(\varepsilon, \hbar\omega) = (2\pi)^{-3} \sum_{f,i} \int' d^3k \delta(\Omega_{fi}) \delta[\varepsilon - \varepsilon_i(\mathbf{k})] |P_{fi}|^2,$$

where the prime on the integration sign denotes that $\varepsilon_i < E_F < \varepsilon_f$. The first delta function selects those transitions with k conservation and the second those with a particular initial energy. $|P_{fi}|^2$ is the square of the momentum matrix element

$$P_{fi} = \langle f, \mathbf{k} | p | i, \mathbf{k} \rangle$$

and weights each transition with its appropriate strength factor.

In order to obtain an expression that may be compared with experimental results, this expression for EDJDOS is multiplied by a transport and escape factor $\mathscr{T}[\varepsilon_f(\mathbf{k}), \mathbf{k}]$. These two steps are considered in more detail below.

A key function in the band theory of solids is the density of states, DOS, given by

$$\rho(\varepsilon) = (2\pi)^{-3} \sum_i d^3k [\varepsilon - \varepsilon_i(\mathbf{k})].$$

It is clear that EDJDOS differs from DOS in that it contains $\delta(\Omega_{fi})$ and $|P_{fi}|^2$. Thus, only under conditions of an unstructured continuum, constant matrix elements, and constant transport and escape, would a photoelectron spectrum be a direct measure of density of states.

There is good evidence that the first condition is fulfilled at sufficiently high photon energies. This is best illustrated by the extensive studies on gold. Shirley (12) has shown that an X-ray photoelectron spectrum, XPS, directly reflects the main features of the d-band densities of state whereas at low photon energies, $\hbar\omega \lesssim 30$ eV, the spectral shape varies with photon energy (11, 13). In this region the experimental spectra reflect structure in the EDJDOS. At energies $\hbar\omega \gtrsim 30 \cdot$eV, the spectral structure positions are essentially invariant (13) and resemble the XPS spectrum. This behavior occurs at higher photon energies, because final states are available for excitation of electrons from all initial states; the continuum is effectively unstructured.

Although in the high-energy range the positions of peaks and shoulders are essentially invariant, the energy distribution curves are not. Freeouf et al (13) suggest that these minor variations are most likely due to band-to-band variations in the matrix elements, although they do suggest other causes. In attempting to correlate experimental results in the high-energy range with theoretical densities of states, attention is focused on structure in the spectra and in the densities of states in order to sidestep the problem of matrix element variation across the band.

Interpretation of spectra in the low-energy range involves calculation of EDJDOS. Williams et al (14) have carried out a calculation on copper using the effective Chodorow one-electron potential. They point out that the steps needed to calculate the photoelectron energy distribution are: (a) calculation of the band energies, (b) evaluation of the interband-momentum matrix elements, (c) **k**-space interpolation of these quantities to a mesh fine enough to permit accurate integration, and (d) evaluation of the **k**-space integrals that define the energy distribution. They obtain band energies and momentum matrix elements directly from the Chodorow potential via the KKR (Korringa-Kohn-Rostoker) method. The zone integration is performed using Janak's generalization of the Gilat-Raubenheimer method. In their treatment of electron transport, the electron was given the group velocity produced by the band calculation, and the mean free path due to electron-electron scattering was obtained from the computed band density of states by the use of Kane's random-**k** model. Their calculated photoelectron energy distributions had approximately the correct shape as functions of initial energy, but this shape changed too rapidly above photon energies of 16.8 eV, which suggested errors in the band energies rather than any major error in the model. Smith (11), in a simpler approach, assumes constant-momentum matrix elements and uses an interpolation scheme of Hodges, Ehrenreich, and Lang, which is fitted for copper to an augmented plane wave (APW) calculation by Burdick. Good correlation was obtained between energy locations of peaks in the experimental spectra and peak loci in the calculated EDJDOS. The direct transition model is regarded as vindicated, and it is suggested that a reliable way of determining the density of states from photoelectron spectra in this low-energy region is to find a band structure whose calculated EDJDOS is consistent with the photoelectron spectra and then compute the density of states from the band structure.

A more direct method has been used by Eastman & Grobman (15). A given initial state can undergo transitions into a number of final states determined by the selection rules of photon energy and crystal momentum. When the photoelectron energy distribution curves are averaged over the photon energy range, which contributes to the optical sum rule, details of the final states and transition-probability effects are largely eliminated, and a replica of the density of occupied states results.

Although sensitivity of the photoelectron spectra in the low-energy range to structure in the conduction band may be viewed as a disadvantage by the experimenter whose interest lies in determination of the density of occupied states, it is evident that this sensitivity may be exploited in an investigation of the unoccupied band structure.

Variations in the Momentum Matrix Elements

Nemoshkalenko et al (16) have calculated the X-ray electron spectra and the density of valence states in copper, silver, and palladium with the aim of elucidating the character of the effect of transition probability on the shape of the photoelectron energy distribution. The wave functions for the valence bands of each metal were found with the aid of the interpolation method developed by Hodges and Ehrenreich, and the electrons in the conduction band were treated as plane waves. The calculated electron spectra for all three metals gave a much better reproduction of the experimental band shape than was achieved by considering only the density of states. The observed differences between the calculated and measured spectra were largely due to neglect of the contribution from s and p electron states in the calculation, since these states determine the electron spectrum near the Fermi level in copper and silver. The chief reason the calculated spectrum differed from the density of states was that electrons with e_g symmetry have higher transition probabilities than those with t_{2g} symmetry.

The majority of the work on experimental determination and theoretical prediction of ionization cross sections has been carried out in gases (10). In this area the dependence of photoionization cross sections on the photoelectron energy has proved particularly useful in aiding assignment of spectra (17, 18). In situations where a tight binding model constitutes an appropriate description of the valence band of a solid, similar empirical observations have been used to decompose the valence band into partial densities of states (19, 20). Some examples of this are discussed below.

Even when variations of transition probabilities are taken into account, interpretation of an XPS spectrum in terms of a density of initial states must be recognized as an approximation. The energy change accompanying the removal of a charged particle from a metal contains a many-body term that is described as the relaxation energy arising from polarization of the electron gas toward (or away from) the resultant hole. Valence electron binding energies are sometimes regarded as having no relaxation contribution because the valence electrons are delocalized in the initial states. However, Ley et al (21) argue that relaxation energies are not strongly dependent on the degree of localization of the hole state and therefore vary little from core- to valence-electron states. Differential relaxation across the valence band is likely to be small in metals; therefore this effect should not cause large discrepancies between initial densities of states and photoemission spectra.

Angular Variation of Photoemission

The theory of angular variation of photoemission was developed initially by Kane (22) and Mahan (23). The potential information in the study of angular distributions is readily appreciated. The kinetic energy, E', of a photoelectron emitted into a vacuum is given by $E' = \hbar^2 K'^2/2m$, where $\mathbf{K}'$ is its wave vector outside the material. If we measure E' and the direction of propagation of a photoelectron, we can obtain values for the three components of the electron's wave vector in a vacuum. If we take the normal to the surface as the z axis, the polar angle θ as the angle between the electron's trajectory and the surface normal, and the azimuthal angle ϕ as the angle of rotation about the normal, the three components are given by

$$K'_x = (2mE'/\hbar^2)^{1/2} \sin\theta \cos\phi,$$
$$K'_y = (2mE'/\hbar^2)^{1/2} \sin\theta \sin\phi,$$
$$K'_z = (2mE'/\hbar^2)^{1/2} \cos\theta.$$

If we could convert these to the three components of **K**, which is the wave vector the electron had inside the material, we could in principle map out portions of the band structure directly from experiment. For primary emission, if refraction of the electron wave on crossing the surface boundary is specular, the components parallel to the surface, K_x and K_y, will be conserved. In a simple free-electron model for the solid, the perpendicular component K_z will be given by

$$\frac{\hbar^2 K_z^2}{2m} = \frac{\hbar^2 K_z'^2}{2m} + W,$$

where W is the inner potential. Detailed calculations of this inner potential have been carried out by Mahan (23) for simple metals but such calculations pose problems in more complex situations.

Applications of this model have been especially fruitful for two-dimensional materials such as the transition metal dichalcogenides (24). These have structures in which covalently bonded sandwiches are loosely coupled together by van der Waals forces, and the energy bands in the direction perpendicular to the sandwiches are expected to be rather flat. The band structure is almost completely specified once it is known in the two dimensions parallel to the crystal layers. The K_z dependence of the spectra and energy bands can be neglected, and the K_x and K_y dependence, which is experimentally determinable, may be used to map two-dimensional energy bands.

Alternatively, for a three-dimensional solid the limitation of the indeterminableness of K_z may be overcome by identifying the same transitions in angular photoemission data from two different crystal surfaces. This approach has been used, for example, in a study of tungsten (25).

An obvious advantage of this experimental approach is that direct comparison with band structure calculations is possible rather than comparison with some derived function such as densities of states.

This paints a rather over-optimistic picture of the state of angular distribution studies. Rowe & Smith (26) have studied the ultraviolet photoelectron spectra (UPS) arising from two different faces, namely the $\langle 100 \rangle$ and the $\langle 111 \rangle$, of a copper single crystal. The spectra from the two faces differ significantly. Some of these differences can be understood in terms of a direct transition model with a specular boundary condition, but calculations predict additional differences that are not observed. For example, for the lowest photon energy used, no d emission is predicted from the $\langle 100 \rangle$ face where abundant d emission is in fact observed. The possibility of contributions from higher "Mahan cones" (23) is considered, but these are found to be insufficient to account for the experimental intensity. Various explanations are suggested; namely, that diffuse refraction occurs, and the specular boundary condition is too restrictive, or that quasi-electron scattering occurs in which the directions of photoelectrons are modified, but their energies are unchanged. Surface photoelectric effect and surface state emission are also considered. In a He-I study on the copper $\langle 001 \rangle$ face (27) it is suggested that the photoemission near E_F in copper is dominated by surface-excited emission: spectral features near E_F are among those changed markedly by oxygen adsorption on the copper surface. Angular-dependence studies on gold (28) can be interpreted within the simple model by the use of the band structure of Christensen and Seraphin.

Electron Escape Depths

These studies highlight the question of whether bulk or surface states are being sampled in a photoemission experiment. Only those ejected electrons that do not lose energy on their passage to the surface will appear in the kinetic energy spectrum at their original characteristic positions. The scattered electrons contribute to the background at lower kinetic energies. The electron escape depth is discussed in terms of the mean free path, λ, which is found to be highly dependent on kinetic energy. The general behavior is a sharp decrease from 10,000 Å to 20–30 Å as the energy increases from 0 to 50 eV, then a fairly flat minimum in the range 50–500 eV, with the lowest escape depth of around 5 Å. Finally an increase occurs with increasing electron energy. There is no universal curve, since the actual position of the curve depends on the material. For further details, two literature compilations of data should be consulted (29, 30) along with a review of measurement techniques (31).

The wave functions of the electrons belonging to the first few atomic layers of a solid may be distorted by the presence of the surface (32). Thus for a short escape depth, the photoelectrons may not provide information characteristic of the bulk material but may be related to the electron structure modified by the surface.

SOME EXPERIMENTAL CONSIDERATIONS

The development of the study of valence bands in solids has been, to a certain extent, a function of the development of monochromatic light sources with a range of photon energies. The original work in photoemission was carried out using spectrometers with lithium fluoride windows, which limit the photon energy to below 12 eV. The development of windowless spectrometers employing differential pumping gave access to highly monochromatic lines such as Ne-I (16.8 eV), Ne-II(26.9 eV), He-I(21.2 eV), and He-II(40.8 eV).

In XPS, commonly used radiation is either $AlK\alpha_{1,2}$ (1486.6 eV) or $MgK\alpha_{1,2}$ (1253.6 eV), but these lines are broad and possess troublesome satellites, which interfere in the study of valence bands. An elegant method has been developed for eliminating the inherent width of X-radiation (33, 34). $AlK\alpha$ radiation is dispersed on a Rowland circle by means of a bent crystal monochromator and impinges on the sample with a spatial energy distribution. The sample is situated so that it is irradiated only by the $AlK\alpha_{1,2}$ radiation, although the full width (~ 0.9 eV) of this line is in fact used. The $AlK\alpha_{3,4}$ satellite radiation is eliminated, together with most of the bremsstrahlung continuum radiation. Within the limits set up by the resolving power of the crystal dispersion system, each point on the sample surface is exposed to monochromatic radiation, whose wavelength varies across the sample. The photoelectrons ejected from a particular energy level will therefore have kinetic energies that depend systematically on the point on the sample surface from which they are emitted. The photoelectrons are passed through a multicomponent lens system designed to focus the image of the sample at the entrance to the analyzer and to magnify the image to match the dispersion of the analyzer. Provided the dispersion of the analyzer is such that it compensates for the energy spectrum of the photoelectron species in question, the electrons will all be focused along one line. Thus the linewidth of the exciting radiation is eliminated without loss of sensitivity. Effectively, resolution of ~ 0.3 eV is achieved.

Synchrotron radiation has a wide spectral range with a smooth intensity distribution. The radiation is also plane-polarized. Coupled with sufficiently flexible monochromators, it gives radiation in the region of <100 eV (35), considerably extending the range of photon energies available. Another advantage of this type of source is that it presents no problems for ultra-high-vacuum work.

Calibration

The binding energy of an electron with respect to the Fermi energy E_B^F, is given by the relationship

$$\hbar\omega = E_B^F + KE + e\phi_{sp} + e\psi.$$

KE is the kinetic energy of the photoelectron, ϕ_{sp} the spectrometer work function, and ψ the potential due to sample charging. If the sample is metallic and is maintained in good electrical contact with the spectrometer, ψ is zero. For semiconductors and insulators, ψ is uncertain, and sample calibration is difficult. The normal procedure is to calibrate by reference to some standard material that is in perfect electric contact with the sample. The most satisfactory calibrant appears to be a submonolayer quantity of metal, normally gold, that can be vacuum-deposited on the surface of the sample (36, 37).

In the study of valence bands, where there seems to be little interest in comparison between different samples, an internal reference is taken. For metals the Fermi edge is normally evident in the photoelectron spectra as a steep edge in the distribution curve, and this is taken as the reference level. In semiconductors and insulators the position of the Fermi energy is uncertain, so the top of the valence band is taken as an arbitrary zero (38, 39).

Sample Preparation

As electron escape depths are rarely in excess of 30 Å for typical electron kinetic energies and may be as small as 5 Å, cleanliness of the sample surface is essential if meaningful results are to be obtained. An adequate vacuum is in the range of 10^{-9}–10^{-10} torr. This is difficult to achieve in the case of windowless discharge lamps but has been realized (40). X-ray and synchrotron sources pose no problems in this respect.

A wide variety of techniques are employed for sample preparation. Volatile samples may be continuously sublimed in order to maintain a fresh surface (41). Cleaving of single crystals under vacuum is an excellent way of producing a reproducible surface. Deposition of a thin film is a good technique for insulators where charging effects may constitute a problem. Argon ion bombardment seems to possess certain disadvantages. A comparison of spectra from palladium samples (42) produced by argon ion bombardment and vacuum evaporation shows the former to give a less detailed spectrum, which suggests an amorphous material. In a study of MoS_2 (43) argon ion bombardment caused preferential sputtering and formation of islands of metallic molybdenum. There exist other instances of chemical damage as a result of this procedure (10) so it should be used with caution. Mechanical scraping in situ is another method that has been used for production of clean surfaces (41, 44). XPS core lines, Auger emission, and low-energy electron diffraction (LEED) have all been used to monitor surface purity.

At low-photon energies, the depth that may be probed below the Fermi level is considerably restricted by the work function of the metal. The work function may be lowered by covering the metal film with a monolayer of cesium (45), a technique known as cesiation. In practice cesiation tends to smear out structure in the spectra, but Smith compensated for this by measuring higher derivative spectra in order to enhance the structure and suppress the smooth background.

SOME CHEMICAL STUDIES

The range and variety of systems for which valence band studies have been undertaken is substantial. For near comprehensive tabulations the reader is referred to the review series *Electronic Structure and Magnetism of Inorganic Compounds* (8–10). The body of results discussed below is necessarily eclectic.

Metals

The XPS spectra of sodium and lithium (46) are rich in plasmon structure. Of the alkali metals, sodium is considered to have the most nearly free electrons and its valence band shows the anticipated $E^{1/2}$ dependence of intensity on energy.

Energy-loss processes are also identified in the XPS spectrum of magnesium (47); the valence band has three identifiable plasmon satellites. A detailed comparison is made with valence band data obtainable from X-ray emission spectra and the $KL_{23}V$ Auger peak. It is argued that valence band spectra obtained by different methods can be compared most directly among states with the same number of core holes.

Traum & Smith have measured the photoemission spectra of Rh, Ir, Ni, Pd, and Pt (45) using clean and cesiated samples in the photon energy range $\hbar\omega =$ 4–11 eV. The results are compared with those obtained previously on Cu, Ag, and Au (11, 48). Traum & Smith favor comparison of the derivative spectra that enhance the spectral structure. The variation of the profile of the spectrum with $\hbar\omega$ displays a consistent pattern from metal to metal. The pattern for one metal is obtained from that of the next by a uniform shifting and stretching of the electron energy scale. The metal d-bandwidths increase when one moves down a column of the periodic table or from right to left along a row. Certain events in the evolution of the spectral profile take place at different photon energies for different metals because of variations in the relative positions of the unoccupied bands.

Using a combined interpolation scheme developed in a previous paper (49), Smith constructs model band structures for Rh, Pd, Ag, Ir, and Pt (50) that include spin-orbit coupling and other relativistic corrections. The EDJDOS are calculated by use of quadratic interpolation and a Monte Carlo sampling of k space. Comparison of the calculated EDJDOS with the measured spectra concentrated on energy locations of structure in the spectra. Relative intensities are affected by the momentum matrix elements and the transport and escape factors, which are not included in the calculation. A strong resemblance is found between the variation of the profile of the EDJDOS with photon energy and the profile changes seen in the experimental photoemission spectra. The structure observed experimentally is weaker than that calculated. In the case of Rh and Pd, no empirical adjustment of the Mattheiss parameters for the d bands was considered necessary. Minor adjustments to the width and positions of the d bands were made for Ag, Ir, Pt, and Au.

Most aspects of the spectra are therefore consistent with direct transitions in a one-electron band model. However, for Ir and Rh (45), the upper edge of the spectrum is predicted to decrease in energy as $\hbar\omega$ is increased from about 5.0 eV, and a forbidden range appears just above E_F until $\hbar\omega = 8.5$ eV. Experimentally, the spectral edge is quite stationary, which suggests nondirect transitions.

XPS spectra were obtained for Rh, Pd, Ag, Ir, Pt, and Au (51) by means of monochromatized Al$K\alpha$ radiation. For the 4d metals there is excellent agreement between the energy positions of peaks in the theoretical densities of states and the XPS valence bands. For the 5d metals, Ir, Pt, and Au, the agreement is reasonable but not as good. The relative intensity of the XPS data in the lower energy region of the d bands is consistently lower than that in the densities of states, indicating appreciable modulation of the spectra by optical transition strengths.

The rare-earth metals provide an interesting case because, in addition to the valence band originating from 5d and 6s states, localized 4f states are present at energies quite near the Fermi level. Baer & Busch (52) have studied these metals under rigorous vacuum conditions, using both monochromatized and unmonochromatized Al$K\alpha$ radiation. The detailed f structure is due to the different final states $4f^{n-1}$, which can be formed by removing one electron from the initial state $4f^n$. In most cases the number of states for a given configuration can be very high. The relative intensities of the final states allowed in photoemission are given by the squared fractional-parentage coefficients (53). The small separation of individual J levels compared with experimental resolution allows a summation over J levels belonging to a term. The separations of the final states in the XPS spectra are slightly larger than those found in optical absorption spectra. This difference is explained by the stronger intra-atomic potential due to the additional positive charge felt by the electrons in the photoemission final state. The spectra are most successfully interpreted by this approach. The high degree of correlation may be exemplified by the spectrum of Sm shown in Figure 1.

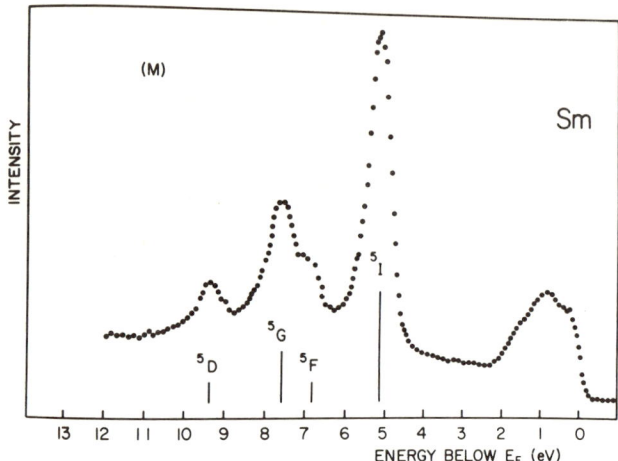

Figure 1 XPS spectrum of Sm measured with monochromatized AlKα radiation. The vertical bars indicate the predicted energies and intensities of the final states. (Reproduced from Reference 52 by kind permission.)

The weak valence bands of these compounds can only be studied with monochromatized radiation since, with unmonochromatized radiation, 4f-band satellites overlap with the valence band. Even so, the intense 4f band tails into the valence band region and distorts the lower part of the valence band. Moderate agreement is found between the band shapes for Gd, Dy, and Er, and those predicted from broadened densities of states based on an APW calculation.

In a study of photoemission from tungsten, Feuerbacher & Egede Christensen (54) have attempted to distinguish between the bulk and surface contributions. Energy-distribution spectra of photoelectrons emitted normal to three single-crystal faces of tungsten have been measured for photon energies between 7.7 and 21.2 eV. The analysis assumes conservation of the momentum component parallel to the crystal surface, so the normal observation angle selects those electrons from the bulk emission with $k_{\parallel}$ zero. The results were interpreted in terms of one-dimensional electronic properties along the symmetry lines in k space that correspond to the emitting crystal faces.

Nonmetallic Elements

Rare-gas solids may be considered prototype insulators. Photoelectron energy distributions for solid Ne, Ar, Kr, and Xe have been measured for $8\,\text{eV} \leq \hbar\omega \leq 30\,\text{eV}$ by use of synchrotron radiation (55). Charging of samples was minimized by using thin films, typically 20–100 Å. The position of the vacuum level was determined, which together with the previously measured energy gap, gave a value for the electron affinity for each gas. The valence bandwidth increased systematically from Ne to Xe, as might be expected from the increasing atomic spin-orbit splitting. In Xe a typical spectrum showed three distinct maxima; the two upper maxima were associated with excitations from the upper valence bands ($j = \frac{3}{2}$) split by the crystal field, and the lower maximum was associated with the lower valence band ($j = \frac{1}{2}$). In the lighter inert gases these maxima were deemed to overlap. Existing band structure calculations were shown to be inadequate for quantitative prediction of the experimental data. In a subsequent paper (56), a self-consistent Hartree-Fock calculation that took into account both polarization and relaxation gave a reasonable account of the Ar spectra, but the lack of relativistic corrections in the calculation led to poor agreement for Kr.

Rowe & Ibach (57) have separated the bulk and surface contributions to the ultraviolet photoemission spectra of silicon. Measurements are reported at a photon energy of 21.2 eV on clean silicon surfaces having $\langle 111 \rangle$ 7 × 7, $\langle 111 \rangle$ 2 × 1, and $\langle 100 \rangle$ 2 × 1 LEED patterns. Since the escape depth of the photoelectrons is near its minimum (5–10 Å) with these photon energies, surface contributions are expected to be important. The surface effects are found to extend over the whole of the valence band; therefore they cannot be separated from the bulk contributions by adsorption studied because the adsorbate states would overlap. Since the surface electronic states are expected to appear at different energies for different LEED structures, an attempt is made to minimize them by averaging over the experimental distributions. As this also produces a "final-state average," the resulting distribution is assumed to be the bulk contribution. It compares well with the bulk density of states calculated by Kane, as well as from X-ray emission data. Surface contributions were localized within small energy ranges by assuming values for the ratio of bulk to surface contributions (three was considered reasonable). Five distinct features were found for the $\langle 111 \rangle$ 7 × 7 surface; three were identified as surface states and two as surface resonances. A dangling-bond state was found at -0.5 eV and a back-bond state at -11.7 eV. By comparison of clean and partially oxidized surfaces of annealed silicon, Rowe (58) identified a surface state 0.1 eV above the valence band edge, which contrasts with the results on cleaved surfaces.

The surface state near the top of the valence bands on the cleaved $\langle 111 \rangle$ face of silicon has been studied by angularly resolved photoemission (59). The intensity of the surface state peak varies dramatically with the polar angle, θ. If conservation of $k_\parallel$ is assumed, the results indicate a surface state band in which the energy decreases with increasing $k_\parallel$. This is consistent with some semiempirical band calculations. However, the surface state appears to split for $\theta > 35°$, which is not predicted by the calculations. Also, there is no agreement on the energy location of the state. This leads to the suggestion that the observed state is not a pure dangling bond but contains a contribution from the back bond. The azimuthal dependence of the surface state emission is quite anisotropic and displays three sharp lobes centered about planes containing the $\langle \bar{2}11 \rangle$ directions. Liebsch (60) has shown that such anisotropies can arise as a consequence of multiple scattering effects of the low-energy electron-diffraction type in the final state of the optical transition that gives rise to the photoelectron. Gadzuk (61), on the other hand, concerns himself with the initial state and proposes that the azimuthal dependence reflects the angular dependence of the atomic orbitals that comprise the initial states, the emission tending to be more intense along the directions in which the lobes of the orbitals point.

In this picture, photoemission from p_z-like dangling bonds would have cylindrical symmetry about the surface normal. However, hybridization with back bonds would predict the observed azimuthal dependence.

Surface states have also been identified in studies on germanium (62–64). Estimates of escape depths of 6–10 Å for photon energy of 21.2 eV lead to the proposal that one third of the emission is due to surface states (64). In a thorough study using synchrotron radiation, Grobman et al (35) use an anisotropic direct transition model to interpret the results, and conclude that for cleaved semiconductors some momentum information survives the photoemission process.

Compounds of the Type $A^N B^{8-N}$

Three groups of workers have published substantial papers on compounds of the type $A^N B^{8-N}$. Ley et al (37) use monochromatized Al$K\alpha$ radiation, Eastman et al (63) use synchrotron radiation in the 15–78 eV range, and Shevchik et al (65) use Ne-I, He-I, He-II, and Al$K\alpha$ radiation. The compounds studied include Ge, GaP, GaAs, GaSb, InP, InAs, InSb, AlSb, ZnO, ZnS, ZnSe, ZnTe, CdS, CdSe, CdTe, and HgTe. There is good agreement between the XPS and UPS data indicating that a bulk density of states is observed in the UPS experiments, with the relative probing depths 4–8 Å at $\hbar\omega = 25$ eV and 15–30 Å at 1487 eV (63). The similarities of the spectra over this wide range of materials is impressive. In general, the spectra show a three-peak structure occasionally with a sharp d-core peak intruding. The upper two bands narrow with increasing ionicity, and the lower valence band splits off from the top two for the heteropolar semiconductors. There is some structure on the least bound peak. Densities of states calculated using the empirical pseudopotential model provide a useful basis for relating features in the photoelectron (PE) spectra to energies of characteristic symmetry points (37, 63). However, experimental bandwidths are greater than those predicted; the discrepancy is largest for the most ionic compounds. The $X_{\alpha\beta}$ orthogonalized plane wave (OPW) method appears to give the best fit (37, 63). Shevchik et al (65) present a simple bond-charge model that simulates realistically the density of states of these germanium and zinc-blende types of semiconductors. Kowalczyk et al (66) base a quantitative scale of ionicity on the XPS studies of these and some $A^N B^{10-N}$ compounds. Studies on lead and tin chalcogenides (67–71) are in qualitative agreement with EPM (empirical-pseudopotential-method), OPW, and APW calculations.

Alkali Halides

In the XPS spectra of the alkali halides (38), the uppermost valence-band peaks strongly resemble core-level peaks, which is understandable in terms of their ionic structure. The lower-lying of the two valence bands shows no structure and is halogen s-like. The upper band, which can be considered to arise from the pure anion p levels, shows definite structure on the low-energy side in the cases of LiF, NaCl, NaBr, NaI, KBr, and KI. This structure is not compatible with a localized ionic model for these salts since the p orbitals are not split by an octahedral crystal field, and the structure separation is too large for spin-orbit splitting. It is therefore concluded that the structure is due to band effects; indeed, there is a smooth progression in the band structure from the group IV elements through III–V and II–VI compounds to these I–VII compounds. For the more ionic compounds KF, KCl, and NaF, only one fairly symmetrical peak is observed in the upper valence band. For the cesium halides (72), no splitting of the valence band is observed for the chloride, that found in the bromide is consistent with bromine spin-orbit splitting, whereas the iodide valence band shows a triplet structure. The higher energy peaks arise from two branches of the $^2P_{3/2}$ level; the lower peak is assigned to the $^2P_{1/2}$ state.

Lapeyre et al (73) have taken advantage of the continuous nature of synchrotron radiation to apply final-state spectroscopy to KCl, which is known to have strong final-state properties. In their experiment they sweep the photon energy, $\hbar\omega$, and simultaneously scan the analyzer to maintain $E_f - \hbar\omega$ constant; E_f refers to the final-state energy of the photoelectrons. This gives a constant initial-state energy spectrum. This is repeated for a series of initial-state energies in order to obtain a family of curves. In this family two types of peaks are recognized: those that occur at the same final-state energy and those that occur at constant $\hbar\omega$. The fixed E_f peaks correspond to high densities of states in the conduction bands. The fixed $\hbar\omega$ band is attributed to electron emission as a result of exciton decay.

Transition Metal Oxides and Halides

Transition metal oxides and halides have been studied both by XPS and UPS (19, 74–78). A prevalent technique for this class of compound is to use the frequency dependence of the photoemission spectra to obtain partial densities of states for the overlapping band-like p states and the localized d states. Eastman & Freeouf (19), for example, have used this approach in the energy range 5–90 eV for NiO, CoO, FeO, MnO, and Cr_2O_3. The photoemission intensities of p and d states have characteristically different dependences on photon energy; the d emission increases relative to the p emission as the photon energy is increased. For the total primary-valence electron emission $N_T(E, \omega)$, they write

$$N_T(E, \omega) = N_p(E, \omega) + N_d(E, \omega),$$

where N_p and N_d are the p- and d-state emission spectra, and E is the electron energy. At sufficiently high photon energies ($\hbar\omega \geq 20$ eV for NiO, etc), N_d and N_p may be factored into $\hbar\omega$-independent spectral shapes, $\bar{N}_d(E)$ and $\bar{N}_p(E)$, and $\hbar\omega$-dependent intensity factors $C_d(\omega)$ and $C_p(\omega)$. This assumes that final-state effects may safely be ignored at these frequencies, so that

$$N_T(E, \omega) = C_d(\omega)\bar{N}_d(E) + C_p(\omega)\bar{N}_p(E).$$

In order to determine $\bar{N}_d(E)$ and $\bar{N}_p(E)$ from a given set of experimental spectra, primary emission is first determined by subtracting secondary emission and multi-electron satellite emission. Separating out the two components is made easier if it may be assumed that only one of the components is responsible for the intensity at any one particular energy; a pair of spectra may then be used to obtain the two partial densities of states. Repetition of this procedure for several independent pairs provides a consistency check.

The partial p- and d-state intensities obtained are shown in Figure 2. As indicated there, the d-state densities were well described by a ligand field and fractional parentage treatment of the $3d^{n-1}$ final states. Agreement may be maximized if we assume a somewhat smaller emission intensity for the e_g relative to the t_{2g} electrons.

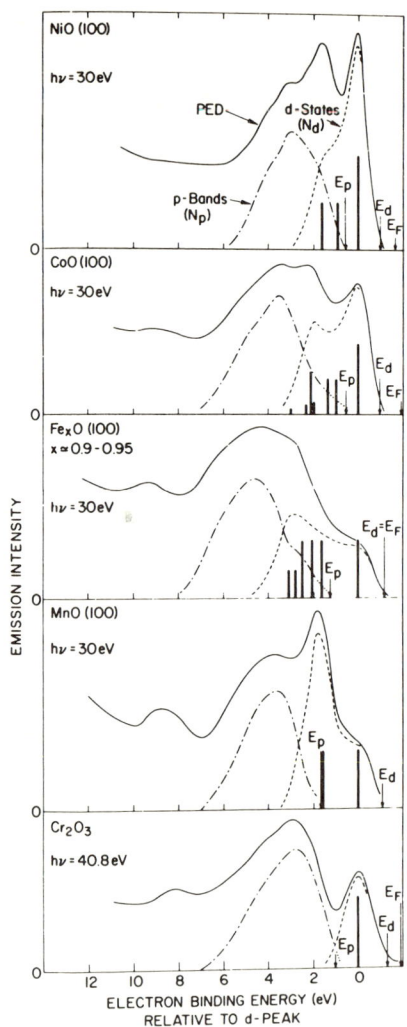

Figure 2 PED's, partial d-state intensities $N_d(E)$, and p-band intensities $N_p(E)$ for NiO, CoO, Fe$_x$O, MnO, and Cr$_2$O$_3$. The vertical lines denote calculated $3d^{n-1}$ final-state ionization potentials. The weak features near 8–10 eV for CoO, Fe$_x$O, MnO, and Cr$_2$O$_3$ are attributed to multielectron satellite peaks. (Reproduced from Reference 19 by kind permission.)

Group IB Halides

The group IB halides may be regarded as members of the $A^N B^{8-N}$ series, but their band structure differs, because the d levels lie higher in energy and thus hybridize with the anion p levels. In the spirit of the tight binding approximation, the valence bands of these compounds are considered to be composed of linear combinations of atomic orbitals.

Studies on the cuprous halides CuCl, CuBr, and CuI have been carried out with a wide variety of photon energies (20, 79–82). Halogen p orbitals may reasonably be expected to undergo cross-section variation with photon energy similar to rare-gas p orbitals. These latter decrease over the range $\hbar\omega = 10$–40 eV, whereas the copper $3d$ cross section remains nearly constant (20, 80). These differences in cross-section variation were used to identify the atomic origins of the cuprous halide valence bands and to achieve decomposition into p and d partial densities of states. In the copper halides the $\Gamma_{12}d$ levels are core-like, whereas the $\Gamma_{15}d$ levels band significantly and lie above the core-like levels. The anion p-like band lies below the $\Gamma_{12}\,d$ levels (80).

In AgI this ordering is reversed (80); the upper band is now p-like, and the core-like $\Gamma_{12}d$ levels lie above the banding $\Gamma_{15}d$ levels. On passing from copper to silver, the d levels are lowered in energy and pass through the anion p band. These trends are reproduced by a calculation using a seven-function basis set of s, p, and d_{xy} orbitals (80).

AgCl and AgBr crystallize in the rock salt structure and thus differ considerably from AgI and the copper halides (zinc blende). Their spectra have been studied at various photon energies (83). The spectra show a sharp peak at the center of the valence band, which decomposition into partial densities of states shows to be mostly d-like. The p density of states shows a gap here. Apart from this region, the p and d functions are mixed almost equally and uniformly throughout the valence band.

Mason (85) assigns the bands in the silver halides by means of the point-charge electrostatic model of Citrin and Thomas. This model predicts that the spacing of the energy levels for a particular ion are independent of crystal environment and identical to those of the free ion. The energies of the ionic states are approximated by the free ion energies as shifted by the Madelung potential. For silver fluoride the first band is identified as Ag$4d$ (4.6 eV), whereas the band at 7 eV is assigned to the fluorine p band. For AgCl and AgBr, the following order is suggested: upper-p state > central d > d–p hybrid.

Silver halides show unusually large and dramatic changes in their photoemission as samples are cooled from 295 to 80°K (84). Since the overlap of the halogen p and silver $4d$ wave functions has a large effect on the energies of the valence bands, and this overlap is dependent upon the separation of the atoms, the energies of the hybridized states will be dependent on this spacing. The vibrations of the lattice thus couple with the electronic structure, and temperature dependence of the former results in temperature dependence of the latter. This is termed *dynamic hybridization* and is most likely to be significant for macromolecular solids such as the silver halides.

Transition Metal Dichalcogenides

Transition metal and B-metal dichalcogenides have proved popular systems for photoemission work. Comparison of XPS and UPS data (86, 87) suggests that at photon energies >26.9 eV, an experimental view of the occupied band structure is obtained. The layered nature of the crystals lends itself to ready cleavage and production of very clean surfaces. The group IVA metals form semiconducting dichalcogenides, and photoemission studies (86–88) show a three-pronged chalcogenide p band widely separated from the s band. In the case of TiS_2 a shoulder near the Fermi energy probably represents the band responsible for the strong n-type conductivity. The widths of the p bands in TiS_2, ZrS_2, and HfS_2 are between 4.5 and 5.0 eV, which is greater than the 3.5 eV suggested by the band structure calculations of Mattheiss.

The group VA compounds are mostly metallic, and many are superconductors. They have one electron in a nonbonding band. This extra band at the Fermi energy is clearly apparent for $1T$-TaS_2 (87, 89). In $2H$-$NbSe_2$ the Se $4p$ valence band overlaps the Nb d band. The 40.8 eV spectrum gives the best resolution of this band (89), but the XPS spectrum gives the clearest indication of where the valence band ends (87). Other forms of TaS_2 show greater d-p overlap (89) than is found for $1T$-TaS_2.

The group VIA dichalcogenides show photoemission thresholds well below the Fermi energy consistent with their semiconducting nature (87, 89). Mattheiss has shown that a hybridization gap in the $2H$ compounds splits the d band into subbands, one of which is completely filled with the two d electrons. The d bands are not completely resolved for these compounds, which suggests that there is no gap between the p and d bands as is suggested by the calculation.

FeS_2 (90, 91), which is a semiconductor, shows a narrow $3d$ band about 1 eV in width. SCF-$X\alpha$ calculations suggest this t_{2g} state is 85% metal-localized and nonbonding. CoS_2(91) shows metallic conductivity, but the band at the Fermi energy, presumably due to the extra e_g electron, merges with the other valence bands. NiS_2 (90, 91) has two e_g electrons and is a p-type semiconductor. The NiS_2 spectra show a shoulder near E_F and a maximum that occurs at -2.2 eV at photon energies greater than 35 eV. In the SCF-$X\alpha$ calculation, the e_g levels are only 59% metal and antibonding; therefore, the associated band would be expected to be broader.

SnS_2 and $SnSe_2$ have also been studied (92). Again, calculated band widths are consistently narrower than experimentally determined values.

Smith & Traum (93) have measured the variation of the photoelectric yield and the photoelectron energy spectra with polar angle, θ, of $1T$-$TaSe_2$, $2H$-$TaSe_2$, and $1T$-TaS_2, by using 10.2 eV photons. Various peaks are found to show a marked θ dependence. In order to plot experimental $\varepsilon(\mathbf{k})$ dispersion curves, the kinetic energy at the peak maximum was determined for each peak. The parallel wave vector $\mathbf{k}_\parallel$ corresponding to each of these values for E, was evaluated from

$$k_\parallel = (k_x^2 + k_y^2)^{1/2} = (2Em/\hbar^2)^{1/2} \sin\theta.$$

The initial-state energy $E - \hbar\omega + \Phi$ was then plotted against $k_\parallel$. Good agreement was found with the APW calculations of Mattheiss if shifts were made in the chalcogen bands.

Williams et al (94, 95) have studied the variation of the photoelectric emission of MoS_2 with polar angle using He-I (94) and $AlK\alpha$ (95) radiation. They point out that the highest energy band that is assigned to d ionization reaches a minimum for $\theta = 55°$ as expected for a d_{z^2} orbital but does not drop to zero, which indicates hybridization with $d_{x^2-y^2}$ and d_{xy} orbitals. Again this is predicted in the band calculation of Mattheiss.

Smith & Traum (24, 93, 96) also discuss the variation of photoelectric yield with azimuthal angle, ϕ, for $1T\text{-}TaSe_2$, $1T\text{-}TaS_2$, and $2H\text{-}TaSe_2$, where the emphasis is on peak intensities rather than positions. By application of an appropriate retarding potential, the Ta d-emission contribution was isolated and measured, along with the total photoemission intensity of the chalcogen p and Ta d bands. The radial plots for both emission intensities are given in Figure 3 for $1T\text{-}TaSe_2$. The radial plot clearly reflects the crystal symmetry.

Two approaches are taken to interpret the d emission-intensity data. The initially more appealing approach assumes that the d photoelectrons are actually created on the Ta sites and are propagated preferentially in directions between the Se atoms. Superposition of the d-emission plot on the two outermost atomic sheets of the crystal structure demonstrates how this accounts for bifurcation of the principal lobes in the d-emission intensity data. Azimuthal dependence of the Ta d-emission intensity for $1T\text{-}TaS_2$ for various polar angles gives added weight to this view.

The other approach follows Gadzuk (61) and emphasizes the properties of the initial state. The final state of the photoelectron is represented by a plane wave $e^{i\mathbf{p}\cdot\mathbf{r}}$, thus deliberately deemphasizing final-state effects. This results in the angular dependence of the photoemission reflecting the angular dependence of $\phi(\mathbf{p})$ where

$$\phi(\mathbf{p}) = \int e^{i\mathbf{p}\cdot\mathbf{r}}\phi(\mathbf{r})d^3r,$$

that is, $\phi(\mathbf{p})$ is simply the momentum representation of the initial-state orbital $\phi(\mathbf{r})$.

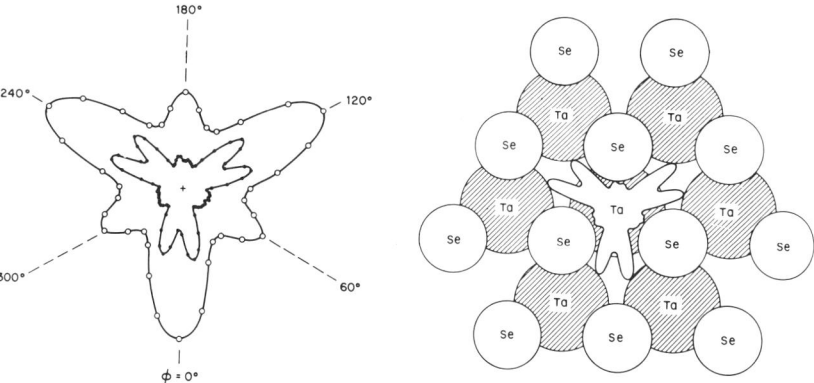

Figure 3 (*a*) Radial plots of the azimuthal dependence at $\theta = 55°$ of the total photoemission intensity (outer curve) and, on a 5× expanded scale, the d-emission intensity (inner curve); (*b*) the experimental Ta d-emission pattern for $\theta = 55°$ superposed on the two uppermost atomic sheets of the real-space crystal structure. (Reproduced from Reference 96 by kind permission.)

A crystal field treatment is successful in predicting the trigonal symmetry of the d-emission intensity, but to account for the bifurcation we invoke the **k**-conservation requirement that optical transitions be confined to certain regions of **k** space.

If an energy range can be found in which it is reasonable to treat the final state as a plane wave, the photoemission intensity is proportional to the momentum density of the initial-state wave function.

The ϕ-dependence of Ta d-emission in $2H$-TaSe$_2$ (24, 93) shows emission symmetry close to sixfold, as expected from the hexagonal symmetry of the crystal; however, there is significant residual threefold symmetry. This is attributed to the finite attenuation length of the photoelectrons, the outermost layer having the most effect on the photoemission. An upper limit of attenuation length of 13 Å is estimated, which is reasonable for photoelectrons of this energy.

Some Rare-Earth Compounds

Spontaneous interconfiguration fluctuations have been detected in thulium monochalcogenides (97). Since photoexcitation takes place in a time short compared to that assumed for the fluctuation (up to 10^6–10^7 times shorter), it is possible to observe an instantaneous picture of the ions in the two valence states. The valence states of the rare-earth ions are well-characterized by photoelectron spectroscopy in that the spectra reflect the accessible terms of the $4f^{n-1}$ final state weighted by the squared fractional parentage coefficients discussed earlier for the rare-earth metals. In a study on TmSe and TmTe, the compound TmSb, in which Tm is strictly trivalent, is used to obtain a characteristic pattern for ionization of a $4f^{12}$ shell. In the spectrum of TmTe, structure is present between 6.4 and 14 eV, which is characteristic of Tm^{3+} ions, but there is also additional structure in the range $E_F - 6.4$ eV, which is identified as Tm^{2+} ions. The signal corresponding to the Tm^{2+} ground state is much smaller in TmSe than in TmTe. In TmTe the two signals are of comparable intensity.

Studies on pure and doped samarium monochalcogenides have been pursued (98–100) because of considerable interest in their peculiar physical properties. For example, SmS undergoes a pressure-induced semiconductor-to-metal phase transition at 6.5 kbar, Sm$_{1-x}$Gd$_x$S crystals undergo an explosive transition upon cooling at atmospheric pressure, and Sm$_{1-x}$Y$_x$S and Sm$_{1-x}$La$_x$S show expansion upon cooling. These properties are ascribed to mixture of two configurations within the ground eigenstate of SmS and its solid solutions; Sm($4f^6$) $\equiv$ Sm^{2+} and Sm($4f^5 d^1_{\text{conduct}}$) $\equiv$ Sm^{3+} are nearly degenerate. Disagreement exists in the literature as to the degree of mixing in pure SmS (98–100).

The metallic doped samarium sulfides studied show clear evidence of both Sm^{2+} and Sm^{3+}. The Sm^{3+}/Sm^{2+} ratio decreases with temperature, which is consistent with the increase of the lattice constant, since Sm^{2+} is larger than Sm^{3+} (99).

In an interesting experiment, Meier et al (101) have studied the spin-polarized photoemission from La-doped EuO in the vacuum ultraviolet. The purpose of the 2% La doping was to eliminate charging effects during illumination. The measurements were made in a magnetic field of 12 kG at $T < 10°$K. The spin polarization of the photoemitted electron beam, P, is defined as

$$P = (N\uparrow - N\downarrow)/(N\uparrow + N\downarrow),$$

where $N\uparrow$ and $N\downarrow$ refer to the number of electrons with magnetic moments parallel and antiparallel to the magnetic field. P was measured as a function of $\hbar\omega$ between 3 and 8 eV and showed a sharp drop between 5.5 and 6 eV. In the region $3 < \hbar\omega < 5$, $P \simeq 0.5$; only electrons from the half-filled $4f$ shell of Eu^{2+} are ionizing. Above 5 eV, P starts to drop. This corresponds to a dip in the photoelectric yield and is attributed to the onset of inelastic scattering of the $4f$ electrons by the $2p$-valence electrons. Around 5.5 eV the valence electrons with highest initial energy are able to escape; they have a large escape depth as they cannot scatter inelastically, and they dominate the photoemission. As a consequence P falls dramatically, and above 6 eV the photocurrent shows zero polarization.

Molecular Solids

Molecular solids are those in which the valence electrons are localized on a molecule or ion; for these k conservation is not important. Where comparison may be made between solid- and vapor-phase spectra of the same material (e.g. 41), the relative energy levels are in good agreement. The valence electronic structure in the solid is dominated by the molecular orbitals, which are relatively unperturbed by condensation. Thus, photoemission provides a means of studying the electronic structure of molecules or ions that may not be obtained in the vapor phase.

The chief difference between vapor- and solid-phase spectra is the broadness of the bands in the latter. Various authors have discussed this question and suggested many causes for this solid-state broadening (102–109).

Several studies of oxyanions have been reported (110–118). The experimental results of all workers on tetraoxoanions are in good agreement. Comparison with the X-ray emission spectra of oxoanions (110, 112) is of great assistance in assigning the XPS spectra. Molecular orbital theory suggests a level scheme for XO_4^{n-} of $1a_1 < 1t_2 < 2a_1 < 2t_2 < 1e, 3t_2, 1t_2$ for the valence region. The two lowest orbitals are mainly O $2s$ in character, with some contribution from the s and p orbitals of the central atom, the $2a_1$ and $2t_2$ levels are of X, s, and p character mixed with oxygen $2s$ and $2p_\sigma$, whereas the uppermost levels, t_1, $3t_2$, and e, are oxygen $2p_\pi$ in character. The first four maxima in the XPS spectra correspond to the levels $1a_1, 1t_2, 2a_1$, and $2t_2$, while the resolution of the next three levels is poor.

Selection rules for K_β emission in these tetrahedral anions only allow transitions from t_2 orbitals, and those with central-atom character will contribute most to the transition moment. Two strong K_β emission lines are observed for all four tetraoxoanions (119–24), and the observed separation is in good agreement with the bands assigned to the $1t_2$ and $2t_2$ orbitals, which have significant X $3p$ character. In a high-resolution K_β-emission spectrum of K_2SO_4 (123), a third weak line, which agrees nicely with the assignments of the $3t_2$ peak, is observed. L II III emission similarly selects orbitals with a_1-$3s$ and t_2, e-$3d$ character. Interestingly, SO_4^{2-}, PO_4^{3-}, and SiO_4^{4-} all show $3d$ character for the upper levels.

The experimental photoemission intensities have been compared with theoretical predictions obtained from atomic cross sections and molecular orbital parentages (112, 113) in conjunction with Gelius' model (18). In general, agreement between the experimental and theoretical line intensities is good.

XPS spectra of transition metal oxoanions (112, 117), VO_4^{3-}, CrO_4^{2-}, and MnO_4^- are much less informative since, in these cases, the levels $2a_1$, $2t_2$, t_1, $3t_2$, and e all ionize in one broad band about 4 eV wide. A similar situation is found in the gas-phase spectrum of the isoelectronic molecules RuO_4 and OsO_4 (125), where the resolution is better, but the total spread of peaks is again 4.5 eV. Again, comparison with the X-ray emission spectrum is fruitful. Group IV tetrahalides have an analogous orbital structure, and their gas-phase PE spectra (126) show a similar pattern with a bunching of the valence levels in the transition metal halides. Studies are also reported on $MoOS_3^{2-}$ and WOS_3^{2-} (111).

Nefedov et al (114) interpret the trends found for series of isoelectronic compounds in terms of a molecular orbital picture. With increasing atomic number of A, the negative charge of the anion in the series AX_n^{x-} decreases, and the s- and p-orbital energies of the atom A are lowered, which brings about a lowering of all the valence levels and the inner level of ligand X. With increasing atomic number, the ionicity of the AX bond decreases, which leads to an increase in a splitting between the $X\sigma$ and $X\pi$ levels. This is enhanced by an increasing difference between the s- and p-orbital energies of atom A. The orbital energy of the valence s state is rapidly lowered with an increase in the atomic number of A.

TCNQ and its salts have been studied because of their interesting conduction properties (41, 127, 128). Lin et al (128) estimate an electron escape depth in TTF-TCNQ of 10 Å and suggest that it is the surface electronic structure that is being probed rather than the bulk, as was assumed in earlier papers (41, 127). Because of the surface sensitivity, Lin et al prepare samples of TCNQ, $Cs_2(TCNQ)_3$, and TTF-TCNQ in situ, and are able to monitor samples in the course of preparation, which is helpful in assignment of bands. Five maxima in the valence band of room temperature TTF-TCNQ films are assigned by comparison with $Cs_2(TCNQ)_3$ to states of TTF^+, $TCNQ^-$, and $TCNQ^\circ$. Thus $TCNQ^\circ$ is shown to be present in the room temperature film, although it is argued that it may only exist on the surface, since the complete Madelung energy is not available there to stabilize the charge transfer. The UPS spectra of TTF-TCNQ films is found to be extremely sensitive to the substrate temperature at deposition and to the film's subsequent thermal history. The spectra at all temperatures show very little electron population at E_F, which is surprising given the metallic electrical behavior of TTF-TCNQ. This result must be due to localization of the states in question; since the escape depth is so small, this may be due to the influence of the surface.

SUMMARY

Photoelectron spectroscopy is able to provide a wealth of information on valence bands in solids. The three-step model provides a good description of bulk photoemission of valence electrons. The spectra may be related to the density of states, whereas angular distribution studies give information on E-k relationships. XPS gives information on intial states whereas UPS, at low $\hbar\omega$, is also capable of giving information on final states.

In many cases, surface emission contributes significantly to photoelectron spectra. This is especially important in gas adsorption studies, which have not been considered in this review.

No attempt has been made in this review to evaluate current methods of calculating valence bands, but it is clear that photoelectron spectroscopy provides experimental criteria that must be met by calculations.

Literature Cited

1. Turner, D. W. 1970. *Ann. Rev. Phys. Chem.* 21:107–28
2. Carlson, T. A. 1975. *Ann. Rev. Phys. Chem.* 26:211–33
3. Turner, D. W., Baker, A. D., Baker, C., Brundle, C. R. 1970. *Molecular Photoelectron Spectroscopy: a Handbook of the He 584 Å Spectra.* London: Wiley. 386 pp.
4. Siegbahn, K., Nordling, C., Fahlman, A., Nordberg, R., Hamrin, K., Hedman, J., Johansson, G., Bergmark, T., Karlsson, S.-E., Lindgren, I., Lindberg, B. 1967. *Electron Spectroscopy for Chemical Analysis—Atomic, Molecular and Solid State Structure Studies by Means of Electron Spectroscopy.* Stockholm: Wicksells Boktryekeri AB. 282 pp.
5. Siegbahn, K., Nordling, C., Johansson, G., Hedman, J., Heden, P. F., Hamrin, K., Gelius, U., Germark, T., Werme, L. O., Manne, R., Boer, Y. 1969. *ESCA Applied to Free Molecules.* Amsterdam & London: North-Holland. 200 pp.
6. Eland, J. H. D. 1974. *Photoelectron Spectroscopy.* New York: Wiley. 239 pp.
7. Carlson, T. A. 1975. *Photoelectron and Auger Spectroscopy.* New York: Plenum. 417 pp.
8. Hamnett, A., Orchard, A. F. 1972. Electronic Structure and Magnetism of Inorgnic Compounds, Spec. Period. Rep., ed. P. Day, 1:1–62. London: Chem. Soc.
9. Evans, S., Orchard, A. F. 1973. See Ref. 8, 2:1–96
10. Hamnett, A., Orchard, A. F. 1974. See Ref. 18, 3:218–415
11. Smith, N. V. 1971. *Phys. Rev. B* 3:1862–78
12. Shirley, D. A. 1972. *Phys. Rev. B* 5:4709–14
13. Freeouf, J., Erbudak, M., Eastman, D. E. 1973. *Solid State Commun.* 13:771–73
14. Williams, A. R., Janak, J. F., Moruzzi, V. L. 1972. *Phys. Rev. Lett.* 28:671–75.
15. Eastman, D. E., Grobman, W. D. 1972. *Phys. Rev. Lett.* 28:1327–30
16. Nemoshkalenko, V. V., Aleshin, V. G., Kucherenko, Y. N., Sheludchenko, L. M. 1975. *J. Electron. Spectrosc. Rel. Phenom.* 6:145–50
17. Price, W. C., Potts, A. W., Streets, D. G. 1972. *Electron Spectroscopy*, ed. D. A. Shirley, pp. 187–98. Amsterdam: North-Holland
18. Gelius, U. See Ref. 17, pp. 311–34
19. Eastman, D. E., Freeouf, J. L. 1975. *Phys. Rev. Lett.* 34:395–98.
20. Goldman, A., Tejeda, J., Shevchik, N. J., Cardona, M. 1974. *Solid State Commun.* 15:1093–95
21. Ley, L., McFreely, F. R., Kowalczyk, S. P., Jenkin, J. G., Shirley, D. A. 1975. *Phys. Rev. B* 11:600–12
22. Kane, E. O. 1964. *Phys. Rev. Lett.* 12:97–98
23. Mahan, G. D. 1970. *Phys. Rev. B* 2:4334–50
24. Smith, N. V., Traum, M. M. 1974. *Surf. Sci.* 45:745–49
25. Turtle, R. R., Callcott, T. A. 1975. *Phys. Rev. Lett.* 34:86–89
26. Rowe, J. E., Smith, N. V. 1974. *Phys. Rev. B* 10:3207–12
27. Lloyd, D. R., Quinn, C. M., Richardson, N. V. 1975. *J. Phys. C.* 8:L371–76
28. Nilsson, P. O., Ilver, L. 1975. *Solid State Commun.* 17:667–71
29. Lindau, I., Spicer, W. E. 1974. *J. Electron. Spectrosc. Rel. Phenom.* 3:409–13
30. Brundle, C. R. 1974. *J. Vac. Sci. Technol.* 11:212–24
31. Powell, C. J. 1974. *Surf. Sci.* 44:29–46
32. Haydock, R., Heine, V., Kelly, M. J., Pendry, J. B. 1972. *Phys. Rev. Lett.* 29:868–71

33. Meleva, A., Moody, R. 1970. Hewlett-Packard Publ.
34. Siegbahn, K., Hammond, D., Fellner-Feldegg, H., Barnett, E. F. 1972. *Science* 176:245-52
35. Grobman, W. D., Eastman, D. E., Freeouf, J. L. 1975. *Phys. Rev. B* 12:4405-33
36. Thomas, J. M., Evans, E. L., Barber, M., Swift, P. 1971. *Trans. Faraday Soc.* 67:1875-86
37. Ley, L., Pollak, R. A., McFeely, F. R., Kowalczyk, S. P., Shirley, D. A. 1974. *Phys. Rev. B* 9:600-21
38. Kowalczyk, S. P., McFeely, F. R., Ley, L. Pollak, R. A. Shirley, D. A. 1974. *Phys. Rev. B* 9:3573-81
39. Eastman, D. E., Freeouf, J. 1973. *Solid State Commun.* 13:1815-18
40. Eastman, D. E. See Ref. 17, pp. 487-514
41. Grobman, W. D., Pollak, R. A., Eastman, D. E., Maas, E. T., Scott, B. A. 1974. *Phys. Rev. Lett.* 32:534-37
42. Hufner, S., Wertheim, G. K., Buchanan, D. N. E. *Chem. Phys. Lett.* 24:527-30
43. Feng, H. C., Chen, J. M. 1974. *J. Phys. C* 7:L75-L78
44. McLachlan, A. D., Jenkin, J. G., Liesegang, J., Lekey, R. C. G. 1974. *J. Electron. Spectrosc. Rel. Phenom.* 3:207-16
45. Traum, M. M., Smith, N. V. 1974. *Phys. Rev. B* 9:1353-64
46. Kowalczyk, S. P., Ley, L., McFeely, F. R., Pollak, R. A., Shirley, D. A. 1973. *Phys. Rev. B* 8:3583-85
47. Ley, L., McFeely, F. R., Kowalczyk, S. P., Jenkin, J. G., Shirley, D. A. 1975. *Phys. Rev. B* 11:600-12
48. Smith, N. V. 1972. *Phys. Rev. B* 5:1192-1202
49. Smith, N. V., Mattheiss, L. F. 1974. *Phys. Rev. B* 9:1341-52
50. Smith, N. V. 1974. *Phys. Rev. B* 9:1365-76
51. Smith, N. V., Wertheim, G. K., Hufner, S., Traum, M. M. 1974. *Phys. Rev. B* 19:3197-3206
52. Baer, Y., Busch, G. 1974. *J. Electron. Spectrosc. Rel. Phenom.* 5:611-26
53. Cox, P. A., Baer, Y., Jørgensen, C. K. 1973. *Chem. Phys. Lett.* 22:433-38
54. Feuerbacher, B., Christensen, N. E. 1974. *Phys. Rev. B* 10:2373-90
55. Schwentner, N., Himpsel, F. J., Saile, V., Skibowski, M., Steinmann, W., Koch, E. E. 1975. *Phys. Rev. Lett.* 34:528-31
56. Kunz, A. B., Mickish, D. J., Mirmira, S. K. V., Shima, T., Himpsel, F. J., Saile, V., Schwentner, N., Koch, E. E. 1975. *Solid State Commun.* 17:761-63
57. Rowe, J. E., Ibach, H. 1974. *Phys. Rev. Lett.* 32:421-24
58. Rowe, J. E. 1974. *Phys. Lett. A* 46:400-2
59. Rowe, J. E., Traum, M. M., Smith, N. V. 1974. *Phys. Rev. Lett.* 33:1333-35
60. Liebsch, A. 1974. *Phys. Rev. Lett.* 32:1203-9
61. Gadzuk, J. W. 1974. *Phys. Rev. B* 10:5030-44
62. Rowe, J. E. 1975. *Phys. Rev. Lett.* 34:398-402
63. Eastman, D. E., Grobman, W. D., Freeouf, J. L., Erbudak, M. 1974. *Phys. Rev. B* 9:3473-88
64. Rowe, J. E. 1975. *Solid State Commun.* 17:673-76
65. Shevchik, N. J., Tejeda, J., Cardona, M. 1974. *Phys. Rev. B* 9:2627-48
66. Kowalczyk, S. P., Ley, L., McFeely, F. R., Shirley, D. A. 1974. *J. Chem. Phys.* 61:2850-56
67. McFeely, F. R., Kowalczyk, S., Ley, L., Pollak, R. A., Shirley, D. A. 1973. *Phys. Rev. B* 7:5228-36
68. Cardona, M., Langer, D. W., Shevchik, N. J., Tejeda, J. 1973. *Phys. Status Solidi B* 58:127-37
69. Abbati, I., Braicovich, L., De Michelis, B. 1974. *J. Phys. C* 7:3661-72
70. Rowe, J. E. 1974. *Appl. Phys. Lett.* 25:576-78
71. Martinez, G., Schlüter, M., Cohen, M. L. 1975. *Phys. Rev. B* 11:651-59
72. Smith, J. A., Pong, W. 1975. *Phys. Rev. B* 12:5931-36
73. Lapeyre, G. J., Anderson, J., Gobby, P. L., Knapp, J. A. 1974. *Phys. Rev. Lett.* 33:1290-93
74. Hüfner, S., Wertheim, G. K. 1973. *Phys. Rev. B* 8:4857-67
75. Kim, K. S. 1974. *Chem. Phys. Lett.* 26:234-39
76. Bishop, S. G., Kemeny, P. C. 1974. *Solid State Commun.* 15:1877-80
77. Sakisaka, Y, Ishii, T., Sagawa, T. 1974. *J. Phys. Soc. Jpn.* 36:1372-76
78. Ishii, T., Kono, S., Suzuki, S., Nagakura, I., Sagawa, T., Kato, R., Watanabe, M., Sato, S. 1975. *Phys. Rev. B* 12:4320-27
79. Ishii, T., Kono, S., Matsukawa, T., Sagawa, T., Kobayasi, T. 1974. *J. Electron. Spectrosc. Rel. Phenom.* 5:559-71
80. Goldmann, A., Tejeda, J., Shevchik, N. J., Cardona, M. 1974. *Phys. Rev. B* 10:4388-4402
81. Kono, S., Ishii, T., Sagawa, T.,

Kobayasi, T. 1972. *Phys. Rev. Lett.* 28:1385–87
82. Kono, S., Ishii, T., Sagawa, T., Kobayasi, T. 1973. *Phys. Rev. B* 8:795–803.
83. Tejeda, J., Shevchik, N. J., Braun, W., Goldmann, A., Cardona, M. 1975. *Phys. Rev. B* 12:1557–66
84. Bauer, R. S., Spicer, W. E. See Ref. 17, pp. 569–74
85. Mason, M. G. 1975. *Phys. Rev. B* 11:5094–5102
86. Shepherd, F. R., Williams, P. M. 1974. *J. Phys. C* 7:4416–26
87. Wertheim, G. K., Di Salvo, F. J., Buchanan, D. N. E. 1973. *Solid State Commun.* 13:1225–28
88. Fischer, D. W. 1973. *Phys. Rev. B* 8:3576–82
89. Shepherd, F. R., Williams, P. M. 1974. *J. Phys. C* 7:4427–40
90. Li, E. K., Johnson, K. H., Eastman, D. E., Freeouf, J. L. 1974. *Phys. Rev. Lett.* 32:470–72
91. Ohsawa, A., Yamamoto, H., Watanabe, H. 1974. *J. Phys. Soc. Jpn.* 37:568
92. Williams, R. H., Murray, R. B., Govan, D. W., Thomas, J. M., Evans, E. L. 1973. *J. Phys. C* 6:3631–42
93. Smith, N. V., Traum, M. M. 1975. *Phys. Rev. B* 11:2087–2108
94. Williams, R. H., Thomas, J. M., Barber, M., Alford, N. 1972. *Chem. Phys. Lett.* 17:142–44
95. Williams, R. H. 1976. *Daresbury Study Weekend Ser.*, ed. I. H. Munro, G. V. Marr, 9:11–22. Warrington, England: S. R. C. Daresbury
96. Traum, M. M., Smith, N. V., Di Salvo, F. J. 1974. *Phys. Rev. Lett.* 32:1241–44
97. Campagna, M., Bucher, E., Wertheim, G. K., Buchanan, D. N. E., Longinotti, L. D. 1974. *Phys. Rev. Lett.* 32:885–89
98. Freeouf, J. L., Eastman, D. E., Grobman, W. D., Holtzberg, F., Torrance, J. B. 1974. *Phys. Rev. Lett.* 33:161–64
99. Pollak, R. A., Holtzberg, F., Freeouf, J. L., Eastman, D. E. 1974. *Phys. Rev. Lett.* 33:820–23
100. Campagna, M., Bucher, E., Wertheim, G. K., Longinotti, L. D. 1974. *Phys. Rev. Lett.* 33:165–68
101. Meier, F., Eib, W., Pierce, D. T. 1975. *Solid State Commun.* 16:1089–91
102. Seki, K., Harada, Y., Ohno, K., Inokuchi, H. 1974. *Bull. Chem. Soc. Jpn.* 47:1608–10
103. Zagrubskii, A. A., Vilesov, F. I. 1973. *Sov. Phys. Solid State* 13:1927–33
104. Kearns, D. R., Calvin, M. 1961. *J. Chem. Phys.* 34:2026–30
105. Sworakowski, J. 1972. *Phys. Status Solidi A* 10:K89–92
106. Vilesov, F. I., Zagrubskii, A. A., Garbuzov, D. Z. 1964. *Sov. Phys. Solid State* 5:1460–64
107. Berglund, C. N., Spicer, W. E. 1964. *Phys. Rev. A* 136:1030–64
108. Belkind, A. I., Bok, J. 1974. *Phys. Status Solidi A* 22:K37–40
109. Ueno, N., Hayasi, Y., Kiyono, S. 1975. *Chem. Phys. Lett.* 35:31–34
110. Connor, J. A., Hillier, I. H., Wood, M. H., Barber, M. 1974. *J. Chem. Soc. Faraday Trans. 2* 70:1040–44
111. Diemann, E., Müller, A. 1974. *Chem. Phys. Lett.* 27:351–54
112. Prins, R. 1974. *J. Chem. Phys.* 61:2580–91
113. Calabrese, A., Hayes, R. G. 1975. *J. Electron. Spectrosc. Rel. Phenom.* 6:1–16
114. Nefedov, V. I., Buslaev, Y. A., Sergushin, N. P., Bayer, L., Kokunov, Y. V., Kovalev, V. V. 1975. *J. Electron. Spectrosc. Rel. Phenom.* 6:221–29
115. Connor, J. A., Hillier, I. H., Saunders, V. R., Barber, M. 1972. *Mol. Phys.* 23:81–90
116. Biloen, P., Prins, R. 1972. *Chem. Phys. Lett.* 16:611–13
117. Connor, J. A., Hillier, I. H., Saunders, V. R., Wood, M. H., Barber, M. 1972. *Mol. Phys.* 24:497–509
118. Prins, R., Novakov, T. 1972. *Chem. Phys. Lett.* 16:86–88
119. Best, P. E. 1968. *J. Chem. Phys.* 49:2797–2805
120. Takahashi, Y. 1969. *Bull. Chem. Soc. Jpn.* 42:3064–72
121. Takahashi, Y. 1971. *Bull. Chem. Soc. Jpn.* 44:587–89
122. Takahashi, Y., 1972. *Bull. Chem. Soc. Jpn.* 45:4–7
123. Urch, D. S. 1969. *J. Chem. Soc. A*, pp. 3026–28
124. Urch, D. S. 1970. *J. Phys. C* 3:1275–91
125. Diemann, E., Müller, A. 1973. *Chem. Phys. Lett.* 19:538–40
126. Green, J. C., Green, M. L. H., Joachim, P. J., Orchard, A. F., Turner, D. W. 1970. *Philos. Trans. R. Soc. London Ser. A* 268:111–30
127. Nielsen, P., Epstein, A. J., Sandman, D. J. 1974 *Solid State Commun.* 15:53–58
128. Lin, S. F., Spicer, W. E., Schechtman, B. H. 1975. *Phys. Rev. B* 12:4184–99

RHEOLOGY AND KINETIC THEORY OF POLYMERIC LIQUIDS

✤ 2644

R. Byron Bird

Chemical Engineering Department and Rheology Research Center, University of Wisconsin, Madison, Wisconsin 53706

INTRODUCTION

Previous reviews of kinetic theory of macromolecules in the *Annual Review of Physical Chemistry* have been concerned primarily with (*a*) equilibrium properties, and departures from equilibrium that include only *linear* viscoelastic phenomena; (*b*) very simple flows, such as simple *shear flows* of the form $v_x = k(t)y$, where $k(t)$ is often taken to be a small-amplitude sinusoidal function; (*c*) the *shear stress* τ_{yx} in shear flows (as opposed to the normal stresses τ_{xx}, τ_{yy}, τ_{zz}); and (*d*) molecular models of polymers that incorporate *linear* or *Hookean springs*. However, they have generally ignored the relation with modern continuum mechanics. These previous reviews have certainly reflected the chief concerns and accomplishments of polymer chemistry investigations during the past two decades, and it is generally acknowledged that many unresolved problems remain in these areas. In this review the four aspects listed above are deemphasized and attention is focused on the following: (*a*) *nonlinear* viscoelastic properties, such as shear-rate-dependent viscosity and normal stress coefficients in steady shear flow and their time-dependent counterparts, (*b*) *large-deformation flows* including viscometric flows, elongational flows, radial flows, and time-dependent flows, (*c*) the *complete stress tensor* expressions from kinetic theory and the measurement of components of the stress tensor other than τ_{yx} in steady shear flow, (*d*) molecular models involving *nonlinear springs* or *constraints* that may be capable of better describing the observed rheological phenomena in terms of internal structural changes, and (*e*) the use of modern *continuum mechanics* to provide a framework for presenting kinetic theory results, analyzing rheological experiments, and solving polymer fluid dynamics problems. Interest in these areas has been steadily growing and reflects the observation of new rheological phenomena, the advances in rheological instrumentation, and the increased activity in computer modeling of polymer processing operations.

Many of the topics discussed here are treated at an introductory level in two recent textbooks by my associates and myself (1, 2). The first deals with the experimental and fluid dynamics aspects, and the second with molecular theory and nonequilibrium statistical mechanics.

NEW BOOKS, REVIEW ARTICLES, AND JOURNALS

Many books have recently appeared that deal primarily with the continuum mechanical treatment of polymer flow. Yamamoto (3) has given a well-organized account of linear and nonlinear viscoelasticity. Astarita & Marrucci (4) have presented continuum mechanics and dimensional analysis, but with little comparison with data on polymeric fluids. Lodge (5) has treated continuum mechanics in terms of body tensors, and discussed applications to rheological measurements. Tomita (6) has emphasized the solution to flow problems with extensive data comparisons. Lodge's influential book on the rheology of elastic liquids has reappeared in Japanese translation (7). Huilgol (8) has included more material on solution to flow problems than do most authors of continuum mechanics texts. Han (9) has attempted to cover continuum mechanics, rheology, molecular theory, and polymer processing in one volume. Walters (10) has summarized the analyses of the myriad devices now available for the measurement of the rheological material functions. The collection of papers edited by Hutton, Pearson & Walters (11) provides a useful overview of the activities in theoretical rheology. Joseph (12) has given analyses of stability problems encountered in rheology. Two additional books are in production: one by Schowalter (13), which will contain some material on suspension rheology, and another by Tadmor & Gogos (14), which will deal with the strategy of process design in the polymer industry. These books have a bearing on the analysis of rheological measuring equipment, data interpretation, and use of laboratory measurements for describing industrial processes. Divergent points of view are expressed by the various authors, as would be expected in a field that is still evolving.

Several recent review articles are relevant to polymer rheologists. Pearson (15) and Petrie & Denn (16) have surveyed the literature on instabilities in polymer flow. Jeffrey & Acrivos (17), Brenner (18, 19), and Batchelor (20) have provided summaries on the rheology of suspensions (much of this material is recommended for workers in the area of kinetic theory of polymer solutions). Bird (21) has summarized and interrelated the most popular "rheological equations of state" (or "constitutive equations") and indicated how they have been used in solving flow problems. The following reviews of the subject at the molecular level have appeared. Williams (22) has summarized the modifications and embellishments on the Rouse theory for dilute polymer solutions. Graessley (23) has surveyed theories, correlations, and data on viscosity and other properties of concentrated solutions and melts. Curtiss, Bird & Hassager (24) have rederived and slightly extended many older polymer solution theories (e.g. those of Kramers, Kirkwood, Rouse, Giesekus, Fixman) by starting with a much more general phase-space kinetic theory.

Several new journals have appeared. The *Journal of Non-Newtonian Fluid Mechanics* began in 1976, and *Nihon Reorojii Gakkai-shi* (*Journal of the Society of Rheology, Japan*) was started in 1973.

The volume of preprints for the International Congress on Rheology in Sweden in 1976 is particularly helpful for getting an overview of current activities in the field (25).

A topic closely related to that of this review is flow birefringence. The reviews of Janeschitz-Kriegl (26), Wales (27), and Gortemaker (27a) are authoritative and contain much information on experimental data.

RHEOLOGICAL EQUATIONS OF STATE

In order to discuss rheology and kinetic theory of polymeric fluids, one begins with the "equations of change," which describe the conservation of mass, momentum, and energy. For an incompressible fluid these are

Continuity, $(\nabla \cdot \mathbf{v}) = 0$;   1.

Motion, $\rho \dfrac{D}{Dt} \mathbf{v} = -[\nabla \cdot \boldsymbol{\pi}] + \rho \mathbf{g}$;   2.

Energy, $\rho \dfrac{D}{Dt} \hat{U} = -(\nabla \cdot \mathbf{q}) - (\boldsymbol{\pi} : \nabla \mathbf{v})$;   3.

in which ρ, $\mathbf{v}$, and $\hat{U}$ are the fluid density, local velocity, and internal energy per unit mass, $\mathbf{g}$ the gravitational acceleration, $\mathbf{q}$ the heat flux vector, and $\boldsymbol{\pi}$ the stress tensor. Note that $(\mathbf{n} \cdot \mathbf{q})\,dS$ gives the heat flow from the negative side of dS to the positive side; similarly $[\mathbf{n} \cdot \boldsymbol{\pi}]\,dS$ is force exerted by the negative-side fluid on the positive-side fluid. We further write

$$\boldsymbol{\pi} = p\boldsymbol{\delta} + \boldsymbol{\tau}, \qquad\qquad 4.$$

thus splitting up the total stress into two parts: an isotropic part containing the pressure p multiplied by the unit tensor $\boldsymbol{\delta}$, and $\boldsymbol{\tau}$, which is that part of the total stress tensor that is zero at equilibrium. For the incompressible Newtonian fluid, one has

$$\boldsymbol{\tau} = -\mu[\nabla \mathbf{v} + (\nabla \mathbf{v})^{\dagger}] = -\mu \dot{\boldsymbol{\gamma}}, \qquad\qquad 5.$$

where the constant μ is the viscosity, † designates the transpose, and $\dot{\boldsymbol{\gamma}}$ is the rate-of-strain tensor. Since polymers and polymer solutions are not Newtonian fluids, the central problem in polymer fluid dynamics is the development of a relation that gives $\boldsymbol{\tau}$ in terms of $\dot{\boldsymbol{\gamma}}$ and/or other kinematic tensors. Such a relation is called a "rheological equation of state" or a "constitutive equation."

In continuum mechanics, considerable success has been achieved by restricting attention to materials called "(rheologically) simple fluids" (5, pp. 250, 254–56). For such fluids one *assumes* that the stress in a particular fluid particle depends only on the kinematic history of that particle alone and not on the kinematic history of neighboring particles. Fortunately this assumption seems to be appropriate for describing most polymeric fluids, and all molecular theories examined to date fall into this category.

It has been shown by Goddard (28) that the stress tensor for a (rheologically) simple fluid can be expanded in a corotational memory-integral expansion, which is shown at the top of Figure 1. This expansion contains an infinite number of

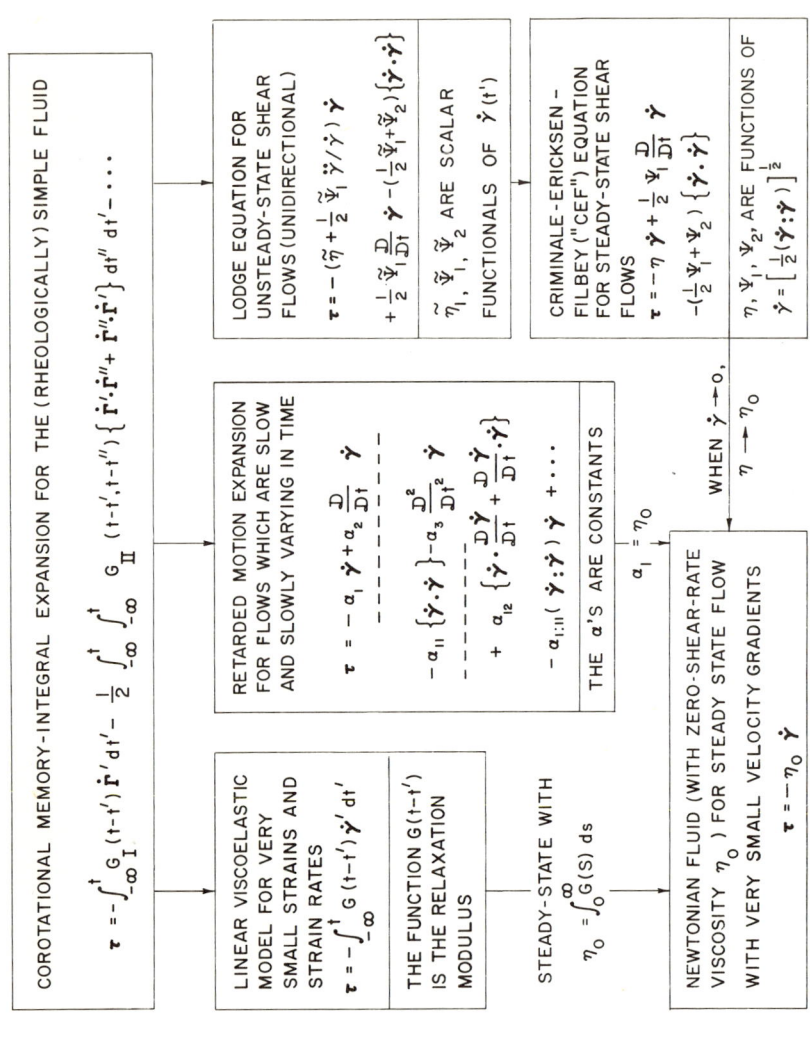

Figure 1 Chart showing the rheological equations of state that can be used as a framework for presenting polymer kinetic-theory results.

Notes:

(a) In the memory-integral expansion $\hat{\mathbf{\Gamma}}' \equiv \hat{\mathbf{\Gamma}}(t, t')$ is the corotating rate-of-strain tensor defined by:

$$\hat{\mathbf{\Gamma}}(t, t') = [\mathbf{\Omega}(t, t') \cdot \dot{\gamma}(t') \cdot \mathbf{\Omega}^\dagger(t, t')].$$

Here $\dot{\gamma} = \nabla \mathbf{v} + (\nabla \mathbf{v})^\dagger$ is the rate-of-strain tensor, and $\mathbf{\Omega}$ is the tensor relating the fixed unit vectors $\boldsymbol{\delta}_1$, $\boldsymbol{\delta}_2$, $\boldsymbol{\delta}_3$ to the mutually orthogonal unit vectors $\tilde{\boldsymbol{\delta}}_1$, $\tilde{\boldsymbol{\delta}}_2$, $\tilde{\boldsymbol{\delta}}_3$ that go along with a fluid particle rotating with the instantaneous local fluid angular velocity:

$$\tilde{\boldsymbol{\delta}}_i = [\boldsymbol{\delta}_i \cdot \mathbf{\Omega}(t, t')], \qquad \mathbf{\Omega}(t, t) = \boldsymbol{\delta},$$

where $\boldsymbol{\delta}$ is the unit tensor. For vanishingly small strains $\hat{\mathbf{\Gamma}}(t, t')$ becomes $\dot{\gamma}(t')$.

(b) $G(t - t')$ in the linear viscoelastic model is identical to $G_I(t - t')$ in the memory-integral expansion.

(c) The α's are related to the G's by

$$\alpha_n = \frac{1}{(n-1)!} \int_0^\infty G_I(s) s^{n-1} \, ds,$$

$$\alpha_{1:1} = \int_0^\infty \int_0^\infty G_{II}(s, s') \, ds' \, ds, \text{ etc.}$$

If only the dashed-underlined terms are included, the equation is called the "second-order fluid"; the "third-order fluid" is obtained by truncating the series after the $\alpha_{1:11}$ term.

(d) The corotational (or Jaumann) derivative is defined by

$$\frac{\mathcal{D}}{\mathcal{D}t} \dot{\gamma} = \frac{\partial}{\partial t} \dot{\gamma} + \{\mathbf{v} \cdot \nabla \dot{\gamma}\} + \tfrac{1}{2} \{\boldsymbol{\omega} \cdot \dot{\gamma} - \dot{\gamma} \cdot \boldsymbol{\omega}\},$$

where $\boldsymbol{\omega} = \nabla \mathbf{v} - (\nabla \mathbf{v})^\dagger$ is the vorticity tensor.

terms, the nth term involving n-fold integrals. In these integrals it is understood that the integration is performed by following the fluid particle at which the stress is to be evaluated. The kernel functions G_I, G_II, G_III, etc, have to be specified for each polymeric fluid. The kinematic tensor $\dot{\Gamma}$ (defined in Figure 1 footnote) is the corotating rate-of-deformation tensor. Goddard patterned his derivation after earlier work of Green & Rivlin (29) and Coleman & Noll (30, 31), who used a different kind of kinematic tensor in their codeformational memory-integral expansion. There is some evidence from kinetic theory that the Goddard corotational memory-integral expansion is to be preferred (32, 33), since it has somewhat better convergence properties for some flows (in particular, steady shear flow). It may be that some other choice of kinematic tensor will lead to a still better memory-integral expansion; the development of Johnson & Segalman suggests this possibility (34).

For many fluid dynamics calculations and kinetic theory developments, the use of the complete memory-integral expansion is either undesirable or impractical. Continuum mechanicians have expended considerable effort to obtain from the complete expansion "specialized" (or "reduced") rheological equations of state that describe accurately the simple fluid behavior in restricted classes of flows; four such specialized equations are shown in Figure 1.

The *general linear viscoelastic model* is obtained by restricting the flow to very small strains (so that $\dot{\Gamma}$ is replaced by $\dot{\gamma}$) and very small rates of strain (so that the double-, triple-, etc, integral terms can be omitted). The objectives in linear viscoelasticity then, are, to measure experimentally and calculate from molecular theory the relaxation modulus $G(t - t')$. This may be done by studying, for example, the small-amplitude oscillatory-motion experiment, in which $v_x = \dot{\gamma} y$, where $\dot{\gamma} = \dot{\gamma}^0 \cos \omega t$; one then measures, or calculates, $\eta'(\omega)$ and $\eta''(\omega)$, which describe the in-phase or out-of-phase contribution to the shear stress required to maintain the motion:

$$\tau_{yx} = -\eta'(\omega)\dot{\gamma}^0 \cos \omega t - \eta''(\omega)\dot{\gamma}^0 \sin \omega t. \qquad 6.$$

There are standard procedures (35) for obtaining $G(t - t')$ from experimental measurements of η' and η'' as functions of frequency. Analyses of experiments used for measuring η' and η'' have been summarized by Walters (10), who includes oscillatory experiments and also the various steady-state eccentric flow systems, such as the "Maxwell orthogonal rheometer."

The *retarded motion expansion* is obtained from the memory-integral expansion by developing $\dot{\Gamma}(t', t)$ in a Taylor series in t' about $t' = t$. If the flow is slow and slowly varying in time, the retarded motion expansion can be useful, since higher-order terms can be discarded. If only the dashed-underlined terms in Figure 1 are retained, one refers to the fluid as the "second-order fluid." For such flows one wants to measure or calculate the lower-order constants, α_1, α_2, α_{11}, etc. Several experiments have been analyzed so that various combinations of these constants can be obtained: flow near a translating and/or rotating sphere (36–39); liquid surface shape near a rotating rod in the liquid (40); free surface shape in flow down a tilted trough (41); radial flow between parallel disks (42, 43).

The *Lodge equation* (5, p. 166, Equation 14) results from restricting the flow to be an unsteady-state unidirectional shear flow. For such flows one has to specify three scalar functionals $\tilde{\eta}$, $\tilde{\Psi}_1$, and $\tilde{\Psi}_2$. To date no attempt has been made to evaluate these functionals from experimental data, although they have been worked out theoretically for rigid dumbbell solutions (44).

The *Criminale-Ericksen-Filbey* (*CEF*) *equation* (45) can be obtained from the Lodge equation by restricting the flows to steady-state (such flows are often called "viscometric"). For such flows we want to measure or compute the three "viscometric functions" $\eta(\dot{\gamma})$, $\Psi_1(\dot{\gamma})$, and $\Psi_2(\dot{\gamma})$. The past two decades have witnessed improved instrumentation and data analysis, as described by Walters (10); such diverse systems as cone-and-plate viscometers, jet thrust, and axial and tangential annular flow have been used. For the steady shear flow $v_x = \dot{\gamma}y$ with $\dot{\gamma}$ a constant, the three material functions η, Ψ_1, and Ψ_2 are related to the stresses thus:

$$\tau_{yx} = -\eta(\dot{\gamma})\dot{\gamma}, \qquad\qquad 7.$$

$$\tau_{xx} - \tau_{yy} = -\Psi_1(\dot{\gamma})\dot{\gamma}^2, \qquad\qquad 8.$$

$$\tau_{yy} - \tau_{zz} = -\Psi_2(\dot{\gamma})\dot{\gamma}^2. \qquad\qquad 9.$$

Both η and Ψ_1 are positive, whereas the best recent measurements seem to indicate that Ψ_2 is negative and that its magnitude (for concentrated solutions and melts) is probably one-tenth to two-fifths that of Ψ_1 (10, p. 89; 46). In many engineering flow and heat-transfer calculations it has been common practice to omit the Ψ_1 and Ψ_2 terms in the CEF equation and to write $\tau = -\eta(\dot{\gamma})\dot{\gamma}$, which is called the *generalized Newtonian fluid*. Many empiricisms have been proposed for $\eta(\gamma)$; the most generally useful is the "power law": $\eta = m\dot{\gamma}^{n-1}$, where m and n are constants.

In addition to the reduced rheological equations of state shown in Figure 1, many empirical equations have been proposed; tabulations of these can be found in several books (1, pp. 444–45; 5, pp. 149–50; 3, pp. 264–78). Particularly useful for exploratory fluid-dynamic studies is the *eight-constant Oldroyd model* (47, 48):

$$\tau + \lambda_1 \frac{\mathscr{D}}{\mathscr{D}t}\tau + \frac{1}{2}\mu_0(\text{tr }\tau)\dot{\gamma} - \frac{1}{2}\mu_1\{\tau\cdot\dot{\gamma} + \dot{\gamma}\cdot\tau\} + \frac{1}{2}v_1(\tau:\dot{\gamma})\delta$$

$$= -\eta_0\left[\dot{\gamma} + \lambda_2\frac{\mathscr{D}}{\mathscr{D}t}\dot{\gamma} - \mu_2\{\dot{\gamma}\cdot\dot{\gamma}\} + \frac{1}{2}v_2(\dot{\gamma}:\dot{\gamma})\delta\right]. \qquad 10.$$

Here η_0 is the zero-shear-rate viscosity, and λ_i, μ_i, and v_i are time constants. This was originally proposed as the most general relationship that (*a*) involves no corotational derivatives higher than the first, (*b*) is linear in the stress tensor, and (*c*) contains all allowable terms quadratic in velocity gradients and in products of the velocity gradients and the stresses. The eight-constant Oldroyd model may also be obtained from the memory-integral expansion by making very special choices for the G's (1, pp. 370–71).

Another empirical rheological equation of state is the *Hand equation* (49),

$$\tau = [\sigma_0 + \sigma_1(\dot{\gamma}:\mathbf{A}) + \sigma_2(\dot{\gamma}:\mathbf{A}^2)]\delta + [\sigma_3 + \sigma_4(\dot{\gamma}:\mathbf{A}) + \sigma_5(\dot{\gamma}:\mathbf{A}^2)]\mathbf{A}_s$$
$$+ \sigma_6\dot{\gamma} + \sigma_7\{\dot{\gamma}\cdot\mathbf{A}\}_s + \sigma_8\{\dot{\gamma}\cdot\mathbf{A}^2\}_s + [\sigma_9 + \sigma_{10}(\dot{\gamma}:\mathbf{A})$$
$$+ \sigma_{11}(\dot{\gamma}:\mathbf{A}^2)]\{\mathbf{A}\cdot\mathbf{A}\}_s, \qquad\qquad 11.$$

in which the "structure tensor" **A** is given by an ancillary equation:

$$\frac{\mathscr{D}}{\mathscr{D}t}\mathbf{A} = [\theta_0 + \theta_1(\dot{\gamma}:\mathbf{A}) + \theta_2(\dot{\gamma}:\mathbf{A}^2)]\delta + [\theta_3 + \theta_4(\dot{\gamma}:\mathbf{A}) + \theta_5(\dot{\gamma}:\mathbf{A}^2)]\mathbf{A}_s$$
$$+ \theta_6\dot{\gamma} + \theta_7\{\dot{\gamma}\cdot\mathbf{A}\}_s + \theta_8\{\dot{\gamma}\cdot\mathbf{A}^2\}_s + [\theta_9 + \theta_{10}(\dot{\gamma}:\mathbf{A})$$
$$+ \theta_{11}(\dot{\gamma}:\mathbf{A}^2)]\{\mathbf{A}\cdot\mathbf{A}\}_s. \qquad 12.$$

Here the subscript s means "symmetrized deviator" and the σ's and θ's are functions of tr $\mathbf{A}$, tr $\mathbf{A}^2$, tr $\mathbf{A}^3$. This equation has attracted attention in connection with the rheology of suspensions and emulsions.

For concentrated solutions and undiluted polymers an interesting modification of Lodge's rubberlike-liquid model has recently been suggested by Acierno et al (50–52), who have done an extensive job of comparing theory with experiment; in their theory they relate the relaxation times to the frictional properties of the entangled chains. A very different approach has been taken by Johnson & Segalman (34); their theory has the advantage of being extended in a systematic way in a memory-integral expansion. In both theories all rheological properties are described in terms of the usual relaxation modulus of linear viscoelasticity plus one additional material constant.

RHEOLOGICAL PHENOMENA

One might wonder whether it is really necessary to use expressions for the stress tensor that are as complicated as those just described. The accumulated evidence from the last decade of experiments and its analysis suggests very strongly that some kind of memory-integral expansion is indeed required for the description and interrelation of the known rheological phenomena. It is evident that polymeric liquids are not Newtonian fluids, and that their flow behavior cannot be fully described by the generalized Newtonian fluid model (a standard empiricism long used by engineers) or by the linear viscoelastic model (widely used by polymer chemists).

These three "classical" models (Newtonian, generalized Newtonian, linear viscoelastic) cannot describe the following polymeric fluid phenomena: (a) a fluid climbing up a rotating rod (53); (b) the secondary flow near a rotating disk, where the fluid near the disk moves radially inward (54); (c) the large-strain recoil of the fluid in a tube when the axial pressure drop is suddenly removed (55); (d) the extrudate swell at a tube exit (53); (e) the continued operation of a siphon when the end of the siphon tube is raised several centimeters above the surface of the liquid (56); (f) the toroidal vortices formed near a centrally located hole at the bottom of a draining cylindrical tank (57); (g) the bulging of the liquid surface in the flow of a liquid down a tilted trough (58); (h) the direction of the secondary flows in the neighborhood of a cylinder oscillating in a plane containing the cylinder axis, in which the fluid is moving away from the plane near the cylinder (59, 60); (i) the increase in the separation between two spheres dropped one after the other into a liquid (61); (j) the abnormally low readings (the "hole-pressure error") of pressure transducers that use holes (62); (k) the production of streamers from a glass rod used to clean the inner surface of a beaker (63); (*l*) secondary flows associated with axial flow in square pipes (64). This list is incomplete, but it serves to emphasize the diversity of flow phenomena observed mainly in concentrated solutions and melts; some of these have been discovered only within the past ten years, and undoubtedly the next decade will produce additional phenomena to amuse the experimenter and baffle the theoretician. The references cited here are ones that have photographs or data presentations and are not necessarily the original publications; additional photographs are assembled in Chapter 3 of (1).

There are also challenging phenomena that are observed when minute quantities of polymers are added to Newtonian solvents: (*a*) "drag reduction," the dramatic lowering of the pressure drop required to pump liquids in turbulent flow (65, 66); (*b*) "vortex inhibition," the suppression of the air core in a cylindrical tank when the fluid is rotating and draining through a centrally located hole (67, 68); (*c*) "heat-transfer reduction" in turbulent flow in tubes (69).

MEASUREMENT OF MATERIAL FUNCTIONS

The recent surge of interest in molecular theories of *non*linear viscoelasticity has largely been due to the development of new rheometric devices and the attendant proliferation of experimental data on "material functions," those rheological functions measured in carefully designed geometric arrangements in which the flow is (hopefully) well understood. Walters' book (10) provides a key to the recent literature.

The search for better methods of measuring the viscometric functions is continuing. High-shear-rate techniques have been summarized and compared by Walters (70). Higashitani & Iwamoto (71) have proposed a new method for determining Ψ_2 from the flow of a film flowing down the inside wall of a vertical cylinder. The hole-pressure error has been used by Lodge, Baird, and Higashitani to develop a "stressmeter" for use as an on-line device for monitoring molecular weight distribution (62, 72, 73). Osmers & Lobo have built an "omegameter" to determine Ψ_1 at high shear rates in a pressure-driven angular annular flow (74). New measurements of Ψ_2 for polymer melts from axial annular flow have been reported by Ehrmann (75). "Shear thickening" of dilute solutions of flexible macromolecules has been observed by Layec-Raphalen & Wolff (76, 77).

Time-dependent shear flows have received increasing attention: stress growth, stress relaxation after step-function shear strain, stress relaxation after cessation of steady shear flow, constrained recoil after removal of stress in a steady shear flow (78–82). Additional measurements on shear flow with small-amplitude oscillations superposed have been reported by Powell & Schwarz (83); they have analyzed their data in terms of the Pipkin-Owen theory. Also the high-frequency limiting behavior of $\eta'(\omega)$ has been further investigated (84, 85). The use of flow birefringence as a tool for studying time-dependent shear flows has been pioneered by Janeschitz-Kriegl's group (85a, 85b). They have verified that for melts, Lodge's rubberlike liquid is valid up to a total shear of unity.

Shear-free flows are of particular interest in kinetic theory developments as well as in processing operations such as fiber spinning and film blowing. Shaw (86) has used a readily available tensile machine for doing elongational stress-growth experiments. Meissner (78) has continued his work on elongational stress growth and recoil of polymer melts. Elongational flows at constant stress and at constant strain rate have been observed for melts by Laun & Münstedt (87), and elongational flows in orifices have been studied for dilute solutions by Balakrishnan & Gordon (88).

A droplet-elongation experiment has been suggested by Hsu & Flumerfelt (89). In addition, Denson and collaborators (90–92) have made contributions to planar and biaxial extension of polymer melts.

An important activity of the past few years has been chronicled in the IUPAC Working Party's report (93) on a comprehensive study of the rheology and processibility of three LDPE (low-density polyethylene) samples.

ANALYSIS OF FLOWS

The fluid-dynamic analysis of viscoelastic flows (or "non-Newtonian fluid dynamics," which uses both the equations of change and also the rheological equation of state) has been the focus of immense effort in the past decade; a recent summary in matrix form classifies the problems that have been solved, according to the geometry of the flow system and the rheological equation of state (21). This activity is of great importance in the development of rheometric devices, the analysis of polymer-processing operations, and the determination (with the help of kinetic theory) of molecular configurations in various flow systems. We mention here only some of the most recent developments.

Brindley, Davies & Walters (94) have analyzed squeezing flow between two parallel disks for a generalized Newtonian fluid with arbitrary viscosity. In two subsequent papers Binding, Davies & Walters (95) analyze the squeeze flow experiment with a superimposed rotation, thereby providing a new method for measuring viscoelastic functions; the authors refer to their new instrument as a "torsional-balance" rheometer. Radial flow between fixed circular disks has been studied experimentally and theoretically (using an Oldroyd five-constant model) by Lee & Williams (96) and by Winter (97); this work is closely related to (42) and (43) for the third-order fluid.

Joseph has analyzed the rod-climbing effect near a torsionally oscillating rod using two terms of a memory-integral expansion (98). Huilgol has determined the upper and lower bounds of the jet diameter in extrudate swell by using the retarded motion expansion to second order (99). The second-order fluid has also been used by Ballal & Rivlin (100) in the flow between eccentric, rotating cylinders.

Experimental determination of velocity profiles continues to be an important activity. Higashitani, Nishio & Hara (101), Whipple (102), and Schowalter & Allen (103) have studied the flow near a die exit. The laser Doppler effect has been used by Busby & MacSporran (104) to examine the flow into a reentrant tube, and by Goulden & MacSporran (105) to observe a jet falling under the influence of gravity.

Tung & Laurence (106) have discussed the appropriate coordinate system for helical flows, and pointed out errors in an earlier publication on screw extruders. Flow of polymers through porous media has been pursued by Park, Hawley & Blanks (107), and by Mena, Rangel & Barboza (108). The motion of a sphere in a cylinder has been studied theoretically and with flow visualization by Sigli & Coutanceau (109). The crossflow of a convected Maxwell model around a cylinder has been compared with experiments by Mizushina & Usui (110). Taylor vortices in tangential annular flow have been studied experimentally and theoretically (retarded motion expansion and Oldroyd model) by Chan Man Fong & Jones (111). Chang & Schowalter (60) have analyzed the acoustical streaming of a polymeric liquid near an oscillating cylinder by use of a corotational Jeffreys model (Oldroyd model with $\mu_0 = \mu_1 = \mu_2 = \nu_1 = \nu_2 = 0$), and report excellent agreement with experimental observations.

Tanner (112) has explored the use of finite-element methods for complex flow problems of industrial interest.

KINETIC THEORY OF DILUTE SOLUTIONS

In statistical mechanics studies on polymer rheology it is standard practice to represent the macromolecule by some kind of ultrasimplified mechanical model—usually made up of beads, rods, and springs. Flory (113) has severely criticized such models by saying: "Irrespective of the particular model chosen, forfeiture of a connection with the actual chemical structure is the inevitable price of adoption of a hypothetical model . . ." and ". . . no matter how faithfully such a model may represent experimental observations, interpretations carried out in its terms are cast in a framework of unreality." Of course, Flory is right. However, we must keep in mind that in the kinetic theory of gases, empirical models have also been used (e.g. the two-parameter Lennard-Jones potential); the model parameters are admittedly not fundamental quantities, but considerable progress has been made in understanding equilibrium and nonequilibrium properties of gases, as well as their interrelations. We can certainly hope that the artificial mechanical models in polymer kinetic theory can fulfill the same role. Another point is that, although very detailed modeling has been successfully undertaken for *equilibrium* properties of isolated molecules, we shall be restricted for some time to very simple models for solving the kinetic theory equations for *nonlinear viscoelastic* properties.

Until relatively recently, kinetic theory derivations and calculations have been performed almost exclusively for shear flows and were usually restricted to small-amplitude motions and small-velocity gradients (linear viscoelasticity). Because of improved knowledge of continuum mechanics, one now knows how to report kinetic theory results in terms of the kernel functions of memory-integral expansions, the constants in the retarded motion expansion, the functionals in the Lodge equation for unsteady shear flows, etc. Table 1 gives a summary of the molecular-model interpretation of continuum mechanics constants, functions, and functionals. We have included a few dilute suspension and emulsion results in order to point out

Table 1 Summary of kinetic theory results and their relation to continuum mechanical equations. (all models are for dilute solutions except Lodge's network model)

Molecular Model	Rheological Equation of State	Investigators	Ref.	Eq. No. in Ref. 2	Remarks
Hookean dumbbells	(a) Oldroyd equation	Giesekus Fris Lodge	114 115 116	10.4–4	A complete rheological equation of state is obtained; it can be written as the first term of the codeformational memory-integral expansion.
	(b) Codeformational memory-integral expansion	—	—	10.4–6	
	(c) Retarded-motion expansion	—	—	10D.2–2	
Finitely extendable nonlinear elastic ("FENE") dumbbells	(a) CEF equation	Warner	117	10.5–24	Tanner (119) has allowed the friction coefficient to increase with dumbbell extension; Christiansen & Bird (120) have given an extensive data comparison.
	(b) Retarded-motion expansion (third order)	Armstrong	118	10.5–18 to 23	
Linear-locked dumbbells	(a) CEF equation	Tanner & Stehrenberger	121	—	Also studied elongational flow.
	(b) An approximate differential rheological equation of state	Tanner	122	—	
Rigid dumbbells	(a) Retarded-motion expansion (third order)	Giesekus Prager	123, 124 125	11.5–29	Bird, Warner & Evans (127) summarized kinetic theory and solutions for a number of special flows. Abdel-Khalik & Bird (128) examined lopsided dumbbells. Stewart & Sørensen (129) made numerical calculations including hydrodynamic interaction. Dumbbells with elliptical beads have been studied by Giesekus (130) and the shish kebab model with equilibrium-averaged hydrodynamic interaction has been used by Kirkwood & Auer (131), Kirkwood & Plock (132), and Kotaka (133).
	(b) Retarded-motion expansion (fourth order)	Armstrong & Bird	126	11D.2–7	
	(c) Codeformational memory-integral expansion (third order)	Armstrong & Bird	126	—	
	(d) Corotational memory-integral expansion (third order)	Abdel-Khalik, Hassager & Bird	32	11.5–21 to 23 11.5–24 to 28	
	(e) Lodge's equation for unsteady shear flows	Bird, Armstrong, Curtiss & Hassager	44	—	
	(f) Oldroyd equation (six-constant)	—	—	11B.9–1 to 6	
Rigid plane polygon (N beads joined by N rigid rods)	Retarded motion expansion	Curtiss, Bird & Hassager Paul & Mazo	24 134	p. 628	Paul & Mazo have given derivation including complete hydrodynamic interaction; their free draining limit does not agree with the Curtiss-Bird-Hassager results.
Rouse freely jointed bead-spring chain (N beads, $N-1$ Hookean springs)	(a) Codeformational memory-integral expansion	Lodge & Wu	135	12.4–18 or 19	A complete rheological equation of state is obtained in the form of the first term of the codeformational memory-integral expansion.
	(b) Corotational memory-integral expansion	Bird, Hassager & Abdel-Khalik	33	—	

Table 1 (*Continued*)

Model		Type of result	Author(s)	Ref.	Eq.	Comments
Kramers freely jointed bead-rod chain (N beads, $N-1$ rigid rods)	(a)	Reiner-Rivlin equation (for steady, irrotational flows) any N	Rivlin	136	—	The results in (a) and (b) are approximate, since the random-walk distribution was used. The $N=3$ chain is a special case of the "once-broken rod" studied by Yu & Stockmayer (139) and Wilemski (139a).
	(b)	Retarded motion expansion (second order), any N	Hassager	137	13.5–11	
	(c)	Retarded motion expansion (second order), $N=3$	Hassager	137	13.5–9 & 10	Hassager's results have been verified by computer simulation by Gottlieb (169, 170).
	(d)	Retarded motion expansion (second order), $N=4, 5$	Curtiss & Bird	138	—	
Arbitrary bead-rod models with no internal potentials	(a)	Linear viscoelastic model	Hassager	140	13.6–12	
	(b)	Linear viscoelastic model (rigid bead-rod model with two principal moments of inertia equal)	Hassager	140	13.6–43	
	(c)	Retarded motion expansion	Curtiss, Bird & Hassager	24	13.14–16 & 17	
Arbitrary bead-rod-spring models		Retarded motion expansion (second order)	Curtiss, Bird & Hassager	24	13.4–16 and 17	This model has been considered further by Curtiss & Hansen (141) who incorporated hydrodynamic interaction.
Ellipsoids		Hand equation	Leal & Hinch	142	—	Leal & Hinch's paper deals with "near spheres." For perturbation calculation of η, Ψ_1, Ψ_2 see Giesekus (130).
Deformable droplets	(a)	Retarded motion expansion	Schowalter, Chaffey & Brenner	143	—	Earlier Oldroyd (145) had obtained the parameters in the Jeffreys linear viscoelastic model for an emulsion.
	(b)	Hand equation	Frankel & Acrivos	144	—	
Deformable elastic spheres		Hand equation	Goddard & Miller	146	—	Earlier Fröhlich & Sack (147) obtained the parameters in the Jeffreys linear viscoelastic model for a suspension of deformable elastic spheres.
Lodge's network theory of a rubberlike liquid		Codeformational memory-integral expansion	Lodge	5, 7, 148	15.3–2	Lodge's network theory leads to the first term of the codeformational memory-integral expansion. Lodge's development is an extension of earlier work by Green & Tobolsky (149).

that investigators in that field feel that the Hand equation (49) is attractive. The review by Barthès-Biesel & Acrivos is particularly helpful (150).

The kinetic theories have been moderately useful in several ways by suggesting possible forms for rheological equations of state:

(a) The molecular origins of the Lodge rubberlike liquid, coupled with its partial successes in describing rheological phenomena, have encouraged many workers to attempt empirical modification of the model; some of these have been moderately successful [e.g. the Carreau model (151) and the model of Acierno et al (50–52)].
(b) The rigid dumbbell results, which have been carried through third order for time-dependent flows, have been used as a probe to explore the relative merits of the corotational and codeformational memory-integral expansions (32); it was concluded that the former was better than the latter for steady shear flow.
(c) Hinch & Leal (152, 153) have suggested that one can extract from the kinetic theory experience some generalizations as to the form rheological equations of state should assume.

In addition, kinetic theory results may be combined with hydrodynamic results to get information about molecular orientation and stretching in complex flows (43, 154). They may also be used to suggest relations among measurable rheological properties (155). Ultimately one would hope to be able to predict differences in flow behavior resulting from differences in structure (chain length, chain stiffness, side groups, etc), but thus far only a start has been made in that direction (24).

Of the models mentioned in Table 1, the simplest are the dumbbell models, useful mainly for pedagogical treatments or exploratory calculations. The FENE ("finitely extendable nonlinear elastic") dumbbell model has a force vs extension expression given by $\mathbf{F}^{(c)} = H\mathbf{R}/[1 - (R/R_0)^2]$, where $\mathbf{F}^{(c)}$ is the connector force, $\mathbf{R}$ is the vector describing the dumbbell configuration, H is a spring constant, and R_0 is the maximum allowable extension of the dumbbell. This model has two time constants: $\lambda_H = \zeta/4H$ and $\lambda_R = \zeta R_0^2/12kT$. The first of these can be related to the zero-shear-rate intrinsic viscosity $[\eta]_0$, molecular weight M, Avogadro's number $\tilde{N}$, Boltzmann's constant k, and absolute temperature T by

$$\lambda_H = \frac{b+5}{b} \frac{[\eta]_0 \eta_s M}{\tilde{N}kT}. \qquad 13.$$

The parameter b (defined by $b = 3\lambda_R/\lambda_H = HR_0^2/kT$) is to be determined from experimental viscosity data; for a wide range of solvents and solutes it has been found (120) that b is in the range from ~ 30 to ~ 150, and increases with molecular weight for a given solute-solvent pair. The FENE dumbbell model seems to be capable of fitting some non-Newtonian viscosity data; it is not appropriate for describing small-amplitude oscillatory behavior (i.e. η' and η'' as functions of frequency), because it does not contain a complete spectrum of relaxation times. If we designate by $\dot{\gamma}_c$ the shear rate at which $\eta(\dot{\gamma})$ has dropped to 0.99 η_0, and by

ω_c the frequency at which $\eta'(\omega)$ has fallen to 0.99 η_0, then (2, 23)

$$\dot{\gamma}_c/\omega_c = [(b+5)(2b+1)/2(4b+17)]^{1/2}. \qquad 14.$$

This ratio, for the values of b cited above, varies from about 3 to almost 6; there are no data on $\dot{\gamma}_c$ and ω_c for any single dilute solution, and hence no experimental check on this relation is yet possible. For concentrated solutions the curve of $\eta(\dot{\gamma})/\eta_0$ always lies to the right of that for $\eta'(\omega)/\eta_0$, and hence values of $\dot{\gamma}_c/\omega_c$ in the range 3–6 are not implausible. Because this very simple model can describe, at least qualitatively, a variety of observed rheological phenomena, it may someday be used for molecular dynamics simulations of complex flows. It is certainly to be preferred over the Hookean dumbbell model, since the latter does not describe the shear-rate dependence of viscosity, and it allows for infinite stretching in elongational flows.

Some investigators [see (22) for list of references] have suggested that inclusion of a dashpot in parallel with the spring could roughly represent the rearrangement process that a polymer undergoes as the ends of a polymer coil are moved farther apart or closer together. However, in the derivations for such "internal viscosity" models, the usual formalism of statistical mechanics has been applied without justification, even though the systems contain nonconservative forces.

With regard to chainlike-models, the papers of Rouse (156), Bueche (157, 158), and Zimm (159) have been very influential. These freely jointed models, with Hookean springs, yield a viscosity that does not drop off with increasing shear rate (even with equilibrium-averaged hydrodynamic interaction) (see 2, Chapter 12). One of the Bueche papers (157), however, gives η as a monotone-decreasing function of $\dot{\gamma}$. This much-cited equation appears to be incorrect, since Bueche apparently formulated his kinetic theory equations in a corotating coordinate frame (which should not be necessary) and then transferred the results to a fixed frame. If this is indeed what he did, then $\eta(\dot{\gamma})$ would have exactly the same functional form as $\eta'(\omega)$ [see 1, p. 340, Equations (A) and (B)]. However, the functions are not quite the same:

$$\left(\begin{array}{c}\text{Summand of Equation 20}\\ \text{of Reference 158}\end{array}\right) = \frac{3\chi^2 n^4 + \chi^4}{n^2(n^4 + \chi^2)^2} \qquad (\chi = \tau_1 \omega), \qquad 15.$$

$$\left(\begin{array}{c}\text{Summand of Equation 5}\\ \text{of Reference 157}\end{array}\right) = \frac{2\chi^2 n^4 + \chi^4}{n^2(n^4 + \chi^2)^2} \qquad (\chi = \tau_1 \dot{\gamma}). \qquad 16.$$

I have not been able to locate the source of this discrepancy, but I agree with Friš (115) and with Lodge & Wu (135) that η is a constant for the bead-spring model with Hookean springs, either without hydrodynamic interaction or with equilibrium-averaged hydrodynamic interaction. The inclusion of the complete hydrodynamic interaction does give a shear-rate dependence of η (160–163), but apparently the "shear-thinning" effect is underestimated (164).

Freely jointed bead-rod chainlike models have been studied much less than bead-spring chains, since the latter are easier to cope with mathematically. A recent series of publications by Fixman and collaborators (165–168) have introduced new ideas and techniques, particularly in regard to linear viscoelasticity. In the first of these

Fixman succeeded in obtaining a general recursion formula for getting the determinant of the metric tensor needed in Kramers' theory for flexible bead-rod chains. The second-order fluid constants for short chains of three (137), four (138), and five (138) beads have been obtained. Of interest here has been the deviation from the constants obtained from the "random-walk distribution." There has been some controversy as to whether, in bead-rod chain problems, one must formulate the kinetic theory by introducing the constraints of constant bond length at the outset, or whether one may begin with bead-spring chains and then "freeze out" the vibrational degrees of freedom to generate bead-rod chain results at the end. The two methods give different results for the second- and higher-order coefficients. A molecular dynamics calculation by Gottlieb (169, 170) strongly suggests that, if constraints are appropriate, they should be introduced at the outset by using generalized coordinates and momenta. Gō & Scheraga (171), however, have arrived at a different conclusion by considering the quantum-mechanical aspects of the problem.

The earlier kinetic theories of Kuhn, Hermans, Kirkwood, Rouse, Bueche, Zimm, etc, were all formulated in the configuration space of a single polymer molecule; Brownian motion was introduced to describe the randomizing thermal forces as the polymer writhes in the sea of solvent molecules. The stress tensor is obtained by elementary pictorial arguments in which the contributions of the connector tensions and bead-momentum transport are adduced. The recent theory of Curtiss, Bird & Hassager (CBH) (24) is formulated in the phase space of the polymer solution and formally accounts for the motions of both polymer and solvent molecules. The "Brownian-motion terms" arise naturally in integrations over momenta, and the stress tensor expression is obtained by specializing the general equation of change and identifying certain integrals as the various contributors to the stress tensor. By introducing well-defined assumptions, the earlier configuration-space theories can be recovered. The original CBH theory did not include hydrodynamic interaction, but complete hydrodynamic interaction has been included by Curtiss & Hansen (141). As an alternative to hydrodynamic interaction calculations using the Oseen tensor, an empirical procedure using hydrodynamic shielding coefficients for the beads has been evolved (172), and the method appears to be attractive for rigid biopolymers (173).

One topic in dilute-solution theory that does not seem to have been studied earlier is the slip effect near a wall. Brunn (174) has obtained an expression for the slip coefficient, but was not able to determine whether the effect is sufficiently large to be measurable.

KINETIC THEORY OF CONCENTRATED SOLUTIONS AND UNDILUTED POLYMERS

As would be expected, the theory of dilute solutions is far better developed than that for concentrated solutions and undiluted polymers. Several approaches are under way in this area, and we mention a few of them.

An expression for the stress tensor in concentrated solutions was first given by Fixman (175); Williams (176, 177) made some calculations of linear and nonlinear properties based on Fixman's equation, but an algebraic error invalidates part of this work (22, footnote † on p. 18). The Fixman equation has been rederived from the phase-space kinetic theory (2, p. 697; 24, p. 91); application of this equation is not easy, since the pair distribution function has to be calculated or estimated.

Because of this, various investigators have sought to apply the dilute-solution models in some approximate way to account for the fact that the polymer molecule finds itself in a cage formed by other polymer molecules and entangled with its neighbors. Hansen, Williams & Shen (178) have developed a modification of the Rouse theory called "the elastically coupled entanglement model," in which entanglement sites are represented as interacting with each other through very weak springs; excellent description of linear viscoelastic phenomena is reported. Wang & Zimm (179) derived a modified hydrodynamic-interaction tensor that accounts approximately for the polymer-polymer interactions in a freely jointed bead-spring chain; their work is also only applicable to the linear viscoelastic regime, but it does predict the transition from "Zimm-like" to "Rouse-like" behavior with increasing concentration.

Still another approach, and by far the most productive thus far, is the network theory (5, 7, 148, 149). This theory results from adapting the theory of rubber elasticity to viscoelastic liquids. Lodge's rubberlike liquid gives the same rheological equation of state as the Rouse and Zimm dilute-solution models, with, of course, different interpretations of the constants. Hence, the viscometric functions are predicted to be constant. Numerous empirical modifications of the Lodge theory have been suggested to correct this deficiency (2, pp. 714–21; 50–52).

CONCLUSIONS

New rheological flow phenomena are still being discovered, and new measurement apparatus and techniques are resulting in an abundance of data on material functions. This activity has stimulated the need for continuum mechanical and molecular explanations. The current feeling among rheologists seems to be that a simple rheological equation of state, capable of describing a wide range of flow phenomena, will not be found. Some kind of memory-integral expansion will be useful for interrelating material functions and for presenting kinetic theory results, but for most fluid dynamics problems, simpler empiricisms will be used for the next decade.

In the kinetic theory of dilute macromolecular solutions, we can get the second-order fluid constants for rather complex models, and we can get the first few kernel functions of a memory-integral expansion only for the simplest of models. Molecular theories for concentrated solutions and undiluted polymers leave much to be desired. So far, all the kinetic theory results for reasonable models seem to fit into the framework of continuum mechanics. Molecular theories have provided limited guidance in suggesting likely forms for the rheological equation of state.

Areas needing increased attention in the future are

(*a*) the rheological behavior of two-phase fluids, one or both phases of which may be viscoelastic; this includes suspensions, emulsions, and foams;

(b) development of devices for measuring steady and unsteady elongational flows; data from such devices are needed for developing and testing rheological equations of state and molecular theories;
(c) analytical and computer solutions of nonviscometric flow problems, particularly for those flows that occur in laboratory measurement devices and in polymer processing operations;
(d) further development of kinetic theories for concentrated solutions and melts, by various approaches (i.e. from the entanglement network approach, the caged molecule approach, the pair-distribution function approach, the molecular dynamics simulation approach);
(e) further calculations for dilute polymer solutions, including use of shielding coefficients for flexible macromolecules, exploration of high shear-rate properties, solution of the differential equation for the configurational distribution function for complex models, analysis of molecular stretching and orientation in complex flows;
(f) development of the dilute-solution theory for thermal conduction, diffusion, and thermal diffusion; also, extension of the classical theories of flow birefringence and other optical phenomena.

NOTES ADDED IN PROOF   Since the manuscript for this review was prepared, two additional books have appeared:

1. Middleman, S. 1977. *Fundamentals of Polymer Processing.* New York: McGraw-Hill. 525 pp. This textbook emphasizes the use of analytical methods for understanding polymer unit operations.
2. Vinogradov, G. V., Malkin, A. Ya. 1977. *Reologiya Polimerov.* Moscow: Khimiya. 438 pp. This book deals primarily with data on material functions; it provides a useful key to the Russian literature.

In addition, the source of the error mentioned in Reference 134 has been identified by C. Y. Mou, working with Professor Mazo, and an erratum to the Paul-Mazo paper will appear in a forthcoming issue of the *Journal of Chemical Physics.*

ACKNOWLEDGMENTS

The author wishes to express his thanks for financial support to the National Science Foundation (grant No. ENG 75-01092) and to the Vilas Trust Fund of the University of Wisconsin. He is also indebted to his colleagues in the Rheology Research Center (Professors J. D. Ferry, M. W. Johnson Jr., A. S. Lodge, and J. L. Schrag) and to his coauthors of *Dynamics of Polymeric Liquids* (Professor C. F. Curtiss, Professor R. C. Armstrong, and Dr. O. Hassager) for assistance and advice during the period of preparation of this review. Special thanks are due to Professor M. C. Williams of the University of California at Berkeley for suggesting many improvements in the manuscript.

Literature Cited

1. Bird, R. B., Armstrong, R. C., Hassager, O. 1977. *Dynamics of Polymeric Liquids. Vol. 1, Fluid Mechanics.* New York: Wiley. 576 pp.
2. Bird, R. B., Hassager, O., Armstrong, R. C., Curtiss, C. F. 1977. *Dynamics of Polymeric Liquids. Vol. 2, Kinetic Theory.* New York: Wiley. 286 pp.
3. Yamamoto, M. 1972. *Buttai no Henkeigaku.* Tokyo: Seibundō Shinkōsha. 373 pp.
4. Astarita, G., Marrucci, G. 1974. *Principles of Non-Newtonian Fluid Mechanics.* New York: McGraw-Hill. 289 pp.
5. Lodge, A. S. 1975. *Body Tensor Fields and Continuum Mechanics.* New York: Academic. 319 pp.
6. Tomita, Y. 1975. *Reorojii: Hisenkei Ryūtai no Rikigaku.* Tokyo: Corona. 444 pp.
7. Lodge, A. S. 1975. *Dansei Ekitai.* (Japanese translation with corrections of *Elastic Liquids* by M. Kurata and K., Osaki.) Kyoto: Yoshioka Shoten. 434 pp.
8. Huilgol, R. R. 1975. *Continuum Mechanics of Viscoelastic Liquids.* New York: Halsted, Wiley. 367 pp.
9. Han, C. D. 1976. *Rheology in Polymer Processing.* New York: Academic. 366 pp.
10. Walters, K. 1976. *Rheometry.* New York: Halsted, Wiley. 278 pp.
11. Hutton, J. F., Pearson, J. R. A., Walters, K., eds. 1974. *Theoretical Rheology.* New York: Halsted, Wiley. 377 pp.
12. Joseph, D. D. 1976. *Stability of Fluid Motions.* Berlin: Springer. 556 pp.
13. Schowalter, W. R. 1977. *Mechanics of Non-Newtonian Fluids.* Oxford: Pergamon.
14. Tadmor, Z., Gogos, C. 1978. *Polymer Processing Principles.* New York: Wiley. In press
15. Pearson, J. R. A. 1976. *Ann. Rev. Fluid Mech.* 8:163–81
16. Petrie, C. J. S., Denn, M. M. 1976. *AIChE J.* 22:209–36
17. Jeffrey, D. J., Acrivos, A. 1976. *AIChE J.* 22:417–32
18. Brenner, H. 1974. *Int. J. Multiphase Flow* 1:195–341
19. Brenner, H. 1972. In *Prog. Heat Mass Transfer* 5:89–129
20. Batchelor, G. K. 1974. *Ann. Rev. Fluid Mech.* 6:227–55
21. Bird, R. B. 1976. *Ann. Rev. Fluid Mech.* 8:13–34
22. Williams, M. C. 1975. *AIChE J.* 21:1–25
23. Graessley, W. W. 1974. *Fortschr. Hochpolym. Forsch.* 16:1–179
24. Curtiss, C. F., Bird, R. B., Hassager, O. 1976. *Adv. Chem. Phys.* 35:31–117. [The plus sign just before the summation sign in the second line of Eq. (17.7) should be a minus sign.]
25. Klason, C., Kubát, J. 1976. *Proc. 7th Int. Congr. Rheol.* Chalmers Univ. Technol., Göteborg, Sweden. 673 pp.
26. Janeschitz-Kriegl, H. 1969. *Fortschr. Hochpolym. Forsch.* 6:170–318
27. Wales, J. L. S. 1976. *The Application of Flow Birefringence to Rheological Studies of Polymer Melts.* Delft, Holland: Delft Univ. Press. 111 pp.
27a. Gortemaker, F. H. 1976. *A Flow Birefringence Study of Stresses in Sheared Polymer Melts.* Meppel, Holland: Kreps REPRO. 142 pp.
28. Goddard, J. D. 1967. *Trans. Soc. Rheol.* 11:381–99. [In Eq. (2.15) change $W^\dagger(\tau)$ to $W^\dagger(t)$.]
29. Green, A. E., Rivlin, R. S. 1957. *Arch. Ration. Mech. Anal.* 1:1–21
30. Coleman, B. D., Noll, W. 1961. *Rev. Mod. Phys.* 33:239–49
31. Coleman, B. D., Noll, W. 1961. *Ann. NY Acad. Sci.* 89:672–74
32. Abdel-Khalik, S. I., Hassager, O., Bird, R. B. 1974. *J. Chem. Phys.* 61:4312–16. [In Eq. 14 replace 1/70 by 1/35.]
33. Bird, R. B., Hassager, O., Abdel-Khalik, S. I. 1974. *AIChE J.* 20:1041–66
34. Johnson, M. W. Jr., Segalman, D. J. 1977. *J. Non-Newtonian Fluid Mech.* 2:255–70
35. Ferry, J. D. 1970. *Viscoelastic Properties of Polymers.* New York: Wiley. 671 pp. 2nd ed.
36. Giesekus, H. 1963. *Rheol. Acta* 3:59–71
37. Walters, K., Waters, N. D. 1963. *Br. J. Appl. Phys.* 14:667–71
38. Walters, K., Waters, N. D. 1964. *Br. J. Appl. Phys.* 15:898–991
39. Walters, K., Waters, N. D. 1964. *Rheol. Acta* 3:312–15
40. Joseph, D. D., Beavers, G. S. 1975. *J. Fluid Mech.* 69:475–511
41. Sturges, L., Joseph, D. D. 1975. *Arch. Ration. Mech. Anal.* 59:359–87
42. Schwarz, W. H., Bruce, C. 1969. *Chem. Eng. Sci.* 24:399–413
43. Co, A., Bird, R. B. 1977. *Appl. Sci. Res.* 32: In press

44. Bird, R. B., Armstrong, R. C., Curtiss, C. F., Hassager, O. 1967. See Ref. 25, pp. 55–57
45. Criminale, W. O. Jr., Ericksen, J. L., Filbey, G. L. Jr. 1958. *Arch. Ration. Mech. Anal.* 1:410–17
46. Pipkin, A. C., Tanner, R. I. 1972. *Mech. Today* 1:262–321
47. Oldroyd, J. G. 1958. *Proc. R. Soc. London Ser. A* 245:278–97
48. Oldroyd, J. G. 1961. *Rheol. Acta* 1:337–44
49. Hand, G. L. 1962. *J. Fluid Mech.* 13:33–46
50. Acierno, D., LaMantia, F. P., Marrucci, G., Titomanlio, G. 1976. *J. Non-Newtonian Fluid Mech.* 1:125–46
51. Acierno, D., LaMantia, F. P., Marrucci, G., Rizzo, G., Titomanlio, G. 1976. *J. Non-Newtonian Fluid Mech.* 1:147–57
52. Marrucci, G., Acierno, D. 1976. See Ref. 25, pp. 538–39
53. Truesdell, C. 1974. *Ann. Rev. Fluid Mech.* 6:111–46
54. Hill, C. T. 1972. *Trans. Soc. Rheol.* 16:213–45
55. Fredrickson, A. G. 1964. *Principles and Applications of Rheology.* Englewood Cliffs, N.J.: Prentice-Hall. 326 pp.
56. James, D. F. 1966. *Nature* 212:754–56
57. Giesekus, H. 1968. *Rheol. Acta* 7:127–38
58. Tanner, R. I. 1970. *Trans. Soc. Rheol.* 14:483–507
59. Chang, C. F., Schowalter, W. R. 1974. *Nature.* 252:686–88. [Erratum: ibid, 253:572.]
60. Chang, C. F., Schowalter, W. R. 1976. See Ref. 25, pp. 668–69
61. Riddle, M. J., Narvaez, C., Bird, R. B. 1977. *J. Non-Newtonian Fluid Mech.* 2:75–87
62. Higashitani, K., Lodge, A. S. 1975. *Trans. Soc. Rheol.* 19:307–35
63. Barnard, B. J. S., Pritchard, W. G. 1974. *Nature* 250:215–16
64. Townsend, P., Walters, K., Waterhouse, W. M. 1976. *J. Non-Newtonian Fluid Mech.* 1:107–23
65. Lumley, J. L. 1973. *Macromol. Rev.* 7:263–90
66. Hoyt, J. W. 1972. *Trans. ASME, J. Basic Eng.* 94D:258–85
67. Balakrishnan, C., Gordon, R. J. 1975. *J. Appl. Polym. Sci.* 19:909–13
68. Chiou, C. S., Gordon, R. J. 1976. See Ref. 25, pp. 283–85
69. Dimant, Y., Poreh, M. 1976. *Adv. Heat Transfer* 12:77–113
70. Walters, K. 1976. See Ref. 25, pp. 147–50
71. Higashitani, K., Iwamoto, K. 1976. See Ref. 25, pp. 222–23
72. Baird, D. G. 1975. *Trans. Soc. Rheol.* 19:147–51
73. Baird, D. G. 1976. *J. Appl. Polym. Sci.* 20:3155–73
74. Osmers, H. R., Lobo, P. F. 1976. *Trans. Soc. Rheol.* 20:239–52
75. Ehrmann, G. 1976. *Rheol. Acta* 15:8–14
76. Layec-Raphalen, M. N., Wolff, C. 1976. *J. Non-Newtonian Fluid Mech.* 1:159–73
77. Wolff, C., Layec-Raphalen, M. N. 1976. See Ref. 25, pp. 504–5
78. Meissner, J. 1976. See Ref. 25, pp. 85–91
79. Osaki, K. 1976. See Ref. 25, pp. 104–9
80. Janeschitz-Kriegl, H., Gortemaker, F. H., Hansen, M. G., deCindio, B. 1976. See Ref. 25, pp. 168–69
81. Chang, K. I., Yoo, S. S., Hartnett, J. P. 1975. *Trans. Soc. Rheol.* 19:155–71
82. Meissner, J. 1975. *Rheol. Acta* 14:201–18
83. Powell, R. L., Schwarz, W. H. 1975. *Trans. Soc. Rheol.* 19:617–43
84. Koh, I.-Y., Birnboim, M. H. 1976. See Ref. 25, pp. 302–3
85. Noordermeer, J. W. M., Kramer, O., Nestler, F. H. M., Schrag, J. L., Ferry, J. D. 1975. *Macromolecules* 8:539–44
85a. Gortemaker, F. H., Hansen, M. G., deCindio, B., Laun, H. M., Janeschitz-Kriegl, H. 1976. *Rheol. Acta* 15:256–67
85b. Gortemaker, F. H., Janeschitz-Kriegl, H., te Nijenhuis, K. 1976. *Rheol. Acta* 15:487–500
86. Shaw, M. T. 1976. See Ref. 25, pp. 304–5
87. Laun, H. M., Münstedt, M. 1976. See Ref. 25, pp. 268–69
88. Balakrishnan, C., Gordon, R. J. 1975. *AIChE J.* 21:1225–27
89. Hsu, J. C., Flumerfelt, R. W. 1975. *Trans. Soc. Rheol.* 19:523–40
90. Denson, C. D., Gallo, R. J. 1971. *Polym. Eng. Sci.* 11:174–76
91. Denson, C. D., Grady, D. L. 1974. *J. Appl. Polym. Sci.* 18:1611–17
92. Denson, C. D. 1976. See Ref. 25, pp. 386–87
93. Meissner, J. 1975. *Pure Appl. Chem.* 42:551–612
94. Brindley, G., Davies, J. M., Walters, K. 1976. *J. Non-Newtonian Fluid Mech.* 1:19–37. [One line after Eq. (45), change "less" to "greater"; the solution

95. Binding, D. M., Davies, J. M., Walters, 1976. *J. Non-Newtonian Fluid Mech.* 1:259–75, 277–86
96. Lee, C. H., Williams, M. C. 1976. *J. Non-Newtonian Fluid Mech.* 1:323–41; 343–55
97. Winter, H. H. 1975. *Polym. Eng. Sci.* 15:460–69
98. Joseph, D. D. 1976. See Ref. 25, pp. 242–43
99. Huilgol, R. R. 1976. See Ref. 25, pp. 342–43
100. Ballal, B. Y., Rivlin, R. S. 1976. *Trans. Soc. Rheol.* 20:65–101
101. Higashitani, K., Nishio, K., Hara, I. 1976. See Ref. 25, pp. 478–79
102. Whipple, B. A. 1974. *Velocity Profiles in Die Swell.* PhD dissertation. Washington Univ., St. Louis
103. Schowalter, W. R., Allen, R. C. Jr. 1975. *Trans. Soc. Rheol.* 19: 129–37
104. Busby, E. T., MacSporran, W. C. 1976. *J. Non-Newtonian Fluid Mech.* 1:71–82
105. Goulden, D. D., MacSporran, W. C. 1976. *J. Non-Newtonian Fluid Mech.* 1:183–98
106. Tung, T. T., Laurence, R. L. 1975. *Polym. Eng. Sci.* 15:401–5
107. Park. H. D., Hawley, M. C., Blanks, R. F. 1975. *Polym. Eng. Sci.* 15:761–73
108. Mena, B., Rangel, C., Barboza, M. 1976. See Ref. 25, pp. 351–53
109. Sigli, D., Coutanceau, M. 1976. See Ref. 25, pp. 372–73
110. Mizushina, T., Usui, H. 1975. *J. Chem. Eng. Jpn.* 8: 393–98
111. Chan Man Fong, C. F., Jones, W. M. 1976. See Ref. 25. pp. 476–77
112. Tanner, R. I. 1976. See Ref. 25, pp. 140–45
113. Flory, P. J. 1969. *Statistical Mechanics of Chain Molecules.* New York: Wiley. 432 pp.
114. Giesekus, H. 1966. *Rheol. Acta* 5:29–36. [In Eq. (9), replace 2 by 4.]
115. Friš, P. 1966. *Czech. J. Phys. B* 16: 563–68
116. Lodge, A. S. 1970. In *Proc. 5th Int. Congr. Rheol.*, ed. S. Onogi, pp. 169–78. Tokyo: Univ. Tokyo Press
117. Warner, H. R. Jr. 1972. *Ind. Eng. Chem. Fundam.* 11:379–87. [Factors of $(b + 7)$ in the denominators of the second and third lines of Eq. (40) on p. 385 should be removed; in Fig. 7, the curves should approach zero at high frequency.]
118. Armstrong, R. C. 1974. *J. Chem. Phys.* 60:724–28
119. Tanner, R. I. 1975. *Trans. Soc. Rheol.* 19:557–82
120. Christiansen, R. L., Bird, R. B. 1977. *J. Non-Newtonian Fluid Mech.* 2: In press
121. Tanner, R. I., Stehrenberger, J. 1971. *J. Chem. Phys.* 55:1958–64. [Erratum: ibid 61:2486; see also Ref. 118, Eq. (23), for discussion of error in Eq. (30).]
122. Tanner, R. I. 1975. *Trans. Soc. Rheol.* 19:37–65
123. Giesekus, H. 1956. *Kolloid Z.* 147–49: 29–45. [Erratum: 1961 *Rheol. Acta* 1:404.]
124. Giesekus, H. 1963. *Rheol. Acta* 3:59–71
125. Prager, S. 1957. *Trans. Soc. Rheol.* 1: 53–62. {D_R is defined as $2kT/b^2\zeta$; the rigid-dumbbell suspension *does* exhibit the Weissenberg effect [statement after Eq. (16) is in error].}
126. Armstrong, R. C., Bird, R. B. 1973. *J. Chem. Phys.* 58:2715–23. [In Eq. (22) replace 27/280 by 27/980.]
127. Bird, R. B., Warner, H. R. Jr., Evans, D. C. 1971. *Fortschr. Hochpolym. Forsch.* 8:1–90. [In Eq. (10.6) replace 15/14 by 4/7; in Eq. (12.9) replace 1/120 by 1/280; on p. 86 replace O. H. Christiansen by O. Hassager.]
128. Abdel-Khalik, S. I., Bird, R. B. 1975. *Appl. Sci. Res.* 30:268–70
129. Stewart, W. E., Sørensen, J. P. 1972. *Trans. Soc. Rheol.* 16:1–13
130. Giesekus, H. 1962. *Rheol. Acta* 2:50–62
131. Kirkwood, J. G., Auer, P. L. 1951. *J. Chem. Phys.* 19:281–83
132. Kirkwood, J. G., Plock, R. J. 1956. *J. Chem. Phys.* 24:665–69. [Serious errors have been corrected by Paul, E. 1969. *J. Chem. Phys.* 51:1271–72.]
133. Kotaka, T. 1959. *J. Chem. Phys.* 30: 1566–45. [In Eq. (6), replace 128/175 by 228/175; in the expression for σ_{xy} in Eq. (3), replace $\sin^2 \theta \sin 2\psi$ by $(3/2)\sin^2 \theta \sin 2\psi$; in Eq. (9), replace 666/1925 by 663/1925.]
134. Paul, E., Mazo, R. M. 1969. *J. Chem. Phys.* 51:1102–7. [These authors get $\tau_{xx} - \tau_{yy} = 0$ and $\tau_{yy} - \tau_{zz} \neq 0$, whereas Curtiss, Bird Å Hassager (24) obtain, in the free-draining limit, $\tau_{xx} - \tau_{yy} \neq 0$ and $\tau_{yy} - \tau_{zz} = 0$ for all bead-rod-spring models at zero shear rate. We feel that the Paul-Mazo work is in error, but the exact source of the difficulty is not yet known.]
135. Lodge, A. S., Wu, Y. 1971. *Rheol. Acta* 10:539–53
136. Rivlin, R. S. 1949. *Trans. Faraday Soc.* 45:739–48. [The random-walk

assumption is introduced in Eq. (3.6); this is in conflict with the rigorous statistical mechanical result in Eq. (14) of Kramers, H. A. 1944. *Physica* 11:1–19.]
137. Hassager, O., 1974. *J. Chem. Phys.* 60:2111–24
138. Curtiss, C. F., Bird, R. B. 1977. *J. Non-Newtonian Fluid Mech.* 2:392–96
139. Yu, H., Stockmayer, W. H. 1967. *J. Chem. Phys.* 47:1369–73. [Errors have been discussed by Hassager (137), Sect. 8.]
139a. Wilemski, G. 1977. *Macromolecules* 10:28–34
140. Hassager, O. 1974. *J. Chem. Phys.* 60:4001–8
141. Curtiss, C. F., Hansen, R. L. 1977. *J. Chem. Phys.* To be published
142. Leal, L. G., Hinch, E. J., 1972. *J. Fluid Mech.* 55:745–65. [In Eq. (6) replace Ω by Φ; on p. 756, in the equation for η', replace 28 by 26.]
143. Schowalter, W. R., Chaffey, C. E., Brenner, H. 1968. *J. Colloid Sci.* 26:152–60
144. Frankel, N. A., Acrivos, A. 1970. *J. Fluid Mech.* 44:65–78
145. Oldroyd, J. G. 1953. *Proc. R. Soc. London Ser. A* 218:122–32
146. Goddard, J. D., Miller, C. 1967. *J. Fluid Mech.* 28:657–73
147. Frölich, H., Sack, R. 1946. *Proc. R. Soc. London Ser. A* 185:415–30
148. Lodge, A. S. 1968. *Rheol. Acta* 7:379–92
149. Green, M. S., Tobolsky, A. V. 1946. *J. Chem. Phys.* 14:80–92
150. Barthès-Biesel, D., Acrivos, A. 1973. *Int. J. Multiphase Flow* 1:1–24
151. Carreau, P. J. 1972. *Trans. Soc. Rheol.* 16:99–127
152. Hinch, E. J., Leal, L. G. 1975. *J. Fluid Mech.* 71:481–95
153. Hinch, E. J., Leal, L. G. 1976. *J. Fluid Mech.* 76:187–208
154. Tadmor, Z. 1974. *J. Appl. Polym. Sci.* 18:1753–72
155. Hassager, O., Bird, R. B. 1972. *J. Chem. Phys.* 56:2498–2501
156. Rouse, P. E. Jr. 1953. *J. Chem. Phys.* 21:1272–80
157. Bueche, R. 1954. *J. Chem. Phys.* 22:1570–76. [Eq. (5) is apparently in error; see fn. 3, p. 592 of Ref. 2.]
158. Bueche, R. 1954. *J. Chem. Phys.* 22:603–609. [Differences between Bueche's results and those of Rouse for η' and η'' have been resolved by DeWames, R. E., Hall, W. R., Shen, M. C. 1961. *J. Chem. Phys.* 46:2781–94.]
159. Zimm, B. H. 1956. *J. Chem. Phys.* 24:269–78. [Errors in this paper (in the expressions for birefringence) have been pointed out by Williams, M. C. 1965. *J. Chem. Phys.* 42:2988–89.]
160. Fixman, M. 1966. *J. Chem. Phys.* 45:785–92, 793–803
161. Pyun, C. W., Fixman, M. 1966. *J. Chem. Phys.* 42:3838–44
162. Pyun, C. W., Fixman, M. 1966. *J. Chem. Phys.* 44:2107–15
163. Stidham, H. D., Fixman, M. 1968. *J. Chem. Phys.* 48:3092–95
164. Yamakawa, H. 1971. *Modern Theory of Polymer Solutions*, p. 317. New York: Harper & Row
165. Fixman, M. 1974. *Proc. Natl. Acad. Sci. USA* 71:3050–53
166. Fixman, M., Kovac, J. 1974. *J. Chem. Phys.* 61:4939–49, 4950–54
167. Kovac, J., Fixman, M. 1975. *J. Chem. Phys.* 63:935–41
168. Fixman, M., Evans, G. T. 1976. *J. Chem. Phys.* 64:3474–81
169. Gottlieb, M., Bird, R. B. 1976. *J. Chem. Phys.* 65:2467–68
170. Gottlieb, M. 1977. *Comput. Chem.* 1: In press
171. Gō, N., Scheraga, H. A. 1976. *Macromolecules* 9:535–42
172. Abdel-Khalik, S. I., Bird, R. B. 1975. *Biopolymers* 14:1915–32
173. Nakajima, H., Wada, Y. 1977. *Biopolymers* 16:875–93
174. Brunn, P. 1976. *Rheol. Acta* 15:23–29
175. Fixman, M. 1965. *J. Chem. Phys.* 42:3831–37. [In Eq. (A16) replace $+\frac{1}{2} + c^2$ by $+\frac{1}{2}c^2$.]
176. Williams, M. C. 1966. *AIChE J.* 12:1064–70. [For errata, see Ref. 22, p. 18.]
177. Williams, M. C. 1967. *AIChE J.* 13:534–39, 955–61. [For errata see Ref. 22, p. 18.]
178. Hansen, D. R., Williams, M. C., Shen, M. 1976. *Macromolecules* 9:345–54
179. Wang, F. W., Zimm, B. H. 1974. *J. Polym. Sci.* 12:1619–37, 1639–48

PICOSECOND SPECTROSCOPY ✶ 2645

Kenneth B. Eisenthal

Department of Chemistry, Columbia University, New York, New York 10027

INTRODUCTION

In the last decade lasers have been developed that can generate light pulses of the order of 10^{-12} sec duration and are capable of peak powers in excess of 10^{13} W, thus enabling the investigation of the interactions of light with matter in a time and power domain not previously possible. Picosecond lasers provide a new and powerful tool for the study of physical and chemical phenomena at the most fundamental level. In chemistry today one of the key issues is the evolution of a system from some initial distribution of energy and geometric structures through various intermediate structures and energy states to some "final" state. The identification and lifetimes of the energy states, geometric structures, and chemical species through which the system rapidly passes is necessary to understand the mechanisms of chemical and physical changes. The competition between various pathways for energy dissipation and structural change determines whether light is emitted or the manifold nonradiative physical and chemical processes dominate the system.

What are some of these key energy-degrading pathways, and how can the unique features of picosecond spectroscopy illuminate some of these basic questions in chemistry? As an example, if we consider a molecule in an excited electronic state, it can dissipate its energy by emitting light, by internal conversion to lower-energy electronic and vibrational states, by flipping a spin and going from a singlet to a triplet state and decaying down the triplet manifold, by intermolecular energy transfer to some other molecule, or by chemical reactions such as unimolecular decomposition or reaction with a neighboring molecule. These processes often occur in the subnanosecond time region. Internal conversion and vibrational relaxation, singlet to triplet intersystem crossing, electron and proton transfer, excited-state structural changes, and collision-induced predissociation often occur in this ultrafast time domain. Picosecond lasers can perturb the system (e.g. by excitation to a given energy state or alignment of molecules by the intense laser electric field). Subsequently, the return to the initial equilibrium state or the development of new structures or chemical species can be measured with picosecond resolution. The monitoring steps are usually spectroscopic and utilize picosecond probe pulses or an ultrafast streak camera to follow absorption, emission, Raman scattering, and changes in the polarization of the probe light as the system evolves.

INTERMOLECULAR ENERGY TRANSFER

The efficiency of intermolecular energy transfer depends on the nature of the intermolecular coupling, the molecular separations and relative orientations, the extent of energy overlap, and the lifetime of the excited donor molecule (1–6). For example a dipole-dipole coupling mechanism can lead to transfer over distances of the order of 100 Å as opposed to an exchange coupling that is short-range and can extend to distances of the order of 10 Å. The donor and acceptor states involved in the transfer process can be excited- and ground-electronic states of various multiplicities containing variable amounts of vibrational excitations. This "diffusion" of energy from some initial distribution of excited and unexcited molecules through a variety of alternative distributions plays a vital role in a variety of physical, chemical, and biological processes (7–11), such as the quenching of the donor fluorescence and the appearing of new emission bands, the initiation of chemical reactions by transfer to a reactive molecule or site, and has been used as an aid in mapping distances between chromophores in biological molecules.

Prior to the development of picosecond lasers, studies of the dynamics of energy transfer were restricted to times of the order of 10^{-8} sec or longer and were generally limited to light-emitting molecules. Thus, the nature of the laws governing the transfer could not be examined in the subnanosecond time scale.

Singlet-Singlet Transfer

One approach to the study of singlet-singlet energy transfer uses the linear polarization of the picosecond exciting light to induce an orientational anisotropy in the distribution of ground and excited molecules (9, 10). Molecules whose transition moments have a large component along the polarization direction of the light are preferentially excited. Thus, the formerly isotropic system becomes anisotropic, i.e. more excited molecules are oriented parallel to the field direction, and hence more unexcited molecules are oriented with their transition moments perpendicular to the field direction. In a rigid environment the induced anisotropy can only relax by unimolecular decay of the excited molecules and by energy transfer. In a system where the donor and acceptor molecules are of the same species, the randomization results from energy transfer between molecules of different orientations. On the other hand, in a mixed system, the anisotropy in donor orientations can also relax by transfer between donor and acceptor molecules regardless of their mutual orientation. The decay of the anisotropy can be monitored by measuring the polarization-dependent change in either the ground- or excited-state donor absorption with time. The ground-state donor absorption is greater for probe light polarized perpendicular to rather than parallel to the polarization of the exciting light because of the relative depletion of ground-state molecules in the "parallel configuration." By probing at successively later times after excitation, the decay of the dichroism due to the continued effects of energy transfer and the unimolecular excited-state decay can be determined. Measurement of the latter quantity in the absence of transfer is then combined with the measured decay of the dichroism to obtain the energy-transfer dynamics.

To determine the nature of the transfer it is advantageous to study a two-component system, i.e. distinct donor and acceptor molecules, versus a one-component system for which the acceptor molecules are the ground-state donor molecules. For the two-component system the acceptor molecules are randomly distributed both in orientation and distance with respect to the anisotropically distributed excited donor; hence the donor decay function can readily be calculated. In the one-component system, although the distribution in distances is random, the distribution in orientations is perturbed by the excitation pulse (11–15). Furthermore, in the two-component system only one transfer step from the donor to acceptor need be considered, assuming that vibrational relaxation in the excited acceptor molecule is rapid compared with excitation transfer. In the one-component system several steps may be necessary before randomization has occurred and would therefore have to be included in any theoretical treatment.

Picosecond experiments with rhodamine 6G as the donor and malachite green as the acceptor showed that the dipole-dipole interaction was a good description of energy transfer up to the earliest time measured, i.e. 20 psec. The critical transfer distance, R_0, was found to be about 53 Å, which is in good agreement with the value calculated from the spectra (48 Å). In experiments of this type one must consider the possibility of stimulated emission caused by the probe pulse and amplified spontaneous-emission processes. For the rhodamine 6G molecule excited-state vibrational relaxation is known to be complete at the time of the earliest measurement (20 psec) (16–18). The power densities used in the excitation pulse must be adjusted to avoid the complications of amplified spontaneous emission. In addition, if the full pulse train is used, rather than a single pulse, the possibility of buildup effects must be considered.

Triplet-Triplet Transfer

Since triplet-triplet energy transfer occurs via an exchange interaction, it is a short-range process and generally occurs between neighboring molecules, perhaps as far as 10–15 Å apart. Most studies in solution are therefore limited by the diffusion of the donor and acceptor molecules to some neighboring or near-neighboring molecular configurations. If the concentration of acceptor molecules is sufficiently high, then the rate of transfer is too rapid to study by conventional flash photolysis methods. Furthermore, earlier studies covered a time domain that was long compared to vibrational relaxation times. Thus the initial and final states of the donor-acceptor pair monitored were thermally equilibrated. Information on the vibronic energy in the excited donor prior to transfer and the distribution of vibronic energy in the ground-state donor and excited acceptor after transfer is thus unobtainable. Since the energy transfer process is dependent on the energy distribution of the interacting states and also on the rate of vibrational relaxation, both intramolecularly as well as intermolecularly, this short-time information is of key importance.

With the application of picosecond laser methods we are now gaining some insight into the role of vibrational energy distribution in triplet excitation transfer processes (19). Some of the donor-acceptor pairs that have been studied include benzophenone as the donor and *cis*-piperylene, *trans*-piperylene, and 1-methylnapthalene as the acceptors. In these investigations the solvent was composed of the acceptor molecules and hence translational molecular diffusion was not the rate-determining step. For some systems rotational motion may be important in satisfying orientational requirements of the donor-acceptor pair for energy transfer (20, 21). The method used for determining the triplet-triplet dynamics was to excite the donor to an excited singlet state with a picosecond pulse and monitor the donor triplet population with a picosecond probe pulse at a wavelength corresponding to a donor triplet-triplet absorption. In this way we obtain both the buildup of the donor triplet due to intersystem crossing from the donor singlet and the decay of the donor triplet due to energy transfer to the acceptor molecule.

A small but definite difference is found in the rate of energy transfer from benzophenone to *cis*- vs *trans*-piperylene with the *cis* form the faster one. The difference in the observed transfer is attributed to the different vibrational overlap functions (Franck-Condon factors) for the two forms. In this interpretation the final vibrational energy distribution would not be the same for the two pairs. In addition the rapid rate of energy transfer observed (10 psec) is in the time domain of vibration relaxation, and thus the transfer might occur from a thermally nonequilibrated triplet benzophenone. This latter point is used in part to explain the slower rate of energy transfer from benzophenone to 1-methylnapthalene (20 psec). However, it is also clear that the differences in the electronic contributions to the transfer can be different for the piperylenes and 1-methylnapthalene, and thus could contribute to the observed differences in transfer rates.

In another type of experiment the dynamics of triplet energy transfer in crystals have been investigated. The fluorescence risetime and decay in a tetracene crystal excited by a picosecond pulse (at 530 nm) has been used to estimate the incoherent hopping rate of triplet excitons (22). Tetracene is a rather interesting system in that one channel for the decay of the singlet exciton involves fission into two triplet excitons. The rate of this process is dependent, in part, on the velocity at which the newly formed triplet excitons separate and thus avoid a geminate recombination process leading to the initial singlet exciton. This triplet energy hopping or transfer from neighbor to neighbor is estimated to be at a rate greater than 10^{13} sec^{-1}. By postulating an average lifetime for the triplet exciton of 100 μsec and an average jump distance of 7 Å, a diffusion length of the order of microns is obtained.

ORIENTATIONAL RELAXATION OF MOLECULES IN LIQUIDS

Although a number of methods have been used to study orientational relaxation processes, the exciting feature of the recent applications of picosecond laser techniques is that rotational motions are measured directly in the time domain. In one method that uses the optical Kerr effect the birefringence induced by an intense picosecond pulse is monitored with an attenuated picosecond pulse as a function of time. This method is most appropriate for studies of pure liquids and highly concentrated mixtures. A second approach monitors the decay of the dichroism induced by picosecond laser excitation of solute molecules present at low concentrations in the solution being investigated. In a related method the decay of the anisotropy induced by picosecond excitation is followed by measurement of the time dependence of the fluorescence polarization. The dichroism and fluorescence polarization decay as the solute molecules rotate and thus transform the orientational distribution from an anisotropic to an isotropic one. The fourth method involves the creation of a transient grating by the intersection of two coherent light pulses in the liquid. Time-resolved measurements can be obtained by monitoring the decay of the induced diffraction pattern with a probe light pulse.

Optical Kerr Effect

The optical Kerr effect results from an intensity-dependent change in the refractive index induced by an intense light pulse propagating through a material. The optical Kerr effect was first observed in a number of liquids using a Q-switched nanosecond ruby laser (23, 24); this followed a theoretical prediction and treatment of the phenomenon (25–29). The use of picosecond lasers to induce a significant nonlinear refractive index in liquids has made it possible to measure the rotational motion of molecules in liquids (30), investigate the short-term nature of optical self-trapping in liquids (31), and study a variety of ultrafast processes with a laser-generated ultrafast light gate (30).

Due to the polarization of the optical field, the change in the refractive index parallel to the beam polarization (assuming a linearly polarized beam) can be different from the change in the perpendicular directions. Therefore an isotropic medium, such as a liquid, can be made anisotropic, and thus birefringent. In a liquid composed of anisotropic molecules, such as carbon disulfide (CS_2), we can view a major contribution to the refractive index difference as resulting from the partial alignment of the CS_2 molecules along the optical field direction. The light pulse induces a dipole in the CS_2 molecules. The induced dipole interacts with the light field and leads to a torque on the dipole that tends to orient the long axis of the CS_2 molecule along the polarization direction of the light field. This induced anisotropy in the orientation of the molecules produces a phase difference as the probe light propagates through the liquid, between the components polarized parallel vs perpendicular to the initial intense picosecond light pulse. Thus by measuring the decay of the birefringence with the probe pulse as a function of time subsequent to the intense alignment pulse, the orientational relaxation can be determined. In addition to rotational motions there can also be an electronic contribution to the nonlinear refractive index that will have a relaxation time far shorter than the picosecond light pulse (30, 32–36). The birefringence due to the electronic part of the nonlinear refractive index will thus only last as long as the initial pulse duration, i.e. it will "instantaneously" follow the excitation pulse in time.

In studies (30, 37, 38) of nitrobenzene the measured decay time of 32 psec, although somewhat lower than the orientational relaxation times obtained from depolarized Rayleigh scattering, 36 psec (39), 39 psec (40), and 50 psec (41), does indicate that the primary contribution to the decay of the optically induced birefringence is orientational in nature. It has been suggested that the lower values found for the time constant of the birefringent decay as compared with the orientational relaxation values obtained from light scattering in a variety of liquids (nitrobenzene, m-nitrotoluene, and various liquid mixtures) arise from the coupling of the orientational motions with shear modes (38). This explanation, although interesting, has not yet been clearly established and requires further study. It has, however, been demonstrated that the rotational motion of nitrobenzene cannot be described by very small angular jumps characteristic of Debye-type rotational diffusion nor by very large jumps. This result has been established by comparing the optically induced birefringence or light-scattering results with dielectric relaxation measurements (37).

Picosecond measurements of liquid CS_2 using a high-repetition-rate rhodamine 6G dye laser yielded a value for the relaxation of the optical Kerr effect of 2.1 psec (42). This is in good agreement with light-scattering results of 1.96 psec (43) and indicates that for CS_2 the decay of the induced birefringence is due to orientational relaxation. Studies of a variety of other liquids including bromobenzene, toluene, iodomethane, and mixtures of CS_2 and CCl_4 indicated fair agreement between the orientational relaxation and the macroscopic viscosity as given by the Debye relation, i.e. the relaxation time scales linearly with the viscosity (37). In mixtures of CS_2 and CCl_4 the scaling appears to be linear, although not for the full concentration range studied. Great care should be exercised in studying mixtures since the contributions of both components to the induced birefringence must be included. Correspondingly, if one seeks to obtain relaxation times from light scattering, the background scattering due to the solvent must also be considered.

Picosecond laser methods have recently been used to induce relaxation processes in liquid crystals (44). This work follows earlier work on liquid crystals that used Q-switched lasers (45). Direct measurement of the orientational relaxation has yielded information about the phase transition and temperature dependence of the viscosity coefficient for the liquid crystal *p*-methoxybenzilidene *p-m*-butylaniline (44).

Induced Dichroism Method

Unlike the induced birefringence method previously described, the induced dichroism method (46) is suitable for studies of solute molecules at low concentrations and can be carried out in any solvent into which the solute can be introduced. The optical Kerr method requires high concentrations, at least several percent of the species of interest, whereas the induced dichroism method is limited to low concentrations of the solute molecules. In the latter method the rotational motions of the individual solute molecules are obtained, whereas in the induced birefringence method the concentration of solute molecules is so high that solute-solute interactions as well as the solute-solvent interactions must be considered. As a further point, the dichroism method is applicable to molecules that have absorptions at frequencies corresponding to the frequency of the picosecond light pulse. The birefringence method can be applied to any liquid that has an optical Kerr constant sufficiently large to yield an induced birefringence. These two methods can be viewed as complementary in the systems amenable to study.

The principal idea of the dichroism method (46) is to induce an anisotropy in the orientational distribution of the excited- and ground-state populations with an intense picosecond pulse in the same way as in the singlet-singlet energy-transfer experiments. The return of the system to an isotropic state is monitored with an attenuated picosecond pulse as a function of time. The difference in absorption of probe light polarized parallel and perpendicular to the polarization of the excitation pulse will decay in time as a function of excited-state lifetime, solute concentration, and solution viscosity. For example, in a fluid environment the anisotropic orientational distribution can transform to an isotropic one via the rotational motion of the molecules. At low concentrations in a highly viscous medium the rotations are frozen out, and the anisotropy decays as the excited molecules return to the ground state. At high concentrations the anisotropic distribution can also decay in time by intermolecular energy transfer between molecules of differing orientations in the way discussed in the energy-transfer section. Clearly, for the study of rotational motions, low concentrations are necessary to avoid the complicating effects of energy transfer.

This method has been used to study the orientational relaxation of rhodamine 6G (46, 47). In particular the effects of solute-solvent hydrogen bonding interactions on the rotational motions of rhodamine 6G in a variety of solvents have been investigated (47). A linear relation between the relaxation times and the solution viscosities was obtained for the series chloroform, formamide, and the alcohols from methanol through octanol. The observed linear scaling, in agreement with the Debye-Stokes-Einstein hydrodynamic model, is surprising since the volumes of the hydrogen-bonded complexes should vary considerably through the series from methanol to octanol. Furthermore one might expect that since the strengths of hydrogen-bonding interactions of rhodamine 6G with chloroform and methanol are different, the relaxation times would not be equal even though the viscosities are the same. Similar arguments can be applied to the results obtained in the liquids formamide and 1-pentanol. To explain the apparent insensitivity of the orientational relaxation times to the volumes of the hydrogen-bonded complexes and the strengths of hydrogen-bonding interactions, it was proposed that the rotational motion of the complex cannot be described as that of a rigid particle. By invoking flexibility in the hydrogen bond as well as a dynamic process of bond formation and dissolution, it was proposed that the rotational motion of the solute could be roughly the same whether it is hydrogen-bonded or not. Thus the hydrodynamic volume was unchanging through the series of liquids studied, and linear scaling of τ_{OR} with η becomes plausible. Furthermore, if the time varying torques experienced by the solute molecule are primarily determined by the solvent-solvent interactions (which also determine the solution viscosity), then the linear scaling of τ_{OR} vs η found in the rhodamine 6G–solvent systems is not surprising. The observed deviation from linearity of τ_{OR} vs η in the liquids 1-decanol and 1-undecanol is probably due to the breakdown of the continuum hydrodynamic model since the solvent molecules are larger than the solute molecule. The sharp deviation observed in the ethyleneglycol solvent is thought to be due to the extensive solvent-solvent aggregation via hydrogen bonding. If the solute does not experience the full frictional effects of this aggregation, τ_{OR} will be faster than the measured viscosity would lead one to expect, which was found to be the case in this system.

Fluorescence Depolarization Method

In the method discussed above, the orientational relaxation is followed by the decay of the induced dichroism, i.e. the polarization dependence of the absorption. If the ground-state absorption is monitored, the decay of the induced anisotropy can be dependent not only on the ground-state molecular rotation but upon that of the excited molecule as well. The time constants for the ground and excited-state rotations can be different. On the other hand, measurement of the excited-state absorption is dependent only on the rotational motions of the excited-state molecule.

Another way to follow the orientational relaxation of excited-state molecules is to measure the time dependence of the fluorescence depolarization. In this case, the decay of the anisotropy in the excited-state population is being measured, i.e. the rotational motion of the molecule in its excited state is obtained. If the excited-state molecules, before emitting light, rotate and thereby change the orientation of their emission dipoles (note that it is the orientation of the emission dipoles that is being monitored), the polarization of the fluorescence is decreased. The time dependence of this loss of fluorescence polarization is a direct measure of the orientational relaxation time of the excited-state molecule. The advantage of the fluorescence depolarization method over the induced dichroism method is that it is easier to monitor fluorescence than absorption. The limiting features of the fluorescence method are that a reasonably good emitter is required and only excited-state rotational motions can be measured.

By using a picosecond-excitation pulse in combination with a streak camera for detection of the fluorescence, the orientational relaxation times of the dyes rose bengal in a series of alcohols and eosin in water have been obtained (48a). The time-dependent fluorescence depolarization indicated that the rotational motions could be described as those of spherical molecules. The hydrodynamic volumes were found to be more than twice the estimated volume of the free molecule. The volume differences and the apparent spherical hydrodynamic shapes are attributed to solvent attachment. In addition, the reorientation times were found to scale only approximately with solvent viscosity.

Transient Grating Method

By using the rather novel approach of a transient grating technique the rotational relaxation times and fluorescence lifetimes of rhodamine 6G in several alcohols have been obtained (48b). Frequency-doubled mode-locked Nd:Yag laser pulses of about 60 psec duration and 530 nm wavelength were used to excite the rhodamine 6G molecules. Attenuated pulses at the same wavelength were used to probe the decay of the induced grating by the time-dependent changes in the probe light diffraction pattern. The grating, which results from strong absorption at the antinodes of the colliding excitation beams, decays as the molecules rotate, i.e. the orientational anisotropy decays; the grating also decays as the excited molecules return to the ground state. The rotational-motion times were found to be in agreement, within experimental error, with the values obtained from the induced dichroism method (47).

PHOTODISSOCIATION AND THE CAGE EFFECT

The chemistry following the dissociation of a molecule in the liquid state is dependent on the relative probabilities of the original fragments recombining (which results in no net chemical change), or escaping each other and subsequently reacting with other fragments or molecules in the medium. In a liquid the original fragments are surrounded by solvent molecules that interfere with their escape. The enhanced probability of recombination of the original fragments, referred to as the cage effect (49), is dependent on the kinetic energy of the fragments and on the nature of the fragment-solvent interactions. To determine the nature of cage-effect reactions it is necessary to obtain information on the early time motions of the fragments since this is the key to the partitioning between geminate (original fragment) and nongeminate processes. Studies of these reactions (50) provide information not only on the chemistry of the reactions but also on the properties of the liquid state with which they are intimately connected.

Prior to picosecond studies (51) of the photodissociation of iodine molecules in solution, the following questions had been unanswered: Did the recombination in the cage take 10^{-12} sec, 10^{-11} sec, or 10^{-10} sec? Or, could the cage be thought of as a static structure? The validity of various theories of the kinetics of recombination in the subnanosecond time domain (50, 52, 53) also remained unconfirmed. The I_2 studies (51) involved excitation of molecular iodine to the $^3\pi_{O^+u}$ state, where the excited molecule undergoes a collisionally induced predissociation and produces a pair of ground-state, $^2P_{3/2}$, iodine atoms. The dissociation of the iodine molecules and the subsequent recombination of the atoms were studied by monitoring the time-dependent population of molecular iodine with an attenuated picosecond laser pulse. It was determined that the lifetime of the "cage," i.e. geminate recombination, was 70 psec in hexadecane and 140 psec in CCl_4. From the time scale of these geminate recombinations, it seems unlikely that a description of the cage effect in terms of a static solvent cage would be physically reasonable. With regard to a theoretical model, it was found that a random flight picture was inadequate in describing the experimentally determined cage kinetics. However, a diffusion-theoretical model (54) that introduced some correlation in the motion of the atoms by treating the atoms as hard spheres (i.e. there is a volume excluded to motion) described the recombination dynamics very nicely. In addition, these experiments provided the first direct observations of the dynamics of a collision-induced predissociation in the liquid state. A rate constant of about 10^{11} sec^{-1} for predissociation from the $^3\pi_{O^+u}$ was obtained. This is about 10^5 larger than the spontaneous predissociation process observed in gas phase I_2 at low pressures.

ELECTRON PHOTOEJECTION AND SOLVATION

Pulse radiolysis, conventional flash photolysis, and picosecond lasers have been used to investigate the dynamics of electron localization in a variety of solvents (55–85). With these methods, information has been obtained on the structure and energies of the solvated electron, ionic aggregates in liquids, and the structure and relaxation properties of the liquid itself.

In picosecond flash-photolysis studies of ionic aggregates (77), such as the sodium salt of tetraphenylethylene dianions (T^{2-}, $2Na^+$), marked differences were observed in the relaxation processes in the solvents tetrahydrofuran and dioxane. The bleaching of the (T^{2-}, $2Na^+$) absorption with a picosecond excitation pulse was found to last 10 psec in dioxane and several nanoseconds in tetrahydrofuran. In the latter, solvent electron ejection occurs. However, the absorption due to the $T^{·-}$ radical that is expected to be generated by the ionization was not observed. Further experiments are necessary to clarify these interesting results.

EXCITED STATE CHARGE-TRANSFER COMPLEXES

In addition to intermolecular energy transfer there is another important class of excited-state interactions that quenches molecular fluorescence. It involves the transfer of charge rather than energy. Charge-transfer interactions not only quench fluorescence, but give rise to a new emission in low dielectric solvents characteristic of the exciplex $(A^- - D^+)^*$, produce ion radicals in high dielectric solvents, provide new pathways for energy degradation, and change the chemistry of the system (86–98).

Just as cage-effect reactions can be used to probe translational motions of atoms and molecules in liquids, the study of electron-transfer reactions can also provide insights into the nature of translational motions in liquids. In the cage-effect experiments, the fragments (e.g. the iodine atoms) generated by the dissociation are not initially uniformly distributed in the solution. In the early time domain the fragments are near each other, and the cage-effect experiments thus yield information on the relative motions of neighboring particles, i.e. within roughly 10 Å of each other. For the excited-state electron-transfer reactions, the excited-state acceptor molecules A* (produced by picosecond-pulse excitation) and the ground-state donor molecules D are initially uniformly distributed in the liquid. These reactions can therefore be used to examine the theoretical treatment of what are commonly referred to as diffusion-controlled reactions. Although the reactants are initially distributed randomly in the liquid, the reaction in the early time domain preferentially depletes the distribution of donor and acceptor molecules that happen to be near each other. The spatial distribution of A* with respect to D is no longer random, i.e. uniform, and the nonequilibrium spatial distribution is thus changing with time. This leads to a rate "constant" that is also changing with time. The reaction cannot be described therefore as a bimolecular reaction with a time-independent rate constant dependent on the diffusion coefficients of D and A*. The kinetics of formation of $(A^- - D^+)^*$ do not follow a simple exponential form but contain transient terms.

Picosecond-laser studies (93) of a hexane solution consisting of anthracene, serving as the acceptor (i.e. excited to S_1 with a picosecond pulse), and ground-state N,N-diethylaniline as the donor showed that a diffusion model that included all transient terms was in excellent agreement with the experimental results. Speculation that all transient terms were not necessary to describe the dynamics for times > 10 psec was found to be incorrect; the full transient description was necessary to conform with the experimental findings. However, at very high concentrations of the donor, 3 M or neat N,N-diethylaniline, no transient behavior was observed. The formation of the exciplex followed an exponential time dependence characteristic of a bimolecular process with a time-independent rate constant of 10^{11} sec^{-1}. At these high donor concentrations the excited anthracene molecules have donor molecules as immediate neighbors and translational motions are not rate-determining.

In addition to the distance requirements for excited-state electron transfer there can also be orientational restrictions on the transfer process. To examine the role of geometry on the dynamics of the electron-transfer process (98), the acceptor anthracene was linked to the donor dimethylaniline, DMA, via three methylene groups, A-$(CH_2)_3$-D, which was initially done by the Weller group (99), Chandross & Thomas (88), and Mataga and co-workers (89). As in the free system, the A moiety was excited with a 347.2 nm pulse, and the electron-transfer step was monitored with a 694.3 nm pulse. The behavior in polar solvents such as acetonitrile and methanol was found to be considerably different than that observed in nonpolar solvents such as hexane. In nonpolar solvents the initial charge-transfer step is rapid and then either levels off or increases very slowly after about 40 psec. However, electron transfer does not occur for all of the A*-$(CH_2)_3$-DMA molecules in the system. This is thought to be due to a distribution of ground-state geometries. It was proposed that molecules that are in the "appropriate" configuration can undergo exciplex formation, whereas molecules that are in the "wrong" configurations (e.g. extended form) cannot achieve the appropriate geometry within the lifetime of the A* moiety (5.5 nsec) to effect electron transfer.

On the other hand, the observation that electron transfer occurs for almost all A*-$(CH_2)_3$-DMA molecules in polar media can be due to favorable molecular configurations for electron transfer in these media. Packing effects due to molecular shapes and sizes as well as strong solvent-solvent interactions could favor the more compact configuration of A-$(CH_2)_3$-D in the polar solvents, thereby minimizing disruption of the solvent structure. Another possibility is the relative shifting of the A*-$(CH_2)_3$-DMA and ·A$^-$-$(CH_2)_3$-DMA$^+$ energy surfaces, leading to an enhanced electron-transfer probability for an "extended" configuration in polar media. In polar media that there is also a fairly rapid decay (probably back-electron transfer), though it is slower than the initial transfer step. This back transfer is not observed in the nonpolar media, at least for times up to 1 nsec. For the polar media the fairly rapid decay process 760 ± 80 psec can involve formation of the acceptor triplet in the back-transfer step, i.e. A·$^-$-$(CH_2)_3$-D·$^+$ to ^{3}A*-$(CH_2)_3$-D. The center-to-center separation in an extended form is about 4–5 Å. From the studies (91, 93) of the free donor and acceptor systems at donor concentrations of 3 M or higher, it is known that the electron transfer is completed in about 20–25 psec. At the distances separating the linked donor and acceptor, one would thus expect the transfer to be completed in this same time period (20–25 psec) rather than the longer times observed in both the polar and nonpolar solvents. It is concluded that the differences are due to the less than favorable geometries achievable in the linked molecule.

INTERNAL CONVERSION AND INTERSYSTEM CROSSING

Internal conversion refers to the nonradiative processes by which a molecule in an excited state converts to a lower energy state of the same multiplicity. Intersystem crossing involves the nonradiative transition between states of differing multiplicity. Picosecond-pulse techniques afford special opportunities for studying internal conversion and intersystem crossing since many organic molecules possess lifetimes of the order of nanoseconds or less because of these ultrafast processes. The interactions responsible for internal conversion and intersystem crossing in molecules are not well understood.

Two picosecond techniques have mainly been used to measure rapid-energy relaxation in large molecules: an absorption technique and a fluorescence emission method. In the former, the sample is prepared with an exciting pulse and the transmission of the sample is probed with a weak interrogation pulse at varying delay times. If a molecule emits sufficient fluorescence, rapid relaxation can also be investigated by observing the emission with either a picosecond-resolution optical gate or a streak camera.

An interesting measurement of rapid internal conversion was carried out (100) on the crystal-violet dye molecule $[(CH_3)_2NC_6H_4]_3C^+$. The structure of this molecule is known to be D_3-propeller-shaped with the phenyl rings rotated 32° from the central plane. The molecule, although exhibiting intense visible absorption bands, is almost completely nonfluorescent with a quantum yield below 10^{-4}. The crystal-violet molecules were excited with an intense 530 nm pulse that promoted the molecules from S_0 to S_1. A very weak interrogating pulse was used to probe the return of the molecules to the ground state. Because the recovery was very rapid and complete within 100 psec, it was concluded that the rapid internal conversion process was being measured. By choosing a series of solvents that covered a viscosity range from 0.01 to 120 P, it was shown that the ground-state recovery time varies as $\eta^{1/3}$, where η is the viscosity of the solvent.

Solvent effects other than viscosity are judged to be relatively unimportant. A model developed (101) for a series of triphenylmethane dyes predicts a viscosity dependence for the quantum yield of $Q = C\eta^{2/3}$. In this model, absorption of light produces a Franck-Condon vertically excited state with the phenyl rings still at a ground-state equilibrium angle, θ_0. The rings then rotate toward a new equilibrium angle, θ, and the nonradiative deactivation of the excited state depends upon $(\theta - \theta_0)^2$. It was further assumed that the radiative rate is independent of θ. A new model is necessary to explain the new data since the present model predicts the same $\eta^{2/3}$ dependence for the lifetime as well as for the quantum yield, whereas the measured dependence for the lifetime is $\eta^{1/3}$.

Further measurements on another triphenylmethane dye, malachite green, have been made using the superior resolution available from a mode-locked CW dye laser (102). With 0.5 psec excitation pulses it was found that the recovery time is only 2.1 psec for malachite green in methanol and that the decay is exponential. By studying the recovery time in a number of solvents it was determined that there was always a long-term recovery, whereas in the higher viscosity solvents, there was also an initial, more rapid, partial recovery. Since the fast initial recovery is in agreement with the S_1 lifetime calculated from the quantum efficiency, it was suggested that on a short time scale, molecules in S_1 rapidly convert to a highly energetic level of S_0, giving rise to a partial recovery of the absorption. Subsequently, this hot distribution in the ground state relaxes, giving rise to the slower rate observed for complete recovery of the absorption. The longer lifetimes were found to depend on the viscosity of the solvent approximately as $\eta^{1/2}$, but the $1/e$ point of the total-recovery curves varies closely as $\eta^{1/3}$ in agreement with the previously described work (100).

Studies (103, 104) of benzophenone and nitronaphthalene have shown that the intersystem-crossing rates for these molecules are solvent dependent. In other investigations on benzophenone, it was found that intersystem crossing was dependent on the wavelength of excitation (105–108). These phenomena may be quite complex, if vibrational relaxation is occurring on a time scale comparable to that of intersystem crossing. The solvent-dependent risetimes indicate that conventional singlet-triplet mixing appears to be an insufficient explanation, because this mechanism is quite sensitive to the spacing between the singlet and triplet states. It has been suggested (104) that the variations in buildup time may be due to the effectiveness of different solvents in relaxing the excited benzophenone molecules vibrationally to or from singlet levels that are strongly coupled to the triplet manifold.

Studies of the internal conversion between excited electronic states of the molecule 4-(1-naphthylmethyl)-benzophenone in a benzene solution have been undertaken (109a). In these experiments the benzophenone molecule and the methylnapthalene molecule are connected by a sigma-type chemical bond so that energy may be transferred from one part of the connected molecule to the other. The benzophenone portion of the molecule is excited to the S_1 state by a pulse at 353 nm, and the triplet formation is probed by the absorption at 530 nm. An initial absorption with a lifetime of 10 psec is accompanied by the development of a much weaker absorption with a decay time greater than the longest delay time used in the experiment. The short-time and long-time components were interpreted as originating from triplet-triplet absorptions from the benzophenone and 1-methylnaphthalene parts of the molecule, respectively. The initial rapid decay of the absorption of the double molecule can be contrasted against a very slow decay of about 1 nsec observed for an equimolar solution of benzophenone and naphthalene in benzene with each solute at 0.2 M to simulate the double molecule experiments at 0.2 M. Thus the contribution to quenching by nearest neighbors is small. The results appear to support a model in which the wave functions of the low-energy states of the double molecule are approximately products of naphthalene- and benzophenone-like single-excitation functions, with the coupling provided by the small interactions between the chromophores.

A mode-locked CW dye laser was used to generate UV (307.5 nm) pulses for excitation of coronene to S_3; 615 nm pulses were also generated to monitor absorption from S_1. With the unusually high time resolution achieved with this system (0.2 psec), the internal conversion from S_3 to S_1 was found to be 2 psec (109b).

A more general extension of the probe technique has been used by several groups to measure internal conversion and intersystem crossing. A pulse at a given "wavelength" excites the sample, but the probe pulse consists of a picosecond continuum that spans the entire visible region and sometimes extends beyond it. These very broad continua, generated by nonlinear optical techniques, were first used (35, 110–112) to monitor inverse Raman spectra, and their use for picosecond flash photolysis experiments was suggested. These continua make it possible to study systematically the picosecond transient behavior of molecules, which are weak emitters.

Experiments using these flash photolysis techniques have helped to resolve discrepancies in the literature on the lifetime of DODCI (3,3'-diethyloxadicarbocyanine iodide) dye (113). In further studies (114) the effect of the solvent on the decay kinetics of bis-(4-dimethylaminodithiobenzil)-Ni(II), or BDN for short, was investigated. BDN is a nickel complex that absorbs in the infrared and is difficult to study by emission spectroscopy because it has a low quantum yield, and the emission probably extends well into the infrared. BDN was excited with a single intense pulse at 1060 nm, and excited-state absorption was then monitored with a picosecond continuum. The excited-state absorption lifetime was measured to be 220 psec for BDN in iodoethane; 3.6 nsec for BDN in 1,2-dichloroethane; 2.6 nsec for BDN in 1,2-dibromoethane; and 9 nsec for BDN in benzene. An external heavy-atom effect apparently leads to the more rapid recovery in the halogenated solvents. Analysis indicates that the excited-state absorption represents the lifetime of a state, or states, in which the $3b_{2g}$ orbital is occupied.

Porphyrin molecules have also been studied by excitation with a 530 nm pulse and the evolution of the absorption spectra probed with picosecond resolution by means of a continuum (115). For octaethylporphinatotin(IV) dichloride [(OEP)SnCl$_2$] the absorption spectrum of the excited singlet state, the decay of the S_1 state, and the growth of the T_1 triplet state have been observed. The S_1 state decays in about 500 psec, and the spectrum represented by T_1 appears in about the same time. By analysis it was found that the quantum yield for triplet formation is 0.8 ± 0.008 for (OEP)SnCl$_2$, and from the known fluorescence quantum yield of about 0.01 the quantum yield for internal conversion from the S_1 to the S_0 state was deduced to be about 0.19. The difference between the absorption spectrum of the first excited singlet state and that of the first excited triplet state in a porphyrin molecule has also been obtained by using these methods. For many of these experiments picosecond flash photolysis is used to locate a number of new absorption bands, and standard techniques are then used to observe changes of optical density at a particularly interesting wavelength.

In other experiments using the picosecond continua, two cyanine dyes, cryptocyanine (1,1'-diethyl-4,4'-carbocyanine iodide) and DTTC (3,3'-diethyl-2,2'-thiatricarbocyanine iodide), were excited with 694.3 nm pulses, and the evolution of transient absorption bands was examined (116). A new absorption band, which decayed in 90 ± 30 psec, was detected in DTTC at 525 nm. Since the ground state was also observed to recover in about this time, the newly observed band was attributed to a transition between excited singlet states. Measurements of the recovery time for the ground state of cryptocyanine were found to be consistent with the results of earlier workers (117).

Picosecond flash photolysis methods have also been used to measure the $S_n \leftarrow S_1$ and $T_n \leftarrow T_1$ absorption spectra of anthracene in solution (118). The anthracene is excited with pulses at 347.2 nm, and a continuum covering the entire region from 390 to 920 nm is used to probe the transient spectra. At short delay times of 250 to 300 psec, a strong absorption band at 600 nm that corresponds to the $S_n \leftarrow S_1$ transition is observed, while at much longer delay times, 4 to 5 nsec, a strong transition is found at 420 nm that is assigned to $T_n \leftarrow T_1$.

Lifetimes of the excited states of a number of transition-metal compounds have been established (119) by use of probe techniques and picosecond flash photolysis. Lifetimes in the subnanosecond range were established for some nonluminescent compounds of iron and ruthenium. For a number of transition-metal complexes the interstate nonradiative processes such as intersystem crossing were found to be extremely fast. A recent improvement for measuring time-resolved absorption spectra in the picosecond range has been developed (120). The system consists of a ruby picosecond laser (694.3 nm) for excitation and a 6 μsec Xe probe flash pulse whose transmission is time-resolved with a streak camera that has a 65 psec resolution capability. With this system, the excited singlet absorption of a number of dyes was obtained. Since there are streak cameras that have better than 5 psec resolution capabilities, the combination of picosecond excitation, Xe probe pulses, and an ultrafast streak camera will make this method increasingly attractive.

Emission Measurements of Internal Conversion and Intersystem Crossing

One of the first applications of the mode-locked laser was its use in measuring the nanosecond fluorescence decay times of dye molecules in solvents. In an early technique the sample was excited with either the fundamental or the second harmonic of pulses generated by a mode-locked ruby laser, and the fluorescence was detected with a planar diode detector used in conjunction with a traveling-wave oscilloscope (121, 122). In these experiments both the ultrashort pumping pulses for minimizing deconvolution problems and the high-intensity picosecond-pulse source were used. Prior to these measurements more conventional schemes used either nanosecond flashlamp pumping or more cumbersome fluorometry techniques. These experiments were still limited, however, to a resolution of about 0.5 nsec.

The first studies in which fluorescence phenomena were detected with "true" picosecond resolution used the optical gate to study the emission from two dyes, DDI (1,1'-diethyl-2,2'-dicarbocyanine iodide) and cryptocyanine, dissolved in methanol (117). The samples were excited with pulses at 530 nm, and the emission was sampled at variable delay times by means of an ultrafast shutter operated by intense pulses at 1060 nm. The fluorescence from DDI in methanol decayed in 14 ± 3 psec, while that of cryptocyanine decayed in 22 ± 4 psec. This ultrafast recovery time in cryptocyanine has since been verified (116, 123, 124). The ultrashort fluorescence decay time demonstrates the rapid internal conversion of the first excited singlet state to the ground state.

Using an optical gate technique, measurements have been obtained for the fluorescence decay time of erythrosin in solution (125–127). For erythrosin in water the measured decay times were found to be 90 psec (125), 110 ± 20 psec (127), and 57 ± 6 psec (126). Since the quantum yield is 0.02 for erythrosin in water, the higher values should probably be preferred. The fluorescence risetime is "instantaneous" as shown in two of the studies (126, 127), although other work (125) mistakenly identified their prompt risetime as a slow one because of the long delay between the peak of the fluorescence curve and the calibrated zero time. These apparent delays originate because of the continuous accumulation of excited-state singlets produced by the wings of the pulse. The fluorescence lifetimes of a number of fluorescein derivatives have also been measured (127). The lifetimes of fluorescein (Fl), eosin ($FlBr_4$), and erythrosin (FlI_4), were measured to be 3.6, 0.9, and 0.11 nsec, respectively. The data is in excellent agreement with quantum-yield predictions, and the decrease in lifetime observed upon addition of heavy halogen atoms has been taken to be consistent with heavy-atom-enhanced intersystem crossing and published triplet quantum yields.

A precautionary note should be added for measurement of fluorescence lifetimes. At high excitation intensities, nonlinear processes such as stimulated emission may occur. Stimulated emission effects were pointed out in two early studies (128, 129a). These effects lead to a nonexponential decay of the fluorescence and a shortening of the lifetime. Other common effects are 1. the formation of transient species when an entire train of pulses is used to excite fluorescence and 2. concentration quenching. A fairly common effect is that leftover triplet states modify the lifetime for singlets produced by later pulses in the train. Such problems have led to a wide range of estimates for the lifetime of DODCI, for example. The correct fluorescence lifetime of about 1.2 nsec has now been firmly established by a number of investigators (42, 113, 129b–132).

The photochemical decomposition of s-tetrazene in benzene has been studied by both excited-state absorption and fluorescence techniques (133). The excited-singlet lifetime was found to be about 450 psec, and a new excited-state transition peaking at 473 nm was observed. In fluorescence experiments (134) on the polymethane dye DTTC in methanol the lifetime from a highly excited singlet state (blue-emitting) was found to be roughly 35 psec.

In a number of experiments a streak camera has been used in conjunction with picosecond excitation to examine fluorescence emission with picosecond resolution. Emission from a number of dyes under mode-locking conditions (135, 136), such as fluorescence from DODCI (132), emission from dye vapors, from photosynthetic samples, and scintillator materials (137–139), has been obtained in this way. In the dye-vapor experiments (137), dimethyl POPOP [1,4-bis-2-(4-methyl-5-phenyloxazolyl) benzene] and perylene molecules were excited in the gaseous phase with picosecond pulses at 353 nm, and then, by using a streak-camera detection technique, the risetime of the fluorescence was found to be ⩽20 psec for dimethyl POPOP and $\leqslant 30$ psec for perylene. Since the vapors were at 300°C, the collision rate between dye molecules could have been no more than 10^8 sec^{-1}. A dye molecule in the liquid phase would collide with the neighboring solvent molecules at a rate of about 10^{12} sec^{-1}. Since fluorescence is commonly observed from the ground vibrational state of the first excited singlet state, the rapid risetime in the vapor was interpreted as direct evidence supporting rapid internal relaxation of the molecules, even in the absence of collisions. The ability of large dye molecules to relax internally, independent of their surroundings, is a manifestation of the fact that such molecules have a large number of vibrational and rotational modes; the probability of mutual interactions among the modes, thus, is very high. These interactions reduce the lifetime of any particular level, because the energy provided by the excitation can be rapidly redistributed over the large ensemble of densely packed levels.

It has been shown that the decay of benzophenone in the vapor phase is non-exponential and in the microsecond range (140, 141). The nonexponential decays have been verified and interpreted in terms of coupling with triplet states (142). These more recent workers indicate that oscillations observed in the tail of the decay previously observed (141) are probably experimental artifacts and not quantum beats.

In another interesting application of picosecond fluorescence techniques, emission originating from upper singlet states has been observed by means of a two-photon absorption technique (143). Ordinarily, these emissions can only be observed in rare cases since the fluorescence is emitted from higher singlet states with a very low quantum efficiency. Two-photon excitation allows the excitation of levels that are spectroscopically forbidden by one-photon excitation. Two-photon excitation also permits a more uniform spatial distribution of excited molecules in a sample that is strongly one-photon-absorbing, e.g. a solid, concentrated solution, or a very intense transition. Furthermore, states that are one-photon-allowed but hidden by stronger overlapping transitions can be amenable to detection by two-photon absorption. In addition, direct-scattering processes that would tend to obscure any weak emission are avoided with two-photon absorption techniques. By irradiating samples with a combination of the wavelengths 1060, 530, and 354 nm, generated from the fundamental or harmonics of the Nd:glass laser, spectra were obtained for the excited states of such dyes as rhodamine 6G perchlorate in 2-propanol, rhodamine B in 2-propanol, and acridine red in 2-propanol. From the intensity of typical spectra, an upper limit of 10^{-4} can be placed on the fluorescence quantum efficiency from the upper states, and the lifetime of these states is estimated to be less than 5 psec.

In addition to the experimental progress due to picosecond lasers, there have been important theoretical advances in recent years in the description of radiationless transitions and optical coupling in complex molecules (144–151).

VIBRATIONAL RELAXATION IN EXCITED ELECTRONIC STATES

With regard to the rapid relaxation of large dye molecules, a number of innovative techniques have been introduced for measuring vibrational relaxation times (16, 17, 153a). In the first technique, molecules are excited from the ground state S_0 to the first excited singlet state S_1 by photons of frequency ω_1, and the molecules then decay to the ground vibrational state of S_1. This decay can be followed by measuring the gain produced by a probe pulse of frequency ω_2, which is usually chosen to correspond with the red end of the fluorescence band. In the second technique, ground-state vibrational relaxation can be monitored by exciting the molecules to S_1, permitting a suitable delay for the molecules to relax to the ground vibrational state of S_1, stimulating these molecules to relax to an upper vibrational state of S_0 by an intense beam at frequency ω_2, and then probing the vibrational relaxation by the return of absorption at ω_1, caused by return of the population to the ground state. Because the pulsewidths were of the order 5 to 6 psec, and the vibrational relaxations were very fast, deconvolution of the pulse-shape function was a difficult matter, but it was concluded that the vibrational relaxation times in the rhodamines were approximately several picoseconds, with an uncertainty also of several picoseconds.

In the other studies (153a) of excited-state vibrational relaxation in rhodamine 6G and rhodamine B in a variety of solvents, the vibrational population lifetimes were found to be 0.5–1 psec, independent of the solvent or its viscosity. Two methods were used that gave consistent results. In one method the nonlinear transmission of an intense picosecond pulse as a function of its intensity was used to calculate the vibrational relaxation time. In the other, the transient population of the initially excited vibrational levels of S_1 was measured with a weak probe pulse. Using subpicosecond pulse excitation (18), it has been possible to apply a previously used technique (16) with greater precision to the study of vibronic relaxation. Rhodamine 6G and rhodamine B molecules were excited in a number of solvents up to a higher electronic state with a subpicosecond pulse (0.9 psec) at 307.7 nm, and the gain of the emission was then probed with a subpicosecond pulse (9.9 psec) at 615 nm. Gain measurements show that the fluorescence emission begins promptly. After deconvolution, the risetime is less than 0.2 psec, which is the resolution of the apparatus. The result is at first somewhat surprising because, in order for the excitation to reach the first excited singlet state, the molecules must relax through an ensemble of levels with a total energy gap of $\sim 14{,}000$ cm^{-1}. Although for these complex molecular systems other possible explanations present themselves, e.g. the stimulated emission cross section of the higher vibronic levels of the first excited singlet state is the same as that from vibrationally relaxed S_1, the following interpretation (18) seems to be indicated: deactivation of the excited state of rhodamine 6G is exceedingly fast.

The rapid deexcitation of an excited electronic state has also been obtained by determining the risetime of the spontaneous fluorescence intensity for a number of dyes (152). Dyes such as rhodamine B, rhodamine 6G, and erythrosin B in such solvents as water, ethanol, and methanol were excited with 530 nm, 10 psec pulses and the emission was detected with an optical shutter. The observed risetime corresponded to a delay of less than 1 psec, consistent with later work (18).

Previously, it was claimed that the time- and frequency-resolved vibrational relaxation in an excited state of rhodamine 6G had been determined (153b). It was observed, upon pumping rhodamine 6G with intense 530 nm pulses of several picoseconds' duration, that intense stimulated emission occurred many picoseconds after the zero time of the apparatus, and it was therefore concluded that the vibrational relaxation time was about 6 psec. It is now clear that the onset of stimulated emission in this experiment is governed more by parameters such as the integral of the excitation pulse-shape function, which is closely related to the number of molecules in the excited state, and by the stimulated emission cross section, and is only indirectly related to any vibrational relaxation time.

Recent experiments carried out in several laboratories (154–156) on azulene in its lowest excited singlet state indicate that the relaxation is $\leqslant 1$ psec. The physical processes contributing to the ultrafast decay time are not clear. The measured relaxation time can be a superposition of vibrational relaxation in S_1, internal conversion to S_0, and intersystem crossing to the triplet manifold.

INTRAMOLECULAR PROTON TRANSFER

One of the elementary processes by which molecules in excited electronic states can relax is by an intermolecular or intramolecular transfer of a proton (157–167). The excited-state proton transfer can generate an anion, e.g. the naphtholate ion (159, 162, 163), or lead to an enol-to-keto isomerization, e.g. salicylic ester (162), or produce the tautomer via a double-proton transfer in the 7-azaindole dimer (164, 165). With these processes, new emissions, new routes for energy degradation, and chemical changes become available to the system.

A picosecond ruby laser was used to measure the proton-transfer kinetics of 2,4-bis-(dimethylamino)-6-(2-hydroxy-5-methylphenyl)-s-triazine in cyclohexane at 298°K (167). The lifetime of the excited singlet (enol form of the molecule) was found to be 6.3×10^{-11} sec, and the rate constant for proton transfer was found to be 1.1×10^{10} sec^{-1}. This proton-transfer rate constant, measured at 298°K, is about a factor of 100 greater than that found for the double-proton transfer in the 7-azaindole dimer measured at 77°K (164, 165). The difference may be due to a quantum mechanical tunneling in the latter case vs going over a potential barrier in the former case.

CONFORMATIONAL CHANGES IN EXCITED ELECTRONIC STATES

The coupling between the electronic systems of identical molecules, one of which is in an excited electronic state whereas the other is in the ground state, can lead to the formation of a transient species, the excimer, whose spectral properties differ markedly from those of the isolated chromophores. Examples of systems composed of two identical parts that are, however, not free, are double molecules such as the biaryls, i.e. biphenyl, binaphthyl, bianthryl, etc. They consist of two planar aromatic molecules connected by a carbon-carbon bond. Unlike the usual excimers that are formed from free molecules, the relative orientations of the two moieties in a double molecule are severely constrained by the bond joining them.

Spectroscopic and X-ray studies (168–173) of one of the members of this series, 1,1'-binaphthyl, have indicated that in the ground state the orientation of the naphthalene moieties is close to 90° and thus weakly interacting, whereas in the excited singlet state in solution at room temperature the moieties are strongly interacting and this indicates a more coplanar configuration. Picosecond dye-laser pulses were used in the recent investigation (174) of the kinetics of the excited-state structural change that results from the internal rotation about the carbon-carbon connecting bond. In addition to the excited-state twisting motion that changes the relative orientation of the naphthalene moieties, the rotational motion was also obtained for the 1,1'-binaphthyl molecule in its equilibrium excited-state configuration. The approach used was to excite 1,1'-binaphthyl with a picosecond light pulse at 307.5 nm and monitor the motion towards coplanarity by the appearance of a strong excited-state absorption at 615 nm, characteristic of the equilibrium "coplanar" excited-state configuration. The experiments were carried out at room temperature in the solvents: benzene, t-butyl benzene, and n-heptanol. The kinetics of the conformational changes were found to be polarization-dependent, consisting of a very rapid part of about 2.5 psec in all of the solvents used and a slower part varying from 11–12 psec in benzene and t-butyl benzene to 22 psec in n-heptanol. The polarization-dependent kinetics were discussed in terms of the shift in energy levels during the motion from the perpendicular to coplanar configurations. As different states come into resonance at the probe wavelength of 615 nm, the polarization of the absorption changes. Although the excited-state structural change, i.e. the internal rotation, does not follow a simple (linear) viscosity dependence, the overall rotation of the molecule in its equilibrium excited-state geometry does scale linearly with viscosity, which suggests a Debye hydrodynamic rotational motion.

VIBRATIONAL RELAXATION IN THE GROUND ELECTRONIC STATE

Vibrational relaxation processes in the condensed phase are extremely rapid, 10^{-10} sec–10^{-12} sec, with liquid nitrogen being one of the notable exceptions. Before the application of picosecond lasers, information on the dynamics of condensed-phase vibrational relaxation was obtained from the measurement of infrared and Raman linewidths. One difficulty with this approach is that the experimentally observed line shape is often determined by a number of different physical processes. These include population (energy) relaxation, dephasing, and isotope effects. If the line is homogeneously broadened, the dephasing time can be obtained from the spectral width. The population lifetime can be obtained from the measured linewidth if the dominant relaxation process is the population decay. In this case the dephasing time and the population lifetime are equal. However, if dephasing processes, e.g. dephasing collisions, are important, the population lifetime (energy relaxation) is hidden by the rapid dephasing processes and cannot be extracted from the linewidth measurements. In such a case, measurements in the time domain are necessary to find the population decay time. If the line is inhomogeneously broadened, neither the dephasing time nor the population decay time is directly related to the observed linewidth. The linewidth is determined by the distribution of closely spaced resonance frequencies, e.g. a Doppler-broadened line in a gas or the overlapping components due to isotope line splittings, and is not determined by the relaxation processes. In this case, time-resolved measurements can yield the various dephasing and population decay times.

A further limitation of the spectral linewidth measurements is that the decay of a vibrational mode into other vibrational modes of the molecule, e.g. stretching into bending, cannot be determined. In a similar way intermolecular vibrational energy transfer is not obtained from linewidth measurements. However, with direct time-resolved measurements, the dephasing times, population decay times, and channels for population decay can be found.

The basic idea of the time measurements is to generate a large excess population in a specified vibrational mode with an intense picosecond light pulse and then probe the relaxation processes with a weak picosecond light pulse (175–178). In the first, and most commonly used, method (179), the large vibrational population is achieved by a stimulated Raman process followed by an anti-Stokes Raman-scattering process with a time-delayed picosecond probe pulse (175–178). If the probe scattering is carried out under phase-matching conditions (coherent Raman scattering), the dephasing time is obtained. If the system is monitored by the spontaneous (incoherent) anti-Stokes Raman scattering of the probe pulse, the population decay is obtained.

Although this method has proved to be a powerful one, it, too, has its shortcomings. Some of the limitations are the selectivity of the stimulated Raman scattering in exciting only a given vibrational mode of the molecule and thus not permitting study of other vibrational modes, the requirement of very high concentrations, the weakness of the anti-Stokes spontaneous scattering, and the difficulty in studying low-frequency vibrations (<1000 cm^{-1}) due to their large Boltzmann populations at room temperature.

To avoid the limitations of the stimulated Raman-scattering approach, a technique has been developed that utilizes tunable infrared picosecond pulses for vibrational excitation (180,181). It has several advantages: it selectively excites different vibrational modes of a molecule, the excitation is a one-photon process rather than the nonlinear stimulated Raman process, and it can be used to study dilute systems. Since the selection rules for infrared absorption are different from those for Raman scattering, the techniques can be viewed as complementary ones. The probe method is the same for both techniques, namely, measurement of the spontaneous anti-Stokes Raman scattering as a function of delay time with an attenuated visible-wavelength picosecond pulse. In studies (180) of ethanol (4×10^{-2} M) and methyl iodide (5×10^{-2} M) in carbon tetrachloride, it was found that intramolecular relaxation of the excited methyl vibrations (2940 cm^{-1} in ethanol and 2950 cm^{-1} and 3050 cm^{-1} in methyl iodide) to neighboring energy states was very rapid (1–2 psec). The decay to lower energy states was found to be much more rapid in methyl iodide than in ethanol.

In an earlier section, I detailed one method for measuring the population decay of ground-state vibrational modes that used a sequence of three picosecond pulses (16, 17). Another technique has recently been developed that combines infrared and visible pulses with fluorescence measurements (181). The infrared pulse, generated by a parametric process at the frequency of interest, is used to excite a vibration of the molecule. A time-delayed visible pulse that cannot excite the molecule to its lowest excited electronic state, S_1, from the $v = 0$ level but can reach S_1 from the vibrationally excited level ($v = 1$) is used to probe the population relaxation. Measurement of the fluorescence from S_1 as a function of the time delay between the visible and infrared pulses gives the vibrational population for that time. By this method (181), the vibrational energy relaxation of the 2970 cm^{-1} asymmetric CH$_3$-stretching mode of coumarin 6 in CCl$_4$ at 295°K was found to be 1.3 psec. The appealing features of this approach include the ability to excite different vibrational modes of the molecule by adjusting the frequency of the infrared pulse, the high sensitivity, and the applicability to dilute systems. However, the method requires molecules that have a reasonable quantum yield of fluorescence. Another factor to be considered is the difficulty in separating two relaxation channels: 1. a channel to neighboring energy states that have different transition probabilities to S_1 (and thus would alter the fluorescence intensity), and 2. a channel to lower-energy vibrational states. In all of these time-domain measurements of ground-state vibrational relaxation, a large excess (relative to thermal) vibrational population must be generated. This is very difficult to achieve at room temperature for the lower frequency vibrations (<1000 cm^{-1}) because of the large Boltzmann populations. For the moment at least, linewidth measurements are necessary for the lower frequency vibrations.

CONCLUDING REMARKS

It is hoped that this review of picosecond spectroscopy has helped the reader perceive not only the diversity of experiments and impact of this young field but also, most importantly, its unusual promise. The applications of picosecond lasers to problems in chemistry, physics, and biology are expanding rapidly. Unfortunately, due to space limitations, I have not discussed the use of these lasers in biology. The enormous gaps in our understanding of the liquid state, molecular motions, physical decay processes, and chemical reactions provide important opportunities for this new field.

ACKNOWLEDGMENTS

I wish to acknowledge support for this work by the National Science Foundation and the US Army Research Office.

Literature Cited

1. Cario, G., Franck, J. 1923. *Z. Phys.* 17:202–12
2. Forster, T. 1948. *Ann. Phys. NY* 2:55–62
3. Forster, T. 1949. *Z. Naturforsch. Teil A* 4:321–34
4. Dexter, D. L. 1953. *J. Chem. Phys.* 21:836
5. Galanin, M. D. 1955. *Sov. Phys. JETP* 28:495–95
6. Terenin, A. N., Ermolaev, V. L. 1952. *Dokl. Akad. Nauk SSSR* 85:547–57
7. Wilkinson, F. 1974. *Advances in Photochemistry*, ed. W. A. Noyes, G. S. Hammond, J. N. Pitts, Jr. 3:241–68. New York: Wiley-Intersci.
8. Bennett, R. G., Kellogg, R. E. 1966. *Progress in Reaction Kinetics*, ed. G. Porter, 4:215–38. London: Pergamon
9. Eisenthal, K. B. 1970. *Chem. Phys. Lett.* 6:155–57
10. Rehm, D., Eisenthal, K. B. 1971. *Chem. Phys. Lett.* 9:387–89
11. Weber, G. 1954. *Trans. Faraday Soc.* 50:552–55
12. Ore, A. 1959. *J. Chem. Phys.* 31:442–43

13. Eisenthal, K. B., Siegel, S. 1964. *J. Chem. Phys.* 41:652–55
14. Eisenthal, K. B., Siegel, S. 1965. *J. Chem. Phys.* 42:2494–2502
15. Steinberg, I. Z. 1968. *J. Chem. Phys.* 48:2411–13
16. Ricard, D., Lowdermilk, H., Ducuing, J. 1972. *Chem. Phys. Lett.* 16:617–21
17. Ricard, D., Ducuing, J. 1975. *J. Chem. Phys.* 62:3616–19
18. Ippen, E. P., Shank, C. V. 1976. Unpublished results
19. Anderson, R. W. Jr., Hochstrasser, R. M., Lutz, H., Scott, G. W. 1974. *J. Chem. Phys.* 61:2500–6
20. Roy, J. K., El-Sayed, M. A. 1964. *J. Chem. Phys.* 40:3442–43
21. Eisenthal, K. B. 1969. *J. Chem. Phys.* 50:3120–21
22. Alfano, R. R., Shapiro, S. L., Pope, M. 1973. *Opt. Commun.* 9:388–91
23. Mayer, G., Gires, F. 1964. *CR Acad. Sci.* 258:2039–43
24. Maker, P. D., Terhune, R. W., Savage, C. M. 1964. *Phys. Rev. Lett.* 12:507–9
25. Buckingham, A. D., Pople, J. A. 1955. *Proc. Phys. Soc. London Sect. A* 68:905–9
26. Buckingham, A. D. 1955. *Proc. Phys. Soc. London Sect. A* 68:910–19
27. Buckingham, A. D. 1956. *Proc. Phys. Soc. London Sect. B* 69:344–49
28. Kielich, S. 1966. *Acta Phys. Pol.* 30:683–707
29. Kielich, S. 1972. *Dielectrics and Related Molecular Processes*, 1:192. London: Chem. Soc.
30. Duguay, M. A., Hansen, J. W. 1969. *Appl. Phys. Lett.* 15:192
31. Shimizu, F., Stoicheff, B. P. 1969. *IEEE J. Quantum Electron.* 5:544–50
32. Fisher, R. A., Kelley, P. L., Gustafson, T. K. 1969. *Appl. Phys. Lett.* 14:140–42
33. Brewer, R. G., Lee, C. H. 1968. *Phys. Rev. Lett.* 21:267–70
34. Veduta, A. P., Kirsanov, B. P. 1968. *Sov. Phys. JETP* 27:736
35. Alfano, R. R., Shapiro, S. L. 1970. *Phys. Rev. Lett.* 24:592–94
36. Sala, K., Richardson, M. C. 1975. *Phys. Rev. A* 12:1036–46
37. Mourou, G., Malley, M. M. 1975. *Opt. Commun.* 13:412–17
38. Ho, P. P., Yu, W., Alfano, R. R. 1976. *Chem. Phys. Lett.* 37:91–96
39. Alms, G. R., Bauer, D. R., Brauman, J. I., Pecora, R. 1973. *J. Chem. Phys.* 59:5310–20
40. Stegeman, G. I. A., Stoicheff, B. P. 1973. *Phys. Rev. A* 7:1160–77
41. Starvanov, V. S., Tirganov, E. V., Fabelinskii, I. L. 1966. *Sov. Phys. JETP* 4:176
42. Ippen, E. P., Shank, C. V. 1975. *Appl. Phys. Lett.* 26:62–63
43. Shapiro, S. L., Broida, H. P. 1967. *Phys. Rev.* 154:129–38
44. Lalanne, J. R. 1975. *Phys. Lett. A* 51:74–76
45. Wong, G. K. L., Shen, Y. R. 1973. *Phys. Rev. Lett.* 30:895–97
46. Eisenthal, K. B., Drexhage, K. H. 1969. *J. Chem. Phys.* 51:5720–21
47. Chuang, T. J., Eisenthal, K. B. 1971. *Chem. Phys. Lett.* 11:368–70
48a. Fleming, G. R., Morris, J. M., Robinson, G. W. 1976. *Chem. Phys.* 17:91–96
48b. Phillion, D. W., Kuizenga, D. J., Siegman, A. E. 1975. *Appl. Phys. Lett.* 27:85–87
49. Franck, J., Rabinowitch, E. 1934. *Trans. Faraday Soc.* 30:120–31
50. Lampe, F. W., Noyes, R. M. 1954. *J. Am. Chem. Soc.* 76:2140–44
51. Chuang, T. J., Hoffman, G. W., Eisenthal, K. B. 1974. *Chem. Phys. Lett.* 25:201–5
52. Noyes, R. M. 1954. *J. Chem. Phys.* 22:1349–59
53. Noyes, R. M. 1956. *J. Am. Chem. Soc.* 78:5486–90
54. Evans, G. T., Fixman, M. 1976. *J. Phys. Chem.* 80:1544–48
55. Bronskill, M. J., Wolff, R. K., Hunt, J. W. 1970. *J. Chem. Phys.* 53:4201–10
56. Matheson, M. S., Mulac, W. A., Rabani, J. 1963. *J. Phys. Chem.* 67:2613–17
57. Grossweiner, L. I., Joschek, H. I. 1965. *Adv. Chem. Ser.* 50:279–96
58. Czapski, G., Schwarz, H. A. 1962. *J. Phys. Chem.* 66:471–74
59. Waltz, W. L., Adamson, A. W., Fleischauer, P. D. 1967. *J. Am. Chem. Soc.* 89:3923–24
60. Bennema, P., Hoytink, G. J., Luprinski, J. M., Oosterhoff, L. J., Sellier, P., Van Voorst, J. D. 1959. *Mol. Phys.* 2:431–35
61. Joussot-Dubien, J., Lesclave, R. 1964. *J. Chem. Phys.* 61:1631
62. Gibbons, W. A., Porter, G., Savadatti, M. I. 1965. *Nature* 206:1355–56
63. Cadogan, K. D., Albrecht, A. C. 1965. *J. Chem. Phys.* 43:2550–52
64. Goldschmidt, C. R., Stein, G. 1970. *Chem. Phys. Lett.* 6:299–303
65. Kenney-Wallace, G., Walker, D. C. 1971. *Ber. Bunsenges. Phys. Chem.* 75:634–37
66. Ottolenghi, M. 1971. *Chem. Phys. Lett.* 12:339–43
67. Shirom, M., Stein, G. 1971. *J. Chem. Phys.* 55:3379–82

68. Jortner, J., Ottolenghi, M., Stein, G. 1963. *J. Am. Chem. Soc.* 85:2712–18
69. Hoytink, G. J., Zandstra, P. J. 1960. *Mol. Phys.* 3:371–89
70. Van Voorst, J. D., Hoytink, G. J. 1965. *J. Chem. Phys.* 42:3995–99
71. Eloranta, J., Linschitz, H. 1963. *J. Chem. Phys.* 38:2214–19
72. Fisher, M., Ramme, G., Claesson, S., Szwarc, M. 1971. *Chem. Phys. Lett.* 9:306–8
73. Makkes van der Deijl, G., Dousma, J., Speiser, S., Kommandeur, J. 1973. *Chem. Phys. Lett.* 20:17–22
74. Rentzepis, P. M., Jones, R. P., Jortner, J. 1972. *Chem. Phys. Lett.* 15:480–82
75. Rentzepis, P. M., Jones, R. P., Jortner, J. 1973. *J. Chem. Phys.* 59:766–73
76. Netzel, T. L., Rentzepis, P. M. 1974. *Chem. Phys. Lett.* 29:337–42
77. Struve, W. S., Netzel, T. L., Rentzepis, P. M., Levin, G. Swarc, M. 1975. *J. Am. Chem. Soc.* 97:3310–13
78. Huppert, D., Struve, W. S., Rentzepis, P. M., Jortner, J. 1975. *J. Chem. Phys.* 63:1205–10
79. Huppert, D., Rentzepis, P. M. 1976. *J. Chem. Phys.* 64:191–96
80. Dorfman, L. M. 1974. *Investigations of Rates and Mechanism of Reactions*, ed. G. G. Hammes, 6:463–519. New York: Wiley-Intersci.
81. Halpern, B., Gommer, R. 1968. *J. Chem. Phys.* 43:1069–71
82. Woolf, M. A., Rayfield, G. W. 1965. *Phys. Rev. Lett.* 15:235–37
83. Jortner, J. 1971. *Ber. Bunsenges. Phys. Chem.* 75:696–714
84. Copeland, D., Kestner, N. R., Jortner, J. 1970. *J. Chem. Phys.* 53:1189–1216
85. Levin, G., Claesson, S., Swarc, M. 1972. *J. Am. Chem. Soc.* 94:8672–76
86. Leonhardt, H., Weller, A. 1963. *Ber. Bunsenges. Phys. Chem.* 67:791–95
87. Weller, A. 1963. *Nobel Symp., 5th*, ed. S. Claesson, p. 413. New York: Wiley-Intersci.
88. Chandross, E. A., Thomas, H. T. 1971. *Chem. Phys. Lett.* 9:393–96
89. Okada, T., Fujita, T., Kubota, M., Masaski, S., Mataga, N., Ide, R., Sakata, Y., Misumi, S. 1972. *Chem. Phys. Lett.* 14:563–68
90. Ottolenghi, M. 1973. *Acc. Chem. Res.* 6:153–60
91. Chuang, T. J., Eisenthal, K. B. 1973. *J. Chem. Phys.* 59:2140–41
92. Chuang, T. J., Cox, R. J., Eisenthal, K. B. 1974. *J. Am. Chem. Soc.* 96:6826–30
93. Chuang, T. J., Eisenthal, K. B. 1975. *J. Chem. Phys.* 62:2213–22
94. Ware, W. R., Novros, J. S. 1966. *J. Phys. Chem.* 70:3246–53
95. Hui, M. H., Ware, W. R. 1976. *J. Am. Chem. Soc.* 98:4712–15; 4718–22
96. Nakashima, N., Mataga, N. 1975. *Chem. Phys. Lett.* 35:487–92
97. Noyes, R. M. 1961. *Prog. React. Kinet.* 1:129–60
98. Gnadig, K., Eisenthal, K. B. 1977. *Chem. Phys. Lett.* 46:339
99. Weller, A. Private communication
100. Madge, D., Windsor, M. W. 1974. *Chem. Phys. Lett.* 24:144–48
101. Förster, T., Hoffmann, G. 1971. *Z. Phys. Chem. Neue Folge* 75:63–67
102. Ippen, E. P., Shank, C. V., Bergman, A. 1976. *Chem. Phys. Lett.* 38:611–14
103. Hochstrasser, R. M., Lutz, H., Scott, G. W. 1974. *Chem. Phys. Lett.* 24:162–67
104. Anderson, R. W., Hochstrasser, R. M., Lutz, H., Scott, G. W. 1974. *Chem. Phys. Lett.* 28:153–57
105. Nitzan, A., Jortner, J., Rentzepis, P. M. 1971. *Chem. Phys. Lett.* 8:445–47
106. Rentzepis, P. M. 1969. *Science* 169:239–47
107. Rentzepis, P. M., Mitschele, C. J. 1970. *Anal. Chem.* 42:20–25
108. Rentzepis, P. M., Busch, G. E. 1972. *Mol. Photochem.* 4:353–67
109a. Anderson, R. W., Hochstrasser, R. M., Lutz, H., Scott, G. W. 1975. *Chem. Phys. Lett.* 32:204–9
109b. Shank, C. V., Ippen, E. P., Teschke, O. Private communication
110. Alfano, R. R., Shapiro, S. L. 1970. *Phys. Rev. Lett.* 24:584–87
111. Alfano, R. R., Shapiro, S. L. 1970. *Phys. Rev. Lett.* 24:1217–20
112. Alfano, R. R., Shapiro, S. L. 1971. *Chem. Phys. Lett.* 8:631–33
113. Magde, D., Windsor, M. W. 1974. *Chem. Phys. Lett.* 27:31–36
114. Magde, D., Bushaw, B. A., Windsor, M. W. 1974. *Chem. Phys. Lett.* 28:263–69
115. Magde, D., Windsor, M. W., Holten, D., Gouterman, M. 1974. *Chem. Phys. Lett.* 29:183–88
116. Tashiro, H., Yajima, T. 1974. *Chem. Phys. Lett.* 25:582–86
117. Duguay, M. A., Hansen, J. W. 1969. *Opt. Commun.* 1:254–56
118. Nakashima, N., Mataga, N. 1975. *Chem. Phys. Lett.* 35:487–92
119. Kirk, A. D., Hoggard, P. E., Porter, G. B., Rockley. M. G., Windsor, M. W. 1976. *Chem. Phys. Lett.* 37:199–203
120. Müller, A., Schulz-Hennig, J., Tashiro, H. 1976. *Z. Phys. Chem. Neue Folge* 101:361

121. Pine, A. S. 1968. *J. Appl. Phys.* 39:106–8
122. Mack, M. E. 1968. *J. Appl. Phys.* 39:2483–85
123. Mourou, G., Busca, G., Denariez-Roberge, M. M. 1971. *Opt. Commun.* 4:40–43
124. Fouassier, J., Lougnot, D., Faure, J. 1975. *Chem. Phys. Lett.* 30:448–50
125. Alfano, R. R., Shapiro, S. L. 1972. *Opt. Commun.* 6:98–101
126. Mourou, G., Malley, M. M. 1974. *Opt. Commun.* 11:282–86
127. Porter, G., Reid, E. S., Tredwell, C. J. 1974. *Chem. Phys. Lett.* 29:469–72
128. Mack, M. E. 1969. *Appl. Phys. Lett.* 15:166–68
129a. Lessing, H. E., Lippert, E., Rapp, W. 1970. *Chem. Phys. Lett.* 7:247–53
129b. Cirkel, H.-J., Ringwelski, L., Schäfer, F. P. 1972. *Z. Phys. Chem. Neue Folge* 81:158–62
130. Shank, C. V., Ippen, E. P. 1973. In *Topics in Applied Physics*, ed. F. P. Schafer, Vol. 1, Chap. 3, p. 141. Berlin, Heidelberg, New York: Springer
131. Shank, C. V., Ippen, E. P. 1975. *Appl. Phys. Lett.* 26:62–63
132. Mialocq. J. C., Boyd, A. W., Jaraudias, J., Sutton, J. 1976. *Chem. Phys. Lett.* 37:236–39
133. Hochstrasser, R. M., King, D. S., Nelson, A. C. 1976. *Chem. Phys. Lett.* 42:8–12
134. Tashiro, H., Yajima, T. 1976. *Chem. Phys. Lett.* 42:553–57
135. Arthurs, E. G., Bradley, D. J., Roddie, A. G. 1973. *Chem. Phys. Lett.* 22:230–34
136. Arthurs, E. G., Bradley, D. J., Puntambekar, P. N., Ruddock, I. S., Glynn, T. J. 1974. *Opt. Commun.* 12:360–63
137. Shapiro, S. L., Hyer, R. C., Campillo, A. J. 1974. *Phys. Rev. Lett.* 33:513–16
138. Shapiro, S. L., Kollman, V. H., Campillo, A. J. 1975. *FEBS Lett.* 54:358–62
139. Campillo, A. J., Hyer, R. C., Shapiro, S. L. 1974. *Nucl. Instrum. Methods* 120:533–40
140. Busch, G. E., Rentzepis, P. M., Jortner, J. 1971. *Chem. Phys. Lett.* 11:437–40
141. Busch, G. E., Rentzepis, P. M., Jortner, J. 1972. *J. Chem. Phys.* 56:361–70
142. Hochstrasser, R. M., Wessel, J. E. 1973. *Chem. Phys. Lett.* 19:156–61
143. Orner, G. C., Topp, M. R. 1975. *Chem. Phys. Lett.* 36:295–300
144. Robinson, G. W., Frosch, R. P. 1964. *J. Chem. Phys.* 38:1187–1203
145. Siebrand, W. 1966. *J. Chem. Phys.* 44:4055–56
146. Henry, B. R., Kasha, M. 1968. *Ann. Rev. Phys. Chem.* 19:161–92
147. Bixon, M., Jortner, J. 1968. *J. Chem. Phys.* 48:715–26
148. Jortner, J., Bixon, M. 1968. *J. Chem. Phys.* 48:2757–66
149. Freed, K., Jortner, J. 1969. *J. Chem. Phys.* 50:2916–27
150. Fischer, S., Schlag, E. W. 1969. *Chem. Phys. Lett.* 4:393–96
151. Spears, K. G., Rice, S. A. 1971. *J. Chem. Phys.* 55:5561–81
152. Mourou, G., Malley, M. M. 1975. *Chem. Phys. Lett.* 32:476–79
153a. Penzkofer, A., Falkenstein, W., Kaiser, W. 1976. *Chem. Phys. Lett.* 44:82–85
153b. Rentzepis, P. M., Topp, M. R., Jones, R. P., Jortner, J. 1970. *Phys. Rev. Lett.* 25:1742–45
154. Wirth, P., Schneider, S., Dörr, F. 1976. *Chem. Phys. Lett.* 42:482–87
155. Heritage, J. P., Penzkofer, A. 1976. *Chem. Phys. Lett.* 44:76–78
156. Ippen, E. P., Shank, C. V. 1976. Private communication
157. Förster, T. 1950. *Z. Electrochem.* 54:42–46; 531–35
158. Weller, A. 1961. *Progress in Reaction Kinetics*, ed. G. Porter, 1:187–214. London: Pergamon
159. Jackson, G., Porter, G. 1961. *Proc. R. Soc. London Ser. A* 260:13–30
160. Mataga, N., Kaifu, Y. 1963. *Mol. Phys.* 7:137–47
161. Godfrey, T. S., Porter, G., Suppan, P. 1965. *Discuss. Faraday Soc.* 39:194–99
162. Beens, H., Grellmann, K. H., Gurr, M., Weller, A. 1965. *Discuss. Faraday Soc.* 39:183–93
163. Ofran, M., Feitelson, J. 1973. *Chem. Phys. Lett.* 19:427–31
164. Ingham, K. C., Ashraf El-Bayoumi, M. 1974. *J. Am. Chem. Soc.* 96:1674–82
165. Ashraf El-Bayoumi, M., Avouris, P., Ware, W. R. 1975. *J. Chem. Phys.* 62:2499–2500
166. Shizuka, H., Matsui, K., Okamura, T., Tanaka, I. 1975. *J. Phys. Chem.* 79:2731–34
167. Shizuka, H., Matsui, K., Hirata, Y., Tanaka, I. 1976. *J. Phys. Chem.* 80:2070
168. Friedel, R. A., Orchin, M., Reggel, L. 1948. *J. Am. Chem. Soc.* 70:199–204
169. Hochstrasser, R. M. 1961. *Can. J. Chem.* 39:459–70
170. Zimmerman, H., Joop, N. 1960. *Ber. Bunsenges. Phys. Chem.* 64:1215–21
171. Brown, W. A. C., Trotter, J., Robertson, J. M. 1961. *Proc. Chem. Soc.*, p. 115
172. Kerr, K. A., Robertson, J. M. 1969. *J. Chem. Soc. B*, pp. 1146–51
173. Post, M. F. M., Langelaar, J., Van Voorst, J. D. W. 1975. *Chem. Phys. Lett.* 32:59–62

174. Shank, C. V., Ippen, E. P., Teschke, O., Eisenthal, K. B. 1976. Unpublished results
175. Laubereau, A., von der Linde, D., Kaiser, W. 1972. *Phys. Rev. Lett.* 28:1162–65
176a. Alfano, R. R., Shapiro, S. L. 1972. *Phys. Rev. Lett.* 29:1655–58
176b. Kaiser, W., Kirschner, L., Laubereau, A. 1973. *Opt. Commun.* 9:182–85
177. Monson, P. R., Patumtevapibal, S., Kaufmann, K. J., Robinson, G. W. 1974. *Chem. Phys. Lett.* 28:312–15
178. Laubereau, A., Kaiser, W. 1975. *Ann. Rev. Phys. Chem.* 26:83–99
179. DeMartini, F., Ducuing, J. 1966. *Phys. Rev. Lett.* 17:117–19
180. Spanner, K., Laubereau, A., Kaiser, W. 1976. *Chem. Phys. Lett.* 44:88–90
181. Laubereau, A., Seilmeier, A., Kaiser, W. 1975. *Chem. Phys. Lett.* 36:232–37

HYDRODYNAMICS IN BIOPHYSICAL CHEMISTRY

✤ 2646

Victor A. Bloomfield

Department of Biochemistry, University of Minnesota, St. Paul, Minnesota 55108

INTRODUCTION

Hydrodynamic methods have been used for many years in biophysical chemistry to determine the molecular weight, size, shape, flexibility, electrical properties, and interactions of biological macromolecules. Recent advances in experimental methodology have made such determinations more precise, rapid, convenient, and broadly applicable. These advances are the subject of this review. The topic has not been reviewed before in this series, and thus there is an enormous backlog of work that might be covered. Emphasis here is placed on work published between 1974 and 1976. For background, the reader is referred to standard monographs (1–8). Space limitations have forced me reluctantly to omit coverage of two areas that should properly be part of this review: hydrodynamic theory and the hydration of biopolymers. Fortunately, these topics have been covered well in other recent reviews (9–13). I have also concentrated on molecular hydrodynamics and omitted topics in physiological hydrodynamics, such as bacterial motility and chemotaxis, that are beginning to attract the attention of biophysical chemists. Even with these omissions, the remaining material is so voluminous as to lead perhaps to an overly broad and superficial treatment. I hope that once the ground has been generally surveyed, subsequent reviews in this series will treat particular areas more critically and in greater depth.

To provide overall perspective and to establish notation that is used later, I list some standard relations between measurable transport properties and molecular parameters. For simplicity, the macromolecule is taken to be a sphere of hydrodynamic radius R_h and molecular weight M, dissolved in a solvent of viscosity η_0 and density ρ at temperature T. For flexible-chain polymers, we find $R_h = KM^{1/2+a}$, where a accounts for deviations from random-coil behavior due to stiffness, excluded volume, or polyelectrolyte effects. Such effects, along with those of nonsphericality for elongated or otherwise asymmetric particles, are reflected in the numerical constants in each equation, which it is the business of theory to calculate in terms of molecular structure. In translational motion, important quantities are the

Translational friction coefficient: $f = 6\pi\eta_0 R_h$,   1.

Translational diffusion coefficient: $D_T = k_B T/f$,   2.

Sedimentation coefficient: $S = M(1 - \bar{v}\rho)/Nf$,   3.

where $\bar{v}$ is the polymer's partial specific volume, k_B is Boltzmann's constant, and N is Avogadro's number. Combination of Equations 2 and 3 leads to the

Svedberg equation: $SNk_B T/D_T(1 - \bar{v}\rho) = M$.   4.

If transport occurs in an electric field, we measure the

Electrophoretic mobility: $\mu = QX/f$,   5.

where Q is the molecular charge and X a function of interaction between the polymer and its ionic environment.

In rotational and shearing motion, measured quantities are the

Rotational relaxation time: $\tau_R = A\eta_0 R_h^3/k_B T$,   6.

Intrinsic viscosity: $[\eta] = BR_h^3/M$,   7.

Viscoelastic relaxation time: $\tau_v = C\eta_0[\eta]M/k_B T$,   8.

where A, B, and C are functions of molecular shape and flexibility. Measurement of a transport coefficient in a solvent of known properties thereby enables determination of molecular structure parameters. Conversely, measurement of a transport coefficient of a known molecule enables determination of the properties, particularly η_0, of the medium in which it is immersed. This approach has been especially useful in studies of membranes and cell interiors.

TRANSLATIONAL MOTION

Quasi-Elastic Laser Light Scattering (QLS)

One of the major experimental advances in biochemical hydrodynamics has been the development of QLS as a rapid, accurate technique for measuring translational diffusion and other dynamical properties of macromolecules in solution. QLS has been the subject of several books (14–17) and numerous general reviews of which only the most recent are listed here (18–21). A more specialized review on signal intensity considerations in optical homodyne detection (22) is also of value. Kam et al (23) have described several novel ways to obtain the autocorrelation function of fluctuating signals, which can be readily implemented using a multichannel signal averager and simple gating circuits.

The most common and straightforward application of QLS to biological macromolecules is the determination of D_T, and its combination with S in the Svedberg equation to determine M. This approach has recently been used to characterize a number of bacterial (24–28), plant (24, 25), and animal (29, 30) viruses, as well as the DNA from fd bacteriophage (31) and calf thymus (32). The papers by Pusey et al (24) and by Jolly & Eisenberg (32) contain particularly useful discussions of experimental procedures and data analysis. Application of QLS to polysaccharides revealed substantial polydispersity (33, 34) and for fractionated samples of dextran enabled the construction of an empirical correlation between D and M (33). This would, however, be expected to vary somewhat with degree of branching. A similar D-M correlation was established for highly sulfated mucopolysaccharides (35); this correlation was then used to demonstrate differences in glucuronic acid substitution of human placental dermatan sulfate at two stages of development.

The majority of QLS studies have been on proteins. McDonnell & Jamieson (36) have proposed two methods for determination of protein molecular weights from diffusion measurements: a D vs M correlation in denaturing solvents where the protein exists as a random coil, and a combination of D with $[\eta]$. The latter procedure is based on the well-known Scheraga-Mandelkern approach. The ratio of hydrodynamic radii determined by D and $[\eta]$ is lower than predicted by theory for impermeable ellipsoids. A similar discrepancy has long been known from $S - [\eta]$ studies, and appears to have been explained by recent theoretical calculations employing porous-sphere hydrodynamics (37). Changes in R_h attendant upon aggregation and conformational transitions have been investigated in a variety of proteins and their complexes, such as antifreeze glycoproteins from an Antarctic fish (38), β-lactoglobulin A (39), glutamate dehydrogenase (40–42), glyceraldehyde-3-phosphate dehydrogenase (43), phycocyanin (44), ribosomes (45), and succinylated aspartate transcarbamylase (46). This last system is of particular interest because a 4% decrease in D in the presence of carbamyl phosphate was detected using an ingenious differential optical mixing spectrometer (47). The thickness of protein films absorbed on polystyrene latex spheres has been measured by the change in D and R_h of these spheres (48); this approach should be generally useful in the study of protein interactions with model membranes in the form of phospholipid vesicles. Determination of S and D for chromatin multimers and interpretation of the frictional coefficients using subunit hydrodynamic theory (49) indicates that chromatin either has a helical superstructure or is a flexible coil with attractive interactions between nucleosome units (50). Individual nucleosomes appear to be separated by spacer regions that contribute about 20% of the frictional resistance of a dimer.

A major complication in QLS analysis of polymeric systems is polydispersity. Considerable theoretical effort has been devoted to procedures for analyzing polydisperse systems. The cumulant expansion analysis of the autocorrelation function devised by Koppel (51) has been successfully applied (52, 53). It has the advantage that the first cumulant, for small particles or in the limit of low scattering angles, yields a well-defined z-average D_T, which when combined with the weight-average S yields $\bar{M}_w$. Extensions to larger, Rayleigh-Debye particles have been carried out (52, 54). In pauci-disperse systems, least-squares fitting to a sum of exponential decays may be more accurate (55–57) than cumulant analysis, given realistic signal-to-noise ratios. Direct Laplace inversion of the correlation function to obtain the distribution function of polymer sizes (58), though attractive in principle, requires data of extremely high precision. If the distribution can be parameterized (e.g. as a two-parameter Schulz distribution), the spectrum or correlation function may be computed and the parameters determined by numerical fitting to the experimental data (59, 60). Aragon & Pecora (59) have carried out such calculations for Rayleigh-Debye particles. Particular application of this approach has been made to polydisperse suspensions of membrane vesicles (61). $\bar{D}_z$ and the z-averaged relative dispersion are obtained from scattering measurements; these may be used to compute number-average radius and dispersion (62).

A commonly encountered type of polydispersity, which does not respond well to analysis by the previously discussed procedures, is one in which a large amount of low-M polymer is present along with a small amount of strongly scattering aggregate. The widely different time scales for decay of the scattering autocorrelation functions of the two components enable separation of their contributions; this has enabled analysis of solutions of collagen (63) and phosphatidylcholine vesicles (64). Where possible, of course, it is desirable to fractionate a complex mixture into its components. This approach has been used, for example, to study molecular weight and size distribution of milk casein micelles containing wild-type (65) and mutant (66) proteins. A clever variation is to band the components at different positions in a density gradient (67) or a capillary tube (68) using preparative ultracentrifugation. The tube may be scanned by scattered light to obtain the band positions, and the distance moved yields the sedimentation coefficients. A second scan using QLS gives D_T for the material in each band. This procedure has been used to characterize ribosomes (67), R17 bacteriophage (68), and ribosomal RNA (68). Lim et al (69) have used a similar combination of band electrophoresis in sucrose density gradients with QLS determination of D_T in each band to characterize bacteriophage T4 tail fiber attachment.

Substantial efforts have been made to understand the concentration dependence of D. Two diffusion coefficients must be distinguished—the tracer diffusion coefficient D_T, given by Equation 1, and the mutual diffusion coefficient D_M, which measures the flow of a species due to a concentration gradient in that species. In the infinite dilution limit, these both become equal to $D_0 = k_B T/f_0$, but they may differ at higher concentrations. QLS measures D_M. It is observed experimentally that at low or moderate concentrations, D_M is generally close to D_0, even though long-range hydrodynamic interactions might be expected to couple motions of separate particles. Altenberger & Deutch (70) rationalized this finding and showed that a concentration dependence of D_M would result only if short-range interparticle forces were added to hydrodynamic interactions. This conclusion has been elaborated by Altenberger (71) and Phillies (72–74), who derived the general expression

$$D_M = (\partial \pi/\partial C)_{P,T}(1 - \phi)/f(1 - G\phi), \qquad 9.$$

where $(\partial \pi/\partial C)_{P,T}$ is the osmotic compressibility of the solute, and equals $k_B T(1 + 8\phi + \ldots)$ for hard spheres. ϕ is the solute volume fraction, and must include hydration (75). The concentration dependence of f has been derived by Batchelor (76, 77) as $f = f_0(1 + 6.55\phi + \ldots)$. G is a hydrodynamic integral related to the cross-diffusion tensor. Taking only first-order terms in ϕ, one obtains $D_M = D_0[1 + (0.45 + G)\phi + \ldots]$, which may be compared with $D_M = D_0(1 + 2\phi)$ obtained by Altenberger & Deutch (70).

Newman et al (78), treating fd DNA as a hard sphere, found the dimensionless diffusion coefficient virial in the same range as predicted by Altenberger & Deutch; but did not observe a variation of D with scattering angle, which was predicted by the same theory. For bovine serum albumin (74, 79) the relation between D_T and D_M predicted by Equations 2 and 9 with $G = 0$ was generally obeyed at concentrations up to 200 g liter^{-1}. D_M varies much less strongly than D_T with C, due to compensating thermodynamic and hydrodynamic nonidealities. For human cyanomethemoglobin (80), the term $(1 - G\phi)$ in Equation 9, with $G = -5.8$, was required to fit the data.

A theory has been constructed (81) that also takes into account hydrodynamic and long-range potential interactions, and computations have been performed for a range of potentials, resulting in rather different predictions. Specifically, for hard spheres, it indicates $D_T = D_M$. The difference between this and the previous theories appears to reside in the method of averaging over direct and hydrodynamic forces (82, 83). Resolution of the dispute is complicated by the fact that D_M at high concentrations for both serum albumin and hemoglobin is different when measured by diaphragm diffusion (84, 85) and QLS (79, 80).

Charge interactions are particularly important between biological polyelectrolytes, and produce substantial variations in D with ionic strength and polyion concentration. Qualitatively, one expects and finds that D will be close to its value at the isoelectric point in solutions of high ionic strength, and will increase due to repulsive interactions as the salt concentration is lowered. This general behavior is predicted by a theory of Stephen (86), but quantitative agreement with experiment on bovine serum albumin (87, 88) is not achieved. This is attributed to the use of the Debye-Hückel approximation, in which it is assumed that the screening length is greater than the interpolyion distance, while in fact the reverse is generally true. Doherty & Benedek (87) observed a substantial variance in the cumulant analysis of their data, which they attributed to polydispersity. Phillies (89) has pointed out that some of this variance may arise from slow fluctuations of the protein charges. Lee & Schurr (90, 91) have studied electrostatic effects on the diffusion of poly(L-lysine HBr). They have constructed a linearized theory of scattering from fluctuations in polyelectrolyte solutions that predicts, under all realizable experimental conditions, a single exponential decay of the correlation function proportional to scattering vector $\mathbf{q}^2$ and to an average D_T dependent on polyion and salt concentrations and diffusion coefficients. Detailed comparison between theory and experiment is made difficult by conformational changes in the polypeptide.

Striking phenomena have been observed in very low salt solutions of charged R17 virus (24, 92, 93) and polystyrene latex spheres (94), where long-range electrostatic interactions appear to produce ordering of the particles. This ordering is reflected in maxima in the scattering intensity as a function of angle, and in concomitant angular variations of the effective particle diffusion coefficient (95). Greatly reduced scattering intensities and diffusion coefficients in low-salt polylysine solutions are also attributed (96) to long-range interionic forces. Theoretical progress in this difficult polyelectrolyte area may be aided by the development of computer simulation techniques for modeling polyelectrolyte solutions (97).

An attractive potential use of QLS is to determine relaxation times of rotational and internal modes of polymers. This field has recently been reviewed by Schurr (98). These intramolecular motions should modulate the scattering dynamics when their scale is comparable to the inverse scattering vector. For flexible random-coil polymers, the generally accepted treatment of dynamics is the normal coordinate approach of Rouse & Zimm. The free-draining Rouse model forms the basis of the spectral calculations by Pecora (99). Büldt (100) has shown how to calculate mean relaxation times for the free-draining, pearl necklace model; while Perico et al (101) have included hydrodynamic interaction and conclude that this enhances the contribution of intramolecular motions relative to translation. While comparison between theory and experiment is made difficult due to requirements for extreme monodispersity of the polymer sample and to problems in unscrambling several exponentially decaying contributions to the correlation function, measurements on polystyrene in cyclohexane and in 2-butanone (102–104) are in good agreement with normal coordinate theories. However, scattering from DNA (32, 105, 106) exhibits quite different behavior. This has led Lee & Schurr (107) to propose a model of unconnected segments moving in a spherical mean-force field. This theory predicts a q^2-dependence of the relaxation rate at both $qR_G \ll 1$ and $qR_G \gg 1$ (R_G is the radius of gyration). The apparent D_T at small qR_G is that for translation of the entire polymer; while at large qR_G, D_T is the sum of macromolecular and segment diffusion coefficients. While this dependence describes much of the data, it is still unclear why the more rigorously based normal coordinate theories fail. At higher polymer concentrations, or with very large DNA molecules, very long relaxation times appear that may be due to anisotropic motion of instantaneously nonspherical molecules crowded by their neighbors (98, 106).

The effect of histones F2A and F1 on the dominant internal mode of DNA has been interpreted in terms of supercoiling (F2A) and uncoiling (F1) (108), but may instead be due to intermolecular aggregations (F1) and local uncoiling (F2A) (109).

Several schemes have been proposed for enhancing the contribution of internal modes to QLS, relative to translational diffusion. Depolarized forward (zero-angle) scattering contains information only on rotational and internal modes; it has been used successfully to study the rigid molecules TMV (110–112) and poly(n-hexylisocyanate) (113), and flexible DNA (114) and polystyrene (113). Bloomfield (115) has proposed to "label" the ends or middle of the chain with a small but strongly scattering molecule and to cancel scattering from the rest of the polymer by refractive index matching. Chemical implementation of this idea has been difficult. Schmitz (116) has suggested application of a sinusoidally oscillating electric field to a random-coil polyelectrolyte. If the applied frequency is greater than the translational Doppler shift, the center of mass will remain stationary while internal modes may be driven. For this concept to succeed, Joule heating problems must be overcome.

Depolarized scattering at finite angles has also been used to study rotation of muscle calcium-binding protein (117) and of lysozyme (117, 118). The equivalence of τ_R for the former protein measured in this way and by ^{13}C NMR implies that the two techniques are sensitive to the same rotational motions and that the α-carbon backbone acts as a rigid body. Applications of depolarized scattering to small molecules, liquid crystals, and nonbiological polymers has been recently reviewed by Bauer et al (119) in this series. A particularly attractive experimental technique appears to be resonance-enhanced depolarized Rayleigh scattering from diphenylpolyenes (120). These may serve as probes of molecular reorientation in membranes, and conjugation to soluble macromolecules may greatly facilitate depolarized scattering spectroscopy.

Kinetics is another area in which QLS holds great promise theoretically, but experimental realization has lagged somewhat. By now it is well known that if a reaction involves a change in polarizability, and/or in D_T in going from reactants to products, and if the reaction rate is on the order of the diffusion rate, $q^2 D_T$, then the reaction should broaden the power spectrum of scattered light or be observable as one or more exponential decays in the autocorrelation function. In practice, however, the magnitudes of these effects can be shown theoretically to be small and the usual difficulties arise in dissecting a sum of exponentials into the individual components. Some potentially favorable cases are discussed by Benbasat & Bloomfield (121). A somewhat different approach is to study macromolecular reactions that are slow compared to the diffusion rate. One then measures the averaged diffusion coefficient as a function of time under conditions where the measurements of $\bar{D}_T$ can be accomplished in a time that is short compared to the half-life of the chemical reaction. This approach was utilized (27) to study the attachment of bacteriophage T4D heads to tails. The reaction appears to be essentially diffusion controlled and, since the virus parts scatter so strongly, the reactants can be diluted to slow down the bimolecular reaction while $\bar{D}_T$ for the system can still be measured with accuracy. This approach should be generally useful for reactions involving large, strongly scattering macromolecules whose translational frictional properties change substantially upon reaction.

Light-scattering studies from hemoglobin solutions have shown a positive intercept of the plot of spectral linewidth versus q^2 (122). This was attributed to hemoglobin dimer-tetramer equilibrium. However, it appears (123) to be due instead to adsorption of the helium-neon laser line by methemoglobin contaminant, and consequent thermal convection.

The combination of QLS with electrophoresis yields an interesting method for simultaneously determining μ and D_T for charged particles in solution (124, 125). In the power spectral mode, the spectrum appears as a Lorentzian, with half-width at half-height proportional to D_T, displaced from zero frequency by an amount proportional to the electrophoretic Doppler shift, which is in turn proportional to μ. This technique affords good precision and much more rapid measurement of mobilities than is possible by traditional moving-boundary techniques. It has by now been applied to a number of biopolymers and their mixtures. Recent applications include calf thymus DNA and tobacco mosaic virus (126), RNA tumor viruses (127), and red and white blood cells (128–130). Resolution among different cell types is possible and it was found (130) that white cells from leukemic patients have lower electrophoretic mobilities than those from normal patients. Progress in electrophoretic light scattering has been slow largely because of difficulties in designing suitable scattering cells and electrodes to overcome problems with reproducible heterodyning, bubble formation, Joule heating, and electrode product contamination. These technical difficulties and approaches to improved design are discussed by a number of workers (131–134).

The usual electrophoretic light-scattering experiment employs heterodyne detection because the phase information characteristic of the electrophoretic Doppler shift disappears in the homodyne mode. Josefowicz & Hallett (135) have designed an ingenious homodyne electrophoretic light-scattering apparatus that employs two parallel laser beams polarized in the same direction. These are focused into the scattering cell by a cylindrical lens and interfere at the focal point to produce a fringe intensity profile. Scattering particles driven through the fringes by the electrophoretic field modulate the scattered light intensity that is detected by digital photoelectron pulse autocorrelation. A combination of electrophoresis in density-gradient-stabilized bands with QLS has been developed by Lim et al (69) and used to study attachment of tail fibers to bacterial viruses. The theory of electrophoretic light scattering has been examined by several workers who have taken ionic interactions into account in a more detailed fashion than in the simple theory generally employed for data analysis. Stephen (136) has extended his polyelectrolyte diffusion theory (86) to the electrophoretic case by employing the Debye-Hückel approximation. Phillies (137) and Friedhoff & Berne (138) have utilized the formalism of irreversible thermodynamics in three-component ionic solutions and have shown that the Doppler shifts are not determined by the mobilities of individual ions but instead depend on more complicated electrochemical interactions. These studies are unlikely to substantially change conclusions based on simple theories for macromolecular polyelectrolytes, however.

In passing from the study of more or less dilute solutions where the scattering macromolecules behave as approximately free particles, one may consider QLS from viscoelastic gels that are crosslinked macromolecular structures swollen by solvent. Such gels are of interest in industrial polymer applications, in gel chromatography and electrophoresis, and in various biological structures such as the eye. Experimental studies of QLS from viscoelastic gels have been carried out by several workers (139–144). The correlation function for scattering from thermally excited longitudinal and transverse displacements of the gel fiber network is predicted (140) to have the form of an exponential decay where the decay rate is a function of the compressional or shear modulus of the gel, the frictional coefficient of the fiber network with respect to the gel liquid, and the scattering angle. Munch et al (144) have constructed a similar theory based on statistical thermodynamics of rubber elasticity rather than macroscopic viscoelastic theory, and have made a molecular interpretation of the various moduli. An alternative theoretical treatment of the scattering spectrum or correlation function from viscoelastic gels is developed in terms of scattering from macromolecules that behave as harmonically bound particles and execute independent Brownian motion about a stationary mean with additional static scattering due to spatial structuring of the macromolecules in the gel (142, 145, 146). Connection between these harmonically bound particle theories and the macroscopic gel theories is as yet unclear. In studying the sol to gel transition in agarose systems Wun et al (147) observed oscillatory behavior of the autocorrelation function that persisted for several hours. This was attributed to convection occurring during microphase separation.

Complexes of the muscle protein F-actin with heavy meromyosin and the myosin fragment S1 show characteristics of gels, and the F-actin does not appear to exhibit free diffusion (145). Carlson (148) has studied intensity fluctuation spectra of coherent light scattered from striated muscle, He observed large-scale fluctuations in position and polarizability at the level of the myofibrillar sarcomere and its structural subunits during steady-state contraction. Fluctuating forces generated by the making and breaking of cross-bridges are shown to account for some of the features of the spectra.

Finally, although it does not depend on scattering, the technique of fluorescence correlation spectroscopy (FCS) has also been devised as a procedure for determining diffusional transport and reaction in bulk solutions and membranes. For reviews see (149, 150). The theory is discussed in detail by Elson & Magde (151), who consider pure diffusion; the binding of a small, rapidly diffusing ligand to a larger, slowly diffusing macromolecule; and unimolecular isomerization. Experimental realization of the technique was achieved in determining D_T of the fluorescent molecule, rhodamine 6G, and the kinetics of the intercalation reaction of macromolecular DNA with the drug ethidium bromide (152). Application of the related technique of fluorescence photobleaching recovery to membrane systems is discussed below. Phillies (153) has considered the information that may be obtained from FCS of a nonideal solution and has elucidated situations where the mutual and tracer diffusion coefficients will be obtained. Koppel (154) has presented a thorough analysis of statistical accuracy in FCS.

Weissman et al (155) have presented an ingenious method for determining molecular weights of large molecules and cells by fluctuation spectroscopy. They monitor spontaneous concentrations of fluctuations in a rotating cell and detect the correlations that have a time lag equal to the period of rotation. The method gave good results when applied to T2 bacteriophage DNA and to replicating *Escherichia coli* DNA and should be extremely useful in studies of very large macromolecules.

Pulsed-Gradient NMR (PGNMR)

Although QLS is the major new experimental method for translational diffusion of biopolymers, PGNMR has emerged as a useful new technique for measuring small-molecule diffusion and has applicability to larger species as well. Cooper et al (156) used PGNMR to measure restricted diffusion of water in animal tissues (red blood cells, plasma, heart, and liver). The measured D_T was larger for short measuring times than for longer ones, because permeability barriers restrict diffusion at larger distances and times. Restrictive lengths from 1.5 to 15 μm are observable by PGNMR; in packed red cells, a length of 2.3 μm was observed and attributed to the red cell membrane. The membrane is not completely impermeable, and the extension of PGNMR theory in inhomogeneous systems to encompass penetrable barriers was developed in a later paper (157). The retardation of diffusion of gases and small molecules by hydrated polymer gels has also been investigated by more traditional techniques (158, 159). PGNMR was used to study diffusion of dimyristoyl-lethecin (DML) vesicles in D_2O solutions (160). Comparison with results obtained by ultracentrifugation and Stokes law calculation indicates that DML diffusion is governed primarily by whole liposome migration. D_T of lecithin in a cubic mesophase with cholate and water is the same as in phospholipid bilayers (161).

Fluorescence Methods

The diffusion-controlled rate of pyrene excited-state dimer (excimer) formation, monitored by the changed excimer fluorescence, has been used to monitor lateral diffusion and phase transitions in membranes and phospholipid vesicles (162, 163). Agreement with experiment is improved if time-dependent rather than steady-state gradients are considered, and if the two-dimensional nature of the diffusion process is taken into account (164, 165).

A promising new technique for measuring lateral mobility of fluorescent particles in two dimensions is fluorescence photobleaching recovery. A small spot on the surface is bleached by a brief, intense laser flash; and the recovery of fluorescence in the spot is monitored as fluorophores diffuse or flow into it from other regions of the system. The basic theory (166) and experimental details (167) have been presented along with results from model membranes and bulk solution and comparison to FCS (149). The technique allows distinction between diffusion and directed flow, and determination of both the mobility coefficient and the fraction of fluorophores that are mobile. A particularly interesting biological application involves the mobility of concanavalin A receptors on myoblasts (168). The method resembles an earlier determination of D_T of visual pigment in photoreceptor disc membrane by bleaching a fraction of the rod outer segment and following the recovery of light absorption (169).

Theoretical understanding of translational diffusion in membranes presents a fundamental problem, for, according to the Stokes paradox, there is no solution of the two-dimensional viscous-flow equations for an infinite membrane. Saffman & Delbrück (170) showed that the paradox could be eliminated in three ways whose relative contribution is unclear: 1. give the membrane a finite size, 2. take account of the viscosity of the bounding liquid, or 3. calculate the mean-square displacement using the Langevin equation with the drag given by unsteady-flow theory. They conclude that, in practice, η_0 measured by translational diffusion should be about one-fourth that measured by rotation. A useful general review of rotational and translational diffusion in membranes has been written by Edidin (171).

Ultracentrifugation

Continuous attempts are being made to improve the specificity, accuracy, and convenience of ultracentrifugation as an analytical (172) and preparative technique. A very useful technique is active-enzyme ultracentrifugation, in which a band of the enzyme, possibly at very low concentrations, is sedimented through a substrate solution; the conversion of substrate to product is detected spectrophotometrically. This method detects only the catalytically active species in a complex mixture. It has been used to determine the state of aggregation of the reactive enzyme complex in pyridine nucleotide transhydrogenase (173), antitumor asparaginase and glutaminase (174), and yeast hexokinase (175). In the last of these studies, the validity of the procedure under conditions of rapid association-dissociation equilibrium was verified. A different type of biological specificity, that of the complex between antigen and antibody, was used by Simpson & Bustin (176) in a technique they called immunosedimentation to determine the histone composition of chromatin subunits. Antibodies to the various histones were prepared, and the formation of complexes detected by an increased sedimentation rate.

Sedimentation in density-stabilized gradients is of widespread utility. Clark (177) has described a computational procedure to obtain $S_{20,w}$ from ultracentrifugation in linear sucrose gradients, taking into account solvent density and viscosity. Koch & Blumberg (178) have developed a theoretical treatment of the distribution of bacteria in density-gradient sedimentation, as a function of variable cell size and mass in asynchronous growth.

Sedimentation studies of very large molecules, such as high-M DNA, are often complicated by vexing anomalies. One of these is the rapid sedimenting-out of solution of a certain percentage of the DNA, depending on concentration and rotor speed; the remainder sediments normally. Molecular entanglement (179) does not appear to be responsible (180). Another suggestion is based on computer simulation of a solution with molecules undergoing sedimentation and diffusion and subject to intermolecular hydrodynamic interaction (181). It is predicted that as D decreases, molecular clusters form in high centrifugal fields. These move through solution in a wave-like fashion, gaining material at the front and losing it at the rear. Convective instabilities also arise in preparative zone centrifugation. Mason (182) has shown that maximum particle loading is greatly overestimated by the simple stability criterion of a total density gradient in the direction of the sedimenting force, and presents specific recommendations for optimum loading. Another type of anomaly is concentration-independent decrease in S at high centrifugal fields for large DNA ($M > 10^8$) (183). This appears to be due to uneven frictional forces on the chain, which are less near the center than at the ends. The ends therefore tend to lag behind the center and cause hydrodynamic shielding to be reduced by increased intersegmental distances (184).

Gel Permeation Chromatography

Gel permeation chromatography has become an important method for characterizing the molecular weight and size (Stokes radius R_h) of proteins and other biopolymers, but careful calibration with known standards is required. Tanford et al (185) have shown how R_h of proteins in detergent solutions can be measured and how to combine this with sedimentation coefficients to determine molecular weights. Such detergent procedures are particularly useful for membrane and lipoproteins, which are sparingly soluble in aqueous solution. When gel chromatography is applied to native, undenatured proteins, anomalous retardation of large, asymmetric particles is observed (186), which may plausibly be attributed to end-on insertion of asymmetric polymers into the gel pores (186, 187). Casassa (188) has examined the theory of gel permeation chromatography and concluded that a characteristic dimension governing the permeation of simple pores is the mean projection of a polymer molecule onto a line. Relations between the mean projection and the hydrodynamic volume, reflected in $[\eta]$, suggest that the product $[\eta]M$ should not be a common calibration factor for elution of all macromolecular species from a gel column. However, this correlation is often found to hold in practice among chemically similar polymers, and even among some rather dissimilar ones such as random-coil polystyrene and helical poly(γ-benzyl-L-glutamate) (189). Small (190) has shown that separation of submicron colloidal particles can occur in flow through beds of solid, nonporous particles.

In these investigations the gel behaves as an inert matrix. More specific separation can be achieved in affinity chromatography, given specific attractive interactions between solute and gel. Denizot & Delaage (191) have applied a statistical theory to affinity chromatography and have found that association and dissociation rate constants in the stopped-flow range may be obtained from analysis of dispersion.

Electrophoresis

Since work by Tiselius in the 1930s, electrophoresis has been an important technique for characterization and separation of biological polyelectrolytes. Its use has vastly increased with the development of procedures for electrophoresis on supporting media such as paper and gels. Among the myriad applications to particular systems that are published each year, we note just a handful of typical ones. Leaback (192) has surveyed techniques of electrophoresis in protein analysis, and a procedure for the analysis of polypeptide molecular weights by electrophoresis in urea has been described (193). The dependence of electrophoretic mobility in polyacrylamide gels on molecular weight of double-stranded RNA (194) and small double- and single-stranded DNA (195) has been determined. Covalently closed circular DNA molecules containing different numbers of superhelical terms can be distinguished and quantitated in agarose gels (196–198). Even particles as large as intact bacteriophage T4D (mol wt $\approx 2 \times 10^8$, length ≈ 3000 Å) can be electrophoresed through dilute polyacrylamide gels (199). On the theoretical side, deKeizer et al (200) have shown that if the relaxation effect is neglected, the electrophoretic mobility of a randomly oriented cylinder is obtained from the arithmetic, rather than the harmonic, mean of the mobilities of the cylinder oriented parallel and perpendicular to the field.

The resolution of electrophoresis may be greatly enhanced by conducting it in the presence of a pH gradient; this is called isoelectric focusing. The theory of transient-state isoelectric focusing has been developed by Weiss et al (201) and the effects of zone load and carrier ampholyte concentration on the kinetics of defocusing and refocusing in sucrose density gradients have been determined (202). Another major experimental advance in electrophoresis, its combination with QLS, has been reviewed above.

Hawley & Mitchell (203, 204) have used electrophoresis to detect intermediates in the reversible denaturation of chymotrypsinogen. Denaturation by high pressures behaves like a simple two-state process, but rapid quenching to low temperature of the mixture formed by thermal denaturation reveals an intermediate species.

Transport in Reacting Systems

An important but difficult area is the analysis of transport in reacting systems (205, 206). Three major cases may be distinguished, depending on the relation of the chemical relaxation time to the duration of the transport experiment (the latter is typically on the order of one to several hours). If the reaction is very slow compared to transport, the system will appear to be essentially nonreacting. Such appeared to be the case in high-speed ultracentrifugal study of the association-dissociation reaction of 30S and 50S *E. coli.* ribosomal subunits (207). However, this is a complicated situation since the sedimentation pattern and average sedimentation coefficient are strongly dependent on rotor speed, and an artifact in the low-speed behavior has been implicated. Moreover, study of the reaction by other techniques indicates that it is rapid on the sedimentation time scale. At the other extreme, where reaction is rapid compared to transport, the classical analysis by Gilbert and co-workers pertains, if diffusion is neglected. Frigon & Timasheff (208) have used Gilbert theory, along with computer simulation, to study the magnesium-induced self-association of calf brain tubulin. Schönert (209) has presented a series solution, including effects of diffusion and various kinds of nonidealities, of the differential equation describing transport of a reversibly reacting substance in an infinite rectangular cell. The solution converges for short times. Monomer-trimer and monomer-dimer-trimer systems are discussed in detail (209).

The third and most difficult case is the intermediate one in which reaction and transport proceed at comparable rates. This is termed kinetic control. An important advance has been achieved by Mitchell (210) for transport in an isomerizing system. Using probabilistic arguments, he has derived exact analytical expressions for the spatial distribution of molecules, with arbitrary values of the diffusion coefficients in the two states. Expressions are obtained for the mean and variance of the distribution, which allow interpretation of experimental results in terms of isomerization rate constants. Consideration is given to both initially equilibrium and nonequilibrium ("jump") conditions. Other theoretical work in kinetically controlled transport systems is based on computer simulation. This includes molecular sieve studies of isomerizing solutes on molecular sieve columns, with experimental comparison to the reversible unfolding of hen egg lysozyme (211); sedimentation of ligand-mediated and -nonmediated dimerizing systems (212); gel chromatography of a monomer-n-mer ($n = 2$ and 4) association-dissociation reaction (213); and gel chromatography of a monomer-dimer-tetramer system (214). In all of these studies, emphasis is on the number of separate boundaries to be expected, and the time range over which kinetic control may be manifested. Consideration of dispersion (diffusion) effects is particularly important in gel chromatography because of the effects of the inert matrix. Claverie & Cohen (215, 216) have presented a general numerical procedure for analysis of sedimentation in interacting systems, based on an integral form of the Lamm equation and the use of a finite-element numerical method. They have applied it to a variety of situations, notably to active-enzyme centrifugation and to concentration-dependent sedimentation.

ROTATIONAL MOTION AND FLEXIBILITY

Advances in instrumentation have brought increasing prominence to methods for measuring rotational diffusion and intramolecular relaxation. For spherical particles, the basic quantity obtained is the rotational-relaxation time τ_R or some closely related rotational diffusion coefficient or frictional coefficient. For nonspherical or flexible particles, more than one relaxation time is expected; each of these will be a function of particle dimensions. The dependence of rotational motion on R_h^3 makes it a much more sensitive indicator of molecular size than translational motion, which depends linearly on R_h. Alternatively, if R_h is known, τ_R may be used to determine the viscosity of the medium.

In all but the simplest rotational-relaxation processes, several exponential decays will be anticipated. Fitting of multiexponential data is a notoriously difficult problem of long standing. Recent interesting proposals are moment index displacement (217), which corrects some systematic instrumental errors, and a method (218) based on the Fourier convolution theorem.

Koenig (219) has corrected an error in the Perrin formula for the rotational diffusion coefficient of the short axes of an ellipsoid of revolution, and Wright & Baxter (220) have presented numerical inversion techniques to obtain the semiaxes of prolate and oblate ellipsoids of revolution from the Perrin equations for translational and rotational diffusion coefficients.

Rotational motion is detected through a variety of optical, spectroscopic, and dielectric probes. It may be produced either by the application of a field, or by spontaneous Brownian motion. We consider the latter cases first.

Fluorescence Anisotropy

Steady-state fluorescence anisotropy measurements have been used for decades, since the fundamental work of Perrin and of Weber, to determine τ_R. Recently, this technique has been employed to study such diverse topics as conformational changes and complexing patterns in histones (221–223), the orientation of myosin cross-bridges in muscle fibers (224, 225), the development of the immune response as reflected in complex formation between fluorescein and antifluorescein antibody (226), changes in electronic properties of dansylgalactoside bound to *lac* carrier protein in "energized" *E. coli* membrane vesicles (227), the rate of renaturation of DNA modified by chloroacetaldehyde (228), the microviscosity of the hydrocarbon region of bovine retinal rod outer-segment disk membrane (229) and of normal and transformed 3T3 cell membranes (230), and phase transitions and fluidity in phospholipid bilayers (231–233). In the bilayer case, results are found to be different for perylene and for diphenylhexatriene, disk- and rod-shaped probes, respectively. Conti (234) has reviewed the use of fluorescent probes in studying nerve membranes.

More direct insight into rotational depolarization of fluorescence is obtainable by nanosecond pulse techniques. Two recent papers have used this method to study complex formation between a small protein subunit and a larger aggregate, taking advantage of the restriction of rotational mobility of the small subunit upon complexation. Wu et al (235) studied complex formation between the σ subunit of

E. coli RNA polymerase and the core polymerase during gene transcription; and Highsmith et al (236) measured thermodynamic parameters for binding of myosin fragment S-1 to F-actin. Weber et al (237) have used the frequency-domain counterpart of nanosecond decay, differential polarized-phase fluorometry (238), to set limits on the fraction of cell-bound fluorophores that change properties upon addition of colicin El (239).

When studying the motion of a small-molecule probe in a high-polymer solution, caution must be exercised in attributing changes in τ_R to changes in solution viscosity, since the local microviscosity may be considerably less than the macroscopic η_0. Mazo (240) has made order-of-magnitude estimates indicating that the decrease in fluorescence depolarization when diphenylanthracene is dissolved in poly(γ-benzyl-L-glutamate) or polystyrene solution is due to short-range interactions rather than long-range hydrodynamic interactions.

Many rotational-relaxation processes, of large molecules or in viscous media, have τ_R much longer than the typical fluorescence lifetime of order 10^{-9} to 10^{-8} sec. Galley (241) has found that buried tryptophans in proteins have phosphorescent lifetimes in the 10^{-3} sec range and are protected from solvent-mediated quenching processes. Phosphorescence anisotropy can therefore be used to monitor protein rotation in the millisecond range. Cherry et al (242, 243) have devised a related procedure, measuring the decay of the dichroism of triplet-triplet absorption following excitation by a pulse of polarized light. By covalent conjugation of the triplet-probe eosin isothiocyanate to proteins, τ_R in the 10^{-6} to 10^{-3} sec range can be determined. It has also been noted (245, 246) that the fluorescence intensity from a solution steadily irradiated by polarized light will be modulated by the rotational motion of the chromophores. For spherical rotors, theory predicts three relaxation times for the fluorescence time autocorrelation function, two of which are independent of fluorescent lifetime and depend only on τ_R. This enormously extends, at least in principle, the time range over which rotational motion can be studied by fluorescence. However, as in other forms of fluorescence correlation spectroscopy, substantial problems in signal-to-noise ratio and photodegradation of the fluorophore must be overcome before experimental implementation is practical.

A constant concern in fluorescence anisotropy measurements is whether the fluorophore moves rigidly with the macromolecule or has independent mobility. Experimentally this is best resolved by several independent measurements of τ_R. Theoretically, Wahl (247) has investigated the effect of reflecting and absorbing barriers on the anisotropy, decay, and quantum yield of chromophores executing rotational Brownian motion on a macromolecular surface. Valeur et al (248, 249) have computed the orientation autocorrelation function for internal bonds of a polymer on a tetrahedral lattice, undergoing three- and four-bond motions. Time-dependent fluorescence polarization studies of polystyrene in solutions (250) confirm the general predictions of the theory.

EPR of Spin Labels

Nitroxide spin labels have been used extensively to probe the fluidity of membranes and to monitor other manifestations of biomolecular rotation. Space limitations and the reviewer's competence preclude an extensive survey of this very active field. The reader is referred to recent books and reviews (251–253) for basic theory and applications. We note here only the important development of saturation transfer EPR by Hyde and co-workers (254, 254a). This extends measurable rotational correlation times from the 3×10^{-7} to 10^{-10} sec range characteristic of linear-response techniques (i.e. rotational frequencies on the order of magnetic anisotropies), to the slower range 10^{-7} to 10^{-3} sec. This has been used (255) to study the motion of myosin subfragment S-1, with a spin label rigidly attached, in intact myosin and in various complexes.

Electrooptics

When orientation is produced by an electric field, the time scale for measurement of rotational motion is greatly extended, but significant difficulties in interpretation of the orienting mechanism may arise. A biological polyelectrolyte may be oriented by interaction of its permanent dipole moment with the E field, by an induced moment (particularly in the ion atmosphere surrounding the polyion), or by some combination of these. In theory these mechanisms may be distinguished by their E dependence, by their saturation behavior at high E (256), and by their behavior in reversing-pulse experiments; but in practice the distinction is often difficult to make. Two monographs on electrooptics have appeared recently (257, 258).

Among the helical proteins and polypeptides, collagen appears to orient primarily by a permanent dipole mechanism (259), while helical poly(L-glutamic acid) mainly shows induced dipole behavior in aqueous solutions (260), as do various myosin fragments (261). Two decays were observed in heavy meromyosin; the faster one is probably related to the reorientation of the S-1 subfragment about the flexible hinge. Studies on paramyosin indicated two forms depending on extraction technique (262). The longer, apparently native, form had a much larger permanent dipole moment at low pH, attributed to excess basic amino acids at one end. Application of an electric field to extremely stretched, skinned muscle fibers increased the intensities of optical diffraction lines; this was attributed to orientation of the permanent dipole of the thin (F-actin) filament (263). Concentrated solutions of rodlike molecules sometimes exhibit anomalous negative birefringence. This is the case for the common strain of tobacco mosaic virus, although at very high fields the steady-state birefringence becomes positive (264). Such behavior is suggestive of a permanent dipole perpendicular to the long axis, although this ought to be impossible on the basis of symmetry.

In globular proteins, where two relaxation times might be expected from the different molecular axes but where only one might previously have been observed, a more rapid transient electric birefringence apparatus has led to the detection of two decays, which can be interpreted in terms of Perrin theory to obtain the axial dimensions. The proteins include bovine serum albumin (265), hemocyanin (266), and rhodopsin solubilized in micelles (267, 268). Two transient electric birefringence studies of protein–sodium dodecyl sulfate complexes (269, 270) are in disagreement with the common model for those complexes, i.e. prolate ellipsoids with constant semi-minor axes and semi-major axes proportional to molecular weight, and also disagree with each other in certain respects.

Double-stranded DNA would be expected to undergo a transition from flexible-chain behavior at high molecular weight to rigid-rod behavior at low molecular weight. This has been observed by electric birefringence in oscillating fields (271, 272). The antiparallel orientation of the two strands would appear to preclude by symmetry a permanent dipole moment, but some experiments (272) suggest the existence of such a moment. Others (273), more plausibly, do not. These experiments on short DNA produced by sonication are complicated by polydispersity that exists even in well-fractionated samples.Electric dichroism has been used to determine the geometry of complexes of DNA with ethidium bromide (274) and with a covalently bound carcinogen analog (275). Dourlent & Hogrel (276) suggest that transient electric dichroism in temperature-jump experiments may be responsible for a fast relaxation process previously attributed to complex formation in DNA-proflavin mixtures.

Greve & Blok (277, 278) have undertaken an interesting transient electric birefringence study of T-even bacteriophage. Fiberless T4 particles and T4B in the absence of tryptophan (tryptophan is required by this strain for infectivity; it causes extension of tail fibers) are hydrodynamically equivalent, with rotational diffusion coefficient $D_{rot} \approx 290$ sec^{-1} for both. Orientation is largely due to a permanent moment, which is about 24,000 D for T4B and 95,000 D for fiberless phage. When fibers are extended, by addition of tryptophan to T4B or by use of T4D, D_{rot} decreases to 133 sec^{-1}, and the dipole moment increases to 2×10^5 D. This enormous increase is due to numerous positive charges on the fibers, which are on the order of 10^3 Å from the center of rotation of the virus. The hydrodynamic changes in T4 and T2 phage attendant on fiber reorientation are fairly accurately predicted by a recent theoretical treatment (279).

A potentially important use of electric birefringence is to study reaction dynamics if the reaction is accompanied by a change in D_{rot} and/or birefringence. Pearlstein (280) has presented an analytical solution to a system of equations describing the decay of birefringence in an isomerizing system (281). The electric field itself, which typically reaches 10 kV cm^{-1}, may itself induce macromolecular reactions, such as helix-coil transitions (282, 283) and aggregation of bacterial flagellar protein filaments (284).

Although birefringence and dichroism have been the most common optical indicators of orientation in electric fields, other properties may also change. These include optical rotation (285) and fluorescence polarization (286). The wavelength dispersion of the dichroism also provides useful information (287). The electric field in the fixed-Q mode of a YAG laser beam has been used to produce orientation birefringence in several biopolymers (288); the field reached 5 kV cm^{-1} in a 200 μsec pulse. At the other extreme, low fields can be used in transient electric birefringence, enhancing sensitivity and accuracy by several orders of magnitude over traditional high-field pulse techniques and minimizing thermal degradation of the sample through the use of a low-powered laser light source, crystal-polarizing optics, signal averaging, and digital data processing (289). This procedure was successfully applied to TMV, and holds promise for other systems as well.

Dielectric Dispersion

Dielectric dispersion also provides information on rotational motion, flexibility, and charge distribution of macromolecules. The wide variation of AC frequency enables detection of multiple relaxation processes, and the low applied voltages in comparison to electric birefringence leads to less thermal denaturation. However, results are often hard to interpret, because of various possible orientation mechanisms: permanent dipole, induced dipole, ion-atmosphere polarization, Maxwell-Wagner, etc. Some of these ambiguities may be resolved by systematically varying the molecular weight, charge, and ionic strength, but polydispersity, aggregation, and basic uncertainties about the behavior of polyelectrolyte solutions have made progress slow.

Aqueous solutions of synthetic polypeptides and other polyelectrolytes typically show two dispersions: one at low frequency, which is molecular weight dependent; and one at high frequency, which is not. This behavior has been observed, for example, for poly(ε,N-succinyl-L-lysine) (290), poly(L-glutamic acid) (291), polymethacrylic acid, and polyacrylic acid (292). This behavior is explained by a theory of van der Touw & Mandel (293) that attributes high-frequency relaxations to fluctuations in bound counterions along small portions of the polymer, while the low-frequency dispersion is determined largely by overall molecular rotation. The static electric permittivity is due to fluctuations in the counterion density over the entire polymer domain. Some other recently constructed theories of dielectric relaxation (294, 295) are restricted to solutions without small ions. A series of once-broken helical polypeptide rods, formed by linking two rigid segments of poly(γ-benzyl-L-glutamate) through a flexible, bivalent initiator, had a dipole moment per residue about half that of the corresponding unbroken helixes, and relaxation times 3 to 4 times lower (296). These results, which are not in good agreement with theory for freely hinged rods, were attributed to hindered rotation of the joint. Dielectric dispersion has also been used to study the very cooperative helix-coil transition of poly(ε-carbobenzoxy-L-lysine) in *m*-cresol (297) and the aggregation of various polypeptides in nonaqueous solvents (298, 299).

Although many dielectric dispersion studies have been made of aqueous proteins, that by Petersen & Cone (300) on rhodopsin solubilized in Triton X-100 appears to be the first on a membrane protein. The dipole moment was determined to be 720 D, which increased upon flash irradiation by about 25 D in a two-step process, the faster of which was related to a change in proton binding.

DNA has been the subject of many dielectric dispersion investigations. A recent one (301), using apparatus that enabled direct detection of the frequency difference spectrum of solution impedance, found a single relaxation in the region 0.2 Hz to 30 kHz. This was at 10 Hz at low DNA and salt concentration, and was attributed to rotation of the DNA coil. At higher concentrations, intermolecular interactions were evident. At higher fields than those commonly employed (about 100 V cm^{-1}), nonlinear electric behavior manifested in a third harmonic of the current was observed (302) and attributed to electrical deformation of the DNA. Binding of aminoacridine dyes lowers the DNA charge and thus the static dielectric constant, while the relaxation time increases owing to lengthening of the molecule upon intercalation (303).

Viscoelasticity and Flow Orientation

Macromolecular rotation and chain flexibility may also be usefully studied by the application of shearing fields, either oscillating or steady. Ferry (304) has written a lucid, brief review of dilute-solution viscoelastic properties of macromolecules. Measurements over a wide frequency range on dilute aqueous biopolymer solutions have been difficult because of, among other reasons, corrosion of the low-loss aluminum alloy resonators by aqueous solvents. Employing a new titanium alloy, Nemoto et al (305) were able to measure the storage and loss shear moduli of dilute TMV solutions in glycerol-water mixtures. Agreement with theory, including the molecular weight dependence of the moduli for aggregated samples, was generally good. Dynamic mechanical studies have also been performed on the melting behavior of reconstituted collagen (306) and on dry elastin at low temperatures (307). In the latter case, a 3.5 Hz resonance at $-71°C$ was attributed to a main-chain relaxation of the pure protein. At the other frequency extreme, in the nsec range, ^{13}C NMR relaxation data were used to detect backbone bond motions in swollen elastin (308). The data support a model of water-swollen elastin as a network of mobile chains, except perhaps near interchain crosslinks. Birefringence measurements on DNA in both oscillatory electric and hydrodynamic fields helped to separate chain flexibility from electrical effects (272). Shear creep measurements on fibrin clots over a time scale from 20 to 10^4 sec showed differences between coarse and fine clots, and between those unligated or ligated by Ca^{2+} and fibrinolygase (309).

The technique of viscoelastic relaxation has been particularly useful in studying very large DNA molecules, since the relaxation time is proportional to $M^{3/2}$ according to Equations 7 and 8. Uhlenhopp et al (310) used the method to show that at least some giant T-even bacteriophage heads were filled with single, multiple-length DNA molecules, thus supporting a "head-full" packaging and cleavage mechanism in phage morphogenesis. Viscoelastometry was also used to measure molecular weights of DNA in the 10^9 to 10^{10} molecular weight range, from *E. coli* bacteria (311) and mouse 3T3 cells (312). The DNA was single-stranded after being released from the cells by alkaline-EDTA-detergent treatment. Quasi-elastic behavior was exhibited by solutions of viral capsid and RNA, produced by alkaline degradation of turnip yellow mosaic virus, at very low shear stresses in a magnetic viscodensimeter (313). The solution structure responsible for this behavior is unclear.

During measurement of viscoelastic retardation times in a concentric cylinder shear apparatus, inward radial migration of large DNA molecules has been observed. Hydrodynamic calculations assuming a free-draining model (314) indicate this migration is due to the curvilinear solvent flow field. Extension to nondraining gaussian polymers (315) shows that the inward radial velocity is proportional to $M^{5/2}$, suggesting a means for separation of DNA molecules according to size. This size dependence is increased for polymers in good solvents, and it is predicted that linear molecules flow inward 8 times faster than do circular molecules of the same M (316).

Birefringence in steady flow has long been used to determine rotational diffusion coefficients of elongated macromolecules. The extinction angle is particularly sensitive to the larger molecules in solution, and the technique is therefore potentially useful in studying polymerization of rodlike structures. It has been used in this way to measure the kinetics of polymerization and depolymerization of microtubules in vitro (317). Orientation in flow can also be detected by other optical properties. A notable recent addition to the list is circular dichroism, which has been used by Chung & Holzwarth (318, 319) to study various conformations of DNA and RNA. The parallel and perpendicular components of the dichroism enable useful checks on CD theory that are not available from isotropic measurements. Linear-flow dichroism is also useful in this regard, and has been used to study DNA–ethidium bromide complexes (320). However, extension and deformation of the chain in flow may complicate interpretation of these measurements.

Simultaneous application of electric and shear fields cause complex effects that have been the subjects of several theoretical investigations (321–323). Whether these will be experimentally useful remains to be seen.

CONCLUSION

The scope of hydrodynamics in biophysical chemistry has become very broad: it has expanded from "traditional" studies of small proteins in dilute solutions to DNA molecules of almost macroscopic size, molecules in membranes, and whole cells. This broadening is due to improved biochemical preparative and purification techniques as well as to more diverse and sensitive physical tools. Laser optics, digital electronics, nanosecond pulse techniques, dynamic spectroscopic probes—these and similar technological advances have strongly complemented the older, but still useful, techniques of ultracentrifugation, electrophoresis, and viscosity. Similar technology transfer may be expected in the future. Areas currently in relatively early stages of development, and in which solid progress is likely to be made in the next few years, include fluctuation spectroscopy, dynamic NMR and ESR spectroscopy, electrooptics (progress here and in electrophoresis will require advances in theoretical understanding of polyelectrolyte solutions), and transport in reacting systems.

ACKNOWLEDGMENTS

Work in my laboratory on biomolecular hydrodynamics has been supported by research grants from NIH (GM17855) and NSF (PCM75-22728).

Literature Cited

1. Tanford, C. 1961. *Physical Chemistry of Macromolecules.* New York: Wiley. 710 pp.
2. Yamakawa, H. 1971. *Modern Theory of Polymer Solutions.* New York: Harper & Row. 419 pp.

3. Walton, A. G., Blackwell, J. 1973. *Biopolymers*. New York: Academic. 604 pp.
4. Morawetz, H. 1975. *Macromolecules in Solution*, Chap. 6. New York: Wiley. 549 pp. 2nd ed.
5. Leach, S. J., ed. 1969–1973. *Physical Principles and Techniques of Protein Chemistry*. New York: Academic. Pt. A, 530 pp.; Pt. B, 491 pp.; Pt. C, 621 pp.
6. Bloomfield, V. A., Crothers, D. M., Tinoco, I. Jr. 1974. *Physical Chemistry of Nucleic Acids*. New York: Harper & Row. 517 pp.
7. Cantoni, G. L., Davies, D. R., eds. 1971. *Procedures in Nucleic Acid Research*, Vol. 2. New York: Harper & Row. 924 pp.
8. Hirs, C. H. W., Timasheff, S. N., eds. 1972, 1973. *Methods in Enzymology*, Vol. 26, 27. New York: Academic. 737 pp., 1063 pp.
9. Yamakawa, H. 1974. *Ann. Rev. Phys. Chem.* 25:179–200
10. Bixon, M. 1976. *Ann. Rev. Phys. Chem.* 27:65–84
11. Bird, R. B. 1977. *Ann. Rev. Phys. Chem.* 28:185–206
12. Cooke, R., Kuntz, I. D. Jr. 1974. *Ann. Rev. Biophys. Bioeng.* 3:95–126
13. Kuntz, I. D. Jr., Kauzmann, W. 1974. *Adv. Protein Chem.* 28:239–345
14. Chu, B. 1974. *Laser Light Scattering*. New York: Academic. 317 pp.
15. Cummins, H. Z., Pike, E. R. eds. 1974. *Photon Correlation and Light-Beating Spectroscopy*. New York: Plenum. 584 pp.
16. Berne, B. J., Pecora, R. 1976. *Dynamic Light Scattering With Applications to Chemistry, Biology, and Physics*. New York: Wiley. 376 pp.
17. Crosignani, B., Diporto, P., Bertolotti, M. 1975. *Statistical Properties of Scattered Light*. New York: Academic. 226 pp.
18. Berne, B. J., Pecora, R. 1974. *Ann. Rev. Phys. Chem.* 25:233–54
19. Carlson, F. D. 1975. *Ann. Rev. Biophys. Bioeng.* 4:243–64
20. Shepherd, I. W. 1975. *Rep. Prog. Phys.* 38:565–620
21. De Maeyer, L., Gnädig, K., Hendrix, J., Saleh, B. 1976. *Q. Rev. Biophys.* 9:83–107
22. Jakeman, E., Oliver, C. J., Pike, E. R. 1975. *Adv. Phys.* 24:349–405
23. Kam, Z., Shore, H. B., Feher, G. 1975. *Rev. Sci. Instrum.* 46:269–77
24. Pusey, P. N., Koppel, D. E., Schaefer, D. W., Camerini-Otero, R. D., Koenig, S. H. 1974. *Biochemistry* 13:952–60
25. Camerini-Otero, R. D., Pusey, P. N., Koppel, D. E., Schaefer, D. W., Franklin, R. M. 1974. *Biochemistry* 13:960–70
26. Dubin, S. B., Benedek, G. B., Bancroft, F. C., Freifelder, D. 1970. *J. Mol. Biol.* 54:547–56
27. Benbasat, J. A., Bloomfield, V. A. 1975. *J. Mol. Biol.* 95:335–57
28. Welch, J. 1976. *The Thermodynamics and Kinetics of the Slow-Fast Transition in Bacteriophage T2L*. PhD thesis. Univ. Minnesota, St. Paul. 237 pp.
29. Salmeen, I., Rimai, L., Liebes, L., Rich, M. A., McCormick, J. J. 1975. *Biochemistry* 14:134–41
30. Hartford, S. L., Lesnow, J. A., Flygare, W. H., MacLeod, R., Reichmann, M. E. 1975. *Proc. Natl. Acad. Sci. USA* 72:1202–5
31. Newman, J., Swinney, H. L., Berkowitz, S. A., Day, L. A. 1974. *Biochemistry* 13:4832–38
32. Jolly, D., Eisenberg, H. 1976. *Biopolymers* 15:61–95
33. Sellen, D. B. 1975. *Polymer* 16:561–64
34. Chen, F. C., Tscharnuter, W., Schmidt, D., Chu, B., Liu, T. Y. 1974. *Biopolymers* 13:2281–92
35. Jamieson, A. M., Lee, T.-Y., Schafer, I. A. 1974. *Biopolymers* 33:2133–46
36. McDonnell, M. E., Jamieson, A. M. 1976. *Biopolymers* 15:1283–99
37. McCammon, J. A., Deutch, J. M., Bloomfield, V. A. 1975. *Biopolymers* 14:2479–87
38. Ahmed, A. I., Feeney, R. E., Osuga, D. T., Yeh, Y. 1975. *J. Biol. Chem.* 250:3344–47
39. Chu, B., Yeh, A., Chen, F. C., Weiner, B. 1975. *Biopolymers* 14:93–109
40. Cohen, R. J., Jedziniak, J. A., Benedek, G. B. 1975. *Proc. R. Soc. London Ser. A* 343:73–88
41. Cohen, R. J., Benedek, G. B. 1976. *J. Mol. Biol.* 108:151–78
42. Cohen, R. J., Jedziniak, J. A., Benedek, G. B. 1976. *J. Mol. Biol.* 108:179–99
43. Gabler, R., Ford, N. C., Westhead, E. W. 1975. *Biophys. J.* 15:747–51
44. Kato, M., Lee, W. I., Eichinger, B. E., Schurr, J. M. 1974. *Biopolymers* 13:2293–2304
45. Gabler, R., Westhead, E. W., Ford, N. C. 1974. *Biophys. J.* 14:528–45
46. Dubin, S. B., Cannell, D. S. 1975. *Biochemistry* 14:192–95
47. Cannell, D. S., Dubin, S. B. 1975. *Rev. Sci. Instrum.* 46:706–12
48. Uzgiris, E. E., Fromageot, H. P. M. 1976. *Biopolymers* 15:257–63

49. Filson, D. P., Bloomfield, V. A. 1968. *Biochim. Biophys. Acta* 155:169–82
50. Shaw, B. R., Schmitz, K. S. 1976. *Biochem. Biophys. Res. Commun.* 73: 224–32
51. Koppel, D. E. 1972. *J. Chem. Phys.* 57: 4814–20
52. Bargeron, C. B. 1974. *J. Chem. Phys.* 61:2134–38
53. Brown, J. C., Pusey, P. N., Dietz, R. 1975. *J. Chem. Phys.* 62:1136–44
54. Brehm, G. A., Bloomfield, V. A. 1975. *Macromolecules* 8:663–65
55. Bargeron, C. B. 1974. *J. Chem. Phys.* 60:2516–19
56. Lee, S. P., Chu, B. 1974. *Appl. Phys. Lett.* 24:261–63
57. Lee, S. P., Chu, B. 1974. *Appl. Phys. Lett.* 24:575–76
58. Tanaka, T. 1975. *Polym. J.* 7:62–71
59. Aragon, S. R., Pecora, R. 1976. *J. Chem. Phys.* 64:2395–2404
60. Raczek, J., Meyerhoff, G. 1976. *Makromol. Chem.* 177:1199–1214
61. Selser, J. C., Yeh, Y., Baskin, R. J. 1976. *Biophys. J.* 16:337–56
62. Selser, J. C., Yeh, Y. 1976. *Biophys. J.* 16:847–48
63. Fletcher, G. C. 1976. *Biopolymers* 15: 2201–17
64. Chen, F. C., Chrzeszczyk, A., Chu, B. 1976. *J. Chem. Phys.* 64:3403–9
65. Dewan, R. K., Chudgar, A., Mead, R., Bloomfield, V. A., Morr, C. V. 1974. *Biochim. Biophys. Acta* 342:313–21
66. Dewan, R. K., Chudgar, A., Bloomfield, V. A., Morr, C. V. 1974. *J. Dairy Sci.* 57:394–98
67. Koppel, D. E. 1974. *Biochemistry* 13: 2712–19
68. Loewenstein, M. A., Birnboim, M. H. 1975. *Biopolymers* 14:419–30
69. Lim, T. K., Baran, G. J., Bloomfield, V. A. 1977. *Biopolymers* 16: In press
70. Altenberger, A. R., Deutch, J. M. 1972. *J. Chem. Phys.* 59:894–98
71. Altenberger, A. R. 1976. *Chem. Phys.* 15:269–77
72. Phillies, G. D. J. 1974. *J. Chem. Phys.* 60:976–82
73. Phillies, G. D. J. 1974. *J. Chem. Phys.* 60:983–89
74. Phillies, G. D. J. 1975. *J. Chem. Phys.* 62:3925–32
75. Alpert, S. S. 1976. *J. Chem. Phys.* 65: 4333–34
76. Batchelor, G. K. 1972. *J. Fluid Mech.* 52:245–68
77. Batchelor, G. K. 1976. *J. Fluid Mech.* 74:1–29
78. Newman, J., Swinney, H. L., Berkowitz, S. A., Day, L. A. 1974. *Biochemistry* 13:4832–38
79. Phillies, G. D. J., Benedek, G. B., Mazer, N. A. 1976. *J. Chem. Phys.* 65: 1883–92
80. Alpert, S. S., Banks, G. 1976. *Biophys. Chem.* 4:287–96
81. Anderson, J. L., Reed, C. C. 1976. *J. Chem. Phys.* 64:3240–50
82. Phillies, G. D. J. 1976. *J. Chem. Phys.* 65:4334–35
83. Anderson, J. L., Reed, C. C. 1976. *J. Chem. Phys.* 65:4336–37
84. Keller, K. H., Canales, E. R., Yum, S. I. 1971. *J. Phys. Chem.* 75:379–87
85. Riveros-Moreno, V., Wittenberg, J. B. 1972. *J. Biol. Chem.* 247:895–901
86. Stephen, M. J. 1971. *J. Chem. Phys.* 55:3878–83
87. Doherty, P., Benedek, G. B. 1974. *J. Chem. Phys.* 61:5426–34
88. Raj, T., Flygare, W. H. 1974. *Biochemistry* 13:3336–40
89. Phillies, G. D. J. 1976. *Macromolecules* 9:447–50
90. Lee, W. I., Schurr, J. M. 1974. *Biopolymers* 13:903–8
91. Lee, W. I., Schurr, J. M. 1975. *J. Polym. Sci. Polym. Phys. Ed.* 13:873–88
92. Pusey, P. N., Schaefer, D. W., Koppel, D. E., Camerini-Otero, R. D., Franklin, R. M. 1972. *J. Phys. Paris Colloq.* 33: 163
93. Schaefer, D. W., Berne, B. J. 1974. *Phys. Rev. Lett.* 32:1110–13
94. Brown, J. C., Pusey, P. N., Goodwin, J. W., Ottewill, R. H. 1975. *J. Phys. A: Math. Nucl. Gen.* 8:664–82
95. Pusey, P. N. 1975. *J. Phys. A: Math. Nucl. Gen.* 8:1433–40
96. Lee, W. I., Schurr, J. M. 1976. *Chem. Phys. Lett.* 38:71–74
97. Ermak, D. L. 1975. *J. Chem. Phys.* 62: 4189–4203
98. Schurr, J. M. 1976. *Q. Rev. Biophys.* 9:109–34
99. Pecora, R. 1968. *J. Chem. Phys.* 49: 1032–35
100. Büldt, G. 1976. *Macromolecules* 9: 606–8
101. Perico, A., Piaggio, P., Cuniberti, C. 1975. *J. Chem. Phys.* 62:2690–95
102. Huang, W.-N., Frederick, J. E. 1974. *Macromolecules* 7:34–39
103. McAdam, J. D. G., King, T. A. 1974. *Chem. Phys. Lett.* 28:90–92
104. McAdam, J. D. G., King, T. A. 1974. *Chem. Phys.* 6:109–16
105. Schmidt, R. L. 1973. *Biopolymers* 12: 1427–30

106. Schmitz, K. S., Pecora, R. 1975. *Biopolymers* 14:521–42
107. Lee, W. I., Schurr, J. M. 1973. *Chem. Phys. Lett.* 23:603–7
108. Wun, K. L., Prins, W. 1975. *Biopolymers* 14:111–17
109. Shaw, B. R., Schmitz, K. S. 1976. *Biopolymers* 15:2313–14
110. Wada, A. 1974. *Biopolymers* 13:237–38
111. Schurr, J. M., Schmitz, K. S. 1973. *Biopolymers* 12:1021–45
112. King, T. A., Knox, A., McAdam, J. D. G. 1973. *Biopolymers* 11:1917–26
113. Han, C. C.-C., Yu, H. 1974. *J. Chem. Phys.* 61:2650–59
114. Schmitz, K. S., Schurr, J. M. 1973. *Biopolymers* 12:1543–64
115. Bloomfield, V. A. 1974. *Macromolecules* 7:846–49
116. Schmitz, K. S. 1976. *Chem. Phys. Lett.* 42:137–40
117. Bauer, D. R., Opella, S. J., Nelson, D. J., Pecora, R. 1975. *J. Am. Chem. Soc.* 97:2580–82
118. Dubin, S. B., Clark, N. A., Benedek, G. B. 1971. *J. Chem. Phys.* 54:5158–64
119. Bauer, D. R., Brauman, J. I., Pecora, R. 1976. *Ann. Rev. Phys. Chem.* 27:443–63
120. Bauer, D. R., Hudson, B., Pecora, R. 1975. *J. Chem. Phys.* 63:588–89
121. Benbasat, J. A., Gloomfield, V. A. 1973. *Macromolecules* 6:593–97
122. Uzgiris, E. E., Golibersuch, D. C. 1974. *Phys. Rev. Lett.* 32:37–40
123. Haas, D. D., Mustacich, R. V., Smith, B. A., Ware, B. R. 1974. *Biochem. Biophys. Res. Commun.* 59:174–80
124. Ware, B. R. 1974. *Adv. Colloid Interface Sci.* 4:1–44
125. Flygare, W. H., Ware, B. R. 1976. *Molecular Electrooptics*, ed. C. T. O'Konski. New York: Dekker. 544 pp.
126. Hartford, S. L., Flygare, W. H. 1975. *Macromolecules* 8:80–83
127. Rimai, L., Salmeen, I., Hart, D., Liebes, L., Rich, M. A., McCormick, J. J. 1975. *Biochemistry* 14:4621–27
128. Uzgiris, E. E., Kaplan, J. H. 1974. *Anal. Biochem.* 60:455–61
129. Uzgiris, E. E., Kaplan, J. H. 1976. *J. Colloid Interface Sci.* 55:148–55
130. Smith, B. A., Ware, B. R., Wiener, R. S. 1976. *Proc. Natl. Acad. Sci. USA* 73:2388–91
131. Uzgiris, E. E. 1974. *Rev. Sci. Instrum.* 45:74–80
132. Uzgiris, E. E., Kaplan, J. H. 1974. *Rev. Sci. Instrum.* 45:120–21
133. Mohan, R., Steiner, R., Kaufmann, R. 1976. *Anal. Biochem.* 79:506–25
134. Haas, D. D., Ware, B. R. 1976. *Anal. Biochem.* 74:175–88
135. Josefowicz, J., Hallett, F. R. 1975. *Appl. Opt.* 14:740–42
136. Stephen, M. J. 1974. *J. Chem. Phys.* 61:1598–99
137. Phillies, G. D. J. 1973. *J. Chem. Phys.* 59:2613–17
138. Friedhoff, L., Berne, B. J. 1976. *Biopolymers* 15:21–28
139. Prins, W., Rimai. L., Chompff. A. J. 1972. *Macromolecules* 5:104–6
140. Tanaka, T., Hocker, L. O., Benedek, G. B. 1973. *J. Chem. Phys.* 59:5151–59
141. McAdam, J. D. G., King, T. A., Knox, A. 1974. *Chem. Phys. Lett.* 26:64–68
142. King, T. A., Knox, A., McAdam, J. D. G. 1974. *J. Polym. Sci. Polym. Symp.* 44:195–202
143. Munch, J. P., Candau, S., Duplessix, R., Picot, C., Benoit, H. 1974. *J. Phys. Paris Lett.* 35:239–41
144. Munch, J. P., Candau, S., Duplessix, R., Picot, C., Herz, S., Benoit, H. 1976. *J. Polym. Sci. Polym. Phys. Ed.* 14:1097–1109
145. Carlson, F. D., Fraser, A. B. 1974. *J. Mol. Biol.* 89:273–81; 1975. 95:139 (correction)
146. Wun, K. L., Carlson, F. D. 1975. *Macromolecules* 8:190–94
147. Wun, K. L., Feke, G. T., Prins, W. 1974. *Discuss. Faraday Soc.* 57:146–55
148. Carlson, F. D. 1975. *Biophys. J.* 15:633–49
149. Elson, E. L., Webb, W. W. 1975. *Ann. Rev. Biophys. Bioeng.* 4:311–34
150. Webb, W. W. 1976. *Q. Rev. Biophys.* 9:49–68
151. Elson, E. L., Magde, D. 1974. *Biopolymers* 13:1–27
152. Magde, D., Elson, E. L., Webb, W. W. 1974. *Biopolymers* 13:29–61
153. Phillies, G. D. J. 1975. *Biopolymers* 14:499–508
154. Koppel, D. E. 1976. *Phys. Rev. A* 10:1938–45
155. Weissman, M., Schindler, H., Feher, G. 1976. *Proc. Natl. Acad. Sci. USA* 73:2776–80
156. Cooper, F. L., Chang, D. B., Young, A. C., Martin, C. J., Ancker-Johnson, B. 1974. *Biophys. J.* 14:161–77
157. Chang, D. B., Cooper, F. L., Young, A. C., Martin, C. J., Ancker-Johnson, B. 1975. *J. Theor. Biol.* 50:285–308
158. Brown, W., Chitumbo, K. 1975. *J. Chem. Soc. Faraday Trans. 1* 71:1–11
159. McCabe, M., Laurent, T. C. 1975. *Biochim. Biophys. Acta* 399:131–38

160. McDonald, G. G., Vanderkooi, J. M. 1975. *Biochemistry* 14:2125–27
161. Lindblom, G., Wennerström, H., Arvidson, G., Lindman, B. 1976. *Biophys. J.* 16:1287–95
162. Soutar, A. K., Pownall, H. J., Hu, A. S., Smith, L. C. 1974. *Biochemistry* 13:2828–36
163. Vanderkooi, J. M., Callis, J. B. 1974. *Biochemistry* 13:4000–6
164. Owen, C. S. 1975. *J. Chem. Phys.* 62:3204–7
165. Vanderkooi, J. M., Fischkoff, S., Andrich, M., Podo, F., Owen, C. S. 1975. *J. Chem. Phys.* 63:3661–66
166. Axelrod, D., Koppel, D. E., Schlessinger, J., Elson, E., Webb, W. W. 1976. *Biophys. J.* 16:1055–69
167. Koppel, D. E., Axelrod, D., Schlessinger, J., Elson, E. L., Webb, W. W. 1976. *Biophys. J.* 16:1315–29
168. Schlessinger, J., Koppel, D. E., Axelrod, D., Jacobson, K., Webb, W. W., Elson, E. L. 1976. *Proc. Natl. Acad. Sci. USA* 73:2409–13
169. Liebman, P. A., Entine, G. 1974. *Science* 185:457–59
170. Saffman, P. G., Delbrück, M. 1975. *Proc. Natl. Acad. Sci. USA* 72:3111–13
171. Edidin, M. 1974. *Ann. Rev. Biophys. Bioeng.* 3:179–201
172. Fujita, H. 1975. *Foundations of Ultracentrifugal Analysis.* New York: Wiley. 459 pp.
173. Widmer, F., Kaplan, N. O. 1976. *Biochemistry* 15:4699–4703
174. Holcenberg, J. S., Teller, D. C., Roberts, J. 1974. *Arch. Biochem. Biophys.* 161:306–12
175. Shill, J. P., Peters, B. A., Neet, K. E. 1974. *Biochemistry* 13:3864–71
176. Simpson, R. T., Bustin, M. 1976. *Biochemistry* 15:4305–12
177. Clark, R. W. 1976. *Biochim. Biophys. Acta* 428:269–74
178. Koch, A. L., Blumberg, G. 1976. *Biophys. J.* 16:389–405
179. Goldstein, B., Zimm, B. H. 1973. *Biopolymers* 12:857–67
180. Schumaker, V. N., Zimm, B. H. 1973. *Biopolymers* 12:869–76
181. Schumaker, V. N., Zimm, B. H. 1973. *Biopolymers* 12:877–94
182. Mason, D. W. 1976. *Biophys. J.* 16:407–16
183. Rubenstein, I., Leighton, S. B. 1974. *Biophys. Chem.* 1:292–99
184. Zimm, B. H. 1974. *Biophys. Chem.* 1:279–91
185. Tanford, C., Nozaki, Y., Reynolds, J. A., Makino, S. 1974. *Biochemistry* 13:2369–76
186. Nozaki, Y., Schechter, N. M., Reynolds, J. A., Tanford, C. 1976. *Biochemistry* 15:3884–90
187. Laurent, T. C., Preston, B. N., Pertoft, H., Gustafson, B., McCabe, M. 1975. *Eur. J. Biochem.* 53:129–36
188. Casassa, E. F. 1976. *Macromolecules* 9:182–85
189. Dawkins, J. V., Hemming, M. 1975. *Polymer* 16:554–60
190. Small, H. 1974. *J. Colloid Interface Sci.* 48:147–61
191. Denizot, F. C., Delaage, M. A. 1975. *Proc. Natl. Acad. Sci. USA* 72:4840–43
192. Leaback, D. H. 1974. *Chem. Br.* 10:376–83
193. Poole, T., Leach, B. S., Fish, W. W. 1974. *Anal. Biochem.* 60:596–607
194. Bozarth, R. F., Harley, E. H. 1976. *Biochim. Biophys. Acta* 432:329–35
195. Maniatis, T., Jeffrey, A., van de Sande, H. 1975. *Biochemistry* 14; 3787–94
196. Depew, R. E., Wang, J. C. 1975. *Proc. Natl. Acad. Sci. USA* 72:4275–79
197. Pulleyblank, D. E., Shure, M., Tang, D., Vinograd, J., Vosberg, P. 1975. *Proc. Natl. Acad. Sci. USA* 72:4280–84
198. Keller, W. 1975. *Proc. Natl. Acad. Sci. USA* 72:4876–80; 1976. 73:2527
199. Childs, J. D., Birnboim, H. C. 1975. *J. Virol.* 16:652–61
200. DeKeizer, A., van der Drift, W. P. J. T., Overbeek, J. T. G. 1975. *Biophys. Chem.* 3:107–8
201. Weiss, G. H., Catsimpoolas, N., Rodbard, D. 1974. *Arch. Biochem. Biophys.* 163:106–12
202. Catsimpoolas, N., Yotis, W. W., Griffith, A. L., Rodbard, D. 1974. *Arch. Biochem. Biophys.* 163:113–21
203. Hawley, S. A., Mitchell, R. M. 1975. *Biochemistry* 14:3257–64
204. Hawley, S. A. 1976. *J. Mol. Biol.* 103:655–57
205. Cann, J. R., Goad, W. B. 1970. *Interacting Macromolecules.* New York: Academic. 249 pp.
206. Nichol, L. W., Winzor, D. J. 1972. *Migration of Interacting Systems.* London: Oxford Univ. Press. 166 pp.
207. Shcherbukhin, V. V., Guermant, C. 1975. *Biophys. Chem.* 3:21–34
208. Frigon, R., Timasheff, S. N. 1975. *Biochemistry* 14:4559–66, 4567–73
209. Schönert, H. 1975. *Biophys. Chem.* 3:161–68
210. Mitchell, R. M. 1976. *Biopolymers* 15:1717–39, 1741–53

211. Halvorson, H. R., Ackers, G. K. 1974. *J. Biol. Chem.* 249:967-73
212. Cann, J. R., Kegeles, G. 1974. *Biochemistry* 13:1868-74
213. Zimmerman, J. K. 1974. *Biochemistry* 13:384-89
214. Zimmerman, J. K. 1975. *Biophys. Chem.* 3:339-44
215. Claverie, J., Dreux, H., Cohen, R. 1975. *Biopolymers* 14:1685-1700, 1701-16
216. Claverie, J. 1976. *Biopolymers* 15:843-57
217. Small, E., Isenberg, I. 1976. *Biopolymers* 15:1093-1100
218. Provencher, S. W. 1976. *Biophys. J.* 16:27-41
219. Koenig, S. H. 1975. *Biopolymers* 14:2421-23
220. Wright, A. K., Baxter, J. E. 1976. *Biophys. J.* 16:931-38
221. D'Anna, J. A., Isenberg, I. 1974. *Biochemistry* 13:2093-98
222. Smerdon, M. J., Isenberg, I. 1974. *Biochemistry* 13:4046-49
223. D'Anna, J. A., Isenberg, I. 1974. *Biochemistry* 13:4987-92, 4992-97
224. Nihei, T., Mendelson, R. A., Botts, J. 1974. *Proc. Natl. Acad. Sci. USA* 71:274-77
225. Tregear, R. T., Mendelson, R. A. 1975. *Biophys. J.* 15:455-67
226. Levison, S. A., Hicks, A. N., Portmann, A. J., Dandliker, W. B. 1975. *Biochemistry* 14:3778-86
227. Schuldiner, S., Spencer, R. D., Weber, G., Weil, R., Kaback, H. R. 1975. *J. Biol. Chem.* 250:8893-96
228. Miller, S. J., Wetmur, J. G. 1974. *Biopolymers* 13:2545-51
229. Stubbs, G. W., Litman, B. J., Barenholz, Y. 1976. *Biochemistry* 15:2766-72
230. Fuchs, P., Parola, A., Robbins, P. W., Blout, E. R. 1975. *Proc. Natl. Acad. Sci. USA* 72:3351-54
231. Shinitzky, M., Barenholz, Y. 1974. *J. Biol. Chem.* 249:2652-57
232. Lentz, B. R., Barenholz, Y., Thompson, T. E. 1976. *Biochemistry* 15:4521-28, 4529-37
233. Jacobson, K., Papahadjopoulos, D. 1975. *Biochemistry* 14:152-61
234. Conti, F. 1974. *Ann. Rev. Biophys. Bioeng.* 4:287-310
235. Wu, C., Yarbrough, L. R., Hillel, Z., Wu, F. Y. 1975. *Proc. Natl. Acad. Sci. USA* 72:3019-23
236. Highsmith, S., Mendelson, R., Morales, M. 1976. *Proc. Natl. Acad. Sci. USA* 73:133-37, 2527
237. Weber, G., Helgerson, S. L., Cramer, W. A., Mitchell, G. W. 1976. *Biochemistry* 15:4429-32
238. Weber, G., Mitchell, G. W. 1976. *Excited States of Biological Molecules*, ed. J. B. Birks, p. 75. London: Wiley
239. Helgerson, S. L., Cramer, W. A., Harris, J. M., Lytle, F. E. 1974. *Biochemistry* 13:3057-61
240. Mazo, R. M. 1976. *Biopolymers* 15:507-12
241. Galley, W. C. 1976. *Nature* 260:554-56
242. Cherry, R. J., Cogoli, A., Oppliger, M., Schneider, G., Semenza, G. 1976. *Biochemistry* 15:3653-56
243. Cherry, R. J., Schneider, G. 1976. *Biochemistry* 15:3657-61
244. Deleted in proof
245. Aragon, S. R., Pecora, R. 1975. *Biopolymers* 14:119-38
246. Ehrenberg, M., Rigler, R. 1974. *Chem. Phys.* 4:390-401
247. Wahl, P. 1975. *Chem. Phys.* 7:210-19, 220-28
248. Valeur, B., Jarry, J. P., Geny, F., Monnerie, L. 1975. *J. Polym. Sci. Polym. Phys. Ed.* 13:667
249. Valeur, B., Monnerie, L., Jarry, J. P. 1975. *J. Polym. Sci. Polym. Phys. Ed.* 13:675
250. Valeur, B., Monnerie, L. 1976. *J. Polym. Sci. Polym. Phys. Ed.* 14:11-27, 29-37
251. Berliner, L. J. 1976. *Spin Labeling, Theory and Application*. New York: Academic. 604 pp.
252. Lasker, S. E., Milvy, P., eds. 1973. *Electron Spin Resonance and Nuclear Magnetic Resonance in Biology and Medicine and Magnetic Resonance in Biological Systems*, *Ann. NY Acad. Sci.* 222:1124 pp.
253. Hyde, J. W. 1974. *Ann. Rev. Phys. Chem.* 25:407-35
254. Smigel, M. D., Dalton, L. R., Hyde, J. S., Dalton, L. A. 1974. *Proc. Natl. Acad. Sci. USA* 71:1925-29
254a. Thomas, D. D., Dalton, L. A., Hyde, J. S. 1976. *J. Chem. Phys.* 65:3006-24
255. Thomas, D. D., Seidel, J. C., Hyde, J. S., Gergely, J. 1975. *Proc. Natl. Acad. Sci. USA* 72:1729-33
256. Kikuchi, K., Yoshioka, K. 1976. *Biopolymers* 15:583-87
257. Fredericq, E., Houssier, C. 1973. *Electric Dichroism and Electric Birefringence*. London: Oxford Univ. Press. 219 pp.
258. O'Konski, C. T., ed. 1976. *Molecular Electrooptics, Vol. 1, Theory and Methods*. New York: Dekker. 544

pp.; *Vol. 2, Applications to Biopolymers.* In press
259. Bernengo, J. C., Roux, B., Herbage, D. 1974. *Biopolymers* 13:641–47
260. Kobayashi, S., Ikegami, A. 1975. *Biopolymers* 14:543–53
261. Kobayashi, S., Totsuka, T. 1975. *Biochim. Biophys. Acta* 376:375–85
262. DeLaney, D. E., Krause, S. 1976. *Macromolecules* 9:455–63
263. Umazume, Y., Fujime, S. 1975. *Biophys. J.* 15:163–80
264. Asai, H., Watanabe, N. 1976. *Biopolymers* 15:383–92
265. Wright, A. K., Thompson, M. R. 1975. *Biophys. J.* 15:137–41
266. Wright, A. K., Fish, W. W. 1975. *Biophys. Chem.* 3:74–82
267. Wright, A. K. 1974. *Biophys. J.* 14:243–45
268. Wright, A. K. 1976. *Biophys. Chem.* 4:199–202
269. Wright, A. K., Thompson, M. R., Miller, R. L. 1975. *Biochemistry* 14:3224–28
270. Rowe, E. S., Steinhardt, J. 1976. *Biochemistry* 15:2579–85
271. Miller, S. J., Wetmur, J. G. 1974. *Biopolymers* 13:115–28
272. Wilkinson, R. S., Thurston, G. B. 1976. *Biopolymers* 15:1555–72
273. Greve, J., de Heij, M. E. 1975. *Biopolymers* 14:2441–43
274. Houssier, C., Hardy, B., Fredericq, E. 1974. *Biopolymers* 13:1141–60
275. Chang, C., Miller, S. J., Wetmur, J. G. 1974. *Biochemistry* 13:2142–48
276. Dourlent, M., Hogrel, J. F. 1976. *Biochemistry* 15:430–36
277. Greve, J., Blok, J. 1973. *Biopolymers* 12:2607–22
278. Greve, J., Blok, J. 1975. *Biopolymers* 14:139–54
279. Garcia de la Torre, J., Bloomfield, V. A. 1977. *Biopolymers.* In press
280. Pearlstein, A. J. 1974. *Biopolymers* 13:2649–51
281. Kobayashi, S. 1971. *Biopolymers* 10:915–22
282. Schwarz, G., Schrader, U. 1975. *Biopolymers* 14:1181–95
283. Kikuchi, K., Yoshioka, K. 1976. *Biopolymers* 15:1669–76
284. Gerber, B. R., Minakata, A. 1975. *J. Mol. Biol.* 92:507–28
285. Jennings, B. R., Baily, E. D. 1973. *J. Polymer Sci. Polym. Symp.* 42:1121–30
286. Weill, G., Sturm, J. 1975. *Biopolymers* 14:2537–53
287. Jennings, B. R., Foweraker, A. R. 1974. *Spectrosc. Lett.* 7:371–75
288. Coles, H. J., Jennings, B. R. 1975. *Biopolymers* 14:2567–75
289. Newman, J., Swinney, H. L. 1976. *Biopolymers* 15:301–15
290. Aiuchi, T., Murai, N., Sugai, S. 1974. *Biopolymers* 13:1499–1510
291. Muller, G., van der Touw, F., Zwolle, S., Mandel, M. 1974. *Biophys. Chem.* 2:242–54
292. van der Touw, F., Mandel, M. 1974. *Biophys. Chem.* 2:231–41
293. van der Touw, F., Mandel, M. 1974. *Biophys. Chem.* 2:218–30
294. South, G. P., Grant, E. H. 1974. *Biopolymers* 13:1777–89
295. Shore, J. E., Zwanzig, R. 1975. *J. Chem. Phys.* 63:5445–58
296. Matsumoto, T., Nishioka, N., Teramoto, A., Fujita, H. 1974. *Macromolecules* 7:824–31
297. Omura, I., Teramoto, A., Fujita, H. 1975. *Macromolecules* 8:284–90
298. Gupta, A. K., Dufour, C., Marchal, E. 1974. *Biopolymers* 13:1293–1308
299. Marchal, E. 1974. *Biopolymers* 13:1309–16
300. Petersen, D. C., Cone, R. A. 1975. *Biophys. J.* 15:1181–1200
301. Sakamoto, M., Kanda, H., Hayakawa, R., Wada, Y. 1976. *Biopolymers* 15:879–92
302. Marion, C., Bernengo, J. C., Hanss, M. 1976. *Biopolymers* 15:373–82
303. Goswami, D. N., das Gupta, N. N. 1974. *Biopolymers* 13:391–400
304. Ferry, J. D. 1973. *Acc. Chem. Res.* 6:60–65
305. Nemoto, N., Schrag, J. L., Ferry, J. D., Fulton, R. W. 1975. *Biopolymers* 14:409–17
306. Nguyen, A., Vu, B. T., Wilkes, G. L. 1974. *Biopolymers* 13:1023–37
307. Pezzin, G., Scandola, M., Gotte, L. 1976. *Biopolymers* 15:283–92
308. Lyerla, J. R. Jr., Torchia, D. A. 1975. *Biochemistry* 14:5175–83
309. Gerth, C., Roberts, W. W., Ferry, J. D. 1974. *Biophys. Chem.* 2:208–17
310. Uhlenhopp, E. L., Zimm, B. H., Cummings, D. J. 1974. *J. Mol. Biol.* 89:689–702
311. Uhlenhopp, E. L., Zimm, B. H. 1975. *Biophys. J.* 15:222–32
312. Uhlenhopp, E. L. 1975. *Biophys. J.* 15:233–37
313. Hodgins, M. G., Hodgins, O. C., Kupke, D. W., Beams, J. W. 1975. *Proc. Natl. Acad. Sci. USA* 72:3501–4

314. Shafer, R. H., Laiken, N., Zimm, B. H. 1974. *Biophys. Chem.* 2:180–84
315. Shafer, R. H. 1974. *Biophys. Chem.* 2:185–88
316. Dill, K., Shafer, R. H. 1976. *Biophys. Chem.* 4:51–54
317. Haga, T., Abe, T., Kurokawa, M. 1974. *FEBS Lett.* 39:291–95
318. Chung, S.-Y., Holzwarth, G. 1975. *J. Mol. Biol.* 92:449–66
319. Chung, S.-Y., Holzwarth, G. 1975. *Biopolymers* 14:1531–45
320. Norden, B., Tjerneld, F. 1976. *Biophys. Chem.* 4:191–98
321. Barisas, B. G. 1974. *Macromolecules* 7:930–33
322. Okagawa, A., Cox, R. G., Mason, S. G. 1974. *J. Colloid Interface Sci.* 47:536–67
323. Okagawa, A., Mason, S. G. 1974. *J. Colloid Interface Sci.* 47:568–87

… 2647

VIBRATIONAL STATE ANALYSIS OF ELECTRONIC-TO-VIBRATIONAL ENERGY TRANSFER PROCESSES[1]

Stephen Lemont[2]
Chemistry Division, Naval Research Laboratory, Washington DC 20375

George W. Flynn
Department of Chemistry and Columbia Radiation Laboratory, Columbia University, New York, NY 10027

INTRODUCTION

The quenching of electronic fluorescence has been studied extensively in a wide variety of atomic and molecular systems over the past several years (1). Nevertheless, most of these quenching experiments have given little information concerning the specific product states into which electronic energy is channeled during collision, the relative amount of electronic energy converted to translational, rotational, and vibrational energy, or the specific rate constants for energy transfer into a given channel. Comprehensive studies of vibration-vibration (V-V) and vibration-translation/rotation (V-T/R) energy transfer phenomena have been made for many molecules over the past few years (2). These studies have provided a rather detailed view of such processes in the ground electronic states of molecules. Partly due to the development of lasers and new experimental techniques and partly due to the increased knowledge of energy transfer processes, there has been renewed interest in the study of electronic-to-vibrational (E-V) energy transfer, which can provide insight into molecular interactions and energy partitioning during collisions. Such studies are of fundamental interest as tests of necessarily approximate scattering theories and as probes of potential surface crossings that play important roles in chemically

---

[1] This work was supported by the Joint Services Electronics Program (US Army, Navy, and Air Force) under Contract DAAB07-74-C-0341, and by the National Science Foundation under Grant MPS 75-04118.

[2] NRC/NRL Postdoctoral Research Associate (September 1976 to the present).

reactive collisions. E-V energy transfer processes can also be used to excite high-lying or infrared inactive vibrational levels in order to provide further insight into V-V and V-T/R relaxation of such states. The possibility exists that preferential excitation of different modes of polyatomic molecules can be achieved by transferring energy from electronic to vibrational degrees of freedom, leading to mode specific chemical reactivity. In addition E-V energy transfer data is of great practical importance in modeling the upper atmosphere, in laser development, and in predicting the efficiency of photochemical reactions. The present review is concerned with experimental investigations of E-V energy transfer in which attempts have been made to analyze vibrational state product distributions or to determine the role of vibrational energy in the transfer process. A brief discussion of experiments probing the inverse event, vibration-to-electronic (V-E) energy transfer, is also given. No attempt has been made to review the extensive theoretical literature, although theoretical models are discussed in relation to experimental data where appropriate. The literature search spans the period from December 1958 through November 1976. No relevant papers were found that had been published prior to 1958. Articles appearing after December 1, 1976 have occasionally been included, but no systematic search of the literature after this date was attempted.

PREVIOUS REVIEWS AND EARLY HISTORY

Earlier reviews (3–8) include, for the most part, descriptions of quenching experiments in which transfer of energy from electronic to vibrational modes was postulated to be the mechanism. Such claims were based in part on the well-established fact that diatomic and polyatomic molecules were much more efficient at quenching electronic excitation than atomic gases of comparable mass (9, 10). In addition, in some quenching experiments involving spin-orbit relaxation of atomic species, the importance of matching the vibrational energy of the quencher to the energy of the electronically excited atom had been demonstrated very early (3–11) and was confirmed more recently (12–22). The observation of resonantly enhanced E-V transfer was in at least partial agreement with an early theoretical model (23) that did not consider crossing or near crossing of the initial and final state potential energy surfaces. For example, both theory (23) and experiment (10) concluded that the order of efficiencies in the quenching of $6^3P_1 \rightarrow 6^3P_0$ mercury radiation by the diatomic molecules N_2, CO, and NO was NO > CO > N_2. The energy of the Hg $6^3P_1 \rightarrow 6^3P_0$ transition is 1770 cm^{-1}, while the vibrational energies of the fundamentals of NO, CO, and N_2 are 1904, 2170, and 2360 cm^{-1}, respectively. In another instance, studies of the relaxation of dissolved singlet molecular oxygen by solvent molecules (24) indicated a strong correlation between the singlet oxygen lifetime and the infrared absorption intensity of the solvent overtone and combination bands that are in near resonance with the $a^1\Delta_g \rightarrow X^3\Sigma_g^-$ transition energies (a lifetime decrease was noted for favorable resonances). However, many experiments (11, 19–22) also demonstrated that quenching efficiency frequently does not correlate with small energy deficit, $\Delta E = \Delta E_{el} - \Delta E_{vib}$, but is instead a function of the tendency of the colliding species to form an intermediate collision complex.

Support for this idea came from another early theoretical model (25) that assumed that quenching occurred via the formation of a quasimolecule. In this picture the potential energy surfaces that describe the interaction $A^* + BC$ and the interaction $A + BC^\dagger$ must intersect (* denotes electronic excitation and † denotes vibrational excitation). In this case the quenching efficiency depends on the energy separation of initial and final states for the quasimolecule instead of the energy deficit at infinite separation as in the theory proposed by Dickens, Linnett & Sovers (23). Thus, a possible alternate explanation for the enhanced quenching efficiency of $Hg^*(6^3P_1)$ by CO over N_2 is the intermediate formation of $Hg\text{-}CO^*$ (N_2 is considerably less "reactive" than CO), which dissociates to ground state $Hg(6^1S_0)$ and vibrationally excited CO (11). An important feature of this model (25) is the prediction of efficient E-V energy transfer even if the energy deficit, $\Delta E = \Delta E_{el} - \Delta E_{vib}$, for the system is large.

Electronic-to-vibrational energy transfer experiments that follow the vibrational state distributions of molecules offer the possibility of discriminating among various theoretical models including the two described above. Such models are very important for the development of a unified description of energy transfer processes that are of prime interest in chemical reaction studies, atmospheric modeling (26–30), and laser development. Only in the last 15 years have such experiments been attempted. In the past 5 years there has been considerable acceleration in experimental efforts of this type, particularly in methods employing laser excitation and time-resolved infrared analysis. The present review concentrates largely on these studies.

E-V ENERGY TRANSFER STUDIES EMPLOYING CLASSICAL METHODS

Experiments Involving Excited Mercury Atoms

The first definite proof of electronic-to-vibrational energy transfer using vibrational state detection was given by Polanyi and co-workers (31–33) who, in studying the quenching of $Hg^*(6^3P_1)$ and $Hg(6^3P_0)$ atoms by CO, observed infrared emission from several CO vibrational levels ($v = 2$–9). The apparatus consisted of a cell containing mercury vapor and CO that was illuminated by several mercury lamps chopped at 780 Hz to produce Hg^*. Steady state CO overtone emission was observed by use of lock-in detection at 780 Hz with an IR spectrometer and PbS detector. From the spectra subsequently obtained under various experimental conditions of gas concentration and illumination intensity, the steady-state vibrational population distribution for CO was obtained; and, from the relative number of molecules N_v in the vibrational states of CO, a set of rate constants k_v and cross sections σ_v^2 for E-V energy transfer via the processes

$$Hg^*(6^3P_{1,0}) + CO \xrightarrow{k_v} Hg(6^1S_0) + CO^\dagger(v) \qquad \qquad 1.$$

were derived. The designation $^3P_{1,0}$ implies the presence of both $^3P_1$ and $^3P_0$ states in the experimental mixture. The major experimental results of this study were that 1. approximately 30–50% of the electronic energy of Hg^* was transferred into vibrational states of CO; 2. the highest vibrational state populated in the E-V transfer was $v = 9$ even though sufficient electronic energy was available to excite $v = 20$; 3. the populations (N_v) for individual CO(v) states appeared to maximize at $v = 2$, fell relatively slowly (a factor of 10) in going from $v = 2$ to $v = 8$, and fell a factor of 200 from $v = 2$ to $v = 9$. This distribution of populations cannot be described by a single temperature. The major problem with this experimental technique, as pointed out by Karl & Polanyi (32), is that rapid V-V relaxation of the type

$$CO(v) + CO(0) \rightleftarrows CO(v - 1) + CO(1) + \Delta E \qquad 2.$$

leads to cascading of energy to low-lying vibrational states (34). Though attempts were made to minimize this problem, subsequent experiments described below indicate that the vibrational populations observed for low-lying states were too high.

In subsequent experiments employing similar methods, sets of relative rate constants for E-V energy transfer were obtained for analogous Hg* + NO (35) and Hg* + HF (36) reactions. NO emission was observed from vibrational levels up to $v = 17$, corresponding to two thirds of the electronic energy converted into vibration, although NO could hypothetically become excited up to $v \simeq 26$ by Hg*. The vibrational population was a slowly decreasing function of v, changing by a factor of 80 in going from $v = 2$ to $v = 17$. Again here, as in the Hg* + CO system, rapid vibrational relaxation (V-V as well as V-T) of NO$^\dagger(v)$ was undoubtedly an important source of error in the experiment (37). In the case of HF, emission was observed from $v = 1$ to $v = 6$. HF can in principle be excited as high as $v = 13$ by Hg*. The relative vibrational state populations were observed to decrease with increasing v as before, but the falloff was much more rapid, a factor of 500 from $v = 1$ to $v = 6$. This phenomenon may be the result of extremely rapid V-V and/or V-T/R relaxation, now known to occur in HF (38), which would quickly cause most of the excited HF molecules to cascade to the lowest vibrational levels. This unusually fast relaxation in HF may account for the large differences between the HF vibrational population distribution and those obtained for CO and NO.

Tsuchiya and co-workers have repeated the Polanyi-type experiments on Hg* + CO (39, 40) and NO (40) using Hg lamps modulated at frequencies from 120–4000 Hz. In these experiments the vibrational fluorescence both in phase and out of phase with the lamp modulation frequency was determined, thus leading to a fluorescence phase shift as a function of modulation frequency. Because relaxation processes are not instantaneous, they contribute to the observed phase shift. By measuring the phase shifts, the vibrational populations were corrected for the effects of V-V relaxation (Equation 2). The resulting population (N_v) and rate constant (k_v) distribution more nearly represents the direct E-V energy transfer probabilities since cascading effects to lower levels are minimized. For the Hg* + CO system, the set of k_v, expressed relative to $k_{v=9} = 1.00$, was determined to be $k_{v=2} = 6.92$, $k_{v=3} = 14.1$, $k_{v=4} = 23.6$, $k_{v=5} = 25.6$, $k_{v=6} = 23.1$, $k_{v=7} = 7.69$, $k_{v=8} = 2.08$, and $k_{v=9} = 1.00$. A similar corrected set of k_v values was also obtained for the Hg* + NO system. These results are significantly different from those uncorrected for relaxation (31–33) especially for the lower vibrational levels.

In order to explain the general features of the Hg* + CO E-V energy transfer, Polanyi and co-workers (33) suggested that the observed vibrational energy distribution, which was clearly nonresonant, might occur via an intermediate compound consisting of an excited Hg atom colliding with a ground state CO molecule to form a short-lived, linear Hg-CO* molecule in some quasi-bound state. The "ground" state of this molecule is repulsive so that Hg and CO separate when a transition is made from the upper to the lower state. In a weak interaction configuration, the potential curves of the initial and final states are expected to run parallel to each other and transition between potential surfaces is difficult, leading to low E-V energy transfer efficiency, unless this process can occur by an alternate mechanism such as resonant energy transfer from the energy donor to the energy acceptor. In a strong interaction configuration, the potential curves are expected to cross and provide a more efficient transition route. If this were the case for the Hg*–CO system, then infrared emission from CO could result because of either or both of the following mechanisms: 1. relaxation of the CO bond from its compressed or extended configuration on the upper potential surface; 2. the effects of relaxation of Hg ↔ CO repulsion as Hg and CO separate on the lower potential surface. If the second process is treated classically as an instantaneous release of repulsive energy between Hg* and CO in a colinear configuration, then conservation of linear momentum places an upper limit on the amount of energy that can go into C–O relative motion (vibration). Since Hg* is very heavy, and if HgCO* is assumed to have Hg at the C end of CO, the fraction of energy released into vibration is approximately $m_O/(m_C + m_O) = 0.57$ (33, 41). This corresponds to $v = 11$ as the highest possible vibrational level of CO that could be excited by Hg*, which is in fair agreement with the experimentally observed value of $v = 9$.

Similar arguments applied to Hg* + NO suggest the highest excited level of NO should be $v \simeq 11$, whereas experimentally $v = 17$ was observed to have significant population (35). In the case of Hg* + HF the observed distribution again did not appear to fill states as energetic as allowed by this simple colinear argument (36). As noted earlier, the HF data may have been seriously affected by V-V and V-T/R relaxation.

While the above model appeared to reproduce at least qualitatively some of the experimental observations, no quantitative prediction could be made for the distribution of E-V rate constants k_v (or populations N_v). Subsequently an impulsive "half-collision" model for E-V energy transfer was developed that quantitatively treated the quenching of electronically excited atoms by collisional transfer to diatomic molecule vibrations (42–46). This theory is an extension of that used to describe vibration-to-translation (V-T) energy transfer processes (47) and was later adapted to explain photofragment vibrational and rotational excitation following photodissociation (48). The treatment involves the use of near-classical collision theory in the high energy limit. Assumptions are that a linear quasidiatomic molecule (e.g. Hg*–CO) in a single excited state is formed upon collision and *suddenly* dissociates into fragments when the excited state crosses the ground repulsive state. The mechanics of vibrational excitation, which may be compared to a hammer striking one end of the CO molecule, are approximated by a forced harmonic oscillator model. The theory predicts that the populations (N_v) of the vibrational states will be described by a Poisson distribution and gives a value for the average energy transferred to the vibrational mode of the diatomic molecule. The corrected CO vibrational population distribution found in the phase shift experiments (39, 40) does in fact approximate a Poisson distribution. The impulsive "half-collision" model also predicts that smaller vibrational spacings will lead to a larger average E-V energy transfer and the population of higher vibrational states. Experimentally (33, 35, 39, 40), NO was found to be more highly excited than CO by collisions with Hg*, which agrees with this prediction.

Although the Hg* + CO and NO deactivation to ground state Hg(6^1S_0) suggests the importance of complex formation and curve-crossing dynamics in E-V energy transfer, other quenching studies appear to indicate that resonance effects can also be important in some cases. For example, the spin-orbit quenching of Hg*

$$\text{Hg*}(6^3P_1) + M \rightarrow \text{Hg*}(6^3P_0) + M^\dagger(v = 1), \qquad 3.$$

has been studied where M = CO or N_2. Both flash kinetic spectroscopy (CO) (49) and phase shift techniques (N_2) (50) were used to show that Equation 3 proceeds with nearly unit efficiency for CO and N_2.

Crossed Beam E-V Studies

A few experiments have been performed using molecular beams. The earliest of these was a study of vibrationally excited N_2 after collisions with electronically excited rubidium (51). The Rb was excited by means of a resonance lamp, the Rb* and N_2 beams were crossed, and $N_2^\dagger$ in the highest energetically allowed vibrational state ($v = 5$) was detected by superelastic electron scattering. Analysis of the energy gained by the electrons showed that in more than 10% of the quenching collisions, $v = 5$ was populated. Unfortunately, due to experimental limitations, electrons of energy below 1.2 eV could not be accurately detected. Thus, the lower vibrational levels of N_2 were not probed.

Additional crossed beam experiments have been performed to study E-V energy transfer from electronically excited sodium atoms to molecules such as H_2, D_2, N_2, O_2, CO, CO_2, N_2O, and C_2H_4 (52, 53). In these experiments, a supersonic Na beam with known mean velocity intersects the molecular beam at right angles. The collision region is defined by a laser beam that is used to excite the sodium. The scattered atoms are velocity analyzed by a mechanical selector and observed by a hot-wire detector. The resulting Na atom velocities may be converted to energies, E, so that the energy, ΔE, transferred to the molecule can be calculated ($\Delta E = E_{\text{in}} + E_{\text{ex}} - E$). E_{in} is the initial relative kinetic energy of the collision partners and E_{ex} is the excitation energy of Na*($3^2P_{1/2,3/2}$). From the resolved energy distribution of scattered Na atoms, the vibrational distribution of quencher molecules may be obtained. The observed energy transfer spectra for H_2, D_2, CO, N_2, and C_2H_4 showed distinct maxima at $v = 2$ for H_2, $v = 3$ for D_2, $v = 3, 4$ for CO, and $v = 3, 4$ for N_2. On the other hand, O_2, CO_2, and N_2O were found to be populated more effectively under resonant conditions. The results were shown to be in agreement with a modified curve-crossing theory, adapted from early similar models (54–58), that assumes an ionic intermediate complex [e.g. $(Na^+ - N_2^-)$].

Although difficult to set up, beam experiments possess some definite advantages over "bulb experiments." First, differential rather than integrated cross sections can be measured; second, results are obtained without recourse to a complex set of rate equations; third, any vaporizable molecules can be studied; and fourth, homonuclear diatomic molecules can be indirectly detected. However, energy transfer processes that show up at $E = E_{in}$ cannot be distinguished from totally elastic scattering processes, which makes it impossible to evaluate data in this region.

Additional E-V Studies Employing Classical Methods

During flash kinetic studies of the reactions of $O^*(2^1D_2)$ atoms with molecular oxygen and ozone, the production of vibrationally excited oxygen molecules was observed (59, 60), through the reaction

$$O^*(2^1D_2) + O_2 \rightarrow O(2^3P_J) + O_2^\dagger. \qquad 4.$$

In these experiments, mixtures of O_3 and O_2 were flash photolyzed to produce $O^*(2^1D_2)$, and vacuum UV absorption spectra of O_2 were subsequently recorded. In this way, the populations of $v = 12$–15 could be monitored (the presence of O_3 interfered with the observation of other vibrational states). The band intensities for $v = 12$–14 were found to increase as O_2 was added to the system. Mixtures of $^{16}O_3$ and $^{18}O_2$ were also flash photolyzed (61), and vibrationally excited $^{16}O^{18}O$ was detected. Furthermore, mass spectrometric analysis of the products revealed that 85% of the $^{18}O_2$ had been converted to $^{16}O^{18}O$. These results illustrate the power of isotopic exchange experiments to distinguish between physical and chemical E-V energy transfer mechanisms. For $O^*(2^1D_2) + O_2$ the quenching undoubtedly proceeds via an O_3^* intermediate. This conclusion is consistent with theoretical treatments (28, 62, 63) that have been used to describe energy transfer in the $O^*(2^1D_2) + N_2$ system. This is an extension of the quenching model for electronically excited alkali metals described earlier (54–58). For $O^* + N_2$ however, the quenching process is assumed to proceed by curve crossing of covalent states (rather than ionic states):

$$O^*(2^1D_2) + N_2 \rightarrow (N_2O^*) \rightarrow O(2^3P_J) + N_2^\dagger. \qquad 5.$$

Flash kinetic spectroscopy was also used to investigate the reaction (64)

$$O^*(2^1D_2) + CO \rightarrow O(2^3P_J) + CO^\dagger(v = 1). \qquad 6.$$

The E-V energy transfer efficiency of this process was estimated to be $9.4 \pm 1.4\%$ in He and $6.0 \pm 1.0\%$ in Ar.

The reaction of $O^*(2^1D_2) + CO$ and $O^*(2^1D_2) + N_2$ have been studied by still a different technique (30). The experimental apparatus consisted of a reaction cell containing O_2/CO mixtures. $O^*(2^1D_2)$ was produced by O_2 photodissociation and the concentration of $CO^\dagger(v = 1)$ was monitored by observing $A^1\pi - X^1\Sigma^+(0 - 1)$

vacuum UV absorption in CO. The E-V energy transfer efficiency for reaction 6 was found to be 40 ± 8%. The reaction

$$O^*(2^1D_2) + N_2 \to O(2^3P_J) + N_2^{\dagger}(v = 1) \qquad 7.$$

was also studied by this technique, but the presence of $N_2(v = 1)$ was monitored by use of CO, which comes into rapid V-V equilibrium with nitrogen via

$$N_2^{\dagger}(v = 1) + CO(v = 0) \rightleftarrows N_2(v = 0) + CO^{\dagger}(v = 1). \qquad 8.$$

The relative efficiency of reaction 7 with respect to reaction 6 was found to be 83 ± 10%. The results for 7 are not in agreement with a theoretical curve-crossing model (28). However, there is excellent agreement with theoretical calculations that assume an N_2O^* complex having a lifetime long enough to allow energy migration to occur (62). Application of RRKM (Rice-Ramsperger-Kassel-Marcus) theory leads to an absolute energy transfer efficiency of 30% as compared to the experimental value of 33 ± 10%.

The results obtained for the $O^*(2^1D_2) + N_2$ system have important aeronomic implications. The N_2 vibrational temperature in the earth's upper atmosphere has been modeled (26, 29), but a large uncertainty has always been introduced into these calculations because of the lack of knowledge of the $O^*(2^1D_2) - N_2$ E-V energy transfer efficiency. Since reaction 7 has an absolute efficiency of 33 ± 10%, it is a major contributor to N_2 vibrational excitation in the earth's upper atmosphere.

Using a similar experimental technique, the overall efficiency of E-V energy transfer in the following reactions has been measured relative to the $O^*(2^1D_2) + CO$ efficiency of 40% (65):

$$CO^*(d^3\Delta)_{v=7} + CO \to 2CO^{\dagger}(v), \qquad 9.$$

$$CO^*(d^3\Delta)_{v=7} + N_2 \to CO^{\dagger}(v) + N_2^{\dagger}(v), \qquad 10.$$

$$CO^*(a^3\Pi)_{v=0} + CO \to 2CO^{\dagger}(v). \qquad 11.$$

$CO(d^3\Delta)$ was generated by use of a Xe flashlamp at 1470 Å and $CO(a^3\Pi)$ was produced by using a pulsed microwave-powered iodine lamp operating at 2062 Å. The efficiencies for reactions 9–11 above, relative to reaction 6, were found to be 62 ± 12%, 96 ± 27%, and 89 ± 24%, respectively.

Other experiments involving electronically excited molecules quenched by molecules include a study of $O_2(^1\Delta_g) + NO$ (66) and $O_2(^1\Sigma_g^+) + H_2O$ or CO_2 (67). Excited O_2 was produced in a microwave discharge and vibrational emission from $NO(v = 2-4)$, $H_2O(v_1$ and $v_3)$, and $CO_2(201 \to 000, 121 \to 000)$ was observed. These authors interpret their results as arising from resonant E-V energy transfer (24, 68) with no significant contribution from curve crossings.

LASER STUDIES OF E-V ENERGY TRANSFER

Since lasers are powerful light sources that can be used to electronically excite atoms, to populate specific vibration-rotation levels of an upper electronic state, or to dissociate molecules, leaving their fragments electronically excited, they have begun to play an increasingly important role in the study of E-V energy transfer processes. Lasers are being used in E-V energy transfer studies both to produce electronically excited states and to probe vibrational levels populated during collisional encounters (69–81). Two different laser excitation techniques, which employ visible (69, 70) and infrared (79) fluorescence detection respectively, were developed almost simultaneously. Due to its simplicity and wide generality, infrared fluorescence has become the method of choice over the more restricted visible fluorescence technique. Infrared laser probing of vibrationally excited states in CO, produced by E-V energy transfer processes, was demonstrated shortly afterward (73–75). These experimental techniques are considered in detail below.

Laser Excitation with Fluorescence-Sensitized Probing

E-V energy transfer from excited alkali metal and mercury atoms to hydrogen, deuterium, and nitrogen has been studied (69, 70). In an initial nonlaser experiment, H_2 vibrational levels were probed by measuring the absorption of the sample in the vacuum UV (H_2 lamp) (69). Atomic species (Na, Rb, Cs, and Hg to $3^2P_{1/2,3/2}$, $5^2P_{1/2,3/2}$, $6^2P_{1/2,3/2}$, and 6^3P_1, respectively) were excited by use of suitable metal vapor resonance lamps. Only H_2 $v = 1$ and 2 were detected, probably because of rapid relaxation of higher levels to $v = 1, 2$. In the case of the $Hg^*(6^3P_1) + H_2$ reaction, there was some evidence that higher H_2 vibrational levels had been populated because the risetime of the H_2 absorption signal was slow at low H_2 pressure (several hundred microseconds) but became more rapid at higher H_2 pressures. $Hg^*(6^3P_1)$ can, in principle, excite H_2 up to $v = 13$. Because of the large energy of $Hg^*(6^3P_1)$, 112.9 kcal mole^{-1}, the excitation of H_2 in collisions with Hg^* could proceed through a mercury-sensitized reaction (71, 72) in which Hg^* and H_2 first react to form HgH:

$$Hg^*(6^3P_1) + H_2 \rightarrow HgH + H, \qquad 12.$$

followed by

$$H + HgH \rightarrow H_2^\dagger + Hg. \qquad 13.$$

Further experimental work will be required to fully elucidate the mechanism of this transfer.

In another experimental arrangement a CW tunable dye laser was chopped (pulse width 50 μsec, with a 1 μsec falltime) and used to excite Na atoms in a sample cell containing Na, Cs, and H_2, D_2 or N_2 (70). Alkali metal vapor pressures were kept low ($\ll 1$ mtorr). E-V energy transfer from Na produces vibrationally excited molecules that are then detected by V-E energy transfer to Cs atoms. The Cs is excited to its $6^2P_{1/2,3/2}$ state, which fluoresces strongly, thus serving as a sensitive probe of vibrationally excited molecules. The decay of the Cs fluorescence signal with time measures the overall rate of deactivation (V-T/R) of the excited molecular vibrational states.

E-V Studies Using IR Laser Probes

Some of the most interesting and important work in the field of E-V energy transfer has been reported recently by Lin and co-workers (73-75). In these experiments the vibrational population distribution of CO(v) was monitored, following E-V excitation, by means of infrared resonance absorption. Figure 1 is a block diagram of the apparatus used in these experiments. Gas mixtures were contained in a long absorption cell. In most cases (73, 74), electronic excitation was achieved using a flash tube; however, when studying E-V energy transfer from Na* to CO, sodium was excited using a tunable flashlamp-pumped dye laser (75). Subsequent to excitation, the vibrational levels of CO(v) ($v = 1-10$) were probed by use of a line-tunable CW CO laser. The CO laser was aligned colinear with the dye laser when the latter was used as the excitation source. The detection system consisted of a suitable IR detector and an oscilloscope or signal averager. In analyzing the data obtained, the initial rates of excitation of each CO vibrational level were extracted from the rising portion of each absorption curve by extrapolating the initial intensity with a power series fit to a function of the form $a + bt + ct^2$, where t is time and the coefficients a, b, c were adjusted to fit the experimental data. This procedure was employed in order to avoid the effects of rapid CO V-V relaxation, which complicates later portions of the signal. From these initial rates, the energy transfer cross sections for each vibrational level can be determined. This technique is clearly one of the two most powerful diagnostic methods available that can provide detailed information on the dynamics of E-V energy transfer processes.

In the initial study that used this method the following reactions were investigated (73):

$$O^*(2^1D_2) + CO \rightarrow O(2^3P_J) + CO^\dagger(v \leq 7), \qquad 14.$$

$$I^*(5^2P_{1/2}) + CO \rightarrow I(5^2P_{3/2}) + CO^\dagger(v \leq 3), \qquad 15.$$

$$Br^*(4^2P_{1/2}) + CO \rightarrow Br(4^2P_{3/2}) + CO^\dagger(v = 1). \qquad 16.$$

O*, I*, and Br* were produced in the reaction tube by flashlamp photodissociation of O_3, CF_3I, and CF_3Br, respectively. In all the reactions above (14-16), CO was found to be excited up to the limits of the available electronic energies of the excited atoms. The efficiency of reaction 14 was determined to be 16%. A comparison between the I* + CO and O*(2^1D_2) + CO systems is of interest. The CO vibrational population distribution produced in collisions with I* is not Boltzmann (73) at

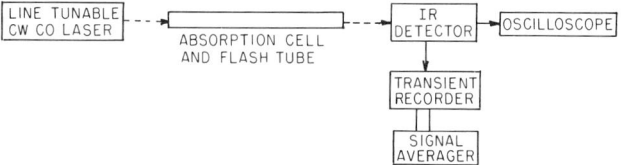

Figure 1 Schematic diagram of the CO laser probing apparatus used by Lin and co-workers (73-75) to study E-V energy transfer.

any temperature, while the distribution produced by O* + CO collisions can be described by a temperature of the order of 8000°K (74). In a later study (74), the $O^*(2^1D_2)$ + CO system was investigated in greater detail, and the complete vibrational population distribution of CO was determined at four different temperatures in the range 246–323°K. The populations N_v were observed to increase with temperature, indicating an increased E-V energy transfer efficiency. An isotope-labeling experiment was also performed using $^{16}O^*(2^1D_2) + C^{18}O$ to yield the results

$$^{16}O^*(2^1D_2) + C^{18}O \begin{array}{c} \nearrow 50\% \; C^{16}O^\dagger(v) + {}^{18}O(2^3P_J) \\ \searrow 50\% \; C^{18}O^\dagger(v) + {}^{16}O(2^3P_J), \end{array} \qquad 17.$$

thus proving unequivocally that the E-V energy transfer process takes place through a CO_2^* intermediate.

A recent theoretical treatment of the analogous $O^*(2^1D_2) + N_2$ system (63), predicts that the vibrational energy of the N_2 product will increase with relative translational energy of the collision partners. This theory treats a model consisting of only one singlet $[O^*(2^1D_2) + N_2]$ and one triplet $[O^*(2^3P_J) + N_2]$ potential energy surface. The reactants move classically, with transitions between surfaces allowed only at a crossing point. The probability of changing surfaces at an intersection is derived from a Landau-Zener approximation. The temperature dependence predicted by the theory is similar to that observed for the $O^*(2^1D_2)$ + CO reaction.

The results of simple statistical calculations based on the RRK and RRKM theories of unimolecular reactions (76) are in reasonable agreement with experimental findings (74) for reaction 14. The observed CO vibrational distribution is very close to an a priori statistical expectation and yields a linear surprisal plot (74, 77).

The vibrational excitation of CO in collisions with $Na^*(3^2P_{1/2,3/2})$ has also been studied (75). CO was found to be excited up to $v = 8$ in this case, with an overall E-V efficiency of 35%. Once again, CO vibrational energy was limited by the electronic excitation energy of Na*. In these experiments Na was promoted to $3^2P_{1/2,3/2}$ by use of a flashlamp-pumped dye laser, and CO was probed as described above. Hassler & Polanyi (78) had previously indicated observation of CO vibrational excitation up to $v = 3$ in this system. The vibrational population distribution observed in the Na* + CO laser study agrees well with those obtained from both the theoretical ionic curve crossing (57) and impulsive models (42) described previously. These results suggest that Na* and CO interact strongly during collisions and that E-V energy transfer in this system is clearly nonresonant.

Laser Excitation by Photofragmentation with IR Fluorescence Detection

Another extremely powerful laser technique has been used to study E-V energy transfer events (79–81). This method, first used by Leone & Wodarczyk (79), appears to be one of the two most promising experimental approaches for obtaining detailed information about specific energy transfer channels [along with the method of Lin (73)]. A block diagram of a typical apparatus is given in Figure 2. Basically, a pulsed visible laser is used to produce electronically excited species, while spontaneous emission in the infrared is employed to monitor the time evolution of the molecular vibrational state populations excited by collision events. IR interference filters are used to achieve spectral resolution, while a wide bandwidth transient recorder/signal averager combination records the time-resolved fluorescence.

The initial experiments using this apparatus can be described by the following set of kinetic equations:

$$Br_2 + h\nu(\text{pulsed laser}) \rightarrow Br(4^2P_{3/2}) + Br^*(4^2P_{1/2}), \quad\quad 18.$$

$$Br^*(4^2P_{1/2}) + HX \xrightarrow{k_0} Br(4^2P_{3/2}) + HX, \quad\quad 19.$$

$$Br^*(4^2P_{1/2}) + HX \xrightarrow{k_1} Br(4^2P_{3/2}) + HX^\dagger, \quad\quad 20.$$

$$Br^*(4^2P_{1/2}) + Br_2 \xrightarrow{k_2} Br(4^2P_{3/2}) + Br_2, \quad\quad 21.$$

$$HX^\dagger + M \xrightarrow{k_M} HX + M. \quad\quad 22.$$

In the above equations HX is ground state ($v = 0$) hydrogen halide (HCl or HBr) and HX† is a vibrationally excited molecule ($v = 1$). The electronic energy of Br*($4^2P_{1/2}$) (3685 cm^{-1}) is sufficient to excite only $v = 1$ of HX. M is any component of the system that quenches HX† according to reaction 22. The rate equations for the system were solved subject to the constraint [HX] $\gg$ [Br], [Br*], [HX†], leading to the result,

$$[Br^*] = [Br^*]_0 \exp(-k_q t), \quad\quad 23.$$

and

$$[HX^\dagger] = \{k_1[Br^*]_0[HX]/(k_q - K_M)\} \times \{\exp(-K_M t) - \exp(-k_q t)\}, \quad\quad 24.$$

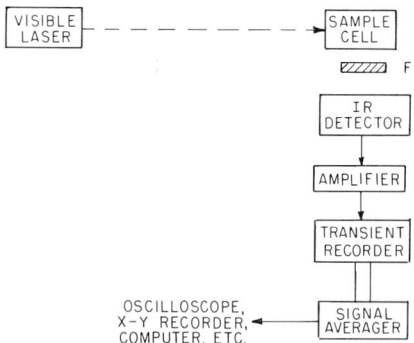

Figure 2 Apparatus used for the study of E-V energy transfer in the experiments involving laser excitation by photofragmentation with IR laser detection. (F = infrared bandpass filter.)

where $[Br^*]_0$ is the initial Br* concentration, $k_q = (k_0 + k_1)[HX] + k_2[Br_2]$, and $K_M = \sum_M k_M[M]$. This solution is valid for $k_q \neq K_M$. According to this treatment, Br* fluorescence decays at the same rate $k_q = 1/\tau$ that characterizes the rise of HX† fluorescence. Division of both sides of the expression for k_q by the total pressure P_{tot}, yields

$$1/(P_{tot}\tau) = (k_0 + k_1) + (k_2 - k_0 - k_1)X_{Br_2}. \qquad 25.$$

Therefore, if $1/P_{tot}\tau$ is plotted against X_{Br_2}, the mole fraction of Br$_2$, the zero intercept gives the total rate coefficient $(k_0 + k_1)$ for quenching of Br* by HX, and the $X_{Br_2} = 1$ intercept gives k_2, the rate coefficient for quenching by Br$_2$.

If the exponential decay of HX† fluorescence is extrapolated to $t = 0$ and divided by the initial Br* intensity, then

$$I(HX^\dagger)_{t=0}/I(Br^*)_{t=0} = \{k_1[HX]/(k_q - K_M)\}\tau_{Br^*}/\tau_{HX\dagger}, \qquad 26.$$

where τ_i is the radiative lifetime of species i and $I(X)_{t=0}$ is the intensity of species X at $t = 0$. Since all quantities on the right-hand side of this equation are known except k_1, the absolute rate coefficient for the E-V process, k_1, can be obtained from a measurement of the ratio of the intensities of HX† and Br* fluorescence. Such measurements are rather difficult due to detector and filter correction problems and thus represent the greatest source of error in obtaining absolute values for k_1.

By use of the above analysis, the E-V energy transfer rates (k_1) for each of the systems studied were found to be 1100 ± 200 msec^{-1} torr^{-1} for (80),

$$Br^* + HF \rightarrow Br + HF^\dagger(v = 1) - 277 \text{ cm}^{-1}; \qquad 27.$$

160 ± 80 msec^{-1} torr^{-1} for (79),

$$Br^* + HCl \rightarrow Br + HCl^\dagger(v = 1) + 799 \text{ cm}^{-1}; \qquad 28.$$

14 ± 7 msec^{-1} torr^{-1} for (79),

$$Br^* + HBr \rightarrow Br + HBr^\dagger(v = 1) + 1126 \text{ cm}^{-1}; \qquad 29.$$

and 570 ± 110 msec^{-1} torr^{-1} for (81),

$$Br^* + HCN(0) \rightarrow Br + HCN^\dagger(001) + 374 \text{ cm}^{-1}. \qquad 30.$$

An earlier study of the Br* + HBr system was also made employing flash kinetic spectroscopy (82). The very fast, nearly gas-kinetic rates measured for HF and HCN may possibly be due to the availability of near-resonant energy transfer channels. Ewing (83), for example, has assumed that for spin-orbit relaxation E-V energy transfer can proceed via a long-range attractive potential interaction. A multipole expansion of the potential is employed with the lowest order nonvanishing term being a dipole-quadrupole interaction. The probability for energy transfer is then just proportional to the product of the molecular dipole transition moment and atomic quadrupole transition moment matrix elements squared.

The detailed kinetic study of E-V transfer from Br* to HCN has been reported (81). In such a system, the greater complexity of vibrational states makes analysis of the data much more difficult. No direct excitation of the (010) bending mode state of HCN was observed, which suggests that the quenching of Br* must populate (001) rather than a combination state such as (011). Vibrational relaxation of HCN(001) was also studied in this experiment.

The detailed kinetics of E-V energy transfer from Br*($4^2P_{1/2}$) to another polyatomic molecule, OCS, have also been studied[3,4] (84). Following Br_2 photodissociation in bromine–carbonyl sulfide mixtures, infrared fluorescence from the v_3(C–O stretch) mode of OCS was monitored at 4.8 μm. By observation of the rate of decay of Br* fluorescence and the filling rate of OCS(v_3) as a function of added OCS, the average number of Br*–OCS collisions required to excite the (001) v_3 state was found to be 110. The most likely filling path appears to be one that proceeds through a combination state with at least one quantum of v_3. In the case of OCS a complete analysis of the system could not have been accomplished without knowledge of V-V energy transfer processes in OCS. The importance of having such data when studying E-V energy transfer in polyatomic systems cannot be overemphasized.

Preliminary work on the Br* + NO system, in which NO fluorescence from $v = 1$ was detected, gives a total Br* quenching rate by NO of approximately 65 msec^{-1} torr^{-1}.[5] However, chemical reaction of Br_2 + NO appears to complicate E-V energy transfer for the Br* + NO system.[6]

The trend in the observed E-V energy transfer rates for Br* to HF, HCl, and HBr certainly supports the idea that resonance plays an important role in these processes. Further support for this view comes from the E-V pumped laser systems described in the next section. For example, the Br*–CO_2 laser has been shown to operate via near resonant transfer from Br*($4^2P_{1/2}$) to the (10°1) state of CO_2 (85). The E-V energy transfer rate for this system was found to be 180 ± 90 msec^{-1} torr^{-1} in an experiment similar to that used to study the Br* + HX systems (86). This rate corresponds to an effectiveness of 40% for quenching collisions that bring about E-V energy transfer.

E-V Lasers

One of the most interesting, practical applications of E-V energy transfer has been the transfer of electronic energy from atomic bromine to CO_2, N_2O, HCN, C_2H_2, H_2O, and NO, accompanied by the observation of gain and stimulated emission from each of these molecular species at previously known or new laser wavelengths (85–90). Each molecule is listed in Table 1 along with its corresponding lasing wavelengths and transition assignments (if known).

These new laser systems have several potential applications. First, by a "sensitization" technique, several molecules that do not interact directly with the electronically excited atom can be made to lase. For example, the N_2O E-V laser is pumped by

[3] S. Lemont, G. W. Flynn. Unpublished work.
[4] C. Wittig. Private communication.
[5] S. Lemont, G. W. Flynn. Unpublished work.
[6] S. Leone. Private communication.

Table 1 List of lasing molecules pumped by E-V energy transfer from $Br^*(4^2P_{1/2})$

Molecule	Laser Wavelength (μm)	Transition
CO_2	4.3	(101)–(100)
CO_2	—	(021)–(020)
CO_2	10.6	(001)–(100)
CO_2	14.1	(101)–(011)
N_2O[a]	10.9	(001)–(100)
HCN	3.85	(001)–(010)
HCN	7.25[b]	(100)–(010)
HCN	8.48	(001)–(100)
C_2H_2	7–8[b]	(00100)–(01000)
H_2O	7.100	(020)–(010)
H_2O	7.294	(020)–(010)
H_2O	7.430	(020)–(010)
H_2O	7.551	(020)–(010)
H_2O	7.710	unassigned
H_2O	16.9	unassigned
NO	5.5	$v = 2-1$

[a] Directly pumped by Br* or indirectly pumped by HCl or CO_2.
[b] These transitions may be due to lasing in water vapor impurities.

V-V transfer from E-V-excited CO_2 or HCl. Such a scheme increases the generality of E-V energy transfer as a laser-pumping mechanism. Second, in the case of the H_2O E-V laser, the large number of vibrational-rotational transitions in the (020)–(010) band, and in the H_2O spectrum in general, suggests the possibility of a tunable laser in the 6–9 μm region. Third, these systems offer the possibility of high-pressure operation. By using CO_2, for example, such a device might provide optical gain and continuous tuning over a range of frequencies.

A particularly interesting feature of E-V lasers is that excitation is mode preferential, and observations of lasing transitions give insight into the exact states involved in the E-V energy transfer process.

Another class of E-V lasers has also recently been developed (91). These are chemical lasers that involve the reaction of Mg, Ca, or Ba with N_2O to form electronically excited metal oxides. E-V transfer from these oxides to high CO_2 energy levels, followed by V-V relaxation to (001) resulting in high gain at 10.6 μm, has been suggested as the pumping mechanism for these laser systems.

V-E ENERGY TRANSFER STUDIES

As mentioned in the introduction, relatively few experiments have been performed that have proved the existence of E-V energy transfer processes. Unfortunately, even less work has been done in studying V-E energy transfer. This may be the result of the difficulty encountered in exciting molecules to specific high vibrational states where they can efficiently transfer their excitation to electronic states. Also, the density of vibrational states at high energy can make the unraveling of such energy transfer processes quite difficult. Up to now, for the most part, V-E processes have been used as experimental tools. For example, as we have already discussed, the experiments of Tsuchiya et al (50) on E-V transfer in Hg* + N_2 mixtures and of Broida and co-workers (70) on vibrational relaxation in H_2 excited by E-V transfer from Na* both used V-E probes of vibrationally excited homonuclear diatomic molecules.

The first direct evidence for the existence of V-E energy transfer processes came with experiments that used the spectrum-line reversal method (92–102), a well-known technique for vibrational relaxation measurements in shock tubes. In these experiments, molecular species that were vibrationally excited by shock waves were found to induce sodium D-line emission. In cases in which reaction with Na atoms could occur, chromium in the form of volatile $Cr(CO)_6$ was used instead, and emission was observed from this atom as well ($^7P_2 \to {}^7S_3$) (95, 96, 101, 102). In these studies, energy transfer to Na or Cr atoms was assumed to occur predominantly from vibrationally excited molecules (e.g. N_2, O_2, CO, and CO_2) so that the excitation temperature of the atom provided a measure of the vibrational temperature of the molecules. Electronic excitation can also be induced by inert gases, but the cross sections for excitation by molecules are considerably larger. The slow rise of sodium excitation temperature behind shock waves has been interpreted as arising from the vibrational relaxation of the particular molecule under investigation. In practice, after molecules were vibrationally excited in the shock tube, both Na D-line emission and absorption were monitored to obtain the ratio $[Na(^2P_{1/2,3/2})]/[Na(^2S_{1/2})]$. Theoretically, this ratio is

$$[Na(^2P_{1/2,3/2})]/[Na(^2S_{1/2})] = \frac{g_P}{g_S} \exp\left(-\frac{hc\omega_d}{kT_g}\right), \qquad 31.$$

where g_i is the degeneracy of state i, ω_d is the D-line transition frequency, and T_g is the reversal temperature, which is assumed to be equal to the vibrational temperature of the molecule. Comparison of the results obtained by this method with those obtained using other techniques conclusively demonstrated that the spectrum-line reversal method was valid (99). In addition, estimates of excitation cross sections in the $N_2^\dagger$ + Na system led to the conclusion that the excitation probability of Na per collision with $N_2^\dagger$ is close to unity, which suggests the importance of near-resonant mechanisms such as

$$Na(3^2S_{1/2}) + N_2^\dagger(v = 8) \to Na(3^2P_{1/2,3/2}) + N_2(v = 0) + 200 \text{ cm}^{-1} \qquad 32.$$

and

$$Na(3^2S_{1/2}) + N_2^\dagger(v = 7) \to Na(3^2P_{1/2,3/2}) + N_2(v = 0) - 1300 \text{ cm}^{-1} \qquad 33.$$

for the excitation process (99). Na D-line emission was also observed when vibrationally excited N_2, produced by a microwave discharge, was mixed with sodium

(103). Addition of N_2O, a known quencher of $N_2^\dagger$, caused a decrease in D-line emission intensity. In the course of this experiment, eight other Na emission lines in addition to the D lines were detected in the range 3100–6110 Å.

The study of V-E energy transfer from N_2 to Na is important from an aeronomic standpoint. For example, it has been proposed that the observed Na D-line emission in Type B red aurora is probably excited by $N_2(v = 8)$ (104).

The first state-by-state analysis of V-E energy transfer from vibrationally excited nitrogen to sodium was done in a crossed-beam experiment (105). This work was later extended to studies of energy transfer from $H_2^\dagger$ and $D_2^\dagger$ (106). A sodium atom beam was crossed with another beam of N_2, H_2, or D_2. The diatomics were vibrationally excited by heating in an oven between 2000 and 3000°K. The degree of excitation of internal energy states was given by a Boltzmann distribution at the oven temperature. Na D-line emission produced at the point of beam crossing was observed with a photomultiplier, and the signal intensity at each diatomic temperature was related to the rate of production of Na*($3^2P_{1/2,3/2}$) to obtain a set of crude rate coefficients. Since the kinetic energy in these experiments is also characterized by the oven temperature, a second measurement was performed to separate out the effects of kinetic energy transfer (translational-to-electronic energy transfer). To separate the internal and kinetic contributions to the observed energy transfer processes, a velocity selector was employed at one temperature, and the resulting distribution of signal intensity vs velocity was used to correct the experimental results. This study demonstrated that N_2 transitions involving $\Delta v = 7$ or 8 quanta were most effective in exciting Na. For $\Delta v = 7$, the N_2 transition causing most efficient energy transfer was found to be $v = 11 \rightarrow 3$. This transition is most nearly resonant with the sodium excitation energy. For H_2 and D_2, similar trends and resonance requirements were observed. For H_2, $\Delta v = 4, 5$ transitions (corresponding to $v = 4 \rightarrow 0$ and $v = 5 \rightarrow 0$) were found to be most efficient in Na* production, and, for D_2, $\Delta v = 6 (v = 6 \rightarrow 0)$ was most effective. Krause et al (106) pointed out that their results agreed well with studies of V-E energy transfer from $N_2^\dagger$ to Na and K atoms (107). By using the principle of microscopic reversibility the cross sections for the inverse process (quenching of Na* by N_2) were determined for each N_2 vibrational state up to $v = 8$. This gives a probability distribution that peaks at $v = 2$. These experimental results are roughly compatible with the ionic curve-crossing theory of Bauer et al (55), which hypothesizes the existence of an $Na^+N_2^-$ intermediate complex in the energy transfer reaction.

Using a shock tube technique, the relative rates of excitation of Na by CO and N_2 were studied for each vibrational level (108, 109). The emission intensity, I_{Na}, of the Na D-lines was expressed as

$$I_{Na} \sim \sum_v k_v X_v, \qquad 34.$$

where k_v is the rate of Na excitation and X_v is the mole fraction of CO or N_2 in vibrational state v. The relative values of k_v were obtained by comparing the observed profile of I_{Na} (the rise of Na* intensity with time) with calculated profiles.

The results of this analysis were interpreted as indicating that the excitation of Na by vibrationally excited CO is most efficient for CO molecules in levels $v = 3$ and 4, rather than level 8, which is resonant with $Na(3^2P_{1/2,3/2})$. (The remaining excitation energy is, of course, drawn from the translational or rotational degrees of freedom that are quite hot.) However, for the case of $N_2^\dagger$ + Na, resonance appeared to be important. The results of both experiments agree qualitatively with the ionic curve-crossing model (55).

The importance of resonance in V-E energy transfer from $N_2^\dagger$ to Na has also been claimed, based on data from a microwave discharge experiment (110). The results of this study were interpreted as supporting the conclusion of the molecular beam study (106) that the reaction

$$Na + N_2^\dagger(v = 11) \rightarrow Na^*(3^2P_{1/2,3/2}) + N_2(v = 3) \qquad 35.$$

is the most efficient V-E energy transfer channel. Despite the apparent agreement between this work (110) and the molecular beam studies (106), two other experiments have been reported (111, 112) that are in serious disagreement with these results. This latter work casts doubt on the validity of the resonant V-E energy transfer argument, at least for the microwave discharge experiments. In the first study (111) active nitrogen was produced via a microwave discharge and mixed with Na in an argon stream. $Na(D_2)$ emission (5890 Å) was observed through an interferometer, and the Na temperature was determined from the Doppler profile. Very strong evidence was found that $N_2(A^3\Sigma_u^+)$ (electronically excited metastable nitrogen) was responsible for collisional excitation of Na via

$$N_2(A) + Na \leftrightarrows N_2 + Na(3^2P) + \Delta E, \qquad 35a.$$

where ΔE is a translational energy release of the order of 6 kcal mole^{-1} (111). Quenching experiments with NO and NH_3 plus the rather large translational energy release mitigate against excitation of Na by resonant collisions with $N_2^\dagger$. Further support for the mechanism of Equation 35a has been obtained in another microwave discharge experiment of similar design (112). These results are not necessarily in conflict with either the molecular beam (106) or shock tube studies (99, 108, 109) since these experiments employ thermal means for excitation of N_2. Thermal production leads to high translational energy but may not yield significant amounts of $N_2(A)$. The beam experiments do indicate that the presence of large relative kinetic energy before collision enhances V-E energy transfer, and the microwave discharge experiments are all done at low relative translational energies.

The only laser V-E energy transfer experiment performed to date was a study of the reaction (113)

$$HF^\dagger(v = 1, J = 5) + Br(4^2P_{3/2}) \rightarrow HF(v = 0, J = 6)$$
$$+ Br^*(4^2P_{1/2}) - 9 \text{ cm}^{-1}. \qquad 36.$$

(The $\Delta J = 1$ transition represents the most strongly allowed HF relaxation channel.) Excitation of HF molecules to $v = 1$ was achieved using an HF chemical laser.

The HF($v = 1$) fluorescence decay signal, initially a single exponential, becomes a double exponential decay upon the addition of Br atoms produced in a microwave discharge. The fast exponential decay is attributed to resonant vibrational-electronic energy exchange as shown in Equation 36. This double exponential behavior is not observed in the deactivation of HF($v = 1$) by O, F, or Cl atoms for which a resonant energy transfer channel is not available. Such double exponential behavior is characteristic of systems where resonant V-V energy transfer is known to occur (e.g. $CO_2 + N_2$ and HF + H_2). The rate for reaction 36 was measured to be 1000 ± 500 msec^{-1} torr^{-1}. This value is in good agreement with the value of 1100 ± 200 msec^{-1} torr^{-1} obtained in studying the reverse E-V process (80).

With the current availability of high-power tunable infrared laser sources, the means exist for exciting a significant fraction of molecules into high vibrational levels via multiphoton absorption. If V-E energy transfer occurs following such excitation the resulting visible (or, optimistically, UV) emission from the electronically excited species could easily be detected by use of a high gain photomultiplier. From a practical viewpoint, V-E processes could be used as a means to populate metastable electronic states, or states not in resonance with existing laser transitions, thus providing new means of studying energy transfer and relaxation processes of excited electronic states. In addition, V-E pumping of atoms or molecules into excited electronic states may lead to population inversion with consequent development of new laser systems, systems that would exchange infrared photon input for visible or UV output.

CONCLUSIONS

Significant progress has been made over the past few years in the study of E-V energy transfer processes that employ vibrational state detection methods. Two laser techniques (73–75, 79–81) appear to be exceptionally promising in their ability to yield large quantities of detailed information regarding such events, and molecular beam studies (51–53) can be expected to grow significantly in importance over the next few years. Nevertheless, the field itself is, at this stage, relatively unexplored and offers rich opportunities for future work. The area of V-E energy transfer appears to be even less explored. Studies in this field, which are certainly related to E-V work, can be expected to expand rapidly, particularly as laser techniques capable of producing high vibrational temperatures are improved and developed.

ACKNOWLEDGMENTS

One of us (S. L.) thanks Dr. Albert B. Harvey for his patience and encouragement during the preparation of this manuscript, and also gratefully acknowledges the award of an NRC/NRL Postdoctoral Research Associateship. We are indebted to the many members of the scientific community who made their results available to us prior to publication. The help of Ms. Patricia Pohlman and Ms. Dorinda Rulle in manuscript preparation and literature search is greatly appreciated.

Literature Cited

1. Steinfeld, J. I. 1970. *Acc. Chem. Res.* 9:313–33 Yardley, J. T. 1974. *Chemical and Biochemical Applications of Lasers*, ed. C. B. Moore, 1:231–47. New York: Academic; Donovan, R. J., Husain, D. 1970. *Chem. Rev.* 70:489–510; Husain, D., Donovan, R. J. 1971. *Adv. Photochem.* 8:1–17; Krause, L. 1966. *Appl. Opt.* 5:1375–87 (These works contain several references concerning the quenching of fluorescence from excited electronic states of atoms and molecules.)
2. Weitz, E., Flynn, G. W. 1974. *Ann. Rev. Phys. Chem.* 25:275–315; Flynn, G. W. 1974. *Chemical and Biochemical Applications of Lasers*, ed. C. B. Moore, 1:163–201. New York: Academic (Additional references may be found in these review articles.)
3. Polanyi, J. C. 1963. *J. Quant. Spectrosc. Radiat. Transfer* 3:471; 1965. *Appl. Opt. Suppl.* 2:109–27
4. Callear, A. B. 1964. *Annu. Rep. Prog. Chem.* 61:48–60
5. Callear, A. B. 1965. *Appl. Opt. Suppl.* 2:145–70
6. Callear, A. B. 1967. *Photochemistry and Reaction Kinetics*, ed. P. G. Ashmore, F. S. Dainton, T. M. Sugden, Chap. 1. London: Cambridge Univ. Press
7. Callear, A. B., Lambert, J. D. 1969. *Comprehensive Chemical Kinetics*, ed. C. H. Bamford, C. F. H. Tipper, Vol. 3, Chap. 1. New York: Elsevier
8. Cundall, R. B. 1969. *Transfer and Storage of Energy by Molecules*, ed. G. M. Bennett, A. M. North, Vol. 1, Chap. 1. New York: Wiley-Interscience
9. Mitchell, A. C. G., Zemansky, M. W. 1961. *Resonance Radiation and Excited Atoms*. London: Cambridge Univ. Press
10. Laidler, K. J. 1955. *The Chemical Kinetics of Excited States*. Oxford: Clarendon
11. Scheer, M. D., Fine, J. 1961. *J. Chem. Phys.* 36:1264–67
12. Donovan, R. J., Little, D. J. 1973. *J. Chem. Soc. Faraday Trans. 2* 69:952–56
13. Butcher, R. J., Donovan, R. J., Strain, R. H. 1974. *J. Chem. Soc. Faraday Trans. 2* 70:1837–46
14. Ewing, J. J., Trainor, D. W., Yatsiv, S. 1974. *J. Chem. Phys.* 61:4433–39
15. Foo, P. D., Wiesenfeld, J. R., Yuen, M. J., Husain, D. 1976. *J. Phys. Chem.* 80:91–97
16. Bevan, M. J., Husain, D. 1976. *J. Phys. Chem.* 80:217–25
17. Wiesenfeld, J. R., Wolk, G. L. 1976. *J. Chem. Phys.* 65:1506–10
18. Pritt, A. T. Jr., Coombe, R. D. 1976. *J. Chem. Phys.* 65:2096–2103
19. Doemeny, L. J., Van Itallie, F. J., Martin, R. M. 1969. *Chem. Phys. Lett.* 4:302–10
20. Van Itallie, F. J., Doemeny, L. J., Martin, R. M. 1972. *J. Chem. Phys.* 56:3689–96
21. Deakin, J. J., Husain, D. 1972. *J. Chem. Soc. Faraday Trans. 2* 68:1603–7
22. Krause, H. F., Datz, S., Johnson, S. G. 1973. *J. Chem. Phys.* 58:367–73
23. Dickens, G., Linnett, J. W., Sovers, O. 1962. *Discuss. Faraday Soc.* 33:52–53
24. Merkel, P. B., Kearns, D. R. 1972. *J. Am. Chem. Soc.* 94:7244–53
25. Bykhovskii, K., Nikitin, E. E. 1964. *Opt. Spectrosc.* 16:111–13
26. Walker, J. C. G., Stolarski, R. S., Nagy, A. F. 1969. *Ann. Geophys.* 25:831–37
27. Bauer, E., Kummler, R. H., Bortner, M. H. 1971. *Appl. Opt.* 10:1861–67
28. Fisher, E. R., Bauer, E. 1972. *J. Chem. Phys.* 57:1966–74
29. Breig, E. L., Brennan, M. E., McNeal, R. J. 1973. *J. Geophys. Res.* 78:1225–31
30. Slanger, T. G., Black, G. 1974. *J. Chem. Phys.* 60:468–77
31. Karl, G., Polanyi, J. C. 1962. *Discuss. Faraday Soc.* 33:93
32. Karl, G., Polanyi, J. C. 1963. *J. Chem. Phys.* 38:271–72
33. Karl, G., Kruus, P., Polanyi, J. C. 1967. *J. Chem. Phys.* 46:224–43
34. Smith, I. W. M., Wittig, C. 1973. *Trans. Faraday Soc.* 69:939–43; Powell, H. T. 1973. *J. Chem. Phys.* 59:4937–42
35. Karl, G., Kruus, P., Polanyi, J. C., Smith, I. W. M. 1967. *J. Chem. Phys.* 46:244–53
36. Heydtmann, H., Polanyi, J. C., Taguchi, R. T. 1971. *Appl. Opt.* 10:1755–59
37. Stephenson, J. C. 1973. *J. Chem. Phys.* 59:1523–31
38. Osgood, R. M. Jr., Sackett, P. B., Javan, A. 1974. *J. Chem. Phys.* 60:1464–84
39. Fushiki, Y., Tsuchiya, S. 1973. *Chem. Phys. Lett.* 22:47–51
40. Horiguchi, H., Fushiki, Y., Tsuchiya,

S. 1976. *Seminar on Molecular Energy Transfer, US-Japan Cooperative Science Program, March 22–24, 1976*, pp. 9–13
41. Suess, H. 1940. *Z. Phys. Chem. Abt. B* 45:312–15
42. Levine, R. D., Bernstein, R. B. 1972. *Chem. Phys. Lett.* 15:1–5
43. Gonzalez, M. A., Karl, G., Watson, P. J. S. 1972. *J. Chem. Phys.* 57:4054–60
44. Simons, J. P., Tasker, P. W. 1973. *Mol. Phys.* 26:1267–73
45. Wilson, A. D., Levine, R. D. 1974. *Mol. Phys.* 37:1197–1201
46. Simons, J. P., Tasker, P. W. 1974. *Ber. Bunsenges. Phys. Chem.* 78:176–79
47. Rapp, D., Kassel, T. 1968. *Chem. Rev.* 69:61–95
48. Holdy, K. E., Klotz, L. C., Wilson, K. R. 1970. *J. Chem. Phys.* 52:4588–95
49. Callear, A. B., Wood, P. M. 1971. *Trans. Faraday Soc.* 67:2862–68
50. Horiguchi, H., Tsuchiya, S. 1975. *J. Chem. Soc. Faraday Trans.* 2 71:1164–72
51. Burrow, P. D., Davidovits, P. 1968. *Phys. Rev. Lett.* 21:1789–91
52. Hertel, I. V., Hofmann, H., Rost, K. J. 1976. *Phys. Rev. Lett.* 36:861–64
53. Hertel, I. V., Hofmann, H., Rost, K. A. 1977. *Chem. Phys. Lett.* 47:163–67
54. Bjerre, A., Nitikin, I. E. 1967. *Chem. Phys. Lett.* 1:179–82
55. Bauer, E., Fisher, E. R., Gilmore, F. R. 1969. *J. Chem. Phys.* 51:4173–81
56. Fisher, E. R., Smith, G. K. 1970. *Chem. Phys. Lett.* 6:438–40
57. Fisher, E. R., Smith, G. K. 1971. *Appl. Opt.* 10:1803–13
58. Andreev, E. A. 1973. *Chem. Phys. Lett.* 23:516–18
59. McCullough, D. W., McGrath, W. D. 1971. *Chem. Phys. Lett.* 12:98–102
60. McCullough, D. W., McGrath, W. D. 1971. *Chem. Phys. Lett.* 8:353–57
61. McCullough, D. W., McGrath, W. D. 1972/1973. *J. Photochem.* 1:241–53
62. Tully, J. C. 1974. *J. Chem. Phys.* 61:61–68
63. Zahr, G. E., Preston, R. K., Miller, W. H. 1975. *J. Chem. Phys.* 62:1127–35
64. Collins, R. J., Husain, D. 1972/1973. *J. Photochem.* 1:481–90
65. Slanger, T. G., Black, G., Fournier, J. 1975. *J. Photochem.* 4:329–39
66. Ogryzlo, E. A., Thrush, B. A. 1973. *Chem. Phys. Lett.* 23:34–36
67. Ogryzlo, E. A., Thrush, B. A. 1974. *Chem. Phys. Lett.* 24:314–16
68. Kear, K., Abrahamson, E. W. 1974/1975. *J. Photochem.* 3:409–12
69. Lee, P. H., Broida, H. P., Braun, W., Herron, J. T. 1973/1974. *J. Photochem.* 2:165–72
70. Jennings, D. A., Braun, W., Broida, H. P. 1973. *J. Chem. Phys.* 59:4305–8
71. Callear, A. B., Hedges, R. E. M. 1970. *Trans. Faraday Soc.* 66:615–27
72. Yang, K. 1967. *J. Am. Chem. Soc.* 89:5344–48
73. Lin, M. C., Shortridge, R. G. 1974. *Chem. Phys. Lett.* 29:42–46
74. Shortridge, R. G., Lin, M. C. 1976. *J. Chem. Phys.* 64:4076–82
75. Hsu, D. S. Y., Lin, M. C. 1976. *Chem. Phys. Lett.* 42:78–83
76. Forst, W. 1973. *Theory of Unimolecular Reactions*. New York: Academic. 220 pp.
77. Bernstein, R. B., Levine, R. D. 1972. *J. Chem. Phys.* 57:434–45; Levine, R. D., Bernstein, R. B. 1974. *Acc. Chem. Res.* 1:393–99; Ben-Shaul, A., Levine, R. D., Bernstein, R. B. 1972. *J. Chem. Phys.* 57:5427–41
78. Hassler, J. C., Polanyi, J. C. 1967. *Discuss. Faraday Soc.* 44:182
79. Leone, S. R., Wodarczyk, F. J. 1974. *J. Chem. Phys.* 60:314–15
79a. Leone, S. R. *Tunable Laser State Selected Energy Transfer and Photochemistry*. PhD thesis. Univ. Calif., Berkeley. 212 pp.
80. Wodarczyk, F. J., Sackett, P. B. 1976. *Chem. Phys.* 12:65–70
81. Hariri, A., Petersen, A. B., Wittig, C. 1976. *J. Chem. Phys.* 65:1872–75
82. Donovan, R. J., Husain, D., Stevenson, C. D. 1970. *Trans. Faraday Soc.* 66:2148–58
83. Ewing, J. J. 1974. *Chem. Phys. Lett.* 29:50–55
84. Lemont, S. 1976. *Laser studies of nanosecond fluorescence and electronic energy transfer*. PhD thesis. Columbia Univ., New York. 110 pp.
85. Petersen, A. B., Wittig, C., Leone, S. R. 1976. *J. Appl. Phys.* 47:1051–54
86. Petersen, A. B., Wittig, C., Leone, S. R. 1975. *Appl. Phys. Lett.* 27:305–7
87. Hariri, A., Wittig, C. 1975. *Bull. Am. Phys. Soc.* 20:1509
88. Petersen, A. B., Leone, S. R., Wittig, C. 1975. *Bull. Am. Phys. Soc.* 20:1510
89. Petersen, A. B., Wittig, C., Leone, S. R. 1976. *Opt. Commun.* 18:125–27
90. Petersen, A. B., Braverman, L. W., Wittig, C. 1977. *J. Appl. Phys.* 48:230–33

91. Benard, D. J. 1975. *Chem. Phys. Lett.* 35:167–71
92. Clouston, J. G., Gaydon, A. G., Glass, I. I. 1958. *Proc. R. Soc. London Ser. A* 248:429–44
93. Clouston, J. G., Gaydon, A. G., Hurle, I. R. 1959. *Proc. R. Soc. London Ser. A* 252:143–57
94. Hurle, I. R., Gaydon, A. G. 1959. *Nature* 184:1858–59
95. Gaydon, A. G., Hurle, I. R. 1961. *Proc. R. Soc. London Ser. A* 262:38–50
96. Gaydon, A. G. 1962. *Energy Transfer in Gases*, ed. R. Stoops, p. 289. New York: Interscience
97. Tsuchiya, S. 1964. *Bull. Chem. Soc. Jpn.* 37:828–42
98. Tsuchiya, S., Kuratani, K. 1964. *Combust. Flame* 8:299
99. Hurle, I. R. 1964. *J. Chem. Phys.* 41:3911–20
100. Hurle, I. R., Russo, A. L. 1965. *J. Chem. Phys.* 43:4434–43
101. Watt, W. S., Bauer, S. H. 1966. *J. Chem. Phys.* 44:2206–15
102. Bradley, J. N. 1969. *Transfer and Storage of Energy in Molecules*, ed. G. M. Burnett, A. M. North, Vol. 1, Chap. 2. New York: Wiley-Interscience
103. Starr, W. L. 1965. *J. Chem. Phys.* 43:73–75
104. Hunten, D. M. 1965. *J. Atmos. Terr. Phys.* 27:583–90
105. Mentall, J. E., Krause, H. F., Fite, W. L. 1967. *Discuss. Faraday Soc.* 44:157–60
106. Krause, H. F., Fricke, J., Fite, W. L. 1972. *J. Chem. Phys.* 56:4593–4605
107. Kalff, P. J. 1971. PhD thesis. Univ. Utrecht, Drukkerij Bronder-Offset, Rotterdam
108. Tsuchiya, S., Suzuki, I. 1969. *J. Chem. Phys.* 51:5725–26
109. Tsuchiya, S., Suzuki, I. H. 1971. *Bull. Chem. Soc. Jpn.* 44:901–7
110. Sadowski, C. M., Schiff, H. I., Chow, G. K. 1972/1973. *J. Photochem.* 1:23–38
111. Gann, R. G., Kaufman, F., Biondi, M. A. 1972. *Chem. Phys. Lett.* 16:380–84
112. Duthler, C. J., Broida, H. P. 1973. *J. Chem. Phys.* 59:167–74
113. Quigley, G. P., Wolga, G. J. 1975. *J. Chem. Phys.* 62:4560–62

TIME DOMAIN REFLECTOMETRY ✼ 2648

Robert H. Cole

Department of Chemistry, Brown University, Providence, Rhode Island 02912

INTRODUCTION

Time domain reflectometry (TDR) is a term used to describe a technique of observing the time-dependent response of a sample of interest after application of a time-dependent electromagnetic field. The response characteristic, whether measured as a current, charge, or other related observable quantity, is thus a real quantity as a function of "real time," as opposed to steady-state measurements of a complex quantity that expresses amplitude and phase relations of response to an alternating sinusoidal (AC) field at specific frequencies. For systems with linear-response characteristics, the two approaches are capable, in principle, of giving the same information differently expressed, with the relation between the two forms being a Laplace transform or its inverse.

An attractive feature of time domain measurements is that a single record can give information over a considerable range of time or frequency, but until recently this was feasible only for times greater than a few microseconds at the least and was done in preference to AC measurements primarily for times greater than several milliseconds (frequencies below, say, 20 Hz). Instrumentation developed in the last few years for generating and observing rapidly changing waveforms, notably tunnel-diode pulse generators and sampling oscilloscopes, now permits measurements ranging from a time resolution of a few picoseconds (1 psec = 10^{-12} sec) to times of several nanoseconds (1 nsec = 10^{-9} sec) or longer. Coverage of the corresponding frequency range from 10^7 to 10^{10} Hz with AC methods has required a considerable amount of complex instrumentation and data processing. As a result, time domain measurements are an attractive alternative for the study of systems that have time-dependent behavior of interest in this range.

In this review, we primarily consider time domain methods and their applications for ranges extending from picoseconds to fractions of a microsecond, which are the ranges that have been subject to the most interest and development using available commercial equipment. For purposes of illustration, we take, for the most part, dielectric response as the quantity of interest, with some reference to aspects of electrical conduction, magnetic properties, and network analysis that can be studied by related methods. Two excellent previous reviews by Suggett (1) and van

Gemert (2) discuss some aspects of the subject in more detail than is possible here and provide examples of earlier results.

Very recently, Clarkson et al (3) wrote a comprehensive discussion of various methods and their applications, with specific examples that supplement the content of this review.

BASIC PRINCIPLES

In most TDR systems, a train of suitably generated fast-rising pulses is applied to a transmission line, usually coaxial with 50 Ω characteristic impedance, and the waveform in the line is observed at some point by a voltage probe connected to a sampling oscilloscope or other data acquisition system. As indicated schematically in Figure 1, a sample of interest is either inserted in or attached to the line with some form of termination, with the resultant changes of the transient waveforms a measure of the response properties of the sample. In an ideal coaxial line, the resultant voltage wave $V(t)$ at the front surface of the sample is the sum of the forward-travelling initial pulse $V_0(t)$ and ensuing reflections, $V(t) = V_0(t) - R(t)$, with a transmitted signal $T(t)$ behind the sample. The relation between $R(t)$ and $T(t)$ to $V_0(t)$ is then dictated by the response characteristic of the sample, propagation in the sample, and boundary conditions.

The most common arrangements for observation, each with its advantages and drawbacks, are approximations to the idealizations:

1. reflected waveform from the sample in the line terminated by a matched section of 50 Ω-impedance for no reflection of the transmitted wave;
2. reflected waveform from a sample used as the termination, i.e. with an open circuit and no current after the sample section;
3. reflected waveform from a short-circuited sample, with no potential difference at the terminal end;
4. transmitted waveform in the section of line behind the sample, terminated by 50 Ω to eliminate reflections from the end of the line.

To illustrate the relation of observed response to sample properties in the four cases, consider a uniform length d of dielectric sample with current $I(t, x)$ or charge

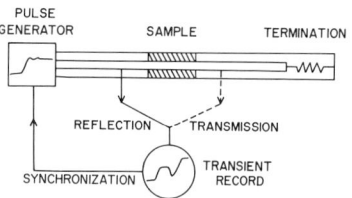

Figure 1 Schematic diagram of instrumentation for time domain measurements.

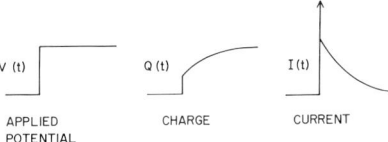

Figure 2 Charge and current of a dielectric sample following application of an ideal step voltage $V(t)$.

$Q(t, x)$ related to potential $V(t, x)$ at a point x by

$$I(t, x) = \frac{\partial Q(t, x)}{\partial xt} = \frac{\partial}{\partial t} C_g \int_0^t dt' \dot{\Phi}(t - t') V(t', x). \qquad 1.$$

A relation of this kind is valid for a dielectric that is linear in the sense of the superposition principle expressed by the time integral or convolution. The parameter C_g is simply the geometric capacitance for unit length of line, while the response function $\Phi(t)$ and its derivative $\dot{\Phi}(t)$ are typically of the form shown in Figure 2, i.e. an abrupt initial change of polarization established virtually instantaneously for the time scale of resolution followed by slower changes, roughly exponential in time, as a result of slower relaxation processes.

The forms of response observed with the four arrangements are shown in Figure 3 for a steplike initial pulse, i.e. a pulse rising rapidly to a plateau with only slight decay over the period of observation. The responses for no sample present are shown as the dashed curves, and the differences between the two indicated by the vertical arrows are in all cases distorted versions of the current-time curve $I(t)$ shown in Figure 2. In the reflection cases, the similarities result from the fact that the differences R, P, and S are directly proportional to the input current to the sample section. The distortion derives from the fact that the input voltages are the algebraic sum

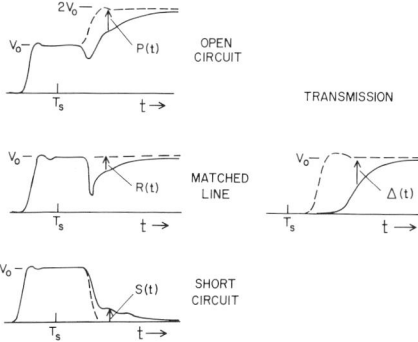

Figure 3 Observed responses for a dielectric sample and common terminations. The dashed curves are voltages for no sample. The time T_s is the time of arrival of the incident pulse at the front surface of the sample.

of incident and reflected waves and, also, the incident pulse is not an ideal step. The curves are further modified by propagation effects in the sample unless its length is very small, as the pulses generated in it necessarily travel more slowly than the speed of light c ($=0.3$ mm psec^{-1} in convenient units). This leads to a delayed arrival of the transmitted pulse and to irregularities in the profiles from successive reflections delayed by intervals greater than $2d/c$ for sample length d. In all cases, however, the integral of the difference curves to infinite time is directly proportional to the static permittivity if the dielectric is nonconducting (effects of finite ohmic conductance are considered later).

To obtain a sample response function $\Phi(t)$ or its equivalent, one needs solutions of the basic electromagnetic equations relating $\Phi(t)$ to the observed waveforms for the appropriate boundary conditions. To obtain real-time solutions one finds that one must solve integrodifferential equations such as Equation 1. Methods for doing this to various degrees of approximation have been developed by Fellner-Feldegg (4), Cole (5), and Chahine & Bose (6) [an error in the last paper has been corrected (7)].

More powerful and generally useful methods of analysis proceed by Laplace transformations of the observed voltage-time curves, the response integral, and the electromagnetic propagation equations. This procedure transforms the convolution integral 1 into a product of transforms and permits solutions for the response by algebraic methods that have long been known and used by electrical engineers, since they are also solutions for steady-state AC behavior. In such solutions, the response function $\Phi(t)$ is replaced by its transform $\phi(i\omega)$, defined as

$$\phi(i\omega) = \mathscr{L}\Phi(t) = \int_0^\infty dt \exp(-i\omega t)\Phi(t). \qquad 2.$$

In the case of dielectrics, one considers a complex permittivity $\varepsilon^*(i\omega) = \varepsilon' - i\varepsilon''$ related to $\phi(i\omega)$ by

$$\varepsilon^*(i\omega) = i\omega\phi(i\omega). \qquad 3.$$

Often, the quantity ε^* is taken as the sum of an "infinite-frequency" value ε_∞, corresponding to the "instantaneous" part of $\Phi(t)$, and a relaxation contribution associated with observable effects of the time dependence of $\Phi(t)$.

The relations in the "frequency domain" give explicit solutions for response, but involve $\varepsilon^*(i\omega)$ implicitly in the arguments of irrational or transcendental functions, necessitating some sort of numerical trial and error or iterative procedure to obtain $\varepsilon^*(i\omega)$ as a function of $\omega = (2\pi)$ (frequency) from numerically evaluated transforms of the incident voltage and response pulses observed in "real time." Fortunately, these necessary tasks are relatively simple even with quite modest computer facilities. It is furthermore often possible to recast the frequency domain solutions as usually given into forms that simplify much of the calculation. These forms also make more evident the nature of the relations of $\varepsilon^*(i\omega)$ to the transforms $v_0(i\omega)$ and $r(i\omega)$, $s(i\omega)$, etc of the observed pulses $V_0(t)$, $R(t)$, $S(t)$, etc, and so express quantitatively the counterpart of the similarities and differences between $\Phi(t)$ and observed responses described at the beginning of this section.

BASIC RELATIONS

In this section, we discuss the essentials of the most commonly used experimental arrangements to show their possibilities and limitations, and give references to more detailed discussions.

Reflection from Sample in a Matched Line

The conventional solution by standard methods (8) gives an expression for the reflection coefficient $S_{\parallel}(\omega)$, defined as the ratio of transform $v_0(i\omega)$ and $r(i\omega)$ of the incident and reflected pulses $V(t)$ and $R(t)$,

$$S_{\parallel}(i\omega) = \frac{r(i\omega)}{v_0(i\omega)} = \rho^* \frac{1 - \exp(-2i\omega d\sqrt{\varepsilon^*}/c)}{1 - \rho^{*2}\exp(-2i\omega d\sqrt{\varepsilon^*}/c)}, \qquad 4.$$

where $\rho^* = (\sqrt{\varepsilon^*} - 1/\sqrt{\varepsilon^*} + 1)$ is the complex reflection coefficient, and d is the sample length. The exponential terms in this expression are the result of the succession of reflections at the two sample surfaces, as is easily appreciated by series expansion of the denominator and the observation that for a nondispersive dielectric with ε^* real, the quantity $2d\sqrt{\varepsilon}/c$ is the time for a round trip in the sample with speed $c/\sqrt{\varepsilon}$, in which case the exponential is simply the transform of an impulse delayed by this time. This formulation is useful for recognizing the accumulated effect of reflections; at the same time one sees problems in solving for ε^*, a complex quantity in general, given $S_{\parallel}(i\omega)$, also a complex quantity, as a function of ω.

One simplification is to omit the exponentials, thereby restricting the range to times before arrival of the first reflection from the back of the sample, i.e. times less than $2d|\varepsilon|^{1/2}/c$. To the extent this direct or single-reflection method is valid, one has merely $\varepsilon^* \cong [(1 + S_{\parallel})/(1 - S_{\parallel})]^2$. The limitations of the approximation are quite severe, as excessively long samples may be required to obtain sufficiently long times or low frequencies, and one must consider the errors in "truncating" the integral transform of $R(t)$ when $R(t)$ is known to a finite time rather than the proper upper limit $t = \infty$ of Equation 2. These problems are discussed in more detail by, among others, Nicolson (9), Gans & Andrews (10), and more generally by Brigham (11).

If all reflections are taken into account, one is faced with solving Equation 4 or its equivalent for ε^*. A direct attack on the problem used by Gestblom & Noreland (12) is to devise a computer algorithm that will find a series of pairs of values ε', ε'' that gives the real part of $S_{\parallel}$ correctly, and a second series that gives the imaginary part; the correct pair is then determined from the intersection of the two loci. Another approach used by Suggett et al (13) proceeds by the Newton-Raphson method of successive approximations. Calling the right side of Equation 4 $F(\varepsilon^*)$, one takes a trial value ε_1^*, obtains a second estimate ε_2^* from the leading term of the Taylor series expansion $S_{\parallel} = F(\varepsilon_1^*) + (\varepsilon_2^* - \varepsilon_1^*)(\partial F/\partial \varepsilon^*)_{\varepsilon^* = \varepsilon_1^*}$, and continues the process until a satisfactory solution is obtained.

The rapidity of convergence to the proper solution evidently depends on how judiciously trial values of ε_1^* can be chosen. Suggett states that usually only a few iterations are necessary if one can, for example, estimate a value of ε_1^* at a low frequency from other measurements, and work to progressively higher frequencies by using preceding results as a guide.

We may remark here that either approach is made possible because *both* real and imaginary parts of $S_{||}$ are calculable from the transient records, a considerable advantage over standing-wave and other steady-state methods that give only absolute values of $S_{||}$ or its equivalent.

A rather different approach with several advantages is to use an alternative form of Equation 4,

$$\varepsilon^* - 1 = (2c/d)\left[\frac{(r/i\omega v_0)}{1 - i\omega(r/i\omega v_0)}\right](x \cot x + i\omega d/c), \qquad 5.$$

where $x = \omega d\sqrt{\varepsilon^*}/c$. This can be obtained by judicious rearrangement of Equation 4, or, more simply and directly, by considering the input admittance of the dielectric section as a four-terminal network terminated by the characteristic conductance G_c of the coaxial line. Although transcendental because ε^* appears in the argument x, this form lends itself to simple interpretation and to simple evaluation of ε^* for $\omega d|\varepsilon|^{1/2}c < 1$. The first bracket takes account of the fact that the voltage at the sample input is not v_0 but $v_0 - r$, while the term $x \cot x$ in the second bracket accounts for propagation effects in the sample section of length d and the term $i\omega d/c$ accounts for termination by G_c [note that $G_c = (C_c/L_c)^{1/2}$, where L_c is the inductance of unit length of the empty line in which the speed of propagation is $c = (L_c C_c)^{-1/2}$]. The simplicity in evaluation arises from the fact that the complicating factor $x \cot x$ has the limiting-value unity for $x \ll 1$ and the series expansion

$$x \cot x = 1 - (\omega d/c)^2 \varepsilon^*/3 - (\omega d/c)^4 \varepsilon^{*2}/45 + \ldots, \qquad 6.$$

is valid for $|x| < \pi$ and rapidly convergent for $|x| < 1$. As a result, Equation 5 can be solved rapidly by iteration. Details are given in a paper by Cole (14); essentially the same results were obtained by Claasen & van Gemert (15) from series expansion of Equation 4.

In first approximation, neglecting terms in powers of $\omega d/c$ and taking $r/v_0 \ll 1$ gives $\varepsilon^* - 1 = (2c/d)(r/i\omega v_0)$, which is the "thin-sample" formula, valid in the limit $d \to 0$, first obtained by Fellner-Feldegg (4). If the incident pulse $V_0(t)$ is approximately of step form with transform $v_0(i\omega) = V_0/i\omega$, where V_0 is the plateau value of $V_0(t)$, one sees that in the thin-sample approximation $\varepsilon^* - 1$ is given directly by the Laplace transform $r(\omega)$ of the reflected pulse. Also, by a Laplace-limit theorem, if incident pulses approach a plateau value V_0 as $t \to \infty$, the static value $\varepsilon_s - 1$ is given by the transform $r(0)$, i.e. the area under the multiple-reflection curve.

The formulas (5) and (6) also make clear an upper frequency limit on the usefulness of the method, as for ε^* real $x \cot x$ has a zero at $x = \omega d\sqrt{\varepsilon}/c = \pi/2$, corresponding to a sample one-quarter wavelength long, and $x \cot x$ changes rapidly near this point, making calculated values of ε^* extremely sensitive to small errors in evaluation of $r(i\omega)/i\omega v_0(i\omega)$. This conclusion has been reached independently by Gestblom & Noreland (12) and by Clarkson & Williams (16) through detailed calculations of the behavior of the reflection coefficient $S_{||}(i\omega)$ or its equivalent.

Reflection from Open-Circuited Sample

In this arrangement, a sample in a length d terminates the coaxial line. As shown in Figure 2, the observed voltage pulse rises to a value $2V_0$, i.e. the voltage-doubling effect from reflection at an open circuit. The difference signal is a measure of the current needed to charge the sample to $2V_0$, and at low frequencies the arrangement corresponds to the simple steady-state procedure of measuring permittivity by placing a sample in a cell at the end of a coaxial line. Neglecting end effects at the back face, the transform $P(i\omega)$ of the difference signal $P(t)$ shown in Figure 2 is related to the transform $v_0(i\omega)$ of the incident pulse by

$$\frac{p(i\omega)}{v_0(i\omega)} = (1 + \rho^*)\frac{1 - \exp(-2i\omega d\sqrt{\varepsilon^*}/c)}{1 - \rho^* \exp(-2i\omega d\sqrt{\varepsilon^*}/c)}. \qquad 7.$$

In this explicit solution for $p(i\omega)$ by the usual method, the relation of ε^* to the observed ratio is not readily apparent. But in this case, network admittance analysis (14) directly yields the very simple result

$$\varepsilon^* = (c/d)\frac{(p/2i\omega v_0)}{1 - i\omega(p/2i\omega v_0)} x \cot x, \qquad 8.$$

also obtainable by rearrangement of Equation 7. The resemblance to Equation 5 for the sample in a matched line is apparent, and much of the discussion of Equation 5 also applies here with a few differences. First, there is no term $i\omega d/c$ added to $x \cot x$ as the sample is not shunted by G_c. The open circuit results in a larger response for a given sample length, by a factor of 4 at low frequencies, and, as before, solutions become impractical for $|x|$ near $\pi/2$. One also notices that for ω small and a steplike pulse, ε^* is directly proportional to the transform $p(i\omega)$ with static permittivity given by the area under the $P(t)$ difference curve.

Advantages of the sample termination method are that in many cases simpler and more readily thermostated cells are possible because the sample is at the end of the coaxial line rather than inserted in it. Disadvantages are that an ideal open circuit is impossible to realize exactly, but corrections for finite stray capacitance at the end of the line or an increased effective electrical length can be made quite simply (14), and that if the permittivity of the sample is very large, e.g. water or aqueous solutions with $\varepsilon \sim 80$, the maximum length for statisfactory results at high frequencies (1 GHz or more) becomes so small (1–2 mm) that a satisfactory cell design is difficult.

The "lumped circuit" method used by Iskander & Stuckly (17, 18) is similar to the procedure just discussed in that the sample of interest is placed at the end of the coaxial signal channel, but differs in that, as the name implies, the sample is assumed to be small enough that propagation effects in it can be neglected. The preceding arguments give an indication of the upper frequency limit for which this would be acceptable; van Gemert (19) has discussed this question, effects of fringing fields, and the like in detail.

Reflection from Short-Circuited Sample

This arrangement is superficially attractive, as an ideal short circuit is closely approximated by simply closing off the coaxial line with a metal disk. The transform $s(i\omega)$ of the difference $S(t)$ between the signals for the short circuit in front of and behind the sample is related to the transform $v_0(i\omega)$ by

$$\frac{s(i\omega)}{v_0(i\omega)} = (1 - \rho^*)\frac{1 - \exp(-2i\omega d\sqrt{\varepsilon^*}/c)}{1 + \rho^* \exp(-2i\omega d\sqrt{\varepsilon^*}/c)}. \qquad 9.$$

Judicious rearrangement of this equation to obtain a counterpart of Equations 5 and 8 for the previous cases or, more directly, use of network analysis gives

$$1 = (c/d)\frac{(s/2i\omega v_0)}{1 - i\omega(s/2i\omega v_0)} x \cot x \qquad 10.$$

for $x = \omega d\sqrt{\varepsilon^*}/c$. This result may seem surprising at first sight, as the permittivity ε^* appears only as the argument of the function $x \cot x$, and at low frequencies for which $x \cot x \simeq 1$ the expression becomes independent of ε^*. The reason for this behavior, and a consequent limit on utility of the method, is that at low frequency one is in effect measuring the difference in *inductance* of the two line lengths (note that $c/d = 1/G_c L_c d$ and $L_c d$ is the inductance of line of length d).

Results for the short-circuited line are not entirely devoid of information about the sample properties at low frequencies, however, as it has been tacitly assumed until now that the sample is nonmagnetic with relative permeability $\mu^* = 1$. If μ^* differs significantly from unity, the sample inductance is $\mu^* L_c d$, the propagation variable x is $\omega d\sqrt{\varepsilon^* \mu^*}/c$, and Equation 10 becomes

$$\mu^* = (c/d)\frac{(s/2i\omega v_0)}{1 - i\omega(s/2i\omega v_0)} x \cot x. \qquad 11.$$

This possibility of using short-circuit TDR measurements for study of magnetic behavior has found some application: Nicolson & Ross (20), for example, have combined short- and open-circuit measurements of the same sample to obtain both μ^* and ε^*, by using the equivalents of Equation 8, with μ^* included in x, and Equation 11. We omit further discussion of such possibilities here on the supposition that few systems likely to be of interest for physical chemistry have permeabilities sufficiently different from one to make the possibilities of much interest.

The use of short-circuit methods for study of dielectric and conductance behavior is rather severely restricted by the requirements on $x = \omega d\sqrt{\varepsilon^*}/c$, hence the frequency must be large enough for $x \cot x$ to differ appreciably from unity but not in a range near $\pi/2$ or an odd multiple of $\pi/2$ (quarter-wave resonance) in which $x \cot x$ is so small that the sample itself is nearly a short-circuit.

There are thus intrinsically limited frequency ranges and corresponding "time windows" for the short-circuit method. It has been used with some success, however, by several investigators (21, 22). Most recently, Bottreau, Dutuit & Moreau (23) have reported success in using measurements for two sample lengths, one twice the other. The reason for this is that one can use the relation cot $2x = (\cot x - \tan x)/2$ to eliminate trigonometric functions and obtain algebraic expressions for ε^*, but there is extra need for caution as one must have both x and $2x$ safely away from the danger zones pointed out above.

Transmission Methods

As the above title indicates, one uses the signal emerging from the sample into the coaxial line rather than that reflected back toward the generator. The simplest arrangement, and the only one we describe, is to terminate the following coaxial line in its characteristic conductance G_c to eliminate further reflections. This alternative has not been used extensively; the principal reason probably is that the time of initial appearance of the signal shown in Figure 2 depends on the permittivity characteristics of the sample, with no indication on the record as to the time of appearance of the signal for no sample present. Proper "time referencing" to establish a common zero of time for the two signals to be analyzed is then hard to establish with sufficient accuracy unless further circuitry or other means is introduced to provide some sort of time reference marker.

Very recently, Gestblom & Noreland (12) have pointed out that the transmission method has two distinct potential advantages over reflection methods. The first is that the expression for the ratio of transmitted-to-incident pulse transforms, from which the permittivity or other information is to be extracted, varies considerably more smoothly with ε^* in ranges where the propagation factor $x \cot x$ in the reflection equations is near zero or infinity. The second is that the delay in arrival of the transmitted signal, which is a measure of $\sqrt{\varepsilon^*} - 1$, can be large enough for otherwise reasonable sample lengths to make uncertainties in the zero of time less serious. In frequency domain language, the phase shifts are large compared to those from the timing error in a realizable marker pulse.

The conventional expression for the ratio of the transform $t(i\omega)$ of the total transmitted pulse $T(t)$ to the transform $v_0(i\omega)$ of the pulse for no sample is

$$\frac{t(i\omega)}{v_0(i\omega)} = (1 - \rho^{*2}) \frac{\exp\left[-i\omega d(\sqrt{\varepsilon^*} - 1)/c\right]}{1 - \rho^{*2} \exp(-2i\omega d\sqrt{\varepsilon^*}/c)}. \qquad 12.$$

In terms of the difference $\delta(i\omega) = v_0(i\omega) - t(i\omega)$ corresponding to the difference $\Delta(t)$ in Figure 2, one has the equivalent expression

$$(\varepsilon^* - 1)\left(\frac{\sin x}{x}\right) + \frac{2c}{i\omega d}\left[\cos x - \cos(\omega d/c)\right] + 2\left[\frac{\sin x}{x} - \frac{\sin(\omega d/c)}{\omega d/c}\right]$$

$$= \frac{2c}{d} \frac{(\delta/i\omega v_0)}{1 - i\omega(\delta/i\omega v_0)} \exp(i\omega d/c). \qquad 13.$$

This is more complicated than the expressions 5 and 8 for the preceding cases, but, like them, helps to clarify relations between $\varepsilon^* - 1$ and the transforms of the observed quantities. For steplike pulses, one sees that at low frequencies $\varepsilon^* - 1$ is directly proportional to $\delta(i\omega)$, as one would expect, with the static value $\varepsilon_s - 1$ proportional to the area of the difference curve $\Delta(t)$. There are no intrinsic difficulties with the solution near $(\omega d/c)\sqrt{\varepsilon^*} \simeq \pi$, but care must be taken with multiple-valued solutions at higher frequencies (e.g. those corresponding to halfwave resonance).

These and other considerations suggest that transmission methods deserve more attention than they have received so far. The problem of time referencing can doubtless be overcome in various ways; Gestblom and Noreland triggered a second reference channel and sampler by the same synchronizing pulse used for the measurements and added a derived output "spike" from the reference to the measuring-channel record.

Finally, direct transmission measurements, i.e. use of only the first signal before arrival of internal reflections in the sample, can also be useful. If reflections are neglected, Equation 12 becomes $t(i\omega)/v_0(i\omega) = (1 - \rho^{*2}) \exp\left[-i\omega d(\sqrt{\varepsilon^*} - 1)/c\right]$, which indicates the possibility, pointed out by Whittingham (24), of combining measurements with different sample lengths. The experimentally determinable ratio for two lengths d_1 and d_2 is just $\exp\left[-i\omega\sqrt{\varepsilon^*}(d_1 - d_2)/c\right]$, permitting explicit solution for $\sqrt{\varepsilon^*}$ and hence ε^* from the (complex) logarithm of the ratio. There are obviously limits on appropriate sample lengths and frequency ranges: if the samples are too short the time window before reflections will be so restricted as to give serious problems with truncation errors, while samples that are too long will produce excessively attenuated, directly transmitted pulses and may also suffer from faster parts of internal reflections overtaking slower parts of the direct pulse. B. Gestblom and E. Noreland (personal communication) have found that, even so, one may use variations with length successfully for relatively small values of ε^*, since ρ^{*2} is then small compared to unity.

PROCEDURES AND PROBLEMS IN USE

The analyses of the several methods just described or variations thereof show a wide range of possibilities for their use; limitations of various kinds, though, depend on the method and shortcomings of available instrumentation as well as the facilities for recording and processing the information. Here we briefly describe some of the practical problems in the use of TDR, with appropriate references to more detailed discussions and solutions in the literature.

Sample Cells

Most commonly, the sample cell is merely a finite length d of coaxial line filled with the unknown sample of interest and suitable spacers to confine the sample. The choice of length is usually a compromise between conflicting requirements. If only the initially reflected or transmitted signal is used, the sample length must be great enough to avoid unwanted reflections out to the maximum time of interest. If a total reflection method is used, the sample must be long enough to produce an adequate difference signal but short enough to keep complications of resonance effects at frequencies above the range of interest. Fortunately, these conditions are not unduly restrictive for a variety of problems and can be relaxed somewhat in some cases, e.g. by varying the ratio of outer-to-inner conductor diameters of the coaxial conductors, or by other geometric changes. Most of the papers already referred to give more or less detailed descriptions of a variety of arrangements.

Pulse Generation and Detection Equipment

Most of the work on TDR reported so far makes use of commercial equipment embodying a tunnel diode or other generator of pulses repeated at a rate of about 10^5 per second, each with an initial rise in about 30 psec or more to a level of about 250 mV, and a decay of about 10% in 0.2 µsec. To obtain a record of the repetitive response picked up by a voltage probe in the coaxial line, a sampling amplifier is used in which a "staircase" scanning voltage determines a sequence of several hundred points in turn at which the voltage of the probe is determined, spread over a time range adjustable from 100 psec to 50 µsec or more. The scanning voltage and hence the time of the points selected can be controlled either manually or by an internal sweep operating at a frequency below 1 kHz. The amplifier and scan voltages can then be displayed on a cathode-ray tube, XY recorder, or other "data-acquisition" device. The circuitry to accomplish all this with various ranges, delays, forms of synchronization, etc, is naturally complex, and we have suggested only those features that need to be appreciated in discussing possibilities and limitations in use.

Pulse Forms and their Transforms

Unless real-time methods of rather limited usefulness are employed, one requires Laplace or one-sided Fourier transforms into the frequency domain of the initial and modified pulses recorded as a function of time. Analytical approximations to the pulse and their transforms are useful for understanding the methods and their use, but numerical methods are usually required to make full use of the available information.

The initial pulses $V_0(t)$ used in TDR are almost always approximations to an ideal step, although impulse-like functions or linearly rising ramp functions have sometimes been used. The Laplace transform of an ideal step with discontinuous rise to a constant value V_0 at $t = 0$ is simply $v_0(i\omega) = V_0/i\omega$. In all the various methods just discussed, one requires only the ratio of a response $r(i\omega)$ to $i\omega v_0(i\omega)$. The latter quantity is just V_0 for the ideal step, so one then has essentially constant Fourier components independent of frequency in determining a permittivity ε^* or its counterpart.

Of course, real pulses have finite risetimes, so it is important to recognize the resulting limitations. If such a pulse is approximated by a linear rise in time T_r to a constant value V_0, the Laplace transform is $v_0(i\omega) = [V_0/(i\omega)^2 T_r][1 - \exp(-i\omega T_r)]$. This is approximately $v_0(i\omega) = V_0/i\omega$ for $\omega T_r \ll 1$ but falls to zero for $\omega T_r = 2\pi$. For the fastest pulses so far available, with T_r about 40 psec, there is thus an increasingly small amount of information from the observed response as the frequency approaches $\omega/2\pi = 25$ GHz, and this is an intrinsic limitation on any TDR methods.

When numerical evaluations of pulse transforms are required, there is usually the fortunate circumstance for most interesting TDR applications that both the incident pulse and response are relatively simple and smooth functions of time (as contrasted, for example, with interferograms from Fourier transform infrared spectroscopy). One can then often use a simple Simpson's rule approximation to a transform

$$F(i\omega) = \int_0^\infty dt\, exp(-i\omega t) F(t) \cong \Delta \sum_{n=1}^{N} exp(-i\omega n\Delta) F(n\Delta), \qquad 14.$$

with a moderate number $N = 30$ to 100 points at evenly spaced intervals Δ. Such a summation, or a more elaborate one if required, necessarily has errors associated with cutting off the summation at a finite N (truncation errors) and using a finite Δ (aliasing errors).

For $F(t)$ a response pulse approaching zero as a limit as $t \to \infty$, (e.g. a negative exponential), truncation usually has a minor effect if total reflection or transmission curves are analyzed; the effect varies roughly as the fraction of the total area that is omitted. For an incident pulse $V_0(t)$ of steplike form, more care must be taken to ensure the proper low-frequency dependence $v_0(i\omega) \sim V_0/i\omega$ of the transform. Samulon's formula is often used (25), which is a modification of Equation 14 in which differences $F(n\Delta + \Delta) - F(n\Delta)$ appear; these go to zero for large n but are subject to error from "noise" in the values. Another, proposed by Nicolson (26) in one form, is to take $V_0(t)$ as the sum of an analytic function with known transform for the long time behavior and the difference, then evaluate the latter by Equation 14.

This is as good a place as any to mention the effect of conduction currents in the sample on the observed response curves and their transforms. If these are ohmic, they can be described explicitly by a "total" complex permittivity $\varepsilon_t^* = \varepsilon^* + \sigma/i\omega\varepsilon_0$, where σ is a specific conductance and ε_0 is a conversion factor to accommodate one's preference in units. Only the total permittivity can be measured, and if σ is taken to be a constant independent of frequency, any dispersion processes of conduction will be reflected in the values of ε^* derived from ε_t^* and assigned values of σ. The presence of a finite low-frequency conductance has the effect of making the response curves decay to a finite value (rather than to zero at long times), with the value determined by simple Ohm's law arguments (14, 27). In obtaining transforms and values of ε_t^*, one must then take precautions similar to those indicated above for step transforms. One should also recognize the further possible effects of electrode polarization or space charge if electrolyte solutions, for example, are of interest, as these will affect the long time behavior as much as in any other method involving electrodes.

Aliasing errors arising from too coarse an interval Δ are largest for higher-frequency components associated with rapidly changing functions of time. The name arises because the use of a finite sum to approximate a transform is really a Fourier series representation that gives false high-frequency components not present in the true transform. The Shannon sampling theorem (28) states that for an interval Δ, no error in the transform results if the true transform has no components at frequencies greater than $f = 1/2\Delta$. Such a property is not to be expected in general, but gives a loose criterion that Δ should be less than $1/2f_c$ for small errors up to a frequency f_c.

The pulse forms from commercial tunnel-diode generators have three kinds of distortion, other than the inevitable finite risetime, which can be reduced in various degrees. The first is a slowly rising and falling signal (~ 1–2 nsec duration) superposed on the baseline and "topline" in the early portions of the tunnel-diode pulse. This arises, as Andrews (29) has pointed out, from leakage of the synchronizing trigger pulse through the tunnel-diode circuit, and can be greatly reduced by feeding the trigger pulse to the tunnel diode via a directional coupler in the coaxial line between the diode and other components. The other artifacts of small oscillations at the initial peak and subsequent decay of tunnel-diode pulses are less easy to reduce, but Andrews (30) has also described a simple pulse circuit giving a much smoother and flatter top (2% drop in 1 μsec) that can be used if the longer risetime (~ 300 psec) is acceptable.

Timing Errors

This source of systematic error becomes increasingly serious at high frequencies. It arises from the fact that if the zeros of time for evaluating the two transforms appearing as a ratio in the various working equations are wrongly chosen with error Δt, there will be a phase factor $\exp(i\omega\Delta t)$ multiplying such ratios as $r(i\omega)/i\omega v_0(i\omega)$ with errors in derived values of complex permittivity $\varepsilon^* = \varepsilon' - i\varepsilon''$ varying with the method, but depending on relative magnitudes of ε' and ε'' as well as on the frequency.

Without special precautions, it is difficult to reduce the error Δt in assigning time zeros to less than, say, one tenth of the risetime of the incident pulse, i.e. to about 5 psec for the fastest tunnel diode pulses available. At a frequency above 1 GHz, the corresponding phase errors $\omega\Delta t > 0.03$ rapidly become serious. Detailed discussions have been given by Nicolson & Ross (20) and by Loeb et al (31).

Various methods and devices have been proposed to reduce timing errors. The crude expedient of extrapolating initially rising parts of the incident and response curves to the baseline by similar convention can hardly be expected to give results better than suggested above. An obvious device is to supply a time marker pulse that is derived by auxiliary circuitry from the synchronizing pulse (12) or tunnel-diode pulse (31). Less direct methods employing a combination of reflection and transmission measurements to provide a time reference for the former have also been used (31).

Unwanted Reflections

In any real system, extraneous signals arise from discontinuities and impedance mismatches in the coaxial line, at the generator, sampling probe circuit, etc, which are superposed on the designed reflection or transmission signals. Their origins are usually easy to identify from the appearance times, especially if these can be changed by using different lengths of coaxial line in the various parts of the system. It is also often possible to eliminate their effects for a particular time range of interest by judicious choice of line lengths to put the offending signal outside this range or "time window," and it is obviously desirable to keep the system as "clean" as possible by using a minimum of circuit components with attention to impedance matching between them.

One source of undesired reflections is the sampling probe circuit, which in commercial units produces asymmetric reflections, i.e. different responses for forward and backward pulses travelling through it. These can appear typically some 400 psec after a reflected pulse and are unavoidable with a given sampling unit. Giese & Tiemann (32) have given examples of their magnitude and effect for a particular sampler, together with a procedure for calibrating and correcting the distortions in the observed pulse record (by observation of the reflection from an essentially ideal short circuit).

Noise and Drift

The various amplifier and scanning circuits in TDR instrumentation must of course have some noise-introducing errors in their outputs. For typical units, signal-channel noise of sampler circuit outputs is largely incoherent and of the order of several millivolts, which, in comparison to the 250 mV amplitude of a tunnel-diode pulse, sets limits on resolution of necessarily smaller difference signals. These can be reduced greatly by signal averaging, in crudest form by the response of an XY recorder to the several hundred points of a manual time sweep, or more effectively by signal averaging for a number of repetitive scans.

Short time fluctuations in time-base circuitry jitter results in distortions of averaged waveforms that are clearly most serious for parts of the waveform changing rapidly in time. In Elliott's analysis (33) of the effects of jitter, he assumed that its presence was equivalent to a low-pass filter in the sampling system; detailed analysis and experimental studies by Gans & Andrews (10) have confirmed the validity of this approach. They found that for tunnel-diode pulses the time distribution is roughly Gaussian, with standard deviation of about 3 psec for a 100 psec time scale, and that its effect is equivalent to that of a Gaussian roll-off low-pass filter with cutoff frequency of 53 GHz. The effect is thus most serious near the high-frequency limit of TDR range.

The preceding discussion is of short-time random noise only. Long time drift of the pulse occurrence times, over periods of seconds or minutes and amounting to 20 psec or more, must also be reckoned with, as it also blurs a recorded trace and acts to nullify improvements by signal averaging over extended periods. Some reduction in time drift, by a factor of 2–4, has been accomplished by Andrews (30) by improved stabilization of bias voltage supplies for tunnel-diode pulse circuits, but much greater improvement by more powerful methods is needed if optimum benefits of signal averaging are to be realized.

Nicolson & Ross (20) have developed a computer-controlled system for time scanning in which the times of observation are automatically maintained in relation to the times of two earlier points, one preceding the pulse and one at a fixed level on the initially rising portion of the pulse, by suitable feedback to the scanning circuits.

Quite recently, Elliott (34) described a very effective control method for reducing time drift in which a small portion of the repeated waveform in the signal channel is fed to the second, synchronous channel of a commercial dual sampler. The ouput of this second channel is then fed to a lock-in amplifier tuned to the scan repetition frequency, and its quadrature output voltage used for feedback control of the bias and, hence, timing of the pulse generator. The effective time drift is reduced to about 0.04 psec by this procedure, permitting effective signal averaging for as long as 15 min with equivalent coherent sampler noise reduced to about 5 μV.

The dramatic improvement made possible by Elliott's method is illustrated by the fact that the smallest detectable capacitance change of about 0.01 pF without feedback (as compared to geometric capacitance of 0.06 pF per millimeter of coaxial line) was reduced by a factor of 200 with feedback and signal averaging for 15 min. As far as I know, such methods have not yet been applied in physical chemical measurements, but the possibilities are obvious.

Data Acquisition and Processing

A wide range of sophistication, and cost, of methods for recording and analyzing TDR data now exists and will doubtless increase. In addition to the basic pulse generation and sampling oscilloscope equipment with a relatively modest cost of $8,000–10,000, one needs at least a simple XY recorder. One can analyze relatively simple response behavior at, say, times greater than 50 psec or frequencies below 1 GHz with useful accuracy using a pocket calculator for real-time analysis (5) or, better, simple Fourier transform programs at a computer facility for frequency analysis. Addition of signal averaging doubles the cost at least, while incorporation of a minicomputer for programmed data acquisition, processing, and print out means an investment of the order of $50,000–60,000. Such complete systems are available commercially; the system developed at the National Bureau of Standards (Boulder) has been described in detail by Gans & Andrews (10).

REPRESENTATIVE RESULTS

Fellner-Feldegg's original paper in 1969 (35) described the first applications of TDR to measurement of dielectric and related properties, and together with more recent papers (4, 36) undoubtedly stimulated much of the work by others to develop and apply increasingly powerful methods.

A favorite subject of study has been the dielectric relaxation of normal aliphatic alcohols from methanol to 1-octanol. These polar liquids have relatively large static permittivities ε_s, in the range 30–10 at room temperature, and steady-state measurements at UHF and microwave frequencies by various workers have established that the principal relaxation by which the permittivity decreases to $\varepsilon_\infty \sim 4.5$–3 is a simple Debye process [$\Phi(t)$ is a negative exponential and $\varepsilon^*(i\omega)$ is a semicircle in the complex plane] with relaxation times in the range 100–1200 psec.

Early determinations of properties of alcohols with TDR were made using direct- or first-reflection methods, and gave rather scattered and inaccurate results, particularly for relaxation times τ, probably as a result of a restricted time "window" of observation and time-referencing errors. [See Suggett (1) for references to this work and representative data.] Subsequent refinements have given more consistent results, particularly after the development of total-reflection and transmission methods, stimulated in large part by Fellner-Feldegg's report (4) of the thin-sample methods. At the present time, data obtained by a variety of TDR methods have given results for several of the alcohols that agree satisfactorily with each other (to a few percent in the parameters ε_s, ε_∞, and τ) and are probably more reliable than many of the results by steady-state measurements. It seems likely that the alcohols will continue to be used to test increasingly refined methods, and may well become the best characterized of all polar liquids. Probably half the references cited in this review give results on one or more alcohols.

Water is the most important dielectric fluid in many respects, both as a pure liquid and as a solvent. Its high static permittivity and short relaxation time ($\varepsilon_s \sim 80$ and $\tau \sim 9$ psec at 25°C) make accurate TDR measurements difficult because time referencing is more critical for the high frequencies of interest, and the high permittivity requires small sample lengths and careful cell design. Loeb et al (31) have, however, obtained very satisfactory data to 10 GHz by taking great care in time referencing, thus demonstrating TDR capabilities when used properly. As Suggett (1) has noted, the satisfactory resolution of these problems has as its reward the virtually infinite resolution in frequency or time of the derivable information, as contrasted with results at a few discrete frequencies in the narrow range obtainable by most steady-state techniques. One is then in a far better position to explore such questions as the number and form of solute relaxation processes.

There have been several studies of aqueous solutions, notably by Suggett and co-workers (37, 38), with such solutes as mono- and disaccharides, amino acids, and peptides, using both direct- and total-reflection techniques. In some of this work it was found that the accuracy of derived results for the relaxation process or processes associated with the solute was improved considerably by taking the difference of time domain responses for the solution and for pure water.

Electrolyte solutions of high permittivity and appreciable conductivity have also been studied successfully. K. Giese (personal communication) has measured solutions of $Mg_2(SO_4)_3$ in water by total reflection and defined a small relaxation process attributable to solute ion pairs below 1 GHz. In unpublished measurements by D. G. Hall in my laboratory, the principal relaxation process with $\tau = 340$ psec in pure sulfuric acid has been determined by total reflection. As H_2SO_4 at the 100% composition point has an ohmic conductance of about 0.01 mho cm^{-1} and static permittivity $\varepsilon_s \sim 85$, these and Giese's results confirm conclusions from theoretical analysis (14, 27) that total-reflection methods can be used successfully to determine dielectric behavior of systems with quite high conductance.

Up to the present, there have been relatively few studies of weakly polar systems such as solutions of polar molecules in a nonpolar solvent. Evidently the reason is that for such systems one is interested in relatively small differences, and if one is limited by the precision of directly recorded response curves large errors can result. Recently, R. G. Campbell, M. Crossley, and L. Glasser (personal communication) have made measurements of solutions of normal aliphatic alcohols at concentrations from 0.3 to 1.0 mole fraction in cyclohexane. Satisfactory results were obtained after making corrections for reflections introduced by the sampler probe and the sample cell; these were determined by use of reference liquids of known permittivity.

Even less work has been done in the study of solid dielectrics or electrolytes, but there appear to be no intrinsic difficulties in application of TDR other than the problems for any measurements of solids, such as electrode contact and polarization effects and formation of voids in the sample.

CONCLUSION

As is doubtless apparent from the preceding discussion, most of the progress to date in time domain reflectometry, or more generally time domain spectroscopy, has been in development of improved methods of analysis and more precise instrumentation, rather than in use of the available techniques as working tools for more or less routine studies.

Proven methods exist at present for study of moderately polar liquids or solids at frequencies from 1 MHz to 3 GHz that use only basic equipment and simple data-processing methods. These can—with sufficient care to reduce timing errors, unwanted signals from impedance mismatches, and the like—be extended to 10 GHz or higher, and, for example, to studies of moderately dilute solutions of polar solutes in nonpolar solvents. It has also been demonstrated that satisfactory results can be obtained for systems with quite high ohmic conductance of the order of 0.02 mho or larger by use of total-reflection methods.

Recent developments strongly suggest that very substantial improvements in the quality of data from available basic instrumentation, and hence much wider ranges of usefulness, can be realized by improvements in associated circuitry to give better time references and stability of time scales, the latter permitting much more effective signal averaging. The feedback control arrangement of Elliott (34) or similar devices should give dramatic improvements in resolution of small effects and be adaptable to differential methods for study of dilute solutions, for example. These more sophisticated methods entail some increase in complexity and cost of the necessary equipment.

Even so, they still appear to be relatively simple and inexpensive means of realizing more fully the intrinsic TDR capability of giving, by a very few experiments that perhaps take as much as a few minutes, detailed information over several decades of time or frequency in regions where other methods are tedious, time consuming, and expensive.

Literature Cited

1. Suggett, A. 1972. *Dielectr. Relat. Mol. Processes* 1:100–20
2. van Gemert, M. J. C. 1973. *Philips Res. Rep.* 28:530–72
3. Clarkson, T. S., Glasser, L., Tuxworth, R. W., Williams, G. 1977. *Adv. Mol. Relaxation Processes.* 10:173–202
4. Fellner-Feldegg, H. 1972. *J. Phys. Chem.* 76:2116–22
5. Cole, R. H. 1975. *J. Phys. Chem.* 79:1459–68
6. Chahine, R., Bose, T. K. 1976. *J. Chem. Phys.* 65:2211–15
7. Chahine, R., Bose, T. K. 1977. *J. Chem. Phys.* 66:1764
8. Johnson, W. C. 1950. *Transmission Lines and Networks.* New York: McGraw. 361 pp.

9. Nicolson, A. M. 1968. *IEEE Trans. Instrum. Meas.* IM-17:395–402
10. Gans, W. L., Andrews, J. R. 1975. *Natl. Bur. Stand. US Tech. Note 672* 165 pp.
11. Brigham, E. O. 1974. *The Fast Fourier Transform.* Englewood Cliffs, NJ: Prentice-Hall
12. Gestblom, B., Noreland, E. 1977. *J. Chem. Phys.* In press
13. Clark, A. H., Quickenden, P. A., Suggett, A. 1974. *J. Chem. Soc. Faraday Trans. 2* 70:1847–62
14. Cole, R. H. 1975. *J. Phys. Chem.* 79:1469–74
15. Claasen, T. A. C. M., van Gemert, M. J. C. 1975. *J. Chem. Phys.* 63:68–73
16. Clarkson, T. S., Williams. G. 1975. *Chem. Phys. Lett.* 34:461–65
17. Iskander, M. F., Stuckly, S. S. 1972. *IEEE Trans. Instrum. Meas.* IM-21:425–31
18. Rzepecka, M. A., Stuckly, S. S. 1975. *IEEE Trans. Instrum. Meas.* IM-24:27–32
19. van Gemert, M. J. C. 1974. *Adv. Mol. Relaxation Processes* 6:123
20. Nicolson, A. M., Ross, G. F. 1970. *IEEE Trans. Instrum. Meas.* IM-90:377–82
21. de Loor, G. P., van Gemert, M. J. C., Gravesteyn, H. 1973. *Chem. Phys. Lett.* 18:295–99
22. Cole, R. H. 1976. *IEEE Trans. Instrum. Meas.* IM-25:371–75
23. Bottreau, A. M., Dutuit, Y., Moreau, J. 1977. *J. Chem. Phys.* In press
24. Whittingham, T. H. 1970. *J. Phys. Chem.* 74:1824
25. Samulon, H. A. 1951. *Proc. IRE* 39:175
26. Nicolson, A. M. 1973. *Electron. Lett.* 9:317–18
27. van Gemert, M. J. C. 1974. *J. Chem. Phys.* 60:3963–74
28. Shannon, C. 1949. *Proc. IRE* 37:10
29. Andrews, J. R. 1975. *IEEE Trans. Instrum. Meas.* IM-21:275–78
30. Andrews, J. R. 1970. *IEEE Trans. Instrum. Meas.* IM-19:171
31. Loeb, H. W., Young, G. M., Quickenden, P. A., Suggett, A. 1971. *Ber. Bunsenges. Phys. Chem.* 75:1155
32. Giese, K., Tiemann, R. 1975. *Adv. Mol. Relaxation Processes* 7:45
33. Elliott, B. J. 1970. *IEEE Trans. Instrum. Mass.* IM-19:391–95
34. Elliott, B. J. 1976. *IEEE Trans. Instrum. Meas.* IM-25:376–79
35. Fellner-Feldegg, H. 1969. *J. Phys. Chem.* 73:616–23
36. Fellner-Feldegg, H., Barnett, E. F. 1970. *J. Phys. Chem.* 74:1962–65
37. Suggett, A. 1975. In *New Techniques in Biophysics and Cell Biology.* New York: Wiley
38. Finer, E. G., Franks, F., Phillips, M. C., Suggett, A. 1975. *Biopolymers.* 14:895

STATISTICAL MECHANICS OF MOLECULAR MOTION IN DENSE FLUIDS

✣ 2649

James T. Hynes[1]

Department of Chemistry, University of Colorado, Boulder, Colorado 80309

INTRODUCTION

In this article we review the theory of molecular motion in dense gases and liquids. Since Berne & Forster reviewed certain aspects of this area earlier in this series (1), there has been extensive activity fuelled by experiment, molecular dynamics (MD) computer simulations, and theoretical advances. There is a large number of reviews of this and related topics (2–14).

Our perspective focuses on the simultaneous importance of both short-time collisional and subsequent collective dynamics for an adequate description of molecular translation and rotation. Enskog and hydrodynamic descriptions provide a good starting point for our discussion.

Enskog theory (15) is applicable to impulsive interaction systems such as hard spheres and rough spheres (respectively, reversal on collision of normal and total relative velocity). In this approach, the (normalized) velocity and angular velocity correlation functions (VCF, AVCF) decay exponentially (15) with respective rates ζ_e/M and $\zeta_{r,e}/I$ given by the Enskog friction constants divided by the mass or moment of inertia. These constants depend only on the binary collision details and the equilibrium pair distribution (df) at contact. The translational and rotational diffusion constants D and D_r are proportional to the time integrals of the CF's; their Enskog values are $D_e = k_B T/\zeta_e$ and $D_{r,e} = k_B T/\zeta_{r,e}$.

The hardsphere MD results of Alder & Wainwright (16) and later work (17) showed that at moderate to high (liquid) densities, small but significant deviations from Enskog behavior for the VCF and deviations of up to 40% for D were observed and attributed to collective effects. The long time $t^{-3/2}$ behavior, in particular, was interpreted in terms of hydrodynamics (16). This unexpected phenomenon led to intensive activity on such "long time tails". This has been admirably reviewed (9).

---

[1] Alfred P. Sloan Fellow, 1975–1977.

The importance of these observations for physical chemists is the appearance, at nearly molecular space and time scales, of hydrodynamic-like behavior.

In a continuum description, the bath for the test particle is governed by the hydrodynamic equations for the flow fields. The particle and fluid are coupled by boundary conditions (BC) at the particle radius R. These are (a) vanishing normal component $\hat{\mathbf{n}} \cdot \mathbf{v}$ of the relative velocity and (b) a condition on the tangential velocity $\mathbf{v}_{tg}$ or the tangential force $(\hat{\mathbf{n}} \cdot \mathbf{\Pi})_{tg}$, where $\mathbf{\Pi}$ is the hydrodynamic stress tensor. The stick BC is $\mathbf{v}_{tg} = 0$; the perfect slip BC is $(\hat{\mathbf{n}} \cdot \mathbf{\Pi})_{tg} = 0$. A more general condition, the slip BC (18),

$$-(\hat{\mathbf{n}} \cdot \mathbf{\Pi})_{tg} = (\beta/R)\mathbf{v}_{tg}, \qquad 1.$$

will be discussed subsequently. The corresponding friction constants ζ_s are the Stokes values (18) $q\pi\eta R$, with $q = 6$ and 4 respectively and η the shear viscosity. With the general Einstein diffusion constant relation $D = k_B T/\zeta$, the Stokes-Einstein result is $D_s = k_B T/q\pi\eta R$. For rotation, the Stokes result (18) for the rotational friction constant ζ_r is $\zeta_{r,s} = 8\pi\eta R^3$ for stick and zero for perfect slip, a much greater sensitivity to BC's than for translation. The rotational diffusion constant D_r is given by the Debye-Stokes relation (19) $D_{r,s} = k_B T/\zeta_{r,s}$.

The importance of collective effects is also suggested by a variety of MD and experimental results.

1. Zwanzig & Bixon (20) have given a fairly successful hydrodynamic model for MD argon.
2. Gass, Alder & Wainwright (17) found that for moderate to high density hard sphere fluids, the Stokes-Einstein result for D with perfect slip held within 10%. For approximately spherical molecules, such results are also found in experiment (8, 21, 22).
3. From the MD data of Alder et al (23) for hard spheres with square well attractions, $D = k_B T/4\pi\eta R$ is fairly well obeyed except at low densities. Attractions (24) simultaneously decrease D and increase η from their hard sphere values.
4. According to Chandler (25), who modelled CCl_4 as a (generalized) rough sphere, torque effects decrease D_e by a factor of two from a hard sphere value. The viscosity increases by a similar factor; perfect slip Stokes-Einstein again holds (22).
5. Further evidence is available from mass-dependence studies. We devote a separate section to these.

For rotation, Kivelson and co-workers (26) have established a strong viscosity dependence for orientational correlation times in ESR experiments. Brauman, Pecora, and co-workers (27) found that such times determined from light scattering satisfied the linear relation

$$\tau = \tau_0 + c\eta, \qquad 2.$$

where c is related to the rotational friction. Hydrodynamic calculations of ζ_r for spheroids by Hu & Zwanzig (28) [and more complex geometries (29)] with the perfect slip condition agree quite well with experiment (27).

Yet hydrodynamics cannot be the complete story. For example, the hard sphere MD results (17, 30) clearly show that D_e remains a significant part of D and that collective effects, while they are important, simply do not dominate. This is also clear from MD experiments on Lennard-Jones systems (31–33). For rotation, the same conclusion follows from the MD results of Levesque, Brot, and co-workers (34, 35) on N_2 and Davis et al (36) and O'Dell & Berne (37) for rough spheres. Returning to translation, we would expect that for equal-sized spheres the appropriate Stokes law would be $\zeta_s = 4\pi\eta(2R)$, since separations less than a diameter are impossible. Finally, it is difficult to believe that no details of the intermolecular forces are important.

The sensitivity of rotation to such questions is counterbalanced by the difficulty of relating AVCF information to measured relaxation times [see, e.g. (38, 39)]. For example, we know of no microscopic derivation of Equation 2.

A very successful approach for the prediction of diffusion constants for real atomic and molecular fluids (rather than correlation functions) was initiated by Dymond & Alder (40), who exploited the idea of an effective or reference hard sphere system. D is constructed as the product of D_e^{ref} and the MD (40) correction factor $D^{\text{ref}}/D_e^{\text{ref}}$ to account for collective effects. An important feature of the hard sphere reference system is that it provides access to the results of the recent significant advances in the theory of equilibrium liquid structure (41) where such systems play a central role. Applications and extensions of these ideas have been given by a number of authors (4, 8, 42, 43). We do not review this approach here. Its success in a wide variety of areas does, however, serve to stress the importance of both microscopic and collective effects, as well as the value of impulsive reference systems as a guide for real molecular fluids.

It will be useful to discuss briefly some aspects of equations of motion for correlation functions and, also, the nature of associated friction kernels. We focus on the VCF; generally similar considerations apply for the AVCF.

It is well known that the VCF $C(t)$ satisfies (7)

$$\partial C(t)/\partial t = -M^{-1} \int_0^t d\tau \zeta(\tau) C(t-\tau). \qquad 3.$$

In Laplace transform language, $C(\varepsilon) = [\varepsilon + M^{-1}\zeta(\varepsilon)]^{-1}$; the friction constant ζ is $\zeta(\varepsilon = 0)$. Here the friction kernel $\zeta(t)$ is $(3k_BT)^{-1}$ times the correlation function of the Mori "random" force, whose dynamics are modified by the presence of a projection operator (PO). This projects out the linear effects of the particle momentum. The general structural features of $\zeta(t)$ for hard spheres are (a) an initial delta function contribution whose amplitude is the Enskog friction ζ_e and (b) a contribution $\delta\zeta$ associated with correlated collisions and collective effects. At low to moderate densities, the latter is negative, due (at least) to vortex motion (17). At high densities, $\delta\zeta$ is, instead, positive for short and intermediate times. This positive correlation implies coherence of retarding effects that tend to reverse the particle motion (backscattering).

For continuous interactions (31–34), there is no really clearcut separation as described above. Here $\zeta(t)$ will have a short-time positive structure as forces remain more or less coherent over a typical interaction time. This will also be true for purely repulsive forces. This is implicit in early modeling studies (44, 45), but most clearly emphasized by Schofield (46), Kushick & Berne (47), and Kim & Chandler (48). Correlated collision effects are also important for continuous interactions, as shown, for example, by MD studies of Levesque et al (31, 32). Their precise magnitudes are difficult to ascertain since no tractable analogues of Enskog results currently exist.

Hynes, Kapral & Weinberg (49) showed by the study of microscopic nonlinear velocity fluctuations that $\zeta(t)$ has an additional decomposition. The numerical effect of such fluctuations can be argued to be small (49, 50), so that $\zeta(t)$ can be well approximated by the contribution of the CF of the random force $\mathbf{F}^\dagger(t)$ first introduced by Mazur & Oppenheim (51). This force has the average effect of the equilibrium bath projected out of its dynamics. This can also be regarded as a first order Sonine polynomial approximation in the velocity (15, 50), which is used in almost all kinetic approaches. Our previous remarks on Enskog and collective separation apply here as well.

The Brownian motion limit (1, 7, 45) is the special case where $\mathbf{F}^\dagger(t)$ approaches the force on a stationary particle, so that $\zeta(t)$ becomes mass independent (as does $D = k_B T/\zeta$). For general mass, the Enskog friction depends on M, due to recoil on collision. All purely (linear) hydrodynamic calculations of $\zeta(t)$ are restricted to the Brownian limit (52) and, hence, M independence. Of course, the VCF will still exhibit M dependence through the M^{-1} factor in Equation 3.

MASS DEPENDENCE

The question of the mass dependence of the translational diffusion constant and velocity correlation function is a useful probe of mechanisms of molecular motion that has only recently received much attention.

Diffusion Constant

For one-component systems (without internal degrees of freedom), D scales like $m^{-1/2}$ (η scales as $m^{1/2}$). This at least implies ηD is independent of m. For test particle diffusion, $D_e \propto \mu^{-1/2}$, with μ as the reduced mass; D_s is mass independent (whatever the boundary condition). At liquid densities, experiments on a series of tracers in N_2 (53), on Ag and Au tracers in Hg (54), and on ^{14}C-substituted benzenes (55) have been interpreted as showing either no mass dependence (53, 54) or one significantly weaker than that predicted by D_e (55). A related conclusion has been reached for computer simulated photodissociation in liquids (56).

In addition to the Brownian limit, strictly mass independent D is also found analytically for a harmonic oscillator in a chain (57), a cluster in such a chain (58), and a cluster of one-dimensional hard rods (59). Approximate mass independence for D is found via MD for clusters in a one-dimensional Lennard-Jones system, although the friction kernel itself only becomes independent for $m/M \lesssim (40)^{-1}$ (60).

The mass dependence of D and the VCF has been studied via MD over a wide range of m/M for test hard spheres in hard sphere fluids by Alder and co-workers (61, 62). A particular aim of these studies was to determine, with progressively lighter test particles, the nature and rate of increase of the importance of backscattering. Decreasing particle momentum progressively leads from persistence of initial momentum direction to reversal in collision with neighboring particles.

The computer results (62) for D/D_e show significant deviations from unity at V/V_0 values of 3, 1.6, and 1.5, with increasingly negative deviations as m/M increases for lighter test particles. At a given density, D/D_e decreases with increasing m/M; thus D increases (but not so rapidly as D_e) due to negative correlations in the VCF. Since D_e itself underestimates D in the absence of correlated collisions (15), the absolute effect is even more pronounced.

We can compare these results with the Enskog and Stokes predictions for D/D_e of invariance with μ and proportionality to $\mu^{1/2}$, respectively. The observed D/D_e values (62) increase with μ but not so strongly as D_s/D_e. For example, at $V/V_0 = 1.5$, a change of m/M from 1 to 10 leads to an decrease in D/D_e by a factor of 2 rather than the Stokes prediction of 3. This is another indication that neither D_e nor D_s completely determines D.

Velocity Correlation Function

The VCF for the hard sphere systems (61, 62) shows more pronounced negative structure as m/M increases; the negative minimum moves initially to shorter times $s = t/t_c$ as m/M goes from 1 to 10. For further increase in m/M, the negative structure gradually (but not monotonically) deepens, but the minimum location (~ 2–3 collision times) does not shift. Alder et al (61, 62) suggest that the first feature is due to easier reversal of the smaller test particle momentum.

Herman & Alder (61) found similar results at lower densities for V/V_0 values of 1.6 and 3. At $V/V_0 = 3$, a change of m/M from 1 to 4 is sufficient to change from velocity persistence to initial backscattering. For heavier test particles ($m/M = 1/4$), positive correlations are stronger (at $V/V_0 = 3$). At $V/V_0 = 1.6$, the heavier particle showed surprisingly strong (though later developed) negative correlations compared to the equal mass case. We know of no theoretical predictions of these results or those for D.

Lorentz Limit

In the limiting case where $m/M \to \infty$ the Lorentz model is approached, where the test particle moves in a sea of fixed scatterers. MD calculations (63) in two and three dimensions show a negative region for the VCF at intermediate and high densities that increases with increasing density. For the Lorentz model, there are, of course, no bath hydrodynamic modes (and no viscosity). Still the single hydrodynamic mode of test particle density can be used to predict some aspects of the behavior of the VCF (64).

Zwanzig & Bishop (65) have investigated a closely related tunnel model by combining exact VCF results for a particle moving in a rigid square with that for a linear hard rod system. The negative correlation arises mainly from the former, while the latter alone determines D. This separation was suggested by MD observations (66, 67) of "rattling" and "slipping" motions at liquid densities. This phenomenon deserves more attention than it has yet received.

Other Systems

Ebbsjö et al (68) have studied, via MD, diffusion in liquid mixtures of isotopes interacting with identical Lennard-Jones potentials. These authors conclude that diffusion constants follow hydrodynamic behavior much more closely than Enskog. Lantelme et al (69) similarly find the change upon isotopic substitution in MD diffusion constants for isotopes is very small for fused NaCl. Both sets of authors note, however, that the corresponding VCF's for different masses differ significantly despite the small effect on the area D. For the NaCl study, the early time behavior of the VCF's has a time dependence that scales with the inverse square root of the mass, a behavior consistent with a kinetic picture. The location of the negative minimum also scales in this fashion but not its depth or the subsequent time variation. These latter are consistent with a collective viewpoint.

ENSKOG AND RELATED KINETIC DESCRIPTIONS

Enskog and related kinetic equations are important for at least two reasons. First, for the VCF and AVCF for dense impulsive systems, Enskog results provide good zeroth order starting points. Secondly, modern kinetic theories of dense systems require as input various space and time dependent correlations that are accessible from such kinetic equations. Finally, these are relevant to real molecular systems to the extent that the effect of repulsive forces and torques on dynamics can be represented by impulsive interactions.

A number of authors have investigated Enskog or generalized (k-dependent) Enskog theory for hard spheres [see, e.g. Sykes (70)]. The general conclusion is that such equations are exact for very short times at any density and that for longer times static correlations are included just as in the regular Enskog equation. In all cases the only such correlation that enters is the contact pair distribution. Résibois & Lebowitz (71) have proposed a generalization of Enskog theory that takes into account binary, ternary, and succeeding dynamical correlations at arbitrary densities. This has been applied by Résibois to the calculation of D and the VCF (72).

Barajas et al (73) pointed out that existing extensions of the Enskog equation to hard sphere mixtures [e.g. Enskog-Thorne theory (15)] lead to violation of the Onsager relations! Such extensions involve particular choices for the evaluation point (e.g. contact, midpoint) for the equilibrium pair df. The ensuing transport coefficients and hydrodynamic equations vary with this choice (73). Van Beijeren & Ernst (74) proposed a modified Enskog equation (free of the above difficulties) that involves the exact local equilibrium pair df as a natural extension of Enskog's original arguments. The resulting Enskog transport coefficients have been given as well as a prescription for relating these to their Boltzmann values.

Dahler & Theodosopulu (11) have reviewed rough sphere and hard nonspherical particle kinetic theory at the Enskog level. Mehaffey & Desai (75) have constructed various space-time correlations at the generalized Enskog level for dense rough sphere fluids.

HYDRODYNAMIC AND RELATED APPROACHES

Correlation Functions via Hydrodynamics

Zwanzig & Bixon (20) first constructed a full hydrodynamic representation for the VCF. The (related) approximations made were the neglect of recoil and test particle motion in computing the friction and the use of hydrodynamics for $\zeta(\varepsilon)$. For the latter, these authors included compressibility and modified the standard hydrodynamic equations by introduction of viscoelasticity, i.e. a nonlocal time relation between stresses and velocity gradients. These were taken to be of the Maxwell form, which vanishes at high frequency. [See also Levesque et al (31, 32)]. Remarkable agreement was found with Rahman's MD results (76) for liquid Ar. Zwanzig & Bixon found that, with D fixed from MD, the results were not very sensitive to a particular choice of boundary condition. A notable discrepancy occurs for short times, where the predicted VCF is linear in t as opposed to the quadratic behavior required for continuous potentials. It is pointed out that this is due to the use of BC's.

Unfortunately, a minor algebraic error cast some doubt on these results. Metiu, Oxtoby & Freed (77) recalculated the corrected VCF with the data used to fit Ar current-current time correlation functions (tcf's) by Ailawadi, Rahman & Zwanzig (78). The agreement is, in fact, improved.

There have been a number of applications of similar character [for references, see (9, 12, 79)]. Rather than discuss these individually, we consider a number of features that have important consequences for the relevance of such calculations for real and MD model systems. For impulsive interactions, the frequency dependent viscosities, e.g. $\eta(\varepsilon)$, must differ qualitatively at high frequencies from their Maxwell form, e.g. $\eta_M(\varepsilon)$, because of short-time collisional transfer effects. With stick boundary conditions, initial slopes of the correlation function are infinite with the use of $\eta(\varepsilon)$; this catastrophe has sometimes been avoided by the incorrect use of $\eta_M(\varepsilon)$. Further, the effects of compressibility are important for short times. Their neglect leads to an incorrect initial value of the VCF; this has been reviewed and analyzed (79). Such effects will also be important for the AVCF for nonspherical hard particles. There is also difficulty with the standard normal velocity BC for hard particles as it leads to an infinite initial time slope rather than the required linear behavior. This BC predicts (52) an instantaneous correlation between test particle and bath velocities that must require some time to establish.

For continuous interactions, viscoelastic properties and the standard boundary conditions must carry the heavy burden of describing the continuous effect of momentum transfer to and from the surroundings. This would seem to be the proper role for some form of generalized BC. Finally, for rotation of nonspherical molecules with continuous potentials, compressibility should be taken into account.

Friction Constants via Hydrodynamics

Hu & Zwanzig (28) numerically determined ζ_r for prolate and oblate spheroids via hydrodynamics, by assuming perfect slip. As with the previously known stick results

(18), ζ_r is linear in η for this BC but is greatly reduced for nearly spherical particles. Youngren & Acrivos (29) calculated ζ_r by a more general numerical method for a benzene model comprised of six hemispheres placed symmetrically on a base. The friction is reduced compared to that for spheroidal modeling. This procedure can evidently be extended to rather complex shapes.

Related Approaches

Kivelson, Kivelson & Oppenheim have considered (80) the relationship of angular velocity and orientational correlation times to D and η. These authors first rewrite the rotational friction in terms of the translational friction

$$\zeta_r = (\langle T^2 \rangle / \langle F^2 \rangle)(\tau_T / \tau_F) \zeta, \qquad 4.$$

where τ_T is the normalized time integral of the Mori torque with a similar definition for τ_F. The viscosity dependence of ζ_r is not derived, but is brought in by the assumption of Stokes' law in the form $\zeta = 6\pi\eta R_e$, where the radius R_e is that required by experiment. With the second assumption that $\tau_T \approx \tau_F$, the result can then be written in the form $\zeta_r = \kappa \zeta_{r,s}$. The correction factor κ then depends only on equilibrium mean square torque and force properties as well as $R_e : \kappa = (3/4 R_e^2)(\langle T^2 \rangle / \langle F^2 \rangle)$. When it is further assumed that rotational diffusion obtains, the ratio r_i of various orientational relaxation times τ_i to their Debye-Stokes values is κ (80).

For the MD results for modelled CO (45), where the normalized friction kernels have been computed, the basic assumption that $\tau_F \approx \tau_T$ seems well justified. For MD-simulated N_2 (34, 35) and for molecules that are only slightly nonspherical, it seems somewhat suspect, particular at high densities where one expects that the VCF goes negative but the AVCF does not.

Kivelson and co-workers (26) have shown in various ESR experiments that κ is a constant for a given solute-solvent system. High pressure NMR studies (8) on pure liquids, however, indicated significant variation of κ [defined by r_2] with density. Fury & Jonas (81) reanalyzed the latter data in the form of Equation 2. With the zero viscosity intercept taken into account, measured κ values were found to be constant. Kowert & Kivelson (26) show that an approximately (numerically) linear behavior of τ_2 with η follows from an expression for κ [again defined by r_2] containing terms in η and η^2 for solute relaxation in a number of binary mixtures.

Hoel & Kivelson (82) considered the separation of κ into contributions due to short ranged repulsive ("hard core") anisotropic interactions and non-hard core interactions. The former are taken into account by the hydrodynamic perfect slip results for spheroids of Hu & Zwanzig (28). The latter are then to be gauged by a parameter s. It was concluded that such a separation is probably not clearcut, since s must also include any deviations from spheroidal geometry and nonhydrodynamic effects.

Hynes, Kapral & Weinberg (83) have considered the rotational motion of a rough sphere in a bath of identical rough spheres. It is argued that, due to kinetic boundary layer effects in the test particle neighborhood, the boundary condition for rotation of a rough sphere is not the hydrodynamic stick condition as often assumed but is rather the slip BC Equation 1. The slip coefficient β reflects the lack of perfect coherence between test particle and solvent motion and gauges the effect of the test particle rotation across the boundary layer into an outer hydrodynamic region. [A somewhat related perspective is given by Kowert & Kivelson (26)]. The exact initial Enskog slope was used to determine β, a procedure that involves extrapolation across an initial time layer during which the slip BC is established.

With this picture, the rotational friction is determined by both the microscopic Enskog friction and the hydrodynamic (i.e. stick BC) friction. Numerical agreement with MD results for D_r and the AVCF is good. It is found that $D_r = D_{r,e} + D_{r,s}$. A further interesting feature is that even though D_r is dominated by $D_{r,e}$, ηD_r is not a strong function of η; over regions of moderate η change, ηD_r appears constant.

MODE COUPLING THEORY

The anomalous behavior of fluid transport coefficients in the critical region first led workers to question the assumption of microscopic lifetimes of transport kernels. These kernels are tcf's of transport fluxes with PO-modified dynamics. The resolution of this difficulty involved the realization that, while linear collective variables or modes had been expunged from these kernels by the Mori PO, important long-lived contributions to the kernel can arise from products of these modes via coupling to the fluxes. The discovery of the long time tail in the VCF and its explanation in hydrodynamic terms by Alder & Wainwright (16) suggested that mode coupling was important for transport even away from the critical region.

Zwanzig (84) has given an extremely lucid account of the general structure and application of mode coupling. The detailed mechanics and applications of mode-coupling calculations, along with aspects of the relationship to kinetic theory and hydrodynamic approaches, particularly with respect to asymptotic time behavior, have been reviewed by Keyes (14), Pomeau & Résibois (9), and Hynes & Deutch (7). Here we focus on those applications of mode coupling theory that attempt to provide reasonably complete descriptions of D or rotational relaxation parameters.

Translational Diffusion

Keyes & Oppenheim (85) approached the calculation of D via the tcf of the density of a tagged particle. If $n_k(t)$ is the kth Fourier component of this density, then one has

$$\frac{\partial}{\partial t}\langle n_{-k}n_k(t)\rangle = -k^2 \int_0^t d\tau D_k(\tau)\langle n_{-k}n_k(t-\tau)\rangle, \qquad 5.$$

and the diffusion constant can be extracted as

$$D = \lim_{k\to 0} \int_0^\infty dt D_k(t). \qquad 6.$$

The diffusion kernel $D_k(t)$ is essentially the tcf of the test particle flux whose dynamics have the linear variable n_k projected out. Keyes & Oppenheim exposed residual collective effects in $D_k(t)$ by explicitly considering coupling of n_k to bilinear products B_k that involve the test particle density and the fluid velocity.

The ensuing expression for D separates into two terms, $D = D_0 + \delta D$, where

$$\delta D = k_B T k_0 / 3\pi^2 \rho (v' + D'). \qquad 7.$$

Here D_0 is a bare diffusion constant and contains both microscopic and collective contributions that are not explicitly accounted for by the effects of the bilinear modes B_k. These latter are exposed in the hydrodynamic term δD. The cutoff wave number k_0 is introduced to approximately account for the fact than even B_k cannot be considered hydrodynamic for large k; k_0 is taken to be less than π/a, with a the particle radius. The coefficients v' and D' are certain bare kinematic viscosity and diffusion coefficients. It is argued (85) that $v' \approx v \gg D'$, so that δD is of the Stokes-Einstein form. Keyes & Oppenheim make the important point that the two contributions D_0 and δD can be comparable.

As important and suggestive as these results are, there remain a number of difficulties. The bare diffusion constant D_0 is not very clearly identified. It is hard to see how D_0 can be, for example, D_e at high densities since $D/D_e < 1$ there. A second and related problem is that k_0 is also somewhat vague; D itself is independent of any imposed cutoff k_0 (used here as a computational tool).

For a large test particle of radius R in a fluid of smaller particles of radius a, the definition of n_k was modified (85) to account for the fact that $n(\mathbf{r}, t)$ is nonvanishing over a finite spatial range, not just at a point. The foregoing analysis (with some important differences arising from finite R) then yields (85, 86)

$$D = k_B T / 5\pi \rho v R = k_B T / 5\pi \eta R, \qquad 8.$$

with the assumptions $v' \approx v \gg D'$. For a large particle, the exclusion of solvent particles from the sphere leads to an effective "geometric" cutoff $\sim \pi/R$ in the calculation when $R \gg a$; k_0 then disappears from the result. It is argued that D_0 vanishes when $R \gg a$.

The above result differs by a factor of $\frac{5}{4}$ from D_s; the origin of this discrepancy is still unclear. It is also curious that the mass of the test particle seems to play no role in the vanishing of D_0. Despite these and other questions, something very like the Stokes-Einstein result has emerged from a mode-coupling treatment.

At the same time, Kapral & Weinberg (87) considered mode-coupling effects on diffusion from the point of view of mutual diffusion in the limit of small concentration of the test particle component. Although D is not given explicitly, their result for equal size particles also has the form of Equation 7. Kapral & Weinberg showed explicitly that, for low density, the bare diffusion constant was the Boltzmann value of D and that the transport denominator of the δD term was $v + D$ in Boltzmann approximation.

Bedeaux & Mazur (88, 89) have developed a phenomenological approach to mode coupling that has the particular virtue of rapidly and clearly yielding general structure and specific predictions without the intricacies of microscopic calculations. Of course at some stage this becomes a limitation. In view of residual cutoff dependences of microscopically based mode-coupling approaches, however, this does not seem to be a serious competitive drawback at present.

Bedeaux & Mazur's calculation of D starts from the conservation law

$$\partial n(\mathbf{r}, t)/\partial t = -\mathbf{V} \cdot \mathbf{j}(\mathbf{r}, t) \qquad 9.$$

for the (low) number density of test particles in the fluid. The current $\mathbf{j}$ is taken to be

$$\mathbf{j}(\mathbf{r}, t) = -D_b[\rho(\mathbf{r}, t)]\mathbf{V}n(\mathbf{r}, t) + \mathbf{v}(\mathbf{r}, t)n(\mathbf{r}, t), \qquad 10.$$

plus a random current $\mathbf{j}_R$ whose average $\overline{\mathbf{j}_R}$ over equilibrium fluid fluctuations vanishes. Here coupling of test particle motion to the fluid occurs by the convective term that involves the fluid velocity field $\mathbf{v}(\mathbf{r}, t)$ and the dependence of the bare diffusion constant D_b on the local fluid density. For small deviations from equilibrium,

$$D_b = D_0 + (\partial D_b/\partial \rho)_{eq}\delta\rho, \qquad 11.$$

where the bare diffusion constant $D_0 = D_b(\rho_{eq})$ is taken to be a constant and $\delta\rho$ is a fluid density fluctuation. It is shown that $\overline{n_k(t)}$ satisfies a generalized diffusion equation identical to Equation 5 with $D_k(t)$ expressed in terms of the fluid fluctuations. When D is evaluated to second order in these fluctuations, Bedeaux & Mazur obtain

$$D = D_0 + D_c + [k_B T k_0/3\pi^2 \rho(v + D_0)]. \qquad 12.$$

A cutoff k_0 again appears, due to the continuum description. The essential new feature in Equation 12 is D_c, which for large particles is a small negative correction (88) proportional to the compressibility. A similar result was found by Lukin et al (90) by a hydrodynamic friction calculation that included nonlinear terms in the fluid density conservation law. Such corrections should be important (90) near one-component critical points where the compressibility diverges. The contribution D_c has not been evaluated for small particles. It would be interesting to know its significance.

In a later paper (89), Bedeaux & Mazur modify their procedures to be directly applicable to a large test particle. By starting from incompressible stick-boundary condition hydrodynamics, it is inferred that Equation 10 is replaced by

$$\mathbf{j}(\mathbf{r}, t) = \mathbf{v}_{\text{eff}}(\mathbf{r}, t)n(\mathbf{r}, t), \qquad 13.$$

so that the bare diffusion constant and random force evidently vanish for a large Brownian particle in continuum description. Here $\mathbf{v}_{\text{eff}}$ is altered from the unperturbed velocity $\mathbf{v}$ owing to the particle's presence, boundary condition, and motion. This modification naturally brings in an effective cutoff in k space; Bedeaux & Mazur ascribe the incorrect numerical factor in Equation 8 to the use of a solely geometric cutoff factor. Bedeaux & Mazur carry out the calculation of D to second order in the fluid fluctuations and recover $D = D_s$.

We have ignored the question of which viscosity and diffusion constants should appear in the denominators of the collective contributions to D. The unspoken consensus, based on higher-order mode and fluctuation expansions, is that $v + D$

should appear. While this rather changes the character of the equation for D, at low densities $D \approx D_e$ and renormalization effects are small, while at high densities $D \ll v$, and D may be dropped from the denominator.

As a final point, we note that mode coupling has not been used to obtain the VCF itself; this would require an evaluation of the bare correlation function whose time integral is D_0.

Rotational Motion

The application of mode-coupling theory to rotation and orientation (O) seems rather more complex than for translation, and only partial results are available.

Garisto & Kapral (91) showed that mode coupling for linear molecules leads to a $t^{-5/2}$ asymptotic behavior for the AVCF while the first rank OCF decays as $t^{-7/2}$. Keyes & Oppenheim (92) also found the latter result. The coefficients of these decays depend on fairly complicated dynamic-coupling parameters. The AVCF result agrees, with respect to time dependence, with a conjecture of Berne (93) based on stick hydrodynamics for a sphere but the OCF result does not agree. This hydrodynamic result depends on the spherical symmetry and an (unproven) generalized Debye equation (93, 94, 95) that implies that OCF's can be found solely from the AVCF. This generalized Debye equation has been shown to follow from the Bedeaux-Mazur formalism applied (95) to a finite-radius spherical particle when the fluid fluctuation expansion is truncated at second order. For a rough sphere, Garisto & Kapral (96) found that the AVCF decays again as $t^{-5/2}$ but with the simple coefficient predicted via hydrodynamic considerations. The origin of this decay in hydrodynamics is the coupling to the fluid transverse velocity, but in mode coupling it is (97) the coupling to transverse spin in the fluid! There is fair agreement on the rough sphere AVCF tail (98) and MD calculations. For spherical particles, Pomeau & Weber (99) argue that the orientation correlation functions decay exponentially for long times even when the AVCF decays as $t^{-5/2}$.

Keyes & Oppenheim (92) used microscopic mode-coupling theory to examine the orientational relaxation time τ_2. They found τ_2^{-1} as a sum of a bare part $(\tau_2^0)^{-1}$ and a collective contribution roughly of the form $A^2 k_0^3 k_B T/\eta$. This could not be clearly reduced to the Debye result $\tau_2^{-1} = 2D_r$, since $(\tau_2^0)^{-1}$ could not be shown to vanish with increasing particle size, and the dynamical coupling coefficient A could not be evaluated (although its relationship to light scattering experiments was suggested). Keyes' & Oppenheim's result is not consistent with the empirical Equation 2. Deutch & Hills (95) have stressed the importance of accounting for finite size effects in microscopic mode coupling.

All of the above clearly emphasizes that the relation between angular velocity, fluid variables, and angular correlations requires considerable elucidation.

RENORMALIZED KINETIC THEORY

Renormalized Kinetic Theory (RKT), as developed by Mazenko (100, 101), is a kinetic description designed to systematically describe dynamics in dense fluids for realistic interactions. It has the potential of bridging the gap between short time Enskog descriptions and long time mode-coupling and hydrodynamic representations. Here we discuss some of its important characteristics and applications to the VCF and AVCF by Furtado, Mazenko & Yip (102) and Mehaffey, Desai & Kapral (97). These latter calculations are similar, in a number of respects, to earlier work on hard spheres by Résibois (72); a comparison has been made (102).

General Features

There are a number of noteworthy features of RKT.

1. The focus is on few-particle distribution and correlation functions. This contrasts with mode-coupling theory, which generally deals with variables and PO's defined in the full phase space.

2. The actual intermolecular forces and torques are systematically renormalized, i.e. the interactions present in the final formulation are spatial and angular gradients of potentials of mean force and torque. This feature introduces the static fluid structure in an essential way. Indeed, equal prominence is given to both static and dynamic properties. This will be important for real molecules in condensed phases where the molecular structure is itself solvent dependent.

3. In low order approximation, the theory makes contact with the Enskog approximation, which we know is a reasonable first approximation for dense systems.

4. In contrast to mode-coupling theory, no wave number cutoffs need be introduced. Instead, integrals over wave vectors are internally "cut off" by various k-dependent static and dynamic factors.

5. The theory is not limited, in principle, to singular potentials, although all applications to date have employed such potentials.

Structure and Analysis

Many important aspects of RKT for single particle motion can be discussed (97, 101, 102) in terms of the VCF. In particular, the translational friction $\zeta(\varepsilon)$ decomposes naturally according to

$$\zeta(\varepsilon) = \zeta_{ge}(\varepsilon) + \zeta_{rc}(\varepsilon). \qquad 14.$$

The first term is the generalized Enskog contribution. In the limit of impulsive collisions it reduces to the Enskog friction; for soft potentials, it is ε-dependent, owing to finite collision duration. All forces are of course renormalized to include the fluid structure. The contribution $\zeta_{rc}(\varepsilon)$ is the recollision term and contains the effects of correlated collisions. The structure of $\zeta_{rc}(\varepsilon)$ is schematically of the form

$$\zeta_{rc} = -BCC_f B^T, \qquad 15.$$

where B and its transpose B^T represent an initial and final correlated collision between the test particle and a bath particle. Here C and C_f represent propagators for the test particle and the bath respectively. The form above is the "ring collision" (97, 101, 102) approximation to the more general result (101).

The propagation between initial and final collisions can be represented in terms of correlations that involve both hydrodynamic and nonhydrodynamic states of the system. Furtado et al (102) have argued that ζ_{rc} can be approximately rewritten as (in time language)

$$\zeta_{rc}(t) = -B[C'C'_f - e^{-(\lambda+\lambda_f)t}C'_0C'_{f0}]B^T, \qquad 16.$$

where $C' = \mathcal{P}C\mathcal{P}$ and $C'_f = \mathcal{P}_f C_f \mathcal{P}_f$ with $\mathcal{P}$ and $\mathcal{P}_f$ as projection operators on the hydrodynamic, or conserved variable, states of the test particle and fluid. This is the mass density for the test particle, while for the fluid these are the densities of mass, momentum, and energy. This retains the hydrodynamic states intact in the propagation and approximates the "microscopic" states through the term that involves the free particle propagators C'_0 and C'_{f0} and microscopic relaxation rates λ and λ_f.

With the above considerations, the recollision friction term assumes the form of a sum of contributions of the general form $\int_0^\infty dk k^2 B(k) B^T(k) \Delta(k, t)$, where the B's are certain Fourier components related to the collision operators and mode coupling, while

$$\Delta(k, t) = C(-k, t)C_f(k, t) - e^{-(\lambda+\lambda_f)t} C_0(-k, t) C_{f0}(k, t) \qquad 17.$$

involves correlations of the system hydrodynamic modes evaluated with Enskog dynamics. As $t \to \infty$, the first terms in Δ will dominate and be given by their low k, hydrodynamic approximation. Just such terms are analyzed in mode-coupling treatments of long time tails. The importance of the RKT results is that couplings to the bath are monitored from their inception to their ultimate decay.

The correlation functions required, e.g. $C(k, t)$, have been constructed, following Résibois (72), from their hydrodynamic limits with k-dependent transport coefficients, and their short time form is determined by time expansion and various sum rules. The two forms are joined by extrapolation formulae (72). In principle, these correlations could also be constructed via generalized hydrodynamics (12).

VCF and AVCF Applications

The preceding considerations have been applied to the calculation of D and the VCF for smooth hard spheres by Furtado et al (102), and the VCF and AVCF for rough spheres by Mehaffey et al (97), who generalized RKT to apply to this system.

Furtado et al included coupling to longitudinal and transverse velocity modes (but not density) to find reasonable agreement with MD results at low and moderate densities. At higher densities, the cage or backscattering effect was obtained, although numerical agreement with MD was not satisfactory. These authors found that the nonhydrodynamic states make important short time contributions and that collisional transfer effects are very important at high densities. This latter point was first stressed by Résibois (72), who connected it with the origin of the backscattering effect.

Mehaffey et al included couplings to all hydrodynamic modes of the rough sphere system. For the VCF, ζ_{rc} is dominated by the coupling to density (positive) and transverse velocity (negative), the former most important at short times with the reverse true at long times. The VCF accordingly exhibits mainly negative deviations from Enskog behavior. The recollision contribution to the rotational friction is generally dominated by the spin angular velocity and the transverse velocity contributions, which are both negative. At short times, the former dominates, while the long time behavior involves a delicate balance of these and a cross-coupling effect. The AVCF exceeds the Enskog value, as found via MD (36, 37).

Both sets of authors find that for moderate and high densities the predicted deviations from Enskog behavior are somewhat too strong and too early. This implies that the initial transmission of correlations into collective modes requires clarification.

BOUNDARY CONDITIONS

The importance of some hydrodynamic component of molecular motion implies the necessity of a deeper consideration of the boundary conditions that provide the coupling of a particle's motion to that of its surroundings. While these are often found from heuristic considerations, this is a dangerous procedure, as illustrated by the work of van Kampen & Oppenheim (103), who discovered a number of nonintuitive boundary effects for a stochastic random walk model.

Although there has been little progress in this area in condensed phases, we include a brief review of related boundary condition studies in view of this area's importance and expected future activity therein.

Nonequilibrium Thermodynamics and Hydrodynamics

Waldmann (104, 105) has shown how the well-known entropy production arguments used to derive linear transport laws in bulk media may be extended to interfaces and boundaries. The total entropy production is shown to have a boundary contribution that can be expressed as a surface integral of various flux-force products. For the case of an interface (between two immiscible media) carrying no flows of energy or momentum, Waldmann constructed local linear relations, valid at the boundary, between the tangential stress and velocity (the slip condition, Equation 1) and heat flux and temperature jump. Within this framework, the boundary conditions can be regarded as linear constitutive relations at the surface. Of course, the values of various slip and jump coefficients are not provided by this approach. If surface flows of energy and momentum are allowed, cross effects appear and are governed by Onsager relations (104).

This nonequilibrium approach has also been used to provide phenomenological boundary conditions for the set of transport and relaxation equations derived for a polyatomic gas in the near continuum regime (106). Boundary conditions have also been obtained by similar methods for mass transfer between phases (105) and particle rotation in a fluid with spin (107).

Bedeaux et al (108) have extended this approach to include singular densities of energy and mass so as to incorporate equilibrium surface phenomena, e.g. surface tension.

Richardson (109) has shown how perfect slip on a sufficiently irregular (macroscopic) surface can lead to the stick BC on a macroscopic scale.

Kinetic Theory Approaches

The preceding macroscopic approach to boundary conditions ignores microscopic details near boundaries and is unable to provide expressions or even magnitudes for slip parameters. For such questions, a more fundamental approach is required.

For low density gases, there has been an enormous amount of effort in these areas. As a number of books (110, 111) on the subject exist, we limit our discussion to a few major points of interest.

The first ingredient for a kinetic approach to boundary conditions for low density gases is a kinetic equation for the gas distribution, usually the linearized Boltzmann equation (LBE) (110, 111). The collision operator in this equation that describes collisions among the gas molecules is often modelled for purposes of tractability. The most popular of these is the BGK model (110, 111).

The second ingredient is a boundary condition on the distribution f that, at this level of description, specifies the interactions of the gas molecules with the boundary (usually a wall). This has the structure $j_- = Kj_+$, where $j_\pm$ denotes incoming and outgoing particle fluxes (with respect to the boundary), and K is a scattering kernel. Conditions for its linear and local character are discussed by Kuščer (112). There have been a few attempts to derive K from microscopic considerations (113), but it is usually modelled, after Maxwell (110), as a linear combination of specular and diffuse reflections, $(1 - \alpha)K_{sp}$ and αK_d, where α is the probability of diffuse scattering, i.e. equilibration with the surface and subsequent Maxwellian emission. Such equilibration is certainly not relevant for molecular test particles but does provide a simple mechanism for tangential velocity change. It should be noted that the above boundary condition on f is stochastic and thus has associated fluctuations (114).

The protoypical steady state BC problem is the Kramers problem of tangential velocity flow past a wall with linear velocity profile at infinity. This was solved exactly by Cercignani (110) and displays previously anticipated features.

1. A kinetic or Knudsen layer a few mean free paths thick exists near the wall. Here the macroscopic hydrodynamic equations are invalid, e.g. stress is no longer proportional to the velocity gradient. Within this layer, the velocity varies rapidly and, in Fourier language, has significant components at high wave numbers.

2. The stress is constant and equal to its bulk hydrodynamic value. The appropriate boundary condition for the velocity in the outer hydrodynamic region is the (back-extrapolated) macroscopic slip velocity v. The stress Π is proportional to *this* velocity. This provides the slip boundary condition $\Pi = \beta v$ with β the slip coefficient. As the mean free path λ decreases, β increases, and the macroscopic stick condition $v = 0$ is approached.

The prototype for time-dependent BC problems is the Rayleigh problem (110) where the bounding wall is set impulsively into motion in its own plane. This, too, has been solved exactly (110). For times that are long compared to a mean free time, the hydrodynamic solution with slip BC applies, with β identical to that for the Kramers problem. The nature of the evolution to this behavior is as yet unclear.

A number of variational and approximation methods for BC problems have been developed (115, 116) that are remarkably successful.

While such exact and approximate results are very instructive, they do not specifically address a number of fundamental questions. How are slip BC's and coefficients to be determined from first principles and what is the extent of their validity? At what stage and in what manner must the hydrodynamic equations themselves be modified along with the hydrodynamic BC's?

Some progress along these lines has been made at the Boltzmann level (117–119). These treatments involve an "inner-outer" analysis via matched asymptotic expansions that recognize the different scales of spatial variation within and outside of kinetic boundary layers. These approaches resolve a long-standing puzzle according to which more boundary conditions seem to be required than are available in a standard Chapman-Enskog analysis.

Scharf (120) has investigated the Stokes' Law problem by including a source term in the LBE to represent the effect of the spherical particle (momentum source). This source is taken to be linear in the flow velocity, localized on the sphere surface and independent of the gas distribution f. This last feature excludes the inclusion of the Knudsen layer. McLennan & Chiu (121) repeated this calculation with a more general expression for the source (still, however, independent of f). With an *assumed* slip BC, these authors found the slip hydrodynamic result $\zeta = \zeta_s[(2\eta + \beta)/(3\eta + \beta)]$ with corrections of order λ/R.

Microscopic Kinetic Theory

Dorfman, van Beijeren & McClure (122) have considered the Stokes problem in low density gases via an LBE approach. Here the distribution is redefined over all space (including the sphere) such that the collision operator $\bar{T}$ for gas particle–sphere collisions is explicit. Thus, one has

$$\partial f/\partial t = -iL_B f + \bar{T}f,\qquad 18.$$

where iL_B is the usual combination of free-streaming and gas collision operator terms. Here the source term $\bar{T}f$ depends on the distribution in contrast with previous treatments. The Maxwell BC was chosen so that $\bar{T}$ splits into specular and diffuse terms.

Dorfman et al analyzed ζ for small λ/R in two ways (122). First a standard Chapman-Enskog solution was attempted outside the sphere with the surface condition $\bar{T}f = 0$ (which generates the boundary conditions). Up to the order in λ/R that generates the usual hydrodynamic equations, the BC's are perfect slip condition for specular reflection ($\alpha = 0$) or the perfect stick BC for completely diffuse reflection ($\alpha = 1$). For small $\alpha[O(\lambda/R)]$, the slip BC Equation 1 is found with β proportional to α. For strongly diffuse scattering $[\alpha = O(1)]$, the Stokes result $\zeta_s = 6\pi\eta R$ is correct up to terms of $O(\lambda/R)$.

The second and more fundamental approach of Dorfman et al involves the splitting of the dynamical propagator for the gas particles (in the absence of the sphere) into (*a*) its projection onto the hydrodynamic variables and (*b*) its complement. This division separates dynamics of a typical gas particle into collective processes involving trajectories in the bath of several mean free paths (with many collisions) and "nonhydrodynamic" processes which involve trajectories with a length of $O(\lambda)$. The complete dynamics is described in terms of these processes and their interruption by collisions with the sphere.

Analysis (122) for small λ/R of this more general formulation (which opens the way for the investigation of boundary layers) reproduces completely the results of the first approach. Dorfman et al state that the dynamical processes responsible for Stokes' Law are those in which gas particles make long excursions between sphere collisions.

Mode-Coupling Considerations

The strict legitimacy of a slip boundary condition or its alternative formulation in terms of a slipping length is very doubtful according to Wolynes (123). The slipping length l is the distance *inside* the boundary at which the hydrodynamic stick BC applies and is finite for finite slip.

Wolynes (123) first shows, in an analysis related to that of van Kampen & Oppenheim (103), that l may be defined in terms of the phase shift $\delta(k)$ that describes a viscous wave vector–dependent eigenmode of a fluid far from the (plane) boundary: $l = \lim_{k \to 0} k^{-1}\delta(k)$. The calculation then proceeds via mode-coupling theory together with a "microscopic" BC applied at the wall. The dominant contribution to $k^{-1}\delta(k)$ is due to sound modes that give a nonanalytic $k^{-1/2}$ dependence; thus no finite slipping length exists! Just as hydrodynamic mode couplings in transport kernels led to long-ranged spatial correlations and vitiated the old assumption of short-ranged character, such couplings here carry the effect of fluctuations near the wall far into the fluid. Sound waves are most effective in this regard. The assumption of a completely local boundary condition then fails; the picture of a localized Knudsen layer is, strictly speaking, invalid. Wolynes (123) notes that the numerical consequences of this divergence are not pronounced and that these divergences are higher order in density than accounted for at the LBE level and so are not observed there. He also suggests that finite radius corrections to Stokes' Law will depend on the k dependence of both the generalized slipping length $l(k) = k^{-1}\delta(k)$ and the fluid shear viscosity; these corrections thus will be nonanalytic. The practical importance of these difficulties remains to be seen. Indeed, it is only very recently that any experimental details of boundary layers have been obtained (124), and these do not extend out sufficiently far in space for divergence effects to be seen.

Literature Cited

1. Berne, B. J., Forster, D. 1971. *Ann. Rev. Phys. Chem.* 22:563–96
2. Alder, B. J. 1973. *Ann. Rev. Phys. Chem.* 24:325–37; Wood, W. W., Erpenbeck, J. J. 1976. *Ann. Rev. Phys. Chem.* 27:319–48
3. Gubbins, K. E. 1973. *Stat. Mech.* 1:194; Deutch, J. M. 1973. In *Transport Phenomena*, ed. J. Kestin, p. 71. New York: AIP
4. Chandler, D. 1974. *Acc. Chem. Res.* 7:246; Eisenthal, K. B. 1975. *Acc. Chem. Res.* 8:118
5. Kivelson, D., Ogan, K. 1974. *Adv. Magn. Reson.* 7:71; Freed, J. H., Pedersen, J. B. 1976. *Adv. Mag. Reson.* 8:1
6. Brot, C. 1975. In *Dielectric and Related Molecular Processes*, ed. M. Davies. 2:1. London: Chem. Soc.
7. Hynes, J. T., Deutch, J. M. 1975. In *Physical Chemistry: An Advanced Trea-*

tise, ed. H. Eyring, D. Henderson, W. Jost. 11B:729
8. Jonas, J. 1975. *Ann. Rev. Phys. Chem.* 26:167–90; Bauer, D. R., Brauman, J. I., Pecora, R. 1976. *Ann. Rev. Phys. Chem.* 27:443–63
9. Pomeau, Y., Résibois, P. 1975. *Phys. Rep.* 19c:64
10. Schofield, P. 1975. See Ref. 3, 2:1
11. Theodosopulu, M., Dahler, J. 1975. *Adv. Chem. Phys.* 31:155
12. Mountain, R. D. 1976. *Adv. Mol. Relaxation Processes* 9:255
13. Steele, W. A. 1976. *Adv. Chem. Phys.* 34:1
14. Keyes, T. 1977. In *Modern Theoretical Chemistry; Statistical Mechanics*, ed. B. J. Berne. New York: Academic. In press
15. Chapman, S., Cowling, T. G. 1970. *The Mathematical Theory of Non-Uniform Gases*. London:Cambridge Univ. Press
16. Alder, B. J., Wainwright, T. E. 1970. *Phys. Rev. A* 1:18
17. Alder, B. J., Gass, D. M., Wainwright, T. E. 1970. *J. Chem. Phys.* 53:3813
18. Happel, J., Brenner, H. 1965. *Low Reynolds Number Hydrodynamics*. Englewood Cliffs, NJ: Prentice-Hall. 553 pp.
19. Debye, P. 1929. *Polar Molecules*. New York: Reinhold. 172 pp.
20. Zwanzig, R., Bixon, M. 1970. *Phys. Rev. A* 2:2002
21. Edward, J. T. 1970. *J. Chem. Educ.* 47:261; McLaughlin, E. 1959. *Trans. Faraday Soc.* 55:28
22. McCool, M., Woolf, L. A. 1972. *J. Chem. Soc. Faraday Trans. I* 68:1971; Van Loef, J. J. 1974. *Physica Utrecht* 75:115
23. Alder, B. J., Alley, W. E., Rigby, M. 1974. *Physica Utrecht* 73:143; Alley, W. E., Alder, B. J. 1975. *J. Chem. Phys.* 63:3764
24. Gosling, E. M., McDonald, I. R., Singer, K. 1973. *Mol. Phys.* 26:1475
25. Chandler, D. 1975. *J. Chem. Phys.* 62:1358
26. Kowert, B., Kivelson, D. 1976. *J. Chem. Phys.* 64:5206
27. Bauer, D. R., Brauman, J. I., Pecora, R. 1974. *J. Am. Chem. Soc.* 96:6840
28. Hu, C.-M., Zwanzig, R. 1974. *J. Chem. Phys.* 60:4354
29. Youngren, G. K., Acrivos, A. 1975. *J. Chem. Phys.* 63:3846
30. Subramanian, G., Levitt, D., Davis, H. T. 1974. *J. Chem. Phys.* 60:591
31. Levesque, D., Verlet, L. 1970. *Phys. Rev. A* 2:2514
32. Levesque, D., Verlet, L., Kürkijasvi, J. 1973. *Phys. Rev. A* 7:1690
33. Jacucci, G., McDonald, I. R. 1975. *Physica A, Utrecht* 80:607
34. Barojas, J., Levesque, D., Quentrec, B. 1973. *Phys. Rev. A*. 7:1092
35. Quentrec, B., Brot, C. 1975. *Phys. Rev. A* 12:272
36. Subramanian, G., Davis, H. T. 1975. *Phys. Rev. A*. 11:1430
37. O'Dell, J., Berne, B. J. 1975. *J. Chem. Phys.* 63:2376
38. Hubbard, P. 1972. *Phys. Rev. A* 6:2421; Cukier, R. I., Lakatos-Lindenberg, K. 1972. *J. Chem. Phys.* 57:3427
39. Kivelson, D. 1974. *Mol. Phys.* 28:321; Evans, G. T. 1976. *J. Chem. Phys.* 65:3030
40. Dymond, J. H., Alder, B. J. 1970. *J. Chem. Phys.* 52:923
41. Andersen, H. C. 1975. *Ann. Rev. Phys. Chem.* 26:145–66; Barker, J. A., Henderson, D. 1976. *Rev. Mod. Phys.* 48:587
42. Gordon, R. G., Armstrong, R. L., Tward, E. 1968. *J. Chem. Phys.* 48:2655; Protopapas, P., Andersen, H. C., Parlee, N. A. D. 1973. *J. Chem. Phys.* 59:15; Dymond, J. H. 1974. *Physica Utrecht* 75:100; Dymond, J. H. 1976. *Physica A* 85: 175
43. Bertucci, S. J., Flygare, W. H. 1975. *J. Chem. Phys.* 63:1; Czworniak, K. J., Andersen, H. C., Pecora, R. 1975. *Chem. Phys.* 11:451; Jones, D. R., Whittenburg, S. L., Wang, C. H. 1976. *J. Chem. Phys.* 65:2033; Carelli, P., DeSantis, A., Modena, I., Ricci, F. P. 1976. *Phys. Rev. A* 13:1131
44. Singwi, K. S., Tosi, S. 1967. *Phys. Rev.* 157:153
45. Berne, B. J. 1971. See Ref. 7, 8B:539
46. Schofield, P. 1973. *Comput. Phys. Commun.* 5:17
47. Kushick, J., Berne, B. J. 1973. *J. Chem. Phys.* 59:3732
48. Kim, K., Chandler, D. 1973. *J. Chem. Phys.* 59:5215
49. Hynes, J. T., Kapral, R., Weinberg, M. 1975. *Physica A, Utrecht* 81:485
50. Cukier, R. I., Hynes, J. T. 1976. *J. Chem. Phys.* 64:2674
51. Mazur, P., Oppenheim, I. 1970. *Physica Utrecht* 50:241
52. Hynes, J. T. 1972. *J. Chem. Phys.* 57:5612; 1973. *J. Chem. Phys.* 59:3459
53. Ricci, F. P. 1967. *Phys. Rev.* 156:184
54. Castleman, A. W. Jr., Conti, J. J. 1970. *Phys. Rev. A* 2:1975
55. Allen, G. G., Dunlop, P. J. 1973. *Phys. Rev. Lett.* 30:316

56. Bunker, D. L., Jacobson, B. S. 1972. *J. Am. Chem. Soc.* 94:1843
57. Silbey, R., Deutch, J. M. 1971. *Phys. Rev. A* 3:2049
58. Rubin, R. J. 1972. *J. Chem. Phys.* 57:312
59. Lewis, J. C., Tjon, J. A. 1976. *Physica A, Utrecht* 85:114
60. Bishop, M., Berne, B. J. 1972. *J. Chem. Phys.* 56:2850
61. Herman, P. T., Alder, B. J. 1972. *J. Chem. Phys.* 56:987
62. Alder, B. J., Alley, W. E., Dymond, J. H. 1974. *J. Chem. Phys.* 61:1415
63. Bruin, C. 1974. *Physica Utrecht* 72:261
64. Ernst, M. H., Weijland, A. 1971. *Phys. Lett. A.* 34:39
65. Zwanzig, R., Bishop, M. 1974. *J. Chem. Phys.* 60:295
66. Alder, B. J., Hoover, W. G., Wainwright, T. E. 1963. *Phys. Rev. Lett.* 11:241
67. Rahman, A. 1966. *J. Chem. Phys.* 45:2585
68. Ebbsjö, I., Schofield, P., Sköld, K., Waller, I. 1974. *J. Phys. C* 7:3891
69. Lantelme, F., Turq, P., Schofield, P. 1976. *Mol. Phys.* 31:1085
70. Sykes, J. 1973. *J. Stat. Phys.* 8:279
71. Résibois, P., Lebowitz, J. L. 1975. *J. Stat. Phys.* 12:483
72. Résibois, P. 1975. *J. Stat. Phys.* 13:393
73. Barajas, L., Garcia-Colin, L. S., Piña, E. 1973. *J. Stat. Phys.* 7:161
74. van Beijeren, H., Ernst, M. H. 1973. *Physica Utrecht* 68:437; 70:225
75. Mehaffey, J. R., Desai, R. C. 1977. *J. Chem. Phys.* In press
76. Rahman, A. 1964. *Phys. Rev.* 136:A405
77. Metiu, H. Oxtoby, D., Freed, J. H. 1977. *Phys. Rev. A.* 15:36
78. Ailawadi, N. K., Rahman, A., Zwanzig, R. 1971. *Phys. Rev. A.* 4:1616
79. Zwanzig, R., Bixon, M. 1975. *J. Fluid Mech.* 69:21
80. Kivelson, D., Kivelson, M. G., Oppenheim, I. 1970. *J. Chem. Phys.* 52:1810
81. Fury, M., Jonas, J. 1976. *J. Chem. Phys.* 65:2206
82. Hoel, D., Kivelson, D. 1975. *J. Chem. Phys.* 62:1323
83. Hynes, J. T., Kapral, R., Weinberg, M. 1977. *Chem. Phys. Lett.* 46:463; 47:575
84. Zwanzig, R. W. 1972. In *Statistical Mechanics*, ed. S. A. Rice, K. F. Freed, J. C. Light. Chicago: Univ. Chicago Press. p. 241
85. Keyes, T., Oppenheim, I. 1973. *Phys. Rev. A.* 8:937
86. Michaels, I. A., Oppenheim, I. 1975. *Physica A, Utrecht* 81:522
87. Kapral, R., Weinberg, M. 1973. *Phys. Rev. A* 8:1008
88. Bedeaux, D., Mazur, P. 1974. *Physica Utrecht* 76:235
89. Bedeaux, D., Mazur, P. 1974. *Physica Utrecht* 78:505
90. Lukin, L., Patashinskii, A., Cherepanova, T. 1968. *Sov. Phys. JETP* 26:933
91. Garisto, F., Kapral, R. 1974. *Phys. Rev. A* 10:309
92. Keyes, T., Oppenheim, I. 1974. *Physica Utrecht* 75:583
93. Berne, B. J. 1972. *J. Chem. Phys.* 56:2164
94. Nee, T. W., Zwanzig, R. 1970. *J. Chem. Phys.* 52:6353
95. Hills, B. P., Deutch, J. M. 1976. *Physica A, Utrecht* 83:401
96. Garisto, F., Kapral, R. 1976. *Phys. Rev. A* 13:1652
97. Mehaffey, J. R., Desai, R. C., Kapral, R. 1977. *J. Chem. Phys.* 66:1665
98. Subramanian, G., Levitt, D. G., Davis, H. T. 1976. *J. Stat. Phys.* 15:1
99. Pomeau, Y., Weber, J. 1976. *J. Chem. Phys.* 65:3616
100. Mazenko, G. F. 1973. *Phys. Rev. A* 7:209, 222
101. Mazenko, G. F. 1974. *Phys. Rev.* 9:360
102. Furtado, P. M., Mazenko, G. F., Yip, S. 1976. *Phys. Rev. A* 14:869
103. van Kampen, N. G., Oppenheim, I. 1972. *J. Math. Phys. NY* 13:842
104. Waldmann, L. 1967. *Z. Naturforsch. Teil A* 22:1269
105. Waldmann, L., Rübsamen, R. 1972. *Z. Naturforsch. Teil A* 27:1025
106. Vestner, H. 1973. *Z. Naturforsch. Teil A* 28:1554
107. Hynes, J. T., Kapral, R., Weinberg, M. 1977. *Physica A., Utrecht* In press
108. Bedeaux, D., Albano, A. M., Mazur, P. 1976. *Physica A, Utrecht* 82:438
109. Richardson, S. 1973. *J. Fluid Mech.* 59:707
110. Cercignani, C. 1969. *Mathematical Methods in Kinetic Theory*. New York: Plenum. 227 pp.
111. Williams, M. M. R. 1971. *Mathematical Methods in Particle Transport Theory*. London: Butterworth. 429 pp.
112. Kuščer, I. 1971. *Surf. Sci.* 25:225
113. Cercignani, C. 1972. *Transp. Theory Stat. Phys.* 2:27
114. Szu, H.-L. 1971. *Contributions to the Kinetic Theory of Dilute Gases*, PhD thesis. Rockefeller Univ., New York. 100 pp.
115. Cercignani, C., Pagani, C. D. 1968. *Phys. Fluids* 11:1395

116. Loyalka, S. 1971. *Phys. Fluids* 14:2291
117. Grad, H. In *Transport Theory, Proc. SIAM-AMS.* 1:269
118. Sone, Y. 1969. In *6th Symp. Rarefied Gas Dyn.* 1:243
119. Ganz, A., Sirovich, L. 1973. *Phys. Fluids* 16:50
120. Scharf, G. 1970. *Phys. Fluids.* 13:848
121. McLennan, J. A., Chiu, S. C. 1974. *Phys. Fluids* 17:1146
122. Dorfman, J. R., van Beijeren, H., McClure, C. F. 1976. *Arch. Mech. Stosow.* 28:333
123. Wolynes, P. G. 1976. *Phys. Rev. A* 13:1235
124. Loyalka, S. 1975. *Phys. Fluids* 18:1666

PHOTODISSOCIATION DYNAMICS OF POLYATOMIC MOLECULES[1]

✦ 2650

William M. Gelbart[2]

Department of Chemistry,[3] University of California, Los Angeles, California 90024

1 INTRODUCTION

Because only one vibrational degree of freedom is involved, the photodissociation of a diatomic molecule (1) is straightforward from the chemical dynamics point of view. 1. If the optically prepared electronic state is repulsive, the two atoms simply fly apart with a relative kinetic energy equal to the difference ($E_{opt} - E_{asymp}^{upper}$) between the exciting light energy E_{opt} and the upper state asymptote E_{asymp}^{upper}. This is of course true as well in the case of a bound initial state, as long as $E_{opt} > E_{asymp}^{upper}$. 2. If it is bound and $E_{opt} < E_{asymp}^{upper}$, the molecule can undergo a radiationless transition into the continuum of nuclear motion states belonging to a lower electronic configuration (hence indirect, or predissociation)—the atoms again separate with a fixed ($E_{opt} - E_{asymp}^{lower}$) translational energy.

For polyatomic molecules, on the other hand, there are at least three (and often many more) coupled vibrational degrees of freedom. A simple description of the fragmentation dynamics is no longer possible in these cases. Even for triatomics, a two-dimensional representation of the potential energy surfaces is incomplete since at least one degree of freedom is suppressed. For larger polyatomics the situation is of course far more awkward. The task of photodissociation theory is to determine which potential energy surfaces are involved—other than the one on which the molecule is prepared by optical excitation—and how the electronic and nuclear motions are coupled to give the observed fragment distributions.

Consider first the case where fragmentation takes place within 10^{-13} sec after the molecule is prepared in an upper electronic state, i.e. the excited state's dissociation threshold (E_{asymp}^{upper}) lies below E_{opt} and no redistribution of vibrational energy is

---

[1] Work supported in part by NSF Grant No. CHE76-01807 and Nato Grant No. 974.
[2] Alfred P. Sloan Foundation Fellow, and Camille and Henry Dreyfus Foundation Teacher-Scholar.
[3] Contribution No. 3779.

necessary for the breakup to occur. These cases are directly analogous to the first diatomic case mentioned in the first paragraph: we refer to them as *direct photodissociations*. Even if the upper state is bound, geometry and/or frequency changes between it and the ground state can lead to optical excitations for which the energy in a particular vibrational mode exceeds its dissociation threshold. H_2O, H_2S, H_2Se, ... are possible examples (2) of this, as are some halogenated methanes, which we discuss in Section 2.1.

When the excited state lifetime against fragmentation is longer than 10^{-12}–10^{-13} sec, the photodissociation must be occurring via either 1. a slow unimolecular reaction in the upper state or 2. a crossing to a lower electronic configuration. In the first case, the molecule is breaking up because of vibrational and/or rotational predissociation, i.e. the continuum belongs to the same electronic state, but either it joins on to a lower limit than that to which the series of optically prepared levels converges (so that vibration-to-vibration energy redistribution is required), or it can be reached only by the "tapping" of rotational excitation. In the second case, a change in electronic state is involved, following which the molecule might either break up immediately (i.e. in less than 10^{-13} sec) or undergo a slow redistribution of vibrational/rotational energy before reaction occurs. We refer to these two latter cases as *direct* and *indirect electronic predissociation*, respectively.

Just as continuous absorption is tha hallmark of molecular spectra showing direct photodissociation, instances of electronic predissociation are fingerprinted by absorption line broadening and weakening of emission. This was first established fifty years ago (3) in NH_3, whose absorption features become diffuse below 2100 Å (where decomposition is observed to set in). The weakening of emission lines is a more sensitive indication of predissociation; it is observable as soon as $k_{diss} \gtrsim k_{rad}$, whereas broadening of absorption features typically requires (for $k_{rad} \approx 10^8$ sec^{-1}, say) that $k_{diss} \gtrsim (100)k_{rad}$ in order for the room temperature Doppler width not to hide everything. An actual breaking-off in emission, while common in diatomics, (1) has been observed in few polyatomics [e.g. HNO (4), C_6H_6 (5)]. Similarly, the diffuseness in absorption features most often sets in gradually. This is due to the multidimensionality of the potential energy surfaces and is discussed in detail by Herzberg (6).

Again, because of the many degrees of freedom involved, selection rules are rather difficult to formulate for polyatomic predissociations. When electronic, vibrational, and rotational motions can be separated, however, there clearly cannot be any change in symmetry of the electronic configuration. As soon as electronic-nuclear interactions are included, "heterogeneous" (2) (electronic symmetry changing) predissociations become possible. These relaxation processes can occur, then, only if particular rotations or vibrations are excited. An A_2'' state of a D_{3h} polyatomic, for example, can be predissociated by an A_1'' state if the molecule is rotating about the principal axis ($\Gamma_{\parallel}^{rot} = A_2'$), and by an E' state if it is rotating about a perpendicular axis ($\Gamma_{\perp}^{rot} = E''$). It cannot be predissociated by A_2' or E'' states (since there are no rotations with proper symmetries) unless it is vibrationally excited. Similarly, excitation of the bending vibration in a linear triatomic makes it possible for a Π state to decay into a Δ configuration, even in the absence of rotation.

Both vibrational and rotational predissociations of electronic ground states prepared by thermal excitation are more commonly referred to as unimolecular decompositions. Following earlier theoretical work from the 1930s (7), Mies & Kraus (8) have explicitly treated these adiabatic (single-potential energy surface) fragmentations as resonance-scattering, and hence dissociative, processes. Unimolecular breakup of "hot" electronic ground states can also correspond to predissociations that occur via a crossing to a lower configuration. Consider, for example, the linear triatomic oxides, which decompose to yield an oxygen atom and a closed-shell diatomic molecule, $XYO(^1\Sigma) \to XY(^1\Sigma) + O$ (9). The lowest dissociation energy D of the $^1\Sigma XYO$ ground state, yielding $XY(^1\Sigma)$ and excited (^1D) atomic oxygen, often exceeds the observed activation energy E_A. There can exist, however, a repulsive triplet state that correlates with $XY(^1\Sigma)$ and ground state (^3P) oxygen. The crossing of the triplet and singlet surfaces is believe to occur well below D, near E_A, and the resulting electronic relaxation has been treated (9, 10) as a predissociation process in the particular case of $N_2O(^1\Sigma) \to N_2(^1\Sigma g) + O(^3P)$. Another dramatic example in which nonadiabaticity is required by symmetry is that of the "metastable" ions reported in mass spectrometric studies of unimolecular decompositions. Consider the breakup of N_2O^+ and H_2S^+ according to $N_2O^+(^2\Pi) \to NO^+(^1\Sigma^+) + N(^4S)$ and $H_2S^+(^2B_1) \to H_2(^1\Sigma_g^+) + S^+(^4S)$. In both cases the fragments involve quartet states, whereas the parent ions are doublets. Thus, instead of a vibrational/rotational predissociation, an electronically nonadiabatic relaxation is driven by the spin orbit coupling. Hence the decomposition is slow enough (it occurs after the ions have been accelerated, but during their flight through the analyzing field) to give weak diffuse peaks at nonintegral mass numbers.

The metastable ions mentioned above comprise the first instance in which time resolution of the molecular decomposition is used to provide information about the dissociation dynamics. It is this aspect of the experimental studies on photofragmentation with which we are primarily concerned. As mentioned earlier, there is a wealth of spectroscopic (e.g. line broadening, background continua) data (2) relating to photodissociations in polyatomic species. We focus instead on those experiments in which a photofragmentation process is explicitly time resolved, and the question of "mechanism" is most directly confronted. More specifically, can the study in question distinguish between the possibilities of direct photodissociation, vibrational/rotational predissociation, and direct versus indirect electronic predissociation? In this light we review in Section 2 the recent time-resolved data acquired on the aryl halides, nitrogen oxides, triatomic cyanides, and formaldehydes.

Theoretical work on photo- and predissociations in polyatomic molecules is in its infancy. Only in the past few years has it appeared in any concerted fashion. "Quasidiatomic" models were developed first, followed by a now extensive series of calculation on collinear triatomic dissociations. This work is reviewed in Section 3. In Section 4 we treat the phenomenological theories that were developed in the late 1960s and early 1970s to describe the dynamics of photofragmentation in larger, nonlinear molecules. Then we focus on the case of formaldehyde, which has been attacked most thoroughly by experiments and calculations, none of which has yielded a clear picture of the predissociation mechanism involved. We conclude in Section 5 by stressing that there is no single example where the actual dynamics of a polyatomic photodissociation have been unambiguously established.

2 TIME-RESOLVED EXPERIMENTAL STUDIES

2.1 Aliphatic Carbonyls and Aryl Halides

Bersohn and co-workers have studied the photodissociation of a large number of aliphatic carbonyls and aryl halides under isolated molecule (collision free) conditions. They determined in particular the angular distribution of the photodissociations. Anisotropy is built into the low-pressure gas-phase samples by exciting the molecules with linearly polarized light. If the dissociation takes place before the molecules have had a chance to rotate, fragments will only be observed along the direction of the transition dipole moment. In the early experiments (11–13) the angular distributions were measured by the "photolysis mapping" technique: the inside of a hemisphere is coated with a metal (e.g. tellurium or cadmium) that is selectively removed by alkyl radicals. In later experiments (14–16), the mirror-lined bulb was replaced by a molecular beam plus quadrupole mass spectrometer, much as in the photofragment studies of Wilson and collaborators described below (Section 2.2).

Let θ be the angle between the polarization of the exciting light and the direction of the detector. Then the angular distribution of fragments is given by (17–19)

$$F(\theta) = \frac{1}{4\pi}[1 + \beta P_2(\cos\theta)], \qquad 1.$$

where P_2 is the usual second Legendre polynomial. The key point is that β can be explicitly related to the symmetry and lifetime (against dissociation) of the optically excited molecular state. If, for example, the molecule breaks up into fragments so fast that no overall rotation has taken place, β takes the simple form $2P_2(\cos\chi)$, where χ is the angle between the transition dipole moment and the direction of dissociation. For $\chi = 0$, say, this gives the maximum possible anisotropy in $F(\theta)$. If, on the other hand, the lifetime τ against dissociation is not short compared to the rotational period $1/\omega$, then the angular distribution will be more uniform. In the case of linear molecules, for example, the anisotropy parameter is given by (17, 19)

$$\beta = 2P_2(\chi)\left\{\frac{1}{4}\left[1 + 3\gamma e^{\gamma}\int_{\gamma}^{\infty} du\, \exp(-u)/u\right]\right\}, \qquad 2.$$

where

$$\gamma \equiv I/8kT\tau^2. \qquad 2a.$$

Here T is the sample's temperature and I is the molecule's principle moment of inertia. The quantity within the braces is a smoothly decreasing function of τ, equal to unity when the lifetime is zero (instantaneous dissociation) and to one quarter when τ becomes very long compared to the rotational period. Values for $\beta(\chi, \tau)$ for nonlinear molecules of various shapes are discussed by Yang & Bersohn (17).

Consider first the photolysis mapping studies of the aliphatic carbonyls. Previous spectroscopic and photochemical work on these compounds had not confronted the question directly: In which direction (relative to the molecular frame of reference), and on what time scale, do the fragments move? From their measured angular distributions, Solomon et al (13) concluded that formaldehyde and acetone decompose via a "perpendicular" transition ($\chi \lesssim 90°$), whereas $\chi \gtrsim 0°$ for acetaldehyde and propionaldehyde. (The transition dipole moment is believed to lie in the molecular plane, perpendicular to the C-O axis.) Furthermore, it is inferred that all the molecules dissociate before they rotate. In the case of formaldehyde, treated in detail in Sections 2.3 and 4.2, these results are argued to imply that the H atom fragment comes off along an axis perpendicular to the molecular plane, and that this takes place on the ground state (A_1) potential energy surface reached by fast internal conversion from the optically prepared singlet (A_2). This out-of-plane motion is excited to a large degree by the $S_1 \rightsquigarrow S_0$ radiationless transition, because the corresponding Franck-Condon factors are made favorable by the nonplanarity of A_2. Furthermore, the molecule cannot dissociate directly in S_1, since this state correlates only with electronically excited fragments that are not energetically accessible under the experimental photolysis conditions ($E_{opt} \lesssim 4.8$ eV). As the alkyl groups in the higher aldehydes and ketones become heavier there will be a point where their mass is large compared to that of the oxygen atom. Then O will do most of the moving out of the molecular plane and the radicals will separate in a parallel fashion, i.e. along the direction of the dipole transition moment (as is observed).

The experiments of Bersohn et al on the aryl halides no longer used the mirror-removal photolysis mapping, but instead used the molecular beam, mass spectrometer technique pioneered by Wilson and his collaborators (20–24). (Instead of the Q-switched laser employed by Wilson et al, they used a chopped dc, high-pressure Hg-Xe arc lamp. This broad band source was convenient for the UV photolysis range of interest, but precludes meaningful velocity analyses of the fragments.) Yang & Bersohn (17) used the appropriate $\beta(\chi, \tau)$ expressions to interpret their observed angular distributions, and deduced lifetimes for four aryl halides: methyl iodide, 0.07 psec; iodobenzene, 0.5 psec; α-iodonaphthalene, 0.9 psec; and 4-iodobiphenyl, 0.6 psec. The analogous aryl bromides show much smaller β values, and it is estimated that their lifetimes against dissociation are probably two orders of magnitude longer. [Note that absolute lifetimes can only be determined if τ is neither too long nor too short compared with $(I/kT)^{1/2}$.]

The lifetime of $\tau \lesssim 0.07$ psec for methyl iodide is of the order of magnitude of the time it takes for a couple of molecular fragments to separate along a rapidly decreasing potential. Thus Dzvonik, Yang & Bersohn (15) conclude that their photolysis of CH_3I is an example of direct photodissociation in which the molecule breaks up "instantaneously" in the optically excited, repulsive state. The order-of-magnitude slower decomposition of phenyl iodide is argued, on the other hand, to involve a direct, electronic predissociation. The optically excited singlet, largely π^* in character, is stable with respect to R-I dissociation; because of large spin orbit coupling on the I atom, however, this state decays rapidly ($\tau \approx 0.6$ psec) into a triplet σ^* configuration that is repulsive in the R-I coordinate. A similar argument is used to explain the comparable ($\tau \lesssim 1$ psec) lifetimes observed for the 1-naphthyl iodide and 4-biphenyl iodide. The optically excited state is the second singlet S_2, and dissociation cannot take place in lower singlets (S_1 or S_0 following internal conversion) since the Franck-Condon factors "force" vibrational energy predominantly into the high-frequency C-H stretching modes, and redistribution of this excitation into the R-I mode is unlikely in less than 1 psec. Thus a triplet σ^* state is postulated in which the molecule is essentially unbound with respect to R-I.

The involvement of a repulsive triplet is also consistent with the dramatic heavy-atom effect observed: the alkyl bromides live two orders of magnitude longer than their corresponding iodides. The repulsive σ^* triplet is still involved, but because of the much smaller spin orbit coupling the decay of the optically excited S_2 is much slower. Yet the ratio of fine structure splittings ($^2P_{3/2}$–$^2P_{1/2}$) of the I and Br atoms is ~ 4.3, rather than 100. Thus Dzvonik et al suggest the possibility of a "detour" through S_1.

2.2 NO_2, $NOCl$, and ICN

Using a monochromatic pulsed laser, Wilson and co-workers (20–24) have carried out measurements of both the angular and velocity distributions of fragments from the photodissociation of NO_2, NOCl, and ICN molecular beams. Except for the light source being monoenergetic, allowing for time-of-flight analyses by the mass spectrometric detector, these experiments are very similar in conception to those discussed above. In the case of ICN, and also NOCl to a lesser extent, the data allow many conclusions to be drawn about the mechanism of the dissociation dynamics.

NO_2 is a notoriously poorly understood molecule, essentially because of the large and numerous intrinsic perturbations that spoil nearly all quantum numbers other than the total energy and angular momentum [see discussion and references in the most recent report of NO_2 ab initio electronic structure calculations (25)]. It is known to photodissociate at energies above 25,000 cm^{-1}. Busch & Wilson (23, p. 3626) have intersected a molecular beam of NO_2 with the linearly polarized, pulsed light from the second harmonic (28,810 cm^{-1}) of a Q-switched ruby laser. The photodissociation fragments, O and NO, that recoil out of the intersection region are detected by a quadrupole mass spectrometer placed a few centimeters away. The distribution of fragments arriving there is analyzed both as a function of time after the laser pulse and as a function of θ, the laboratory angle of recoil relative to the light's polarization direction. The measured $F(\theta)$ suggests that the excited molecule breaks up in $\lesssim 0.2$ psec, consistent with the estimate made by Gaedtke & Troe (26) from their measurements of the pressure dependence of the quantum yield at 27,320 cm^{-1} for the disappearance of NO_2. The translational energy of the fragments is found on the average to be $\sim 60\%$ of the available energy; its distribution shows a pair of peaks, corresponding to nearly equal probability of recoil with NO in the $v = 0$ and $v = 1$ vibrational states. The positions and widths of the peaks indicate further that there is a broad range of rotational excitation.

Busch & Wilson (23, p. 3626) have attempted to rationalize the above data in terms of two simple models. First, they ("with reservation") apply statistical models to the dissociation, i.e. introduce the assumption that every state of a critical configuration (e.g. a bent triatomic with one N–O bond extended and the other associated with the nascent NO fragment) is equally probable. The statistical assumption is also applied, alternatively, to the separated fragments. But in both cases, the predicted vibrational energy distribution peaks much too predominantly at $v = 0$. Also, the measured lifetime against dissociation is on the short side, suggesting that the usual randomization postulates are inappropriate. Second, Busch and Wilson then discuss several "direct," dynamical models according to which the molecule breaks apart in a simple vibrational motion. They distinguish between the interfragment effects, in which the changes in potential energy surface in going from initial to final configurations involve mainly forces and torques between the fragments, and intrafragment effects, in which the changes are mainly within the fragments (e.g. bond length changes). (This distinction, as discussed in Section 3.1, plays an important role in current theories of collinear triatomic dissociations.) Busch & Wilson noted that a combination of intra- and interfragment direct models can reproduce many translational and vibrational energy distributions, including those measured. Thus they observed (23, p. 3626): "if pressed to make a choice in the face of inconclusive evidence, we would describe the photodissociation of NO_2 at 28,810 cm^{-1} as excitation to a state with high probability of vibrational excitation in the nascent NO fragment, followed by dissociation rapid enough to prevent statistical equilibrium." It is indeed a testament to the fractiousness of NO_2 that Busch & Wilson were forced to conclude further in a later paper (23, p. 3638), detailing the angular distribution of fragments, that the single-potential energy-surface direct mechanism considered above is no more likely than an electronic predissociation involving the lower $^2B_1$ or ground $^2A_1$ states. Too much obviously remains unknown about the potential energy surfaces involved and the interaction of their vibrational/rotational structures.

Other recent studies of NO_2 photodissociation, which do not directly time resolve the fragmentation dynamics, include the works of Lee & Uselman (27), Jones & Bayes (28), Levy and co-workers (29), and Creel & Ross (30).

Wilson and co-workers have used their photofragment spectroscopy technique to study the photolysis of NOCl and ICN, and in both these cases the conclusions reached about "mechanisms" are on better footing than those for NO_2. Busch & Wilson (23, p. 3655) have photodissociated NOCl with doubled ruby (28,810 cm^{-1}) light and found that $\sim 70\%$ of the energy in excess of that ($\sim 13,000$ cm^{-1}) required to break the ON–Cl bond goes into translational energy of the recoiling fragments. Enough internal energy remains for the NO to be vibrationally excited with several quanta, but this distribution is not measured directly. From the very high angular anisotropy observed, it is inferred that the lifetime against dissociation is short compared to the rotational period, being at most ~ 0.08 psec. From this very short lifetime it follows that the molecular breakup must correspond to a direct photodissociation in the optical excited state (whose symmetry is inferred to be A', since the angular distribution peaks along the direction of the linear polarized laser light, implying that the transition dipole moment probably lies close to the N–Cl bond). Unpublished classical trajectory calculations by Okuda & Wilson (31) assume that all of the available energy ($E_{\text{opt}} - E_{\text{asymp}}^{\text{upper}}$) appears at the beginning of the dissociation as a repulsive potential between the N and Cl atoms. This assumption prescribes the quasidiatomic model that Goodeve & Katz (32) have used to interpret their visible and UV absorption spectra of NOCl, and that Holdy, Klotz & Wilson (33) and others (34) have used to treat the ICN photolysis. The potential between N and Cl is assumed further to fall off fast with increasing R_{N-Cl} so that the force force between the two atoms is "nearly" impulsive on the time scale scale of rotation, but not sharp enough to vibrationally excite the stiff No bond. If the ONCl bond angle is estimated to be $\sim 113°$, and an approximate N-Cl repulsive potential is inferred from the continuous absorption spectra, this "modified impulsive" model (23, pp. 3626, 3655; 32) predicts that 66% of the available energy will go into translation, in reasonable agreement with the measured value of 70% ± 3%. The classical trajectory calculations show further that most of the remaining energy goes into rotation rather than vibration of the NO fragment. Obviously it would be useful to check this prediction directly. Basco & Norrish (35), for example, exciting into a different continuum than that corresponding to Busch and Wilson's A state, have measured the NO vibrational state populations following flash photolysis of NOCl. They find considerable ($v \lesssim 11$) vibrational excitation, which has been confirmed by Pollack (36) for the photodissociation of NOCl under similar conditions.

In the case of ICN's photofragment spectrum, Ling & Wilson (37) have found two peaks in the internal energy distribution, both with angular distributions peaking parallel to the light polarization vector. This result contradicts previous spectral assignments (38) that ascribed predominantly perpendicular transitions to the absorption region probed by Ling and Wilson's quadrupled neodymium laser (37,550 cm^{-1}). The possible internal state possibilities consistent with the two-peak structure of the measured energy distributions are discussed in terms of two models for the dissociation process. The first assumes that each peak arises from a separate ICN excited state. The second, which we treat further in Section 3.1 where we discuss the theory of Simons & Tasker (40), assumes that all the CN fragments are initially formed in their excited $^2\Pi_i$ state from which some are electronically quenched during recoil from their I atom partners. Baronavski & McDonald (39), however, have concluded from their attempts to observe chemiluminescence from CN ($\tilde{A}^2\Pi v' = 0$, or $v' = 1 \to \tilde{X}^2\Sigma v'' = 0$, or $v'' = 1$) following laser photolysis at 37,550 cm^{-1} that the CN fragments are virtually all formed in their ground electronic state. They also find a vibrational state distribution that is essentially Boltzmann, in marked contrast to that suggested by Ling & Wilson (37). Clearly, further studies of the I and CN fragments (e.g. $5^2P_{3/2}$ vs $5^2P_{1/2}$ iodine atom populations) are desirable to pin down the actual photodissociation mechanism. This is still more imperative in the NOCl case—we discussed above only the single-potential energy surface, direct photodissociation models but, as Busch & Wilson (23, p. 3655) point out, nothing in their data precludes the possibility of a fast vibrational/rotational predissociation in the optically excited state.

2.3 *Formaldehyde*

As our third and last example of time-resolved photodissociation studies of polyatomic molecules, we discuss the case of formaldehyde. The photochemistry of H_2CO in the low-pressure gas phase has been reviewed by Moore and his co-workers (41–43). McQuigg & Calvert (44) had demonstrated earlier that there are several important processes following optical excitation of the first excited singlet: $H_2CO(S_1, v) \to H_2 + CO$, $\to H + HCO$, and $\to H_2CO(S_0, v) + hv$. That is, following absorption to S_1, formaldehyde is observed to yield either molecular or radical (at shorter wavelengths) products; there is also a small quantum yield for fluorescence. The high-resolution electronic spectrum in the S_1 region has been well studied. Vibrational and rotational assignments for the S_1 state are essentially complete (45–47), with the fundamental frequencies and structural constants established. The origins of the S_1 and T_1 states are 28,188 and 25,194 cm^{-1}, respectively (2), and Brand, Stevens & Liu (48) have analyzed partially the S_1-T_1 perturbations. Especially relevant to our discussion of the photodissociation "mechanism" are the correlation diagrams for the low-lying electronic states. S_0 correlates with both molecular and radical products in their ground electronic configurations. The reaction $H_2CO(S_1) \to H_2 + CO$ is practically thermoneutral ($\Delta E \approx -140$ cm^{-1}) but is believed (49, 50) to involve an energy barrier whose height is comparable with E_{S_1}. As far as the radical dissociation is concerned, the ground state surface probably (51) rises smoothly to $E(H + HCO) \approx E_{S_1}$. The first triplet T_1, on the other hand, correlates only with the radical products. Finally, S_1 gives only electronically excited HCO at much higher energies ($\sim 40,000$ cm^{-1}). For this reason it can be inferred that an electronic predissociation is necessary when formaldehyde is optically prepared with $E_{opt} \gtrsim E_{S_1}$.

To achieve single-vibronic-level excitation of S_1 formaldehyde, Yeung & Moore (41, 52) constructed a tunable UV source, with a bandwidth of $\leqslant 1$ Å, based on the summation frequency of a pulsed (7 nsec) ruby laser and a tunable dye laser. Miller & Lee (53) have also reported fluorescence lifetimes for individual vibrational levels in S_1, using a D_2 flashlamp with comparable bandwidth. There are discrepancies between the results of these two laboratories, but the data can be summarized as follows. The fluorescence lifetimes drop from ~ 100–200 nsec to ~ 10–20 nsec as the excess vibrational energy in H_2CO-S_1 is increased from a few hundred to a few thousand cm^{-1}. [These results are consistent with a single τ_F measured in this region (3371 Å excitation from a N_2 laser) by Sakurai, Capelle & Broida (54).] The corresponding fluorescence quantum yields are all smaller than 0.04 (53), implying that the measured τ_F values differ by only a few percent from the actual nonradiative lifetimes. For D_2CO (41) the lifetimes are about twenty times longer and decrease faster with excess energy. The lifetimes for both isotopes do not depend too strongly on the particular normal mode combination excited and are essentially independent of rotational quantum number.

To test the possibility that the measured $\tau_F \approx \tau_{nonrad}$ values correspond to electronic predissociation of formaldehyde, Houston & Moore (43) have monitored, as a function of initial vibronic levels, the appearance rate, relative yield, and vibrational distribution of the CO formed from short-pulse (~ 10 nsec) photoexcitation of H_2CO. Collisional deactivation of CO ($v \gtrsim 1$) by formaldehyde was also studied, using three different techniques. In the first the fluorescence decay of the CO ($v = 1$) produced from H_2CO photolysis was monitored as a function of the original H_2CO pressures. The second technique was similar to the first, but the CO ($v = 1$) decay was monitored by the change in absorption of the $1 \to 2$ line of a CO probe laser. In the third, the Q-switched CO laser was used to directly excite CO and the resulting CO fluorescence was monitored as a function of H_2CO pressure. All three techniques gave the same result: CO ($v = 1$) is collisionally relaxed by H_2CO with a rate of $1.31 \pm 0.11 \times 10^3$ sec^{-1} torr^{-1} ($\sim 1/7500$ gas kinetic). The appearance rate of CO product following 3371 Å excitation (overlapping the $2_0^1 4_0^1$ vibronic level of H_2CO) was measured as a function of pressure by time resolving the absorption of either the $1 \to 2$ or $0 \to 1$ line of the CO laser. The appearance rates were found to vary linearly with pressure, extrapolating to zero in the limit of zero pressure. The slopes of the plots of k_{app} versus P_{H_2CO} were found to be $1.65 \pm 0.12 \times 10^6$ sec^{-1} torr^{-1} ($\sim 1/10$ gas kinetic) for both CO ($v = 1$) and CO ($v = 0$). The rates measured at the lowest pressure studied (~ 0.1 torr) were as small as $\sim 0.26 \times 10^6$ sec^{-1}.

Thus it appears that, upon exciting H_2CO at 0.1 torr and a couple thousand cm^{-1} above its S_1 origin, CO fragments do not "grow in" until ~ 5 μsec, even though the electronic excitation (fluorescence) decays in $\lesssim 0.05$ μsec. As we discuss in detail in Section 4, there exists no explanation of these results, In the meantime, Houston and Moore have suggested that the time lag between the decay of S_1 and the formation of CO be thought of in terms of an "intermediate state," for which they consider two candidates: the first triplet, $T_1(^3A_2)$, and the "hot" ground state, S_0^*. The additional observations for which either of these mechanisms must account include the relative yields of product, and its vibrational distribution, as a function of initial vibronic level and of collisional partner and pressure. Since the measured appearance rates are always more than ten times faster than the observed collisional deactivation rates of CO ($v \gtrsim 1$), the vibrational distributions (measured by tuning the CO probe laser to successively high $v \to v + 1$ absorptions) can be assumed to be those of the nascent product. Also, McQuigg & Calvert (44), from their measurement of the extent of isotopic exchange (i.e. HD formation) following photolysis of H_2CO/D_2CO mixtures, have concluded that the molecular quantum yield at 3371 Å is ~ 0.8. Lewis, Tang & Lee (55), from their studies of HNO chemiluminescence following photolysis of H_2CO mixed with NO, estimate that ϕH_2 is even larger at this wavelength. Thus, one cannot explain away the time lag observed by Houston and Moore by ascribing it to radical recombination.

The lower triplet state appears an obvious candidate for the long-lived intermediate since it readily explains many of the properties observed for the collision-induced dissociation. Miller & Lee (56) have found, for example, that triplet benzene ($^3B_{1u}$)-sensitized decomposition of H_2CO in the gas phase (a few torr) gives an appreciable yield of $H_2 + CO$. Also, Luntz & Maxson (57) have time resolved the phosphorescence decay of the T_1 state of D_2CO excited directly by a dye laser. The lifetime observed is 16.9 ± 0.5 μsec, independent of the vibronic level initially pumped. The pressures used were high enough so that phosphorescence is always coming from the thermally relaxed state. The absence of emission in H_2CO was taken to imply a substantially shorter triplet lifetime than in D_2CO. Comparison with the D_2CO emission intensity yields an upper limit of ~ 0.4 μsec for $T_1(H_2CO)$. Since $\tau_{ph} \ll \tau_{rad}^{T_1} \approx 2$ msec, Luntz and Maxson conclude that T_1 must undergo a fast predissociation via the ground state. The triplet has been further implicated in the formaldehyde photolysis because of the dramatic way in which both O_2 and NO are observed (43) to enhance the product yield. However, the self-quenching of $T_1(D_2CO)$ measured by Luntz & Maxson (57) is much slower in rate than the collision-induced production of CO monitored by Houston & Moore (43). More serious is the fact that it is difficult to account simultaneously for both the fast $S_1 \rightsquigarrow T_1$ intersystem crossing and the slow $T_1 \rightsquigarrow$ products reaction. The same problem arises of course in the case of the ground state ($S_1 \rightsquigarrow S_0 \rightsquigarrow$ Products) mechanism; we return to this question in Section 4.

3 THEORY: COLLINEAR TRIATOMICS

3.1 Optical Excitation

The earliest of recent theories of collinear triatomic photodissociation have come to be known as the "quasidiatomic" description. The central assumption of the quasidiatomic model is that the reaction coordinate—the A–B separation, say, in the reaction ABC → A + BC—is the same as one of the parent molecule's normal modes. That is, the ABC normal modes are taken to be the A–B and B–C "bond oscillators." If, for example, the linear triatomic is initially in its ground vibrational state, light absorption (in the case of direct photodissociation) or electronic relaxation (direct predissociation) is assumed to produce an excitation of the A–B bond on a repulsive surface, while the B–C oscillator remains bound but is in general excited because of a change in its bond length. The fact that the final surface is repulsive with respect to A–B means that the BC oscillator can be further excited by the recoiling A atom. These two mechanisms for the excitation of the BC diatomic have been christened intra- and interfragment, respectively.

The first developments of the quasidiatomic model appear to have included either intra- or interfragment effects, but not both simultaneously. Caplan & Child (58), for example, treated a formal generalization of the "golden rule," which leads to expressions for both the spectroscopic linewidth Γ_n of a predissociating level and the corresponding fragment internal-state distribution $I_n(m)$. For the ABC → A + BC reaction mentioned above, it is found that (58)

$$\Gamma_n = \sum_m \left[\int dQ_{AB} \, dQ_{BC} \phi_n(\mathbf{Q}) V(\mathbf{Q}) \xi_m(Q_{AB}) N_m(Q_{BC}) \right]^2, \qquad 3.$$

where $\mathbf{Q} \equiv (Q_{AB}$ and $Q_{BC})$ denotes the two bond coordinates, ϕ_n is the initial bound vibrational state of the parent triatomic, V is the interaction energy driving the predissociation (assumed here to be direct electronic), and N_m is the wavefunction describing the BC diatomic excited with m quanta. ξ_m is the continuum state of the recoiling fragments when BC is in its mth state. Similarly, Caplan and Child show that the distribution $I_n(m)$ of the fragment's final internal states is proportional to the mth term in the sum defining Γ_n in Equation 3. Note that all interfragment effects are neglected; couplings of nuclear motions on the final A-B-C potential energy surface are assumed to have no effect on the BC excitation determined by the mth term in Equation 3. Also, the change in normal modes between the initial (ABC) and final (A + BC) states has not been taken into account.

The role of the recoiling A atom in exciting the diatomic fragment has been treated in detail by a number of investigators, assuming both statistical (59, 60) and impulsive (33, 61–64) dynamics. The impulsive model treats the system as having undergone a "half-collision" following its preparation at the classical turning point of an imagined full collision. That is, the repulsive potential is suddenly switched on at the midpoint of the hypothetical full encounter, the initial state of the BC oscillator is taken into account, and the ensuing forced-oscillator problem is solved semiclassically. At first, all three calculations were carried out (61–64) under the assumptions that the potential function of BC remains identical to that in the isolated fragment and that it is initially unexcited. Recall that in both the direct photo- and direct predissociations, the fast change in electronic state brought about by light absorption or radiationless transition results in a sudden release in potential energy as the molecule breaks up. It is this fact that stands behind Caplan and Child's use of generalized Franck-Condon factors in determining $I_n(m)$. When the Q dependence of their "electronic factor" V is suppressed, the mth term in Equation 3 becomes simply the square of the vibrational overlap integral between initial (n_{AB}, n_{BC}) and final (E_{A-BC}, m_{BC}) states of the bond oscillators. Mitchell & Simons (65) has evaluated these Franck-Condon factors earlier by using empirical bond order–bond length relationships to determine the change in BC potential. Simons & Tasker (40) were the first to introduce these (intrafragment) effects in conjunction with the half-collision description of the A-BC final state (interfragment) dynamics. Instead of referencing the impulsive model calculation to an initial $n_{BC} = 0$ state, it is assumed that light absorption or radiationless transition into the repulsive electronic state prepares the diatomic fragment BC in a range of vibrational states, with their distribution given by the appropriate Franck-Condon factors. Good agreement is found between theory and experiment for the quenching—through formation of a triplet exiplex, ^{3}(HgCO)—of Hg(6^3P_1) by CO; the crucial effect is argued to be the diatomic's change in bond length as the triplet complex intersystem crosses to a repulsive singlet.

A proper treatment of the normal mode change accompanying the molecular breakup into fragments was first provided by Band & Freed (66–70) (see also 71). They start with an expression for the transition rate from (n_1, n_2) to (E, m), which is essentially the same as that given by the mth term in Equation 3. However, the "normal modes" are not taken to be the same in the initial and final states. n_1 and n_2 refer now to the quantum numbers for the vibrational coordinates Q_1 and Q_2, which are no longer the AB and BC bond stretches, but rather the actual normal coordinates (symmetric and asymmetric stretches) for the parent triatomic: $\mathbf{Q} \equiv Q_1, Q_2$. In the final state, the normal coordinates are different: $\bar{\mathbf{Q}} \equiv Q_{BC}, Q_{rxn}$. The "old" and "new" coordinates are related to each other through a linear transformation, i.e. $\mathbf{Q} \equiv \mathbf{C} \cdot \bar{\mathbf{Q}} + \mathbf{b}$. The elements of the 2×2 matrix $\mathbf{C}$ are completely determined (72) by the atomic masses, equilibrium bond lengths, and harmonic frequencies for the molecule and its fragments; $\mathbf{b}$ incorporates the effects of bond length changes. As a result of $\mathbf{Q} \neq \bar{\mathbf{Q}}$, the integrations in Equation 3 are no longer factored into a product of one-dimensional integrals. Band & Freed (66–70) have shown how these two-dimensional vibrational overlaps can be transformed into a sum of one-dimensional integrals, which latter can be identified with an "effective oscillator" (66) whose equilibrium position and frequency are functions of the equilibrium geometries, atomic masses, and normal mode force constants for ABC and BC. [In the case where both fragments are *polyatomic*, but with rotations suppressed, reduction of the Franck-Condon overlaps to two-dimensional integrals has also been treated (70)]. It remains only to choose the repulsive potential describing the recoiling fragments, e. g. $V(Q_{rxn}) = V_0 \exp(-DQ_{rxn})$.

Band & Freed treated first (66) the (assumed) direct photodissociation of HCN into H + CN ($B^2\Sigma^+$), choosing V_0 and D to fit the experimental data of Mele & Okabe (73) on the relative vibrational populations of the CN fragment. All the remaining constants (i.e. bond lengths and harmonic frequencies) are obtained from spectroscopic data (1, 2) as mentioned above. Similar calculations were carried out for ICN → I($P_{3/2}$) + CN($B^2\Sigma^+$), for which the repulsive potential has added to it an attractive part (67). The $V(Q_{rxn})$ fit to obtain agreement with experiment at a given exciting-light energy is then used to predict relative vibrational populations of CN at other incident wavelengths.

Berry (74) has also treated the HCN dissociation under the assumption that interfragment effects are negligible. Instead of taking into account the actual change in normal coordinates, he defines the initial conditions for the triatomic in terms of a localized diatomic oscillator (CN), which is "clothed" by its interaction with the neighboring atom (H). Transition to the repulsive potential (via light absorption or radiationless decay) strips the oscillator of its "clothes" suddenly; its properties (internuclear separation, dissociation energy, and force constant) become those of the separated diatomic fragment. Thus the dissociation rate becomes proportional to the Franck-Condon factor for initial and final oscillator states. The bound-free contribution to the transition probability—accounted for explicitly through additional, bound-free Franck-Condon factors in the theories described above—is treated statistically by Berry through a density-of-states function (74). The one-dimensional vibrational overlap integrals are evaluated using Morse wavefunctions for the initial and final oscillator states. The final oscillator (separated diatom) properties are spectroscopically well known. As for the initial oscillator, its force constant is calculated via Badger's rule (75) from the known internuclear separation in the triatomic, and its dissociation energy from Pauling and Johnston's rules (76). Berry (74) has also explained the vibrational distributions observed in the dissociations of CO_2 and in the electronic quenching of $Hg^*(6^3P_1)$ by CN and CO. By comparing the initial diatom internuclear separations that are necessary to give agreement with experiment with those of the ground electronic state of the triatomic he infers whether the reaction proceeds via direct photo- (versus pre-)dissociation. Consider, for example,

$CO^2 \xrightarrow{h\nu} CO^{*\dagger}(a^3\Pi, 0) + 0$ (Ref. 77),

$CO^{*\dagger}(d^3\Delta_i, v) + 0$,

$CO^{*\dagger}(a'^3\Sigma^+, v) + 0$ (Ref. 78).

Best fits to the $CO^{*\dagger}(a^3\Pi)$ and $CO^{*\dagger}(a'^3\Sigma^+)$ vibrational distributions are obtained for an initial internuclear separation of ~ 1.12 Å. Since the CO bond length is 1.16 Å in the ground electronic state of CO_2, it is concluded that these reactions correspond to direct photodissociation via the optically excited, repulsive state of CO_2. The $CO^{*\dagger}(d^3\Delta_i, v)$ data, on the other hand, are best fit by an initial internuclear separation of ~ 1.23 Å, suggesting that $CO^{*\dagger}(d^3\Delta_i)$ is formed through a predissociation, i.e. the upper CO_2 state is "crossed" by a lower, repulsive configuration near $R_{CO} \approx 1.23$ Å. In all three cases the final CO fragment corresponds to an appreciably stretched CO bond, consistent with all the $CO^{*\dagger}(v)$ distributions showing large population inversions (79).

Atabek and co-workers (80, 81), following up the earlier work of Mukamel & Jortner (82, 83) and Atabek, Beswick & Lefebvre (84), have argued that it is not sufficient to describe the photodissociation of collinear triatomics in terms of an intrafragment mechanism alone. Instead, it is necessary to consider the effects of the forces between receding fragments in order to determine how they might alter the final oscillator vibrational distributions predicted by the two-dimensional Franck-Condon factors described above. Atabek et al have treated collinear triatomic dissociations from a unified quantum mechanical point of view that 1. simultaneously includes *inter-* and *intra*fragment effects and 2. accounts for the change $(Q_1, Q_2) \rightarrow (Q_{BC}, Q_{rxn})$ in the normal coordinates. From a thorough numerical study of the ICN and HCN dissociations, they concluded that 1. the vibrational distribution of the CN fragment becomes broader with increasing excess energy; 2. the vibrational distribution is very sensitive to (a) the parameters characterizing the repulsive potential, (b) the change in CN bond length, and (c) the initial state (n_1, n_2) of the bound triatomic; and 3. the *inter*fragment, final-state interactions ["intercontinuum couplings" (80)] drastically modify the population of high vibrational levels of the CN fragment. In their recent review Freed & Band (70) report roughly equal contribution of inter- and intrafragment effects to CO excitation in the CO_2 photodissociation; for HCN the interfragment effects are found to be significant only for certain of the potential curve choices.

Atabek et al (80) have also attempted to fit the observed vibrational distributions by varying the two parameters characterizing the repulsive potential $Ae^{-\alpha Z}$. In the case of HCN they find a marked difference between their preexponential factor A and that used by Band & Freed (66, 67). This difference in repulsive potential makes the conclusions reached about the importance of final-state interactions uncertain. Also, Band & Freed (68) have predicted a many-orders-of-magnitude larger $H \rightarrow D$ isotope effect on the vibrational distributions and absolute dissociation rates. Atabek et al have argued that this result is a pathological result associated with Band and Freed's repulsive potential (i.e. the factor A) being ~ 100 times too large. Direct measurements of the DCN versus ICN photodissociation should be of considerable help in resolving this controversy.

Finally, the collinear theory of Atabek et al has been modified (80) to approximately take into account the different number of rotational states available in each vibrational channel (85–87). Morse, Band & Freed (88) have taken an alternative approach in which the bending vibration and overall rotations of the linear molecule are included explicitly in the description of the initial state. (General features of the relationship between vibrational, rotational, and translational distributions in triatomic dissociations had been discussed earlier by Florida & Rice (89).] The results for ICN have been discussed in detail (80, 88). Clearly, the investigation of rotational distributions and their relationship to molecular geometry, vibrational excitation, and final-state interactions is important in clarifying further the mechanism of triatomic dissociations. Atabek et al (80) for example, believe that all of the photofragmentation data on ICN and possibly HCN in the range 7.2–10.6 eV correspond to predissociation from Rydberg states. The ICN excited states are linear (38), whereas the low-lying electronic configurations of HCN have a bent geometry (2). Further, less ad hoc theories of the rotational degrees of freedom and interfragment effects are necessary before surer conclusions can be reached about the identity of "initial" and "final" states in these photofragmentations.

3.2 Thermal Excitation

The dynamics of thermal fragmentation of collinear triatomics have been treated in detail only for $N_2O(^1\Sigma^+) \rightarrow N_2(^1\Sigma_g^+) + O(^3P)$, first by Gilbert & Ross (10) and independently by Gebelein & Jortner (9). Gilbert and Ross introduced a stepladder model in which each vibrational level of the N_2O ground state has associated with it a rate of collisional exchange with neighboring levels, and a "golden rule" rate of intramolecular decay into the repulsive electronic state that crosses the ground state. The "reactant" (bound N_2O) and "product" ($N_2 + O$) state wavefunctions are written as products of electronic and vibrational factors in the usual way, and the corresponding matrix elements of the intramolecular interaction (assumed here to be spin orbit coupling) are written in the Condon approximation (90) as $\lambda_{\text{spin orbit}}$ times vibrational overlap integrals. $\lambda_{\text{spin orbit}}$ is estimated to be $\lesssim 100$ cm^{-1} for the ground state triplet crossing; recall from Section 1 that the singlet ground state cannot correlate with triplet products. The collisional exchange rates for the N_2O levels are calculated from SSH theory (91, 92). Since so little is known about the two potential energy surfaces involved (especially the triplet), Gilbert and Ross pursued two complementary, but crudely simplified, approximations. In both, anharmonicity was neglected—except in the NO bond—and the repulsive state was assumed to be linear, with no allowance made for rotation.

In the first, one-dimensional approximation, a simple quadratic potential $(\frac{1}{2})k_{\text{NO}}\bar{R}_{\text{NO}}^2$ was used for the ground state: $R_{\text{NO}}^{\text{eq}}$ was taken to be the observed bond length in N_2O, and k_{NO} was set equal to the spring constant in a diagonal N_2O force field. For the excited state corresponding to product, the repulsive potential was written as $D_e + B \exp(-C\bar{R}_{\text{NO}})$, where D_e is the thermochemical dissociation energy and B was chosen to fit, at $\bar{R}_{\text{NO}} \equiv R_{\text{NO}} - R_{\text{NO}}^{\text{eq}} = 0$, the weak maximum in the UV absorption spectrum; the remaining constant C was used as an adjustable parameter to reproduce the experimental activation energy. In a complementary description, the ground state potential was taken to be quadratic in the normal coordinates for the in- and out-of-phase stretches Q_1 and Q_2: $\frac{1}{2}a_1 Q_1^2 + \frac{1}{2}a_2^2 Q_2$, with a_1 and a_2 determined from the well-known fundamental frequencies. The repulsive potential was as before, with the addition of an independent quadratic term in the NN stretch: $D_e + B \exp(-C\bar{R}_{\text{NO}}) + \frac{1}{2}k_{\text{NN}}\bar{R}_{\text{NN}}^2$, with k_{NN} and $R_{\text{NN}}^{\text{eq}}$ determined from the properties of the ground state N_2 fragment.

Solving approximately the master equations for collisional and intramolecular energy exchange within the above models, Gilbert and Ross have found that all the observed features of the $N_2O(^1\Sigma^+) \to N_2(^1\Sigma_g^+) + O(^3P)$ thermal fragmentation can be qualitatively reproduced. In particular, the frequency factor and activation energy characterizing the high-pressure unimolecular rate constant can be made to agree arbitrarily closely with experiment for reasonable choices of $\lambda_{\text{spin orbit}}$, B, and C. The description of the transition region from low-pressure (bimolecular) to high-pressure (unimolecular) kinetics is less satisfactory. As Gilbert & Ross (10) have emphasized, however, their object is not to reproduce observed data, but rather to explore a new interpretation of thermal fragmentations in terms of nonadiabatic, intramolecular processes.

Gebelein & Jortner (9) have pursued further this analogy between unimolecular decomposition reactions and the theory of nonadiabatic "radiationless transitions" (e.g. predissociation). They concern themselves only with the high-pressure, thermally averaged rate constant given by

$$k = \sum_\alpha \exp(-E_{g\alpha}/k_B T) W_{g\alpha} / \sum_\alpha \exp(-E_{g\alpha}/k_B T). \qquad 4.$$

Here $E_{g\alpha}$ is the energy of the αth vibrational level of the N_2O ground electronic state and

$$W_{g\alpha} = \frac{2\pi}{\hbar} \lambda_{\text{spin orbit}}^2 \left| \int d\mathbf{R} \chi_{g\alpha}(\mathbf{R}) \chi_T(\mathbf{R}) \right|^2 \qquad 5.$$

is its rate of decay into the repulsive triplet (T); χ_T is the energy-normalized continuum wavefunction describing the triplet state with energy $E_{g\alpha}$. As in the Gilbert & Ross (10) treatment, the molecular wavefunctions have been written as products of electronic and vibrational parts, the Condon approximation has been invoked to separate out the average electronic interaction $\lambda_{\text{spin orbit}}$, and first-order perturbation theory has been assumed valid so that the "golden rule" result (Equation 5) is obtained. Gebelein & Jortner (9) carry out one-dimensional calculations of Equations 4 and 5, which are essentially the same as those of Gilbert & Ross (10) except that they use WKB vibrational wavefunctions (93) and use Br^{-n} rather than $B \exp(-Cr)$ repulsive potentials. Also, they manipulate the general expression for the high-pressure unimolecular rate constant to yield an analytical result that illustrates the physical basis for the increase with temperature of the apparent activation energy. Gebelein and Jortner have also considered two-dimensional calculations, in which they argue that the bond stretches are more appropriate than the normal modes (used by Gilbert and Ross) in describing the separable ground state surface and in predicting the vibrational distribution of the N_2 fragment. Again, though, these differences cannot be resolved until it becomes possible to treat more realistically the effects of both final-state interactions and rotational degrees of freedom.

The work on N_2O described above cannot be placed directly within the context of our earlier discussion of collinear triatomic predissociations since it refers to initial (bound) states that are extremely "hot," having been prepared by collisional activation rather than by Franck-Condon radiative transitions. Nevertheless it can be thought of as a theory that includes only intrafragment effects, with the initial coordinates referenced alternatively to bond stretches (9) and normal mode (10) in the bond triatomic. Further discussion relevant to the mechanism of N_2O dissociation is given in the work of Chutjian & Segal (94) and Zahr et al and Tully (95) on N_2O photoelectron spectra and N_2 quenching of $O(^1D)$, respectively.

I should also mention here the quasidiatomic picture suggested by Lawetz, Siebrand & Orlandi (96) to describe the possible (97, 98) predissociation of the first singlet state of benzene. Using the earlier "local" CH mode arguments of Henry & Siebrand (99), they propose that the excess energy dependence of the $C_6H_6(S_1) \rightarrow C_6H_5\cdot + H$ process can be explained by the vibrational quantum number dependence of the bound-continuum Franck-Condon factor for a single CH stretch. Morse oscillator wavefunctions were used for the S_1 levels, and WKB approximations for the S_0 states; the relevant anharmonicities and bond length and frequency changes were estimated from electronic and vibrational overtone spectra. It is concluded that the predissociation rate associated with an excess energy of ~ 3000 cm^{-1} is only ~ 10 times greater than the internal conversion rate at the S_1 origin. Thus the several orders of magnitude increase (97, 98) in the observed S_1 radiationless decay rate cannot be accounted for in this way.

4 THEORY: NONLINEAR POLYATOMICS

The 1968 paper by Rice, McLaughlin & Jortner (100) launched a great deal of interest in the formal theory of molecular photodissociation processes. Inspired by the earlier work of Mies & Kraus (8) and Peters (101) on collisionally activated unimolecular reaction and direct predissociation, respectively, it focused attention on the intermediate states linking the photochemical preparation and the final products. The key idea involved simulating the dynamics of a molecular photodissociation by solving for the time-dependent, quantum mechanical behavior of a model system whose spectrum consisted of 1. a discrete vibrational state ϕ_s belonging to an electronically excited configuration, which is directly populated by light absorption; 2. a manifold of bound vibrational levels $\{\phi_l\}$ belonging to a lower electronic configuration—these are uniformly distributed with spacing ε, and each is coupled to ϕ_s through an interaction matrix element v_1; 3. a single (energy normalized, say) continuum ξ_E, representing the dissociating molecular states, coupled to each ϕ_l through the energy-independent v_2. Kay & Rice (102) showed that under certain limiting conditions, the dynamics of this model could be solved for analytically, with the following results: 1. the decay of ϕ_s is exponential, but the corresponding rate constant is given by $[2\pi v_1^2/\hbar(\varepsilon + \pi^2 v_2^2)]$ instead of the usual "golden rule" $2\pi v_1^2/\hbar\varepsilon$; 2. the overall decay is nonsequential, i.e. $\{\phi_l\}$ and ξ_E are populated simultaneously ("in parallel") rather than consecutively. Similar results were found by Lefebvre & Beswick (103), who replaced the $\{\phi_l\}$ manifold by a true continuum, and by Nitzan, Jortner & Berne (104), who included an arbitrary number of continua.

The origin of nonsequential photodissociation dynamics was first explored critically by Kay (105). Nitzan, Jortner & Rentzepis (106) had earlier studied a model similar to that of Rice, McLaughlin, and Jortner except that a large number of continua were assumed to exist and each intermediate manifold level ϕ_l was coupled to "its own" continuum. In this case the initial state ϕ_s decays with the familiar $2\pi v_1^2/\hbar\varepsilon$ rate, independent of the continua couplings, and the overall decay is sequential—the probability of $\{\phi_l\}$ first increases and then falls off as the continua "grow in"—and the populations for the ϕ_s, $\{\phi_l\}$, and continua states obey the usual kinetic equations for consecutive decay. Kay (105) has argued (see also 102, 103) that the difference between the nonsequential and sequential results is due to quantum mechanical interference effects that arise in the RMJ (100) but not in the NJR (106) scheme. In both models, interaction with a fragmentation continuum causes each ϕ_l to be unstable and hence to acquire a width $\Gamma = 2\pi v_2^2$. Interference effects arise when there is overlap of the widths of two or more states ϕ_l that interact with the same continuum, i.e. these states become indirectly coupled to one another through a mutual continuum and hence can no longer decay independently. This is asserted to be the origin of the nonsequential behavior of the RMJ model. Arguing the physical unreasonableness of a separate continuum for each ϕ_l, Kay (105) has extended the original RMJ scheme to include both the existence of many continua and interference effects. "Quasi"-sequential behavior is found, whose details depend on the discrete energy level spacing and the various interaction matrix elements; the sequential and nonsequential behaviors are contained as limiting cases of the general theory. Heller & Rice (107) have bridged much of the gap between the NJR and RMJ-Kay models by getting the assumption of constant coupling (v_1, v_2) between ϕ_s, $\{\phi_l\}$, and ξ_E. They take the $\phi_s - \phi_l$ interaction matrix elements to be random, and show that this washes out the interference effects arising from the ϕ_l's being coupled to the same continuum; the overall decay is sequential, as if each ϕ_l were coupled to its own continuum. Thus it is concluded, by means of other (e.g. perturbation theory) points of view as well, that the unphysical aspect of the RMJ-Kay models is their assumption of constant coupling.

Formal theories of the kind discussed above, aside from providing a phenomenology for electronic relaxation and vibrational energy redistribution in polyatomic photodissociations, have been used in a few instances to interpret specific data. Evans et al (108), for example, have measured fluorescence lifetimes and quantum yields for individual vibrational levels of the first excited singlet states of chloro- and bromoacetylene. Because the absorbed photon energies are $\sim 10,000$ cm^{-1} greater than the C–X bond strengths, it is assumed that photofragmentation occurs. A model in which ϕ_s is directly coupled to $\{\phi_l\}$ and ξ_E, with no $\{\phi_l\} - \xi_E$ interaction, is used to rationalize the observed increase (HCCCl) and constancy (HCCBr) in fluorescence lifetimes. Direct dissociation of the prepared S_1 state into ground state fragments is not ruled out by electronic symmetry correlations, but is argued (since this unimolecular breakup would have to be slower than the observed fluorescence rate of $\lesssim 10^8$ sec) to be less likely than predissociation through S_0. Intersystem crossing to T_1 cannot be the rate-determining step since the ratio of S_1 lifetimes for HCCCl and HCCBr is not compatible with that calculated from the known spin orbit coupling constants. On the basis of the observed lifetimes, angular momentum conservation, and known molecular geometry and frequency changes, Evans et al (108) argue that HCCX must dissociate from a nearly linear configuration; that the reactive degree of freedom, the

CX stretch, is the poorest acceptor of excitation in the internal conversion; and that the energy exchange between this vibration and the remaining modes is slow compared with the RRKM (109, 110) rate. Hase & Sloane (109) have performed classical trajectory calculations that simulate the ground state unimolecular breakup of HCCCl. They obtain $\sim 10^{-12}$ sec as a lower limit to the vibrational dissociation time in S_0, in direct contrast to Evans et al's third conclusion above. Very probably it is the $S_1 \rightsquigarrow S_0$ internal conversion that is the rate-determining step; arguments (108) that the $S_1 - S_0$ interaction matrix elements would have to be unrealistically small are not strong. Further resolution of the photofragmentation mechanism clearly awaits new experimental data (e.g. time-dependent monitoring of fragments) and theoretical work (e.g. calculation of $S_1 \rightsquigarrow S_0$ decay rates).

Gillispie & Lim (111) have used a RMJ-Kay model to treat the S_1 radiationless decay in formaldehyde. As discussed briefly in Section 2.3, and in detail below, the photoexcitation of $H_2CO - S_1$ involves molecular fragmentation. Solving the $\phi_s\{v_{sl}\}\{\phi_l\}\{v_l(E)\}\xi_E$ model for slowly varying $\{v_{sl}\}$ and $\{v_l(E)\}$, they found that the S_1 decay rate could be expressed approximately [for $v_{sl}^2 < (E_s - E_l)^2 + \Gamma_l^2 \approx (E_s - E_l)^2$] as

$$k_{S_1} = \frac{2\pi}{\hbar} \sum_l \frac{v_{sl}^2 \Gamma_l}{(E_s - E_l)^2}. \qquad 6.$$

Here $\Gamma_l = 2\pi v_l^2(E)$ is the width of ϕ_l due to its interaction with ξ_E. [A similar result had been obtained in a different physical context by Nitzan & Jortner (112).] Gillispie & Lim (111) then argued that the observed isotope and excess energy dependences of k_{S_1} can be rationalized by the isotope and excess energy dependences of the unimolecular dissociation rates $\Gamma_l/\hbar$ to which k_{S_1} has been shown (cf Equation 6) to be essentially proportional. This nonsequential mechanism offers a distinctly different explanation of the fluorescence lifetime data than that given by Yeung & Moore (42)—see below—in which the widths Γ_l never appear explicitly in the S_1 decay rate.

Recall that Yeung & Moore (41) and Miller & Lee (53) had found that H_2CO, excited to its first singlet with one quantum in the out-of-plane wag (v_4), decays with a lifetime of $\sim 100-200$ nsec. Excitation of levels with more quanta in v_4 and v_2 (the CO stretch) gives rise to shorter lifetimes — τ_{obs} decreases by more than an order of magnitude for excess vibrational energies of ~ 5000 cm^{-1}. D_2CO decays about twenty times slower (than H_2CO), but its increase (two orders of magnitude) in relative rates is greater. Yeung & Moore (42) assumed that the overall photodissociation occurs in two steps: the first step, internal conversion ($S_1 \rightsquigarrow S_0$), is slow and rate determining; the second, unimolecular decomposition on the ground state surface, is fast. Thus the measured $k_{obs} \equiv 1/\tau_{obs}$ corresponds to the $S_1 \rightsquigarrow S_0$ radiationless decay rate k_{IC}. It is assumed further that the dissociation continua need not enter explicitly; instead they are invoked implicitly to push the H_2CO internal conversion into the "large molecule" ("statistical") limit (e.g. see 113). k_{IC} is written accordingly as

$$k_{IC} = \frac{2\pi}{\hbar} \sum_{v_f} |\langle S_1 v_i | H' | S_0 v_f \rangle|^2 \delta(E_{S_0 v_f} - E_{S_1 v_i}), \qquad 7.$$

where v_i and v_f denote the initial and final sets, respectively, of vibrational quantum numbers. H' is the nuclear kinetic energy that gives rise to the coupling between the $S_1 v_i$ and $S_0 v_f$ product (Born/Oppenheimer) wavefunctions: $H' = -(\hbar^2/2) \sum_k \partial^2/\partial Q_k^2$, where $\{Q_k\}$ are the usual mass-weighted normal coordinates. Substituting this form for H' into Equation 7, and retaining only the first ("cross") term in

$$H' \psi_{el} \chi_{vib} = -\hbar^2 \sum_k \frac{\partial \psi_{el}}{\partial Q_k} \frac{\partial \chi_{vib}}{\partial Q_k} - \frac{\hbar^2}{2} \sum_k \chi_{vib} \frac{\partial^2 \psi_{el}}{\partial Q_k^2}, \qquad 8.$$

k_{IC} can be expressed as

$$k_{IC} = \frac{2\pi}{\hbar} \frac{\hbar^4}{\Delta} \sum_{v_f}' \left| \left\langle \chi_{S_1 v_i}(Q) \left| \left\langle \psi_{S_1} \left| \sum_k \frac{\partial}{\partial Q_k} \right| \psi_{S_0} \right\rangle \frac{\partial}{\partial Q_k} \right| \chi_{S_0 v_f}(Q) \right\rangle \right|^2. \qquad 9.$$

Note that the sum in Equation 7 over energy-conserving Dirac delta functions has been replaced by a sum over all states $S_0 v_f$ whose energy lies within a range Δ of $E_{S_1 v_i}$.

Yeung and Moore evaluate the electronic factor $\langle \psi_{S_1}(qQ) | \sum_k (\partial/\partial Q_k) | \psi_{S_0}(qQ) \rangle$ appearing in Equation 9 by using Herzberg-Teller theory [see (114) and references therein] to display the nuclear coordinate dependence of ψ_{S_1} and ψ_{S_0}. The magnitudes of most of the expansion coefficients are treated as adjustable parameters in the calculation of the absolute internal conversion rates. The vibrational wavefunctions, $\chi_{S_1 v_i}$ and $\chi_{S_0 v_f}$, are written as products of harmonic oscillator states whose properties are known from infrared and electronic spectra. Finally, Δ is chosen to give the best agreement between calculated and measured relative rates; if it is too small or too big the vibrational quantum number dependence of k_{IC} is too strong or too weak, respectively. In this way the observed data on k_{IC} vs excess energy is accounted for, and—with no further adjustment of parameters—the isotope ($H_2CO \rightarrow D_2CO$) effects are well reproduced.

There are several reasons why the Yeung/Moore calculations fall short of providing a theory of the formaldehyde photodissociation dynamics:

1. The fact that the molecule breaks up as part of the $S_1 \rightsquigarrow S_0$ electronic relaxation process means that the S_1 nonradiative decays cannot be described as simple transitions between bound states.

2. Similarly, because formaldehyde is not a "large molecule," the "golden rule" rate expression given by Equation 7 is not appropriate; introducing the restricted sum $\sum_{v_f}'$, and treating the energy range Δ as an adjustable parameter, obscures the basic physics.

3. The Herzberg-Teller expansions for the electronic wavefunctions are treated semiempirically, so that a priori estimates of absolute internal conversion rates are not possible.

4. More importantly, the assumption of a fast, S_0 unimolecular breakup following a slow, rate-determining $S_1 \rightsquigarrow S_0$ decay is not consistent with the subsequent experimental work of Houston & Moore (43) in which CO products are not seen until microseconds after the optical excitation.

Morokuma et al (49–51) have performed extensive, ab initio, SCF-CI and MCSCF calculations on the S_0, S_1, and T_1 potential energy surfaces of formaldehyde. To study the $H_2CO \to H + HCO$ process (51) they stretched one of the CH bonds by small increments, reoptimizing the energy at each step with respect to the remaining five coordinates; a minimal Slater basis was used to construct several hundred CI wavefunctions. This procedure correctly accounted for the known CO bond lengthening upon electronic excitation, the nonplanarity of the S_1 and T_1 states, and the bent structure of the ground state HCO radical, etc. S_0 and T_1 correlate with ground state radical products, whereas S_1 can only give excited HCO. The lowest energy path to $H + HCO$ is planar for S_0 and nonplanar for T_1 (and S_1), e.g. at the saddle point of the T_1 surface the stretched CH bond is almost perpendicular to the HCO plane. Furthermore, as the CH bond is stretched, the S_0 energy increases monotonically, the energy barrier being as high as $\lesssim 4.4$ eV ($\lesssim 33,000$ cm^{-1}). Calculations (49, 50) on the $H_2CO \to H_2 + CO$ process (which must proceed through S_0) have also focused on locating the minimum energy barrier on the ground state surface. The minimal basis SCF-CI (49) results were qualitatively confirmed by more recent (50), extended basis set MCSCF computations. For each of many values of the distance D' between the carbon atom and the H_2 center of mass, the energy is optimized with respect to the remaining five coordinates (R_{CO}, R_{HH}, and three angles). This procedure is carried out for planar symmetric (C_{2v}), nonplanar symmetric (C'_s), planar asymmetric (C_s), and nonplanar asymmetric (C_1) geometries. Only the C_s path gives an energy barrier ($\lesssim 4.5$ eV) close to the observed threshold of 3.7 eV (44). 1. The strong asymmetry of the path (at the saddle point the angle between the D' vector and CO axis is $\sim 50°$, and that between the H_2 and CO axes is $\sim 70°$); 2. the fact that the saddle point H–H and C–O bond lengths are much closer to H_2CO's than to the diatomic products; and 3. the thermoneutrality ($\Delta E \approx 0.3$ eV) of the overall reaction suggest that the fragments will be highly rotationally and vibrationally excited. All of these potential energy surface "facts" should be taken into explicit account by any "first principles" theory of the formaldehyde photodissociation.

Elert, Heller & Gelbart (115) have recently attempted a description of the simultaneous electronic relaxation and vibrational dissociation (to molecular products) in formaldehyde. They start by making certain simplifying assumptions: the possible role of the T_1 state is neglected for the $H_2CO(S_1) \leadsto H_2 + CO$ process; rotational degrees of freedom are not included; and the initial and final vibrational (normal mode) coordinates are taken to be the same. We discussed earlier the several shortcomings associated with each of these assumptions, and the various ways in which they can be improved upon. The starting idea, however, is to unravel in as simple a fashion as possible the competition between electronic and vibrational energy redistribution processes in determining the dynamics of the overall photodissociation. Elert et al (115) begin by considering the following basis as providing a complete set of states for the excited formaldehyde molecule in the region of S_1's origin: 1. the initially prepared vibrational level ϕ_s in S_1; 2. all S_0 normal-mode combinations and overtones $\{\phi_l\}$ that correspond to less than D ($\sim 25,000$ cm^{-1}, say) in the in-plane symmetric H–C–H bend (v_3); and 3. all continuum states $\{\xi_E^{(n)}\}$ in which

$E > D$ in v_3. Here v_3 has been taken to be the reaction coordinate, even though we know that the CH stretches must be involved as well. Each bound S_0 state is written as a product of harmonic oscillator wavefunctions, while in the continuum wavefunctions the v_3 factor is replaced by the appropriate unbound state of a Morse potential. ϕ_s interacts with $\{\phi_l\}$ and $\{\xi_E^{(n)}\}$ through the nuclear kinetic energy, while the S_0 states, $\{\phi_l\}$ and $\{\xi_E^{(n)}\}$, are coupled through the anharmonic ground-electronic-state potential energy surface $E_{S_0}(Q)$.

Elert et al (115) solve for the time development of the initial state ϕ_s by using the damping matrix ("effective Hamiltonian") formalism [e.g. see (116) and references therein]. The probability $P_s(t)$ of being in ϕ_s at time t, for example, is given by

$$P_s(t) = \left| \sum_k O_{sk}^2 \exp\left(-\frac{i}{\hbar}\Lambda_k t\right) \right|^2, \qquad 10.$$

where Λ_k and O_{sk} denote the kth eigenvalue and the ϕ_s component in the kth eigenvector of the matrix representing the effective hamiltonian $\mathbf{H}_{\text{eff}}$ in the $\{\phi_s, \phi_l\}$ basis:

$$\mathbf{H}_{\text{eff}} \equiv \mathbf{H}_{\text{mol}} - \frac{i}{2}\mathbf{\Gamma}. \qquad 11.$$

H_{mol} is the full molecular energy including both $E_{S_0}(\mathbf{Q})$ and $T(Q)$, the nuclear kinetic energy. Γ is the (in general, nondiagonal) "damping matrix" defined by

$$(\Gamma)_{ij} = 2\pi \sum_n \langle \phi_i | H_{\text{mol}} | \xi^{(n)} \rangle \langle \xi^{(n)} | H_{\text{mol}} | \phi_j \rangle, \qquad 12.$$

with $i, k = s$ or l. Equations similar to Equation 10, but with $O_{sk}^2 \to O_{sk}O_{lk}$, hold for each of the $P_l(t)$ values; P_{diss}, the probability of the molecule flying apart in any of the dissociation channels, is given by $1 - P_s(t) - \sum_l P_l(t)$. Elert et al (115) compute each of the matrix elements involving $T(Q)$ by expanding the electronic wavefunctions $\psi_{S_0}(qQ)$ and $\psi_{S_1}(qQ)$ in powers of $\{Q_i\}$, through second order in Herzberg-Teller theory (114); for the $\psi_{S_i \neq 1,0}(q, Q_{\text{eq}}^{S_1})$ values they use the LCAO-MO-SCF results of Aung, Pitzer & Chan (117). The matrix elements involving $E_{S_0}(Q)$ are calculated using the ab initio potential energy surface of Jaffe, Hayes & Morokuma (49, 50). The resulting $\mathbf{H}_{\text{eff}}$ matrix is diagonalized numerically, and plots are generated for $P_s(t)$, $P_l(t)$, and $P_{\text{diss}}(t)$. Successively more bound states are included in the computation until convergence is achieved in the dynamics, e.g. the function $P_s(t)$ is constant to within a few percent at any time. Collisional interactions never appear since the calculations are intended to explain zero-pressure experimental data.

Taking the ground state dissociation threshold, D, to lie just below the S_1 origin, Elert et al (115) find that $P_s(t)$ falls off on a time scale of 10^{-7} sec. Within about the same time, $P_{\text{diss}}(t)$ grows in, i.e. there is no time lag between decay of electronic excitation ("internal conversion") and appearance of $H_2 + CO$ products. This τ_{S_1} (for one quantum in the v_4 mode) is in good agreement with the measured values (41, 53) of 100–200 nsec; similarly, the calculated decreases in τ_{S_1} upon further vibrational excitation in S_1 agree with those observed experimentally. The lifetime lengthening upon deuteration is underestimated.

It is not clear how the above results would be changed by corrections arising from rotation, the T_1 state, or a more realistic choice of reaction coordinate. It is almost certain, however, that—as long as the S_0 fragmentation threshold is assumed to lie below the optical excitation energy—dissociation will take place in the absence of collisions and with no appreciable time lag. Recall, however, that the experimentally observed (43) CO appearance rate was directly proportional to the pressure, extrapolating to zero at zero pressure. This is suggestive of a collisionally activated dissociation, consistent with the theoretical estimates (49, 50) of $D_{S_0} \gtrsim 4.3$ eV > $E_{opt} \lesssim 4.0$ eV. But, if indeed $D_{S_0} > E_{opt}$, then collisions would also be necessary to explain the fact that S_1 (for $E_{S_1} < D$) decays exponentially with a ~ 100 nsec lifetime. This is because, without collisions, the S_0 states are not broadened and are too widely spaced (even with anharmonicity included) to provide a dissipative manifold into which S_1 might decay. Numerical calculations (115) suggest that gas-kinetic collisional deactivation at 0.1 torr of the "hot" S_0 states is necessary to explain the S_1 lifetimes in the $E_{S_1} < D_{S_0}$ case. And yet the exponential fluorescence decays measured by Yeung & Moore (41) at ~ 0.1 torr are believed to involve no collisional effects. Experiments at still lower pressures would be helpful in determining whether or not the "collision-free" limit of the S_1 decay has been reached. If indeed $E_{S_1} < D$, and collisional deactivation of the S_0 states accounts for the S_1 decay, then the time lag observed for dissociation could be explained by a thermal activation process. Otherwise, it is quite unclear what is happening.

5 CONCLUSION

It is remarkable that, after the considerable theoretical and experimental work reviewed above, so little is understood about photodissociation of polyatomic molecules. There is not a single example, in fact, where the actual "mechanism" (dynamics) of a polyatomic photodissociation has been unambiguously established. Even in the case of collinear triatomics (see Sections 2 and 3), where isotope effects and fragment vibrational/rotational distributions have been accounted for in detail, the identity of the electronic states involved can only be guessed. For example, do the XCN initial (bound) and final (repulsive) states correspond to the ground state S_0 and the optically prepared S_1—and hence to a direct photodissociation—or to an electronic predissociation involving S_1 and a lower-lying configuration? Similarly, in the case of the aryl halides (Section 2), does the postulated σ^* triplet (15) really exist and, if so, does it have the properties necessary to explain the observed lifetime data? What is the nature of the electronic states involved in the photodissociation of NO_2 (20–30), the halogenated acetylenes (108), and substituted ketones (13) studied respectively by photofragment spectroscopy, single vibronic level fluorescence, and photolysis mapping? Finally, in the case of formaldehyde, how is the T_1 state involved and what are the actual dissociation thresholds for radical and molecular products?

Questions such as the above will only be answered by more thorough theoretical investigations of the relevant electronic structure. The details of the photodissociation mechanism can then be established by further time-resolved experiments at low ("collision-free") pressures. It is clear that a more concerted cooperation is necessary between investigators studying electronic structure, on the one hand, and vibrational/rotational dynamics, on the other. It will also be important to persist in probing further the specific examples mentioned in this review. We must resist the temptation of hopping from one system to the next in a series of incomplete experimental or theoretical attacks that flaunt sophisticated technique but quit before the elementary question of mechanism is answered.

NOTE ADDED IN PROOF   Pack (118) has recently offered a theoretical study of the photodissociation of symmetric triatomics in which he shows that—even when the upper electronic configuration is unbound—the absorption spectrum can show diffuse vibrational bands reminiscent of predissociating states. The possibility of these "bobsledding oscillations" on repulsive surfaces, and their resulting spectral structures, had been suggested earlier by Hall et al (119).

ACKNOWLEDGMENTS

It is a pleasure to acknowledge innumerable discussions on formaldehyde with my collaborators, M. L. Elert and D. F. Heller. For helpful conversations on photodissociation I am also grateful to M. J. Berry, K. F. Freed, E. J. Heller, J. Jortner, E. K. C. Lee, J. Michl, C. B. Moore, S. Leach, A. Chutjian, and S. A. Rice.

Literature Cited

1. Herzberg, G. 1950. *Spectra of Diatomic Molecules.* New York: van Nostrand
2. Herzberg, G. 1966. *Electronic Spectra of Polyatomic Molecules.* New York: van Nostrand Reinhold
3. Bonnhoeffer, K. F., Farkas, L. 1928. *Z. Phys. Chem. Abt. A* 134:337
4. Clement, M. J. Y., Ramsay, D. A. 1961. *Can. J. Phys.* 39:205
5. See, for example, the discussion in (97, 98)
6. Herzberg, G. 1966. See Ref. 2, pp. 466–69
7. Rice, O. K. 1933. *J. Chem. Phys.* 1:375; Rosen, N. 1933. *J. Chem. Phys.* 1:319; Langer, R. M. 1929. *Phys. Rev.* 34:92
8. Mies, F. M., Kraus, M. 1966. *J. Chem. Phys.* 45:4455; 1969. 51:787, 798
9. Gebelein, H., Jortner, J. 1972. *Theor. Chim. Acta* 25:143
10. Gilbert, R. G., Ross, I. G. 1971. *Aust. J. Chem.* 24:1541
11. Solomon, J. 1967. *J. Chem. Phys.* 47:889
12. Jonah, C., Chandra, P., Bersohn, R. 1971. *J. Chem. Phys.* 55:1903
13. Solomon, J., Jonah, C., Chandra, P., Bersohn, R. 1971. *J. Chem. Phys.* 55:1908
14. Dzvonik, M. J., Yang, S. C. 1974. *Rev. Sci. Instrum.* 45:750
15. Dzvonik, M. J., Yang, S., Bersohn, R. 1974. *J. Chem. Phys.* 61:4408
16. Kawasaki, M., Lee, S. J., Bersohn, R. 1975. *J. Chem. Phys.* 63:809
17. Yang, S. C., Bersohn, R. 1974. *J. Chem. Phys.* 61:4400
18. Zare, R. 1972. *Mol. Photochem.* 4:1
19. Jonah, C. 1971. *J. Chem. Phys.* 55:1915
20. Busch, G., Cornelius, J. R., Mahoney, R. T., Morse, R. I., Schlosser, D., Wilson, K. 1970. *Rev. Sci. Instrum.* 41:1066
21. Oldman, R. J., Sander, R. K., Wilson, K. 1971. *J. Chem. Phys.* 54:4127
22. Busch, G., Mahoney, R. T., Morse, R. I., Wilson, K. 1969. *J. Chem. Phys.* 51:837
23. Busch, G., Wilson, K. 1972. *J. Chem. Phys.* 56:3626, 3638, 3655
24. Riley, S., Wilson, K. 1972. *Discuss. Faraday Soc.* 53:133
25. Jackels, C. F., Davidson, E. R. 1976. *J. Chem. Phys.* 65:2941
26. Gaedtke, H., Troe, J. 1970. *Z. Naturforsch. Teil A* 25:789
27. Lee, E. K. C., Uselman, W. M. 1972. *Faraday Discuss. Chem. Soc.* 53:125

28. Jones, I. T. N., Bayes, K. D. 1973. *J. Chem. Phys.* 59:4836
29. Solarz, R., Butler, S., Levy, D. H. 1973. *J. Chem. Phys.* 58:5172; Butler, S., Kahler, C., Levy, D. H. 1975. *J. Chem. Phys.* 62:815
30. Creel, C. L., Ross, J. 1976. *J. Chem. Phys.* 64:3560
31. Okuda, M., Wilson, K. R. Unpublished results quoted in (23)
32. Goodeve, C. F., Katz, S. 1939. *Proc. R. Soc. London Ser. A* 172:432
33. Holdy, K. E., Klotz, L. C., Wilson, K. R. 1970. *J. Chem. Phys.* 52:4588
34. See discussion and references given in Section 3.1
35. Basco, N., Norrish, R. G. W. 1962. *Proc. R. Soc. London Ser. A* 268:291
36. Pollack, M. A. 1966. *Appl. Phys. Lett.* 9:94
37. Ling, J. H., Wilson, K. R. 1975. *J. Chem. Phys.* 63:101
38. King, G. W., Richardson, A. W. 1966. *J. Mol. Spectrosc.* 21:339; Rabalais, J. W., McDonald, J. M., Scherr, V., McGlynn, S. P. 1971. *Chem. Rev.* 71:73
39. Baronavski, A. P., McDonald, J. R. 1977. *Chem. Phys. Lett.* 45:172
40. Simons, J. P., Tasker, P. W. 1973. *Mol. Phys.* 26:1267
41. Yeung, E. S., Moore, C. B. 1973. *J. Chem. Phys.* 58:3988
42. Yeung, E. S., Moore, C. B. 1974. *J. Chem. Phys.* 60:2139
43. Houston, P. L., Moore, C. B. 1976. *J. Chem. Phys.* 65:757
44. McQuigg, R. D., Calvert, J. G. 1969. *J. Am. Chem. Soc.* 91:1590
45. Job, V. A., Sethuraman, V., Innes, K. K. 1969. *J. Mol. Spectrosc.* 30:365; 1970. 33:189
46. Jones, V. T., Coon, J. B. 1969. *J. Mol. Spectrosc.* 31:137
47. Callomon, J. H., Innes, K. K. 1963. *J. Mol. Spectrosc.* 10:166
48. Brand, J. C. D., Stevens, C. G. 1973. *J. Chem. Phys.* 58:3331; Brand, J. C. D., Liu, D. S. 1974. *J. Phys. Chem.* 78:2270
49. Jaffe, R. L., Hayes, D. M., Morokuma, K. 1974. *J. Chem. Phys.* 60:5108
50. Jaffe, R. L., Morokuma, K. 1976. *J. Chem. Phys.* 64:4881
51. Hayes, D. M., Morokuma, K. 1972. *Chem. Phys. Lett.* 12:539
52. Yeung, E. S., Moore, C. B. 1971. *J. Am. Chem. Soc.* 93:2059
53. Miller, R. G., Lee, E. K. C. 1975. *Chem. Phys. Lett.* 33:104
54. Sakurai, K., Capelle, G., Broida, H. P. 1971. *J. Chem. Phys.* 54:1412
55. Lewis, R. S., Tang, K. Y., Lee, E. K. C. 1976. *J. Chem. Phys.* 65:2910
56. Miller, R. G., Lee, E. K. C. 1974. *Chem. Phys. Lett.* 27:475
57. Luntz, A. C., Maxson, V. T. 1974. *Chem. Phys. Lett.* 26:553
58. Caplan, C. E., Child, M. S. 1972. *Mol. Phys.* 23:249
59. Light, J. C. 1969. *Faraday Discuss. Chem. Soc.* 44:14
60. Riley, S. J., Wilson, K. R. 1972. *Faraday Discuss. Chem. Soc.* 53:132
61. Holdy, K. E., Klotz, L. C., Wilson, K. R. 1970. *J. Chem. Phys.* 52:4588
62. Shapiro, M., Levine, R. D. 1970. *Chem. Phys. Lett.* 5:499
63. Heidrich, F. E., Wilson, K. R., Rapp, D. 1971. *J. Chem. Phys.* 54:3885
64. Gonzalez, M. A., Karl, G., Watson, P. J. S. 1972. *J. Chem. Phys.* 57:4054
65. Mitchell, R. C., Simons, J. P. 1967. *Faraday Discuss. Chem. Soc.* 44:208
66. Band, Y. B., Freed, K. F. 1974. *Chem. Phys. Lett.* 28:328
67. Band, Y. B., Freed, K. F. 1975. *J. Chem. Phys.* 63:3382
68. Band, Y. B., Freed, K. F. 1975. *J. Chem. Phys.* 63:4479; Band, Y. B., Freed, K. F. 1976. *J. Chem. Phys.* 64:4329
69. Band, Y. B., Freed, K. F. *Energy Distribution in Selected Fragment Vibrations in Dissociation Processes in Polyatomic Molecules.* Preprint
70. Freed, K. F., Band, Y. B. 1977. "Product Energy Distributions in the Dissociation of Polyatomic Molecules." In *Excited States*, ed. E. C. Lim. New York: Academic. In press
71. Abgrall, H., Fiquet-Fayard, F. 1974. *J. Chem. Phys.* 60:11; Fiquet-Fayard, F., Sizun, M., Abgrall, H. 1976. *Chem. Phys. Lett.* 37:72
72. See, for example, (70, 81)
73. Mele, A., Okabe, H. 1969. *J. Chem. Phys.* 51:4798
74. Berry, M. J. 1974. *Chem. Phys. Lett.* 29:329
75. Badger, R. M. 1934. *J. Chem. Phys.* 2:128; 1935. 3:710
76. Johnston, H. S. 1966. *Gas Phase Reaction Rate Theory* Chap. 4. New York: Ronald Press
77. Lawrence, G. M. 1972. *J. Chem. Phys.* 56:3435
78. Lee, L. C., Judge, D. L. 1973. *Can. J. Phys.* 51:378
79. Berry, M. J. 1974. *Chem. Phys. Lett.* 29:323
80. Atabek, O., Beswick, J. A., Lefebvre, R., Mukamel, S., Jortner, J. 1976. *J. Chem. Phys.* 65:4035

81. Atabek, O., Beswick, J. A., Lefebvre, R., Mukamel, S., Jortner, J. 1976. *Mol. Phys.* 31:1
82. Mukamel, S., Jortner, J. 1974. *J. Chem. Phys.* 60:4760
83. Mukamel, S., Jortner, J. 1976. *J. Chem. Phys.* 65:3735
84. Atabek, O., Beswick, J. A., Lefebvre, R. 1975. *Chem. Phys. Lett.* 32:28; 33:228
85. Bernstein, R. B., Levine, R. D. 1974. *Chem. Phys. Lett.* 29:314
86. Baer, M. 1975. *J. Chem. Phys.* 62:4545
87. Kinsey, J. L. 1971. *J. Chem. Phys.* 54:1206
88. Morse, M. D., Freed, K. F., Band, Y. *Rotational Distributions in Photodissociation: Application to ICN.* Preprint
89. Florida, D., Rice, S. A. 1975. *Chem. Phys. Lett.* 33:207
90. Condon, E. U. 1928. *Phys. Rev.* 32:858
91. Slawsky, Z. I., Schwartz, R. N., Herzfeld, K. F. 1972. *J. Chem. Phys.* 20:1591; 1954. 22:767
92. Stretton, J. L. 1965. *Trans. Faraday Soc.* 61:1053
93. Child, M. S. 1970. *J. Mol. Spectrosc.* 33:487
94. Chutjian, A., Segal, G. 1972. *J. Chem. Phys.* 57:3069
95. Zahr, G. E., Preston, R. K., Miller, W. H. 1975. *J. Chem. Phys.* 62:1127; Tully, J. C. 1974. *J. Chem. Phys.* 61:61
96. Lawetz, V., Siebrand, W., Orlandi, G. 1972. *Chem. Phys. Lett.* 16:448
97. Parmenter, C. S. 1972. *Adv. Chem. Phys.* 22:365
98. Callomon, J. H., Lopez-Delgado, R., Parkin, J. E. 1972. *Chem. Phys. Lett.* 13:125
99. Henry, B. R., Siebrand, W. 1968. *J. Chem. Phys.* 49:5369
100. Rice, S. A., McLaughlin, I., Jortner, J. 1968. *J. Chem. Phys.* 49:2756
101. Peters, D. 1964. *J. Chem. Phys.* 41:1046
102. Kay, K. G., Rice, S. A. 1972. *J. Chem. Phys.* 57:3041
103. Lefebvre, R., Beswick, J. A. 1972. *Mol. Phys.* 23:1223
104. Nitzan, A., Jortner, J., Berne, B. J. 1973. *Mol. Phys.* 26:281
105. Kay, K. G. 1974. *J. Chem. Phys.* 60:2370
106. Nitzan, A., Jortner, J., Rentzepis, P. M. 1971. *Mol. Phys.* 22:585
107. Heller, E. J., Rice, S. A. 1974. *J. Chem. Phys.* 61:936
108. Evans, K., Heller, D. F., Rice, S. A., Scheps, R. 1973. *J. Chem. Soc. Faraday Trans.* 2 69:856
109. Hase, W. L., Sloane, C. S. 1976. *J. Chem. Phys.* 64:2256
110. Robinson, P. J., Holbrook, K. A. 1972. *Unimolecular Reactions.* New York: Wiley
111. Gillispie, G. D., Lim, E. C. *Radiationless Decay from the S_1 and T_1 States of Formaldehyde.* Preprint
112. Nitzan, A., Jortner, J. 1973. *Theor. Chim. Acta* 29:97
113. Avouris, P., Gelbart, W. M., El-Sayed, M. A. 1977. *Chem. Rev.* In press
114. Roche, M., Jaffé, H. H. 1974. *J. Chem. Phys.* 60:1193
115. Elert, M. L., Heller, D. F., Gelbart, W. M. Unpublished data
116. Freed, K. F. 1976. *Topics in Applied Physics*
117. Aung, S., Pitzer, R. M., Chan, S. I. 1966. *J. Chem. Phys.* 45:3457
118. Pack, R. T. 1977. *J. Chem. Phys.* 65:4765
119. Hall, R. I., Chutjian, A., Trajmar, S. 1973. *J. Phys. B* 6:L264

LASER-INDUCED FLUORESCENCE ✣ 2651

James L. Kinsey

Department of Chemistry, Massachusetts Institute of Technology, Cambridge, Massachusetts 02139

INTRODUCTION

The discovery of the laser and its subsequent development have profoundly affected the course of experimental research in practically every area of science. The laser's unique capacity to deliver large amounts of energy within extraordinarily small ranges of spectral width and directionality has made possible many kinds of measurement that would otherwise have been quite impossible. Wavelength tunability has proved to be an especially important advance that has not only revolutionized spectroscopy but has also brought the full grandeur of spectroscopic detail into many new areas.

Applications of lasers to various aspects of chemistry, physics, and biology have been the subjects of several recent review articles and monographs (1–7). This article concentrates on a single technique, laser-induced fluorescence (LIF), as a tool for the detection of minute amounts of atoms or molecules in specific quantum states. Applications of LIF to purely spectroscopic questions are not covered unless there is some direct bearing on the main subject matter. The literature survey for this review was completed in January 1977.

Although LIF is still in its infancy as a means for studying molecular collisions, interest in this technique is growing rapidly. Dye lasers with spectral widths in the range 1–10 MHz, which have become commercially available, offer an appealing wealth of resolvable detail. Unlike many other high-resolution methods, LIF even affords a favorable efficiency for detection. Fairbank, Hänsch & Schawlow (8) have recently demonstrated the ability for quantitative detection of sodium atoms by LIF at densities as low as 10^2 atoms cm^{-3}. Intensity estimates for more usual conditions show that LIF compares well in detection efficiency with mass spectrometric detection (9). Densities in the range of 10^4 cm^{-3} have successfully been detected (10). As it may be expected that technology will advance in the laser field more rapidly than in competing means of detection, the trend of such comparisons is likely to favor LIF in the future.

The use of LIF for determining the relative populations of individual quantum states was first realized in two papers from the research group of R. N. Zare (11, 12) that demonstrated the utility of such measurements for investigation of single-collision chemical reactions (11) and for molecular beam diagnostics (12). Early progress in these areas was surveyed in 1974 by Zare & Dagdigian (9).

The basic idea of LIF is simple. Light from a tunable laser is impinged on the sample to be investigated. As the frequency of the laser is changed, molecules within the irradiated portion of the sample will be excited to fluorescence whenever the spectral envelope of the laser overlaps an absorption line of the molecule. Since the only molecules so excited are those in the specific lower level(s) of the corresponding transition, the excitation is selective for this state or group of states. Hence, the "excitation spectrum" obtained by recording the fluorescent intensity as a function of wavelength reflects the relative populations of these different states. The resolution achieved depends on the nature of the absorption bands investigated and the spectral characteristics of the laser.

To be a viable candidate for LIF, a molecule or atom must, first of all, have appreciably populated states from which absorption lines within the wavelength span of the laser can originate. Further, the spectrum of interest must be sufficiently analyzed to permit unambiguous identification of the levels participating at any wavelength. The number of species meeting these requirements is enormous, as reference to various collections of spectroscopic data will demonstrate (13–17). The number of possibilities is somewhat reduced by the additional need for accurate relative line strengths, whose determination becomes a nontrivial part of the experimental task if they are not previously available. Some molecules will be unsuitable because of poor fluorescence yields owing to predissociation in the excited state. The number of atoms or molecules meeting all these criteria is large and destined to increase as the accessible wavelength regions expand and as more careful spectroscopic studies are reported.

MOLECULAR BEAMS

As it is sensible to ask questions about internal-state distributions only when such distributions depart from equilibrium, many of the studies in which LIF is employed involve molecular beams, although some of the examples were carried out in low-pressure static gases. LIF has been used both as a probe of product distributions in molecular reactions and as a diagnostic technique for studying the properties of a single beam. Molecular beams provide a highly desirable environment for detailed studies of the dynamics of elementary bimolecular collisions. The low densities of colliding species guarantee that the reactions investigated are single-collision processes and that the observations are made with products unrelaxed from their nascent distributions. The relative velocities, internal states, and orientation of reactants can be controlled to varying degrees.

These virtues, along with rapidly increasing technical capabilities, have produced a booming effort in the molecular beam field, which has been treated in several recent reviews on the subject (18–22).

Reactive Scattering

One of the main thrusts in studies of reactive collisions has been towards greater levels of specificity in the preparation of reactants and analysis of products. Prior to the appearance of LIF, the main sources of information on product-state distributions were infrared and visible chemiluminescence. Some such data could be obtained in molecular beam studies by electric or magnetic deflection or by electric resonance (see review articles cited above for discussions of these methods). In one kinematically favorable case it was possible to infer partial internal-state information from velocity-angle maps (23).

Bernstein & Faist (24) have compared molecular beam velocity-angle maps to directly resolved product-state distributions (as in LIF) from an information-theoretic point of view. The information content of internal-state distributions, even without observation of angular distributions, was found to be superior to that of velocity-angle maps in many cases. When measurements of angular distributions are combined with product-state determinations, LIF provides substantially greater information content.

Several possible geometries that have been used in molecular beam studies of reactions with LIF are illustrated schematically in Figure 1. In the "beam-beam total" arrangement, the region illuminated by the laser coincides with the scattering region where the two beams cross. Such measurements yield the *total* relative populations of the individual states, integrated over all laboratory scattering angles. As the work of Bernstein & Faist indicates, the information content of the angular distributions is generally not as great as that of the internal-state distribution. At the present stage of theoretical development, it is also true that internal-state distributions can be more directly related to potential-surface features of the process than can angular distributions.

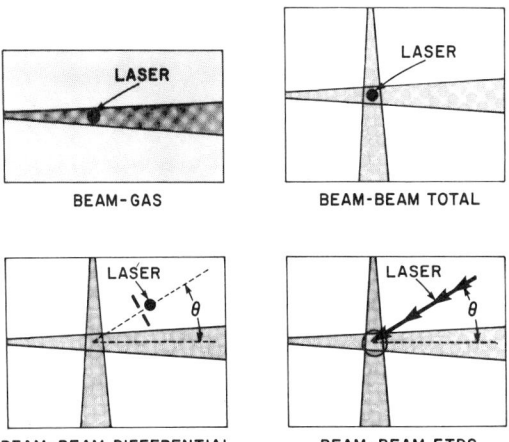

Figure 1 Schematic representation of four experimental configurations for laser-induced fluorescence studies.

The intensity of fluorescence is proportional to the *number* density of absorbers in the illuminated region, but the rate of formation of products is related to the *flux* density. Hence, LIF "total" measurements cannot yield relative rates without the adoption of some assumption as to the velocity distribution. The simplest possibility is to assume that all the molecules have the same magnitude of the laboratory velocity, in which case no correction is required. This can be quite a good approximation if the product molecule detected by LIF is much more massive than its partner, since its velocity will be constrained to lie near the centroid velocity, and all molecules will have a distribution of laboratory velocities roughly equal to the centroid distribution. In other situations, some additional knowledge of the distribution of laboratory velocities is needed for a reliable interpretation of the LIF intensities in terms of relative rates. For reactions whose products are an atom and a diatomic molecule, determination of the internal state of the diatomic establishes the magnitude of the center-of-mass (CM) velocity to within limits set by the spread in initial conditions. In that case, the angular distribution is the remaining unknown, and the conversion may be rather insensitive to reasonable assumptions about it.

Two crossed-beam configurations for measurement of differential cross sections (angular distributions) are indicated in Figure 1. In the first, labeled "beam-beam differential," the region illuminated by the laser beam is displaced from the reaction zone, causing fluorescence to be excited only in those molecules whose velocities are so directed as to carry them into the angular region where the light passes. Variation of the angle of the detector with respect to the colliding beams then allows direct measurement of the angular distribution. Ambiguity as to the direction of the product velocities is thus eliminated, and the magnitude of the velocity can often be determined by "triangulation." When such simple cases do not obtain, or greater precision is required, velocity analysis of the product molecules will be necessary. No such measurements have yet been reported for reactive scattering studies using LIF, but techniques for accomplishing velocity analyses by time-of-flight (TOF) methods (25–28) and Doppler analysis (26, 29) have been reported.

The fourth configuration indicated in Figure 1 is an as yet unimplemented technique that has been proposed for obtaining velocity-angle distributions by LIF (30). The basis of this method is the demonstration that the set of Doppler profiles as functions of the angle of illumination of a sample can be directly inverted into the three-dimensional velocity distribution of the sample. Thus, measurement of the Doppler profiles for fluorescence at a sufficient number of angles is fully equivalent to determination of the velocity-angle distribution. The proposed method, called Fourier transform Doppler spectroscopy (FTDS), affords a potential gain typically in excess of 10^4 in the rate of signal acquisition over other methods for achieving the velocity-angle distribution. This is because *every* molecule in the sample contributes intensity to some part of the Doppler profile no matter what the angle of incidence, whereas only the small fraction of molecules with velocities in a given direction will enter a remote detector located in that direction. Although the technical difficulties of implementing a full-blown application of FTDS are nontrivial at present, the enormity of this signal-gathering advantage makes it an attractive possibility. Moreover, FTDS analysis suggests new and inviting possibilities for measurements in beam-gas configurations, even for measurements in static gases, capable of yielding far more detailed information than was previously thought possible for these experimental situations.

Nozzle Beam Diagnostics

The growth of molecular beam studies of scattering processes is due in large part to the remarkable properties of supersonic beams extracted from hydrodynamic expansions (31–33). Such sources can produce extremely high flux densities in combination with narrow velocity spreads and selective population of the lowest internal states. By "seeding" heavy species in a light carrier, translational energies can be selected over a wide range extending much beyond that of thermal sources. Beams formed from mixtures can also serve to prepare exotic species such as "van der Waals molecules" (34–36).

LIF analyses of the internal-state distributions and velocity distributions in such beams are useful in characterizing the details of beams to be employed in collision studies, and in their own right, since they indicate the collision processes occurring within the beam in the process of its formation.

ANALYSIS OF FLUORESCENT INTENSITIES

The intensity of the fluorescence observed at any wavelength of the laser will be proportional to the population of the state or states from which fluorescence is excited in the monitored region. To extract the populations from the data, however, we require the factors of proportionality, or at least knowledge of how they vary through the spectral range investigated. In the most general case, the relationship between intensity and population can be astonishingly complex. Fortunately, simpler expressions can be derived for many ordinary experimental situations.

The Breit Formula

Unlike simple absorption or emission spectroscopy, fluorescence has a nontrivial polarization dependence, even for an isotropic sample. The general principles underlying polarization and interference effects in fluorescence were established by the early work of Breit (37) on the quantum theory of dispersion of radiation. Subsequent rederivation and interpretation have helped to clarify the results (38–42).

For most LIF measurements in the absence of external fields, "level-crossing" phenomena such as the Hänle effect can be ignored. In that case, excitation from initial level **a** to intermediate level **b** by photons of frequency and polarization v_i and μ, followed by emission of photons of frequency and polarization v_f and μ' to final state **c**, produces an instantaneous rate of fluorescence I (photons $\times$ s^{-1}) given by the following Breit formula:

$$I(\mathbf{a},\mathbf{b},\mathbf{c};\mu\mu') = N_a \rho(v_i,\mu) v_f^3 K(v_f,\mu') S(\mathbf{a},\mathbf{b}) S(\mathbf{b},\mathbf{c})$$
$$\times \sum_{M_b M_{b'}} (-1)^{M_b - M_{b'}} \mathscr{B}^\mu(J_a J_b; M_b M_{b'}) \mathscr{A}^{\mu'}(J_b J_c; M_{b'} M_b), \qquad 1.$$

where N_a is the number of molecules in level **a**, ρ is the radiation density in the incident field, and K is an "apparatus constant" containing the photodetector sensitivity and the like. $S(\mathbf{a}, \mathbf{b})$ and $S(\mathbf{b}, \mathbf{c})$ are the line strengths of the indicated transitions (43–46).

For diatomic molecules, the labels specifying a state **a** include the electronic state (n_a), the vibrational state (v_a), the total angular momentum, and its laboratory-space z-component $(J_a$ and $M_a)$. Any additional labels required for unambiguous specification of the state **a** are denoted by the symbol l_a, which is suppressed where it is not essential for clarity. The general labels **a**, **b**, and **c** are retained for the *set* of quantum numbers for the initial, intermediate, and final state, respectively. In this notation, the definitions of the matrices $\mathscr{A}$ and $\mathscr{B}$ appearing in Equation 1 are:

$$\mathscr{B}^\mu(J_a J_b; M_b M_{b'}) = \sum_{M_a} f(M_a) \begin{pmatrix} J_a & 1 & J_b \\ -M_a & \mu & M_b \end{pmatrix} \begin{pmatrix} J_a & 1 & J_b \\ -M_a & \mu & M_{b'} \end{pmatrix}, \qquad 2.$$

$$\mathscr{A}^{\mu'}(J_b J_c; M_b M_{b'}) = \sum_{M_c} \begin{pmatrix} J_b & 1 & J_c \\ -M_b & \mu' & M_c \end{pmatrix} \begin{pmatrix} J_b & 1 & J_c \\ -M_{b'} & \mu' & M_c \end{pmatrix}, \qquad 3.$$

where $f(M_a)$ is the fraction in magnetic sublevel M_a of state **a**. The coordinate system is defined so that the z-direction is the axis of quantization for the M_a states. For linear polarization along z ($\mu = 0$), for right-circular ($\mu = +1$), or for left-circular ($\mu = -1$) polarization, the quantities in parentheses on the right-hand sides of Equations 2 and 3 are Wigner 3j symbols. For the other two possibilities of linear polarization ($\mu = x$ or $\mu = y$), the quantities that look like 3j symbols are to be interpreted as follows:

$$\begin{pmatrix} J_1 & 1 & J_2 \\ M_1 & x & M_2 \end{pmatrix} = 2^{1/2} \begin{pmatrix} J_1 & 1 & J_2 \\ M_1 & -1 & M_2 \end{pmatrix} - 2^{1/2} \begin{pmatrix} J_1 & 1 & J_2 \\ M_1 & 1 & M_2 \end{pmatrix}, \qquad 4.$$

$$\begin{pmatrix} J_1 & 1 & J_2 \\ M_1 & y & M_2 \end{pmatrix} = 2^{-1/2} \begin{pmatrix} J_1 & 1 & J_2 \\ M_1 & -1 & M_2 \end{pmatrix} + 2^{-1/2} \begin{pmatrix} J_1 & 1 & J_2 \\ M_1 & 1 & M_2 \end{pmatrix}. \qquad 5.$$

Various choices for μ' present so many possibilities that each experimental situation is best treated individually, forming the $\mathscr{A}$ and $\mathscr{B}$ "monitoring operators" (40) that are appropriate for it.

If the detector collects the total fluorescence into all polarization states, and if the M_a sublevels are equally populated, the sums in Equations 1–3 can easily be evaluated. The result, which turns out to be independent of the polarization of the incident light, is given by:

$$I(\mathbf{a}, \mathbf{b}, \mathbf{c}) = K'(v_f)\rho(v_i)N_a B(\mathbf{a}, \mathbf{b})A(\mathbf{b}, \mathbf{c}), \qquad 6.$$

where $B(\mathbf{a}, \mathbf{b})$ and $A(\mathbf{b}, \mathbf{c})$ are the Einstein coefficients for absorption and spontaneous emission for the transitions indicated. This simplified expression has been the basis of most analyses to date, even though the two underlying assumptions were imperfectly justified in some instances. There has been one study of the M_a dependence of the distribution (47), which is discussed in a later section.

Unresolved Initial or Final States

It is not usual in LIF measurements to disperse the fluorescence so that the quantity measured is actually the sum of the intensity over all final states **c**. If the detector efficiency is not significantly dependent on frequency over the range of frequencies emitted from a given intermediate state, Equation 6 becomes even simpler:

$$I(\mathbf{a}, \mathbf{b}) = K'(\bar{v}_f)\rho(v_i)N_a B(\mathbf{a}, \mathbf{b})\tau^{-1}(\mathbf{b}),\qquad 7.$$

where $\tau(\mathbf{b})$ is the radiative lifetime of the intermediate state.

Most of the lasers employed up to now for LIF studies have been pulsed devices having spectral widths on the order of 30–60 GHz. Excitation spectra of molecules with rotational spacings smaller than this appear as unresolved bands. In this event, Equation 1 can reasonably be summed over all the values J_c, l_c, J_b, l_b, J_a, and l_a that belong to the *vibrational* transition ($n_a v_a \rightarrow n_b v_b \rightarrow n_c v_c$). If the line strengths S are then assumed to be given by:

$$S(\mathbf{a}, \mathbf{b}) = R_e^2(n_a, n_b)q(n_a v_a, n_b v_b)\mathscr{S}(J_a l_a; J_b l_b),\qquad 8.$$

where R_e is the electronic transition moment, q the Franck-Condon factor, and $\mathscr{S}$ the general Hönl-London factor. These summations lead to the following simple formula:

$$I(n_a v_a, n_b v_b) = N(n_a v_a)\rho(\bar{v}_i)R_e^2(n_a, n_b)q(n_a v_a, n_b v_b)$$
$$\times \sum n_c v_c K''[\bar{v}_f(n_b v_b, n_c v_c)]R_e^2(n_b, n_c)q(n_b v_b, n_c v_c).\qquad 9.$$

All polarization dependence has disappeared from Equation 9 by virtue of the sums without requiring any assumption about polarization of the incident light, polarization sensitivity of the detector, or isotropy of the distribution.

Beyond the Breit Formula

Apart from possible failure of some of the assumptions that lead to simpler forms of the Breit formula, there are several areas where caution is advised. Since the Breit formula is an instantaneous relationship, the times of excitation and observation must be taken into account. In most instances this will be a straightforward but essential aspect of the analysis (48).

More serious problems can arise at high levels of excitation. The Breit formula was obtained from a first-order treatment and thus ignores the possibility for saturation or optical pumping (49, 50).

The importance of saturation or optical pumping can be estimated from knowledge of the laser intensity and the absolute line strength. However, it is essential to know the spectral and temporal characteristics in some detail for this judgment to be made confidently. Many LIF studies have used the Hänsch-designed (51) nitrogen-laser-pumped dye laser, which produces a spectral output ranging smoothly over a bandwidth of 30–60 GHz. Other kinds of lasers, however, may have pronounced mode structure in the output, i.e. the power may be distributed in a number of sharp spikes located at cavity modes of the dye laser rather than in a smooth envelope. Hence, estimates of saturation or optical pumping based on the *average* power density over the spectral envelope could be seriously in error. There could, for example, be strong saturation within each of the cavity modes with no excitation in between them. Unless the spectral characteristics of the laser are well understood, the best test for saturation and/or optical pumping is evidence that the observed fluorescence varies linearly with incident power.

Finally, we note that the Breit formula only contains *populations*. There is no guarantee that molecules formed in a chemical reaction have not been prepared in coherent superpositions of M-states. This would require analysis in terms of the molecular density matrix. We can also envision incident light and detection methods that require description of the incoming and/or fluorescent radiation in terms of density matrices. The full analysis in terms of density matrices has not been carried out.

CHEMICAL REACTIONS STUDIED BY LIF

Table 1 summarizes all the reactions for which LIF studies have been reported as of this writing—eighteen in total. It is interesting to reflect that, at a similar age, the field of molecular beam reactive scattering had yielded a smaller harvest by an order

Table 1 Summary of chemical reactions studied by LIF[a]

Reaction	Configuration[b]	Comments	References
1. Ba + O_2 → **BaO**(v) + O	B	vib. band heads, rot. struct. almost resolved	11
Ba + O_2 → **BaO**(v, J) + O	A	rot. vib. distributions	52
2. Ba + CO_2 → **BaO**(v, J) + CO	A	rot. vib. distributions	52
3. Ba + SO_2 → **BaO**(v, J) + SO	A	rot. vib. distributions	53
4. Ba + HF → **BaF**(v, T_{rot}) + H	A, B	vib. state dist., plus envelope of rot. bands	54
Ba + $HF^\dagger(v')$ → **BaF**(v) + H	A	state-to-state	55
5. Ba + HCl → **BaCl**(v, T_{rot}) + H	A, B	⎰ vib. state dist., plus	54
Ba + DCl → **BaCl**(v, T_{rot}) + D	A, B	⎱ envelope of rot. bands	
6. Ba + HBr → **BaBr**(v) + H	A, B	vib. state dist.	54
7. Ba + HI → **BaI**(v) + H	A, B	vib. state dist.	54
8. Ba + CH_3I → **BaI**(v) + CH_3	A	vib. state dist.	56
9. Ba + CH_2I_2 → **BaI**(v) + CH_2I	A	vib. state dist.	56
10. Ba + CCl_4 → **BaCl**(v) + CCl_3	B	vib. state dist.	57
11. Ba + BrCN → **BaCN** + Br	A	rad. lifetime of C electronic states	58
Ba + BrCN → **BaBr**(v) + **CN**(v, T_{rot})	A	both products observed	58
12. Ba + LiCl → **BaCl**(v) + Li	B	angular and vib. dist.	10
13. Ba + KCl → **BaCl**(v) + K	B	ang. and vib. dist., width of rot. dist.	59
14. Ca + NaCl → **CaCl**(v) + Na	B	ang. and vib. dists.	60
15. Ca + BrCN → **CaCN** + Br	A	rad. lifetime of $A-B$ and C electronic states	58
16. Sr + BrCN → **SrCN** + Br	A	rad. lifetime of $A-B$	58
→ **SrBr**(v) + **CN**(v, T_{rot})	A	both products obs.	58
17. Al + O_2 → **AlO**(v, J) + O	A	rot. vib. dist.	61
18. H + NO_2 → **OH**(v, J) + NO	B	rot. vib. dist.	48, 62

[a] The product species detected by LIF is indicated by boldface type.
[b] "Beam-gas" and "beam-beam" configurations are indicated, respectively, by A or B.

of magnitude (63). If the expansion of the offshoot parallels that of the parent, a reviewer in this field five years from now will have a formidable literature to cover.

Except for the final two entries in the table, all the reactions studied have been reactions of alkaline-earth metals, mainly barium atoms. A surprising diversity of chemistry and detailed dynamical behavior is exhibited nonetheless. Experimental conditions and levels of refinement of measurement are also diverse, ranging from beam-gas configurations to crossed beams with angular detection, and from unresolved vibrational bands to fully resolved individual lines.

Reactions Producing Metal Oxides

Reactions yielding the oxides BaO and AlO have provided some of the most detailed data on internal-state distributions to be derived from LIF (11, 52, 53, 61). Both BaO and AlO have well-analyzed band systems in readily accessible spectral regions with known Franck-Condon factors. Figure 2 shows the AlO excitation spectrum obtained for the reaction Al + O_2 → AlO + O (61). This spectrum is typical of those exhibiting both vibrational and rotational structure.

Although individual rotational-state peaks are clearly discernible in these spectra, it is not possible to assume the heights of these peaks to be simply proportional to the populations of the corresponding rotational states for several reasons. The spectral width of the laser is too broad to resolve individual lines near the band heads. Also, different branches or bands may have tails extending into other branches or bands, giving a "background" fluorescence that is difficult to eliminate. In the case of AlO, for which the $B^2\Sigma^+ - X^2\Sigma^+$ system is used, there is the additional complication of spin splittings in the spectrum growing increasingly resolvable for higher rotational states with a resulting distribution of the intensity between two lines.

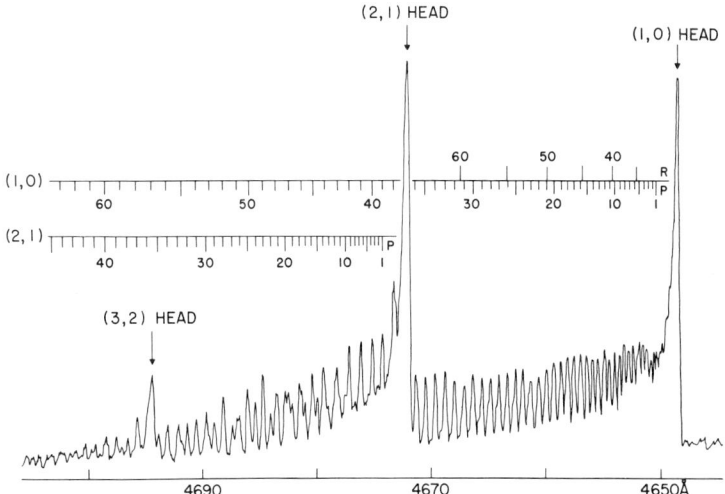

Figure 2 AlO excitation spectrum of the $B^2\Sigma^+ - X^2\Sigma^+ \Delta v = 1$ sequence for the Al + O_2 reaction (61), showing resolved rotational structure. Figure provided by R. N. Zare.

The following successful method has been developed for avoiding these difficulties: From a rough set of populations estimated from the spectrum, a simulated spectrum is generated, by use of the known positions and strengths of the lines. This spectrum is "smeared" over the laser bandwidth and compared with the experimental spectrum. The populations are then iteratively adjusted until satisfactory agreement between the observed and simulated spectrum is achieved.

Analysis of the data for these reactions is further complicated, however, in the conversion from number densities to flux densities. None of the reaction products is strongly constrained by kinematics to lie near the centroid. In the case of Ba + O_2, "conventional" molecular beam studies (64) indicate that the reaction proceeds through a long-lived intermediate complex. The characteristic angular distribution for such complexes can therefore be used to circumvent the difficulty. Unfortunately, no such data are available for Ba + CO_2, Ba + SO_2, or Al + O_2. For those reactions it was necessary to introduce additional assumptions regarding the product's laboratory velocities in the data analysis. The uncertainty in interpretation introduced by these assumptions is difficult to assess quantitatively, but it probably does not affect qualitative trends.

The Ba + O_2 reaction, the first to be studied by LIF (11), turned out upon careful reexamination (52) to involve such high levels of rotational excitation as to prevent use of the iterative analysis described above. Significant populations persisted into levels far beyond those for which reliable spectroscopic data exist. It was nonetheless possible to estimate the rotational-state and vibrational-state populations for this reaction by a more complicated procedure. More than half the available energy was estimated to appear as rotational excitation for products in the $v = 0$ vibrational state. The overall distribution of products showed that about two thirds of the total energy is internal excitation of BaO.

In contrast, the BaO from Ba + CO_2 (52) was found predominantly in low vibrational and rotational states. Despite these substantial differences, both processes seem to occur through an intermediate complex. The internal-state distributions for both reactions are extremely well fit by the phase-space theory (PST) of Pechukas, Light & Rankin (65). The modified transition state theory (TST) of Safron et al (66), on the other hand, failed to agree with the observations either for "loose" or "tight" complexes. There appears to be some disagreement in this respect, since it has been claimed that TST for a "loose" complex is equivalent to PST. There are two extremely useful appendices in (52) that summarize the general features of these two theories.

Smith & Zare (53) concluded that the spin-forbidden reaction Ba(^1S) + SO_2(^1A) → BaO($^1\Sigma^+$) + SO($^3\Sigma^-$) was a direct process, in part on the basis of a tentative report (67) of a strongly forward-peaked angular distribution of the BaO product. Subsequent reinvestigation of the crossed-beam reaction by the same group, however, has yielded compelling evidence of a long-lived intermediate complex (68) in the form of the observation of forwards-backwards symmetry in the CM angular distribution. Less vibrational and rotational excitation of BaO is seen than in the products from Ba + CO_2, but the reaction exothermicity is also smaller. In the absence of computed PST distributions, it is difficult to assess how "nonstatistical" the distribution is. The fact that the spin-forbidden reaction Ba + SO_2 goes at several times the rate of spin-allowed Ba + CO$_2$ is substantiated by both the LIF (53) and crossed-beam (68) studies.

The tentative assignment of Al + O$_2$ as a direct process was made on the basis of disagreement between the observed vibrational and rotational distributions with those computed for PST (61). The discrepancy between the experimental rotational-state distribution for $v = 0$ and $v = 1$ and PST calculations at a single average energy was much diminished, however, by a repeated calculation incorporating an average over initial energies. A similar energy-averaged PST calculation of the vibrational-state distribution gave excellent results for $v = 1$ relative to $v = 0$ but then fell significantly more rapidly with increasing vibrational level than did the experimental distribution. None of the levels for $v \geqslant 2$ has a large population in either the observed or computed distributions, however, so that it is unclear how much these small vibrational populations could be affected by uncertainties in the number density/flux density correction or in other assumptions employed in the analysis. Since the discrepancy with PST is not gross, the decision as to whether Al + O$_2$ is a direct or a complex process should probably be held in abeyance until angular distributions become available.

Interesting by-products of Al + O$_2$ (61, 69) and Ba + O$_2$ (52) studies were improved bounds for the AlO and BaO bond strengths, respectively.

Electron-Jump Reactions: Ba + RX → BaX + R

Barium halides (BaX) can be detected via LIF by their $C^2\Pi$–$X^2\Sigma^+$ emission in the neighborhood of 500 nm. Since the C and X potential curves are virtually identical, the spectra are dominated by the $\Delta v = 0$ sequence. These sequences for the barium halides have extremely small spacings between the bands, and rotational states cannot be resolved with the lasers used. Moreover, the spectroscopic parameters are only poorly known. For BaF and BaCl, sufficient spectroscopic data were available to allow the calculation of Morse-Franck-Condon factors for use in the analysis of LIF spectra (54). This was not possible in the case of BaBr or BaI. The data for BaBr (54) were analyzed by assuming the Franck-Condon factors were constant within a given Δv sequence. This approximation was also used for BaCl in (57). The $\Delta v = 0$ sequence is so predominant in the spectrum of BaI, at least for small v, that the approximation $q(v', v'') = \delta_{v', v''}$ was adopted for its Franck-Condon factors (54, 56). Figure 3 shows the BaI excitation spectrum for the reaction Ba + CH$_3$I (56), which is typical of LIF spectra without resolved rotational structure.

Because of the great disparity between the masses of the two products in the reactions of Ba atoms with hydrogen halogen particles (54), the laboratory velocity of the BaX product is kinematically restricted to a small region near the center-of-mass velocity for any accessible internal state. Such restrictions are highly advantageous in LIF since the number density/flux density conversion becomes trivial. The conversion is less straightforward for the reactions of Ba with CH$_3$I, CH$_2$I$_2$, or CCl$_4$ (56, 57), which have no significant kinematic constraints. Corrections for these reactions are fortunately facilitated by the availability of data on angular distributions from conventional molecular beam studies in each case (64, 70). The crossed-beam studies did not include velocity analysis, however, so the risk of significant errors in

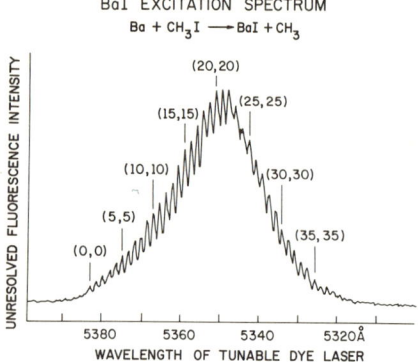

Figure 3 BaI excitation spectrum of the $C^2\Pi_{3/2} - X^2\Sigma^+ \Delta v = 0$ sequence for the Ba + CH_3I reaction (56). Rotational structure is not resolved in this spectrum. Figure provided by R. N. Zare.

the estimates of translational energies (71) must be borne in mind. Schmidt, Siegel & Schultz (57) also discuss the importance of averaging over initial beam conditions.

Table 2 summarizes the conclusions drawn on energy disposal in the reactions of Ba with various RX molecules (54, 56, 57). Analogous reactions of alkali metals with RX have been extensively discussed in terms of an electron-jump ("harpoon") model (63), according to which a sudden transition from ionic to covalent character occurs in the course of reactive collisions. The basic parameters in the electron-jump model are the ionization potential of the metal and the electron affinity of RX. Since the ionization potential of Ba(5.210 eV) lies between those of Li(5.390 eV) and Na(5.138 eV), it is reasonable to examine the Ba atom reactions in terms of the same model.

All the Ba + HX reactions exhibit broad vibrational distribution in the products, reminiscent of the broad translational-energy distribution found in the products of K + HBr, the only alkali-atom plus hydrogen-halide reaction for which such data are available.

Table 2 Energy disposal in Ba + RX → BaX + R reactions

Reaction	Total energy of products (kJ mol^{-1})	$\bar{f}_V$	$\bar{f}_R$	$\bar{f}_T$	$\bar{f}_{int}(R)$	Reference for LIF study
Ba + HF → BaF + H	53	0.12	0.13[a]	0.75	—	54
Ba + HCl → BaCl + H	46	0.28	0.18[a]	0.54	—	54
Ba + DCl → BaCl + D	41	0.29	0.20[a]	0.51	—	54
Ba + HBr → BaBr + H	59	0.36	—	—	—	54
Ba + CCl$_4$ → BaCl + CCl$_3$	165	0.75	0.04[a]	0.06[b]	0.15[c]	57

[a] Based on the width and shape of the envelope of unresolved branches.
[b] Obtained by difference.
[c] Based on the assumption that $E_{int}(CCl_4)$ is the same as found in dissociative electron attachment (73).

Reactions with polyatomic halides have more possibilities for disposal of the energy. LIF results suggest substantial internal excitation of the CCl_3 product resulting from Ba + CCl_4 (57). Initial interpretation of the data for Ba + CH_3I and Ba + CH_2I_2 (56) indicated even greater internal excitation of the R group. However, these conclusions are now in the process of revision following recent work of R. C. Estler of Columbia University showing that the previously assumed BaI dissociation energy was too high by about 120 kJ mol^{-1} (R. N. Zare, private communication). Although this change will affect all the quantitative analyses in reactions producing BaI, the qualitative trend of increasing internal excitation with increasing degree of halogenation still appears to hold up. This is also borne out by the results on Ba + CCl_4 (57), which are unaffected by the erroneous BaI dissociation energy. Analogies between alkaline earth metals and alkali metals have proved extremely instructive in the past. It will be interesting to see how well the revised analyses of the Ba + CH_3I and Ba + CH_2I_2 reactions fit into the scheme of the electron-jump model and Herschbach's photodissociation model (72) for internal excitation of polyatomic reaction products.

M + BrCN Reactions

Pasternack & Dagdigian (58) report LIF studies of the reactions Ca, Sr, and Ba with BrCN. The Ba + BrCN reaction was also studied in crossed beams by Mims, Lin & Herm (74). These reactions are interesting because there are two possible sets of products MBr + CN and MCN + Br. The cyanyl radical CN, the metal halide, and the metal cyanide are all observable by LIF; all were detected for the reactions with Ba and Sr (58). The branching ratio for the rates favors BaCN over BaBr by a factor of 25–100, and the corresponding ratio is at least a factor of 10 greater for SrCN over SrBr. No attempt was made to assess relative vibrational populations for the bromides because of the overlapping, unresolved MCN spectra, which were observed for the first time in these studies. It was possible to distinguish the fluorescence of MBr from that of MCN by lifetime discrimination, however. BaBr was observed in levels up to $v = 10$, and SrBr up to $v = 3$ in this manner. Only the $v = 0$ state of the CN radical was observed. The rotational distributions for the $v = 0$ vibrational state of CN were fit to Boltzmann distributions with $T_{rot}(CN) \sim 1250$ K for the Ba reaction and $T_{rot}(CN) \sim 750$ K for the Sr reaction. Radiative lifetimes were directly determined for two electronic states of CaCN and SrCN and for one electronic state of BaCN.

Angular Distribution of Reaction Products by LIF

Separation of the laser excitation zone from the reaction zone permits the selective detection of molecules reactively scattered in a given direction. LIF measurements of angular distributions in specific vibrational states have been reported for the reactions Ba + LiCl → BaCl + Li (10), Ba + KCl → BaCl + K (59), and Ca + NaCl (60).

Kinematic constraint of the detected product and the broad distribution of initial conditions in the Ba + LiCl reaction prevented the recovery of significant information on its center-of-mass angular distribution. This initial study, however, demonstrated the feasibility of such experiments. It was estimated that densities approximately 10^5 cm^{-3} were detected in the most populated level ($v = 0$). Signals well above the noise level could be seen for states with at least a factor of ten smaller intensity.

Ba + KCl is far more favorable kinematically and more exothermic than the Ba + LiCl reaction ($\Delta H = -40 \pm 1$ kJ mol^{-1} for Ba + KCl; $\Delta H = 4 \pm 12$ kJ mol^{-1} for Ba + LiCl). Since individual rotational states were not resolved, the rotational populations were fit by Boltzmann distributions depending on both vibrational state and laboratory scattering angle. The rotational temperatures, believed to be good to within ± 400 K, indicated large amounts of rotational excitation (59). The laboratory distributions in scattering angle and vibrational state were then simulated by computer averaging of an assumed CM cross section over initial conditions. The CM cross section was assumed to be the product of an angular factor characteristic of an "osculating complex" and a vibrational-state-dependent velocity distribution.

The forward-backward asymmetry of the angular function used in the successful simulation indicated an intermediate complex of mean lifetime roughly equal to one half of a rotational period. No evidence was reported for any vibrational-state dependence of the lifetime. In combination with estimates of the total angular momentum and the moments of inertia of the complex, this led to an estimate of approximately 1.5 ps for the lifetime of the intermediate (59). If RRKM theory (75) is then used to estimate the well depth in the potential surface that would give such a lifetime, a value of about 75 kJ mol^{-1} is obtained. This value compares reasonably with well depths of analogous alkali metal systems.

Use of statistical theories is suspect, however. The total vibrational distribution, obtained by coalescing the excitation and reaction zones, was poorly fit by PST, as was the rotational distribution. PST predicts a vibrational-state distribution that falls off monotonically from $v = 0$, whereas the actual distribution peaks near $v = 7$.

Dagdigian (60) has looked at the CaCl product from Ca + NaCl, using both the $B^2\Sigma^+ - X^2\Sigma^+$ and $A^2\Pi - X^2\Sigma^+$ systems. The $B-X$ system is superior for the determination of vibrational-state distributions. Since band heads occur at widely different values of N in the $^2\Pi_{1/2} - ^2\Sigma^+$ and $^2\Pi_{3/2} - ^2\Sigma^+$ subsystems of the $A-X$ system, it is particularly well suited for estimating the extent of rotational excitation. Lacking resolved rotational lines, Dagdigian fit the rotational distributions to a Boltzmann form. A value 2500 K $\pm$ 500 K was found for T_{rot}, which was independent of scattering angle and vibrational state to within the limits of error. The possibility of errors originating in the nonphysical high-energy "tail" of the Boltzmann distribution was tested by a comparison fit to a "linear surprisal" functional form (76–79). No significant defects were found in the use of the simpler Boltzmann form.

At each scattering angle, the vibrational intensities were estimated by areas under peaks at appropriate band heads, then adjusted until the computed and experimental spectra agreed. Because of the near coincidence of the P_2 head of a given (v, v) band with the P_1 head of the $(v - 1, v - 1)$ band, the intensities reflect $N_v + N_{v-1}$ rather than just N_v.

The product angular distributions were least-squares–fitted to an expansion in Legendre polynomials in the CM coordinate system (80) by use of a computer program that averaged over the (measured) incident beam velocity distributions and spread in beam intersection angles. For $v = 2, 3$ the CM angular distribution peaks at 0° and 180° and is nearly symmetric at about 90°, which strongly suggests a long-lived intermediate complex. The CM angular distribution for $v = 0, 1$ also peaks at 0° and 180°, but the backward peak is only about 65% as high as the forward peak, as would be expected for a complex whose lifetime was comparable to a rotational period.

The overall vibrational-state distribution for the product is in good agreement with PST. PST appears to overestimate the rotational excitation, although it is not clear that the discrepancy between PST and experiment is outside the experimental error limits.

State-to-State Reaction Rates:
$Ba + HF(v = 0, 1) \rightarrow BaF(v = 0-12) + H$

Pruett & Zare (55) have pushed LIF investigation of reaction dynamics close to the ultimate limit in their work on the reaction $Ba + HF \rightarrow BaF + H$. In 1973 the same reaction was studied by use of a thermal HF beam (54). The HF molecules are essentially all in the $v = 0$ vibrational state in such a beam. In the new study the HF beam is pumped by a pulsed HF chemical laser to produce a substantial population of excited $HF^\dagger(v = 1)$. Figure 4 is a schematic representation of the apparatus used. The HF laser was forced to oscillate on a single line, $P(2)$ of the 1–0 transition, which gave resonant excitation at the peak of the Boltzmann rotational distribution.

For both the thermal and the laser-excited HF reactant, relative vibrational populations of the BaCl product were determined by LIF, through use of a "beam-gas" configuration. Since only a small fraction of the HF beam is excited by the laser, the reaction is dominated by $HF(v = 0)$ even when excitation is present. Therefore, it is necessary in the analysis to estimate how many HF molecules actually were pumped to the vibrationally excited state. The radiative lifetime of $HF(v = 1)$ is long enough and the pressures in the experiment low enough that relaxation back to the ground vibrational state by radiative or collisional processes could be neglected.

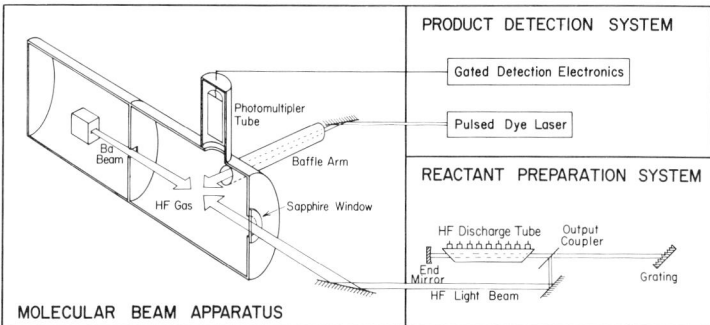

Figure 4 Schematic of the experimental arrangement for the state-to-state investigation of Pruett & Zare (55). Figure provided by R. N. Zare.

Calculation of the fraction excited was complicated by a number of experimental factors that are extremely difficult to evaluate. However, a reasonable estimate of $\beta \geqslant 0.01 \pm 0.005$ was adopted. Figure 5 shows the resulting values of the relative reaction rates as derived from assuming $\beta = 0.01$. The effect of assuming $\beta = 0.002$ is also shown in Figure 5. The extrapolated intensities into $v = 0, 1, 2$ of BaCl from $HF^\dagger(v = 1)$, which could not be obtained from the data, are also shown.

The total product energy in the thermal reaction is 52.7 kJ mol^{-1} of which about 12% is found as vibrational excitation of BaF. In the reaction with HF($v = 1$), about 60% of the additional 46.0 kJ mol^{-1} of energy is retained as vibrational excitation in the products. High retention of vibrational excitation of reactants in the vibrational modes of products has also been observed in a number of infrared chemiluminescence studies (81–83). Blackwell, Polanyi & Sloan (83) have recently reported nearly complete retention of reactant vibrational excitation in the products of $OH^\dagger + Cl \rightarrow HCl^\dagger + O$.

Pruett & Zare (55) suggest two quantitatively different kinds of potential surface that might produce the effects they have seen. One is a smoothly varying surface, such as an LEPS potential, with most of the energy release in the product channel and only slight curvature of the minimum energy path in the region of energy release. The second, which these authors favor, has abruptly changing features, brought about by an electronic switch from covalent to ionic bonding in the course of the reaction (electron-jump model). A narrow constriction in the product channel would, in this picture, tend to force the products into low vibrational states. Vibrationally excited reagents would be able to enter this channel somewhat earlier on the surface and receive part of the repulsive force in a direction that promotes vibrational excitation of products.

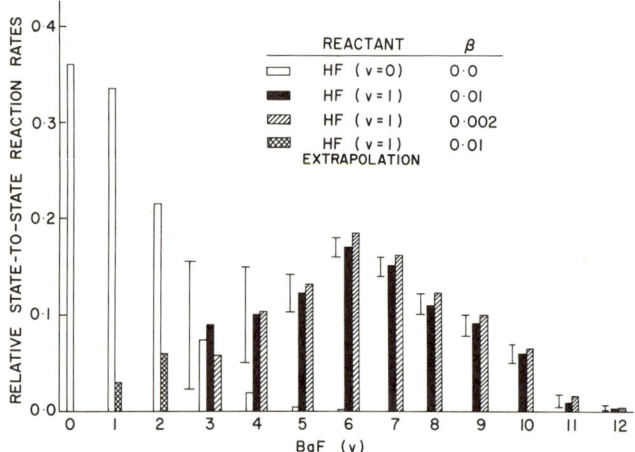

Figure 5 Bar graph showing the state-to-state reaction rates for Ba + HF($v = 0, 1$) → BaF($v = 0$–12) + H. Two calculations are shown for Ba + HF($v = 1$), corresponding to different assumed fractional excitation of HF. Figure provided by R. N. Zare.

Fully Resolved LIF Spectra: $H + NO_2 \rightarrow OH(v, J, N) + NO$

The reaction $H + NO_2 \rightarrow OH + NO$ has been extensively studied by various means, and it continues to present surprises to the investigator. OH is an ideal candidate for LIF because of the well-known and thoroughly documented $A^2\Pi-X^2\Sigma$ spectroscopic system near 300 nm (84) with carefully studied line strengths (85). Since OH is a hydride, it has much larger rotational spacings than the other molecules studied so far by LIF. Accordingly, it has proved possible to resolve individual lines in all the 12 sub-branches that occur for a $^2\Pi-^2\Sigma$ transition (48). Figure 6 shows a part of the LIF spectrum of OH products of $H + NO_2$ recently observed by Silver et al (48).

The conventional view of this reaction, which is widely used as a source of OH radicals for gas kinetic studies, is that OH is produced by it essentially only in $v = 0$ (86, 87). A series of more recent studies, including the LIF work, have shown that $OH(v = 0)$ is actually not even the major product. Haberland, Rohmer & Schmidt's (88) work on the molecular beam angular-velocity distribution showed that only about 24% of the total product energy was translational. The data from these measurements were useful in the number density/flux density conversion in the LIF analysis.

Polanyi & Sloan (89) studied the $H + NO_2$ reaction by infrared chemiluminescence under "arrested relaxation" conditions. They found considerable rotational excitation of OH in the $v = 1, 2, 3$ vibrational states ($v = 0$ cannot be seen in infrared chemiluminescence). From the trend of their vibrational populations, they estimated

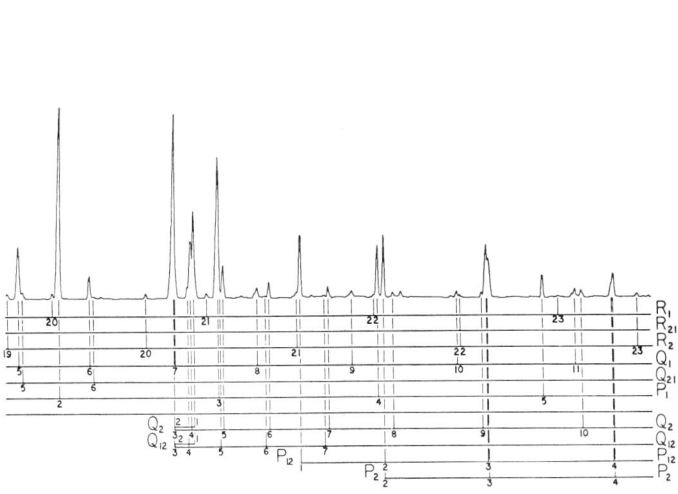

Figure 6 Partial OH excitation spectrum of the $A^2\Sigma - X^2\Pi(0-0)$ sequence for $H + NO_2 \rightarrow OH + NO$ (48). Rotational subbranches are indicated in the notation of Crosswhite & Dieke (84).

about equal rates for the foundation of OH($v = 0$) and OH($v = 1$). Preliminary LIF results of Brophy, Silver & Kinsey (62) gave a lower limit of about 0.17 for $N(v = 1)/N(v = 0)$. The later crossed-beam LIF studies by the same group (48), however, revealed a much higher value $N(v = 1)/N(v = 0) = 1.3 \pm 0.3$, in excellent agreement with the estimate of Polanyi & Sloan. Vibrational levels about $v = 1$ could not be observed by LIF because of predissociation in the $A^2\Sigma$ electronic state. Hence, both LIF and the infrared chemiluminescence are required to produce the full story. Altogether, it is estimated that 40% of the total energy is in vibrational excitation of OH.

Rotational-state populations for OH($v = 0$) were subjected to information-theoretic analysis (77–79). Figure 7 shows the rotational "surprisal" I, defined by:

$$I(f_R|E, f_v) = -\ln\left[P(f_R|E, f_v)/P^o(f_R E, f_v)\right],\qquad 16.$$

where f_R and f_v are, respectively, the fraction of energy in rotational and vibrational excitation of OH. $P(f_R|E, f_v)$ is the experimental probability of finding the value of f_R, given a total energy E and fraction f_v vibrational excitation. P^o is a "prior" microcanonical distribution (90). As Figure 7 shows, the data exhibited a "linear surprisal" behavior with a slope $\lambda = dI/df_R = -2.69$. Features of the H + NO$_2$ surface or surfaces that might lead to the observed behavior are discussed in (48).

The data used in the analysis by (48) were essentially all taken from the R$_2$ branch since this branch was fairly free of coincidences and remained in a relatively short region of wavelengths because of band-head formation. Restriction to R$_2$-branch data prevented testing of possible differences in population of the two spin-states or

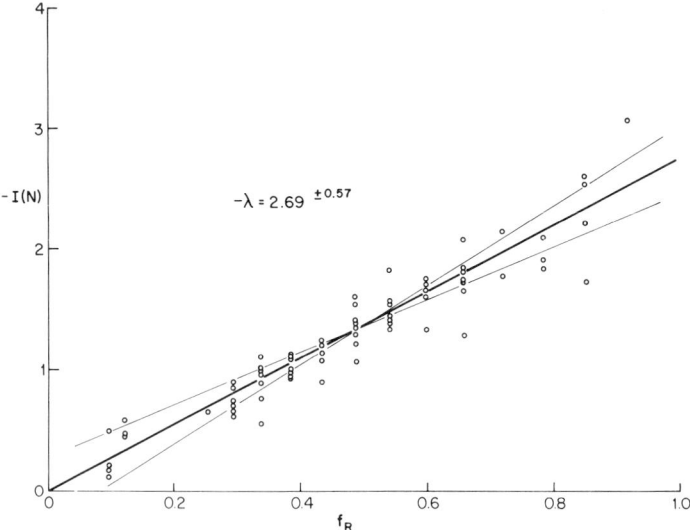

Figure 7 Rotational surprisal plot for the $v = 0$ vibrational state of OH formed in H + NO$_2 \to$ OH + NO (48). The slope $\lambda = -2.69$ corresponds to a negative rotational temperature, $T_{rot} \approx -2940$ K.

members of a Λ-doublet. Reexamination of the spectra at slightly higher resolution should make all the branches useful and may reveal more surprises. Ter Meulen et al (91) have recently found population inversion in the Λ-doublets for low rotational states. It should be interesting to see whether this inversion persists to higher states.

INTERNAL-STATE DISTRIBUTIONS OF PHOTODISSOCIATION PRODUCTS

Another important area in which LIF measurements are useful for measuring internal-state distributions is photodissociation. Jackson, Cody & Sabety-Dzvonik have published a series of LIF investigations of CN radicals produced in vacuum ultraviolet photodissociation of C_2N_2 (92–95), ClCN (96), ICN (97), C_4N_2 (98), C_2HCN (99), and HCN (99). Their measurements were made in a static gas cell at pressures ranging from a few millitorr to a few torr, using the $B^2\Sigma^+-X^2\Sigma^+$ system for the LIF measurements.

Some of the studies were complicated by the possibility that either $CN(X^2\Sigma^+)$ or $CN(A^2\Pi)$, or both, were possible products of the photodissociation. This was checked for by varying the pressure of buffer gas and delay time for detection following the photodissociating flash because of the well-known efficient collisional transfer from $CN(A^2\Pi, v = 0)$ into nearly degenerate $CN(X^2\Sigma^+, v = 4)$ levels.

A summary of the principal findings is contained in Table 3. The photodissociation of C_2N_2 is particularly interesting (95). Adiabatic electronic-state correlation diagrams suggest that the products should be one molecule each of $CN(X)$ and $CN(A)$,

Table 3 Internal-state distributions of $CN(X^2\Sigma^+)$ produced by vacuum ultraviolet photodissociation

Parent Compound	λ	Product of $CN(A^2\Pi)$	v	$T_{rot}(v)$	Relative Population	Comments	References
C_2N_2	160 nm	Equal amts. of X, A	0	1400 K	—	$T_{vib} \sim 2750$ K; 95% in $v = 0, 1$	92–95
—	—	—	1	1100 K	—	Data correct for cascade from A state	—
C_4N_2	160 nm	no A	0	1400 K	0.76	—	98
—	—	—	1	1200 K	0.17	—	—
—	—	—	2	—	0.07	—	—
CH_2CN	⩾145 nm	no A	0	1368 K	0.63	—	99
—	—	—	1	1130 K	0.25	—	—
—	—	—	2	—	0.11	—	—
—	⩾135 nm	small amt. of A	0	1822 K	0.55	—	99
—	—	—	1	1667 K	0.27	—	—
—	—	—	2	—	0.18	—	—
ClCN	⩾160 nm	negl. A	0	—	0.52	Non-Boltz. dist. even at low p, short delays	96
—	—	—	1	—	0.34	—	—
—	—	—	2	—	0.14	Init. rot. dist. for $v = 0$ peaks at $N \sim 66$–70	—
ICN	⩾220 nm	mostly X	—	—	—	$T_{vib} \approx 750$ K from N_1/N_0. Rot. dists. fit by sum of 2 Boltz. dists., with $T_{rot} \approx 400$ K and 2700 K	97
—	⩾145 nm	mostly A	—	—	—	$T_{vib} \sim 1450$ K in A-state	97
CH_3CN	⩾135 nm	mostly A	—	—	—	—	99
HCN	—	—	—	—	—	—	99

a conclusion that is supported by relaxation results obtained by varying the buffer-gas pressure and delay time. The authors also argue that linearity of the ground and excited electronic states of C_2N_2 involved in the photodissociation should require equal and oppositely directed angular momenta in the two CN fragments. This assumption was used with Franck-Condon factors for the CN(A–X) system to correct for radiational cascading of the CN(A) into CN(X).

MOLECULAR BEAM DIAGNOSTICS

LIF has proved a powerful tool for determining properties of supersonic nozzle beams. The earliest measurements of this kind, by Sinha, Schultz & Zare (12), combined white-light fluorescence and LIF excited by different (nontunable) lines of an argon-ion laser to establish the internal-state distribution of alkali dimers in nozzle beams. The Na_2 molecules in a beam formed by expansion through a nozzle of 0.5 mm in diameter of sodium vapor at a stagnation pressure of 50 torr were found to have a vibrational temperature of 153 ± 5 K and a rotational temperature of 55 ± 10 K, both much cooler than the nozzle temperature 920 K. Similar cooling in vibrational and rotational degrees was found at other expansion conditions as well as in supersonic beams of K_2. No evidence was found either for Na_2 or K_2 "vibrationally hot" dimers expected to be found in a stage of the expansion where vibrational relaxation has been frozen.

Later, the alignment of Na_2 molecules in a nozzle beam was investigated by Sinha, Caldwell & Zare (47) by measuring the degree of polarization of laser-excited fluorescence as a function of both the stagnation pressure and the angle between the direction of polarization and that of the molecular beam. From these data, it was possible to determine the coefficients of P_2 and P_4 (relative to P_0) in the Legendre expansion of the probability density function in the variable $\cos\theta$, defined by $\cos\theta = M/[J(J+1)]^{1/2}$. The measurements relied on a coincidence between an absorption line of Na_2 and one of the lines of the argon-ion laser. The results showed that $n(\cos\theta) = 1 - 0.203 P_2(\cos\theta) - 0.14 P_4(\cos\theta)$ for the $v = 3$, $J = 43$ state of $Na_2(X^1\Sigma_g^+)$. A classical model that qualitatively reproduced the observations was based on collisions between hard spheres and hard ellipsoids.

Dagdigian (100) investigated the rotational-state distributions of LiH and NaH in their ground vibrational states in a supersonic beam. The alkali hydride molecules were formed in a source that contained molten alkali metal and several torr of H_2 gas. The vapor in this chamber expanded through a hole 0.055 cm in diameter to form the beam. At a source pressure and temperature of 35 torr and 1250 K, the LiH beam had $T_{rot} = 600 \pm 100$ K. The conditions for NaH were 60 torr and 925 K for the source, 230 ± 50 K for T_{rot}. The exit hole was kept about 100 K hotter than the source temperature in both instances.

Recently, two different methods of velocity analysis (26, 28) employing a tunable laser were used in a study (26) that revealed a strong internal-state dependence in the velocity distributions of Na_2 molecules in nozzle beams. The mean velocity was found to decrease with increasing internal excitation. Higher internal levels exhibited greater ratios of transverse to parallel velocities. This indicates different radial distributions for different internal states.

MISCELLANEOUS APPLICATIONS OF LIF

This section is intended to draw attention to a few additional thought-provoking applications of LIF to various phenomena. No attempt at completeness or fairness of coverage of these topics is claimed.

LIF is a natural method for determining radiative lifetimes of individual vibration-rotation states by direct observation of the decay of fluorescence. Since absolute determinations are notoriously difficult for quantities related to oscillator strengths, such data are extremely welcome. A partial list of molecules and electronic states for which LIF investigations of radiative lifetimes have been reported includes: AlO($B^2\Sigma^+$) (61); the $A^2\Pi$, $B^2\Sigma^+$, and C^2 states of alkaline earth monohalides (101); the B and C states of alkaline earth cyanides (58); the $A'\Sigma^+$ state of ^{7}LiH (100); ^{23}NaH (100) and KH (102); OH($A^2\Sigma^+$) (103–106); SD($A^2\Sigma^+$) (103); C$_2$($d^3\Pi g$) (107); and CN($B^2\Sigma^+$) (108).

Another noteworthy application is the LIF detection of atmospheric OH (109, 110). The direct observation in a molecular beam of internal conversion from the excited singlet state S_1 to vibrationally excited ground-state singlet S_0 in pentacene (111) by LIF is the most straightforward demonstration to date of a radiationless process in individual molecules. The demonstrated ability to detect "sub-part-per-trillion" quantities of organic solutes (112), especially biologically interesting ones, is an application of LIF that is certain to find many uses.

CONCLUSIONS

The impact of laser-induced fluorescence on microscopic studies of collision dynamics has already been felt. In combination with allied and complementary techniques such as molecular beams, infrared and visible chemiluminescence, and laser excitation of reagents, LIF brings the study of molecular reactions to levels of detail only dreamed of a few years ago. The addition of velocity analysis to LIF techniques is an important improvement that can be expected in collisional studies in the near future.

Use of polarization measurements in LIF to probe the anisotropy of molecular distributions is a vast area thus far broached only by the work of Sinha, Caldwell & Zare in Na$_2$ beams (47). These properties contain a great deal of information about reaction dynamics that is not available from any "scalar" measurement (113, 114). It is likely that such measurements will appear before long.

With the rapid advancement of laser technology and the growth of interest in phenomena for which internal-state analyses are pertinent, it is apparent that LIF will continue to occupy an important place in the arsenal of experimental methods, and fluorescence will continue to flourish.

ACKNOWLEDGMENTS

This work was supported in part by grants from the National Science Foundation and the Office of Naval Research.

I am grateful to the many people who sent reprints and preprints of their work and especially to R. N. Zare for helpful conversation and permission to use several of his figures.

Literature Cited

1. Moore, C. B. 1971. *Ann. Rev. Phys. Chem.* 22:387–428
2. Schäfer, F. P., ed. 1973. *Topics in Applied Physics, Vol. 1, Dye Lasers.* New York: Springer. 285 pp.
3. Moore, C. B., ed. 1974. *Chemical and Biochemical Applications of Lasers,* Vol. 1. New York: Academic. 398 pp.
4. Weitz, E., Flynn, G. 1974. *Ann. Rev. Phys. Chem.* 25:275–315
5. Berry, M. J. 1975. *Ann. Rev. Phys. Chem.* 26:259–88
6. Shimoda, K., ed. 1976. *Topics in Applied Physics, Vol. 13, High Resolution Laser Spectroscopy.* New York: Springer. 378 pp.
7. Walther, H., ed. 1976. *Topics in Applied Physics, Vol. 2, Laser Spectroscopy.* New York: Springer. 383 pp.
8. Fairbank, W. M. Jr., Hänsch, T. W., Schawlow, A. L. 1975. *J. Opt. Soc. Am.* 65:199–204
9. Zare, R. N., Dagdigian, P. J. 1974. *Science* 185:739–47
10. Dagdigian, P. J., Zare, R. N. 1974. *J. Chem. Phys.* 61:2464–65
11. Schultz, A., Cruse, H. W., Zare, R. N. 1972. *J. Chem. Phys.* 57:1354–55
12. Sinha, M. P., Schultz, A., Zare, R. N. 1973. *J. Chem. Phys.* 58:549–56
13. Herzberg, G. 1950. *Spectra of Diatomic Molecules.* New York: Van Nostrand. 658 pp.
14. Herzberg, G. 1966. *Electronic Spectra and Electronic Structure of Polyatomic Molecules.* New York: Van Nostrand. 745 pp.
15. Moore, C. E. 1949, 1952, 1958. *Atomic Energy Levels as Derived from the Analyses of Optical Spectra, Natl. Bur. Stand. Circular 467.* Washington DC: GPO
16. Rosen, B., ed. 1970. *Selected Constants—Spectroscopic Data Relative to Diatomic Molecules.* Oxford: Pergamon. 515 pp.
17. Suchard, S. N., ed. 1975. *Spectroscopic Data,* Vol. 1. New York: Plenum; Suchard, S. N., Melzer, J. E., eds. 1976. *Spectroscopic Data,* Vol. 2. New York: Plenum. 585 pp.
18. Kinsey, J. L. 1972. In *Chemical Kinetics,* ed. J. C. Polanyi, pp. 173–212. Oxford: Butterworth–Med. Tech. Press
19. The Chemical Society. 1973. *Faraday Discuss. Chem. Soc., Vol. 55, Molecular Beam Scattering.* 410 pp.
20. Toennies, J. P. 1973. In *Physical Chemistry, an Advanced Treatise, Vol. 6A, Kinetics of Gas Reactions,* ed. W. Jost, pp. 228–381. New York: Academic
21. Farrar, J. M., Lee, Y. T. 1974. *Ann. Rev. Phys. Chem.* 25:357–86
22. Grice, R. 1975. *Adv. Chem. Phys.* 30:247–312
23. Schafer, T. P., Siska, P. E., Parson, J. M., Tully, F. P., Wong, Y. C., Lee, Y. T. 1970. *J. Chem. Phys.* 53:3385
24. Bernstein, R. B., Faist, M. B. 1976. *J. Chem. Phys.* 65:5436–44
25. Gaily, T. D., Rosner, S. D., Holt, R. A. 1976. *Rev. Sci. Instrum.* 47:143–45
26. Bergmann, K., Hefter, U., Hering, P. 1976. *J. Chem. Phys.* 65:488–90
27. Pasternack, L., Dagdigian, P. J. 1977. *Rev. Sci. Instrum.* 48:226–28
28. Bergmann, K., Demtröder, W., Hering, P. 1975. *Appl. Phys.* 8:65–70
29. Hertel, I. V., Hofmann, H., Rost, K. A. 1975. *J. Phys. E* 8:1023–26
30. Kinsey, J. L. 1977. *J. Chem. Phys.* 66:2560–65
31. Anderson, J. B., Andres, R. P., Fenn, J. 1966. *Adv. Chem. Phys.* 10:275
32. Anderson, J. B. 1974. In *Molecular Beams and Low Density Gasdynamics,* ed. P. P. Wegener. New York: Dekker
33. Pauly, H., Toennies, J. P. 1968. In *Methods in Experimental Physics,* ed. B. Bederson, W. L. Fite. 7:227–340. New York: Academic
34. Klemperer, W. 1974. *Ber. Bunsenges. Phys. Chem.* 78:128–34
35. Freeman, R. R., Mattison, E. M., Pritchard, D. E., Kleppner, D. 1974. *Phys. Rev. Lett.* 33:397
36. Smalley, R. E., Levy, D. H., Wharton, L. 1976. *J. Chem. Phys.* 64:3266–76
37. Breit, G. 1933. *Rev. Mod. Phys.* 5:91–140
38. Franken, P. A. 1961. *Phys. Rev.* 121:508–12
39. Rose, M. E., Carovillano, R. L. 1961. *Phys. Rev.* 122:1185–94
40. Carver, T. R., Partridge, R. B. 1966. *Am. J. Phys.* 34:339–50
41. Zare, R. N. 1966. *J. Chem. Phys.* 45:4510–18
42. Alexander, M. H., Dagdigian, P. J., DePristo, A. E. 1977. *J. Chem. Phys.* 66:59–66
43. Tatum, J. B. 1967. *Astrophys. J. Suppl. Ser.* 16:21–55
44. Zare, R. N. 1972. In *Molecular Spectroscopy. Modern Research,* ed. K. N. Rao, C. W. Mathews, pp. 207–22. New York: Academic. 422 pp.
45. Schadee, A. 1967. *J. Quant. Spectrosc. Radiat. Transfer* 7:169–83
46. Allen, C. W. 1973. *Astrophysical Quantities.* London: Athlone. 310 pp.

47. Sinha, M. P., Caldwell, C. D., Zare, R. N. 1974. *J. Chem. Phys.* 61:491–503
48. Silver, J. A., Dimpfl, W. L., Brophy, J. H., Kinsey, J. L. 1976. *J. Chem. Phys.* 65:1811–22
49. Drullinger, R. E., Zare, R. N. 1969. *J. Chem. Phys.* 51:5532–42
50. Drullinger, R. E., Zare, R. N. 1973. *J. Chem. Phys.* 59:4225–34
51. Hänsch, T. W. 1972. *Appl. Opt.* 11:895–98
52. Dagdigian, P. J., Cruse, H. W., Schultz, A., Zare, R. N. 1974. *J. Chem. Phys.* 61:4450–65
53. Smith, G. P., Zare, R. N. 1975. *J. Am. Chem. Soc.* 97:1985–86
54. Cruse, H. W., Dagdigian, P. J., Zare, R. N. 1973. *Faraday Discuss. Chem. Soc.* 55:277–92
55. Pruett, J. G., Zare, R. N. 1976. *J. Chem. Phys.* 64:1774–83
56. Dagdigian, P. J., Cruse, H. W., Zare, R. N. 1976. *Chem. Phys.* 15:249
57. Schmidt, W., Siegel, A., Schultz, A. 1976. *Chem. Phys.* 16:161–73
58. Pasternack, L., Dagdigian, P. J. 1976. *J. Chem. Phys.* 65:1320–26
59. Smith, G. P., Zare, R. N. 1976. *J. Chem. Phys.* 64:2632–40
60. Dagdigian, P. J. 1977. *Chem. Phys.* In press
61. Dagdigian, P. J., Cruse, H. W., Zare, R. N. 1975. *J. Chem. Phys.* 62:1824–33
62. Brophy, J. H., Silver, J. A., Kinsey, J. L. 1975. *J. Chem. Phys.* 62:3820–22
63. Herschbach, D. R. 1966. *Adv. Chem. Phys.* 10:319–96
64. Dixon, D. A., Parrish, D. D., Herschbach, D. R. 1973. *Faraday Discuss. Chem. Soc.* 55:385–87
65. Pechukas, P., Light, J. C., Rankin, C. 1966. *J. Chem. Phys.* 44:794
66. Safron, S. A., Weinstein, N. D., Herschbach, D. R., Tully, J. C. 1972. *Chem. Phys. Lett.* 12:564
67. Herm, R. R., Lin, S.-M., Mims, C. A. 1973. *J. Phys. Chem.* 77:2931
68. Behrens, R. Jr., Freedman, A., Herm, R. R., Parr, T. P. 1976. *J. Am. Chem. Soc.* 98:294–96
69. Zare, R. N. 1974. *Ber. Bunsenges. Phys. Chem.* 78:153–57
70. Lin, S.-M., Mims, C. A., Herm, R. R. 1973. *J. Phys. Chem.* 77:569
71. Rulis, A. M., Bernstein, R. B. 1972. *J. Chem. Phys.* 57:5497
72. Herschbach, D. R. 1973. *Faraday Discuss. Chem. Soc.* 55:233–51
73. DeCorpo, J. J., Bafos, D. A., Franklin, J. L. 1970. *J. Chem. Phys.* 54:1592
74. Mims, C. A., Lin, S.-M., Herm, R. R. 1973. *J. Chem. Phys.* 58:1983
75. Miller, W. B., Safron, S. A., Herschbach, D. R. 1972. *J. Chem. Phys.* 56:3581
76. Ben-Shaul, A., Levine, R. D., Bernstein, R. B. 1974. *J. Chem. Phys.* 57:5427
77. Levine, R. D., Bernstein, R. B. 1974. *Molecular Reaction Dynamics.* New York: Oxford Univ. Press. 250 pp.
78. Levine, R. D., Bernstein, R. B. 1975. In *Modern Theoretical Chemistry,* ed. W. H. Miller, Vol. 3, Part B, pp. 323–60. New York: Plenum
79. Levine, R. D., Johnson, B. R., Bernstein, R. B. 1973. *Chem. Phys. Lett.* 19:1
80. Warnock, T. T., Bernstein, R. B. 1968. *J. Chem. Phys.* 49:1878; 1969. Erratum, *J. Chem. Phys.* 51:4682
81. Ding, A. M. G., Kirsch, L. J., Perry, D. S., Polanyi, J. C., Schreiber, J. L. 1973. *Faraday Discuss. Chem. Soc.* 55:252–76
82. Polanyi, J. C., Schreiber, J. L. 1977. *Faraday Discuss. Chem. Soc.* 62:In press
83. Blackwell, B. A., Polanyi, J. C., Sloan, J. J. 1977. *Faraday Discuss. Chem. Soc.* 62:In press
84. Crosswhite, H. M., Dieke, G. H. 1962. *J. Quant. Spectrosc. Radiat. Transfer* 2:97
85. Crosley, D. R., Lengel, R. K. 1975. *J. Quant. Spectrosc. Radiat. Transfer* 15:579
86. DelGreco, F. P., Kaufman, F. 1961. *J. Chem. Phys.* 35:1895
87. Anderson, J. G., Margitan, J. J., Kaufman, F. 1974. *J. Chem. Phys.* 60:3310
88. Haberland, H., Rohmer, P., Schmidt, K. S. 1974. *Chem. Phys.* 5:298
89. Polanyi, J. C., Sloan, J. J. 1975. *Int. J. Chem. Kinet. Symp.* 1:51–59
90. Kinsey, J. L. 1971. *J. Chem. Phys.* 54:1206
91. Ter Meulen, J. J., Meerts, W. L., van Mierlo, G. W. M., Dymanus, A. 1976. *Phys. Rev. Lett.* 36:1031–34
92. Jackson, W. M. 1973. *J. Chem. Phys.* 59:960–61
93. Jackson, W. M. 1974. *Ber. Bunsenges. Phys. Chem.* 78:190–91
94. Jackson, W. M., Cody, R. J. 1974. *J. Chem. Phys.* 61:4183–85
95. Cody, R. J., Sabety-Dzvonik, M. J., Jackson, W. M. 1977. *J. Chem. Phys.* 66:2145–72
96. Sabety-Dzvonik, M. J., Cody, R. J. 1976. *J. Chem. Phys.* 64:4794–96
97. Sabety-Dzvonik, M. J., Cody, R. J. 1977. *J. Chem. Phys.* 66:125–35

98. Sabety-Dzvonik, M. J., Cody, R. J., Jackson, W. M. 1976. *Chem. Phys. Lett.* 44:131–34
99. Cody, R. J., Sabety-Dzvonik, M. J. 1977. In *Proc. 12th Informal Conf. Photochem., Natl. Bur. Stand.* In press
100. Dagdigian, P. J. 1976. *J. Chem. Phys.* 64:2609–15
101. Dagdigian, P. J., Cruse, H. W., Zare, R. N. 1974. *J. Chem. Phys.* 60:2330–39
102. Cruse, H. W., Zare, R. N. 1974. *J. Chem. Phys.* 60:1182
103. Becker, K. H., Capelle, G., Haacks, D., Tatarczyk, T. 1974. *Ber. Bunsenges. Phys. Chem.* 78:1157–60
104. Brophy, J. H., Silver, J. A., Kinsey, J. L. 1974. *Chem. Phys. Lett.* 28:418–21
105. Hogan, P., Davis, D. D. 1974. *Chem. Phys. Lett.* 29:555–57
106. German, K. H. 1975. *J. Chem. Phys.* 62:2584–87
107. Tatarczyk, T., Fink, E. H., Becker, K. H. 1976. *Chem. Phys. Lett.* 40:126–30
108. Jackson, W. M. 1974. *J. Chem. Phys.* 61:4177–82
109. Baardsen, E. L., Terhune, R. W. 1972. *Appl. Phys. Lett.* 21:209–11
110. Wang, C. C., Davis, L. I., Wu, C. H., Japar, S., Niki, A., Weinstock, B. 1975. *Science* 189:797
111. Sander, R. K., Soep, B., Zare, R. N. 1976. *J. Chem. Phys.* 64:1242–43
112. Bradley, A. B., Zare, R. N. 1976. *J. Am. Chem. Soc.* 98:620–21
113. Case, D. A., Herschbach, D. R. 1975. *Mol. Phys.* 30:1537
114. Case, D. A., Herschbach, D. R. 1976. *J. Chem. Phys.* 64:4212

LIQUIDS OF LINEAR MOLECULES: COMPUTER SIMULATION AND THEORY

✤ 2652

W. B. Streett

Science Research Laboratory, United States Military Academy, West Point, NY 10996

K. E. Gubbins

School of Chemical Engineering, Cornell University, Ithaca, NY 14853

INTRODUCTION

Until about 1970, the literature on computer simulations and equilibrium theory of liquids dealt almost exclusively with simple liquids, that is, liquids composed of spherical molecules with forces acting between the molecular centers. These topics have been the subject of numerous reviews (1–15). Simple liquids are now fairly well understood, mainly as a result of balanced progress in theory, computer simulation, and experiment over the past twenty years. Since the beginning of this decade, liquid state studies have passed beyond their preoccupation with the structure of simple liquids, and have begun to come to grips with other problems.

A variety of liquid state problems have recently been attacked by computer simulation methods: rough sphere fluids (17), nonequilibrium molecular dynamics (18), melting and freezing (19–23, 27), molten salts (24–26), the gas-liquid interface (28–31), fluids of nonspherical molecules (7, 16, 32–72, 104), and others. See Wood & Erpenbeck (1) for more complete references.

Simulations of molecular liquids have proceeded along three fronts: (*a*) simulations of water and aqueous solutions (32–46); (*b*) other simulations of real liquids (56–63, 72, 186, 204); and (*c*) simulations of model liquids (47–55, 64–71, 73). The work on water has been reviewed by Stillinger (74), but the remaining literature in this field has not been comprehensively reviewed. Apart from simulations of ammonia (41), benzene (49, 63), and *n*-butane (57), the literature on the remaining two topics deals almost exclusively with linear molecules—molecules in which the nuclei lie on a straight line—that have an axis of rotational symmetry.

Related developments in the equilibrium theory of molecular liquids have taken place in three areas: (*a*) intermolecular potential functions (13, 75–89); (*b*) theory of the molecular distribution functions (90–101); and (c) approximate theories for estimating physical properties (104–154).

This review is limited to computer simulations of linear molecules and to related developments in intermolecular potential functions and equilibrium statistical mechanics. We further restrict our attention to isotropic, homogeneous, dense fluids. Considerable effort has been devoted to theory and simulations of the dynamic properties of these systems; however, space limitations preclude coverage of this topic.

The division of simulation studies between real and model liquids is a matter of approach rather than substantive difference. All simulations deal with model liquids, inasmuch as they are carried out for systems of molecules of known geometry, that obey precisely defined intermolecular potential functions, and, in most cases, classical mechanics. In studies designed to simulate real liquids, results are compared to experiment, and molecular models and intermolecular potential functions are designed, perhaps modified, to produce the best agreement. This approach is typified by the work of Rahman, Stillinger, and others on water (32–46) and the work of Levesque, Quentrec, Weis, Cheung, and others (56, 58–62, 72, 186) on nitrogen and oxygen. In studies of model liquids, on the other hand, the model is usually chosen for simplicity, and the goal is to understand the structure and properties of hypothetical fluids composed of highly idealized molecules. This approach is typified by work on ellipsoids of revolution (48, 49), prolate spherocylinders (47, 51, 52), hard diatomics (also called hard dumbbells) (50, 53, 55), generalized Stockmayer models (64–71), and atom-atom models of the soft sphere type (54, 56, 58–62, 72, 186, 204).

In many cases the potential models used in simulations also form the basis of approximate solutions to the equations of statistical mechanical theory, and the simulation results then provide unambiguous tests of the mathematical approximations and simplifications that are used. It is in this context that one can view computer simulations as experiments on precisely defined, hypothetical fluids. This seems to us a perfectly reasonable view. Stillinger, however, argues (74) that Monte Carlo and molecular dynamics studies ought to be classified as computer theory. The distinction is clearly in the eye of the beholder.

INTERMOLECULAR POTENTIAL ENERGY MODELS

Although the intermolecular pair potential is now quite accurately known for inert gas (monatomic) molecules, much less is known about potentials for diatomic or polyatomic molecules. As a consequence, computer simulation studies for molecular fluids have assumed pairwise additivity, and have used simplified forms for the pair potential model itself. Both intermolecular and interatomic potentials have been used.

For linear molecules in their ground electronic and vibrational states, the intermolecular pair potential $u(r\omega_1\omega_2)$ depends only on the magnitude of the vector $\mathbf{r}$ (here $r \equiv |\mathbf{r}|$) from the center of molecule 1 to the center of molecule 2, and on the molecular orientations ω_1 and ω_2; here $\omega_i = \theta_i\phi_i$ are the orientations referred to some arbitrary coordinate frame (Figure 1a). Unless otherwise stated, we shall choose

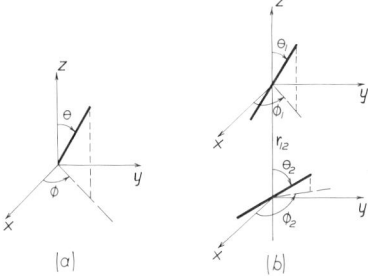

Figure 1 (*a*) The orientation of a line in space can be described by two Euler angles: the polar angle θ and the azimuthal angle ϕ. (*b*) The relative orientation of two molecules in a system in which the line of centers, r_{12}, is chosen to be the polar axis. Since the location in the xy plane of the reference line for measuring the azimuthal angle, ϕ_i, is arbitrary, the relative orientation of molecules 1 and 2 is a function only of the difference $\phi = \phi_1 - \phi_2$.

the polar axis to lie along r (the intermolecular frame) so that u will depend on r, θ_1, θ_2, and ϕ, where $\phi = \phi_1 - \phi_2$ (Figure 1*b*).

Intermolecular Potentials for Rigid Molecules

The intermolecular potential may be classified as multipolar (direct electrostatic), induction, or dispersion at long range, and as charge overlap at short range. The potential models that have been used to simulate systems of rigid molecules may be classed as follows.

GENERALIZED STOCKMAYER MODELS The total potential is written as a sum of parts,

$$u(r\omega_1\omega_2) = u_0(r) + u_{\text{mult}}(r\omega_1\omega_2) + u_{\text{ind}}(r\omega_1\omega_2) + u_{\text{dis}}(r\omega_1\omega_2) + u_{\text{ov}}(r\omega_1\omega_2), \qquad 1.$$

where u_{mult}, u_{ind}, u_{dis}, and u_{ov} are multipolar, induction, anisotropic dispersion, and anisotropic overlap terms, respectively, and u_0 is a spherically symmetric potential that is usually taken to be the Lennard-Jones model:

$$u_0(r) = 4\varepsilon[(\sigma/r)^{12} - (\sigma/r)^6]. \qquad 2.$$

The multipolar, induction, dispersion, and overlap terms are usually approximated by the first few terms in a spherical harmonic expansion. The equations for these terms are given in several reviews (13, 75–81). Computer simulation studies have for the most part been directed towards multipolar forces (dipole-dipole and quadrupole-quadrupole) and anisotropic overlap forces. The dipole-dipole and quadrupole-quadrupole potentials are

$$u_{\mu\mu} = \frac{\mu^2}{r^3}(s_a s_b c - 2 c_a c_b) \qquad 3.$$

and

$$u_{QQ} = \frac{3Q^3}{4r^5}(1 - 5c_a^2 - 5c_b^2 + 17c_a^2 c_b^2 + 2s_a^2 s_b^2 c^2 - 16 s_a s_b c_a c_b c), \qquad 4.$$

respectively, where $s_i = \sin\theta_i$, $c_i = \cos\theta_i$, $c = \cos\phi$, and μ and Q are dipole and quadrupole moments. If the molecular charge distribution is regarded as discrete, the first few multipole moments are given for linear molecules by

$$\mu = \sum_i q_i z_i, \qquad 5.$$

$$Q = \sum_i q_i z_i^2, \qquad 6.$$

$$\Omega = \sum_i q_i z_i^3, \qquad 7.$$

$$\Phi = \sum_i q_i z_i^4, \qquad 8.$$

where q_i is the charge located at z_i, Ω is the octopole moment, and Φ is the hexadecapole moment. [Here the multipole moments are defined as in Stogryn & Stogryn (82).] For symmetrical linear molecules (N_2, Br_2, CO_2, C_2H_2, etc) the odd moments (dipole, octopole, etc) vanish, but in general, for unsymmetrical molecules (HCl, CO, etc) none of the moments vanish (Figure 2). The first nonvanishing multipole moment (dipole or quadrupole for unsymmetric or symmetric molecules, respectively) is independent of the choice of molecular center, but higher moments are strongly affected by this choice. Experimental values of the moments are usually referred to the center of mass (82).

The anistropic overlap forces may be approximated by the first few terms in a spherical harmonic expansion. For symmetrical linear molecules this gives (13, 79, 80)

$$u_{ov} = 4\varepsilon\left(\frac{\sigma}{r}\right)^{12}\delta_2(3\cos^2\theta_1 + 3\cos^2\theta_2 - 2), \qquad 9.$$

while for unsymmetric linear molecules one has

$$u_{ov} = 4\varepsilon\left(\frac{\sigma}{r}\right)^{12}\delta_1(\cos\theta_1 - \cos\theta_2), \qquad 10.$$

where δ_1 and δ_2 are dimensionless overlap parameters that must lie within the ranges $-0.5 \leqslant \delta_1 \leqslant 0.5$ and $-0.25 \leqslant \delta_2 \leqslant 0.5$.

Plots of pair potentials of this form are shown in Figures 6b and 7b as functions of distance for several relative orientations.

Figure 2 Discrete charge models of (*a*) symmetrical and (*b*) unsymmetrical molecules.

SITE-SITE OR ATOM-ATOM MODELS Interactions are assumed to occur between sites in the molecules (e.g. the nuclei) and the total potential is a sum of the site-site interactions (83). Thus we have

$$u(r\omega_1\omega_2) = \sum_{\alpha\beta} u_{\alpha\beta}(r_{\alpha\beta}), \qquad 11.$$

where α and β are sites in two different molecules, $r_{\alpha\beta}$ is the site-site distance, and the sum is over all sites in each of the two molecules. The site-site potential has usually been taken to be the Lennard-Jones model. This potential is shown for several relative orientations in Figure 6a. The sites may also consist of charges, placed so as to reproduce the experimental dipole and/or quadrupole moment.

KIHARA CORE MODEL The potential is taken to be

$$u(r\omega_1\omega_2) = 4\varepsilon\left[\left(\frac{\rho_0}{\rho}\right)^{12} - \left(\frac{\rho_0}{\rho}\right)^6\right], \qquad 12.$$

where $\rho = \rho(\omega_1\omega_2)$ is the minimum distance between impenetrable molecular cores (84, 85). The core may take any shape as long as it is a convex body.

GAUSSIAN OVERLAP MODEL Two interacting spheroids of fixed position and orientation obey a repulsive potential that is directly proportional to the intermolecular overlap volume (86, 87). The mass distribution of each molecule is given by one or more spheroidal Gaussian distributions. The potential then consists of terms of the form

$$u(r\omega_1\omega_2) = 4\varepsilon(\omega_1\omega_2)\left[\left(\frac{\sigma(\omega_1\omega_2)}{r}\right)^{12} - \left(\frac{\sigma(\omega_1\omega_2)}{r}\right)^6\right], \qquad 13.$$

where, for two identical ellipsoids of revolution, the orientation dependence of ε and σ is given by

$$\varepsilon(\omega_1\omega_2) = \varepsilon(1 - X^2 \cos \gamma_{12})^{-1/2}, \qquad 14.$$

$$\sigma(\omega_1\omega_2) = \sigma\left[1 - X\frac{\cos^2\theta_1 + \cos^2\theta_2 - 2X\cos\theta_1\cos\theta_2\cos\gamma_{12}}{1 - X^2\cos^2\gamma_{12}}\right]^{-1/2}, \qquad 15.$$

where γ_{12} is the angle between the symmetry axes of the ellipsoids and X is a shape parameter given by

$$X = \frac{\kappa^2 - 1}{\kappa^2 + 1}, \qquad 16.$$

where κ is the ratio of the lengths of the major and minor axes of the ellipsoid.

OTHER MODELS Various refinements of the above models have been proposed. Thus the Kihara core model has been modified by adding multipole and polarizability terms (85), as has the site-site model for benzene (88).

Each of the above models is highly idealized, and each suffers from certain disadvantages. The generalized Stockmayer potential has the correct form at long range, and is convenient to use for theoretical calculations. Its chief limitation seems to be that for molecules with strongly anisotropic cores the overlap potential cannot be adequately represented by the first few terms of a spherical harmonic expansion. The site-site model is better in this respect, but does not give the correct potential at long range, unless multipolar and polarizability terms are included. This model is convenient for computer simulation, but inconvenient for theoretical work. The site-site model becomes impractical for molecules with many sites; for example, 144 interactions are included in the pair potential for benzene if the atomic nuclei are taken as the sites. In some cases it may be possible to cope with this problem by taking clusters of atoms, rather than a single nucleus, to be the centers of force. The Kihara core model, in addition to giving the long-range potential incorrectly, is limited to convex core shapes, and calculation of the minimum distance between cores is often a complicated matter. The Gaussian overlap model avoids many of the difficulties of the site-site and Kihara models, however it has not yet been extensively studied. MacRury et al (89) have recently reported a comparison of the site-site, Kihara, and Gaussian overlap models for nitrogen, carbon dioxide, and benzene. The three models give equally good fits to the second virial and solid data for these molecules.

Multipolar forces have a large effect on both the structure and equilibrium properties of fluids, and it is reasonable to ask if the multipole series converges rapidly enough to give useful results at separation distances typical of liquids. Figure 3 compares potentials that contain both the full electrostatic interaction (the sum of Coulomb interactions between point charges) and the ideal quadrupole-quadrupole interaction of Equation 4, for a symmetrical linear molecule that has a charge distribution of the type shown in Figure 2a. In spite of the large quadrupole moment, the two results are seen to agree very well for the most probable T-shaped orientation, and are quite good for other orientations. For the T orientation this good agreement holds even when the charge separation is increased to $L^* = 0.8$. Figure 4 shows a similar plot for a dipolar fluid based on the charge distribution of Figure 2b. Again, the molecular center is taken to be the geometric center (the midpoint of the internuclear bond) so that $Q = 0$ (cf Equation 6). The agreement between the exact electrostatic result and the multipole series truncated at the first contributing term is still quite good, although not as good as for the quadrupole case. The agreement becomes poorer, however, as (a) the molecular center is moved away from the

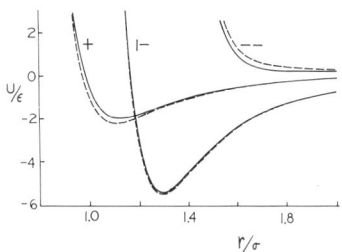

Figure 3 Intermolecular potential curves for symmetrical linear molecules interacting with a diatomic Lennard-Jones, atom-atom potential (with elongation, $L^* = L/\sigma = 0.5471$) plus electrostatic interaction due to the charge distribution of Figure 2a. The charge separation is $L^* = 0.5471$, $Q^* = Q/(\varepsilon\sigma^5)^{1/2} = 2$, where ε and σ are site-site Lennard-Jones parameters, and the molecular center is taken to be the geometric center. Dashed curves are for the Lennard-Jones atom-atom plus full electrostatic potential; full curves are for the Lennard-Jones atom-atom plus quadrupole-quadrupole potential. The curves are for $+$ = crossed ($\theta_1 = \theta_2 = \phi = \pi/2$), $\vdash$ = tee ($\theta_1 = \pi/2$, $\theta_2 = \phi = 0$), and $--$ = end-to-end ($\theta_1 = \theta_2 = \phi = 0$) orientations.

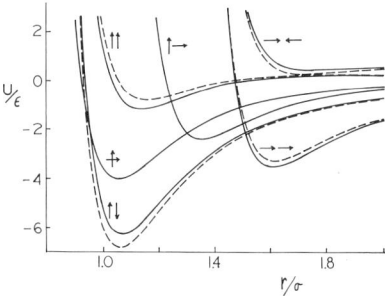

Figure 4 Intermolecular potential curves for unsymmetrical molecules interacting with a diatomic Lennard-Jones, atom-atom potential ($L^* = 0.5471$) plus electrostatic interaction due to the charge distribution of Figure 2b. The charges are at the Lennard-Jones sites (charge separation = $L^* = 0.5471$) and the molecular center is chosen to be the geometric center; the reduced multipole moments are $\mu^* = \mu/(\varepsilon\sigma^3)^{1/2} = 2$, $Q^* = 0$. Dashed curves include the full electrostatic potential, full curves include the dipole-dipole term only. Arrows represent dipole orientations.

geometric center of the molecule, and (b) the separation between the charges is increased.

Central Force Models

A second approach to intermolecular forces in molecular fluids is through the use of central force models that consist of both inter- and intramolecular potentials. In its most general form (40, 41), the central force model consists of spherically symmetric, molecule-independent, interatomic potential functions, designed not only to produce atom combinations of the correct composition and geometry, but also to give the correct intermolecular forces. Models of this form have been used in computer simulations of water (40) and ammonia (41). The central force method has also been used in a form consisting of separate inter- and intramolecular potentials (7), and in a form in which intramolecular bond lengths and angles are accounted for by mathematical constraints imposed on the equations of motion through the use of Lagrange multipliers (205). These potentials are not only more realistic physically, since they permit studies of molecules with internal degrees of freedom, but they are sometimes more efficient for use in molecular dynamics because they reduce the combined translational and rotational motion of molecules to simple translational motion of the component atoms. They are discussed further in the section on molecular dynamics. It remains to be seen whether potentials of this type can provide reasonable approximations to real intermolecular potential functions. The early results (40, 41) are promising.

STATISTICAL MECHANICS

In this section we briefly review topics in the statistical mechanics of molecular liquids that are pertinent to computer simulation studies. The following paragraph introduces the molecular distribution functions; subsequent paragraphs summarize the equations relating physical properties to the distribution functions, and outline the theories that have been compared with simulation results. The canonical ensemble is used throughout (fixed N, V, T). We restrict our attention to isotropic, homogeneous, dense fluids of linear molecules, and we further assume that vibrational motions within these molecules are dynamically and statistically independent of the center of mass and rotational motions.

Molecular Distribution Functions

The angular correlation function of order h, $g(\mathbf{r}^h\omega^h)$, is proportional to the probability density for observing any set of h molecules in the configuration $(\mathbf{r}^h\omega^h)$, where $(\mathbf{r}^h\omega^h)$ is a shorthand for $(\mathbf{r}_1\mathbf{r}_2\ldots\mathbf{r}_h\omega_1\omega_2\ldots\omega_h)$. It is given by

$$g(\mathbf{r}^h\omega^h) = \frac{\Omega^h}{\rho^h}\frac{N!}{(N-h)!}\int d\mathbf{r}^{N-h}\,d\omega^{N-h}\exp\left[-\beta\mathcal{U}(r^N\omega^N)\right]/Z_c, \qquad 17.$$

where $d\omega = \sin\theta\,d\theta\,d\phi$, $\Omega = \int d\omega = 4\pi$, $\rho = N/V$ is the number density, $\beta = 1/kT$, $\mathcal{U}(r^N\omega^N)$ is the total intermolecular potential energy, and

$$Z_c = \int d\mathbf{r}^N\,d\omega^N\exp\left[-\beta\mathcal{U}(r^N\omega^N)\right] \qquad 18.$$

is the configurational partition function (13). The notation $d\mathbf{r}^{N-h}\,d\omega^{N-h}$ means $d\mathbf{r}_{h+1}\,d\mathbf{r}_{h+2}\ldots d\mathbf{r}_N\,d\omega_{h+1}\,d\omega_{h+2}\ldots d\omega_N$. For gases $g(\mathbf{r}^h\omega^h) = 1$, and its departure from unity in other cases is a measure of the effect of intermolecular forces on the molecular distribution. The angular pair correlation function, $g(r\omega_1\omega_2) = g(r\,\theta_1\theta_2\phi)$, where r is the center-to-center distance, plays a central role in the pairwise additivity theory of liquids of linear molecules.

Correlation functions for molecular centers, irrespective of orientations, are obtained by averaging the corresponding angular function over orientations. These functions, sometimes called partial correlation functions, are defined by

$$g(r^h) = \frac{1}{\Omega^h}\int d\omega^h g(\mathbf{r}^h\omega^h) \equiv \langle g(\mathbf{r}^h\omega^h)\rangle_{\omega^h}, \qquad 19.$$

where the notation

$$\langle(\ldots)\rangle_{\omega^h} = \frac{1}{\Omega^h}\int d\omega_1\cdots d\omega_h(\ldots) \qquad 20.$$

is used to represent an unweighted average over orientations. The centers pair correlation function, $g(r)$, is termed the radial distribution function.

One often introduces other pair distribution functions that, while giving information equivalent to $g(r\omega_1\omega_2)$, are more convenient for some purposes. These include the direct correlation function, the site-site correlation functions, and the spherical harmonic coefficients $g_{ll'm}(r)$.

The direct correlation function $c(r\omega_1\omega_2)$ is defined by

$$g(r\omega_1\omega_2) - 1 = c(r\omega_1\omega_2) + \rho\int d\mathbf{r}_3\langle c(r_{13}\omega_1\omega_3)[g(r_{32}\omega_3\omega_2) - 1]\rangle_{\omega_3}. \qquad 21.$$

which is the generalization of the Ornstein-Zernike equation to nonspherical molecules; the range of $c(r\omega_1\omega_2)$ is approximately that of the pair potential.

Another type of partial correlation that is independent of orientation is the site-site pair correlation function $g_{\alpha\beta}(r_{\alpha\beta})$. It is proportional to the probability of finding the β-site of a molecule at a distance $r_{\alpha\beta}$ from the α-site of a different molecule. The sites may be the nuclei (as in the theory of neutron diffraction) or may be intermolecular force sites at arbitrary locations within the molecules. The site-site correlation function is given by (79, 91)

$$g_{\alpha\beta}(r_{\alpha\beta}) = \langle g(|\mathbf{r}_{\alpha\beta} + \mathbf{r}_{c\alpha 1} - \mathbf{r}_{c\beta 2}|, \omega_1\omega_2)\rangle_{\omega_1\omega_2}, \qquad 22.$$

where $\mathbf{r}_{c\alpha i}$ is the vector from the center of molecule i to the α-site, so that $(\mathbf{r}_{\alpha\beta} + \mathbf{r}_{c\alpha 1} - \mathbf{r}_{c\beta 2}) = \mathbf{r}_{12}$. The physical interpretation of Equation 22 is that $g_{\alpha\beta}(r_{\alpha\beta})$ is obtained by averaging $g(r_{12}\omega_1\omega_2)$ over the orientations ω_1 and ω_2, keeping $r_{\alpha\beta}$ fixed. We note that the centers pair correlation function $g(r)$ is a special case of the site-site function. The function $g(r\omega_1\omega_2)$ generally contains more information than the set of $g_{\alpha\beta}(r_{\alpha\beta})$ functions.

The angular pair correlation function can be expanded in spherical harmonics:

$$g(r\omega_1\omega_2) = 4\pi \sum_{ll'm} g_{ll'm}(r) Y_{lm}(\omega_1) Y_{l'\bar{m}}(\omega_2), \qquad 23.$$

where Y_{lm} is a spherical harmonic (92), $g_{ll'm}(r)$ is the harmonic coefficient, a function of distance r, and $\bar{m} \equiv -m$. Unless otherwise noted, r is the distance between molecular centers. It is convenient to introduce the factor 4π on the right-hand side of Equation 23, as then the leading harmonic coefficient, $g_{000}(r)$, is equal to the radial distribution function $g(r)$. The distance r can be any uniquely defined intermolecular distance. If it is taken to be a particular site-site distance, $r_{\alpha\beta}$, then $g_{\alpha\beta}(r_{\alpha\beta}) = g_{000}(r_{\alpha\beta})$; that is, the site-site functions can be viewed as the leading terms in spherical harmonic expansions based on distances other than the center-center distance.

Explicit expressions for the harmonic coefficients are obtained by exploiting the orthogonality properties of spherical harmonics. If Equation 23 is multiplied by the complex conjugate of the harmonic of a particular rank (i.e. a particular set of the indices $ll'm$), and integrated over all angular space, terms containing all other ranks are zero, and one has

$$g_{ll'm}(r) = \frac{1}{4\pi} \int\int g(r\omega_1\omega_2) Y^*_{lm}(\omega_1) Y^*_{l'\bar{m}}(\omega_2) \, d\omega_1 \, d\omega_2. \qquad 24.$$

If Equation 24 is divided by $\int\int g(r\omega_1\omega_2) \, d\omega_1 \, d\omega_2$, the right-hand side is the formal definition of the ensemble average of the product $Y^*_{lm}(\omega_1) Y^*_{l'\bar{m}}(\omega_2)$. Since $\int\int g(r\omega_1\omega_2) \, d\omega_1 \, d\omega_2 = 16\pi^2 g_{000}(r)$, it follows (53) that

$$g_{ll'm}(r) = 4\pi g_{000}(r) \langle Y^*_{lm}(\omega_1) Y^*_{l'\bar{m}}(\omega_2)\rangle_{\omega_1\omega_2}, \qquad 24a.$$

where the average is taken in a spherical shell of thickness Δr, with midpoint at r. Equation 24a provides a powerful, efficient means of calculating the $g_{ll'm}$ in computer simulations (53, 54).

Expressions for Physical Properties

The configurational Helmholtz free energy is related to intermolecular forces by

$$A = -kT \ln Z. \qquad 25.$$

Expressions for the configurational internal energy and pressure may be obtained from Equation 25 by using $U = (\partial \beta A/\partial \beta)$ and $P = -(\partial A/\partial V)$. Assuming pairwise additivity, this gives

$$U = \left\langle \sum_{i<j} u(r_{ij}\omega_i\omega_j) \right\rangle = \tfrac{1}{2}\rho N \int d\mathbf{r} \langle u(r\omega_1\omega_2) g(r\omega_1\omega_2) \rangle_{\omega_1\omega_2}, \qquad 26.$$

and

$$P = \rho kT - \frac{1}{3V} \left\langle \sum_{i<j} r_{ij} \frac{\partial u(r_{ij}\omega_i\omega_j)}{\partial r_{ij}} \right\rangle$$

$$= \rho kT - \tfrac{1}{6}\rho^2 \int d\mathbf{r} \left\langle r \frac{\partial u(r\omega_1\omega_2)}{\partial r} g(r\omega_1\omega_2) \right\rangle_{\omega_1\omega_2}. \qquad 27.$$

The configurational specific heat is given by

$$C_v = \frac{1}{kT^2} \left[\left\langle \left\{ \sum_{i<j} u(r_{ij}\omega_i\omega_j) \right\}^2 \right\rangle - \left\langle \sum_{i<j} u(r_{ij}\omega_i\omega_j) \right\rangle^2 \right]$$

$$= \frac{\partial U}{\partial T} = \tfrac{1}{2}\rho N \int d\mathbf{r} \left\langle u(r\omega_1\omega_2) \frac{\partial g(r\omega_1\omega_2)}{\partial T} \right\rangle_{\omega_1\omega_2}. \qquad 28.$$

In each of the above three equations, the first form of the expression gives the property in terms of the average over a canonical ensemble, represented by $\langle \ldots \rangle$, while the second form is the expression in terms of the angular pair correlation function; $\langle \ldots \rangle_{\omega_1\omega_2}$ is defined in Equation 20.

For a system of two fluid phases (e.g. gas/liquid or isotropic liquid/liquid-crystal phases) with a plane interface lying in the xy plane, the interfacial tension γ is given by (93, 94)

$$\gamma = \frac{1}{2S} \left\langle \sum_{i<j} \left[r_{ij} \frac{\partial u(r_{ij}\omega_i\omega_j)}{\partial r_{ij}} - 3z_{ij} \frac{\partial u(r_{ij}\omega_i\omega_j)}{\partial z_{ij}} \right] \right\rangle$$

$$= \frac{\Omega^2}{4} \int_{-\infty}^{\infty} dz_1 \int d\mathbf{r} \left\langle f(z_1 r_{12}\omega_1\omega_2) \left\{ r_{12} \frac{\partial u(r_{12}\omega_1\omega_2)}{\partial r_{12}} \right. \right.$$

$$\left. \left. - 3z_{12} \frac{\partial u(r_{12}\omega_1\omega_2)}{\partial z_{12}} \right\} \right\rangle_{\omega_1\omega_2}, \qquad 29.$$

where S is the surface area and $f(zr_{12}\omega_1\omega_2)$ is the inhomogeneous pair distribution function. In the Fowler-Kirkwood-Buff model, in which an abrupt transition layer is assumed, this formula reduces to

$$\gamma^F = \frac{1}{16V}\left\langle \sum_{i<j} r_{ij}^2 \frac{\partial u(r_{ij}\omega_i\omega_j)}{\partial r_{ij}} \right\rangle$$

$$= \frac{1}{32}\rho^2 \int \mathrm{d}\mathbf{r}\, r^2 \left\langle \frac{\partial u(r\omega_1\omega_2)}{\partial r} g(r\omega_1\omega_2) \right\rangle_{\omega_1\omega_2}, \quad 30.$$

while the superficial internal energy u_s is given in this model by (94)

$$u_S^F = -\frac{1}{4V}\left\langle \sum_{i<j} r_{ij} u(r_{ij}\omega_i\omega_j) \right\rangle$$

$$= -\tfrac{1}{8}\rho^2 \int \mathrm{d}\mathbf{r}\, r \left\langle u(r\omega_1\omega_2)g(r\omega_1\omega_2) \right\rangle_{\omega_1\omega_2}. \quad 31.$$

In Equation 29 the averaging is over the states for a two phase system, whereas in Equations 30 and 31 it is over the homogeneous liquid phase only; thus in the latter equations both ρ and $g(r\omega_1\omega_2)$ are the homogeneous liquid values.

The structure factor $S(q)$ for molecular liquids, as determined by neutron diffraction, is given by a sum of intramolecular and intermolecular parts (79, 95–97),

$$S(q) = S_{\text{intra}}(q) + S_{\text{inter}}(q), \quad 32.$$

where

$$S_{\text{intra}}(q) = \left(\sum_{\alpha} \bar{b}_\alpha\right)^{-2} \sum_{\alpha\beta} \bar{b}_\alpha \bar{b}_\beta \frac{\sin(qr_{\alpha\beta})}{qr_{\alpha\beta}}, \quad 33.$$

and

$$S_{\text{inter}}(q) = \frac{2}{N}\left\langle \sum_{i<j} \exp(i\mathbf{q}\cdot\mathbf{r}_{ij}) F(\mathbf{q}\omega_i) F(\mathbf{q}\omega_j)^* \right\rangle$$

$$= \rho \int \mathrm{d}\mathbf{r}\, \exp(i\mathbf{q}\cdot\mathbf{r}) \langle g(r\omega_1\omega_2) F(\mathbf{q}\omega_1) F(\mathbf{q}\omega_2)^* \rangle_{\omega_1\omega_2}. \quad 34.$$

In these equations $\bar{b}_\alpha$ is the mean coherent scattering length for nucleus α, $r_{\alpha\beta}$ is the internuclear $\alpha\beta$ distance for a given molecule, and $F(\mathbf{q},\omega_i)$ is proportional to the scattering amplitude for molecule i in the orientation ω_i and is given by

$$F(\mathbf{q}\omega_i) = \left(\sum_{\alpha} \bar{b}_\alpha\right)^{-1} \sum_{\alpha} \bar{b}_\alpha e^{i\mathbf{q}\cdot\mathbf{r}_{\alpha i}}. \quad 35.$$

If we assume that the electron density distribution is rigidly coupled to the nucleus, the X-ray scattering formulae are the same as Equations 32 to 35, provided we interpret the indices α, β to denote electrons rather than nuclei. (Some authors subtract the forward scattering term from Equation 34; the function $g(r\omega_1\omega_2)$ is then replaced by $g(r\omega_1\omega_2) - 1$.)

The mean-squared torque $\langle \tau_1^2 \rangle$ and mean-squared force $\langle F_1^2 \rangle$ on molecule one are given by (66, 98)

$$\langle \tau_1^2 \rangle = \frac{2}{N} kT \left\langle \sum_{i<j} \nabla_{\omega_1}^2 u(r_{ij}\omega_i\omega_j) \right\rangle$$

$$= \rho kT \int d\mathbf{r} \langle \nabla_{\omega_1}^2 u(r\omega_1\omega_2) g(r\omega_1\omega_2) \rangle_{\omega_1\omega_2}, \qquad 36.$$

and

$$\langle F_1^2 \rangle = \frac{2}{N} kT \left\langle \sum_{i<j} \nabla_1^2 u(r\omega_i\omega_j) \right\rangle$$

$$= \rho kT \int d\mathbf{r} \langle \nabla_1^2 u(r\omega_1\omega_2) g(r\omega_1\omega_2) \rangle_{\omega_1\omega_2}, \qquad 37.$$

where $\nabla_{\omega_1}^2 u(r\omega_1\omega_2)$ is the angular Laplacian and $\nabla_1^2 u(r\omega_1\omega_2) = (2/r)\partial u(r\omega_1\omega_2)/\partial r + \partial^2 u(r\omega_1\omega_2)/\partial r^2$ is the radial Laplacian.

Finally, the Kirkwood theory (99–101) for the dielectric and Kerr constants introduces the angular correlation parameters G_l, where

$$G_l = \frac{2}{N} \left\langle \sum_{i<j} P_l(\cos \gamma_{ij}) \right\rangle$$

$$= \rho \int d\mathbf{r} \langle P_l(\cos \gamma) g(r\omega_1\omega_2) \rangle_{\omega_1\omega_2}. \qquad 38.$$

Here, γ_{ij} is the angle between the axes of molecules i and j, and P_l is the lth Legendre polynominal: $P_1(x) = x$, $P_2(x) = \frac{1}{2}(3x^2 - 1)$, etc. The above equations express equilibrium properties in terms of averages over a canonical ensemble, and it is usual to assume that computer simulations carried out in other ensembles will give thermodynamically equivalent results. Lebowitz et al (102) have discussed the analysis of simulation results from various ensembles. The assumption of thermodynamic equivalence appears to hold provided N is large, with the exception of the specific heat, for which Equation 28 requires modification in other ensembles (102, 103).

Approximate Theories

Theroretical evaluations of the equations given in the two proceeding sections are based on approximations for either the correlation functions or the partition function. Such approximations may be classified as (a) integral equation methods, (b) scaled particle theory, or (c) perturbation theories (13, 104). Each of these is discussed briefly below.

Integral equation methods include the Percus-Yevick (PY) approximation, the mean spherical approximation (MSA), and the reference interaction site model (RISM). Each of these involve a correction for the direct correlation function, which is substituted into the Ornstein-Zernike equation. The resulting integral equation is then solved for the pair correlation function [$g(r\omega_1\omega_2)$ in the case of PY and MSA, the site-site $g_{\alpha\beta}$ in the case of RISM]. The PY approximation is

$$c(r\omega_1\omega_2) = g(r\omega_1\omega_2)\{1 - \exp[\beta u(r\omega_1\omega_2)]\} \qquad 39.$$

and provides (105) a consistent way of requiring that $c(r\omega_1\omega_2)$ has the same range as $u(r\omega_1\omega_2)$. Chen & Steele (106) and Morrison (107) have solved the PY equation for site-site hard sphere and Lennard-Jones potential models, respectively, by expanding $g(r\omega_1\omega_2)$ in spherical harmonics and truncating the resulting set of coupled equations. The MSA theory is applicable to fluids with hard-core potentials of the type

$$u(r\omega_1\omega_2) = \infty, \qquad r < \sigma,$$
$$ = u_t(r\omega_1\omega_2), \qquad r > \sigma, \qquad 40.$$

where u_t is the potential "tail." In the MSA c is approximated by (108):

$$c(r\omega_1\omega_2) = -\beta u_t(r\omega_1\omega_2), \qquad r > \sigma, \qquad 41.$$

while for $r < \sigma$

$$g(r\omega_1\omega_2) = 0, \qquad r < \sigma. \qquad 42.$$

The MSA integral equation is obtained by substituting Equations 41 and 42 into Equation 21. Wertheim (109) has solved the MSA exactly for hard spheres plus dipole-dipole interaction. His work has been extended to higher multipoles (110–112), short-range angle-dependent overlap forces (111, 112), and mixtures (113, 114). Generalizations of the MSA have also been given (115, 116). The RISM theory was proposed by Chandler & Andersen (117) for fluids in which the molecules are represented as fused hard spheres, so that the potential is of the form of Equation 11. A site-site direct correlation function $c_{\alpha\beta}(r_{\alpha\beta})$ is defined by means of a set of equations analogous to Equation 21 (the Ornstein-Zernike equation). These equations are then solved by introducing the RISM closure approximation:

$$c_{\alpha\beta}(r_{\alpha\beta}) = 0, \qquad r_{\alpha\beta} > \sigma_{\alpha\beta}, \qquad 43.$$

together with

$$g_{\alpha\beta}(r_{\alpha\beta}) = 0, \qquad r_{\alpha\beta} < \sigma_{\alpha\beta}, \qquad 44.$$

where $\sigma_{\alpha\beta}$ is the hard-core diameter between atoms α and β. In this model the sites are the centers of the hard spheres. In the limit when only one hard sphere is involved, the RISM reduces to the PY compressibility approximation for hard spheres, while for the general case it is an obvious extension of PY theory. The derivation of the RISM has been discussed in the literature (118–121), and it has been applied to the calculation of both the site-site correlation functions and the static structure factor for several linear, axially symmetric and tetrahedral molecules (122–124).

Scaled particle theory provides a method of calculating the free energy change on adding a rigid solute molecule to a system that already contains N rigid molecules. It has been extended to fluids of rigid convex bodies by Gibbons (125, 126). His result for the equation of state is

$$\frac{\beta P}{\rho} = \frac{1}{1-Y} + \frac{B^2 C}{3\rho(1-Y)^3} + \frac{AB}{\rho(1-Y)^2}, \qquad 45.$$

where $A = \rho \bar{R}$, $B = \rho S$, $C = \rho \bar{R}^2$, and $Y = \rho v$; here $\bar{R}$, S, and v are the mean radius, surface area, and volume of the convex particle, respectively (127). Kihara

(127) and Gibbons (125, 126) have given expressions for $\bar{R}$, S, and v for various shapes. Scaled particle theory suffers from two restrictions. First, it is limited to convex shapes; however, a semiempirical extension to nonconvex shapes has been proposed (128). Second, it provides expressions only for the macroscopic thermodynamic properties, and gives little information about the distribution functions [although it does yield the average pair correlation function at contact (129)]. This latter restriction prevents the use of scaled particle theory to calculate reference properties in perturbation treatments.

In thermodynamic perturbation theory the properties of the real system, in which the pair potential is u, are expanded about the values for a reference system, in which the pair potential is u_0. Pople (130) and Gubbins & Gray (131) considered the potential

$$u(r\omega_1\omega_2;\alpha) = u_0(r) + \alpha u_a(r\omega_1\omega_2), \qquad 46.$$

where

$$u_0(r) = \langle u(r\omega_1\omega_2)\rangle_{\omega_1\omega_2} \qquad 47.$$

and

$$u_a(r\omega_1\omega_2) = u(r\omega_1\omega_2) - u_0(r), \qquad 48.$$

so that $\langle u_a(r\omega_1\omega_2)\rangle_{\omega_1\omega_2}$ vanishes. From Equation 46 it is seen that $u(r\omega_1\omega_2;1) = u(r\omega_1\omega_2)$ and $u(r\omega_1\omega_2;0) = u_0(r)$. Expanding $g(r\omega_1\omega_2)$ in powers of α and subsequently putting $\alpha = 1$ gives, to first order,

$$g(r_{12}\omega_1\omega_2) = g_0(r_{12}) - \beta u_a(r_{12}\omega_1\omega_2)g_0(r_{12})$$
$$- \beta\rho \int dr_3 g_0(r_{12}r_{13}r_{23})\langle u_a(r_{13}\omega_1\omega_3) + u_a(r_{23}\omega_2\omega_3)\rangle_{\omega_3}. \qquad 49.$$

The second-order term in this series has also been evaluated (131). The term involving the triplet correlation function in Equation 49 vanishes for anisotropic intermolecular potentials that involve spherical harmonics of order $l \neq 0$ in the orientation ω_3 (e.g. multipolar interactions). The corresponding expansion for the Helmholtz free energy is (13)

$$A = A_0 + A_2 + A_3 + \ldots, \qquad 50.$$

where A_2 and A_3 are the second- and third-order perturbation terms (the first vanishes with this choice of reference potential). For multipole-like ($l \neq 0$) potentials, A_2 and A_3 are given by

$$A_2 = -\frac{1}{4}\beta\rho^2 \int d\mathbf{r}_1\, d\mathbf{r}_2 \langle u_a(r_{12}\omega_1\omega_2)^2\rangle_{\omega_1\omega_2} g_0(r_{12}), \qquad 51.$$

$$A_3 = \frac{1}{12}\beta^2\rho^2 \int d\mathbf{r}_1\, d\mathbf{r}_2 \langle u_a(r_{12}\omega_1\omega_2)^3\rangle_{\omega_1\omega_2} g_0(r_{12})$$
$$+ \frac{1}{6}\beta^2\rho^3 \int d\mathbf{r}_1\, d\mathbf{r}_2\, d\mathbf{r}_3$$
$$\times \langle u_a(r_{12}\omega_1\omega_2)u_a(r_{13}\omega_1\omega_3)u_a(r_{23}\omega_2\omega_3)\rangle_{\omega_1\omega_2\omega_3} g_0(r_{12}r_{13}r_{23}). \qquad 52.$$

Twu (132, 133) has given the corresponding expressions for a general u_a. Explicit expressions for the various terms in A_2 and A_3 have also been obtained (134–136) by carrying out the orientation averages in Equations 51 and 52. The rather slow convergence of Equation 50 for strong multipole strengths led Stell et al (137) to suggest the following simple Padé approximation for the free energy,

$$A = A_0 + A_2 \left[\frac{1}{1 - (A_3/A_2)} \right]. \qquad 53.$$

Similar Padé approximants have been proposed for the mean squared torque (138) and for the surface tension (139). Høye & Stell (140) have proposed an expression for $g(r\omega_1\omega_2)$ for polar fluids that reproduces the Padé equation, Equation 52, for the free energy. Henderson & Gray (141) have also presented a perturbation theory for the pair correlation function $g(r\omega_1\omega_2)$, which is based on an expansion of the direct correlation function; a mean field approximation is used for the anisotropic contribution to $c(r\omega_1\omega_2)$. Madden & Fitts (142, 143) have evaluated higher order terms in the series for $g(r)$ (obtained from Equation 49 by averaging over orientations) using both Percus-Yevick and hypernetted chain approximations. They found poor convergence for strong dipole and quadrupole forces. However, a Padé approximant for $g(r)$ gave good results. Perturbation expansions based on the Pople reference potential of Equation 47 have also been used to study the effects of induction forces (144, 145), triplet potentials (146, 147), the interfacial density-orientation profile (148), and the angular correlation parameters that occur in the dielectric and Kerr constants (149).

Although perturbation expansions based on the Pople reference potential of Equation 47 have been the most extensively studied, several alternate choices of reference potentials have been proposed. A method of building part of the anisotropic potential into the reference fluid, while keeping the reference molecules isotropic, has been proposed by Perram & White (150). They define an average angle-dependent potential $\bar{u}_a(r_{12})$ by

$$\exp\left[-\beta \bar{u}_a(r)\right] = \langle \exp\left[-\beta u_a(r\omega_1\omega_2)\right] \rangle_{\omega_1\omega_2}, \qquad 54.$$

where $u_a(r\omega_1\omega_2)$ is defined by Equation 48. The reference potential $u_0(r)$ is taken to be

$$u_0(r) = \langle u(r\omega_1\omega_2) \rangle_{\omega_1\omega_2} + \bar{u}_a(r), \qquad 55.$$

and $u(r\omega_1\omega_2; \alpha)$ is defined by

$$u(r\omega_1\omega_2; \alpha) = u_0(r) - \frac{1}{\beta} \ln\left[1 + \alpha f(r\omega_1\omega_2)\right], \qquad 56.$$

where

$$f(r\omega_1\omega_2) = \exp\left\{-\beta[u_a(r\omega_1\omega_2) - \bar{u}_a(r)]\right\} - 1. \qquad 57.$$

Examination of Equation 56 shows that $u(r\omega_1\omega_2; 1) = u(r\omega_1\omega_2)$, and $u(r\omega_1\omega_2; 0) = u_0(r)$. The expansion of $g(r\omega_1\omega_2)$ to first order then gives

$$g(r_{12}\omega_1\omega_2) = g_0(r)[1 + f(r_{12}\omega_1\omega_2)]$$
$$+ \rho \int d\mathbf{r}_3 g_0(r_{12}r_{13}r_{23}) \langle f(r_{13}\omega_1\omega_3) + f(r_{23}\omega_2\omega_3) \rangle_{\omega_3}. \qquad 58.$$

In view of the relatively poor convergence of the above expansions for strongly anisotropic potentials, some authors (151, 152) have suggested using a reference fluid of hard nonspherical molecules interacting with

$$u_0(r\omega_1\omega_2) = 0, \quad r > d(\omega_1\omega_2)$$
$$= \infty, \quad r < d(\omega_1\omega_2), \qquad 59.$$

where $d(\omega_1\omega_2)$ is the shortest distance of approach of the centers for two molecules with orientations ω_1 and ω_2. In the theory of Mo & Gubbins (151) the potential $u(r\omega_1\omega_2)$ is divided into repulsive and attractive regions, and these are treated by separate expansions as in the Weeks-Chandler-Anderson theory (99). When first order terms are retained in each of these expansions the resulting equation for the free energy is [the corresponding equation for $g(r\omega_1\omega_2)$ is given in 151]:

$$A = A_0 + 2\pi\rho N \int_0^\infty dr r^2 \langle u_p(r\omega_1\omega_2) g_0(r\omega_1\omega_2)$$
$$\times \exp\{-\beta[u_{\text{rep}}(r\omega_1\omega_2) - u_0(r\omega_1\omega_2)]\}\rangle_{\omega_1\omega_2}, \qquad 60.$$

where $d(\omega_1\omega_2)$ has been chosen to satisfy

$$\int_0^\infty dr r^2 \left(\exp\{-\beta[u_{\text{rep}}(r\omega_1\omega_2) - u_0(r\omega_1\omega_2)]\} - 1\right) g_0(r\omega_1\omega_2) = 0 \qquad 61.$$

and

$$u_{\text{rep}}(r\omega_1\omega_2) = u(r\omega_1\omega_2) + \varepsilon(\omega_1\omega_2), \quad r < r_{\min}(\omega_1\omega_2)$$
$$= 0, \quad r > r_{\min}(\omega_1\omega_2), \qquad 62.$$

$$u_p(r\omega_1\omega_2) = -\varepsilon(\omega_1\omega_2), \quad r < r_{\min}(\omega_1\omega_2)$$
$$= u(r\omega_1\omega_2), \quad r > r_{\min}(\omega_1\omega_2). \qquad 63.$$

Here $\varepsilon(\omega_1\omega_2)$ and $r_{\min}(\omega_1\omega_2)$ are the magnitude and separation distance, respectively, of $u(r\omega_1\omega_2)$ at the minimum of the pair potential (for fixed $\omega_1\omega_2$). The theory of Sandler (152) differs from the above in that a single expansion is made about the reference fluid with potential u_0. He considers the case of a site-site hard sphere potential for u_0, with a multipolar u_p.

Two other perturbation expansions have been proposed for molecules that interact with a purely repulsive potential. These are the theories of Sung & Chandler (153) and of Bellemans (154). The former applies for an arbitrary repulsive potential, while the latter is specifically for rigid molecules with potentials as in Equation 59.

Comparisons of these theories with computer simulation results are given in a later section. However, it is possible to make some general remarks at this stage concerning the limitations of the various approaches. The PY theory is attractive in principle, but severe numerical difficulties arise in solving the integral equation. This problem is less severe for RISM. However, both RISM and scaled particle theory are restricted (at least, at present) to intermolecular potentials that are "hard", i.e. of the form of Equation 59. The RISM may well prove useful as a means for calculating reference properties in a perturbation scheme. Such is not the case for scaled particle theory, because it does not provide a theory for the pair correlation function. Of the perturbation theories, that based on the Pople reference is easy to apply for potentials of the generalized Stockmayer type, Equation 1, particularly if u_0 is the hard sphere or Lennard-Jones potential. The Perram-White expansion is, in general, more difficult to apply because the functional form of u_0 (and hence g_0, etc) changes with the form of u_a. Both of these approaches have difficulty in accounting for significantly nonspherical overlap potentials (e.g. site-site models for large elongations) and fail for hard molecules. The use of a nonspherical reference potential overcomes these problems to a large extent, but at the cost of greatly increased numerical effort in calculating results.

SIMULATION METHODS

Computer simulations of fluids have been carried out by Monte Carlo (155) and molecular dynamics (156) methods. The Monte Carlo method employs random numbers and is based on the ensemble concepts of Gibbs, while the molecular dynamics method proceeds from the solution of the Newtonian equations of motion of the particles in a many-body system and is based on the time-averaging concepts of Boltzmann and Maxwell. Each has its own advantages. Molecular dynamics treats real-time motion and can be used to study both equilibrium and time-dependent properties, whereas the Monte Carlo method is based on Markov-chain ensemble averaging, and is suitable for calculating only static equilibrium properties. On the other hand, the Monte Carlo method can be cast in several different ensembles, including the canonical ensemble (N, V, T constant), the isothermal-isobaric ensemble (N, P, T constant), and the grand canonical ensemble (T, V, μ constant), while the molecular dynamics method is limited to the microcanonical ensemble (N, V, E constant). (Here N, V, T, P, μ, E represent the number of particles, volume, temperature, pressure, chemical potential, and energy of the system.) The (N, P, T) ensemble is particularly appropriate to the study of mixtures (164), while the (T, V, μ) ensemble can be used to calculate free energies (161–163). Variations of the standard Monte Carlo methods make it possible to calculate free energy differences (68, 157–160). For detailed descriptions of simulation methods for simple fluids, see the papers by Wood and co-workers (5, 165, 166) on the Monte Carlo method, and those of Rahman (167) and Verlet (168) on molecular dynamics. The following paragraphs describe how these methods are extended to linear molecules.

Five coordinates are required to specify the position and orientation of a linear molecule; those most commonly used are the Cartesian coordinates of the center of mass and the two Euler angles that define the orientation of the axis (Figure 1a). Other coordinate choices are possible, such as the coordinates of a fixed point and the three-direction cosines of the axis, or the six Cartesian coordinates of two fixed points on the axis (e.g. the nuclei in a diatomic); however, these coordinates are not all mutually independent.

Monte Carlo Methods for Linear Molecules

The extension of Monte Carlo methods to linear molecules is straightforward; however, there are minor pitfalls to be avoided. The mechanics of the Monte Carlo method consists of making small, provisional, random displacements of particles, and accepting them with a probability $\exp(-\Delta E/kT)$, where ΔE is the energy change associated with the proposed move. For spherical particles, the random move is accomplished by changing the center of mass coordinates from (x, y, z) to $(x + R_1 A, y + R_2 A, z + R_3 A)$, where R_1, R_2, R_3 are random numbers in the interval $(-1, 1)$, and A is an adjustable parameter. This parameter is important because it determines the rate of acceptance of proposed moves, which in turn determines the rate of sampling of configurational space and the rate of convergence of the ensemble averages. If the acceptance rate is too high or too low the ensemble averages converge slowly, and computing time is wasted. Even though Wood & Jacobson (166) found evidence that the optimum acceptance rate for spherical particles is in the range 0.1–0.3, values close to 0.5 have almost always been used. In practice, A is usually adjusted during the course of the simulation to maintain the desired rate of acceptance.

Random displacement of a linear particle requires changes in all five independent coordinates. This has usually been accomplished by combined center of mass and rotational displacements. Rotational displacements can be made in different ways, some of which are ill suited to Monte Carlo purposes. Small displacements in the Euler angles (Figure 1a) is a poor choice, for example, because changes in ϕ become meaningless as θ goes to zero, and the allowed displacement becomes a function of the existing orientation. This can lead to the so-called "bottleneck effect," in which the simulation becomes artificially trapped in a small region of configurational space, thus producing results applicable to a long-lived metastable state rather than a state at thermodynamic equilibrium. This effect may have been responsible for the spurious liquid-solid phase transition first reported in a fluid of prolate spherocylinders, and later shown by Vieillard Baron (52) to be an artifact of the simulation.

Few authors give full details of how rotational displacements are made. Watts and co-workers (46, 63) use a rotation in the interval $(-10°, +10°)$ about a randomly chosen axis. Jansoone (67) describes a method based on choosing a random point on a unit sphere, which insures that all displacements are equally probable from all orientations. In these methods, a second parameter, B, is introduced to limit the change in a single orientational coordinate in any one move. No systematic studies of optimum choices for the parameters A & B appear to have been carried out. Watts and co-workers (46, 63) apparently adjust A and B to produce equal numbers of rejections on the basis of rotational and translational displacements. Jansoone (67) adjusts both A and B to produce an acceptance rate of 74% for center of mass displacement and 71% for orientational displacement, with the latter decision made only after a center of mass displacement has been accepted. Vieillard-Baron (52) separates rotational and translational displacements as follows: in the first "step," an attempt is made to give the center of mass of each particle a translational displacement, selecting particles at random, and accepting or rejecting each move in the normal way; in the second "step," rotational displacements are attempted for each particle in the same way. Comparisons of this method with simultaneous translation and rotation reportedly showed them to be about equally efficient in sampling configurational space. Streett & Tildesley (53) describe a method of simultaneous translation and rotation, applicable to linear molecules, which depends on a single parameter A and is independent of the existing orientation.

Two areas in the methodology of Monte Carlo simulations appear to merit further study: (a) the dependence of the rate of convergence of ensemble averages on the parameters A and B; and (b) the optimum number of particles to displace in one move. To our knowledge, no reports of Monte Carlo simulations have been published in which more than one particle has been displaced in each move, although in principle any number between 1 and N can be displaced. Simultaneous displacement of several particles may well result in more efficient sampling of configurational space and more rapid convergence.

Molecular Dynamics for Linear Molecules

The extension of molecular dynamics to polyatomic fluids has been effected by two fundamentally different methods: (a) the central force method, which treats the motions of individual atoms interacting through spherically symmetric potentials; and (b) the rigid particle method, which treats the molecules as rigid, nonspherical rotors.

THE CENTRAL FORCE METHOD   In one version of this method, individual atoms interact through different inter- and intramolecular potentials. In the work of Berne & Harp (7), on CO and N_2, a simple harmonic potential has been used for intramolecular interactions and a Stockmayer potential (spherical Lennard-Jones plus dipole-dipole) for intermolecular interactions. The two atoms in each molecule are permanently bound together by a strong (harmonic) intramolecular potential that holds the internuclear distance within narrow limits, giving results that are in fact very close to those for rigid rotors. The intramolecular potential is therefore merely a device to simplify the computations by resolving the molecular motions into pure translational motions of the individual atoms. A more general central force method has been used by Rahman et al (40) for water and by McDonald & Klein (41) for ammonia. In this case separate central force potentials are used for each of the three types of atom-atom interactions: O-O, O-H, and H-H interactions in water, and N-N, N-H, and H-H interactions in ammonia. No distinction is made between inter- and intramolecular forces; hence the potentials must not only produce molecules of the proper composition and geometry, but also appropriate intermolecular interactions. Vibration and dissociation can be studied. The central force methods reduce the molecular dynamics to translational motions of spherical particles, thereby avoiding the lengthy and sometimes troublesome calculations associated with the solution of equations of rotational motion. One disadvantage is that intramolecular vibrations may take place on a shorter time scale, necessitating shorter time steps; however Rahman et al (40) report that for water models this is more than offset by the increased speed of computation.

RIGID PARTICLE METHODS   This method was first used by Rahman & Stillinger (33) for water and by Barojas et al (56) for N_2. Both groups used the coupled Newton-Lagrange equations (170), which describe translational motion in terms of the Cartesian center-of-mass coordinates and rotational motion in terms of the Euler angles for a body-fixed axis. In the general case (33) all three Euler angles are used, while for linear molecules (56) only two are required. Here again, the use of Euler angles has a serious drawback: one or more of the equations of rotational motion contains a factor $(\sin \theta)^{-1}$ (see Figure 1a) that becomes infinite when θ is zero or π. The equations are mathematically well behaved, because the full terms containing this factor go to zero as θ goes to zero or π; however, it introduces a numerical instability into the algorithm used to solve the differential equations. Barojas et al (56) circumvent this problem by redefining the Euler angles whenever θ becomes smaller than $\pi/10$ or greater than $9\pi/10$. (The range of θ is 0 to π.) If the polar and azimuthal angles, θ and ϕ, are initially measured from the space-fixed z and x axes respectively, then the new polar angle θ' is measured from, say, the x axis and the new azimuthal angle ϕ' from the z axis [see Figure 2 in (56)]. When used with a predictor-corrector numerical method of the type due to Norsdieck (171, 172) these coordinate transformations are especially lengthy, since not only the angles but also several of their time derivatives must be transformed. The method is cumbersome, but adds little to computing time because the transformations are relatively infrequent.

Cheung (60, 173), by casting the problem in vector differential form and taking advantage of certain unique features of the dynamics of idealized linear molecules, devised a simple and elegant method for solving the equations of motion. It is not only free of numerical instabilities but also highly efficient for computation. Cheung's method depends upon the fact that in systems of idealized linear molecules, all centers of force lie on the molecular axis, and in consequence all vectors representing the instantaneous torque and the time derivatives of angular position for a molecule are always perpendicular to its axis. Rotation around the axis does not occur. This makes it possible to express the rotational motion as a first order differential equation for angular velocity, instead of the second order equation for angular position used by Barojas et al (56). At each step in the simulation the orientation of each molecular axis is calculated as the direction of the vector cross product of its angular velocity and angular acceleration vectors, since these are both perpendicular to its axis. For complete details see Cheung (60, 173). This method is easier to program and capable of greater computational efficiency than the method described by Barojas.

Evans (174) has examined the problem of rotational motion of rigid molecules of arbitrary shape and has cast the general problem in a form that is both computationally efficient and free of numerical instabilities. He has shown that these requirements are met if the orientations of molecules in space are described by quarternion parameters—nonlinear functions of the conventional Euler angles—first introduced by Euler in 1776 (175).

Computing Requirements

A major barrier to progress in computer simulations of molecular fluids is the requirement for massive amounts of computing time. The time required to simulate molecular liquids, even diatomics, usually exceeds that required for simple liquids by about an order of magnitude. Stillinger (74) has pointed out that with presently available computers and simulation methods it would take $\sim 3 \times 10^8$ years to simulate one second of real time in a molecular dynamics study of several hundred model water molecules. Fortunately, many of the equilibrium- and time-dependent properties of interest can be estimated from simulated real times of the order of

10^{-12} to 10^{-10} sec, which require computing times of the order of a few minutes to a few hours. With the high cost of time on modern fast computers, simulations continue to be expensive, in many instances prohibitively so.

In both Monte Carlo and molecular dynamics simulations, the bulk of the computing time—usually 95% or more—is used in evaluating pair interactions: calculating relative orientations, potential energies, forces, and torques. The complexity of these calculations is a function mainly of the form of the inter- and intramolecular potential functions, hence the further development of simulation methods is intimately bound up—as, indeed, is liquid theory as a whole—with the development of these functions. In this regard, it would appear that the central force molecular dynamics methods used in conjunction with molecule-independent interatomic potentials (40, 41) offer great potential for future development, not only because they are physically more realistic than rigid body models, but also because they eliminate the need to deal explicitly with equations of rotational motion. Simulations of rigid polyatomic molecules will also continue to be important, and Evans' recommended use of quarternion parameters (174) to facilitate calculations of rotational motion should be exploited. The intermediate method of Berne & Harp (7), in which an essentially artificial intramolecular potential is used as a device to reduce the motion of polyatomic molecules to the Cartesian motions of their component atoms, deserves further study.

An important aspect of molecular dynamics calculations is the choice of a numerical method for solving the equations of motion. For model systems with continuous potential functions, most workers have followed the lead of Verlet (168), who used a simple, central-difference predictor method, or Rahman (167), who used a higher order predictor-corrector method (172, 176). Beeman (185) has examined the relative merits of several methods of this type. Berne & Harp (7) suggest that the Runge-Kutta-Gill method (177) may be superior to the multistep methods in some cases. In general, the computing time spent in solving the differential equations of motion is trivially small compared to the time spent in evaluating and summing pair interactions, so the choice of a particular method is dictated mainly by considerations of stability (which determines the maximum length of the time step) and storage requirements. In terms of these criteria, Berne & Harp (7) present evidence in favor of Runge-Kutta methods; however, we are unaware of other examples of their use in simulations of dense fluids.

Several ingenious programming devices have led to significant savings in computing time. Perhaps the best known is the use of "neighbor lists," generally attributed to Verlet and co-workers, in which a list is kept, for each molecule, of the neighboring molecules that are close enough to be counted when summing pair interactions. These lists are updated periodically. In large systems ($N \gtrsim 150$) with small potential cutoff distances, this reduces the number of pair interactions that must be examined in each step, thereby expediting the single most time-consuming operation. This device is particularly useful with short-range potentials, such as the hard sphere potential or the repulsive part of the Lennard-Jones potential (178). Other techniques for speeding up molecular dynamics simulations have been reported (179–182). Another recent innovation is the separation of the neighbors that interact with a central particle into inner and outer groups, coupled with the use of different time steps to predict the time evolution of the force and torque exerted on the central particle by these two groups. A similar technique has been used by Ahmad & Cohen (183) for molecular dynamics studies of the time evolution of galaxies, where the "particles" comprising the system are stars, and the forces gravitational. G. Saville of London University (private communication) developed this method independently and used it in molecular dynamics simulations of the gas-liquid interface in spherical Lennard-Jones models. Saville's method can be outlined briefly as follows. If the cutoff distance for the potential energy function is r_c, the force and torque on each particle are calculated in two parts: one due to the inner neighbors, lying within a distance ar_c, and another due to the outer neighbors, lying at distances between ar_c and r_c, where $a \approx 0.5$. The simulation is carried forward in blocks of n steps, where $n \approx 10$. During the first two steps in each block, the forces and torques due to the two groups of neighbors are calculated in the usual manner, and all interactions at distances up to r_c are counted. During the remaining $n - 2$ steps, the force and torque contributions of the inner neighbors are calculated in the usual way; however, the *net* contributions of the outer neighbors are calculated by linear extrapolation of the values calculated in the first two steps. The same group of inner neighbors must be used throughout a single block. Saville has used this method with $n = 10$ and has found that the results are essentially identical to those obtained by conventional methods. One of us (W.B.S.) has used it successfully in molecular dynamics simulations of systems of 256 rigid diatomic molecules interacting via Lennard-Jones atom-atom and quadrupole-quadrupole potentials; the computing speed increased by a factor of about 2.5. This method, like the neighbor lists of Verlet, attacks the heart of the computing speed problem by reducing the number of pair interactions that must be explicitly calculated in each time step.

Once the molecular model and the mathematical methods have been selected, the controlling factor in simulations is programming logic. Those contemplating entry into this field should be wary of the advice of professional programmers. A popular philosophy of programming holds that the ideal program is the one with the smallest number of statements. This can lead to programs that are aesthetically pleasing but highly inefficient for lengthy, repetitive computations. Speed of execution should be the dominant consideration. If programming is done in Fortran, for example, one should learn how the computer converts statements into assembly language as well as the relative amounts of time required for different basic operations; one can then program for maximum efficiency of execution. A carefully structured Fortran program can easily reach at least 80% of the efficiency of an original assembly language program, with only a fraction of the programming effort.

ORIENTATIONAL STRUCTURE

A relatively simple method of categorizing phases in terms of gross orientational ordering is through the use of orientational order parameters of the form $P_l(\mathbf{h}_i \cdot \mathbf{h}_j)$ and $P_l(\mathbf{h}_i \cdot \mathbf{n})$, where P_l denotes the lth Legendre polynomial, $\mathbf{h}_i$ is a unit vector in the direction of the axis of molecule i, and $\mathbf{n}$ is a space fixed unit vector (Equation 38). Ensemble averages of these functions take on different characteristic values for isotropic liquids and ordered phases. They have been used, for example, to distinguish between isotropic and nematic liquid phases (49, 50, 52, 67).

The detailed orientational structure of liquids of linear molecules is described by the hierarchy of angular correlation functions defined in Equation 17. Our interest here is in the second order function, $g(r\omega_1\omega_2)$, the angular pair correlation function. In terms of the angles shown in Figure 1b, it is written $g(r\theta_1\theta_2\phi)$, where $\phi = \phi_1 - \phi_2$. In contrast to the case of spherical molecules, for which the pair correlation function depends only on distance, the angular pair correlation function is a multidimensional function of distance and three angles; consequently it is more complex and thus more difficult to calculate or interpret.

Computer Calculations of Angular Pair Correlation Functions

Several methods for calculating angular pair correlation functions from computer simulations have been developed, and each provides a different way of describing and interpreting orientational structure. These are outlined below.

DIRECT CALCULATION FROM MULTIDIMENSIONAL HISTOGRAMS The most direct method of calculating $g(r\theta_1\theta_2\phi)$ for linear molecules is an extension of the method used to calculate $g(r)$ for spherical molecules: radial and angular space are divided into suitable increments and a multidimensional histogram is constructed by counting the numbers of molecular pairs that fall in each element. This method is expensive in both computing time and storage requirements; it has found only limited use. It has been used by Quentrec & Brot (59) with radial intervals of 0.05 and 0.1σ and angular intervals of $\pi/12$, and by Cheung & Powles (60) with radial intervals of 0.05σ and angular intervals of $\pi/16$. In the former case the histogram for a single state contains 8640 elements and in the latter 13,750. These divisions are probably too coarse to reveal complete details of orientational structure. Even if one has the computing resources to use this method, it is still impractical because the mass of numerical data it produces is awkward to work with and difficult to interpret or compare to theory and experiment.

PARTIAL PAIR CORRELATION FUNCTIONS We define a partial pair correlation function to be one that describes the dependence of g on a limited number of variables, taken as an unweighted average over the remaining variables. The radial distribution function (Equation 19) and the site-site pair correlation function (Equation 22) are examples of this type, in which all angular dependence is integrated out. They give the dependence of g on distance as an unweighted average over all orientations. In practice, these radial functions are calculated exactly as for spherical molecules by constructing a two-dimensional histogram of the numbers of molecular pairs separated by distances $r_k \pm \delta r/2$, where k indicates the kth element of the histogram. Radial distribution functions and site-site functions have almost always been calculated in simulations of linear molecules (53, 54, 56, 59, 60, 62, 64–66, 68–70, 186, 190, 191). A closely related function is the static structure factor (Equations 32–34) calculated in simulations of liquids (60–62, 186, 204). Another class of partial correlation functions has the form $g(r\omega)$, and describes the dependence of g on distance and one angle as an unweighted average over the remaining angles. These have been used by Barojas et al (56) and Quentrec & Brot (59). The latter, for example, define and calculate the following partial correlation functions (59):

$$g_1(r\theta_i) = \frac{1}{4\pi} \int_0^{2\pi} d\phi_j \int_0^{\pi} g(r\theta_i\theta_j\phi_j) \sin\theta_j \, d\theta_j, \qquad 64.$$

$$g_2(r\theta_j) = \frac{1}{4\pi} \int_0^{2\pi} d\phi_j \int_0^{\pi} g(r\theta_i\theta_j\phi_j) \sin\theta_i \, d\theta_i. \qquad 65.$$

(In this instance, relative orientation is defined in a coordinate system in which the polar axis lies along the axis of molecule i, hence the angles in Equations 64 and 65 are not the same as those defined in Figure 1b.) These definitions reduce the number of independent variables, making the functions more manageable and hence more useful in visualizing and interpreting angular order; however, this is done at the expense of losing information contained in the complete function. The partial function g_1, for example, describes the probability of finding the center of any molecule j at a distance r from the center of molecule i, along an axis r that makes an angle θ_i with the axis of i. It is independent of the orientation of j. Barojas et al (56) define a similar partial correlation function, in which r is the site-site distance and the angles are measured relative to r as the polar axis. In practice, these functions are calculated from three-dimensional histograms, constructed by sorting pairs of particles according to the values of r and one angle, without regard to the remaining angles.

CONDITIONAL PAIR CORRELATION FUNCTIONS   We define a conditional pair correlation function to be one that describes the dependence of g on a limited number of variables, for fixed values of the remaining variables. The most common example is a function $g_\omega(r)$ that gives the radial dependence of g for fixed relative orientation (indicated by the subscript ω). For linear molecules, conditional functions describing "regular" orientations, such as T-shaped, parallel, and end-to-end, are widely used (53, 54, 60, 62, 64–66, 186–190, 191, 204). Examples are shown in Figures 5–7. These functions can be calculated directly by constructing two-dimensional histograms based on the numbers of molecular pairs with relative orientations close to the orientations of particular interest. A more efficient means of obtaining $g_\omega(r)$ is through the use of spherical harmonic expansions.

SPHERICAL HARMONIC EXPANSIONS   A compact, efficient method for calculating the complete angular pair correlation function $g(r\theta_i\theta_j\phi)$, in computer simulations, has been developed by Streett & Tildesley (53), who calculate the coefficients $g_{ll'm}(r)$ in the spherical harmonic expansion (Equation 23) as ensemble averages of the products of two spherical harmonics (Equation 24a). When these coefficients are known, Equation 23 can be used to calculate any of the partial or conditional pair correlation functions of interest, within limits imposed by the convergence of the series. As many as 22 coefficients have been calculated in simulations of diatomics (53, 54) at a modest cost in computing time, and this number has been found more than adequate to provide detailed, quantitative information about orientational order in diatomics of low-to-moderate elongation ($L/\sigma \lesssim 0.65$). Detailed studies of the convergence of the series have been reported (54).

The harmonic coefficients are functions only of distance, hence each is calculated from a separate two-dimensional histogram, which gives the value of the product of two spherical harmonics, averaged over all pairs, as a function of distance. The result is a table of $g_{ll'm}$ vs r. For a radial interval of 0.05σ extending to 3σ, a table of the first 20 coefficients contains about 900 entries—significantly fewer than the thousands of entries required when the method of multidimensional histograms is used.

Compared to the methods described above, the spherical harmonics method possesses decided advantages in computational efficiency, compactness, and flexibility. A further advantage arises from the fact that much of the statistical mechanical theory of diatomic fluids can be cast in terms of these expansions (187–189). Furthermore, a number of experimentally observable properties, such as dielectric and Kerr constants (184) and depolarized light scattering (189), can be estimated from integrals over discrete subsets of the $g_{ll'm}$.

In some cases (60, 62, 186, 204) only a few of the leading coefficients have been calculated; however, Streett & Tildesley (53, 54) and Haile and co-workers (190, 191) have calculated 16 or more coefficients for a variety of model systems. The latter two groups have made extensive use of the spherical harmonic expansions to calculate conditional correlation functions of the form $g_\omega(r)$. In addition, Haile (190) has produced a family of three-dimensional, conditional correlation functions showing the dependence of g on distance and one angle, for fixed values of the other two. The model systems studied by Haile are governed by generalized Stockmayer potentials (Equation 1).

Descriptions of Orientational Structure

Examples of conditional correlation functions of the form $g_\omega(r)$ are shown in Figures 5–7. Figure 5 shows $g_\omega(r)$ for the T orientation in a model diatomic liquids, calculated from different numbers of terms in the harmonic series. Convergence is excellent for $r > L + \sigma$ ($\sim 1.6\sigma$ in this case), but is slower at close approach. Figures 6

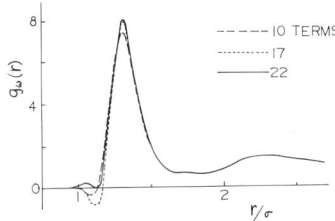

Figure 5 The conditional pair correlation function for the T-shaped orientation, calculated from different numbers of terms in the harmonic series (Equation 23), from molecular dynamics simulations of homonuclear diatomics obeying Lennard-Jones atom-atom plus quadrupole-quadrupole potentials (54). The parameters in this case are: $L^* = L/\sigma = 0.5471$; $T^* = kT/\varepsilon = 2.20$; $\rho^* = \rho\sigma_e^3 = 0.911$ (where σ_e is the diameter of a sphere having a volume equal to the diatomic); $Q^* = Q/(\varepsilon\sigma^5)^{1/2} = 2.0$.

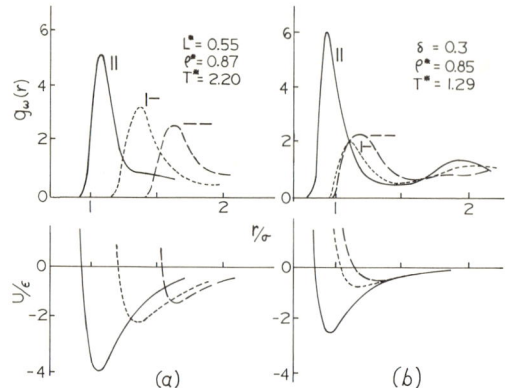

Figure 6 Conditional pair correlation functions and intermolecular potential functions for several orientations, calculated for two different model systems: (a) diatomic Lennard-Jones (atom-atom) model (54) and (b) spherical Lennard-Jones, plus anisotropic overlap represented by first two contributing terms in spherical harmonic expansion of the potential (Equation 9) (190).

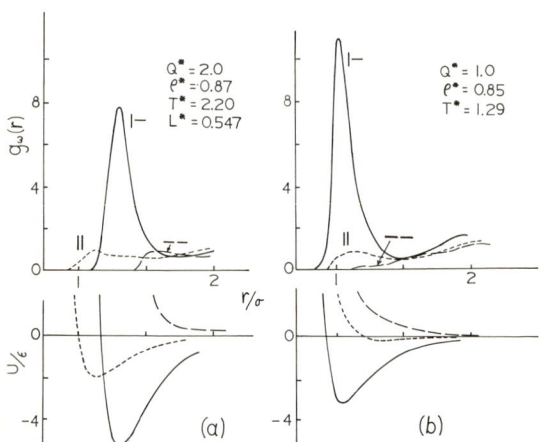

Figure 7 Conditional pair correlation functions for the models described in Figure 6, but with quadrupole-quadrupole interactions included. The quadrupole-quadrupole interaction deepens the potential well for the T-shaped orientation, and lessens it for the other two shown here. As the quadrupole strength is increased, the probability of T-shaped nearest neighbor orientations increases, at the expense of other orientations (54, 190, 191).

and 7 show $g_\omega(r)$, together with $u_\omega(r)$, the conditional intermolecular potential function, for three different orientations in several model systems. Figures 6a and 7a are from the molecular dynamics simulations of homonuclear diatomics by Streett & Tildesley (54) for Lennard-Jones atom-atom and quadrupole-quadrupole potentials, while Figures 6b and 7b are from the molecular dynamics simulations of Haile and co-workers (190, 191) for spherical Lennard-Jones plus anisotropic overlap and quadrupole-quadrupole potentials (Equations 1 and 9). Figures 6 and 7 show, respectively, results for potentials without and with quadrupole-quadrupole terms.

Some general features of orientational structure in systems of molecules governed by generalized Stockmayer potentials and atom-atom and quadrupole-quadrupole potentials, are summarized below. (These models might be expected to be reasonably good approximations to nitrogen, oxygen, and bromine, for example.)

1. Orientational ordering is short ranged, extending in most cases to distances of only two or three atom diameters. Increasing density increases the degree of angular ordering, but has little effect on its range. Increasing the elongation of the molecule governed by atom-atom potentials increases both the degree and the range of angular ordering. Changes in temperature as large as ε/k (where ε is the Lennard-Jones energy parameter and k is Boltzmann's constant) produce only minor changes; orientational order is weakly dependent on temperature.

2. In systems governed by atom-atom potentials of the hard sphere or Lennard-Jones type, angular ordering can be described as a separation of different relative orientations (T-shape, parallel, skewed, etc) according to the distances at which they are most likely to occur (Figure 6a), with little change in the overall frequency of occurrence of any particular orientation in comparison to a uniform random distribution. In other words, all orientations are about equally probable over distances equal to or greater than the range of significant ordering, but different orientations are most likely to occur at different distances in that range. (Statements in Reference 53 concerning a preference for T-shaped orientations in systems of hard diatomics are incorrect, as pointed out in Reference 54.) For the generalized Stockmayer potential, on the other hand, there is relatively little separation of different relative orientations, but there are marked differences in the peak probabilities of these orientations (Figure 6b).

3. The addition of a quadrupole-quadrupole term to either an atom-atom potential function or a spherical Lennard-Jones function produces dramatic changes in orientational structure, resulting in a strong preference for T-shaped orientations and suppression of others (Figure 7). This increase in T-orientations is reflected in changes in the site-site functions (54) and static structure factors (186).

4. Orientational structure in atom-atom models is virtually unchanged by the substitution of the repulsive part of the Lennard-Jones potential for the full potential (54). It therefore appears that orientational structure is determined mainly by the steep repulsive part of the potential function, as in the case of simple liquids (12). (In Figures 6 and 7, peak probabilities for the conditional distribution functions are closely correlated with the minima in the corresponding potential energy curves;

however, it is not the minima that are important, but the steep negative slopes in the potential functions, representing strong repulsive forces, which begin at these distances.)

COMPARISON WITH EXPERIMENT

Apart from work on water, relatively few comparisons have been made between simulation results and real molecular liquids. For linear molecules, the most extensive comparisons have been made for nitrogen (56, 60–62, 186), and for fluorine, chlorine, bromine, and carbon dioxide (204). Comparisons have been made for the static structure factor and for the following equilibrium thermodynamic properties (as functions of density and temperature): pressure, internal energy, constant volume specific heat, specific heat ratio, adiabatic compressibility, and mean-squared force and torque.

Nitrogen

Molecular dynamics simulations of nitrogen have been reported for several models, including Lennard-Jones atom-atom (56, 60–62), atom-atom plus quadrupole-quadrupole (186), and generalized Stockmayer (7) potentials.

THERMODYNAMIC PROPERTIES   Barojas et al (56) calculated pressures and internal energies for nine temperature density points along two isochores, and found agreement to within a few percent with experimental data (192). They used fixed values of the parameters ε and σ (193). Cheung & Powles (60) used the same potential to perform simulations for thirteen temperature-density points, and calculated the optimum values of ε and σ (having used a fixed bond length $L^* = 0.3292$) by fitting the computer results to experimental equation of state and internal energy data (192, 194, 195). They also found agreement between simulation and experiment to within a few percent. In addition, good agreement ($\sim 5\%$) with experiment was found for derived values of the thermal pressure coefficient, and fair agreement ($\sim 10\%$) for constant volume specific heat. Computed values of the mean-squared torque were found to be higher by a factor of two than those estimated from light-scattering data.

Subsequently, Cheung & Powles (186) reported molecular dynamics simulations of nitrogen using atom-atom plus quadrupole-quadrupole potentials. In this case the pressure, internal energy, constant volume specific heat, thermal pressure coefficient, and adiabatic compressibility were all computed in the simulation and found to agree closely with experimental values. However, the agreement is only marginally better than that obtained with the atom-atom potential alone. The mean squared torque increased by about $10-20\%$ with addition of the quadrupole-quadrupole interaction; however, uncertainties in estimates of this property from light-scattering data are such that the only reasonable conclusion is that the computed values are of the correct order of magnitude. The second virial coefficients calculated with this potential show poorer agreement with experiment than those calculated from the atom-atom potential alone.

STRUCTURE FACTOR   Cheung & Powles (60, 62, 186), Weis & Levesque (61), and Barojas et al (56) calculated the static structure factor for nitrogen by use of the Lennard-Jones atom-atom model and compared it to experimental neutron and X-ray-scattering data. The agreement is good in every case; however, the results were found to differ only trivially from those obtained for spherical molecules. This led Weis & Levesque (61) to conclude that little can be learned about the anisotropic part of the nitrogen potential from structure factor computations and neutron scattering. Cheung & Powles, however, found that the addition of a quadrupole-quadrupole term to the intermolecular potential (186) produces small but significant changes, bringing the calculated structure factor notably closer to experiment.

ORIENTATION CORRELATION   Cheung & Powles (60, 186) have calculated the orientational correlation parameter G_2 of Equation 38. This parameter is related to the intensity of the depolarized Rayleigh or Rahman spectrum. A value of $+0.10$ was found for the atom-atom model and a value of -0.12 for the atom-atom plus quadrupole-quadrupole model. Both values lie within the uncertainty of the experimental value (196) of 0.1 ± 0.2, so the results are inconclusive.

TIME-DEPENDENT PROPERTIES   We note in passing that translational and reorientational correlation functions have been calculated (56, 59–62, 186) and compared to experiment. For atom-atom potentials the results are qualitatively similar to experimental results for nitrogen, but quantitative agreement is less satisfactory than in the case of equilibrium properties.

Other Fluids

Cheung (62) has fitted the simulation data originally obtained for nitrogen (56) to oxygen. The equilibrium properties agree with experiment almost as well as in the case of nitrogen.

Singer and co-workers (204) have carried out molecular dynamics calculations for systems of 108 and 256 molecules interacting through Lennard-Jones atom-atom potentials with elongations $L^* = L/\sigma$ between 1.5 and 0.8, and have compared the results to experimental data for fluorine, chlorine, bromine, and carbon dioxide. They have obtained thermodynamically consistent expressions for the energy and pressure as functions of density for each system. They have examined orientational order by means of atom-atom distribution functions, leading terms in the spherical harmonics expansion of $g(r\theta_1\theta_2\phi)$, and several conditional correlation functions. With ε and σ values determined by fitting energy and temperature data along the coexistence line, they obtain agreement between experimental and computed thermodynamic properties that they characterize as very good for fluorine, quite good for chlorine and bromine, and poor for carbon dioxide. However they note that the resulting ε and σ values differ appreciably from those reported in the literature. Their calculated values for the mean-squared torque for liquid chlorine are in poor agreement with experimental estimates.

The site-site distribution functions for atom-atom (54, 204) and quadrupole-quadrupole models (54) have been compared to that obtained from neutron and

X-ray scattering experiments on bromine. Although uncertainties in the experimental data rule out quantitative comparisons, Streett & Tildesley (54) conclude that the addition of a moderately strong quadrupole to the atom-atom potential results in significantly better qualitative agreement than that obtained with the atom-atom potential alone. This is consistent with the findings of Cheung & Powles (186) for nitrogen.

The composition of a vapor of an isotopic mixture differs from that of the liquid with which it is in equilibrium. This phenomenon has no explanation in classical statistical mechanics but is accounted for in the first order quantum correction to the partition function. In molecular fluids this correction depends on the mean-square force and torque in the liquid of the isotopically unsubstituted molecule. Since these quantities are difficult to measure directly and are sensitive to the anisotropic nature of the potential, the experimentally accessible isotope separation factor is a sensitive test of the adequacy of a model intermolecular potential function. The isotope separation factor for nitrogen has recently been calculated from molecular dynamics simulation results for the Lennard-Jones atom-atom model by Thompson et al (206), and found to be in excellent agreement with experiment.

Finally, Gray et al (66) compared the mean-squared force calculated from Monte Carlo simulations with an experimental value for carbon monoxide. The simulation results are based on several alternative forms of the generalized Stockmayer potential, which include dipole, quadrupole, and anisotropic overlap forces. The quadrupolar force is found to produce a much larger effect on mean-squared force than either the dipole or anisotropic overlap forces. The calculated mean squared force is smaller than the experimental value by $\sim 25\%$.

COMPARISON WITH THEORY

Correlation Functions

Few of the theories outlined in the section on statistical mechanics have been tested against computer-calculated correlation functions, mainly because of numerical difficulties in the theoretical calculations.

The RISM equations provide an efficient means of calculating site-site pair correlation functions for models consisting of fused hard spheres. Chandler et al (197) have compared RISM predictions of these functions to those calculated by Streett and co-workers from Monte Carlo simulations for homonuclear (53) and heteronuclear (197a) hard diatomics. The two-site and three-site versions of the RISM equations give slightly different results (197). A typical comparison is shown in Figure 8 for $g(r_{\alpha\beta})$ for a heteronuclear diatomic of moderate asymmetry and low elongation. The agreement is poor at close approach, but excellent at $r \gtrsim 2\sigma$. RISM will clearly be useful for calculating these functions at large distances, which are beyond the reach of simulations based on a few hundred molecules.

The Gubbins-Gray perturbation expansion for $g(r\omega_1\omega_2)$, Equation 49, has been tested against Monte Carlo results for a generalized Stockmayer model (cf Equation 1) in which u_0 is the Lennard-Jones model and the anisotropic potential is either a

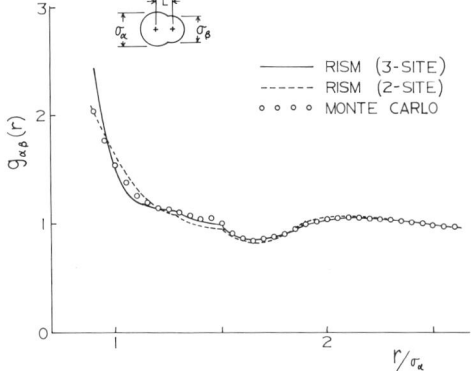

Figure 8 Comparison of site-site distribution function for unlike atoms, $g_{\alpha\beta}(r)$, calculated from Monte Carlo simulations of heteronuclear hard diatomics, with those calculated from the 2-site and 3-site RISM equations (197, 197a). The model parameters in this case are: $\sigma_\beta/\sigma_\alpha = 0.80$; $\rho\sigma_e^3 = 0.9$, and $L^* = L/\sigma_\alpha = 0.48$. ($\sigma_e$ is the diameter of a sphere having a volume equal to that of the diatomic, and ρ is the number density.)

dipole-dipole or a quadrupole-quadrupole term (65, 191). A typical result is shown in Figure 9. It is seen that the theory is only valid for rather weak quadrupole moments, up to about $Q^* = Q/(\varepsilon\sigma^5)^{1/2}$ values of 0.3 or 0.4. Similar results are obtained for dipolar forces.

The Sung-Chandler theory (153) supposes that the liquid structure is determined by the repulsive part of the potential and then predicts that the function $g(r\omega_1\omega_2)$

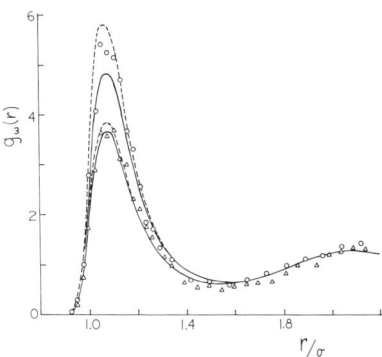

Figure 9 Test of Gubbins-Gray perturbation expansion, Equation 49, for molecules interacting with a potential $u_0 + u_a$ where u_0 is the Lennard-Jones model and u_a is the quadrupole-quadrupole interaction, for the (most probable) T orientation at $\rho^* = 0.800$, $T^* = 0.719$. Lines are theory, points are Monte Carlo data; upper line and circles are for $Q^* = Q/(\varepsilon\sigma^5)^{1/2} = 0.5$, lower line and triangles for $Q^* = 0.3$. [From Wang et al (65), by courtesy of *Chemical Physics Letters*.]

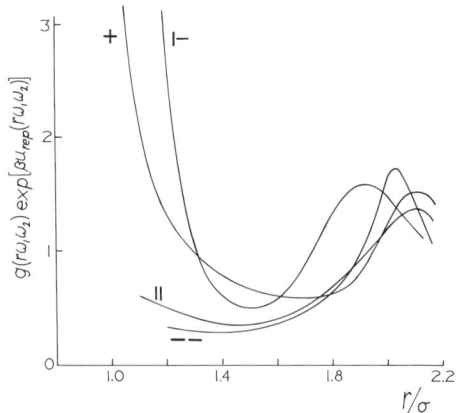

Figure 10 Monte Carlo results for the function $g(r\omega_1\omega_2) \exp[\beta u_{\text{rep}}(r\omega_1\omega_2)]$ for the state condition and potential of Figure 9, with $Q^* = 1$. [From Wang et al (65), by courtesy of *Chemical Physics Letters*.]

$\exp[\beta u_{\text{rep}}(r\omega_1\omega_2)]$ is independent of orientations; here u_{rep} is given by Equation 63. Figure 10 shows a test of this prediction for quadrupolar molecules. The prediction is seen to break down, especially at small separations. However, Steele & Sandler (198) found good results when comparing this theory with PY theory for diatomic hard sphere site-site potentials, suggesting that the expansion may work well for anisotropic repulsive forces.

Comparisons with computer simulation results for the center correlation function $g(r)$ have been carried out for the Gubbins-Gray (64), Perram-White (143, 150, 199), Sung-Chandler (153), and Madden-Fitts (142, 143) theories. It is known, however, that the centers pair correlation function is not very sensitive to the anisotropic forces.

Bulk Thermodynamic Properties

The scaled particle theory equation of state, Equation 45, has been tested against Monte Carlo data for a dense fluid of hard spherocylinders by both Few & Rigby (51) and Vieillard-Baron (52). The scaled particle theory pressures are slightly too high, but the agreement is generally good.

Figure 11 shows a test of the Pople expansion and of the Padé approximant to it (cf Equations 50 and 53) for a generalized Stockmayer potential model. Even when the third-order perturbation term is included, the expansion is seen to fail for moderately strong anisotropic strengths. The Padé approximant gives excellent results for both the dipole and quadrupole cases, however. For the anisotropic overlap potential of Equation 9 the agreement is good only in the range $-0.2 < \delta_2 < 0.35$. Similar tests of the Padé approximant have been reported by McDonald (200), Patey & Valleau (69), and Verlet & Weis (201) for the dipole case, and by Patey & Valleau (70) for the quadrupole case. Table 1 compares the Padé approx-

LIQUIDS OF LINEAR MOLECULES 405

Table 1 Contribution of dipole-dipole forces to thermodynamic properties for a fluid of hard spheres with embedded point dipoles at $\rho/\rho_{cp} = 0.59$, where ρ_{cp} is the close-packed density [from Patey & Valleau (92), by courtesy of the American Institute of Physics]

T_μ^{*a}	μ/D (T = 298.15 K, d = 3.0 Å)	$-U_a/NkT$			$C_{v,a}/Nk$			$\beta P_a/\rho$		
		MSA[b]	Padé[c]	MC	MSA	Padé	MC	MSA	Padé	MC
17.781	0.25	0.0036	0.0060	0.0064 ± 0.0004	0.0035	0.0061	0.0057 + 0.0003	0.0018	0.0045	−0.03
1.976	0.75	0.226	0.368	0.348 ± 0.006	0.168	0.308	0.238 ± 0.011	0.102	0.265	0.58
0.711	1.25	1.20	1.87	1.79 ± 0.01	0.631	1.160	0.935 ± 0.038	0.468	1.26	1.35
0.363	1.75	3.21	4.64	4.80 ± 0.04	1.288	2.075	2.068 ± 0.083	1.09	2.92	3.08

[a] $T_\mu^* \equiv d^3 kT/\mu^2$, where d is hard sphere diameter.
[b] Mean spherical approximation.
[c] From Equation 53, by application of the usual thermodynamic relations.

imant of Equation 53 with Monte Carlo results (69) for hard spheres with embedded point dipoles. The Padé gives excellent results at both high and low dipole moments, and gives moderately good agreement at intermediate μ values. The MSA is seen to give poor results even for small dipole moments.

Table 2 shows a test of the Mo-Gubbins theory, which uses a nonspherical reference potential for a Stockmayer fluid. The properties of the reference fluid were

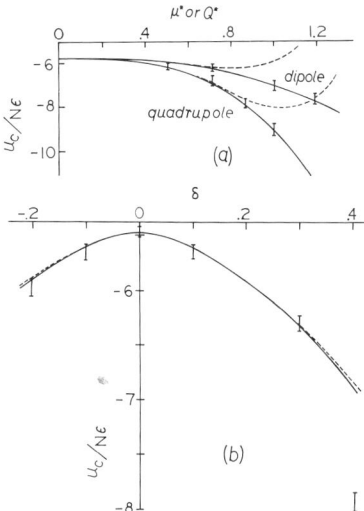

Figure 11 Comparison of Pople expansion to third order (*dashed curves*) and Padé approximant to the series (*solid curves*) with computer simulation data (64, 65, 191). u_c is the configurational contribution to the internal energy. The potential is of the form $u_0 + u_a$ where u_0 is the Lennard-Jones model. u_a is the dipole-dipole or quadrupole-quadrupole potential in figure (*a*), and is the anisotropic overlap model of Equation 9 in figure (*b*). Here $\mu^* = \mu(\varepsilon\sigma^3)^{1/2}$, $Q^* = Q/(\varepsilon\sigma^5)^{1/2}$. [From Twu et al (202).]

Table 2 Test of perturbation expansion about a nonspherical reference potential, Equation 60, for the anisotropic contributions to the internal energy of a Stockmayer fluid, $\mu^* = 1$. The Monte Carlo data are from McDonald (200), and the estimated accuracy is ± 0.03 [from Mo & Gubbins, courtesy of American Institute of Physics (151)]

$\rho^* = \rho\sigma^3$	$T^* = kT/\varepsilon$	$(U_a/N\varepsilon)_{\text{calc}}$	$(U_a/N\varepsilon)_{\text{MC}}$
0.5	1.35	−0.50	−0.53
0.6	1.35	−0.58	−0.60
0.7	1.35	−0.71	−0.70
0.8	1.35	−0.80	−0.82
0.8	2.74	−0.61	−0.57
0.8	1.35	−0.80	−0.82
0.8	1.35	−0.83	−0.86

obtained by using the Bellemans theory (154). The theory gives good results at this dipole strength but breaks down at high values because the Bellemans expansion no longer converges. The conclusions are similar when u_a is a quadrupole-quadrupole term.

Other Macroscopic Properties

Padé approximants to the perturbation expansions for the mean-squared torque (138) and the surface tension (139) have been proposed. The equation for the mean-squared torque has been tested against Monte Carlo results (138); the agreement is about as good as that shown in Figure 10 for internal energy. By introducing the Fowler approximation, the Padé expression for surface tension has also been tested against Monte Carlo data for a Stockmayer fluid (71). The results are excellent at liquid-like densities but slightly poorer at low densities.

CONCLUSIONS

Since the beginning of this decade, computer simulations and theory of liquids have broken free of their earlier preoccupation with simple liquids, and have begun to come to grips with liquids of more complex molecules. Only a few tentative steps have been taken thus far; the situation in this field is now about as it was for simple liquids in the mid-1960s when the exciting work of Rahman and Verlet for Lennard-Jones fluids first appeared. Future progress will depend, as it did in the case of simple liquids, on balanced, mutually supportive progress in simulation, theory, and experiment. There are two pressing needs at this point: (a) better intermolecular potential models; and (b) experiments that yield accurate values of properties that are sensitive to the anisotropic part of the potential and that can be calculated from simulation and theory.

Literature Cited

1. Wood, W. W., Erpenbeck, J. J. 1976. *Ann. Rev. Phys. Chem.* 27:319-48
2. Barker, J. A., Henderson, D. 1976. *Rev. Mod. Phys.* 48:587-671
3. Alder, B. J. 1973. *Ann. Rev. Phys. Chem.* 24:325-37
4. Alder, B. J. 1968. In *Physics of Simple Liquids*, ed. H. N. V. Temperley, J. S. Rowlinson, G. S. Rushbrooke, Chap. 4, pp. 79-114. Amsterdam: North-Holland
5. Wood, W. W. 1968. See Ref. 4, Chap. 5, pp. 115-230
6. Berne, B. J., Forster, D. 1971. *Ann. Rev. Phys. Chem.* 22:563-96
7. Berne, B. J., Harp, G. D. 1970. *Adv. Chem. Phys.* 17:63-227
8. McDonald, I. R., Singer, K. 1970. *Q. Rev. Chem. Soc.* 24:238-62
9. Egelstaff, P. A. 1973. *Ann. Rev. Phys. Chem.* 24:159-87
10. Barker, J. A., Henderson, D. 1972. *Ann. Rev. Phys. Chem.* 23:439-84
11. Andersen, H. C., Chandler, D., Weeks, J. D. 1976. *Adv. Chem. Phys.* 14:105-56
12. Andersen, H. C. 1975. *Ann. Rev. Phys. Chem.* 26:145-66
13. Croxton, C. A. 1974. *Liquid State Physics*. Chap. 5. London: Cambridge Univ. Press. pp. 190-271
14. McDonald, I. R. 1973. *Statistical Mechanics*, ed. K. Singer, Vol. 1, Chap. 3, pp. 134-93. London: Chem. Soc.
15. Henderson, D. 1974. *Ann. Rev. Phys. Chem.* 25:361-83
16. Egelstaff, P. A., Gray, C. G., Gubbins, K. E. 1975. In *Molecular Structure and Properties, International Review of Science, Physical Chemistry*, ed. A. D. Buckingham, Vol. 2. London: Butterworth. 316 pp.
17. O'Dell, J., Berne, B. J. 1975. *J. Chem. Phys.* 63:2376-94
18. Hoover, W. G., Ashurst, W. T. 1975. In *Theoretical Chemistry: Advances and Perspectives*, ed. H. Eyring, D. Henderson. New York: Academic. pp. 2-52
19. Hansen, J. P., Verlet, L. 1969. *Phys. Rev.* 184:151-61
20. Streett, W. B., Raveche, H. J., Mountain, R. D. 1974. *J. Chem. Phys.* 61:1960-69
21. Raveche, H. J., Mountain, R. D., Streett, W. B. 1974. *J. Chem. Phys.* 61:1970-84
22. Cotterill, R. M. J., Jensen, E. J., Kristensen, W. D. 1973. *Phys. Lett. A* 44:127-28
23. Cotterill, R. M. J., Kristensen, W. D., Jensen, E. J. 1974. *Philos. Mag.* 30:245-63
24. Hansen, J. P., McDonald, I. R. 1975. *Phys. Rev. A* 11:2111-23
25. Lewis, J. W. E., Singer, K. 1975. *J. Chem. Soc. Faraday Trans. 2* 71:301-11
26. Jacucci, G., Klein, M. L., McDonald, I. R. 1975. *J. Phys. Paris* 36:97-100
27. Mandell, M. J., McTaque, J. P., Rahman, A. 1976. *J. Chem. Phys.* 64:3699-3702
28. Opitz, A. C. L. 1974. *Phys. Lett. A* 47:439-40
29. Liu, K. S. 1974. *J. Chem. Phys.* 60:4226-30
30. Lee, J. K., Barker, J. A., Pound, G. M. 1974. *J. Chem. Phys.* 60:1976-80
31. Chapela, G. A., Saville, G., Rowlinson, J. S. 1975. *Faraday Discuss. Chem. Soc.* No. 59, pp. 22-28
32. Barker, J. A., Watts, R. O. 1973. *Mol. Phys.* 26:789-92
33. Rahman, A., Stillinger, F. H. 1971. *J. Chem. Phys.* 55:3336-59
34. Stillinger, F. H., Rahman, A. 1972. *J. Chem. Phys.* 57:1281-92
35. Stillinger, F. H., Rahman, A. 1974. *J. Chem. Phys.* 60:1545-57
36. Sarkisov, G. N., Dashevsky, V. G., Malenkov, G. G. 1974. *Mol. Phys.* 27:1249-69
37. Kistenmacher, H., Popkie, H., Clementi, E., Watts, R. O. 1974. *J. Chem. Phys.* 60:4455-65
38. Dashevsky, V. G., Sarkinsov, G. N. 1974. *Mol. Phys.* 27:1271-90
39. Watts, R. O., Clementi, E., Fromm, J. 1973. *J. Am. Chem. Soc.* 95:7943-48
40. Rahman, A., Stillinger, F. H., Lemberg, J. 1975. *J. Chem. Phys.* 63:5223-30
41. McDonald, I. R., Klein, M. L. 1976. *J. Chem. Phys.* 64:4790-91
42. Heinzinger, K., Vogel, P. C. 1974. *Z. Naturforsch. Teil A* 29:1164-71
43. Vogel, P. C., Heinzinger, K. 1975. *Z. Naturforsch. Teil A* 30:789-96
44. Fromm, J., Clementi, E., Watts, R. O. 1975. *J. Chem. Phys.* 62:1388-98
45. Lie, G. C., Clementi, E. 1975. *J. Chem. Phys.* 62:2195-99
46. Watts, R. O. 1974. *Mol. Phys.* 28:1069-83
47. Rigby, M. 1970. *J. Chem. Phys.* 53:1021-23
48. Freasier, B. C., Bearman, R. J. 1976. *Mol. Phys.* 32:551-54
49. Kushick, J., Berne, B. J. 1976. *J. Chem. Phys.* 64:1362-67

50. Freasier, B. C., Jolly, D., Bearman, R. J. 1976. *Mol. Phys.* 31:255–63
51. Few, G. A., Rigby, M. 1973. *Chem. Phys. Lett.* 20:433–35
52. Vieillard-Baron, J. 1974. *Mol. Phys.* 28:809–18
53. Streett, W. B., Tildesley, D. J. 1976. *Proc. R. Soc. London Ser. A* 348:485–510
54. Streett, W. B., Tildesley, D. J. 1977. *Proc. R. Soc. London Ser. A* 355:239
55. Aviram, I., Tildesley, D. J., Streett, W. B. 1977. *Mol. Phys.* In press
56. Barojas, J., Levesque, D., Quentrec, B. 1973. *Phys. Rev. A* 7:1092–1105
57. Ryckaert, J. P., Bellemans, A. 1975. *Chem. Phys. Lett.* 30:123–25
58. O'Brien, E. F. 1973. *Mol. Phys.* 26:453–72
59. Quentrec, B., Brot, C. 1975. *Phys. Rev. A* 12:272–81
60. Cheung, P. S. Y., Powles, J. G. 1975. *Mol. Phys.* 30:921–49
61. Weis, J. J., Levesque, D. 1976. *Phys. Rev. A* 13:450–57
62. Cheung, P. S. Y. 1976. Private communication
63. Evans, D. J., Watts, R. O. 1976. *Mol. Phys.* 32:93–100
64. Wang, S. S., Gray, C. G., Egelstaff, P. A., Gubbins, K. E. 1973. *Chem. Phys. Lett.* 21:123–26
65. Wang, S. S., Egelstaff, P. A., Gray, C. G., Gubbins, K. E. 1974. *Chem. Phys. Lett.* 24:453–56
66. Gray, C. G., Wang, S. S., Gubbins, K. E. 1974. *Chem. Phys. Lett.* 26:610–12
67. Jansoone, V. M. 1974. *Chem. Phys.* 3:78–86
68. Patey, G. N., Valleau, J. P. 1973. *Chem. Phys. Lett.* 21:297–300
69. Patey, G. N., Valleau, J. P. 1974. *J. Chem. Phys.* 61:534–40
70. Patey, G. N., Valleau, J. P. 1976. *J. Chem. Phys.* 64:170–86
71. Gubbins, K. E., Haile, J. M., McDonald, I. R. 1977. *J. Chem. Phys.* 66:364–65
72. Levesque, D., Weis, J. J. 1975. *Phys. Rev. A* 12:2584–86
73. Aviram, I., Tildesley, D. J. 1977. *Mol. Phys.* In press
74. Stillinger, F. H. 1975. *Adv. Chem. Phys.* 31:1–101
75. Buckingham, A. D. 1967. *Adv. Chem. Phys.* 12:107–42
76. Buckingham, A. D., Utting, B. D. 1970. *Ann. Rev. Phys. Chem.* 21:287–316
77. Gray, C. G. 1976. *Can. J. Phys.* 54:505–12
78. Gray, C. G., Van Kranendonk, J. 1966. *Can. J. Phys.* 44:2411–30
79. Gubbins, K. E., Gray, C. G., Egelstaff, P. A., Ananth, M. S. 1973. *Mol. Phys.* 25:1353–75
80. Ananth, M. S., Gubbins, K. E., Gray, C. G. 1974. *Mol. Phys.* 28:1005–30
81. Armstrong, R. L., Blumenfeld, S. M., Gray, C. G. 1968. *Can. J. Phys.* 46:1331–40
82. Stogryn, D. E., Stogryn, A. P. 1966. *Mol. Phys.* 11:371–93
83. Sweet, J. R., Steele, W. A. 1967. *J. Chem. Phys.* 47:3029–35
84. Kihara, T. 1963. *Adv. Chem. Phys.* 5:147–88
85. Koide, A., Kihara, T. 1974. *Chem. Phys.* 5:34–48
86. Berne, B. J., Pechukas, P. 1972. *J. Chem. Phys.* 56:4213–16
87. Kushick, J., Berne, B. J. 1976. *J. Chem. Phys.* 64:1362–67
88. Evans, D. J., Watts, R. O. 1975. *Mol. Phys.* 29:777–85
89. MacRury, T. B., Steele, W. A., Berne, B. J. 1976. *J. Chem. Phys.* 64:1288–99
90. Workman, H., Fixman, M. 1973. *J. Chem. Phys.* 58:5024–30
91. Chandler, D., Andersen, H. C. 1972. *J. Chem. Phys.* 57:1930–37
92. Rose, M. E. 1957. *Elementary Theory of Angular Momentum*. New York: Wiley. 214 pp.
93. Gray, C. G., Gubbins, K. E. 1975. *Mol. Phys.* 30:179–92
94. Gubbins, K. E., Haile, J. M. 1977. In *Improved Oil Recovery by Surfactant and Polymer Flooding*, ed. D. O. Shah, R. S. Schechter. New York: Academic. In press
95. Steele, W. A., Pecora, R. 1965. *J. Chem. Phys.* 42:1863–71
96. Sears, V. F. 1966. *Can. J. Phys.* 44:1297, 1299–1311
97. Narten, A. H. 1972. *J. Chem. Phys.* 56:5681–87
98. Twu, C. H., Gray, C. G., Gubbins, K. E. 1974. *Mol. Phys.* 27:1601–12
99. Kirkwood, J. G. 1939. *J. Chem. Phys.* 7:911–19
100. Buckingham, A. D. 1967. *Discuss. Faraday Soc.* 43:205–11
101. Buckingham A. D., Graham, C. 1971. *Mol. Phys.* 22:335–40
102. Lebowitz, J. L., Percus, J. K., Verlet, L. 1967. *Phys. Rev.* 153:250–54
103. Hanley, H. J. M., Watts, R. O. 1975. *Aust. J. Phys.* 28:315–24
104. Gray, C. G. 1975. *Statistical Mechanics*, ed. K. Singer, 2:300–23. London: Chem. Soc.

105. Rowlinson, J. S. 1967. *Discuss. Faraday Soc.* 43:243–47
106. Chen, Y. D., Steele, W. A. 1971. *J. Chem. Phys.* 54:703–10
107. Morrison, P. F. 1972. PhD thesis. Calif. Inst. Technol.
108. Lebowitz, J. L., Percus, J. K. 1966. *Phys. Rev.* 144:251–58
109. Wertheim, M. S. 1971. *J. Chem. Phys.* 55:4291–98
110. Blum, L. 1973. *J. Chem. Phys.* 58:3295–3303
111. MacInnes, D. A., Farquhar, I. E. 1975. *Mol. Phys.* 30:457–67
112. MacInnes, D. A., Farquhar, I. E. 1975. *Mol. Phys.* 30:889–98
113. Adelman, S. A., Deutch, J. M. 1973. *J. Chem. Phys.* 59:3971–80
114. Isbister, D., Bearman, R. J. 1974. *Mol. Phys.* 28:1297–1304; 32:597–603
115. Høye, J. S., Lebowitz, J. L., Stell, G. 1974. *J. Chem. Phys.* 61:3253–60
116. Wertheim, M. S. 1973. *Mol. Phys.* 20:1425–44
117. Chandler, D., Andersen, H. C. 1972. *J. Chem. Phys.* 57:1930–37
118. Chandler, D., 1973. *J. Chem. Phys.* 59:2742–46
119. Landanyi, B. M., Chandler, D. 1975. *J. Chem. Phys.* 62:4308–24
120. MacInnes, D. A., Farquhar, I. E. 1976. *J. Chem. Phys.* 64:1481–84
121. Chandler, D. 1976. *Mol. Phys.* 31:1213–23
122. Lowden, L. J., Chandler, D. 1973. *J. Chem. Phys.* 59:6587–95; 62:4246
123. Lowden, L. J., Chandler, D. 1975. *J. Chem. Phys.* 61:5228–41
124. Hsu, C. S., Chandler, D., Lowden, L. J. 1976. *Chem. Phys.* 14:213–28
125. Gibbons, R. M. 1969. *Mol. Phys.* 17:81–86
126. Gibbons, R. M. 1970. *Mol. Phys.* 18:809–16
127. Kihara, T. 1970. *Physical Chemistry. An Advanced Treatise. Vol. 5, Valency*, ed. H. Eyring. pp. 663–716. New York: Academic
128. Rigby, M. 1976. *Mol. Phys.* 32:575–78
129. Boublik, T. 1974. *Mol. Phys.* 27:1415–27
130. Pople, J. A. 1954. *Proc. R. Soc. London Ser. A* 221:498–507
131. Gubbins, K. E., Gray, C. G. 1972. *Mol. Phys.* 23:187–91
132. Twu, C. H. 1976. PhD thesis. Univ. Florida. 212 pp.
133. Gray, C. G., Gubbins, K. E., Twu, C. H. To be published
134. Flytzani-Stephanopoulos, M., Gubbins, K. E., Gray, C. G. 1975. *Mol. Phys.* 30:1649–76
135. Rasaiah, J. C., Stell, G. 1974. *Chem. Phys. Lett.* 25:519–22
136. Rasaiah, J. C., Larsen, B., Stell, G. 1975. *J. Chem. Phys.* 63:722–33
137. Stell, G., Rasaiah, J. C., Narang, H. 1974. *Mol. Phys.* 27:1393–1414
138. Twu, C. H., Gubbins, K. E., Gray, C. G. 1975. *Mol. Phys.* 30:1607–10
139. Haile, J. M., Gray, C. G., Gubbins, K. E. 1976. *J. Chem. Phys.* 64:2569–78
140. Høye, J. S., Stell, G. 1975. *J. Chem. Phys.* 63:5342–47
141. Henderson, R. L., Gray, C. G. 1977. *Mol. Phys.* In press
142. Madden, W. G., Fitts, D. D. 1974. *Chem. Phys. Lett.* 28:427–29
143. Madden, W. G., Fitts, D. D. 1976. *Mol. Phys.* 31:1923–28
144. Chambers, M. V., McDonald, I. R. 1975. *Mol. Phys.* 29:1053–62
145. Patey, G. N., Valleau, J. P. 1976. *Chem. Phys. Lett.* 42:407–10
146. Singh, Y. 1975. *Mol. Phys.* 29:155–66
147. Shukula, K. P., Ram, J., Singh, Y. 1976. *Mol. Phys.* 31:873–82
148. Haile, J. M., Gubbins, K. E., Gray, C. G. 1976. *J. Chem. Phys.* 64:1852–53
149. Gray, C. G., Gubbins, K. E. 1975. *Mol. Phys.* 30:1481–87
150. Perram, J. W., White, L. R. 1974. *Mol. Phys.* 28:527–33
151. Mo, K. C., Gubbins, K. E. 1974. *Chem. Phys. Lett.* 27:144–48; 1975. *J. Chem. Phys.* 63:1490–98
152. Sandler, S. I. 1974. *Mol. Phys.* 28:1207–23
153. Sung, S., Chandler, D. 1972. *J. Chem. Phys.* 56:4989–94
154. Bellemans, A. 1968. *Phys. Rev. Lett.* 21:527–29
155. Metropolis, N., Rosenbluth, A. W., Rosenbluth, M. N., Teller, A. H., Teller, E. 1953. *J. Chem. Phys.* 21:1087–92
156. Alder, B. J., Wainwright, T. E. 1959. *J. Chem. Phys.* 31:459–66
157. McDonald, I. R., Singer, K. 1967. *J. Chem. Phys.* 47:4766–72
158. Torrie, G. M., Valleau, J. P., Bain, A. 1973. *J. Chem. Phys.* 58:5479–83
159. Torrie, G. M., Valleau, J. P., Bain, A. 1974. *Chem. Phys. Lett.* 28:578–81
160. Bennett, C. H. 1975. *Diffusion in Solids: Recent Developments*, ed. A. S. Nowick, J. J. Burton. New York: Academic
161. Adams, D. J. 1974. *Mol. Phys.* 28:1241–52
162. Adams, D. J. 1975. *Mol. Phys.* 29:307–11

163. Rowley, L. A., Nicholson, D., Parsonage, N. G. 1975. *J. Comput. Phys.* 17:401–14
164. McDonald, I. R. 1969. *Chem. Phys. Lett.* 3:241–43
165. Wood, W. W., Parker, F. R. 1957. *J. Chem. Phys.* 27:720–33
166. Wood, W. W., Jacobson, J. D. 1959. *Proc. Western Joint Computer Conf.*, San Francisco
167. Rahman, A. 1964. *Phys. Rev. A* 136:405–11
168. Verlet, L. 1967. *Phys. Rev.* 159:98
169. Deleted in proof
170. Goldstein, H. 1953. *Classical Mechanics*, pp. 156–58. Cambridge, Mass: Addison Wesley
171. Gear, C. W. 1971. *Numerical Initial Value Problems in Ordering Differential Equations*, pp. 148–50. Englewood Cliffs, NJ: Prentice-Hall
172. Norsdieck, A. 1962. *Math. Comp.* 16:22–49
173. Cheung, P. S. Y. 1976. *Chem. Phys. Lett.* 40:19
174. Evans, D. J. 1977. *Mol. Phys.* In press
175. Euler, L. 1776. *Novi. Comment. Petrop.* 20:208–38
176. Gear, C. W. 1966. *Argonne Nat. Lab. Rep. ANL-7126*
177. Ralston, A., Wilf, H. 1966. *Mathematical Methods for Digital Computers.* New York: Wiley
178. Weeks, J. D., Chandler, D., Andersen, H. C. 1971. *J. Chem. Phys.* 54:5237–47
179. Schofield, P. 1973. *Comput. Phys. Comm.* 5:17–23
180. Erpenbeck, J. J., Wood, W. W. 1966. In *Statistical Mechanics*, Part B, ed. B. J. Berne. New York: Plenum. 238 pp.
181. Quentrec, B., Brot, C. 1973. *J. Comput. Phys.* 13:430–32
182. Bennett, C. H. 1975. *J. Comput. Phys.* 19:267–79
183. Ahmad, A., Cohen, L. 1973. *J. Comput. Phys.* 12:389–402
184. Buckingham, A. D. 1967. *Discuss. Faraday Soc.* 43:205–11
185. Beeman, D. 1976. *J. Comput. Phys.* 20:130–39
186. Cheung, P. S. Y., Powles, J. G. 1976. *Mol. Phys.* 32:1383–1405
187. Steele, W. A. 1963. *J. Chem. Phys.* 39:3197–3208
188. Sweet, J. R., Steele, W. A. 1967. *J. Chem. Phys.* 47:3022–28
189. Ladanyi, B. M., Keyes, T. 1977. *J. Chem. Phys. Mol. Phys.* 33:1063–70
190. Haile, J. M. 1976. PhD thesis. Univ. Florida. 376 pp.
191. Haile, J. M., Gubbins, K. E., Streett, W. B., Gray, C. G. To be published
192. Van Itterbeek, A., Verbeke, O. 1960. *Physica* 26:931–38
193. Laufer, J. C. 1969. PhD thesis. Princeton Univ.
194. Goldman, K., Scrase, N. G. 1969. *Physica Utrecht* 44:555–86
195. Rowlinson, J. S. 1969. *Liquids & Liquid Mixtures*, Chap. 8. London: Butterworth. 371 pp. 2nd ed.
196. Bruining, J., Clarke, J. H. R. 1976. *Mol. Phys.* 31:1425–46
197. Chandler, D., Hsu, C. S., Streett, W. B. 1977. *J. Chem. Phys.* 66:5231–34
197a. Streett, W. B., Tildesley, D. J. 1977. *J. Chem. Phys.* In press
198. Steele, W. A., Sandler, S. I. 1974. *J. Chem. Phys.* 61:1315–25
199. Smith, W. R., Madden, W. G., Fitts, D. D. 1975. *Chem. Phys. Lett.* 36:195–98
200. McDonald, I. R. 1974. *J. Phys. C* 7:1225–96
201. Verlet, L., Weis, J. J. 1974. *Mol. Phys.* 28:665–82
202. Twu, C. H., Gubbins, K. E., Gray, C. G. 1977. To be published
203. Deleted in proof
204. Singer, K., Taylor, A., Singer, J. V. L. 1977. *Mol. Phys.* 33:1757–95
205. Ryckaert, J. P., Ciccotti, G., Berendsen. J. C. 1977. *J. Comput. Phys.* 23:327–41
206. Thompson, S. M., Tildesley, D. J., Streett, W. B. 1976. *Mol. Phys.* 32:711–19

DECORATED LATTICE-GAS MODELS OF CRITICAL PHENOMENA IN FLUIDS AND FLUID MIXTURES

✣ 2653

John C. Wheeler

Chemistry Department, University of California, San Diego, La Jolla, California 92093

INTRODUCTION

The purpose of this article is to review the use of decorated lattice models in the study of critical phenomena in fluids and fluid mixtures. The study of critical phenomena at phase transitions has undergone a remarkable renaissance during the past twenty years. It is the subject of numerous monographs and reviews (1–29) and a massive literature that is itself the subject of a bibliographic volume (30). It has seen several major developments in the theoretical ideas used to understand it and is presently an extremely active area of study. Any attempt at a comprehensive review of the field here would be both futile (because of the limited space available) and superfluous (because of the many fine surveys and reviews that exist, both at the introductory and technical levels). Rather than attempt to provide a cursory overview of the entire field, I have chosen to focus on one theoretical tool that has not been adequately reviewed elsewhere. This tool has played a particularly important role in our understanding of critical phenomena in fluids and fluid mixtures, systems that are of special interest to chemists.

The remainder of the introduction gives a brief historical resumé of the major developments in our understanding of static critical phenomena in fluids and fluid mixtures and attempts to place in perspective the contribution of decorated lattice models. The following section is devoted to a brief introduction to the simple lattice-gas, Ising model in order to establish the terminology and notation needed for the discussion of decorated lattice models that follows. The possibilities for further extensions and the inherent limitations of the decorated lattice models are discussed in a concluding section.

The resurgence of interest in critical phenomena in fluids and fluid mixtures has been due to the discovery that a fluid near its critical point behaves in a manner fundamentally different from that predicted by all simple, analytic equations of state, of which the van der Waals equation (31) may be considered a prototype. These "classical" equations of state either arise from a mean-field approximation, or they are empirical formulas based entirely on analytic functions. They are unanimous in making a class of predictions about the nature of the liquid-vapor critical point that are qualitatively incorrect. As the critical point of a pure fluid is approached, they predict that the density difference $(n_l - n_g)$ between the coexisting liquid and vapor phases vanishes proportionally to the square root of the deviation, $(T_c - T)$, of the temperature T from the critical temperature T_c. Experimentally, the coexistence curve is more nearly cubic than parabolic:

$$n_l - n_g \approx B(T_c - T)^\beta \qquad (\beta \approx \tfrac{1}{3}). \qquad \qquad 1.$$

Along the critical isotherm, $(T = T_c)$, the deviation $(p - p_c)$ of the pressure from its critical value is predicted to vary as the cube of the corresponding density difference: $(n - n_c)^3$. Instead, the critical isotherm is much flatter:

$$(p - p_c) \approx D(n - n_c)^\delta \qquad \delta \approx (4\text{–}5). \qquad \qquad 2.$$

The isothermal compressibility, $\kappa_T = n^{-1}(\partial n/\partial p)_T$, which measures density fluctuations, is predicted to diverge proportionally to $|T - T_c|^{-1}$ as the critical point is approached, either from above T_c along the critical isochore, $n = n_c$, or from below T_c in the liquid or vapor phase. In fact, the divergence is more rapid:

$$K_T \approx C^+(T - T_c)^{-\gamma} \qquad (T > T_c, n = n_c)$$
$$K_T \approx C^-(T_c - T)^{-\gamma'} \qquad (T < T_c, n = n_l, n_g) \qquad \qquad 3.$$
$$(\gamma \approx \gamma' \approx 1.2).$$

The classical theories predict that the constant-volume heat capacity exhibits a finite jump discontinuity at T_c along the critical isochore, whereas experiments indicate that, in fact, C_v diverges to infinity as a small power of $|T - T_c|^{-1}$:

$$C_v \approx (A^+/\alpha)[(T - T_c)^{-\alpha} - 1] \qquad (T > T_c, n = n_c)$$
$$\approx (A^-/\alpha)[(T_c - T)^{-\alpha'} - 1] \qquad (T < T_c, n = n_c) \qquad \qquad 4.$$
$$(\alpha \approx \alpha' \approx 0.1).$$

(The functional form employed in Equation 4 is convenient because it includes the possibility of a logarithmic divergence in the limiting case $\alpha = \alpha' = 0$.)

Analogous singularities occur at critical points in ferromagnets, antiferromagnets, superfluid ^{4}He, liquid mixtures, substitutional alloys, and in certain structural transitions in solids. Similar discrepancies exist for these systems between the exponents predicted by classical theories and those found experimentally.

For more than eighty years following the discovery of the liquid-vapor critical point by Andrews in 1869 (32) it was widely accepted that critical points could be adequately described by simple analytic equations of state such as that of van der Waals. Although there were early experimental indications that the coexistence curve was more nearly cubic than parabolic (33, 34), and although the exact solution of the two-dimensional Ising model by Onsager in 1944 (35) and subsequent developments (36–38) demonstrated conclusively that the classical predictions were incorrect in two dimensions, it was not until roughly the decade between 1955 and 1965 that a determined and conclusive test was made of the mean field predictions for three-dimensional systems. Painstaking experimental investigations of pure fluids established the nonclassical exponent values given in Equations 1–4. Analogous experiments in magnets, alloys, liquid mixtures, and superfluid ^{4}He revealed strikingly similar behavior in corresponding properties with critical exponents very similar to those for fluids. Extensive theoretical analysis by series expansion methods of the three-dimensional lattice gas or spin-$\frac{1}{2}$ Ising model showed conclusively that this model—which serves as a model for pure fluids, liquid mixtures, and alloys as well as ferro- and antiferromagnets—exhibits nonclassical exponents with values close to those found for fluids. These (and subsequent) analyses established that the exponents for the lattice-gas, Ising model can be expressed quite accurately by the simple rational forms

$$\alpha = \alpha' = \tfrac{1}{8}, \qquad \beta = \tfrac{5}{16}, \qquad \gamma = \gamma' = \tfrac{5}{4}, \qquad \delta = 5. \qquad\qquad 5.$$

Moreover, the functional forms of the coexistence curve and the pressure, isothermal compressibility, and specific heat along the critical isochore and coexistence curve could be obtained, at *all temperatures*, "essentially exactly" in the sense that a graph of the functions used to model them could almost surely be superimposed upon the exact results to within the accuracy of any line drawing. Useful reviews of the work of this period can be found in References 1–4.

While a pure fluid possesses a coexistence *curve* in the $p-T$ plane, which ends at a unique critical *point*, the addition of a second component introduces another degree of freedom, and a binary fluid may have a coexistence *surface* bounded by a critical *curve*, often called the plait-point curve. The geometric and topological structure of the coexistence surface and critical curve. can be extremely complex (8, 39), which adds a richness to the possibilities for critical behavior in a binary fluid that is not possible in a pure fluid. Critical end points, critical double points, and critical azeotropy are three examples of critical behavior not found in pure fluids but exhibited by binary mixtures. In a ternary fluid mixture the third component adds still another degree of freedom, and the critical locus becomes a surface, adding still more possibilities for variation in critical behavior. In addition, entirely new kinds of critical phenomena can occur that are impossible with fewer components. For example, in a ternary mixture, three fluid phases may become simultaneously critical at a tricritical point (40–43). With the addition of a fourth or fifth component, still other interesting types of critical phenomena are possible (44). Moreover, in fluid mixtures there arises the possibility of observing phase transitions within the interface between two fluid phases (45).

Classical theories of fluid mixtures that describe phase equilibrium and critical phenomena by means of analytic equations of state, such as the multicomponent van der Waals equation of state (46) and the quasichemical approximation (47), have existed for many years. (For reviews of these methods for lattice models see References 48, 49.) As might be expected by analogy to the pure fluid case, however, they give misleading results near critical points. For liquid mixtures, in which density fluctuations play only a minor role, the spin-$\frac{1}{2}$ Ising model provides an adequate theoretical model, and experimental investigations are relatively straightforward. By 1965 it was already well established, both on the basis of Ising model results and direct experimental investigations (3, 4), that liquid mixtures exhibit nonclassical critical behavior analogous to that in pure fluids and ferromagnets.

In contrast, for fluid mixtures in which density variations play a significant role (for example, at liquid-vapor critical lines), theoretical and experimental investigations of critical phenomena are greatly complicated by the additional independent variable introduced by the second component. Experimentally, purification of components and precise determination of composition present difficulties beyond those encountered for pure fluids. Gravitational effects are more difficult to analyze and correct for in mixtures; long equilibration times for establishment of diffusional equilibrium make the experiments even more tedious than those on pure fluids. Determination of the properties of coexisting phases requires sampling of composition as well as density—a significant complication—and the specification of, and adherence to, a well-defined path of approach to the critical locus presents an additional difficulty. Theoretically, the direct generalization of the lattice-gas model to mixtures encounters similar difficulties. The presence of additional components greatly increases the difficulties in obtaining series expansions for these models. The presence of a critical curve and coexistence surface seriously complicates the analysis of the resulting series. The absence of obvious symmetries in realistic models of fluid mixtures makes series analysis vastly more difficult than for the simple lattice-gas, Ising model. Because of these difficulties, neither experimental measurements nor direct analysis of lattice models for fluid mixtures have ever reached the high level of precision obtained for the simple lattice-gas model for fluids.[1]

It was therefore a major breakthrough in the understanding of critical phenomena in mixtures when it was shown (50) that *decorated lattice-gas models* of fluid mixtures could be constructed that were not seriously restricted by the symmetry inherent in the simple lattice-gas, Ising model but could nevertheless be mapped onto it exactly. As a result the very complete and precise information gained about that model, through exact results in two dimensions and exhaustive series analysis studies in three, could immediately be applied to the seemingly much more complicated problems in mixtures. Similar decorated lattice models had earlier proved useful in understanding complex magnetic systems such as ferrimagnets (51) and the behavior of an antiferrimagnet in a uniform magnetic field (52). A review of the use of decorated lattice models in spin systems has been given by Syozi (53).

During the past decade, decorated lattice-gas models have served as a valuable guide in dealing with critical phenomena in multicomponent fluids. They provide a detailed microscopic model, which is essentially exactly soluble, for a variety of interesting phenomena such as "gas-gas equilibrium" (54, 55), critical mixing in ternary solutions (50, 54–57), maximum upper critical solution temperatures (56), critical azeotropy and critical anomalies in dilute solutions (58), lower critical solution points (59), and critical double points (60). Perhaps even more importantly,

---

[1] The extensive, careful measurements of Meyer et al on He^3-He^4 mixtures (144) and of Chu (145) are indicative of the limits of precision so far attainable in experiments.

they have served as a useful guide and testing ground in constructing more general thermodynamic arguments about behavior to be expected near multicomponent critical points, such as the theory of renormalization of exponents by Fisher (61) and the geometric approach of Griffiths & Wheeler (62).

Decorated lattice models have also been useful in understanding liquid-vapor asymmetries in pure fluids in terms of the lattice-gas Ising model, which is symmetric in liquid and vapor. Here, too, they have served both as explicit microscopic models for predicting new behavior (63–67) and understanding observed behavior (68, 69) in fluids, and also as a guide in constructing a more general thermodynamic formulation (70, 71) of the expected behavior.

Finally, although they do not fall directly into the subject matter of this review, mention should be made of uses of decorated lattice models to study complicated systems away from critical points. Three interesting examples are the studies of phase diagrams in metal-hydride systems by Hall & Stell (72), the modeling of the density maximum in water by Bell & Sallouta (73), and of aqueous solutions of nonelectrolytes by Sallouta & Bell (148).

Since 1965 a number of significant advances in our understanding of magnets and pure fluids have taken place, which have added to our understanding of mixtures as well. The introduction of the *homogeneous function*, or *scaling* equation of state (74–77) provided a powerful new tool in the analysis of experimental data. It predicted that, in terms of appropriately scaled variables, the entire two-dimensional neighborhood of the critical point could be described by a function of only a single variable, thus collapsing a whole set of isotherms or isochores into a single curve. In addition, it predicted that a set of inequalities (78–80) among the critical exponents required by thermodynamics should be satisfied as equalities. Thus, the exponents $\alpha, \beta, \gamma, \delta$ defined by Equations 1–4 were predicted to satisfy the relations:

$$\alpha + 2\beta + \gamma = 2, \quad \beta(\delta - 1) = \gamma$$
$$\alpha = \alpha', \quad \gamma = \gamma'. \qquad 6.$$

The dramatic data collapse predicted by scaling was strikingly confirmed during the period between 1965 and 1970 (81–83). The scaling idea was soon recast in a variety of forms (84–86) and generalized to deal with dynamic critical phenomena (87), transport properties (88), critical phenomena in mixtures (89, 90), lack of symmetry in pure fluids (71, 91, 92), and quite general types of critical phenomena (93–96).

During this same period another important concept evolved—that of *universality* of critical behavior and the notion of *universality classes*. (It is much harder to assign credit precisely for the idea of universality than for that of scaling. Many of the ideas developed gradually and had been stated in one form or another early in the development of modern critical phenomena. For two explicit formulations of the ideas, see References 97, 98.) While it was known that critical exponents depended upon the *spatial dimensionality* (Onsager's solution of the spin-$\frac{1}{2}$ Ising model in two dimensions gave $\beta = \frac{1}{8}$ and $\alpha = 0$, and subsequent work gave $\gamma = \frac{7}{4}, \delta = 15$), it was found, in both two and three dimensions, that the exponents were independent of the lattice type. They were found to depend on the *symmetry* of the Hamiltonian (the Heisenberg model in which all three components of spin enter on equal footing was found to have different critical exponents than the Ising model), but not upon spin quantum number, range of interactions, degree of anisotropy, or spatial non-uniformity of interactions. Thus the concept developed that, while there might be a number of different *universality classes* that differed in their spatial dimensionality or symmetries of the Hamiltonian, all members of the same class should nevertheless exhibit similar critical behavior.

Beginning in 1971 with a brilliant series of papers by Wilson and co-workers (99–101, 21), the *renormalization group approach* has developed into an extremely powerful tool for understanding critical phenomena. It provides a microscopic explanation of the origin of power laws, nonclassical critical exponents, scaling, and universality (21, 102). In addition, it provides a practical alternative to standard series expansion methods for directly calculating the values of critical exponents and the forms of scaling functions (100, 101, 103–107). Numerous reviews and introductions to the renormalization group approach are available (13, 14, 20–22, 24, 27, 108), and it will not be discussed in any detail here. One essential point of interest is that the assumption of an analytic free energy implicit in mean field theories is replaced with the alternative assumption that a suitable transformation of the underlying Hamiltonian of the system in an appropriate general space of Hamiltonians is *analytic* in the vicinity of its *fixed points*. From this assumption, together with the assumed existence of the fixed points and the nature of the transformation, follow the existence of power laws, homogeneous function equations of state, and universality of critical behavior.

Because of the extremely general terms in which the renormalization group approach is couched, it can in principle describe critical behavior in the most complex fluid mixtures as well as in the simpler pure fluids and magnets. Indeed, the renormalization group approach has been extremely valuable in sorting out the complex network of *crossovers* that can occur between the various *universality classes* in spin systems as parameters in the Hamiltonian are varied that change the symmetry, range of interaction, and spin or spatial dimensionality (20). However, it is often difficult to determine the appropriate mapping of the detailed interactions between the various molecular species of a fluid mixture into the parameters that characterize one of the very general Hamiltonian spaces appropriate to renormalization group calculations. Moreover, the global properties of the transformations, critical surfaces, and fixed point structure in this space is itself a very challenging problem within renormalization group theory. As a result, it seems likely that the ideas involved in decorated lattice models will continue to be useful in extending to seemingly more complex systems the applicability of those models for which essentially exact results are available.

THE LATTICE-GAS ISING MODEL

Certainly the most important single model for the development of modern theories of critical phenomena has been the Ising model of ferromagnetism and its many transcriptions. It is the only model of ordinary critical phenomena for which an exact solution is available in more than one dimension. The exact solution in two dimensions showed conclusively that mean field theories are inadequate at the critical point and suggested many of the current ideas in critical phenomena, e.g. power laws, scaling, and universality. Moreover, it is sufficiently simple that extensive series analysis can be used to obtain an accurate picture of its behavior in three dimensions even though no exact solution is available. In addition, it can be reinterpreted as a model for a wide variety of other transitions, thus helping to explain why so many apparently unrelated transitions share a common description.

As a model of magnetism, the spin-$\frac{1}{2}$ Ising model consists of an array of "spins," σ_i located on the C sites of a regular lattice, such as the square planar lattice shown in Figure 1. The spins are scalar variables that may take on only two values, plus or minus one, corresponding to the spin pointing *up* or *down*. The energy of a given configuration of up and down spins is given by

$$E = -J(N_{\uparrow\uparrow} + N_{\downarrow\downarrow} - N_{\uparrow\downarrow}) - H(N_\uparrow - N_\downarrow), \qquad 7.$$

where J is a coupling constant, positive for ferromagnetism and negative for antiferromagnetism, H is the magnetic field multiplied by the magnetic moment per spin and has dimensions of energy, where $N_{\uparrow\uparrow}$, $N_{\downarrow\downarrow}$, and $N_{\uparrow\downarrow}$ denote the number of nearest neighbor pairs of spins that are up-up, down-down, and up-down, respectively, while $N_\uparrow$ and $N_\downarrow$ denote the number of up and down spins. The partition function for the system is then given by

$$Z_I = \Sigma' \exp(-\beta E), \qquad 8.$$

where Σ' denotes the sum over every assignment of "up" or "down" to each spin, and $\beta = 1/kT$ where k is Boltzmann's constant and T is the absolute temperature. The free energy, f, magnetization, m, and energy, e, per spin are given, in the thermodynamic limit, by

$$f(T, H) = \lim_{C \to \infty} -\frac{kT}{C} \ln Z_I,$$

$$m(T, H) = \lim_{C \to \infty} \frac{\langle N_\uparrow - N_\downarrow \rangle}{C} = (\partial f/\partial H)_T, \qquad 9.$$

$$e(T, H) = \lim_{C \to \infty} J \frac{\langle N_{\uparrow\downarrow} - N_{\uparrow\uparrow} - N_{\downarrow\downarrow} \rangle}{C} = \left(\frac{\partial \beta f}{\partial \beta}\right)_H,$$

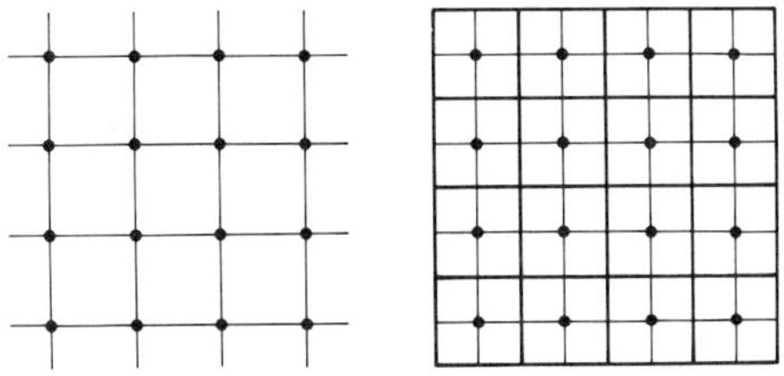

Figure 1 (a) Sites of a square-planar lattice. The lattice vertices are the locations of spins that point either "up" or "down." (b) Cells of the corresponding lattice gas, centered on the sites of the underlying lattice. Particles are free to move from cell to cell, but the pair potential depends only upon which cells the particles are in.

where $\langle \ \rangle$ denotes the statistical mechanical average. The Ising model can be generalized to higher spin and to contain longer-ranged two-spin interactions as well as four-spin, six-spin, etc interactions. Such generalizations, however, destroy its exactly soluble nature in two dimensions and make series analysis in three dimensions much more difficult.

Note that if, in the partition function sum, each "up" spin is replaced by a "down" and vice-versa, and if, simultaneously, H is replaced by $-H$, each term remains unchanged. Since to every configuration of up and down spins in the sum there is a corresponding configuration with the spins reversed, it follows that Z, and hence f, is symmetric in the field H. This means that any phase transition that occurs at nonzero H must occur at both $\pm |H|$. For ferromagnetic Ising models the free energy is in fact analytic in H except at $H = 0$, so that the transition must occur at $H = 0$, if at all (109, 110).

The Ising model can be reinterpreted as a model of a pure fluid by centering a cell of volume v_0 on each site of the underlying lattice so that the entire volume is filled, and every point in the volume is in exactly one cell. This is illustrated for the square planar lattice in Figure 1b. In the case of the square planar or simple cubic lattices the cells are themselves squares or cubes, but more generally they are the Wigner-Seitz (111) cells of the lattice. Each molecule is free to move throughout the entire volume, and the cells serve only to provide a coordinate system for determining the potential energy of a configuration of particles. The pair potential $\phi(\mathbf{r}_i, \mathbf{r}_j)$ between molecules i and j located at $\mathbf{r}_i$ and $\mathbf{r}_j$ is defined to be

$$\phi(\mathbf{r}_i, \mathbf{r}_j) = +\infty \quad (\mathbf{r}_i, \mathbf{r}_j \text{ in same cell}),$$
$$= -\varepsilon \quad (\mathbf{r}_i, \mathbf{r}_j \text{ in adjacent cells}),$$
$$= 0 \quad (\text{otherwise}). \qquad 10.$$

This corresponds to a hard-core repulsion and a short-ranged attraction. The Hamiltonian is then given by

$$H = \sum_{i=1}^{N} (\mathbf{p}_i^2/2m) + \sum_{i<j} \phi(\mathbf{r}_i, \mathbf{r}_j), \qquad 11.$$

with m the particle mass and $\mathbf{p}_i$ the momentum. The (classical) grand partition function is

$$\Xi(T, \mu, V) = \sum_{N=0}^{\infty} (N! h^{dN})^{-1} \iint d\mathbf{p}_1 \cdots d\mathbf{r}_N \exp(\beta \mu N - \beta H), \qquad 12.$$

where μ is the chemical potential, d is the spacial dimensionality, and the integral is over all values of the momentum components from $-\infty$ to $+\infty$ and over all values of $\mathbf{r}$ in the container volume. The thermodynamic pressure, $p(\mu, T)$ is given by the limit

$$\beta p = \lim_{V \to \infty} \frac{1}{V} \ln \Xi(T, \mu, V). \qquad 13.$$

The momentum integrations in Equation 12 can be carried out immediately in the usual manner (112). Moreover, in the spatial integrations the energy is unchanged by motions of a particle within a given cell so that, for a given assignment of particles to cells, the spatial integrations over the cells can be carried out to give a factor of v_0^N. The hard core prevents multiple occupancy of any cell, so the grand partition function can be rewritten

$$\Xi = \Sigma' z^N \exp(\beta \varepsilon N_{11}), \qquad 14.$$

where the sum Σ' is now over all assignments of "filled" or "empty" to each cell, where N is the total number of occupied cells, N_{11} is the number of nearest neighbor pairs of cells in which both cells are occupied, and where z is the (dimensionless) activity defined, in d spatial dimensions by

$$z = v_0 (2\pi mkT/h^2)^{d/2} \exp(\beta \mu). \qquad 15.$$

The analogy with the Ising ferromagnet is now obvious. If we make the identification of "filled" cells with "down" spins on the underlying lattice sites and "empty" cells with "up" spins, then an exact transcription between the models is possible. To complete the transcription we use the lattice identities

$$qN_1 = 2N_{11} + N_{01},$$
$$qN_0 = 2N_{00} + N_{01}, \qquad 16.$$

where q is the coordination number of the underlying lattice (the number of nearest neighbors of any site or cell), where $N_1 \equiv N$ and $N_0 \equiv C - N$ are the number of filled and empty cells, and where N_{01} and N_{00} are defined analogously to N_{11} and correspond to $N_{\uparrow\downarrow}$ and $N_{\downarrow\downarrow}$ in the Ising ferromagnet. The identities are simply a statement that the total number of neighbors of filled (or empty) cells may be counted by counting the number of bonds of each type. By use of these identities the Ising and lattice-gas partition functions can be rewritten in the form

$$Z_I = \exp(-\beta CH - qC\beta J/2) \Sigma' \exp(-2\beta H N_\downarrow - 2\beta J N_{\uparrow\downarrow}). \qquad 17.$$

$$\Xi = \Sigma'[z \exp(q\beta\varepsilon/2)]^N \exp(-\beta\varepsilon N_{01}/2). \qquad 18.$$

Except for the innocuous prefactor in Z_I, which is a simple analytic function of H and T, the two expressions are identical. The identifications

$$\varepsilon = 4J, \qquad z \exp(q\beta\varepsilon/2) = \exp(2\beta H) \qquad 19.$$

allow a precise map of results from either model to the other. If we can evaluate the sum

$$X = \Sigma' \lambda^N \xi^N_{01} \qquad 20.$$

and the corresponding thermodynamic potential

$$\pi^*(\lambda, \xi) = \lim_{C \to \infty} C^{-1} \ln X(\lambda, \xi, C), \qquad 21.$$

we will have all of the properties of both models.

In particular, the particle and potential energy densities of the lattice gas, along with the magnetization and energy per spin of the Ising model can be obtained straightforwardly from the lattice identities in Equation 16 and the derivatives of $\pi^*(\lambda, \xi)$:

$$\rho(\lambda, \xi) \equiv (\partial \pi^*/\partial \ln \lambda)_\xi = \lim_{C \to \infty} \frac{\langle N \rangle}{C}$$

$$\omega(\lambda, \xi) \equiv (\partial \pi^*/\partial \ln \xi)_\lambda = \lim_{C \to \infty} \frac{\langle N_{01} \rangle}{C}.$$
22.

The second derivatives of $\pi^*(\lambda, \xi)$ with respect to $\ln \lambda$ and $\ln \xi$ yield susceptibilities, compressibilities, heat capacities, etc.

Moreover, the symmetry of the Ising ferromagnetic free energy in H, together with its known analyticity for $H \neq 0$, implies immediately that the coexistence curve of the lattice gas in the activity-temperature plane is given by the equation

$$\lambda = z \exp(q\beta\varepsilon/2) = 1.$$
23.

The *up-down* symmetry of the Ising magnet, when H is replaced by $-H$, implies a corresponding *particle-hole* symmetry for the fluid upon replacing λ in Equation 23 with λ^{-1}. One consequence of this symmetry is that the densities of coexisting liquid and vapor must satisfy the identity

$$\rho_l + \rho_g = 2\rho_c = 1,$$
24.

where $\rho = Nv_0/V$ is the fraction of cells occupied. This exact symmetry of the coexistence curve is obviously an artificial feature of the model and not to be expected in real fluids. Rather surprisingly, however, the prediction made by Equation 23—that the coexistence curve is analytic in the chemical potential-temperature plane—seems to agree rather well with the experimental results on pure fluids (113, 114), and served as an essential ingredient in Widom's (74) formulation of scaling.

If, instead of labeling each cell as "filled" or "empty," one considers them as filled with either "component A" or "component B," one has a model of a binary liquid mixture or substitutional alloy. In this case the potential energy might be written, with an obvious notation,

$$V = \varepsilon_{AA} N_{AA} + \varepsilon_{BB} N_{BB} + \varepsilon_{AB} N_{AB}.$$
25.

With the lattice identities (Equation 16) this can be rewritten

$$V = [\varepsilon_{AB} - (\varepsilon_{AA} + \varepsilon_{BB})/2]N_{AB} + [q(\varepsilon_{AA} - \varepsilon_{BB})/2]N_A + \tfrac{1}{2}qC\varepsilon_{BB}.$$
26.

The appropriate partition function for a system in which N_A can vary, but $N_A + N_B$ is fixed, is (115)

$$Y\left(\frac{z_A}{z_B}, T, C\right) = \Sigma' \left(\frac{z_A}{z_B}\right)^{N_A} \exp(-\beta V),$$
27.

where the sum Σ' is now interpreted to be over every assignment of A or B to each cell. By use of Equation 16 this can be rewritten, in terms of the partition function $X(\lambda, \xi, C)$ in Equation 20, as

$$Y = \exp(-qC\varepsilon_{BB}/2)X(\lambda, \xi, C), \qquad 28.$$

where

$$\xi = \exp[(\varepsilon_{AB} - (\varepsilon_{AA} + \varepsilon_{BB})/2)/kT],$$
$$\lambda = (z_A/z_B)\exp[q(\varepsilon_{AA} - \varepsilon_{BB})/2kT]. \qquad 29.$$

A feature common to all of the models above is that each cell or lattice site has precisely two possible states of occupancy. This feature is essential for the exact solution in two dimensions and greatly facilitates the ease of series analysis in three dimensions. No models with more than two states of occupancy or with longer-ranged forces than nearest-neighbor have yet been studied with the care and thoroughness of the nearest-neighbor spin-$\frac{1}{2}$ Ising model. Thus, the generalization of the simple lattice-gas, Ising model to multicomponent systems by allowing each cell to have more than two states of occupancy does *not* provide a model for which the properties are known. It is this gap that the decorated lattice models help to close. They provide a way of constructing models of these more complex systems that can be mapped exactly onto the spin-$\frac{1}{2}$ Ising model.

Before turning to the decorated lattice models, we summarize briefly the steps in obtaining "essentially exact" representations or fitting functions for the properties of the lattice-gas Ising model in three dimensions. The thermodynamic potential, $\pi^*(\lambda, \xi)$, in Equation 21, and its derivatives with respect to λ and ξ are expanded in power series at low and high temperatures in the variables ξ and $(1-\xi)/(1+\xi)$, respectively, for $\lambda = 1$, or in powers of λ for general ξ. The resulting series are then analyzed to obtain estimates of the location of the critical point, ξ_c, the critical exponents $\alpha, \beta, \gamma, \delta$ and amplitudes A, B, C, D in Equations 1–4, and critical constants such as π_c and w_c. For a pedagogical introduction to these techniques see Reference 1. For more extensive technical reviews, see References 3, 116, 117. Once the critical exponents, amplitudes, and constants have been estimated, they are combined with the series expansions to produce fitting functions for π and its derivatives that agree with the estimated behavior in the critical region and also with the series expansions at high or low temperature. A variety of techniques have been used to accomplish this (117). In the method of Scesney (118), which has thus far been most used for decorated lattice problems, the critical amplitudes A, B, C in Equations 3, 1, 2 are replaced by ratios of polynomials in ξ (Padé approximants) that are determined by the requirement that the fitted function have the correct low- or high-temperature series expansion about $\xi = 0$ or 1. The reliability of such methods can be tested by the consistency of the results when different choices of model functions are used. Such tests indicate that the thermodynamic properties of interest can be fitted at *all* temperatures as accurately as could be distinguished in any reproduced line drawing, and sufficiently accurately to answer most questions concerning the properties of decorated lattice-gas models. It is in this sense that we refer to the three-dimensional spin-$\frac{1}{2}$ Ising model as an *essentially exactly soluble model*.

DECORATED LATTICE MODELS

In this section we review the various decorated lattice models of fluids and fluid mixtures that have been used in studying critical phenomena in these systems. A variety of different interpretations and methods of solution of the same basic model have been employed in applying the decorated lattice model idea to different problems. The various interpretations are well worth learning, for they make the model applicable to a wide variety of physical situations. It appears, however, that regardless of interpretation, there is a single easiest way to reduce the various models to the simple lattice-gas, Ising model that amounts to always working in an ensemble with the fewest possible constraints. One purpose of this section is to illustrate the great similarity between, and simplicity of, these models when they are treated in this way.

Mixtures: Widom's Model

A decorated lattice model for describing binary and ternary fluid mixtures was first introduced by Widom (50), who treated it in the microcanonical ensemble. The model was subsequently extended and analyzed by Clark, Neece, and Wheeler (54–56, 58) in the grand canonical ensemble. The essential idea is to augment the primary cells, which serve as the coordinate system for the molecules of the simple lattice gas (component 1), by a second set of cells, each centered on the midpoint of a bond joining the centers of two adjacent primary cells. These secondary cells serve as the coordinate system for the molecules of the second component. The secondary cells, like the primary cells, fill all of the space in the container, and each secondary cell overlaps only with the two primary cells between which it lies. A few primary and secondary cells for a square planar lattice are shown in Figure 2. The same procedure can be carried out in three dimensions. For a simple cubic lattice, for example, the secondary cells are square-based bipyramids.

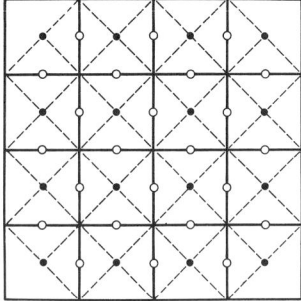

Figure 2 Primary and secondary cells of Widom's decorated lattice model. The primary cells (solid lines) are centered on the sites of the underlying square planar lattice (solid circles). The secondary cells (dashed lines) are centered on the midpoints of the bonds of the underlying lattice (open circles).

As in the simple lattice-gas model, all particles are free to move anywhere in the container volume. The cells serve only to determine the pair potential between different molecules. Two molecules of type 1 contribute energy $+\infty$ if they are in the same primary cell, $-\varepsilon$ if in adjacent primary cells, and zero otherwise, just as in the simple lattice gas. Two molecules of type 2 contribute $+\infty$ to the energy if they are in the same secondary cell, but zero otherwise. A molecule of type 1 and a molecule of type 2 contribute energy ϕ if the primary cell containing the 1 overlaps the secondary cell containing the 2, and zero otherwise. The restriction that there be no interaction between molecules of type 2 in different secondary cells is essential if the model is to be mapped exactly into the spin-$\frac{1}{2}$ Ising model. It has the unfortunate consequence that pure component 2 cannot itself undergo a liquid-vapor transition. The model is thus most appropriate for discussing binary mixtures in which the liquid-vapor critical point of one component lies far below that of the other. Nevertheless, the model is sufficiently rich that it can exhibit such interesting phenomena as critical azeotropy (58) and a maximum upper critical solution temperature (56).

There is no compelling reason why the energy of a type 2 molecule should be the sum of interactions with its neighboring cells. One might imagine, for example, that a type 2 in a secondary cell that overlaps with a single filled primary cell could adjust its position so as to experience a negative (favorable) energy, while a type 2 "squeezed" between two filled primary cells would contribute a positive (unfavorable) energy. One can generalize the model without increased difficulty (54) to allow energies $\psi_{00}, \psi_{01}, \psi_{11}$ for a molecule of type 2 that lies between two empty, one empty and one filled, or two filled primary cells. The model can also be interpreted as a three-component liquid mixture if empty primary cells are interpreted as being filled with a third component in analogy with the way in which the simple lattice gas may be interpreted as a binary liquid solution. Within the language of a two-component fluid it is most natural to choose $\psi_{00} = 0$, but when the model is interpreted as a three-component liquid mixture (54, 55), this is not necessary.

Widom (50) first investigated the model for the case $\varepsilon = 0$, $\phi = +\infty$—that is, when the only interaction beyond exclusion of multiple occupancy is a hard core repulsion between types 1 and 2. Widom treated the model in a microcanonical formulation and found a coexistence curve and plait point that he analyzed to show the existence of nonclassical critical exponents in mixtures. Clark (56) considered a generalization of the model with ϕ finite and positive, but $\varepsilon = 0$, and found a coexistence dome with a maximum critical solution temperature. Neece (54) considered the case $\varepsilon > 0$, $\phi = +\infty$ in the grand canonical formulation, both as a binary and a ternary mixture. This version of the model provides a good qualitative description of certain types of "gas-gas equilibrium" such as that found in the system He-Xe (119). Subsequently, Clark & Neece (55) considered the model with various combinations of ε and ψ_{ij} all finite and produced a variety of shapes for the coexistence surface. Wheeler (58) considered the binary case with $\varepsilon > 0$ and ϕ finite and found that the model is capable of exhibiting critical azeotropy when $\phi < 0$. This version of the model also proved useful in understanding the properties of critical phenomena in very dilute solutions.

The derivation given below of the grand partition function for this model is different (and I believe simpler) than that in any of the treatments above. Let

$$x = e^{-\varepsilon/kT}, \qquad \eta_{ij} = \exp(-\psi_{ij}/kT), \qquad i,j = 0, 1, \qquad (30)$$

and let z_1 and z_2 be the dimensionless activities defined by

$$z_1 = v_0(2\pi m_1 kT/h^2)^{d/2} \exp(\mu_1/kT),$$
$$z_2 = (qv_0/2)(2\pi m_2 kT/h^2)^{d/2} \exp(\mu_2/kT), \qquad (31)$$

where q is the coordination number, d is the dimensionality, v_0 is the volume of a primary cell, and $2v_0/q$ the volume of a secondary cell.

The grand partition function of the decorated lattice gas model can then be written as

$$\Xi = \Sigma^{(1)}\Sigma^{(2)} z_1^{N_1} x^{N_{11}} z_2^{N_2} \exp(-E^{(2)}/kT), \qquad (32)$$

where the sums $\Sigma^{(1)}$ and $\Sigma^{(2)}$ are over all assignments of "filled" or "empty" to every primary and secondary cell, respectively, where $E^{(2)}$ is the sum of all primary-secondary interactions, and where N_1 and N_2 are the number of filled primary and filled secondary cells and N_{11} is the number of filled-filled nearest neighbor pairs of primary cells. The crucial feature of the decorated lattice model that allows it to be reduced to the Ising model is that, once a specific configuration of the primary cells is chosen, the sum over filled and empty for each secondary cell can be evaluated in closed form. Thus, each secondary cell that lies between empty-empty, empty-filled, or filled-filled primary cell pairs contributes, respectively, to the partition function a factor

$$1 + z_2\eta_{00} \quad \text{(empty-empty)},$$
$$1 + z_2\eta_{01} \quad \text{(empty-filled)}, \qquad (33)$$
$$1 + z_2\eta_{11} \quad \text{(filled-filled)}.$$

Thus, the total grand partition function may be rewritten as

$$\Xi = \Sigma^{(1)} z_1^{N_1} x^{N_{11}} (1 + z_2\eta_{00})^{N_{00}} (1 + z_2\eta_{01})^{N_{01}} (1 + z_2\eta_{11})^{N_{11}}, \qquad (34)$$

where N_{00} and N_{01} are the number of empty-empty and empty-filled nearest neighbor primary cells in any term of the sum $\Sigma^{(1)}$. By use of the lattice identities in Equation 16 we can rewrite the grand partition function in terms of the partition function $X(\lambda, \xi, C)$ in Equation 20:

$$\Xi = (1 + z_2\eta_{00})^{qC/2} X(\lambda, \xi, C), \qquad (35)$$

where

$$\lambda = z_1 x^{q/2} \left(\frac{1 + z_2\eta_{11}}{1 + z_2\eta_{00}}\right)^{q/2},$$

$$\xi = \frac{x^{-1/2}(1 + z_2\eta_{01})}{[(1 + z_2\eta_{00})(1 + z_2\eta_{11})]^{1/2}}. \qquad (36)$$

This constitutes an exact correspondence between the decorated lattice-gas model and the lattice-gas Ising model. Once the details of the phase transition and critical point of the spin-$\frac{1}{2}$ Ising model are known, they can be mapped analytically to the behavior of this model. For example, the coexistence surface in this model in z_1-z_2-T space is given by the condition (cf Equation 23) $\lambda = 1$ in Equation 36, and the critical curve by the additional constraint $\xi = \xi_c = \exp(-2J/kT_c)$. The grand canonical potential,

$$\pi(z_1, z_2, x) = \lim_{C \to \infty} \frac{1}{C} \ln \Xi, \qquad 37.$$

is given by the equation:

$$\pi(z_1, z_2, x) = \tfrac{1}{2} q \ln(1 + z_2 \eta_{00}) + \pi^*(\lambda, \xi), \qquad 38.$$

where $\pi^*(\lambda, \xi)$ is the grand canonical potential for the one-component lattice defined in Equation 21. All thermodynamic properties can be obtained from $\pi(z_1, z_2, x)$ by differentiation. For example, the number densities $\rho_1 = N_1 v_0/V$ and $\rho_2 = N_2 q v_0/2V$ are given by

$$\rho_1 = (\partial \pi / \partial \ln z_1)_{z_2, x} = \rho(\lambda, \xi),$$

$$\rho_2 = (\partial \pi / \partial \ln z_2)_{z_1, x} = \frac{z_2 \eta_{00}}{1 + z_2 \eta_{00}} + \rho(\lambda, \xi) \left(\frac{\partial \ln \lambda}{\partial \ln z_2} \right)_{z_1, x}$$

$$+ \omega(\lambda, \xi) \left(\frac{\partial \ln \xi}{\partial \ln z_2} \right)_{z_1, x}, \qquad 39.$$

where $\rho(\lambda, \xi)$ are defined in Equation 22. Note that the coefficients of $\rho(\lambda, \xi)$ and $\omega(\lambda, \xi)$ are analytic functions of z_1, z_2, and x. Similarly, the (configurational) internal energy is given by

$$U = \left(\frac{\partial \pi}{\partial 1/kT} \right)_{z_1, z_2} = \left(\frac{z_2 \eta_{00}}{1 + z_2 \eta_{00}} \right) \psi_{00} + \rho(\lambda, \xi) \left(\frac{\partial \ln \lambda}{\partial 1/kT} \right)_{z_1, z_2}$$

$$+ \omega(\lambda, \xi) \left(\frac{\partial \ln \xi}{\partial 1/kT} \right)_{z_1, z_2}. \qquad 40.$$

Derivatives in which densities are held constant may be obtained as ratios of determinants of the derivatives of π^* by straightforward manipulation of Jacobians (120). For example, the specific heat at constant volume, $C_{v,x} = (\partial U/\partial T)_{\rho_1, \rho_2}$, is given by

$$-kT^2 C_V = \left(\frac{\partial U}{\partial 1/kT} \right)_{\rho_1, \rho_2} = \frac{\partial(U, \rho_1, \rho_2)}{\partial(1/kT, \ln z_1, \ln z_2)} \bigg/ \frac{\partial(\rho_1, \rho_2)}{\partial(\ln z_1, \ln z_2)}. \qquad 41.$$

In this way, Clark (56) showed that $C_{v,x}$ is finite along the critical line but diverges when the critical line passes through an extremum in temperature. The bounded character of $C_{v,x}$ was shown to be a consequence of thermodynamic constraints by Wheeler & Griffiths (121), and the behavior near an extremum in T_c was valuable in suggesting a geometric approach to critical-point thermodynamics (62). In a similar way, Wheeler (58) used this decorated lattice model to understand seemingly paradoxical anomalies in partial molar volumes and energies at critical points in very dilute solutions.

Pure Fluid Models

Decorated lattice models have also been useful in studying the liquid-vapor critical point in pure fluids. They provide a way to obtain essentially exact results for a model of the liquid-vapor transition that does not have the particle-hole symmetry of the simple, lattice-gas, Ising model. That symmetry requires that the *coexistence curve diameter*, ρ_d, defined by (cf Equation 24)

$$\rho_d = (\rho_l + \rho_g)/2, \qquad 42.$$

is constant, equal to $\rho_c = \frac{1}{2}$, for the lattice gas. For real fluids the coexistence curve diameter is known to be a function of temperature. It is predicted by mean-field theories to be linear in T near T_c, i.e.

$$\frac{d\rho_d}{dT} \to \text{const}, \qquad (T \to T_c). \qquad 43.$$

This prediction, which is the *law of the rectilinear diameter*, appeared to be borne out by the many careful experiments on critical behavior and was (and still is) often used to estimate the critical density from coexistence curve data.

In 1970, however, Widom & Rowlinson (122) discovered a continuum model of pure fluids, the penetrable sphere model, which lacks the particle-hole symmetry of the lattice gas. This model exhibits a singularity in the diameter of the coexistence curve at the critical point that is related to the (expected) singularity in the constant volume heat capacity of the model. Widom & Rowlinson showed that the penetrable sphere model was related to a model of a two-component fluid with an obvious symmetry and used this relationship to derive a number of properties of the model. While the penetrable sphere model has many attractive features, it suffers from the disadvantage that calculation of series expansions for its properties is much harder than for the lattice models. As a result, although a rigorous proof of the *existence* of a phase transition in the model has been given (123), and a few virial coefficients have been obtained for this and related models (124–126), no detailed analysis of the thermodynamic behavior of the sort available for the lattice-gas Ising model has yet been possible. Consequently, a decorated lattice-gas model of the liquid vapor transition that lacks particle-hole symmetry and exhibits a singular coexistence curve diameter is of some interest. Two such models have been discovered, both by Mermin (63, 64). These models served to emphasize the generality of the singularity to be expected in the coexistence curve diameter, and they motivated very general geometric arguments (70) for the universal presence of this singularity in real fluids. They also suggested other singularities (65) arising from lack of particle-hole symmetry in real fluids, and served as guides in constructing generalized scaling hypotheses for pure fluids (71, 92) that maintained much of the simplicity of the original symmetric scaling hypothesis but were nevertheless sufficiently general to allow for the singularities arising from lack of particle-hole symmetry.

WIDOM'S MODEL AS A PURE FLUID (MERMIN'S BAR MODEL) Widom's version of the decorated lattice model with $\varepsilon = 0$ and $\phi = +\infty$ can also be interpreted as a model of a pure fluid that has hard-core short-ranged repulsions, a special kind of many-body attraction, and lacks particle-hole symmetry (63). The development of Mermin's "bar model" given here is intended to emphasize its true lattice-gas character and its relationship both to Widom's decorated lattice-gas model and to the penetrable sphere model. While the language used here appears somewhat different from that originally employed by Mermin, the models are mathematically identical.

Consider, first, Widom's two-component model with $\varepsilon = 0$ and $\phi = +\infty$ so that molecules of type 1 and 2 cannot occupy overlapping primary and secondary cells. The grand partition function of this two-component model, $\Xi^{(2)}(z_1, z_2, C)$, is given in Equation 32. Summing over the secondary cells for each primary cell configuration yields (cf Equations 34–36):

$$\Xi^{(2)}(z_1, z_2, C) = \Sigma^{(1)} z_1^{N_1} (1 + z_2)^{N_{00}} = (1 + z_2)^{qC/2} X(\lambda, \xi), \qquad 44.$$

where

$$\lambda = z_1 (1 + z_2)^{-q/2},$$
$$\xi = (1 + z_2)^{-1/2}. \qquad 45.$$

Alternatively, we may first sum over the *primary* cells for each secondary cell configuration. A primary cell must be empty unless all q secondary cells that overlap it are empty, in which case it may be either empty or filled. Let N_q be the number of such primary cells in any secondary cell configuration. Also, let

$$z' = 1/z_2, \qquad N' = qC/2 - N_2. \qquad 46.$$

Then the grand partition function may also be written in the form

$$\Xi^{(2)}(z_1, z_2, C) = \Sigma^{(2)} z_2^{N_2} (1 + z_1)^{N_q}$$
$$= (z')^{-qC/2} \Sigma^{(2)} (z')^{N'} (1 + z_1)^{N_q}. \qquad 47.$$

Now consider a one-component model defined as follows. The particles are free to move throughout the volume, and their potential energy is determined by the secondary cells that they occupy. No two particles can occupy the same secondary cell. In addition, there is a q-body interaction potential (where q is the coordination number of the primary lattice) that contributes energy $-\varepsilon$ whenever the q secondary cells that overlap a given primary cell are all filled and zero otherwise. Reference to Figure 3 will be helpful in understanding this potential. The solid square cells are to be thought of as the *secondary* cells of Figure 2, rotated through 45°. The black circles are the centers of the primary cells (shown here as dashed lines). Note that the four-body interaction only acts at half of the possible vertices—those that are centers, not corners, of the primary cells. This feature of the model is essential

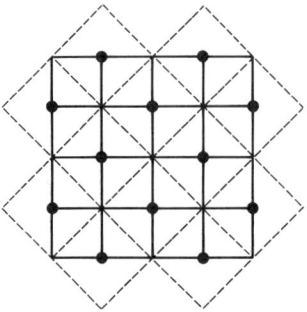

Figure 3 Cells for a one-component decorated lattice gas with q-body interactions (Mermin's bar model). The solid square cells are the *secondary* cells of Figure 2 rotated through 45°. The dashed square cells are the primary cells of Figure 2, centered at the solid circles. Particles in solid cells interact only if the q ($q = 4$ for this lattice) cells adjacent to a solid circle are all filled.

for the transcription to the simple lattice gas and is analogous to the ban on interactions between secondary cells in the two-component decorated lattice model. In the simple cubic case the secondary cells meet at two kinds of vertices: those at which six secondary cells meet, which are at the centers of primary cells and carry a six-body interaction, and those at which twelve secondary cells meet, which are at the corners of the cubic primary cells and carry no interaction.

Let

$$z' = (2v_0/q)(2\pi mkT/h^2)^{d/2} \exp(\mu/kT) \qquad 48.$$

be the activity of this one-component gas and, for any assignment of "filled" or "empty" to each secondary cell, let N' be the number of filled secondary cells and let N_q be the number of primary cells with all q overlapping secondary cells filled. (We are thus establishing a correspondence between particles in this fluid and "holes" or empty secondary cells in Widom's binary model.) Setting $x = \exp(\varepsilon/kT)$, the grand partition function of the pure fluid model, $\Xi^{(1)}(z', x, C)$, is given by

$$\Xi^{(1)}(z', x, C) = \Sigma^{(2)}(z')^{N'} x^{N_q}. \qquad 49.$$

Comparison with Equations 44–47 gives

$$\begin{aligned}\Xi^{(1)}(z', x, C) &= (z')^{qC/2} \Xi^{(2)}(x - 1, 1/z') \\ &= (1 + z')^{qC/2} X(\lambda, \xi, C),\end{aligned} \qquad 50.$$

where

$$\begin{aligned}\lambda &= (x - 1)\xi^q, \\ \xi &= [z'/(1 + z')]^{1/2}.\end{aligned} \qquad 51.$$

Thus, the grand partition function of this new one-component lattice gas with hard-core repulsions and q-body attractions can be expressed analytically in terms of that for the simple lattice gas. The density $\rho' = \langle N' \rangle / \tfrac{1}{2} q C$, and the energy density $e' = -\varepsilon \langle N_q \rangle / C$ of the new model are obtained by the chain rule of differentiation in terms of the density $\rho(\lambda, \xi)$ and (symmetrized) energy density $\omega(\lambda, \xi)$ of the lattice gas; they are found to be:

$$\rho'(z', x) = z'(1 + z')^{-1} + (1 + z')^{-1}[\rho(\lambda, \xi) + \omega(\lambda, \xi)/q], \qquad 52.$$

$$e'(z', x) = -\varepsilon x(x - 1)^{-1} \rho(\lambda, \xi).$$

The coexistence curve diameter is given by

$$\rho'_d(x) = (x - 1)^{-2/q} + [1 - (x - 1)^{-2/q}][\tfrac{1}{2} + \omega(1, (x - 1)^{-1/q})/q]. \qquad 53.$$

Its derivative, $d\rho'_d(x)/dT$ clearly diverges proportionally to the heat capacity at constant volume of the simple lattice gas at the critical point. It is a straightforward exercise in partial differentiation to verify that the constant volume heat capacity of this model also diverges proportionally to $(\partial \omega / \partial \xi)_\lambda$ at the critical point.

MERMIN'S DECORATED LATTICE MODEL A second interpretation of the decorated lattice-gas model, also introduced by Mermin (64) to study pure fluids, has proved very useful in describing both pure fluids and mixtures. In this interpretation one imagines that the primary cells shrink while the lattice spacing remains constant, and that the space left over is filled by inserting between each pair of primary cells a secondary cell of shape and size so that the entire volume is exactly filled by the primary and secondary cells together. This process is illustrated for the square lattice in Figure 4. It can also be carried out for three-dimensional lattices. For the simple cubic lattice, the secondary cell has the gemlike shape shown in Figure 5.

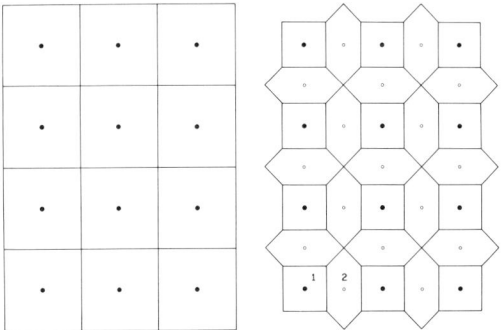

Figure 4 Cells of the simple lattice-gas model (*a*), centered on sites of a square planar lattice (solid circles), and the corresponding primary and secondary cells of the Mermin decorated lattice-gas model (*b*). The primary cells (1) shrink, and the remaining space is filled with secondary cells (2) centered on the midpoints of bonds of the underlying lattice (open circles).

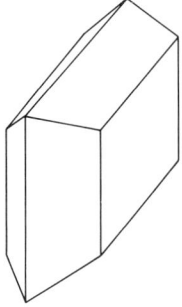

Figure 5 Shape of the secondary cell for the Mermin decorated lattice-gas model on a simple cubic lattice.

Particles in this model are free to move throughout the volume, and the potential energy is determined by the occupancy of the primary and secondary cells. A hard-core repulsion forbids multiple occupancy of either primary or secondary cells. Two particles in nearest neighbor primary cells contribute an energy $-\eta\varepsilon$, while particles in adjacent primary-secondary cell pairs contribute energy $-\lambda\varepsilon$. There is (as usual in these models) *no* interaction between particles in adjacent secondary cells. While the interpretation here is somewhat different than in Widom's model, the mathematics is identical. It is convenient and always possible, although not necessary, to choose the volumes of primary and secondary cells to be equal. In that case, if v_0 is the volume of a cell of either kind, if z is the dimensionless activity defined by Equation 15, and if $x = \exp(\varepsilon/kT)$, then the grand partition function $\Xi^{(1)}(z, x, C)$, can be written down immediately, in the notation of Equations 32–36, as

$$\Xi^{(1)}(z, x, C) = \Sigma^{(1)}\Sigma^{(2)} z^{N_1} x^{\eta N_{11}}(1 + z)^{N_{00}}(1 + zx^\lambda)^{N_{01}}(1 + zx^{2\lambda})^{N_{11}}$$
$$= (1 + z)^{qC/2} X(\lambda, \xi, C), \qquad 54.$$

where

$$\lambda = zx^{q\eta/2}\left(\frac{1 + zx^{2\lambda}}{1 + z}\right)^{q/2}$$
$$\xi = \frac{x^{-\eta/2}(1 + zx^\lambda)}{[(1 + z)(1 + zx^{2\lambda})]^{1/2}}. \qquad 55.$$

Again, as in the "bar model" and Widom's two-component model, the grand partition function of the system can be obtained exactly in terms of that for the simple lattice-gas model as a function of λ and ξ where λ and ξ are, in turn, analytic functions of the independent field variables of the system of interest. As before, the density and energy density of the model fluid can be obtained by differentiation in terms of the density and (symmetrized) energy density of the simple lattice-gas or Ising model. As in the case of the "bar model," one finds that both C_v and $d\rho_d/dT$ diverge proportionally to the constant volume heat capacity of the simple lattice gas at the critical point.

The model was originally introduced by Mermin as an example of a lattice gas with pair interactions that violated the law of the rectilinear diameter. Subsequently, it was shown by Mulholland & Rehr (67) that for negative values of (η/λ) the model possessed a rich structure of phase transitions and critical points, including the possibility of three first-order transition curves, a critical double point, and a maxithermal point. One interesting feature of the model, first noted by Rehr & Mermin (66), is that for a certain (negative) value of the ratio η/λ, the model *as a whole* regains particle-hole symmetry, but the liquid-vapor coexistence curve nevertheless has a singular diameter. This is possible because the entire coexistence curve is mapped into another high-density coexistence curve rather than into itself. Only when there is but a single coexistence curve does particle-hole symmetry guarantee the law of the rectilinear diameter.

A comparison of the model with real fluids was made by Zollweg & Mulholland (68), using approximate equations of state for the simple lattice-gas model based on series expansions. They concluded that with appropriate (positive) values of η/λ, the diameter could be fit satisfactorily away from the critical point, but the data very close to the critical point $[(T_c - T)/T_c < 7 \times 10^{-3}]$ was not sufficiently precise to determine whether or not $d\rho_d/dT$ diverged. Subsequently Mulholland, Zollweg & Levelt Sengers (69), using a similar analysis, have established correlations between behavior away from the critical point and expected visibility of asymptotic behavior.

Although the arguments for the singularity in ρ_d are extremely compelling, and such a singularity must surely exist in all real fluids, it is nevertheless exceedingly hard to see experimentally and, as yet, no completely convincing experimental demonstration of its presence has been given.[2]

Mixtures: Mermin's Model and Hybrids

The Mermin interpretation of the decorated lattice-gas model and hybrids of the Mermin and Widom interpretations have proved useful in describing mixtures as well as pure fluids.

CRITICAL DOUBLE POINTS   Bartis & Hall (60) used a hybrid model to study critical double points in "gas-gas equilibria." The phenomenon of interest is illustrated in Figure 6, where various possible pressure-temperature critical curves for decorated lattice models are illustrated. The curves of type a, c, and d can easily be obtained with Widom's model (58; J. C. Wheeler, unpublished). A more interesting case is curve b, in which the critical curve passes through a minimum temperature. The coexistence curve at this temperature in the P–x plane is sketched in Figure 7. The system exhibits a *critical double point* at which two apparently separate coexistence curves have a common critical point.

---

[2] Perhaps the most compelling experimental evidence for the existence of the singularity in pure fluids is that of Weiner, Langley & Ford (146). Similar evidence in a binary mixture has been found by Gopal et al (147).

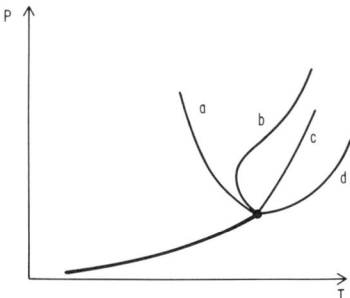

Figure 6 Coexistence curve (heavy solid line) and possible critical curves for binary systems (schematic). (Curves of type *a*, *c*, and *d* can be produced from Widom's decorated lattice-gas model. Curve *b*, which exhibits a minimum critical temperature, leads to a critical double point. This behavior was produced by Bartis and Hall, who used a hybrid of the Widom and Mermin models.

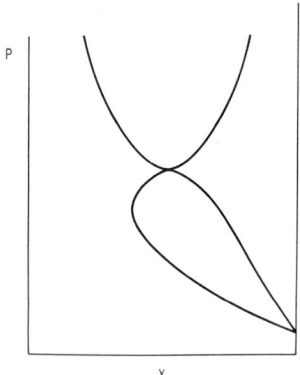

Figure 7 Pressure-composition diagram at fixed temperature (schematic) of a system exhibiting a *critical double point*. Classical theories predict that the coexistence curves should meet with nonzero slope, i.e. $\beta_D = 1$. Modern theory predicts an exponent $\beta_D = 2\beta$ with $\beta < \frac{1}{2}$, so that the curves meet with zero slope, as shown, but with infinite curvature.

An issue of some dispute in the past has been the shape of the coexistence curves in Figure 7 at the critical double point—in particular, the exponent β_D with which the composition difference vanishes as $p \to p_c$:

$$x^{(\alpha)} - x^{(\beta)} \approx B_D |p - p_c|^{\beta_D}, \quad (T = T_D). \qquad 56.$$

Classical (mean field) theories predict that β_D should be exactly unity so that the coexistence curves would have the appearance of two straight lines crossing at the double point. Both from exact analysis of their model and by a more general geometric argument, Bartis & Hall (60) found that β_D was twice the exponent β that characterizes the coexistence curves of the simple lattice gas and the mixture at typical isothermal or isobaric slices:

$$\beta_D = 2\beta. \qquad 57.$$

Thus, for mean field theories, for which $\beta = \frac{1}{2}$, β_D is indeed unity; but in real fluid mixtures and exact treatments of statistical mechanical models, where β is closer to $\frac{1}{3}$, β_D will be less than unity, and the coexistence curves should meet as in Figure 7.

In the model of Bartis & Hall, molecules of type 1 can occupy either primary or secondary cells of the Mermin variety, while molecules of type 2 can occupy only secondary cells of the Widom variety. Thus, either kind of particle can move freely throughout the entire volume, but the interactions can be of a more interesting structure than in the original Widom model. Multiple occupancy of either primary or secondary cells is forbidden. (This includes a proscription against the simultaneous occupancy by particles of type 1 and 2, respectively, of the Mermin and Widom secondary cells between the same two primary cells.) Two particles of type 1 contribute a potential energy $-\varepsilon$ if they are in adjacent primary cells, $-\psi_1$ if they are in an adjacent primary-secondary cell pair and zero otherwise. A particle of type 1 and a particle of type 2 contribute energy $-\psi_2$ if they are in an overlapping primary-secondary cell pair and zero otherwise. There is no interaction between molecules of type 2 other than the hard-core exclusion of multiple occupancy. (The model can easily be generalized to include nonadditive interactions of particles in secondary cells with the adjacent primary cells, but Bartis & Hall found this unnecessary for the existence of a critical double point.) If we let

$$x = \exp(\varepsilon/kT), \qquad \eta_1 = \exp(\psi_1/kT), \qquad \eta_2 = \exp(\psi_2/kT), \qquad 58.$$

then the grand partition function for this model may be expressed exactly in terms of that for the simple lattice gas in the following way. Given any configuration of the primary cells, a particular nearest-neighbor primary pair of cells may have no particle between them, a particle of type 1 in a Mermin secondary cell or a particle of type 2 in a Widom secondary cell. Thus, the grand partition function may be written, in the notation of Equations 32–36 as

$$\begin{aligned}\Xi(z_1, z_2, x, C) &= \Sigma^{(1)} z_1^{N_1} x^{N_{11}} (1 + z_1 + z_2)^{N_{00}} \\ &\quad \times (1 + z_1\eta_1 + z_2\eta_2)^{N_{01}} (1 + z_1\eta_1^2 + z_2\eta_2^2)^{N_{11}} \\ &= (1 + z_1 + z_2)^{qC/2} X(\lambda, \xi, C),\end{aligned} \qquad 59.$$

where

$$\lambda = z_1 x^{q/2} \left[\frac{1 + z_1\eta_1^2 + z_2\eta_2^2}{1 + z_1 + z_2}\right]^{q/2}$$

$$\xi = x^{1/2} \frac{(1 + z_1\eta_1 + z_2\eta_2)}{[(1 + z_1 + z_2)(1 + z_1\eta_1^2 + z_2\eta_2^2)]^{1/2}}. \qquad 60.$$

The coexistence surface and thermodynamic properties follow as for the models above. One significant difference is that here both ρ_1 and ρ_2 contain contributions from the energy of the reference Ising model. This is because ξ depends on z_1 as well as on z_2 and x. Bartis & Hall (60) found that a minimum in the critical temperature could be obtained only when ε was negative and ψ_1 positive, that is, when the primary-primary cell interaction between particles of type 1 was repulsive and the primary-secondary cell interaction was attractive. The necessity for this rather artificial pair potential (dubbed the "mermaid potential" because it is attractive on the average, but has a repulsive tail) is unfortunate, but with it Bartis and Hall were able to obtain coexistence curves in the P–x plane that were qualitatively quite similar to those observed near the critical double point in Ne-Kr mixtures by Schouten & Trappeniers (127). It is possible that a generalization of the model to include either Mermin primary and secondary cells of different size or nonadditive potentials between primary and secondary cells might eliminate the necessity for the unrealistic mermaid potential.

LOWER CRITICAL SOLUTION POINTS   Mermin's decorated lattice model has also been used to model lower critical solution points and closed-loop T-x coexistence curves in liquid mixtures (59). The more common phenomenon of an *upper* critical solution temperature, above which two components are miscible in all proportions but below which phase separation occurs, can easily be understood in terms of a competition between a positive energy of mixing, which favors phase separation and dominates at low temperature, and the positive entropy of mixing, which favors miscibility and dominates at high temperature. There are also mixtures that exhibit a *lower critical solution temperature*, above which phase separation occurs and below which the components are miscible in all proportions. Lower critical solution points are often found in conjunction with upper critical solution points to give closed-loop coexistence curves such as that for the system *m*-toluidine-glycerine (128) shown in Figure 8.

A theoretical understanding of these systems is somewhat more difficult than of those with only an upper critical solution point. A possible mechanism for this behavior based on orientation-dependent potentials was suggested by Hirschfelder, Stevenson & Eyring (129) in 1937. Barker & Fock (130) gave a concrete realization of this suggestion in 1953 in the form of a lattice solution with orientation-dependent interaction energies. Their model is sufficiently difficult that it could not be solved exactly or treated easily by series expansions. They examined its behavior in the quasi-chemical approximation and found that it gave closed-loop coexistence curves for appropriate choices of the interactions. Wheeler (59) has constructed a decorated lattice-gas analog of the Barker-Fock model using a decorated lattice of the Mermin type in which each cell (primary or secondary) is filled with a molecule, either of type 1 or 2. Each molecule has ω possible orientations. The only interactions occur between molecules in adjacent primary-secondary cell pairs. The interactions depend on the orientation of the molecules in secondary cells but not upon the orientation of molecules in primary cells. In this way the essential requirement is met that primary cells have only two energetically distinct states of occupation, and the model can be exactly transcribed to the simple lattice-gas model. Following Barker & Fock, the 1–1 and 2–2 interaction is taken as zero, independent of orientation. The energy of a 1 and a 2 in an adjacent primary-secondary cell pair depends upon the orientation of the molecule in the secondary cell: if its active site is pointed towards the unlike molecule, then the energy is U_2 (<0); if it is pointed in any of the ($\omega - 1$) other possible directions, then the energy is U_1 (>0).

The grand partition function appropriate to a lattice solution model in which all cells are occupied is given by (115)

$$Y(\zeta, T, C) = \Sigma'' \zeta^{N_1} \exp(-E'/kT), \qquad 61.$$

where $\zeta = z_1/z_2$ is the activity ratio, E' the energy of a configuration, and N_1 the number of molecules of type 1; the sum Σ'' runs over all assignments of "1" or "2" to each cell (primary or secondary) and over all orientational configurations of the molecules within the cells. As is always the case in decorated lattice models, the contribution of a secondary cell to the grand partition function depends only on the state of occupation (in this case "1" or "2") of the adjacent primary cells. Let Q_{ijk} ($i, j, k = 1, 2$) be the partition function of a particle of type j in a secondary cell between neighboring primary cells containing particles of type i and k. Then the grand partition for the model may be written (cf Equations 20, 32–36):

$$Y(\zeta, T, C) = \omega^C \Sigma' \zeta^N{}'(Q_{222} + \zeta Q_{212})^{N_{00}}(Q_{121} + \zeta Q_{111})^{N_{11}} \times (Q_{122} + \zeta Q_{112})^{N_{01}}$$
$$= \omega^C (Q_{222} + \zeta Q_{212})^{qC/2} X(\lambda, \xi, C), \qquad 62.$$

where

$$\lambda = \zeta \left[\frac{Q_{121} + \zeta Q_{111}}{Q_{222} + \zeta Q_{212}} \right]^{q/2} \qquad 63.$$

$$\xi = \frac{Q_{122} + \zeta Q_{112}}{[(Q_{222} + \zeta Q_{212})(Q_{121} + \zeta Q_{111})]^{1/2}}.$$

If we define

$$\eta_1 = \exp(-U_1/kT), \qquad \eta_2 = \exp(-U_2/kT), \qquad 64.$$

then the partition functions Q_{ijk} for the model defined above are

$$Q_{111} = Q_{222} = \omega,$$
$$Q_{121} = Q_{212} = 2\eta_1\eta_2 + (\omega - 2)\eta_1^2, \qquad 65.$$
$$Q_{122} = Q_{112} = \eta_2 + (\omega - 1)\eta_1.$$

Because of the 1–2 symmetry of the model, apparent in Equation 65, the coexistence curve condition, $\lambda = 1$, is satisfied by $\zeta = 1$. Provided the ratio $|U_2|/U_1$ is sufficiently small and ω sufficiently large, ξ decreases as T decreases from $+\infty$, passes through a minimum value less than ζ_c, and then increases, diverging to $+\infty$ as $T \to 0$. Thus there is a range of temperatures where phase separation occurs, bounded above and below by critical points where $\zeta = \zeta_c$.

This decorated lattice model is a distortion of the more realistic model originally proposed by Barker & Fock. Nevertheless, it does exhibit closed loop coexistence curves with nonclassical critical exponents at both upper and lower critical temperatures, for choices of interaction parameters analogous to those of Barker &

Fock. Moreover, the coexistence curves from this model are in much better agreement over the *entire* temperature range than are those obtained by Barker & Fock for their model in the quasi-chemical approximation. This is illustrated in Figure 8, which shows the coexistence curve for the system *m*-toluidine-glycerine (128) together with the decorated lattice model and Barker-Fock curves for the same ratio of upper to lower (absolute) critical solution temperature.

The reason for this improved agreement is almost surely that in the decorated lattice model, one can use an essentially exact coexistence curve for the reference model, based on series expansions in three dimensions, while Barker & Fock were forced to use a mean-field type treatment of their model. It seems certain that an *exact* solution of the Barker-Fock model in three dimensions would give greatly improved coexistence curves—perhaps better than those from the decorated lattice model. This is an interesting example of how mean field treatments—even rather sophisticated versions such as the quasi-chemical approximation—can give results that are misleading not only in the immediate vicinity of the critical point, but over the entire coexistence curve.

One reason why closed-loop coexistence curves are so sensitive to approximations of the mean-field type (at least in the decorated lattice model) is that $\xi(T)$ passes through a very shallow minimum at which ξ is still very close to ξ_c. Consequently the nonclassical exponent $\beta < \frac{1}{2}$ in the reference system reflects itself not only in the degree of the coexistence curve at the critical solution points, but also in the width of the entire curve. No such simple measure of the "effective" temperature is known in the original Barker-Fock model, but it seems likely that there also, for physically realistic choices of $|U_2|/U_1$, the system is in some sense "close to critical"

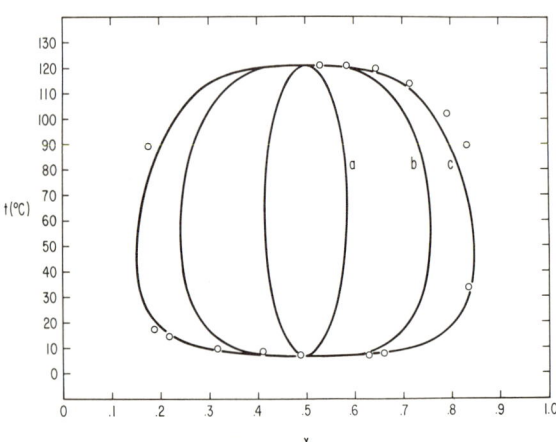

Figure 8 Lower critical solution point and closed loop coexistence curve for the system *m*-toluidine-glycerine. The open circles are the experimental data of Parvatiker & McEwan (128). Curve (*a*) is the quasi-chemical approximation to the Barker-Fock model. Curve (*b*) is the "exact" result for the decorated lattice model on a simple cubic lattice with $\omega = 6$. Curve (*c*) is the same decorated lattice model with $\omega = 5000$.

over the entire coexistence curve and therefore sensitive to any approximation in the statistical mechanics.

The decorated lattice model described above can easily be generalized in a number of ways. Larger values of ω than employed in Reference 59 correspond more realistically to the highly directional hydrogen bonds formed in many systems with closed-loop coexistence curves and also give (J. C. Wheeler, in preparation) more realistic values of $|U_2|/U_1$ and better coexistence curves, as illustrated in Figure 8. The 1–2 symmetry can be broken, and directional bonds between molecules of the same type included in order to give an adequate description of closed-loop curves in aqueous systems (G. Andersen, J. C. Wheeler, in preparation) where the curves are highly asymmetric.

MODEL WITH A PHASE TRANSITION IN BOTH PURE COMPONENTS All of the two-component fluid mixture models considered thus far have one feature in common: component 2 has no liquid-vapor coexistence curve or critical point. Thus, the simplest form of liquid-vapor critical line in a binary system—a smooth curve extending from one pure fluid critical point to the other—is not represented. In the Widom interpretation and in hybrids such as that used by Bartis & Hall (60) this is an inevitable consequence of the requirement that there be no interaction between secondary cells. However, by use of the Mermin interpretation, it is possible (J. C. Wheeler, G. Andersen, in preparation) to construct a decorated lattice-gas model that gives a simple liquid-vapor critical line extending from one pure component to the other and is nevertheless exactly reducible to the simple lattice gas.

Consider the Mermin decorated lattice in Figure 4 (or its counterpart in three dimensions). This lattice serves as a coordinate system for both component 1 and component 2. Multiple occupancy of any cell is excluded by an infinite hard-core potential. Thus any cell, primary or secondary, may be empty, filled with a 1, or filled with a 2. In order that there be only two energetically distinguishable states of a primary cell, the energy is taken to depend only on whether a primary cell is "filled" (with either a 1 or a 2) or "empty". Two adjacent "filled" primary cells contribute an energy $-\varepsilon$ as in the usual lattice gas. The only distinction between components 1 and 2 is in their strength as a secondary-cell "glue" that binds the filled primary cells. A molecule of type 1 in a secondary cell between an empty-filled or filled-filled primary cell pair contributes energy ψ_1 or ψ_{11}, respectively, while a particle of type 2 contributes energy ψ_2 or ψ_{22}. There are no interactions between secondary cells.

If we let $x = \exp(\varepsilon/kt)$ and define

$$\eta_1 = \exp(-\psi_1/kT) \qquad \eta_2 = \exp(-\psi_2/kT)$$
$$\eta_{11} = \exp(-\psi_{11}/kT) \qquad \eta_{22} = \exp(-\psi_{22}/kT), \qquad 66.$$

the grand partition function for this model can be written immediately as (cf Equations 20, 32–36)

$$\Xi(z_1, z_2, T, C) = \Sigma^{(1)}(z_1 + z_2)^{N_1} x^{N_{11}} (1 + z_1 + z_2)^{N_{00}}$$
$$\times (1 + z_1\eta_1 + z_2\eta_2)^{N_{01}} (1 + z_1\eta_{11} + z_2\eta_{22})^{N_{11}}$$
$$= (1 + z_1 + z_2)^{qC/2} X(\lambda, \xi, C), \qquad 67.$$

where

$$\lambda = (z_1 + z_2)x^{q/2}\left[\frac{1 + z_1\eta_{11} + z_2\eta_{22}}{1 + z_1 + z_2}\right]^{q/2}, \qquad 68.$$

$$\xi = \frac{x^{1/2}(1 + z_1\eta_1 + z_2\eta_2)}{[(1 + z_1 + z_2)(1 + z_1\eta_{11} + z_2\eta_{22})]^{1/2}}.$$

If we let

$$z = z_1 + z_2, \qquad \zeta = z_1/(z_1 + z_2) \qquad (0 < \zeta < 1), \qquad 69.$$

these equations can be rewritten

$$\lambda = zx^{1/2}\left[\frac{1 + z\eta^{11}(\zeta)}{1 + z}\right]^{q/2} \qquad 70.$$

$$\xi = \frac{x^{1/2}[1 + z\eta^{1}(\zeta)]}{\{(1 + z)[1 + z\eta^{11}(\zeta)]\}^{1/2}},$$

where

$$\eta^{1}(\zeta) = \zeta\eta_1 + (1 - \zeta)\eta_2, \qquad 71.$$
$$\eta^{11}(\zeta) = \zeta\eta_{11} + (1 - \zeta)\eta_{22}.$$

These are of essentially the same form, at any fixed ζ, as Equation 55 for the Mermin model of a pure fluid. They show clearly how the coexistence surface and critical curve extend smoothly from one pure component to the other as ζ is varied between 0 and 1. Very simple surfaces such as that of the ^{3}He-^{4}He system, in which the critical curve is nearly a straight line, are easily reproduced by this model. It is also capable of exhibiting negative critical azeotropy. A serious price has been paid, however, in order to obtain a phase transition in both pure components. There is no longer any way of introducing specific interactions between unlike components. Thus, one cannot adjust the energy parameters so as to favor or disfavor 1–2 contacts relative to 1–1 or 2–2 contacts. As a consequence, positive azeotropy does not exist for the model, and the critical curve for the model cannot be adjusted in shape at will. A rather artificial extension of the model does exhibit positive azeotropy, but the restriction on possible interactions remains a serious limitation to the general utility of this model. It is perhaps more useful as an illustration of the many ways in which the restrictions on solvable decorated lattice models can be met.

EXTENSIONS AND LIMITATIONS

The possibilities for new decorated lattice models have by no means been exhausted. As the last two examples illustrate, the energy may depend in a rather complicated way on many possible states of occupation of the secondary cells, and the primary cells may have more than two physically distinct states of occupation provided that they have only two *energetically* distinguishable states. Moreover, the interpretation of Widom's model as a pure fluid (Mermin's bar model) shows that the ban on interactions between secondary cells is not absolute, provided that they are of an appropriate nature.

In addition, *multiple occupancy* of primary and secondary cells can be incorporated. Secondary cells can be allowed to have any degree of multiple occupancy with arbitrary interactions between particles in the same secondary cells and with primary-secondary interactions that depend on the occupancy of the secondary cell. Multiply occupied secondary sites have been used with interesting results by Scesney (118) in a model of a mobile electron ferromagnet, and by Eggarter (131) to produce multiple spin transitions. Multiple occupancy of secondary cells can be used (G. Andersen and J. C. Wheeler, unpublished) to produce positive critical azeotropy. Primary cells may be multiply occupied provided that the interaction of a primary cell with other cells depends only on whether it is "filled" or "empty" (with however many particles). Griffiths (132) pointed out, before the discovery of the singular coexistence curve diameter, that multiple occupancy of cells of the simple lattice gas was sufficient to break the particle-hole symmetry of the lattice gas. It does not, however, lead to a singular diameter.

Despite the great versatility of the decorated lattice models in providing a microscopic description of a wide variety of critical phenomena, there are some severe limitations on the method that restrict the kinds of phase transitions and critical phenomena that can be studied. These limitations stem from the nature of the phase transition in the underlying one-component lattice gas or spin-$\frac{1}{2}$ Ising model, which is that it has only two distinct, coexisting phases. As a result, it seems unlikely that decorated lattice-gas models based on *this* underlying model can be constructed that adequately describe three- or four-phase equilibrium, critical end points, or tricritical points in fluid mixtures. To see the difficulty more clearly, consider how we might try to produce a decorated lattice-gas model with three or more coexisting phases. Since the underlying model has at most two phases, we must produce several-phase equilibrium by an intersection of two or more solutions of the equation $\lambda(z_1, z_2, T) = 1$. This is by no means impossible. Such an intersection occurs in the decorated lattice model of a ferrimagnet (133) and in Mermin's decorated lattice model of a pure fluid when nonadditive primary-secondary interactions are allowed (J. C. Wheeler, G. Andersen, unpublished). Such a point, however, is very different from an ordinary quadruple point in a typical fluid mixture. At any such multiple point there is a single set of field variables (e.g. activities, temperature, magnetic field) (62) in all of the coexisting phases. Any density in the decorated lattice model (number density, energy, or entropy density, etc) can be written in the form of a linear combination of densities of the underlying model with coefficients that depend only on fields, as in Equations 39 and 40. Thus, there are only two possible sets of values for the densities of the coexisting phases. The properties of the four coexisting "phases" at this special "quadruple" point are equal in pairs. In fact, by introducing separate activities (or magnetic fields for the ferrimagnet) for primary and secondary cells (sites), one can see that in the augmented variable space there is no quadruple point at all, but merely a curving two-phase coexistence surface that intersects the physical variable plane in two intersecting curves.

So far as we are aware, no one has yet produced a decorated lattice model in which three, rather than four, phases are in equilibrium. Even if such a model is produced, however, two of the phases must become identical in all of their properties at the triple point, as argued above. It would thus be a quite different triple point than that found in real fluid mixtures.

Similarly, no possibility of a critical end point exists for such a model. A critical end point requires that one first-order phase transition be at its critical point while another is not. In the decorated lattice-gas model, however, there is only a single value of the effective temperature variable ξ at the multiple point. Thus, if ξ is at the critical value for one coexistence curve at the multiple point, then it is necessarily at the critical value for all coexistence curves meeting there.

The limitation to two distinct coexisting phases is a restriction imposed by the underlying nearest-neighbor spin-$\frac{1}{2}$ Ising model. Ising models with longer-ranged interactions or spin greater than $\frac{1}{2}$ are capable of exhibiting multiple-phase equilibrium, and higher-order critical points (134–143). Decorated lattice models based on these underlying models could be constructed that would possess triple points, critical end points, tricritical points, etc.

To date, none of these more complex models have been studied beyond mean field approximations with anything approaching the detail with which the nearest-neighbor spin-$\frac{1}{2}$ Ising model has been studied. They are a topic of intense current interest, however, and when sufficiently exhaustive studies of these models are made—either by series expansions or by renormalization group techniques—that their properties are essentially exactly known, they will then serve as useful reference systems for a whole new family of potentially far richer and more rewarding decorated lattice models of multicomponent systems.

NOTE ADDED IN PROOF   Another interesting application of decorated lattice models, which came to our attention only after this review was completed, is the "interstitial model" of aqueous solutions of Bell & Sallouta (148). The model may be thought of as follows. Water molecules occupy either primary or secondary cells of a Mermin decorated lattice model, and solute molecules occupy the secondary cells of a Widom decorated lattice model, as in the model for critical double points of Bartis & Hall (60), discussed above. As in that model, all cells may be either filled or empty, and there is an infinite repulsion between water and solute molecules in overlapping Mermin and Widom secondary cells. The hydrogen bond in water is modeled by an energy that favors an empty decorated cell between two filled primary cells. No "hydrogen bond" is formed between water molecules in adjacent primary cells if either the Mermin or Widom secondary cell between them is occupied. This leads to a density maximum in pure water at low temperatures (73). The model is sufficiently rich to produce critical azeotropy and a variety of interesting coexistence surfaces, as well as to model the behavior of aqueous solutions at low temperatures.

ACKNOWLEDGMENTS

Support during the preparation of this review by National Science Foundation Grant CHE-75-20624, by Grant No. RR-00757 from the Division of Research

Resources, National Institutes of Health, and by an Alfred P. Sloan Foundation Fellowship is gratefully acknowledged. In addition, the author wishes to thank J. McGinness for technical assistance in its preparation.

Literature Cited

1. Fisher, M. E. 1965. In *Lectures in Theoretical Physics*, 7c:1–159. Boulder: Univ. Colo. Press. 488 pp.
2. Green, M. S., Sengers, J. V., eds. 1966. *Critical Phenomena, Proc. Conf., Wash., D. C. 1965*. Nat. Bur. Stand., Misc. pub. 273. 242 pp.
3. Fisher, M. E. 1967. *Rep. Prog. Phys.* 30:615–730
4. Heller, P. 1967. *Rep. Prog. Phys.* 30:731–826
5. Sengers, J. V, Sengers, A. L. 1968. *Chem. Eng. News* 46:104–18
6. Kac, M. 1968. In *Fundamental Problems in Statistical Mechanics*, ed. E. G. D. Cohen, 2:71–105. Amsterdam: North Holland. 338 pp.
7. Kastelyn, P. W. 1968. See Ref. 6, pp. 30–70
8. Schneider, G. M. 1970. *Adv. Chem. Phys.* 17:1–42
9. Stanley, H. E. 1971. *Introduction to Phase Transitions and Critical Phenomena*. New York, Oxford: Oxford Univ. Press. 308 pp.
10. Green, M. S., ed. 1971. *Critical Phenomena, Proceedings of the Enrico Fermi Summer School of Physics, Varenna*. London, New York: Academic. 580 pp.
11. Domb, C., Green, M. S., eds. 1972. *Phase Transitions and Critical Phenomena*, Vols. 1, 2. London, New York: Academic. 506 pp., 518 pp.
12. Schneider, G. M., ed. 1972. *Ber. Bunsenges. Phys. Chem.* 76:179–361
13. Ma, S. K. 1973. *Rev. Mod. Phys.* 45:589–614
14. Gunton, J. D., Green, M. S. eds. *Renormalization Group in Critical Phenomena and Quantum Field Theory: Proc. Conf., Temple Univ., Philadelphia, Pa., 1973*. 200 pp.
15. Domb, C., Green, M. S., eds. 1974. *Phase Transitions and Critical Phenomena*, Vol 3. London, New York: Academic. 694 pp.
16. Levelt Sengers, J. M. H. 1974. In *Proc. Van der Waals Centen. Conf. Stat. Mech.*, ed. C. Prins, pp. 73–106. Amsterdam: North-Holland. 260 pp.
17. Widom, B. 1974. See Ref. 16, pp. 107–18
18. Wilson, K. G. 1974. See Ref. 16, pp. 119–28
19. Griffiths, R. B. 1974. See Ref. 16, pp. 174–83
20. Fisher, M. E. 1974. *Rev. Mod. Phys.* 46:597–616
21. Wilson, K. G., Kogut, J. 1974. *Phys. Rep.* 12C:76–199
22. Widom, B. 1975. See Ref. 6, 3:1–45
23. Cohen, E. G. D. 1975. See Ref. 22, pp. 47–79
24. van Leeuwen, J. M. J. 1975. See Ref. 22, pp. 81–101
25. Kasteleyn, P. W. 1975. See Ref. 22, pp. 103–55
26. Wilson, K. G. 1975. *Rev. Mod. Phys.* 47:773–840
27. Ma, S. 1976. *Modern Theory of Critical Phenomena*. Reading, Mass: Benjamin. 561 pp.
28. Levelt Sengers, J. M. H. 1976. *Physica* 82A:319–51
29. Domb, C., Green, M. S., eds. 1976. *Phase Transitions and Critical Phenomena*, Vols 5a, 5b, 6. London, New York: Academic, 420 pp., 412 pp., 575 pp.
30. Stanley, H. E., ed. 1973. *Cooperative Phenomena Near Phase Transitions*. Cambridge, Mass.: MIT Press. 308 pp.
31. Van der Waals, J. D. 1873. *Die Kontinuität des gasförmingen und flussigen Zustandes I*. Leipzig: Barth.
32. Andrews, T. 1869. *Philos. Trans. R. Soc. London* 159:575–90
33. Levelt Sengers, J. M. H. 1974. See Ref. 16, pp. 78–81
34. Levelt Sengers, J. M. H. 1976. See Ref. 28, pp. 332–34
35. Onsager, L. 1944. *Phys. Rev.* 65:117–49
36. Kaufman, B. 1949. *Phys. Rev.* 76:1232–43
37. Kaufman, B., Onsager, L. 1949. *Phys. Rev.* 76:1244–52
38. Yang, C. N. 1952. *Phys. Rev.* 85:808–16
39. Rowlinson, J. S. 1969. *Liquids and Liquid Mixtures*, Chap. 6. London: Butterworth. 371 pp. 2nd ed.
40. Widom, B. 1973. *J. Phys. Chem.* 77:2196–2200

41. Griffiths, R. B., Widom, B. 1973. *Phys. Rev. A* 8:2173–75
42. Griffiths, R. B. 1974. *J. Chem. Phys.* 60:195–206
43. Lang, J. C. Jr., Widom, B. 1975. *Physica* 81A:190–213
44. Griffiths, R. B. 1975. *Phys. Rev. B* 12:345–55
45. Widom, B. 1975. *Phys. Rev. Lett.* 34:999–1001
46. Van der Waals, J. D. 1900. *Die Kontinuität des gasförmingen und flüssingen Zustandes II.* Leipzig: Barth.
47. Guggenheim, E. A. 1952. *Mixtures.* London: Oxford Univ. Press. 270 pp.
48. Domb, C. 1960. *Adv. Phys.* 9:149–361
49. Burley, D. M. 1972. See Ref. 11, Vol. 2, pp. 329–74
50. Widom, B. 1967. *J. Chem. Phys.* 46:3324–33
51. Syozi, I., Nakano, H. 1955. *Prog. Theor. Phys.* 13:69–78
52. Fisher, M. E. 1960. *Proc. R. Soc. London Ser. A* 254:66–85
53. Syozi, I. 1972. See Ref. 11, Vol. 1, Chap. 7, pp. 269–329
54. Neece, G. 1967. *J. Chem. Phys.* 47:4112–16
55. Clark, R. K., Neece, G. A. 1968. *J. Chem. Phys.* 48:2575–82
56. Clark, R. K. 1968. *J. Chem. Phys.* 48:741–8
57. Wheeler, J. C., Widom, B. 1968. *J. Am. Chem. Soc.* 90:3064–70
58. Wheeler, J. C. 1972. *Ber. Bunsenges. Phys. Chem.* 76:308–18
59. Wheeler, J. C. 1975. *J. Chem. Phys.* 62:433–39
60. Bartis, J. T., Hall, C. K. 1975. *Physica* 78:1–21
61. Fisher, M. E. 1968. *Phys. Rev.* 176:257–72
62. Griffiths, R. B., Wheeler, J. C. 1970. *Phys. Rev. A* 2:1047–64
63. Mermin, N. D. 1971. *Phys. Rev. Lett.* 26:169–72
64. Mermin, N. D. 1971. *Phys. Rev. Lett.* 26:957–59
65. Mermin, N. D., Rehr, J. J. 1971. *Phys. Rev. A* 4:2408–10
66. Rehr, J. J., Mermin, N. D. 1973. *Phys. Rev. A* 7:379–80
67. Mulholland, G. W., Rehr, J. J. 1974. *J. Chem. Phys.* 60:1297–1306
68. Zollweg, J. A., Mulholland, G. W. 1972. *J. Chem. Phys.* 57:1021–25
69. Mulholland, G. W., Zollweg, J. A., Levelt Sengers, J. M. H. 1975. *J. Chem. Phys.* 62:2535–49
70. Mermin, N. D., Rehr, J. J. 1971. *Phys. Rev. Lett.* 26:1155–56
71. Rehr, J. J., Mermin, N. D. 1973. *Phys. Rev. A* 8:472–80
72. Hall, C. K., Stell, G. 1975. *Phys. Rev. B* 11:224–38
73. Bell, G. M., Sallouta, H. 1975. *Mol. Phys.* 29:1621–37
74. Widom, B. 1965. *J. Chem. Phys.* 43:3892–97, 3898–3905
75. Domb, C., Hunter, D. L. 1965. *Proc. Phys. Soc. London* 86:1147–51
76. Kadanoff, L. P. 1966. *Physics* 2:263–72
77. Patashinskii, A. Z., Pokrovskii, V. L. 1966. *Sov. Phys. JETP* 23:292–97
78. Rushbrooke, G. S. 1963. *J. Chem. Phys.* 39:842–43
79. Griffiths, R. B. 1965. *J. Chem. Phys.* 43:1958–68
80. Griffiths, R. B. 1972. See Ref. 11, Vol. 1, pp. 7–109. (See especially pp. 98–105.)
81. Green, M. S., Vicentini-Missoni, M., Levelt Sengers, J. M. H. 1967. *Phys. Rev. Lett.* 18:1113–17
82. Vicentini-Missoni, M. 1971. See Ref. 10, pp. 157–87
83. Vicentini-Missoni, M. 1972. See Ref. 11, Vol. 2, pp. 39–78
84. Griffiths, R. B. 1967. *Phys. Rev.* 158:176–87
85. Josephson, B. D. 1969. *J. Phys. C* 2:1113–15
86. Schofield, P. 1969. *Phys. Rev. Lett.* 22:606–8
87. Halperin, B. I., Hohenberg, P. C. 1969. *Phys. Rev.* 177:952–70
88. Sengers, J. V., Keyes, P. H. 1971. *Phys. Rev. Lett.* 26:70–73
89. Saam, W. F. 1970. *Phys. Rev. A* 2:1461–66
90. Griffiths, R. B., Leung, S. S. 1973. *Phys. Rev. A* 8:2670–83
91. Green, M. S., Cooper, M. J., Levelt Sengers, J. M. H. 1971. *Phys. Rev. Lett.* 26:492–94
92. Widom, B., Stillinger, F. H. 1973. *J. Chem. Phys.* 58:616–25
93. Fisher, M. E. 1971. See Ref. 10, pp. 1–99
94. Stanley, H. E., Hankey, A., Lee, M. H. 1971. See Ref. 10, pp. 237–64
95. Hankey, A., Stanley, H. E. 1972. *Phys. Rev. B* 6:3515–42
96. Fisher, M. E. 1973. In *Collective Properties of Physical Systems,* ed. B. Lundquist, S. Lundquist. Stockholm: Nobel Foundation. 271 pp.
97. Griffiths, R. B. 1970. *Phys. Rev. Lett.* 24:1479–82
98. Kadanoff, L. 1971. See Ref. 10, pp. 100–17

99. Wilson, K. G. 1971. *Phys. Rev. B* 4: 3174–83, 3184–3205
100. Wilson, K. G. 1972. *Phys. Rev. Lett.* 28:548–51
101. Wilson, K. G., Fisher, M. E. 1972. *Phys. Rev. Lett.* 28:240–43
102. Wegner, F. 1972. *Phys. Rev. B* 5: 4529–36
103. Brezin, E., Wallace, D. J., Wilson, K. G. 1973. *Phys. Rev. B* 7:232–39
104. Brezin, E., Le Guillou, J. C., Zinn-Justin, J., Nickel, B. G. 1973. *Phys. Lett. A* 44:227–28
105. Niemeijer, T., van Leeuwen, J. M. J. 1974. *Physica* 71:17–40
106. Kadanoff, L. P., Houghton, A., Yalabik, M. 1976. *J. Stat. Phys.* 14:171–203
107. Baker, G. A., Nickel, B. G., Green, M. S., Meiron, D. I. 1976. *Phys. Rev. Lett.* 36:1351–54
108. See Ref. 29, Vol. 6
109. Yang, C. N., Lee, T. D. 1952. *Phys. Rev.* 87:404–9
110. Griffiths, R. B. 1969. *J. Math. Phys. NY* 10:1559–65
111. Ziman, J. M. 1965. *Principles of the Theory of Solids*, pp. 1–6. London, New York: Cambridge Univ. Press. 360 pp.
112. Hill, T. L. 1956. *Statistical Mechanics*, p. 123. New York: McGraw. 432 pp.
113. Wallace, B. Jr., Meyer, H. 1970. *Phys. Rev. A* 2:1563–75
114. Kierstead, H. A. 1971. *Phys. Rev. A* 3:329–39
115. Hill, T. L. 1956. See Ref. 112, pp. 293–94
116. Domb, C. 1974. See Ref. 15, pp. 357–484
117. Gaunt, D. S., Guttman, A. J. 1974. See Ref. 15, pp. 181–243
118. Scesney, P. E. 1970. *Phys. Rev. B* 1: 2274–88
119. Arons, J. D., Diepen, G. A. M. 1966. *J. Chem. Phys.* 44:2322–30
120. Callen, H. B. 1960. *Thermodynamics*, p. 128. New York: Wiley. 376 pp.
121. Wheeler, J. C., Griffiths, R. B. 1968. *Phys. Rev.* 170:249–56
122. Widom, B., Rowlinson, J. S. 1970. *J. Chem. Phys.* 52:1670–84
123. Ruelle, D. 1971. *Phys. Rev. Lett.* 27:1040–41
124. Straley, J. P., Cotter, M. A., Lie, T-J., Widom, B. 1972. *J. Chem. Phys.* 57:4484–92
125. Stillinger, F. H. Jr., Helfand, E. 1964. *J. Chem. Phys.* 41:2495–2502
126. Helfand, E., Stillinger, F. H. Jr. 1968. *J. Chem. Phys.* 49:1232–42
127. Trappeniers, N. J., Schouten, J. A. 1974. *Physica* 73:527–38, 539–45, 546–55
128. Parvatiker, R. R., McEwan, B. C. 1924. *J. Chem. Soc.* 125:1484–92
129. Hirschfelder, J., Stevenson, D., Eyring. H. 1937. *J. Chem. Phys.* 5:896–912
130. Barker, J. A., Fock, W. 1953. *Discuss. Faraday Soc.* 15:188–95
131. Fradkin, E. H., Eggarter, T. P. 1976. *Phys. Rev. A* 14:495–99
132. Griffiths, R. B., private communication
133. Bell, G. M. 1974. *J. Phys. C* 7:1189–1205
134. Meijering, J. L. 1950. *Philips Res. Rep.* 5:333–56
135. Meijering, J. L. 1951. *Philips Res. Rep.* 6:183–210
136. Blume, M., Emery, V. J., Griffiths, R. B. 1971. *Phys. Rev. A* 4:1071–77
137. Harbus, F., Stanley, H. E. 1973. *Phys. Rev. B* 8:1156–67
138. Wortis, M., Harbus, F., Stanley, H. E. 1975. *Phys. Rev. B* 11:2689–92
139. Lajzerowicz, J., Sivardière, J. 1975. *Phys. Rev. A* 11:2079–89
140. Sivardière, J., Lajzerowicz, J. 1975. *Phys. Rev. A* 11:2090–2100, 2101–10
141. Krinsky, S., Mukamel, D. 1975. *Phys. Rev. B* 11:399–410; 12:211–15
142. Berker, A. N., Wortis, M. 1976. *Phys. Rev. B* 14:4946–63
143. Burkhardt, T. W. 1976. *Phys. Rev. B* 14:1196–1201
144. Wallace, B. Jr., Meyer, H. 1972. *Phys. Rev. A* 5:953–64; Wallace, B. Jr., Harris, J., Meyer, H. 1972. *Phys. Rev. A* 5:964–67; Brown, G. R., Meyer, H. 1972. *Phys. Rev. A* 6:364–77
145. Chu, B., Lin, S. L. 1974. *J. Chem. Phys.* 61:5132–46
146. Weiner, J., Langley, K. H., Ford, N. C. 1974. *Phys. Rev. Lett.* 32:879–81
147. Gopal, E. S. R., Ramachandra, R., Chandrasekhar, P., Govindarajan, K., Subramanyan, S. V. 1974. *Phys. Rev. Lett.* 32:284–86
148. Sallouta, H. A., Bell, G. M. 1976. *Mol. Phys.* 32:839–55

ION THERMOCHEMISTRY AND SOLVATION FROM GAS PHASE ION EQUILIBRIA

✧ 2654

P. Kebarle

Chemistry Department, University of Alberta, Edmonton, Canada T6G 2G2

INTRODUCTION

Recently it has become possible to measure ion-molecule equilibria in the gas phase. The resulting vast quantity of thermochemical information on organic and inorganic ions is of great value not only to situations where gaseous ions are involved but also to the much wider area of ions in condensed phases. Thus, for example, it is possible for the first time to examine quantitatively the energetics of important classes of reactions like Brönsted and Lewis acids and bases in the absence of solvent.

The present article presents thermochemical data for organic and inorganic ions obtained by measurement of ion-molecule reaction equilibria in the gas phase. A brief history of the development in ion-equilibrium measurements is followed by a brief examination of apparatus and methods. The available thermochemical data are organized according to the reactions whose equilibria were measured: (a) proton transfer equilibria with positive ions: $B_1H^+ + B_2 = B_1 + B_2H^+$, which lead to data on the proton affinities of the bases B; (b) proton transfer in negative ions: $A_1^- + A_2H = A_1H + A_2^-$, which lead to dilute gas phase acidities of the acids AH and to $D(A-H) - EA(A)$, where D and EA are bond dissociation energy and electron affinity; (c) hydride ion transfer in positive ions $R_1^+ + R_2H = R_1H + R_2^+$ leading to hydride ion affinities of carbonium ions R^+ and heats of formation of carbonium ions R^+; (d) association and clustering reactions of positive or negative ions $I^\pm$ with molecules: $I^\pm + nM = (IM_n)^\pm$, which provide data on the stabilities of ion-molecule complexes, Lewis acidities and basicities, strong hydrogen-bonding energies, and ionic solvation in the gas phase.

DEVELOPMENT OF THE ION EQUILIBRIUM METHOD

Results from ion-equilibria measurements in solution recorded in the form of acid-base dissociation constants, instability constants of ion-ligand complexes, solubility products, and electrochemical reduction potentials represent one of the most important reservoirs of numerical data available to the chemist. The measurement and recording of such data began in the early 1900s. Similarly valuable information was potentially available from measurement of ion-molecule equilibria in the dilute gas phase. However, because ions in the gas phase become discharged either on collision with the wall of the apparatus or on collision with an ion of opposite charge, the study of gas phase ion equilibria requires special techniques that were only developed relatively recently. The methods emerged from the general field of mass spectrometric studies of ion-molecule reactions in the gas phase. For recent reviews of ion-molecule reactions see Ausloos (1), Franklin (2, 3), McDaniel (4), Beauchamp (5), and Ferguson (6).

In the pressure range above 1 torr and typically at 4–10 torr, which is "high" by mass spectrometric standards, discharge of the ions on the walls of the ion source is slow because the ions must diffuse through the gas. Positive-negative charge recombination is also slow when low ion densities are used. On the other hand, the high concentration of neutral reactant increases the ion-molecule reaction rates to the point where they may be many orders of magnitude faster than the ion discharge rates. In ion sources that are free of external electrical fields these conditions lead to the establishment of true thermal ion-molecule reaction equilibria. The accidental observation of the equilibria $H^+(H_2O)_{n-1} + H_2O = H^+(H_2O)_n$ by Kebarle (7), and recognition of the general power of the method led, in the period 1964–1972, to the first systematic studies of ion-molecule equilibria (7–24). These studies were executed in mass spectrometric apparatus specially designed for high ion source pressure conditions and variable ion source temperatures so that not only $\Delta G° = -RT \ln K$ could be obtained but also $\Delta H°$ and $\Delta S°$ by means of van't Hoff plots of the equilibrium constants K obtained at different temperatures.

A very important expansion of the ion-equilibria studies occurred in 1971 when McIver, Bowers & Aue (25) reported the first proton transfer equilibria measured with a pulsed electron beam, trapped ion cell, ion cyclotron resonance (ICR) spectrometer. Soon afterwards Beauchamp, McIver & Taft (26) reported proton transfer equilibria measurements also done with ICR instruments. Independently of the ICR work, and at the same time, proton transfer equilibria measurements had also been initiated in Alberta with the high-pressure mass spectrometric technique (27). Since then equilibrium measurements with the flowing afterglow technique have also been reported by Ferguson & Fehsenfeld (28) and particularly Bohme (29). The measurement of hydride transfer equilibria was introduced by Field (30), who used pulsed high-pressure mass spectrometry. The increase of the ion thermochemical data based on ion-equilibria measurements has been extremely rapid in recent years, and even more data may be expected in the near future.

CONDITIONS AND APPARATUS FOR MEASUREMENT OF ION EQUILIBRIA

The thermodynamic conditions for the measurement of gas phase ion equilibria have been considered recently (31). We restate them briefly, using the proton transfer reaction $A_1^- + A_2H = A_1H + A_2^-$ as an example:

(a) The reactants and products must be in thermal equilibrium with their common surroundings (carrier gas or walls of the reaction vessel). Since the neutrals AH are generally stable and in vast concentration excess they will be in thermal equilibrium with the gas and walls. However, the ions may be nonthermal because of energy imparted to them in the primary ionization, energy gained due to applied electric potentials, or energy contained in ionic products due to exothermicity of the reaction by which the ion was produced.

(b) The path(s) that reactively couple the ions engaged in the equilibrium must be appreciably faster than all other processes affecting the concentration of the ions.

(c) Sufficient time must be given for the system to reach equilibrium.

In practice, conditions b and c require that the kinetics of the reaction system be well understood.

The varying apparatuses used achieve conditions a–c in somewhat different ways as detailed below.

Three different types of apparatus have been successfully applied to the measurements of ion equilibria. These are mass spectrometry that utilizes a pulsed electron beam and a high-pressure ion source, the flowing afterglow, and the pulsed electron beam, trapped ion ICR spectrometer. The mass spectrometer that employs a pulsed electron beam and a high-pressure ion source, which is presently used in my laboratory, has been described in greater detail elsewhere (32). It consists of a temperature-controlled ($-190°$–$650°C$) reaction chamber in which a suitable reaction mixture, at 0.5–10 torr total pressure is irradiated by short electron pulses. The ions created by a given electron pulse react and reach equilibrium as they diffuse towards the walls of the ion source. Ions diffusing to the vicinity of a very narrow slit (3 × 0.010 mm) are carried by gas flow into an evacuated region pumped by a high-capacity pumping system. Here the ions are captured by electric fields, accelerated and subjected to magnetic or quadrupole mass analysis. Collection of an ion of given mass in a multichannel scaler gives the time dependence of the ion after the electron pulse. Since the concentration of the neutral molecules is constant, the intensity ratios of ions after they reach equilibrium become constant. The ion-thermalization condition is met by having the ions suffer many collisions with an inert third gas like CH_4 before they experience a reactive collision. No electric fields are present in the ion source. A pulsed electron beam, high-pressure mass spectrometer of design similar to that described above is also being successfully used in Field's laboratory (33).

Positive ions of metals like K^+, Cs^+, Sr^+, etc cannot be successfully produced by electron beam irradiation of molecules. However, they may be obtained for use in high-pressure ion sources by thermionic emission from a filament painted with a suitable salt containing the desired ion. Apparatus designed for ion-equilibria measurements involving such ions has been described by Kebarle (18, 20, 34). Apparatus of somewhat similar design has also been developed by Castleman (35–38).

The flowing afterglow apparatus (Ferguson & Fehsenfeld 39, 40, Bohme 41) is a fast-flow system. The carrier gas, most commonly helium, passes with very high flow velocity ($\bar{v} \approx 10^4$ cm sec^{-1}) through the reactor tube, which is 8 cm in diameter and 1 m long. The pressure is typically 0.5 torr. The ions are obtained by electron impact.

The electrons originate from a heated filament mounted upstream in the tube. Suitable gases may be added upstream of the filament in order to obtain the desired primary ions. The neutral reactant gases are added through one or more nozzles downstream of the filament. Some distance d downstream of the neutral reactant inlet, the ions are sampled by bleeding a small part of the gas through a small orifice into an evacuated region. Mounted in the evacuated region is a quadrupole mass spectrometer with which the ions are detected. The reaction time $t = d/\bar{v}$ is a few milliseconds. It remains constant in rate constant and equilibrium runs. Since the linear velocity at a given cross section of the tube is not constant but a function of the radius (viscous flow), the effective reaction time is obtained by integration over the flow profile. The method has been highly developed and has proved itself as an excellent source of rate constants. Equilibrium constants can be obtained either by separate measurements of the forward and reverse rate constants or by admitting sufficient reactant such that the reaction reaches equilibrium in a time shorter than the effective reaction time (Bohme 41). Experiments at different reaction temperatures are possible but the fast flow and low gas pressure require rather elaborate arrangements (40) to insure uniform gas temperature.

A drawback common to all systems described so far is connected with the fact that the ion concentrations are not determined in situ but by bleeding of gas through a sampling leak into mass analyzers located in a vacuum. Possible ion sample adulteration has been discussed (10, 42). Particularly serious is cluster breakup in the region immediately past the sampling orifice. Since the ions are already accelerated by electric fields, dissociation may occur on collision of the ions with neutral molecules (42). For example, a proton-held dimer like $(NH_3HNH_3)^+$ may be collisionally dissociated to $NH_4^+ + NH_3$. Collisional dissociation is of particular concern in the measurement of association equilibria like $NH_4^+ + NH_3 = (NH_3HNH_3)^+$. However, it may also affect the measurement of other equilibria. Thus in the proton transfer equilibria, $B_1H^+ + B_2 = B_1 + B_2H^+$, appreciable concentration of dimers $(BHB)^+$ may also be present at equilibrium. Their collisional breakup outside the ion source leads to nonequilibrium B^+ ions, which will affect the measurement of the proton transfer reaction. Cluster breakup is avoided by reducing the size of the sampling leak.

The pulsed, trapped ion ICR spectrometer developed by McIver (43) for the determination of rate constants and long reaction sequences proved most useful for the determination of ion equilibria (25). The mode of operation particularly deviates from conventional ICR instrumentation (5) by the introduction of electron pulsing and ion trapping. The ions are produced in a rectangular cell ($1 \times 1 \times 3.5$ inch) by a short (0.1 msec) electron pulse. The gas pressure is typically 10^{-6} torr. The ions are trapped by the magnetic field H in the z direction, and in the x and y direction by small static voltages (~ 1 V) applied to the side plates of the cell. The ions circle in the xy plane with their cyclotron frequency $\omega_c = qH/mc$, where m and q are mass and charge. After a desired reaction time t, the concentration of ions with a given m/q is determined by pulsing the magnetic field H to a value such that the resonance frequency ω_c is equal to the applied RF frequency of a marginal oscillator. A somewhat different pulsed and trapped ICR arrangement has also been described by McMahon & Beauchamp (44).

Concerning the ion-equilibria determinations, the ICR spectrometer has the unique advantage of detecting the ion concentrations in situ and does not depend on a sampling leak. Because of efficient ion trapping, equilibrium can be observed, for second-order reactions, even though the pressures of the reactants may be as low as 10^{-6} torr.

The presence of trapping voltages and the very low pressures utilized in the ICR method have caused some concern as to whether the equilibria measured correspond to truly thermal processes. Recently this point has been examined (45, 46) and several experimental tests have been given that demonstrate the thermal nature of the equilibrium results.

For various reasons (46), it becomes difficult to measure equilibria in which the second-order rate constants for the forward and reverse reaction differ by more than a factor of 100. This corresponds to a maximum free energy change of $\Delta G° = 300\,R \ln 100 \approx 2.5$ kcal mole^{-1} (45, 46). Free energy differences that are larger by more than a factor of three can be measured with the high-pressure method (27). This makes the high-pressure method more suitable for the rapid coverage of a new group of compounds spread over a wide $\Delta G°$ gap.

Adduct or clustering reactions that are third order are too slow, and the equilibria cannot be measured by the ICR technique.

Up to the present, ICR spectrometers have only been operated at room temperature. This means that experimental determination of only $\Delta G°$ values has been possible, but not of $\Delta H°$ values. However, there is no fundamental cause that restricts the operation to a single temperature. In fact, a variable temperature, pulsed and trapped ICR spectrometer has been constructed, and its performance will be reported in the near future (M. T. Bowers, private communication).

Each of the three methods described above has its unique advantages. The considerable abundance of ion thermochemical data obtained from ion-equilibria measurements is due to important contributions from each method.

PROTON TRANSFER EQUILIBRIA WITH POSITIVE IONS: PROTON AFFINITIES

Proton transfer is probably the most important class of reactions from the standpoint of ion thermochemistry. Excellent recent reviews by Taft (47) and Bohme (41) are available. Nevertheless the present section should be useful, particularly since numerous new measurements have been published since completion of the above work.

The general reaction involving proton transfer in positive ions is given by Equation 1:

$$B_1H^+ + B_2 = B_1 + B_2H^+, \qquad \qquad 1.$$

The free energy change is obtained from Equation 2, where K is given by Equation 3:

$$-RT \ln K = \Delta G_T°, \qquad \qquad 2.$$

$$K_1 = \frac{[B_1][B_2H^+]}{[B_2][B_1H^+]}. \qquad \qquad 3.$$

Most of the equilibrium measurements have been done at a single temperature: 300°K by ICR (McIver, Beauchamp, Taft, Bowers, Aue), 300°K by flowing afterglow (Bohme), and 600°K by pulsed high-pressure mass spectrometry (Kebarle). The first two types of instruments were not provided with facility for temperature variation. The high-pressure mass spectrometry measurements had to be done most often at the higher temperature because the adduct-forming Reaction 4 is fast at high pressure.

$$BH^+ + B + M = BHB^+ + M. \qquad 4.$$

Bases like water, alcohols, and amines form strongly hydrogen-bonded adducts BHB^+, and at room temperature the Equilibrium 4 is shifted almost completely to the right. At 600°K the Equilibrium 4 is shifted to the left, and the Equilibria 1 and 4 can be observed simultaneously (27, 48).

The temperature dependence of some proton transfer equilibria has been measured by high-pressure mass spectrometry. van't Hoff plots of the results (27, 48, 49) have shown that the entropy changes ΔS_1 are generally small, as could have been expected because of cancellation of entropy terms in Reaction 1. van't Hoff plots could be obtained over a relatively wide temperature range for proton transfer equilibria involving protonated benzenes (49). The entropy changes in these reactions were found to be essentially identical to the entropy change calculated by only considering changes in rotational symmetry numbers σ. This entropy change for Reaction 1 is given by Equation 5:

$$\Delta S°_{rot\,sym} = R \ln \frac{\sigma(B_1H^+)\sigma(B_2)}{\sigma(B_2H^+)\sigma(B_1)}. \qquad 5.$$

As is shown below, other experimental results also indicate that a very good estimate of the total entropy change for most proton transfer reactions involving the neutral bases B can be obtained by considering $\Delta S°_{rot\,sym}$ only.

The proton affinity of the base B_2, $PA(B_2)$, defined as the enthalpy change for the gas phase reaction $B_2H^+ = B_2 + H^+$ can be obtained from ΔH_1 and Equation 6 if $PA(B_1)$ is known:

$$\Delta H_1 = PA(B_1) - PA(B_2). \qquad 6.$$

In practice the gas phase basicities and proton affinities are most often measured systematically by obtaining a continuous "ladder" of overlapping proton transfer free energy changes (Figure 1). The multiplicity of the thermodynamic cycles increases the confidence in the results obtained. In general the consistency of the multiple cycles is within 0.2–0.5 kcal mole^{-1} for a chain of compounds with total $\Delta G°_1$ difference of some 10 kcal mole^{-1}. Corresponding ladders of ΔH_1 values can be obtained from the approximation $\Delta G°_1 = \Delta H°_1 - T\Delta S°_{rot\,sym}$. The availability of only one (external) standard value $PA(B_0)$ allows the determination of all the other proton affinities.

The available proton affinities can be divided into three groups: compounds of low PA, from H_2 to H_2O; compounds of medium PA, from water to ammonia; and compounds with high PA, above ammonia.

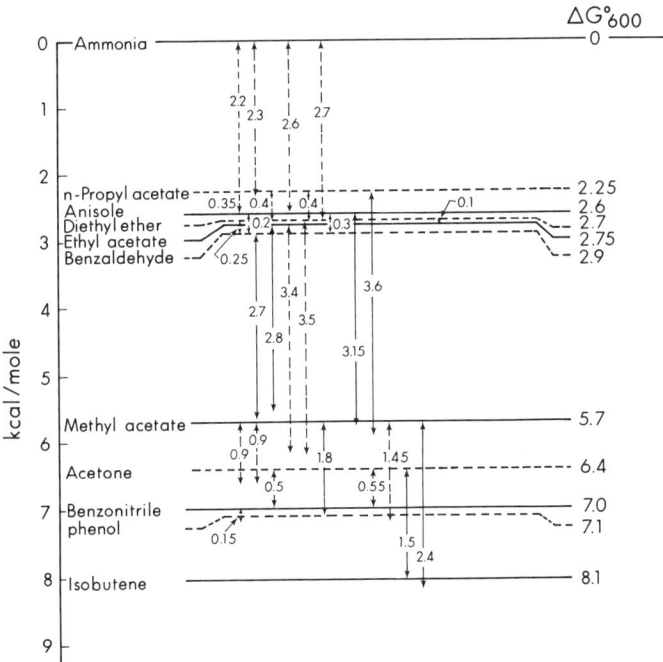

Figure 1 "Ladder" of redundant determinations connecting external standard iso-butene to ammonia. [From Lau, Yamdagni, and Kebarle (49, 55).] Numbers between arrows correspond to ΔG°_{600} for reaction $B_1H^+ + B_2 = B_1 + B_2H^+$, where B_1 is the base appearing on top of the arrow. Dashed arrows connect to dashed lower lines. ΔG°_{600} given on right side of figure is for reaction $NH_4^+ + B = NH_3 + BH^+$.

PROTON AFFINITIES OF COMPOUNDS BETWEEN H_2 AND H_2O

The proton affinities of the first group of compounds, contributed mostly by Bohme (41, 50), are summarized in Table 1. The external standard $PA(H_2) = 101$ kcal mole^{-1} was used (50), which is based on a theoretical calculation of the total energy of H_3^+ by Duben & Lowe (51). Some cross checks with other external values also exist. Thus $PA(CH_4) = 128.2$ kcal mole^{-1} is close to a determination $PA(CH_4) \approx 127$ kcal mole^{-1} by Chupka & Berkowitz (52) based on appearance potentials and thermochemical inferences. Bohme's value for ethane $PA(C_2H_6) = 140.4$ kcal mole^{-1} (see Table 1) is close to the value 139.7 kcal mole^{-1} obtained by Hiraoka & Kebarle (53) from a study of the equilibrium $C_2H_5^+ + H_2 = C_2H_7^+$.

He, Ne, and Ar have proton affinities that are lower than that of H_2. Values of 42, 40, and 90 kcal mole^{-1}, respectively, have been quoted by Beauchamp (54). These determinations were not made by ion equilibria. A variety of compounds have proton affinities that fall between those of C_2H_6 and H_2O. Three such groups

Table 1 Proton affinities of neutral molecules B—Compounds between H_2 and H_2O

B	$PA(B)$ (kcal mole^{-1})	Ref.
H_2^a	101	50, 51
O_2	101.05	41
Kr	101.4	50
N_2	113.7	41
Xe	114.2	41
CO_2	126.8	41
CH_4	128.2	41
N_2O	135.6	41
CO	139.0	41
C_2H_6(A)	131.9[b]	53
C_2H_6(B)	139.7[b], 140.4	41, 53

[a] Primary standard (50, 51).
[b] Two isomers are possible in the protonation of C_2H_6. The less stable isomer $C_2H_7^+$(A) behaves as a very loosely bonded $C_2H_5^+ \cdot H_2$. For evidence that $C_2H_7^+$(B) is C–C protonated ethane see Hiraoka & Kebarle (53).

are the hydrogen halides, the alkyl halides, and the paraffins. Equilibrium values for the hydrogen and alkyl halides are not available, but values based on other determinations can be found in Beauchamp (53). Proton affinities of alkanes higher than ethane have not been determined by proton transfer because the protonated compounds are unstable and decompose to an alkyl ion and H_2 or a paraffin. Some insights into the structures and stabilities of protonated alkanes can be obtained from condensation equilibria like $C_2H_5^+ + CH_4 = C_3H_9^+$. These are discussed in the last section of this paper.

PROTON AFFINITIES OF COMPOUNDS BETWEEN H_2O AND NH_3

Many important classes of compounds have proton affinities in the intermediate range, i.e. between those of water and ammonia. The important groups are π donor bases like olefins and aromatic compounds, and oxygen n donor bases like alcohols, ethers, aldehydes, ketones, acids, and esters. Recently the range between water and ammonia was covered by a complete ladder of proton transfer equilibria (Yamdagni & Kebarle 55), see Figure 1. An even more comprehensive ladder covering the same range will be published shortly (McIver, Taft & Beauchamp 45). The information obtained from the above two publications and two recent equilibrium studies of substituted benzenes (49, 56) is summarized in Tables 2 and 3. Table 2 compares data from the ICR ladder (45) obtained at $300°K$ ($p \sim 10^{-5}$ torr) with values from the high-pressure measurements (55) obtained at $600°K$. The enthalpy changes $\Delta H°_{300}$ and $\Delta H°_{600}$ were calculated from the corresponding $\Delta G°$ changes and $\Delta S_{\text{rot sym}}$

(see Equation 5). In the absence of experimental error and entropy contributions other than rotational symmetry, one expects $\Delta H°_{300} \approx \Delta H°_{600}$. Inspection of Table 2 shows that $\Delta G°_{300}$ values are already rather close to $\Delta G°_{600}$. However, the ΔH values are generally much closer. The table thus illustrates the following points: entropy changes are small, a good estimate of the (small) total entropy change is obtained by considering only $\Delta S_{\text{rot sym}}$, and the experimental data from proton transfer equilibrium obtained by two very different methods at very different conditions are in good agreement.

Table 3 gives the proton affinities obtained by proton equilibria measurements (45, 47–49, 55, 56). The compounds are arranged in groups: alcohols, ethers, etc. The absolute PA in Tables 2 and 3 are based on the external standard $PA(i - C_4H_8) = 194.06$ kcal mole^{-1}. This value is based on the ionization potential of t-butyl by Lossing (57) and the bond dissociation energy of $D(t - C_4H_9 - H)$ by Tsang (58) that leads to $\Delta H_f(t - C_4H_9^+) = 169$ kcal mole^{-1}. This, combined with $\Delta H_f(H^+) = 367.1$ kcal mole^{-1} (59), leads to $PA(i - C_4H_8) = 194.06$ kcal mole^1. $\Delta G_{600} = 8.05$ kcal mole^{-1} was obtained in Figure 1 by the combined data of Lau (49) and Yamdagni (55) for the reaction $NH_4^+ + i$-butene $= NH_3 + t - C_4H_9^+$. The corresponding $\Delta H = 8.35$ kcal mole^{-1}. This leads to $PA(NH_3) = 194 + 8.35 = 202.35$ kcal mole^{-1}.

Table 2 Comparison of $\Delta G°$ and $\Delta H°$ for gas phase proton transfer in reaction $NH_4^+ + B = NH_3 + BH^+$ obtained by two different methods[a]

B	$\Delta G°_{300}$[b]	$\Delta G°_{600}$[c]	$\Delta H°_{300}$[d]	$\Delta H°_{600}$[d]
NH_3	0	0	0	0
EtOAc	3.4	2.7	4.2	4.3
Et_2O	3.7	2.6	4.9	5.0
MeOAc	6.1	5.3	6.9	6.9
Me_2CO	7.2	6.1	8.4	8.5
$Me_2C=CH_2$	8.6	7.7	8.8	8.1
HCO_2n-Pr	9.2	9.0	10.0	10.6
HCO_2Et	10.3	9.5	11.1	11.1
Me_2O	11.0	10.6	12.2	13.0
EtCHO	13.5	12.8	14.3	14.4
HCO_2Me	13.8	12.9	14.6	14.5
MeCN	14.5	13.3	15.3	14.9
EtOH	14.8	13.9	15.6	15.5
McCHO	16.5	13.3	17.3	16.9
MeOH	19.3	18.6	20.1	20.2
HCO_2H	21.8	22.6	22.2	23.2
H_2S	27.8	28.7	28.4	29.9
H_2O	31.4	31.8	32.0	33.0

[a] In kcal mole^{-1}. Adapted from (45).
[b] Pulsed electron beam ICR (45) measured at 300°K.
[c] Pulsed electron beam high-pressure mass spectrometry (55) measured at 600°K.
[d] Enthalpy changes calculated from corresponding $\Delta G°$ by assuming $\Delta S = \Delta S_{\text{rot sym}} = \ln(\sigma NH_4^+ \sigma B/\sigma NH_3 \sigma BH^+)$, where σ are rotational symmetry numbers.

Table 3 Proton affinities of compounds with basicities between those of water and ammonia[a]

Alcohols and ethers		Aldehydes and ketones		Acids and esters		Substituted benzenes		Thiophenes, mercaptans, nitriles	
						Substituent			
HOH	169.3[b], 170.3	H_2CO	174.6	CF_3COOH	170.3[b]	Cl	179[c]	H_2S	173.9
CF_3CH_2OH	172.2	CH_3CHO	185, 185.4[b]	HCOOH	178.1[b]	F	178.8[c]	MeSH	185.5
CHF_2CH_2OH	177.9	EtCHO	188, 187.9[b]	$CH_2ClCOOH$	182.4[b]	H	180.1[c]	EtSH	189.3
CCl_3CH_2OH	179	n-PrCHO	189.7	CH_3COOH	187.4[b]	Me	188.5[c]	i-PrSH	192.0
CH_3OH	182.1[b], 182.2	n-BuCHO	190.6	C_2H_5COOH	189.5[b]	Et	189.5[c], 187.3[d]	t-BuSH	194.6
C_2H_5OH	186.8[b], 186.7	i-PrCHO	190.6	$HCOOCH_3$	187.8[b], 187.9	n-Prop	188.3[d]	Me_2S	197.6
CH_3OCH_3	190.1, 189.3[b]	$(CH_3)_2CO$	193.9, 193.8[b]	$HCOOC_2H_5$	191.2[b], 191.2	i-Prop	188.7[d]	EtSMe	200.7
	192.5			$HCOOn$-Pr	191.6[b], 192.3	n-But	188.7[d]	Et_2S	202.6
$CH_3OC_2H_5$	194			$HCOOn$-Bu	192.5	t-But	188.9[d]	$(n$-Pr$)_2$S	204.0
THF	196.4			CF_3CO_2Me	180.3	NO_2	191.1[c]	MeCN	187.0
THP	197.1			CF_3CO_2Et	183.2	OH	193.5[c]	EtCN	190.2
$C_2H_5OC_2H_5$	197.4, 197.3[b]			CF_3CO_2n-Pr	184.2	CN	193.6[c]		
				CF_3CO_2n-Bu	184.4	CHO	197.6[c]		
				CH_3CO_2Me	195.4[b], 195.4	OMe	197.9[c]		
				CH_3CO_2Et	198[b], 198	NH_2	207.8[c]		
				CH_3CO_2n-Pr	198.5[b]				

[a] All proton affinities given in kcal mole^{-1}. Absolute values based on $PA(NH_3) = 202.3$ kcal mole^{-1} and ΔH for proton transfer from NH_4^+. Values without superscript from Beauchamp, McIver & Taft (45).
[b] Yamdagni & Kebarle (55).
[c] Lau & Kebarle (49).
[d] McIver (56).

The availability of precise basicity data like those in Tables 2 and 3 permits the examination of substituent effects on alkylation, site of protonation (49), resonance, hyperconjugative effects, etc. Comprehensive treatments of the above topics are available (45, 47).

BASICITIES OF COMPOUNDS WITH PROTON AFFINITIES HIGHER THAN THAT OF AMMONIA

The compounds whose PA is greater than that of NH_3 are almost exclusively nitrogen bases, although there are some exceptions. Proton transfer equilibria involving nitrogen bases were the first to be studied (25–27) and therefore there are presently a very large number of determinations available. These are summarized in Table 4. The proton affinities are normalized to the value of ammonia $PA(NH_3) = 202.3$ kcal mole^{-1}, which was discussed in the previous section. The data in Table 4 are based exclusively on work by Taft, McIver and co-workers (47). Although numerous determinations have also been done by Aue & Bowers (25, 46) and some by Kebarle & Yamdagni (27, 60), these are generally in agreement with the values of Taft (47). Since the data are numerous it was felt that only one consistent set should be presented in the review.

The availability of the gas phase basicity data for the aliphatic amines, substituted anilines, and pyridines has led to several examinations of the solvent effects on the basicities of these compounds. Arnett et al (61) have made a complete thermodynamic analysis of the "anomalous" order of alkyl-amine basicities in aqueous solution. These results have shown that the large stabilizing effect of alkyl substitution on the gaseous onium ions is almost cancelled by an equally large decrease in the free energy of solvation of these ions. Therefore the aqueous basicity changes with alkyl substitution are small and sensitive to small changes in other properties. An extension of such Born-type cycle analysis to fluorosulfonic acid solutions was also made by Arnett (62). Elaborations of the interpretation were presented in two review articles (Arnett 63, 64).

Aue et al (65, 66) have also compared the gaseous and aqueous basicity of alkyl amines. The analysis (66), apart from Born-type cycles, also attempts to rationalize the observed changes on the basis of extra-thermodynamic assumptions. In particular, the enthalpy of solvation of the alkyl ammonium ions $\Delta H_s^{\circ}(BH^+)$, whose magnitude can be obtained from the cycles, is expressed as a sum of two terms, a "neutral solvation term" and an "electrostatic term." The neutral term is assumed equal to the known enthalpy of solvation of the neutral bases $\Delta H_s^{\circ}(B)$. This allows an evaluation of the "electrostatic" terms. The electrostatic terms are correlated to the charge density of the ions with use of the Born equation. Hydrogen bonding to the ions is not explicitly considered.

Reversal of gaseous and aqueous acidities often occurs when the "substituents" are bulky hydrocarbon groups and the basic functional group is small. On the other hand, substituents of large bases like aniline or pyridine lead (67, 68) to linearly related gaseous and solution basicity changes. In such linear relationships the

Table 4 Proton affinities of compounds with basicities higher than that of ammonia[a]

Amines			Pyridines				Anilines				Miscellaneous Compounds		
	(ΔG°_{300})[a]	(ΔH°_{300})[b]	(PA)[c]	Substituent	(ΔG°_{300})[a]	(ΔH°_{300})[b]	(PA)[c]	Substituent	(ΔG°_{300})[a]	(ΔH°_{300})[b]	(PA)[c]		(ΔG°_{300})[a]	(ΔH°_{300})[b]	(PA)[c]
NH_3	0	0	202.3	2-CN	4.1	3.3	205.6	p-F	4.3	4.1	206.4	DMSO	7.2	6.0	208.2
$MeNHCH_2CN$	2.4	2.0	204.2	4-NO_2	4.7	3.9	206.2	p-Cl	4.8	4.6	206.9	DMF	7.5	6.7	209.0
$H_2N(CH_2)_2CN$	2.4	2.2	204.4	3-CN	5.3	4.5	206.8	H	6.7	6.5	208.8	DMA	12.4	11.6	213.9
$F_2CHCH_2NH_2$	3.8	3.6	205.6	4-CN	6.1	5.3	207.6	N,N-Me_2	18.6	17.8	220.1	LiOH	—	—	~240.7[d]
CF_3CH_2NHMe	5.9	5.5	207.8	2-F	6.9	6.1	208.4					NaOH	—	—	~247.6[d]
$CF_3CH_2CH_2NH_2$	6.9	6.7	209.0	2-CF_3	7.5	6.7	209.0					KOH	—	—	~262.6[d]
$NCCH_2NMe_2$	7.2	6.4	208.7	3-CF_3	8.5	7.7	210					CsOH	—	—	~269.2[d]
$FCH_2CH_2NH_2$	8.1	7.9	210.2	4-CF_3	8.8	8.0	210.3								
$MeNH_2$	9.2	9.0	211.3	3-F	10.0	9.2	211.5								
$CF_3(CH_2)_3NH_2$	10.2	10.0	211.3	2-Cl	10.3	9.5	211.8								
$CF_3CH_2NMe_2$	11.0	10.2	213.5	3-Cl	10.8	10.0	212.3								
$CH_2=CHCH_2NH_2$	11.6	11.4	213.7	4-F	12.7	11.9	214.2								
$(HC≡CCH_2)_2NH$	11.9	11.5	213.8	4-Cl	13.5	12.7	215.0								
$C_2H_5NH_2$	11.9	11.7	214.0	H	16.6	15.8	218.1								
n-$PrNH_2$	13.4	13.2	215.5	2-MeO	17.7	16.9	219.2								
n-$BuNH_2$	13.9	13.7	216.0	3-Me	19.3	18.5	220.8								
i-$PrNH_2$	14.4	14.2	216.5	3-MeO	19.6	18.8	221.1								
Neopentyl NH_2	15.1	14.9	217.2												
Me_2NH	16.0	15.6	217.9												
c-$C_6H_{11}NH_2$	18.4	16.2	218.5												
t-$BuNH_2$	16.7	16.5	218.8												
MeEtNH	18.5	18.1	220.4												
$(H_2C=CHCH_2)_2NH$	19.9	19.5	221.8												
Me_3N	20.6	19.8	222.1												
Et_2NH	20.8	20.4	222.7												
Piperidine	21.2	20.8	223.1												
Et i-PrNH	22.3	21.9	224.2												
(n-$Pr)_2NH$	22.8	22.4	224.7												
Me_2Et N	23.0	22.2	224.5												
(i-$Pr)_2NH$	24.7	24.3	226.6												
(i-Pr)NMe_2	25.2	24.4	226.7												
Et_3N	27.5	26.7	229.0												
(t-Bu)$_2NH$	28.0	27.6	229.9												
(n-$Pr)_3N$	29.4	28.6	230.9												

[a] ΔG°_{300} for reaction $BH^+ + NH_3 = B + NH_4^+$ from Taft (47).
[b] ΔH°_{300} for reaction as in footnote a and $\Delta S^\circ = R\ln(\sigma_{NH_3}\sigma_{BH^+})/(\sigma_B\sigma_{NH_4^+})$.
[c] Proton affinity from ΔH°_{300} and $PA(NH_3) = 202.3$ kcal mole^{-1}.
[d] From $\Delta H^\circ_{0,1}$ for reaction $M^+ + OH_2 = MOH_2^+$ and where M^+ = alkali ion and thermochemical data (16).

aqueous basicities are found to be considerably attenuated. For example the attenuation factor observed (68) for the free energies of protonation of the substituted pyridine is ~ 2.

As mentioned earlier, the entropy change for gas phase proton transfer reactions is small and can be quite accurately estimated by consideration of the rotational symmetry numbers. However, an important exception occurs when one of the bases has two basic groups such that internal cyclization can occur after protonation. The loss of internal freedom on cyclization can lead to substantial entropy loss. Determination of the temperature dependence of the equilibrium constant K_1, which involves $B_1H^+ = Me_3NH^+$ and $B_2 = \alpha,\omega$-diamino propane, pentane, and heptane, obtained with high-pressure mass spectrometry experiments (60), led to measured entropy losses of about 20 eu. The proton affinities of the diamines were found (60) to be considerably larger than those of the monoamines of similar structure, which must again be a consequence of the internal hydrogen bond. High basicities resulting from proton-bridged cyclization of two basic groups have also been observed in ICR experiments (69, 70).

The lowest ionization potential of nitrogen bases generally involves removal of one of the lone pair electrons on the nitrogen. Substituents that decrease this ionization potential may be expected to increase the gas phase basicity since electronic charge is withdrawn from the nitrogen in both cases. This correlation has been confirmed experimentally (46, 71, 72).

The availability of proton affinities of the bases B, together with their lowest ionization potentials, permit the evaluation of the hydrogen affinity of B^+. The hydrogen affinity is defined as the enthalpy change of the reaction $BH^+ = B^+ + H$. Since the proton affinities and the ionization potentials are related, the hydrogen affinities are found to follow similar trends (45, 46, 71).

The ionization energies of the core 1s electrons of the heteroatom of the bases B measured with ESCA apparatus were also found to be linearly related to the proton affinities. This relationship was first examined by Shirley (73). Their results showed that the correlation was close to 1:1, i.e. the increase of PA was nearly the same as the decrease of the core 1s ionization potential. While the 1:1 correlation is not rigorously obeyed, the quality of the linear relationship in related bases is often so good that it permits determination of the site of protonation (74). Thus protonation in anisole or benzaldehyde may be on the ring or on the oxygen of the substituent group. A linear correlation of 1s oxygen core electron energies of oxygen bases with the corresponding proton affinities showed that the PA of benzaldehyde fitted on the line, while that of anisole was too high (75). This meant that benzaldehyde was substituent protonated, while anisole was ring protonated.

The availability of exact proton affinities has also been of considerable significance to theoretical quantum mechanical calculations. The energy changes ΔE for proton transfer processes like $NH_3^+ + MeNH_2 = NH_3 + MeNH_3^+$, calculated by evaluating the total energy E of each reactant by the minimal basis set STO-3G method (76), were found to be very close to the enthalpy changes for the same reactions. The accuracy of the ΔE values results from cancellation of errors in the evaluations of the total energies E of the reactants since the proton transfer reactions are isodesmic processes (77), i.e. the number and the nature of the bonds remains constant in the reaction. The results from such theoretical calculations have provided a very useful contribution to the understanding of the underlying electronic effects (78). Also, since the site of protonation is exactly known in the theoretical calculation, comparison of experimental energy changes with theoretical ΔE allows one to establish the most basic site of protonation. For example, a recent equilibrium study of the proton affinities of substituted benzenes (49) showed that the proton affinities for N- or ring-protonation of aniline must be very close. Recent theoretical calculations (79) could establish that the ion obtained on protonation of the amino group is more stable, but only by 1–3 kcal mole^{-1}. Another example of a combination of experimental proton affinities and theoretical ΔE calculations is a study of the Baker-Nathan effect (80) that showed that the effect is not an intrinsic molecular property but is caused by solvation.

PROTON TRANSFER TO NEGATIVE IONS—INTRINSIC ACIDITIES OF AH

The proton transfer equilibria involving negative ions are represented by Equation 7:

$$A_1^- + A_2H = A_1H + A_2^-. \qquad 7.$$

The available literature data on gas phase acidities are given in Table 5. The vast majority of the data on aliphatic carboxylic acids, benzoic acids, phenols, carbon, and nitrogen acids were obtained by high-pressure mass spectrometry (81–87). Some measurements by ICR on substituted phenols and alcohols were made by McIver (88–90).

The enthalpy change for Reaction 7 can be related to the bond dissociation energies D(H-A) and the electron affinities EA(A) by Equation 8:

$$\Delta H_7 = D(A_2 - H) - EA(A_2) - D(A_1 - H) + EA(A_1) \qquad 8.$$

The proton affinities of A$^-$ PA(A$^-$) are connected to ΔH_7 by Equations 9 and 10, where I_p(H) = 313.6 kcal mole^{-1} is the ionization potential of hydrogen:

$$\Delta H_7 = PA(A_2^-) - PA(A_1^-), \qquad 9.$$

$$PA(A^-) = D(A\text{-}H) - EA(A) + I_p(H). \qquad 10.$$

To obtain absolute data from ΔH_7, the D(A–H) – EA(A) for at least one of the acids involved in the ladder of equilibria must be known. The standard acid selected (81–85) was HCl since the values D(H–Cl) = 103 kcal mole^{-1} (59) and EA(Cl) = 83 kcal mole^{-1} (91) are well established. The proton transfer reaction involving HCl as a standard is given by Equation 11:

$$Cl^- + HA = HCl + A^-. \qquad 11.$$

The $\Delta G_{11}^\circ(600^\circ K)$ obtained directly from the equilibrium measurements (81–86) are shown in Table 5.

As explained in the preceding section, most of the proton transfer equilibria were measured at a single temperature: 300°K by ICR and 600°K by high-pressure

ION THERMOCHEMISTRY 459

Table 5 Acidities in dilute gas phase from proton transfer equilibria measurements[a]

AH	(ΔG^-)[b]	$(D-EA)$[c]	Ref.	AH	(ΔG^-)[b]	$(D-EA)$[c]	Ref.
CH_3OH	—	63.2	89	Trifluoroacetone	33.2	36.7	84
C_2H_5OH	—	61.3	89	Phenol[d]	33.2	36.2	85
$i\text{-}C_3H_7OH$	—	59.9	89	Diethyl malonate	32.4	34.7	84
$t\text{-}C_4H_9OH$	—	59.1	89	Acetic acid	31.7	34.9	81
$(CH_3)_3CCH_2OH$	—	57.8	89	Acetyl urea	31.1	34.9	84
Acetonitrile	55.6	59.9	84, 86	Propionic acid	30.5	33.7	81–83
DMSO	57.3	61.0	86	Diacetamide	30.2	33.2	84
Acetone	52.5	56.4	84, 86	n-Butyric acid	29.7	32.9	81–83
2-Butanone	51.9	54.6	84, 86	γ-Chlorobutyric acid	28.6	31.8	81–83
3-Pentanone	51.2	54.6	84, 86	Formic acid	28.4	31.6	81–83
3-Methyl, 2-Butanone	51.0	54.0	84, 86	Acetyl acetone	27.8	30.1	84
Acetaldehyde	50.1	53.4	86	Succinimide	27.6	31.6	84
Propionaldehyde	49.9	52.8	86	β-Chlorobutyric acid	25.1	28.3	81–83
Dimethyl sulfone	49.0	52.8	86	Benzoyl acetone	24.1	26.3	84
Diphenyl methane	46.1	49.1	84	β-Chloropropionic acid	24.0	27.2	81–83
1,4-Pentadiene	45.7	47.5	84	Benzoic acid[d]	23.1	26.9	85
Acetophenone	45.8	49.2	84, 86	Fluoroacetic acid	21.1	24.0	81–83
Propiophenone	44.7	47.6	84, 86	α-Chlorobutyric acid	21.1	23.9	81–83
Pyrrole	41.6	45.6	84	Dimedone	20.3	25.3	84
Nitromethane	41.1	44.0	86	α-Chloropropionic acid	20.0	22.8	81–83
Nitroethane	41.0	43.7	86	HCl	20.0	20.0	59, 91
2-Nitropropane	40.9	43.0	86	Chloroacetic acid	19.0	21.8	81–83
Cyclopentadiene	38.8	41.9	84	Bromoacetic acid	17.8	20.4	81–83
p-Nitrotoluene	36.8	39.5	86	Malonitrile	17.4	22.4	84
H_2S	36.2	38.4	81–83	Barbituric acid	16.5	21.5	84
Phenylacetone	36.0	38.2	84	Difluoroacetic acid	13.9	16.4	81–83
Acetanilide	35.1	39.2	84	Trifluoroacetyl acetone	12.6	14.9	84
Fluorene	34.7	39.7	84	Dichloroacetic acid	12.3	14.8	81–83
Phenylacetonitrile	34.8	38.3	84	Trifluoroacetic acid	6.9	9.1	81–83
Acetonyl acetate	33.3	36.7	84				

[a] All values in kcal mole^{-1}. $D(H\text{-}Cl) - EA(Cl) = 20$ kcal mole^{-1} (59, 91) used as primary standard.
[b] ΔG^- corresponds to $\Delta G^\circ = \Delta G^\circ_{11} + 20$ kcal mole^{-1}, where ΔG°_{11} is free energy change for proton transfer Reaction 11, $Cl^- + HA = HCl + A^-$, measured generally at 600°K. ΔG° corresponds to $D(HA) - EA(A)$ if the assumption is made that $\Delta G^\circ_{11} = \Delta H^\circ_{11}$ since $D(H\text{-}Cl) - EA(Cl) = 20$ kcal mole^{-1} and EA stand for bond dissociation energy and electron affinity.
[c] $D(H\text{-}A) - EA(A)$ obtained from $\Delta H^\circ_{11} = \Delta G^\circ_{11} + T\Delta S^\circ_{11}$. ΔS°_{11} is evaluated by considering symmetry number changes, changes of internal rotation, and changes of moments of inertia (86). The proton affinity of $A^- = D(H\text{-}A) - EA(A) + I_p(H) = D(H\text{-}A) - EA(A) + 313.6$ kcal mole^{-1}.
[d] Values for gas phase acidities and $D(AH) - EA(A)$ of many monosubstituted phenols and benzoic acids available in (85).

mass spectrometry. A few measurements of the temperature dependence of the equilibrium constant K_7 showed (81–85) that ΔS is generally small. Although it was recognized (82, 84) that in Reaction 7 significant entropy changes may occur because of the creation and disappearance of internal rotations and rotational barriers, the assumption $\Delta G_T^\circ = \Delta H_T^\circ$ was generally made (81–85) in the presentation of the data. Recent reexamination of the entropy changes has shown (86) that this assumption was not warranted in all cases. In particular, Reaction 11 involves a large entropy increase due to the appearance of two rotational degrees of freedom in the change from Cl$^-$ to HCl. This rotational change is generally not cancelled by the change HA → A$^-$ since the AH involved are most often large molecules so that the moments of inertia of HA and A$^-$ are essentially the same. However, changes of rotational symmetry numbers, disappearance of internal rotations (for example, internal rotation around O–C bond in phenol), and changes of internal rotational barriers occur. The combined effects of these lead to a $T\Delta S$ term for Reaction 11 at 600°K that may be as large as 6 kcal mole^{-1} but is generally in the 2–3 kcal mole^{-1} range (86).

The values for $D(A\text{-}H) - EA(A)$ given in Table 5 were obtained from ΔG_{11}° and estimates of ΔS_{11}° based on consideration of rotational changes involving not only symmetry numbers but also changes of moments of inertia and barriers for internal rotations (86). Since the barriers for internal rotation in the ions A$^-$ had to be estimated (86), the accuracy of the D–EA values is in some cases no better than ± 2 kcal mole^{-1}.

A comparison between the $D(A\text{–}H) - EA(A)$ data from the acid equilibria and from separate determinations of $D(A - H)$ and $EA(A)$ available in the literature is given in Table 6. Most electron affinities used are due to recent measurements of photodetachment thresholds of A$^-$ by Brauman (92–94a,b,h). The interpretation of the thresholds for polyatomic species and the extraction of an electron affinity is not a simple, straightforward matter (92–94a,b,h). Nevertheless, these results are probably accurate to ± 1.5 kcal mole^{-1}. The literature bond dissociation energies are judged to have a similar accuracy, i.e. ± 2 kcal mole^{-1}. Inspection of Table 6 shows that the $D(A\text{–}H) - EA(A)$ data obtained from the equilibria are generally about 1 kcal mole^{-1} higher than the literature data. This is as good an agreement

Table 6 Comparison between $D(A\text{-}H) - EA(A)$ from ion equilibria measurements and thermochemical data[a]

AH	$D(A\text{-}H)$[c]	Ref.	$EA(A)$[d]	Ref.	$D(A\text{-}H) - EA(A)$ literature	$D(A\text{-}H) - EA(A)$ equilibria
HCl	103	59	83	91	20	20[b]
H$_2$S	91.1	59	53.5	94c	37.6	38.4
Phenol	89.0	94d	54.5	93	34.5	36.2
c-Pentadiene	81.2	94e	42.4	92	38.8	41.9
2-Butanone	92.3	94f	38.6	94b	53.7	54.6
Acetone	98.0	94f	40.6	94b	57.4	56.2
CH$_3$CN	92.9	94g	34.8	94h	58.1	59.3

[a] All values in kcal mole^{-1}.
[b] Primary standard used in equilibria measurements.
[c] Error in literature bond dissociation energy determinations generally ± 2 kcal mole^{-1} except for HCl, which is accurately known.
[d] Error in literature electron affinity generally ± 1.5 kcal mole^{-1} except for Cl and HS, which are accurately known.

as could be expected in view of the uncertainties of the literature data and the entropy estimates made in the equilibrium determinations.

The results obtained by McIver (89) on the alcohol acidities are shown in Table 5. The absolute $D(HA) - EA(A)$ values were obtained (89) by use of HF as external standard:

$$F^- + HA = HF + A^-. \qquad 12.$$

ΔH_{12} was obtained by estimating ΔS_{12} with the assumption that $S°(RO^-) = S°(RF)$, since the entropies of the isoelectronic alkyl fluorides RF are available in the literature.

The acidities of the aliphatic alcohols are seen to increase in the order MeOH < EtOH < i-PrOH, i.e. with increasing size of the alkyl substituent. The same order was observed earlier in qualitative gas phase acidity measurements by Brauman & Blair (95). This order was unexpected by organic chemists since the reverse order occurs in solution. Brauman (95) explained that the stabilization of RO^- by the polarizability of the group R is the important effect in the gas phase; the acidities in solution must be affected by adverse changes in the solvation of RO^- with increase of R.

The acidities of the aliphatic acids increase in the order MeCOOH < EtCOOH < BuCOOH, i.e. with increasing size of the hydrocarbon substituent (Table 5 and Reference 81). This order is the same as the order of alcohol acidities and should be due to the same reason, namely the stabilization of the negative ion by the increasing polarizability of the alkyl groups. Since formic acid is more acidic than butyric acid, it evidently does not fit in the above order. This reversal has been explained (81, 96) in terms of the permanent dipole moment effect of the alkyl substituent on sp^2 hybridized systems.

The halo substituents of aliphatic acids have a strong acidifying effect (81, 82 and Table 5). This is due to stabilization of the acid anion by the electron-withdrawing halogen atom. The substituent effect in the gas phase is about 4 times larger than in solution and is a ΔH effect rather than the ΔS effect observed in solution (81, 82). The order F > Cl > Br, observed in solution, is reversed in the gas phase (81, 82, 87).

Examination of the substituent effects in benzoic acids and phenols has shown that good linear correlations are obtained with the aqueous acidities and substituent σ values (85). However, the substituent effects in aqueous solution are attenuated by a factor of ~10 for benzoic acids and ~7 for phenols (85). In the gas phase the substituents affect the enthalpy, while in aqueous solution the substituent induces an entropy change. The reasons for this can be explained in terms of the gas phase data (85). The gas phase acidities of substituted benzoic acids and phenols were used by Arnett (98) for the construction of Born-type cycles by which the substituent effects can be dissected into intrinsic molecular effects and solvation effects on the anions A^- and neutral acids AH. Among other things these results show that the effect of acidifying substituents in the gas phase (ΔH) is almost completely cancelled in protic solvents by an adverse change of the enthalpy of solvation of A^-. Gas phase studies of the hydrogen bonds between phenolate anions and protic solvents

(Cumming, French, and Kebarle, to be published) show that this decrease in solvation enthalpy is a consequence of the decrease in hydrogen bonding between the anion and protic solvent molecules.

Only a few of the bond dissociation energies of substituted benzoic acids and phenols are known. However, because the substituent effect changes the electron affinity much more than the dissociation energy, estimates of the electron affinities are possible by assuming that the bond energies remain approximately constant (85).

The acidities of a number of carbon and nitrogen acids measured by McMahon & Kebarle (84) are given in Table 5. Inspection of the table shows that a number of these compounds have surprisingly high gas phase acidities. Thus barbituric acid and malononitrile are stronger gas phase acids than HCl, and diacetamide, acetylurea, diethylmalonate, dimedone, and succinimide are stronger than acetic acid. Yet these carbon and nitrogen acids are so weak in aqueous solution that their acidity cannot be measured directly. The reason for this drastic difference in behavior lies in the fact that the anions of the carbon and nitrogen acids achieve their stability not through a strongly electronegative atom but through charge delocalization into a relatively large molecule. The solvation of such anions is very poor, particularly in protic solvents (84).

The acidities of carbon and nitrogen acids can be measured in aprotic solvents like DMSO. Bordwell (97) has made a comparison of the acidities in the gas phase and in DMSO; the acidity orders are shown to be very similar. Furthermore, the attenuation in solution is small (factor ~ 2) and is reflected mainly in the enthalpy and not the entropy change. Arnett (99) has used the gas phase data (84, 85) and the DMSO data of Bordwell (97) for an evaluation of the relative enthalpies of solvation of A^- in DMSO.

The gas phase acidities of carbon and nitrogen acids allow an evaluation of the electron affinities $EA(A)$ for cases where $D(A - H)$ is known or can be estimated. Such data can be found in (84).

As mentioned earlier, theoretical STO-3G calculations for the energy change in the proton transfer Reaction 1, involving positive ions, have been very successful. However proton transfer in Reaction 7 involves negative ions, which are electron-crowded systems. The negative ions are stable, relative to the radical and the electron, mostly because of electron correlation. Therefore, it might have been expected that the simple STO-3G calculations cannot be applied to the isodesmic Reaction 7. However, Radom (100) has shown recently that the closed shell anions involved in the isodesmic proton transfer Reaction 7 can be handled, provided one uses a closed shell self-consistent field procedure so that a description of the anion in terms of a radical and an electron, even though energetically favorable, is not possible. The calculated ΔE_7 were found in good agreement with the experimental $\Delta G_7^\circ \approx \Delta H_7^\circ$.

THE OCCURRENCE-NONOCCURRENCE (BRACKETING) TECHNIQUE

The availability of a large number of closely spaced proton affinities and gas phase acidities obtained by proton transfer equilibria permits the determination of quite accurate basicities or acidities with the qualitative bracketing technique (101). The bracketing technique was widely used (5, 95) before proton transfer equilibria measurements were initiated and is still of great utility. For a proton transfer reaction like $BH^+ + B_x = B + B_xH^+$, instead of determining the equilibrium constant K it might be easier to establish which is larger, the forward or the reverse rate constant. This can be done by observing which protonated base is increasing with time. Since $K = k_f/k_r$ such a determination establishes the sign of the $\Delta G°$. The proton affinity of the unknown B_x can be established by bracketing it between a stronger and a weaker base B of known proton affinity. The bracketing technique is particularly useful in cases where B_xH^+ is unstable or engages in some side reaction.

HYDRIDE ION TRANSFER EQUILIBRIA, HYDRIDE AND HALIDE ION AFFINITIES, AND STABILITIES OF CARBONIUM IONS

The hydride ion equilibria

$$R_1^+ + R_2H = R_1H + R_2^+, \qquad 13.$$

where R is a hydrocarbon radical, have been studied by Field (30, 33, 101a–104) with a pulsed high-pressure mass spectrometer. Both the temperature dependence of the kinetics (30, 101a, 102) and the temperature dependence of the equilibria (30, 33, 103, 104) were determined. van't Hoff plots of the Equilibria 13 led to both $\Delta H_{13}°$ and $\Delta S_{13}°$. In most cases $\Delta S_{13}°$ was found to be less than 3 eu. $\Delta H_{13}°$ is related to the heats of formation $\Delta H_f(R^+)$, ionization potentials $I_p(R)$, and the hydride affinities of R^+, $D(R^+-H^-)$, by Equations 14–16:

$$\Delta H_{13} = D(R_2-H) + I_p(R_2) - D(R_1-H) - I_p(R_1), \qquad 14.$$

$$\Delta H_{13} = \Delta H_f(R_2^+) - \Delta H_f(R_2H) - \Delta H_f(R_1^+) + \Delta H_f(R_1H), \qquad 15.$$

$$\Delta H_{13} = D(R_2^+-H^-) - D(R_1^+-H^-). \qquad 16.$$

In order to obtain absolute values, $D(R-H) + I_p(R)$, must be known for at least one compound, and the other compounds must be connected to this standard by a ladder of equilibria.

The thermochemical results obtained by Field are summarized in Table 7. The identity of the ion R^+ was deduced from the known structure of the neutral RH used and the assumption that the lost hydride ion originates from the position that leads to the most stable R_2^+, without any rearrangement. Since the Equilibria 11 involved ΔG changes of only a few kcal mole^{-1}, this assumption is probably justified. NMR experiments in superacids by Olah (105) and Hogeveen (106) have shown that many tert-carbocations rearrange rapidly, even at $-20°C$, equilibrating to mixtures of isomeric ions. However, as Solomon & Field (103) have pointed out, ionic solvation probably reduces the free energy differences between the ions in solution. Therefore the solution observations of rapid rearrangement probably do not hold for the gas phase. The data in Table 7 complement the important work by Lossing (57, 107, 108) on the ionization potentials of alkyl free radicals based on electron impact measurements. Unfortunately, very few cross checks between the two sets of data are available. Of the three cases shown in Table 7, the agreement for two is within less than 1 kcal

Table 7 Heats of formation of carbonium ions R^+ and hydride ion affinities of R^+, i.e. $D(R^+-H^-)$ obtained from equilibria $R_1^+ + R_2H = R_1H + R_2^+$

R^+	$\Delta H_f(R^+)$[a]	$D(R^+ - H^-)$[b]	Ref.
(isopropyl cation)	169.1[c]	235.9[c]	c
(t-butyl cation)	161.1, 161[d]	232.5, 232.5[d]	33
(2-pentyl cation)	155.3	231.1	33
(2-hexyl cation)	155.6	232.1	33
(t-pentyl cation)	153.2	230.5	33
(2-methyl-2-pentyl cation)	151.5, 150[d]	232.7, 231[d]	103
(2,4-dimethyl-2-pentyl cation)	148.7	231.2	103
(3-pentyl cation)	150.6	230.5	103
(2,3-dimethyl-2-butyl cation)	144.5	228.1	103
(cyclopentyl cation)	192.6, 197[d]	245.8	104
(methylcyclopentyl cation)	169.3	229.5	104
(norbornyl cation, numbered)	187.3	234.4	104
(2-methylnorbornyl cation)	~174	228.4	104

[a] In kcal mole^{-1}, obtained by combining ΔH_{i1}° with external standard $\Delta H_f(t\text{-}C_4H_9^+) = 169.1$ kcal mole^{-1} and known $(\Delta H_f(RH))$.
[b] In kcal mole^{-1} obtained from $\Delta H_f(R^+)$, $\Delta H_f(RH)$, and $\Delta H_f(H^-) = 34.7$ kcal mole^{-1} (59).
[c] External standard (Lossing 57, Tsang 58).
[d] Data obtained by electron impact measurements of the ionization potentials of the corresponding free radicals R (Lossing 107, 108).

mole^{-1}, but for cyclopentyl$^+$ Solomon & Field (104) have obtained a heat of formation that is some 4 kcal mole^{-1} lower than that of Lossing (108).

Most useful for the examination of carbonium ion stability are the hydride ion affinities of R$^+$, i.e. the heterolytic bond energies $D(R^+-H^-)$. Since most of the ions in Table 7 are tertiary, the results illustrate the effects of branching and chain length of the alkyl substituents. Thus maximum stabilization is obtained by substituents showing maximum branching at the carbon α to the carbonium ion center.

The results of Solomon & Field (104) dealing with the norbornyl cation are included in Table 7. The hydride ion affinity of the ion produced by hydride abstraction from norbornane was found to be 234.4 kcal mole^{-1}. If one makes the justifiable assumption that the abstracted hydride ion comes from the position that leads to the carbonium ion with highest stability, the ion formed will be the norbornyl cation, whose electronic stabilization has been the central point of the classical-nonclassical ion controversy (104). The measured hydride affinity is very close to that of t-butyl$^+$ and is 10 kcal lower than that of the secondary ion cyclopentyl$^+$. The fact that the norbornyl cation is so much more stable than the secondary cyclopentyl ion argues for the nonclassical structure shown in Table 7 (104).

Beauchamp (109–111) has used the halide ion transfer Reaction 17 for the evaluation of the stabilities of carbonium ions R$^+$:

$$R_1X + R_2^+ = R_1^+ + R_2X. \qquad 17.$$

In principle, the approach, which precedes Field's hydride ion studies, is the same. However, thermochemical data on the neutral bromides are often lacking and must be estimated (111). Also, no examples of the establishment of equilibria for these systems have been given (109–111). Therefore, while the method provides valuable data on a variety of carbonium ions, these are of a more qualitative nature.

Hydride ion affinities and heats of formation of R$^+$ can also be obtained from the proton affinities of corresponding π bases. For example, the proton transfer equilibrium $C_3H_6 + BH^+ = $ iso-$C_3H_7^+ + B$, in which a double bond is protonated, provides the proton affinity of C_3H_6 *and* the heat of formation and hydride affinity of iso-$C_3H_7^+$. The method is equally applicable to polyunsaturated and aromatic systems. For example, the proton affinity of benzene that leads to the heat of formation of the benzenium ion can be used for the evaluation of the hydride affinities of benzenium leading to 1,3- and 1,4-cyclohexadiene, provided that the heats of formation of the neutral dienes are known. Hydride ion affinities of RH$^+$, obtained via the proton affinities of the π bases, have been summarized by Taft & Beauchamp (45).

ASSOCIATION AND CLUSTERING REACTIONS

The association and clustering equilibria that have been studied can be summarized by Equations 18 and 19:

$$I^\pm + M = (IM)^\pm \qquad (0, 1), \qquad 18.$$
$$(IM_{n-1})^\pm + M = (IM_n)^\pm \qquad (n-1, n), \qquad 19.$$
$$(IM_n)^\pm + B = (IM_{n-1}B)^\pm + M. \qquad 20.$$

The reaction (0, 1) is, of course, the first step in a clustering sequence ($n - 1, n$). As we see later, the bonding in the 0, 1 step often may be significantly different from that in the subsequent steps. Reaction 20 represents an exchange equilibrium.

The data obtained from measurement of Equilibria 18–20 are summarized in Tables 8–10. In nearly all cases the temperature dependence of the equilibria was also studied so that not only $\Delta G°$ but also $\Delta H°$ and $\Delta S°$ were obtained. The majority of the results originate from high-pressure equilibria measurements by Kebarle (112–129), Field (130, 131), Conway (132–134), Phelps (135), and Ferguson (136, 137).

Table 8 Enthalpy changes for dissociation of ion hydrates $M^{\pm}(H_2O)_n = M^{\pm}(H_2O)_{n-1} + H_2O$

Ion M$^+$	n, n−1 1, 0	2, 1	3, 2	4, 3	5, 4	6, 5	Ref.	Ion M$^-$	n, n−1 1, 0	2, 1	3, 2	4, 3	5, 4	Ref.
Li$^+$	34	26	21	16	14	12	20	F$^-$	23.3	16.6	13.7	13.5	13.2	21
Na$^+$	24	20	16	14	12	11	20	Cl$^-$	13.1	12.7	11.7	11.1	—	21
K$^+$	16.9	16	13	12	11	10	20, 34	Br$^-$	12.6	12.3	11.5	10.9	—	21
Rb+	16	14	12	11	10	—	20	I$^-$	10.2	9.8	9.4	—	—.	21
Cs$^+$	14	12	11	10.6	—	—	20							
Pb$^+$	22.4	16.9	12.2	10.8	10	9.6	35	OH$^-$	25	16.4	15	14.2	14.1	117
Sr$^+$	34.5	30.5	25.7	22.3	20.6	18.3	38							
H$_3$O$^+$	32	20	17	15	13	12	11, 32, 141	CN$^-$	13.8	—	—	—	—	117
NH$_4^+$	17.2	14.7	13.4	12.2	9.7	—	112	O$_2^-$	18.4	17.2	15.4	—	—	21
H$_3$S$^+$	~18	—	—	—	—	—	113, 141							
NO$^+$	18.5	16.1	13.5	—	—	—	115	NO$_2^-$	14.3	12.9	10.4	—	—	117
NO$_2^+$	~21.2	—	—	—	—	—	136							
CH$_5^+$	66	—	—	—	—	—	48, 116	NO$_3^-$	12.4	—	—	—	—	117
C$_2$H$_5^+$	37	—	—	—	—	—	48, 116							
C$_2$H$_5$OH$_2^+$	24	19.2	14.2	12.5	—	13.1	118							
s-C$_3$H$_7^+$	22.8	—	—	—	—	—	48, 116							
t-C$_4$H$_9^+$	11.2	—	17.7	14	—	—	48, 116							
CH$_3$CO$^+$	19	—	—	—	—	—	48, 116							
CH$_3$CH$_2$CO$^+$	17.4	—	16.4	12.7	—	—	48, 116							
Val H$^+$	19.3	—	—	—	—	—	130							
Prol H$^+$	18.9	—	—	—	—	—	130							

Table 9 Entropy changes for dissociation of ion hydrates: $M^{\pm}(H_2O)_n = M^{\pm}(H_2O)_{n-1} + H_2O$

Ion M$^+$	n, n−1 1, 0	2, 1	3, 2	4, 3	5, 4	6, 5	Ref.	Ion M$^-$	n, n−1 1, 0	2, 1	3, 2	4, 3	5, 4	Ref.
Li$^+$	23	21	25	30	31	32	20	F$^-$	17.4	18.7	20.4	37	30.7	21
Na$^+$	22	22	22	25	28	26	20	Cl$^-$	16.5	20.8	23.2	25.8	—	21
K$^+$	19.9	24	23	25	25	26	20, 34	Br$^-$	18.9	23	25	27	—	21
Rb$^+$	21	22	24	25	25	—	20	I$^-$	16.3	19.0	21.3	—	—	21
Cs$^+$	19.4	22	24	25	—	—	20							
Pb$^+$	35.5	25.3	20.2	20.8	22.3	23.6	35	OH$^-$	20.8	21	25	25	33	117
Sr$^+$	31.1	28.1	28.6	28.2	30.5	30.5	38	CN$^-$	15.8	—	—	—	—	117
H$_3$O$^+$	24.4	22	27	32	30	20	11, 33	O$_2^-$	20	25	28	—	—	21
NH$_4^+$	20	22	25	27	22	—	112	NO$_2^-$	21	23.7	21.2	—	—	117
H$_3$S$^+$	19	—	—	—	—	—	113, 141	NO$_3^-$	19.1	—	—	—	—	117
NO$^+$	23	25.5	—	—	—	—	115							
C$_2$H$_5$OH$_2^+$	26	28	26	26	—	28	118							
t-C$_4$H$_9^+$	22	—	29	28.7	—	—	48, 116							
CH$_3$CH$_2$CO$^+$	—	—	24	24	—	—	48, 116							
Val H$^+$	36.3	—	—	—	—	—	130							
Prol H$^+$	36.8	—	—	—	—	—	130							

Table 10 Enthalpy and entropy changes for cluster dissociation[a, b]

Bond	ΔH°	ΔS°	Ref.	Bond	ΔH°	ΔS°	Ref.
K^+-OH_2	16.9	19.9	34	OH^--HCl	70.1	—	117
K^+-NH_3^c	17.8	20.0	34	OH^--HBr	79.3	—	117
K^+-$O(CH_3)_2$	20.8	24.8	34	OH^--HNO_3	80.1	—	117
K^+-$O(C_2H_5)_2$	22.3	24.7	34	O_2^--CH_3OH^c	19.1	22	121
K^+-NH_2CH_3	19.1	21.8	34	O_2^--CH_3CN^c	16.4	17.4	121
K^+-$NH(CH_3)_2$	19.5	21.4	34	NO_3^--HNO_3	>26.4	—	137
K^+-$N(CH_3)_3$	20.0	23.4	34	NO_3^--SO_2^m	>28	—	137
K^+-$NH_2(nC_3H_7)$	21.8	25.5	34	$NO_3^-SO_2^n$	~22.5	—	137
K^+-(aniline)	22.8	23.7	34	H_3^+-H_2^c	9.6	24.6	125
K^+-(pyridine)	20.7	18.6	34	CH_3^+-H_2^d	40	—	53
K^+-(acetonitrile)^c	24.4	21.5	34	$C_2H_5^+$-H_2^e	4	—	53
K^+-$CH_2NH_2\cdot CH_2NH_2^c$	25.7	22.3	119	$C_2H_5^+$-H_2^f	11.8	—	53
K^+-$CH_3OCH_2\cdot CH_2OCH_3^c$	30.8	26.8	119	sec-$C_3H_7^+$-H_2	~2.5	—	53
NH_4^+-NH_3^c	25	26	112	CH_5^+-CH_4^c	7.4	20.8	42
$CH_3OH_2^+$-CH_3OH^c	33.1	30.5	123, 124	$C_2H_5^+$-CH_4^g	6.6	23.4	127
$(CH_3)_2OH^+$-$O(CH_3)_2^c$	30.7	29.6	123, 124	HCO^+-H_2^h	3.9	20.5	128
Cl^--HCl^c	23.7	23.5	120	HCO^+-CO^c	11.7	20.9	129, 131
Cl^--CH_3OH^c	~14.2	~15	121	CO^+-CO	>25.4	—	131
Cl-t-C_4H_9OH	19.2	27	122	HN_2^+-N_2^c	16	24	131
Cl^--$(CH_3CN)^c$	~13.4	~12.2	121	N_2^+-N_2	22.8	19.5	23
Cl^--$CHCl_3$	~15.2	~14.8	122	HO_2^+-O_2^c	20	27	129
Cl^--Phenol	27.4	25	114	O_2^+-O_2^c	9.6	20.6	15, 132, 133^e
Cl^--$HCOOH$	~28	~25	114	HO_2^+-H_2	12.5	22	129
Cl^--SO_2^l	~22	—	—	HN_2^+-H_2	7.2	22.6	129
OH^--HF	44	—	122	O_2^--O_2	13.6	—	134
OH^--HNO_2	65	—	117	O_2^--CO_2	18.4	—	135

[a] Thermodynamic quantities for dissociation reaction $A^+ - B = A^+ + B$. It should be noted that $A^+ - B$ is actually $(AB)^+$, i.e. the positive charge of the complex $(AB)^+$ is not necessarily located on A.

[b] ΔH° values given in kcal mole^{-1}, ΔS° values in cal degree^{-1} for standard state 1 atm. $\Delta G_T^\circ = \Delta H^\circ - T\Delta S^\circ$. Equilibrium constant in (atm^{-1}) is obtained from $-\Delta G^\circ = RT \ln K$.

[c] Information for ΔH and ΔS also of higher clusters is available in quoted reference.

[d] Calculated from proton affinity $PA(CH_4) = 127$ kcal mole^{-1} (see section on Proton Affinities) and $\Delta H_f(CH_3^+) = 261$ kcal mole^{-1} (57).

[e] Dissociating species believed to correspond to $\left(CH_3CH_2{<}{}^H_H\right)^+$.

[f] Dissociating species believed to correspond to $\left(CH_3 \overset{H}{-\!\!-\!\!-} CH_3\right)^+$.

[g] Dissociating species believed to correspond to $\left(CH_3-CH_2\overset{H}{-\!\!-\!\!-}CH_3\right)^+$.

[h] $HCOH_2^+$ species believed (116) to be the methoxy cation CH_3-O^+.

[k] Estimate (137) based on $\Delta H = 19.1$ for NO_3^--H_2O and $-\Delta G_{300} \geq 7.3$ for reaction

$N_3O^-H_2O + HNO_3 = NO_3^-HNO_3 + H_2O.$ Assumption $\Delta G_{300}^\circ \approx \Delta H.$

[l] Estimate based on $\Delta H = 16.5$ for Cl^--H_2O (Table 8) and $\Delta G_{300}^\circ = -6$ for reaction

$Cl^-H_2O + SO_2 = Cl^-SO_2 + H_2O$ (137).

[m] Estimate based on $\Delta H = 21$ for NO_2^--H_2O (Table 8) and $-\Delta G_{300}^\circ > 7$ kcal mole^{-1} for reaction

$NO_2^-H_2O + SO_2 = NO_2^-$-$SO_2 + H_2O$ (see 137)

[n] Estimate based on $\Delta H = 19.1$ for NO_3^--H_2O (Table 1) and $-\Delta G_{300}^\circ = 3.4$ (see 137).

High pressure is required for Equilibria 18 and 19 because these reactions are exothermic processes where the reaction product requires third-body stabilization. At pressures below ~ 10 torr these reactions are most often third order and the forward and reverse reaction rates become too slow for equilibrium to be reached at total pressures in the millitorr range. Therefore such equilibria have not been measured by ICR. The flowing afterglow measurements (136, 137) were done only at room temperature so that only the $\Delta G°_{300}$ values are available. The exchange reactions 19 do not require third-body stabilization, and their equilibria can be measured also by ICR as has been done by Beauchamp (138).

Comparisons of association equilibria data obtained in different laboratories and by different methods are given in Table 11. The apparatus used is high-pressure mass spectrometry, Futrell (139); flowing afterglow, Ferguson (137, 140); pulsed high-pressure mass spectrometry, Field (141) and Kebarle (21, 32, 112). The agreement observed in the $\Delta G°$ results is generally within 1 kcal mole^{-1}. The agreement in the ΔH results is also within 1–3 kcal mole^{-1}. The errors in the $\Delta G°$ and $\Delta H°$ values obtained from association equilibria are larger than those for, say, the proton transfer reactions. This is due to experimental effects, i.e. cluster dissociation in the sampling process (10, 42). A spectacular disagreement between the experimental results of Kebarle (11, 32) and Field (142–144) for the reaction $H_3O^+ + H_2O = H_5O_2^+$ has been cleared up. Field's (141) most recent measurements of this system, done with a pulsed instrument, have given results in agreement with those of Kebarle (11, 32).

A good portion of the thermochemical data of Tables 8–10 has been discussed in three review articles by Kebarle (145–147). The remainder of the present article is devoted to a brief, up-to-date summary of the significance of the results.

The results in Tables 8–10 are best systematized by considering the nature of the ion involved in the association reaction (0, 1). The noble gas–type alkali ions Li^+-Cs^+ represent a series where considerable experimental data and theoretical results are available. Ab initio SCF-MO calculations on the alkali ion hydrates by Diercksen (148) and Clementi (149) have shown that the bonding, particularly in the larger ions like K^+, is almost purely electrostatic. Thus Mulliken electron population analysis

Table 11 Comparison of results for clustering reactions obtained by different workers

Reaction	$\Delta G°_{300}$ kcal mole^{-1a}	ΔH kcal mole^{-1}
$NH_4(NH_3)_3^+ = NH_4(NH_3)_2^+ + NH_3$	6.1 (112), 6.0 (139), 6.5 (140)[a]	—
$NH_4(NH_3)_4^+ = NH_4(NH_3)_3^+ + NH_3$	3.7 (112), 3.4 (139), 3.4 (140)[a]	—
$OH(H_2O)_3^- = OH(H_2O)_2^- + H_2O$	7.7 (21), 8.2 (137)[a]	—
$NO_2(H_2O)^- = NO_2^- + H_2O$	8.0 (117), 8.0 (137)[a]	—
$NO_3(H_2O)^- = NO_3^- + H_2O$	6.8 (117), 6.8 (137)[a]	—
$H_9O_4^+ = H_7O_3^+ + H_2O$	9.3 (32), 9.9 (141)	17.5 (32), 16 (141)
$H_7O_3^+ = H_5O_2^+ + H_2O$	13.0 (32), 15 (141)	19.5 (32), 21 (141)
$H_5O_2^+ = H_3O^+ + H_2O$	24.3 (32), 23 (141)	31.6 (32), 33 (141)

[a] Results marked with superscript a obtained by flowing afterglow measurements (Ferguson 137, 140); remaining results obtained by high-pressure mass spectrometry from three different groups: Kebarle (21, 32, 112), Field (141), and Futrell (139).

showed (149) a charge transfer of only 0.018, 0.013, and 0.004 electron from the water molecule to Li$^+$, Na$^+$, and K$^+$, respectively. The calculations also provided binding energies that were in good agreement with the experimental data (20, 34). Good accuracy of the calculated binding energies may be expected for these systems since the number and nature of electron pairs in the complex and the separated cation and water molecules are conserved.

Ab initio SCF-MO calculations for the association (0, 1) involving the noble gas-type halide ions X$^-$ by Diercksen (150) and Clementi (151) also provide important insights. Both calculations show that structure I, involving a linear hydrogen bond, is more stable than structure II:

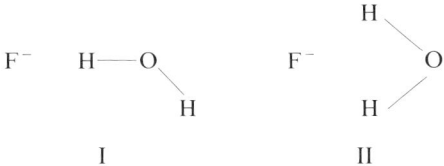

However, as the radius of X$^-$ increases, the energy difference between I and II decreases (151). The calculated binding energies (150, 151) were in agreement with the experimental $\Delta H_{0,1}$ values (21). Mulliken electron population analysis also showed (151) that in this case there is relatively little electron transfer. The maximum electron transfer occurring in structure I, with the smallest ion F$^-$, was 0.05 electrons transferred from F$^-$ to H$_2$O. Therefore the hydrogen bond in X$^-$HOH is essentially electrostatic.

Experimental studies of the stepwise hydrations ($n - 1, n$) of the alkali and halide ions, Table 8, have shown (20, 21, 147) that in the gas phase symmetrical inner hydrate structures that have n even, like Na$^+$(H$_2$O)$_4$, are not significantly more stable than, say, Na$^+$(H$_2$O)$_5$. Theoretical calculations (152) of such hydrates, which provide predictions of the geometries, have also been made.

The stepwise hydration gas phase results for the alkali and halide ions have been used as support to the total single-ion enthalpies of hydration based on the measurements of Randles (20, 21).

Davidson & Kebarle (34) studied the temperature dependence of (0, 1) equilibria involving the potassium ion and a number of polar compounds like methanol, amines, and ethers (see Table 10). Since quantum mechanical calculations were not available, improved classical electrostatic calculations of the binding energies were made (34). Instead of using the point dipole approximation, the net atomic charges of the molecules obtained in LCAO calculations were used. This and other improvements led to calculated values in good agreement with the experimental measurements (34). Similar calculations for Cl$^-$ and the molecules H$_2$O and CH$_3$CN also gave good agreement with experiment (153). The experimental enthalpies for the aprotic solvent molecule CH$_3$CN and the electrostatic calculations predict considerably stronger binding energies to positive ions than to negative ions; see (153) and Table 10. This is a consequence of the charge distribution in the acetonitrile molecule, which consists of a negative charge concentrated on the accessible N atom and a positive charge diffusely distributed over the bulky CCH_3 group (153). This charge distribution is characteristic for all aprotic solvent molecules.

The alkali ion-molecule interactions can be formally considered as a Lewis acid (alkali ion) base interaction. However, since the interaction is electrostatic and there is no electron transfer, electron pair donation does not occur and the application of the Lewis acid pair concept does not appear useful.

Beauchamp (138) has measured the exchange Equilibria 21, which involves a large variety of molecules B:

$$Li^+B_1 + B_2 = LiB_2 + B_1. \qquad 21.$$

Since the experiments were done at a single temperature, only ΔG_{21}° were obtained. The binding energies were calculated by estimating ΔS_{21}° and normalizing to the known (20) $\Delta H_{1,0}$ for Li^+H_2O.

Recent results of Castleman (38) on the hydration of the Sr^+ ion provide an interesting contrast to the result for the alkali ions. Sr^+ has one electron more than Rb^+. This electron is in a $5s$ orbital and there are vacant $4d$ orbitals nearby. As can be seen in Table 8, the binding energies of the Sr^+ hydrates are very much larger than those for Rb^+. This effect persists right up to the last hydrate measured.

The positively charged Brönsted acids H_3O^+, $CH_3OH_2^+$, NH_4^+, $CH_3NH_3^+$, etc represent a large, important group for which considerable information from gas phase equilibria is available. The proton hydrates $H_{2n+1}O_n^+$ (11, 32, 147) (Tables 8 and 9) were found to have binding energies that decreased gradually with increase of n. The Eigen structure $H_3O^+(H_2O)_3$ was found to have very prominent stability. Theoretical ab initio MO calculations of $H_5O_2^+$ by Dierksen (154) and Allen (155) have predicted binding energies in agreement with the experimental results. Less accurate CNDO calculations for the higher hydrates are also available (Ehrenson 156).

The binding energies for monohydrates of nitrogen and oxygen BH^+ ions like H_3O^+, $C_2H_5OH_2^+$, NH_4^+, and $(CH_3)_3NH^+$ are found (Table 8) to decrease in the above order, which is also the order in which the Brönsted acidity of the BH^+ acids decreases (see Tables 1-6). This very useful systematization is part of a more general rule for mixed proton-held dimers $(B_1HB_2)^+$ and $(A_1HA_2)^-$, involving oxygen and nitrogen acids and bases, which states that the strength of the bond B_1H^+–B_2 increases with the Brönsted acidity of B_1H^+ and the basicity of B_2, and in A_1^-–HA_2 with the basicity of A_1^- and the acidity of HA_2 (60, 117, 122).

The higher hydrates of H_3O^+, NH_4^+, $(CH_3)_3NH^+$, OH^-, and NO_3^- show a fairly regular decrease of binding energies with increasing water content (Table 8). However, for $C_2H_5OH_2^+$ the binding energies pass through a minimum (118, Table 8). The binding energies beyond the minimum become similar to those for the proton hydrates $H_{n+1}O_n^+$. These results are believed (118) to derive from the expulsion of the C_2H_5OH to the periphery of the cluster $C_2H_5OH \cdot (H_{2n+1}O_n)^+$. In effect this is a deprotonation of the base C_2H_5OH. Thus while C_2H_5OH is a strong base in the absence of water, it becomes weak, compared with H_2O, when n becomes large. Similar behavior is expected for other oxygen bases that have bulky hydrophobic groups.

The reactions of the alkyl ions $C_nH_{2n+1}^+$ and other carbonium ions like the acylium ions $C_nH_{2n+1}CO^+$, and benzenium ions $C_6H_7^+$, which have an empty bonding orbital, with n donor bases like water, alcohols, ammonia, etc may be expected to constitute true Lewis acid-base reactions in which an electron pair bond is formed. Recent measurement (116) of the equilibrium

$$t\text{-}C_4H_9^+ + OH_2 = t\text{-}C_4H_9OH_2^+ \qquad 22.$$

has shown that the association product has the same heat of formation as protonated t-butanol. This shows that the product of the hydration is protonated t-butanol, which was expected on the basis of the Lewis acid-base consideration and reactions in solution. It could also be shown that the acylium ions on hydration in the gas phase lead to the corresponding protonated acids (116). The energies released on hydration of the s-$C_3H_7^+$, t-$C_4H_9^+$, and CH_3CO^+ at 23, 11.2, and 19 kcal mole^{-1}, respectively (Table 8), are of the same magnitude as the energies for the hydration of Na^+, H_3O^+, NH_4^+, etc where only electrostatic forces or hydrogen bonds are involved.

The monohydration of NO^+ and NO_2^+ in Reactions 23 and 24 can be also considered as a Lewis acid-base reaction where an electron pair bond is formed:

$$NO^+ + OH_2 = [H_2ONO^+] \rightarrow (HO)_2N^+, \qquad 23.$$

$$NO_2^+ + OH_2 = [H_2ONO_2^+] \rightarrow (HO)_2NO^+. \qquad 24.$$

The expected products are protonated nitrous and nitric acid. The conclusion that the hydrates of NO^+ and NO_2^+ were the protonated acids was also reached by Fehsenfeld et al (136) on the basis of reactivity considerations.

The reactions of the alkyl cations with H_2 and CH_4, as, for example,

$$C_2H_5^+ + H_2 = C_2H_7^+, \qquad 25.$$

$$C_2H_5^+ + CH_4 = C_3H_9^+, \qquad 26.$$

may be considered as processes involving interaction of the Lewis acid $C_2H_5^+$ with the σ Lewis bases H_2 and CH_4. It could be shown (53) that Reaction 25 leads to two $C_2H_7^+$ isomers, one a weakly bonded $C_2H_5^+ \cdot H_2$ probably of structure

$$CH_3CH_2 - \!\!\!\!<^{H^+}_{H}$$

and the other a more strongly bonded $C_2H_7^+$ that could have the structure

$$CH_3 \overset{H}{\wedge} CH_3^+,$$

i.e. three centre-bonded C–C protonated ethane (126). The product of Reaction 26 was assigned (53, 127) the structure

$$C_2H_5 \overset{H}{\wedge} CH_3^+,$$

i.e. C–C protonated propane. As $-\Delta H_{26}$ is quite low (Table 10), protonated propane is stable only at low temperatures. Some isomers of the same type were also observed for the higher protonated alkanes, all of which are unstable and dissociate by loss of either H_2 or a paraffin molecule (53).

Another somewhat similar reaction is

$$HCO^+ + H_2 = H_3CO^+. \quad\quad\quad 27.$$

Measurement of the Equilibrium 27 at low temperature allowed the determination of the enthalpy change $\Delta H_{27} = -3.9$ kcal mole^{-1} (128). This led to a heat of formation for H_3CO^+ that was much higher than that of protonated formaldehyde. If one considers Reaction 27 to be an electrophilic attack of the carbonium ion on the bond in H_2, the resulting species may be expected to be

$$H_2^+\!\!-\!\!\underset{H}{\overset{H}{C}}\!\!=\!\!O \longleftrightarrow H\!\!-\!\!\underset{H}{\overset{H}{C}}\!\!-\!\!O^+,$$

which should be the methoxy cation. It has been shown (116) that the heat of formation of H_3CO^+ derived from ΔH_{27} is very close to that of the methoxy cation.

The bond energy obtained from equilibrium measurements (125) for the $(H_3–H_2)^+$ species is 9.6 kcal mole^{-1} (Table 10). Theoretical treatments of this molecule (see references in 125), generally predict somewhat lower energy; however, the theoretical bond energies are not reliable in this case. The predicted geometry is that of an elongated H_3^+ triangle in which the H at the apex interacts with the σ bond of the H_2 molecule. The H_2 axis is perpendicular to the plane of the triangle. The equilibria results (125) indicate that the $H_3^+(H_2)_3$ might be an equilateral triangle with three H_2 molecules facing the three corners of the H_3^+.

The equilibria leading to the species $CH_5^+(CH_4)_n$ have also been studied (42); see Table 10. The binding energy in $CH_5^+CH_4$ at 7.4 kcal mole^{-1} is somewhat lower than that in $H_3^+ \cdot H_2$. Binding energies of the higher clusters are available and structures for the higher clusters have been suggested (42).

The binding energies for the species N_4^+, O_4^+, and O_4^- are shown in Table 10. Conway has considered the bonding in these species (157, 158). Since the polarizabilities of N_2 and O_2 are very similar, and since it is found that $NO^+–N_2$ and $O_2^+–N_2$ have much lower binding energies than $O_2^+–O_2$ and $N_2^+–N_2$, the bonding in the N_4^+, O_4^+, and O_4^- species cannot be due to ion-induced dipole interactions only. Furthermore, ion-induced dipole interactions are too weak to account for the observed binding energies, which are in the 10–20 kcal mole^{-1} range. The bonding must come about by the sharing of one electron by the two molecules. Simple Hückel MO may be used (157) to predict that the orbital containing the electron is π_g type. Conway has also used an improved semiempirical CNDO approach to obtain angles and distances between the two interacting molecules as well as binding energies (157). The higher clusters have been treated by classical electrostatic methods (158).

The bonding and structures of several species in Table 10, like $(CO)_2^+$, HN_4^+, HO_4^+, $H(CO)_2^+$, and $(HN_2N_2)^+$, have not received much attention so far.

CONCLUSIONS

The gas phase ion equilibrium method initiated some 11 years ago has experienced rapid growth and has become one of the principal sources of thermochemical information for ions. Redundant thermodynamic cycles allow the achievement of very good internal consistency. The method provides, for the first time, abundant and accurate information for many organic ions of interest. It is not an exaggeration to say that the method initiates a new era in organic ion energetics. The measurements of ion-solvent molecule interactions in the gas phase provide valuable insights into the solvation of ions and thus form a bridge from ion behavior in the gas phase to the behavior in solution.

Literature Cited

1. Ausloos, P., ed. 1975. *Interactions Between Ions and Molecules.* New York: Plenum
2. Franklin, J. L., ed. 1972. *Ion Molecule Reactions.* New York: Plenum
3. Franklin, J. L., Harland, P. W. 1974. *Ann. Rev. Phys. Chem.* 25:485–526
4. McDaniel, E. V., Cermak, V., Dalgarno, A., Ferguson, E. E., Friedman, L. 1970. *Ion Molecule Reactions.* New York: Wiley-Interscience.
5. Beauchamp, J. L. 1971. *Ann. Rev. Phys. Chem.* 22:527–61
6. Ferguson, E. E. 1975. *Ann. Rev. Phys. Chem.* 26:17–38
7. Kebarle, P., Godbole, E. W. 1963. *J. Chem. Phys.* 39:1131
8. Kebarle, P., Hogg, A. M. 1965. *J. Chem. Phys.* 42:798
9. Hogg, A. M., Kebarle, P. 1965. *J. Chem. Phys.* 43:449
10. Hogg, A. M., Haynes, R. N., Kebarle, P. 1966. *J. Am. Chem. Soc.* 88:28
11. Kebarle, P., Searles, S. K., Zolla, A., Scarborough, J., Arshadi, M. 1967. *J. Am. Chem. Soc.* 89:6393
12. Searles, S. K., Kebarle, P. 1968. *J. Phys. Chem.* 72:742
13. Kebarle, P., Haynes, R. N., Collins, J. G. 1967. *J. Am. Chem. Soc.* 89:5753
14. Kebarle, P., Arshadi, M., Scarborough, J. 1968. *J. Chem. Phys.* 49:817
15. Durden, D. A., Kebarle, P., Good, A. 1969. *J. Chem. Phys.* 50:805
16. Searles, S. K., Dzidic, I., Kebarle, P. 1969. *J. Am. Chem. Soc.* 91:2810
17. Kebarle, P., Arshadi, M., Scarborough, J. 1969. *J. Chem. Phys.* 50:1049
18. Searles, S. K., Kebarle, P. 1969. *Can. J. Chem.* 47:2620
19. Good, A., Durden, D. A., Kebarle, P. 1970. *J. Chem. Phys.* 52:212, 222
20. Dzidic, I., Kebarle, P. 1970. *J. Phys. Chem.* 74:1466
21. Arshadi, M., Yamdagni, R., Kebarle, P. 1970. *J. Phys. Chem.* 74:1475
22. Arshadi, M., Kebarle, P. 1970. *J. Phys. Chem.* 74:1483
23. Payzant, J. D., Kebarle, P. 1970. *J. Chem. Phys.* 53:4723
24. Kebarle, P. 1970. *J. Chem. Phys.* 53:2129
25. Bowers, M. T., Aue, D. H., Webb, H. M., McIver, R. T. 1971. *J. Am. Chem. Soc.* 93:4314
26. Henderson, W. G., Taagepera, D., Holtz, D., McIver, R. T., Beauchamp, J. L., Taft, R. W. 1972. *J. Am. Chem. Soc.* 94:4728
27. Briggs, J. P., Yamdagni, R., Kebarle, P. 1972. *J. Am. Chem. Soc.* 94:5128
28. Fehsenfeld, F. C., Ferguson, E. E. 1973. *J. Chem. Phys.* 59:6272
29. Bohme, D. K., Hemsworth, R. S., Rundle, H. W., Schiff, H. I. 1973. *J. Chem. Phys.* 58:3504
30. Solomon, J. J., Meot-Ner, M., Field, F. H. 1974. *J. Am. Chem. Soc.* 96:3727
31. Kebarle, P. 1974. See Ref. 1, p. 549
32. Cunningham, A. J., Payzant, J. D., Kebarle, P. 1972. *J. Am. Chem. Soc.* 94:7627
33. Solomon, J. J., Field, F. H. 1975. *J. Am. Chem. Soc.* 97:2625
34. Davidson, W. R., Kebarle, P. 1976. *J. Am. Chem. Soc.* 98:6125
35. Tang, I. N., Castleman, A. W. 1972. *J. Chem. Phys.* 57:3638
36. Tang, I. N., Castleman, A. W. 1974. *J. Chem. Phys.* 60:3981
37. Tang, I. N., Castleman, A. W. 1975. *J. Chem. Phys.* 62:4576
38. Tang, I. N., Lian, M. S., Castleman, A. W. 1976. *J. Chem. Phys.* 65:4022
39. Ferguson, E. E., Fehsenfeld, F. C., Schmeltekopf, A. L. 1969. *Adv. At. Mol. Phys.* 5:1

40. Dunkin, F. C., Fehsenfeld, F. C., Schmeltekopf, A. L., Ferguson, E. E. 1968. *J. Chem. Phys.* 49:1365; Fehsenfeld, F. C. 1975. *Int. J. Mass Spectrom. Ion Phys.* 16:151
41. Bohme, D. K. 1974. See Ref. 31, p. 489
42. Hiraoka, K., Kebarle, P. 1975. *J. Am. Chem. Soc.* 97:4179
43. McIver, R. T. 1970. *Rev. Sci. Instrum.* 41:555
44. McMahon, T. B., Beauchamp, J. L. 1972. *Rev. Sci. Instrum.* 43:509
45. Wolf, J. F., Staley, R. H., Koppel, I., Taagepera, M., McIver, R. T., Beauchamp, J. L., Taft, R. W. 1977. *J. Am. Chem. Soc.* 99: In press
46. Aue, D. H., Webb, H. M., Bowers, M. T. 1976. *J. Am. Chem. Soc.* 98:311
47. Taft, R. W. 1975. In *Proton Transfer Reactions*, ed. E. F. Caldin, V. Gold. London: Chapman & Hall. p. 31
48. Yamdagni, R., Kebarle, P. 1973. *J. Am. Chem. Soc.* 95:3504
49. Lau, Y. K., Kebarle, P. 1976. *J. Am. Chem. Soc.* 98:7452
50. Payzant, J. D., Schiff, H. I., Bohme, D. K. 1975. *J. Chem. Phys.* 63:149
51. Duben, A. J., Lowe, J. P. 1971. *J. Chem. Phys.* 55:4270
52. Chupka, W. A., Berkowitz, J. 1971. *J. Chem. Phys.* 54:4256
53. Hiraoka, K., Kebarle, P. 1976. *J. Am. Chem. Soc.* 98:6119
54. Beauchamp, J. L. See Ref. 31, p. 413
55. Yamdagni, R., Kebarle, P. 1976. *J. Am. Chem. Soc.* 98:1320
56. Hehre, W. J., McIver, R. T., Pople, J. A., Schleyer, P.v.R. 1974. *J. Am. Chem. Soc.* 96:7162
57. Lossing, F. P., Semeluk, G. P. 1970. *Can. J. Chem.* 48:955
58. Tsang, W. 1972, *J. Phys. Chem.* 76:143
59. *Selected values of chemical thermodynamic properties.* 1968. Natl. Bur. Stand. Tech. Note 270–3
60. Yamdagni, R., Kebarle, P. 1973. *J. Am. Chem. Soc.* 95:3504
61. Arnett, E. M., Jones, F. M. III, Taagepera, M., Henderson, W. G., Beauchamp, J. L., Holtz, D., Taft, R. W. 1972. *J. Am. Chem. Soc.* 94:4724
62. Arnett, E. M., Wolf, J. F. 1975. *J. Am. Chem. Soc.* 97:3262
63. Arnett, E. M., 1975. See Ref. 47, p. 79
64. Arnett, E. M. 1973. *Acc. Chem. Res.* 6:404
65. Aue, D. H., Webb, H. M., Bowers, M. T. 1976. *J. Am. Chem. Soc.* 94:4728
66. Aue, D. H., Webb, H. M., Bower, S. M. T. 1976. *J. Am. Chem. Soc.* 98:318
67. Taft, R. W., Taagepera, M., Summerhays, K. D., Mitsky, J. 1973. *J. Am. Chem. Soc.* 95:3811
68. Aue, D. H., Webb, H. M., Bowers, M. T., Liotta, C. L., Alexander, C. J., Hopkins, H. P. Jr. 1976. *J. Am. Chem. Soc.* 98:854
69. Aue, D. H., Webb, H. M., Bowers, M. T. 1973. *J. Am. Chem. Soc.* 95:2699
70. Morton, T. H., Beauchamp, J. L. 1972. *J. Am. Chem. Soc.* 94:3671
71. Aue, D. H., Webb, H. M., Bowers, M. T. 1972. *J. Am. Chem. Soc.* 94:4726
72. Aue, D. H., Webb, H. M., Bowers, M. T. 1975. *J. Am. Chem. Soc.* 97:4136
73. Martin, R. L., Shirley, D. A. 1974. *J. Am. Chem.* 17:5299
74. Carroll, T. X., Smith, S. R., Thomas, T. D. 1975. *J. Am. Chem. Soc.* 97:659
75. Benoit, F. M., Harrison, A. G. 1977. *J. Am. Chem. Soc.* 99:3980
76. Hehre, W. J., Stewart, R. F., Pople, J. A. 1969. *J. Chem. Phys.* 51:2657
77. Hehre, W. J., Ditchfield, R., Radom, L., Pople, J. A. 1970. *J. Am. Chem. Soc.* 92:4796
78. Devlin, J. L. III, Wolf, J. F., Taft, R. W., Hehre, W. J. 1976. *J. Am. Chem. Soc.* 98:1990
79. Pollack, S. K., Devlin, J. L., Summerhays, K. D., Taft, R. W., Hehre, W. J. 1977. *J. Am. Chem. Soc.* 99:4583
80. Hehre, W. J., McIver, R. T., Pople, J. A., Schleyer, P.v.R. 1974. *J. Am. Chem. Soc.* 96:7162
81. Yamdagni, R., Kebarle, P. 1973. *J. Am. Chem. Soc.* 95:4050
82. Hiraoka, K., Yamdagni, R., Kebarle, P. 1973. *J. Am. Chem. Soc.* 95:6833
83. McMahon, T. B., Kebarle, P. 1974. *J. Am. Chem. Soc.* 96:5940
84. McMahon, T. B., Kebarle, P. 1976. *J. Am. Chem. Soc.* 98:3399
85. McMahon, T. B., Kebarle, P. 1974. *J. Am. Chem. Soc.* 96:4035; 1977. *J. Am. Chem. Soc.* 99:2222
86. Cumming, J. B., Kebarle, P. 1978. *Can. J. Chem.* 56: In press
87. Yamdagni, R., Kebarle, P. 1974. *Can. J. Chem.* 52:861
88. McIver, R. T., Silvers, J. H. 1973. *J. Am. Chem. Soc.* 95:8462
89. McIver, R. T., Miller, J. S. 1974. *J. Am. Chem. Soc.* 96:4323
90. Arnett, E. M., Small, R. T., McIver, R. T., Miller, J. S. 1974. *J. Am. Chem. Soc.* 96:5640
91. Berry, R. S., Reimann, C. W. 1963. *J. Chem. Phys.* 38:1540
92. Richardson, J. H., Stephenson, L. M., Brauman, J. I. 1973. *J. Chem. Phys.* 59:5068

93. Zimmerman, A. H., Brauman, J. I. 1977. *J. Chem. Phys.* In press
94a. Richardson, J. H., Stephenson, L. M., Brauman, J. I. 1975. *J. Chem. Phys.* 62:1580
94b. Zimmerman, A. H., Reed, W. J., Brauman, J. I. 1977. *J. Am. Chem. Soc.* In press
94c. Steiner, B. 1968. *J. Chem. Phys.* 49:5097
94d. Fine, D. H., Westmore, J. B. 1969. *Chem. Commun.* p. 273
94e. Furuyama, S., Golden, D. M., Benson, S. W. 1971. *Int. J. Chem. Kinet.* 3:237
94f. Solly, R. K., Golden, D. M., Benson, S. W. 1970. *Int. J. Chem. Kinet.* 2:11, 381
94g. King, K. D., Goddard, R. D. 1975. *Int. J. Chem. Kinet.* 7:837
94h. Zimmerman, J. H., Brauman, J. I. 1977. *J. Am. Chem. Soc.* 99:3565
95. Brauman, J. I., Blair, L. K. 1970. *J. Am. Chem. Soc.* 92:5986
96. Brauman, J. I., Blair, L. K. 1971. *J. Am. Chem. Soc.* 93:4315
97. Bordwell, F. G., Bartmess, J. E., Drucker, G. E., Margolin, Z., Matthews, W. S. 1975. *J. Am. Chem. Soc.* 97:3226
98. Arnett, E. M., Small, L. E., Oancea, D., Johnston, D. 1976. *J. Am. Chem. Soc.* 98:7346
99. Arnett, E. M., Johnston, D. E., Small, L. E. 1975. *J. Am. Chem. Soc.* 97:5598
100. Radom, L. 1974. *J. Chem. Soc. Chem. Commun.*, pp. 403–
101. Tal'rose, V. L. 1962. *Pure Appl. Chem.* 5:455; Long, J. W., Munson, B. 1973. *J. Am. Chem. Soc.* 95:2427; Long, J. W., Franklin, J. L. 1974. *J. Am. Chem. Soc.* 96:2320
101a. Meot-Ner, M., Field, F. H. 1975. *J. Am. Chem. Soc.* 97:2014
102. Meot-Ner, M., Field, F. H. 1976. *J. Chem. Phys.* 64:277
103. Meot-Ner, M., Solomon, J. J., Field, F. H. 1976. *J. Am. Chem. Soc.* 98:1025
104. Solomon, J. J., Field, F. H. 1976. *J. Am. Chem. Soc.* 98:1567
105. Olah, G. A., Pittman, C. U. 1966. *Adv. Phys. Org. Chem.* 4:305
106. Brouwer, D. M., Hogeveen, H. 1972. *Prog. Phys. Org. Chem.* 9:179
107. Lossing, F. P., Maccoll, A. 1976. *Can. J. Chem.* 54:990
108. Lossing, F. P., Traeger, J. C. 1975. *J. Am. Chem. Soc.* 97:1579
109. McMahon, T. B., Blint, R. J., Ridge, D. P., Beauchamp, J. L. 1972. *J. Am. Chem. Soc.* 94:8934
110. Blint, R. J., McMahon, T. B., Beauchamp, J. L. 1974. *J. Am. Chem. Soc.* 96:1269
111. Wieting, R. D., Staley, R. H., Beauchamp, J. L. 1974. *J. Am. Chem. Soc.* 96:7553
112. Payzant, J. D., Cunningham, A. J., Kebarle, P. 1973. *Can. J. Chem.* 51:3242
113. Hiraoka, K., Kebarle, P. 1977. *Can. J. Chem.* 55:24
114. French, M. A., Kebarle, P. 1977. Unpublished work
115. French, M. A., Hills, L. P., Kebarle, P. 1973. *Can. J. Chem.* 51:456
116. Hiraoka, K., Kebarle, P. 1977. *J. Am. Chem. Soc.* 99:360, 366
117. Payzant, J. D., Yamdagni, R., Kebarle, P. 1971. *Can. J. Chem.* 49:3309
118. Sensharma, D. K., Kebarle, P. 1978. *Can. J. Chem.* In press
119. Davidson, W. R., Kebarle, P. 1976. *Can. J. Chem.* 54:2594
120. Yamdagni, R., Kebarle, P. 1974. *Can. J. Chem.* 52:2449
121. Yamdagni, R., Payzant, J. D., Kebarle, P. 1973. *Can. J. Chem.* 51:2507; Yamdagni, R., Kebarle, P. 1972. *J. Am. Chem. Soc.* 94:2940
122. Yamdagni, R., Kebarle, P. 1971. *J. Am. Chem. Soc.* 53:7139
123. Grimsrud, E. P., Kebarle, P. 1973. *J. Am. Chem. Soc.* 95:7939
124. Hiraoka, K., Grimsrud, E. P., Kebarle, P. 1974. *J. Am. Chem. Soc.* 96:3359
125. Hiraoka, K., Kebarle, P. 1975. *J. Chem. Phys.* 62:2267
126. Kebarle, P., Yamdagni, R., Hiraoka, K., McMahon, T. B. 1976. *Int. J. Mass Spectrom. Ion Phys.* 19:71
127. Hiraoka, K., Kebarle, P. 1975. *Can. J. Phys.* 53:970
128. Hiraoka, K., Kebarle, P. 1975. *J. Chem. Phys.* 63:1685
129. Hiraoka, K., Kebarle, P. 1978. *J. Am. Chem. Soc.* 100: In press
130. Meot-Ner, M., Field, F. H. 1974. *J. Am. Chem. Soc.* 96:3168
131. Meot-Ner, M., Field, F. H. 1974. *J. Chem. Phys.* 61:3742
132. Conway, D. C., Janik, G. S. 1970. *J. Chem. Phys.* 53·1859
133. Yang, J. H., Conway, D. C. 1964. *J. Chem. Phys.* 40:1725; 1965. 43:2500
134. Conway, D. C., Nesbitt, L. E. 1968. *J. Chem. Phys.* 48:509
135. Pack, J. L., Phelps, A. V. 1966. *J. Chem. Phys.* 45:4316
136. Fehsenfeld, F. C., Howard, C. J., Schmeltekopf, A. L. 1975. *J. Chem. Phys.* 63:2835

137. Fehsenfeld, F. C., Ferguson, E. E. 1974. *J. Chem. Phys.* 61:3181
138. Staley, R. H., Beauchamp, J. L. 1975. *J. Am. Chem. Soc.* 97:5920
139. Arshadi, M. R., Futrell, J. H. 1974. *J. Phys. Chem.* 78:1482
140. Fehsenfeld, F. C. Ferguson, E. E. 1973. *J. Chem. Phys.* 59:6272
141. Meot-Ner, M., Field, F. H., 1977. *J. Am. Chem. Soc.* 99:998
142. Beggs, D. P., Field, F. H. 1971. *J. Am. Chem. Soc.* 93:1567
143. Field, F. H., Beggs, D. P. 1971. *J. Am. Chem. Soc.* 93:1576
144. Bennett, S. L., Field, F. H. 1972. *J. Am. Chem. Soc.* 94:5186
145. Kebarle, P. 1972. In *Ions and Ion Pairs*, ed. M. Szwarc. New York: Wiley. p. 27
146. Kebarle, P. 1972. See Ref. 2, p. 315
147. Kebarle, P. 1974. In *Modern Aspects of Electrochemistry*, ed. B. E. Conway, M. Brockris, 9:1. New York: Plenum
148. Diercksen, G. H. F., Kraemer, W. P. 1972. *Theor. Chim. Acta* 23:387
149. Clementi, E., Popkie, H. 1972. *J. Chem. Phys.* 57:1077; Kistenmacher, H., Popkie, H., Clementi, E. 1973. *J. Chem. Phys.* 58:1689; 1974. 59:5892
150. Diercksen, G. H. F., Kraemer, W. P. 1970. *Chem. Phys. Lett.* 5:570
151. Kistenmacher, H., Popkie, H., Clementi, E. 1973. *J. Chem. Phys.* 58:5627
152. Kistenmacher, H., Popkie, H., Clementi, E. 1974. *J. Chem. Phys.* 61:799
153. Davidson, W. R., Kebarle, P. 1976. *J. Am. Chem. Soc.* 98:6133
154. Kraemer, W. P., Diercksen, G. H. F. 1970. *Chem. Phys. Lett.* 5:463
155. Kollman, P. A., Allen, L. C. 1970. *J. Am. Chem. Soc.* 92:6101
156. Newton, M. D., Ehrenson, S. 1971. *J. Am. Chem. Soc.* 53:4971
157. Conway, D. C. 1969. *J. Chem. Phys.* 50:3864
158. Conway, D. C. 1970. *J. Chem. Phys.* 52:2689

SURFACE SCATTERING ❖ 2655

Sylvia T. Ceyer and Gabor A. Somorjai

Department of Chemistry and Materials and Molecular Research Division, Lawrence Berkeley Laboratory, University of California, Berkeley, California 94720

INTRODUCTION

The molecular beam–surface scattering technique is a powerful method for detailed investigation of the dynamics of gas–surface interactions. Its main thrust is to explore on the atomic scale such macroscopic gas–surface processes as catalysis, corrosion, adsorption, desorption, and energy accomodation. It also provides a tool for surface structure analysis.

Three basic characteristics of the beam–surface scattering experiment allow the interaction details to be probed. First, a well-collimated flow of atoms or molecules of known energy allows precise control of gas concentration at the surface. Secondly, there are no collisions of the incident particles or the outgoing interaction products with surfaces other than the one studied. Lastly, the orientation of the macroscopic surface or any ordered microstructure on the surface is fixed, relative to the incident particles.

An example of a beam surface scattering apparatus, among several others previously described (1, 2), is shown in Figure 1 (W. J. Siekhaus, S. T. Ceyer, Y. T. Lee, and G. A. Somorjai, unpublished). Differential pumping of the two source chambers with enough pumping speed to accomodate nozzle sources allows two directed beams with fluxes that are orders of magnitude above the background to impinge on the experimental surface. The surface must be clean and its structure well characterized. Thus, an ion gun for cleaning is included, along with an electron gun for Auger analysis and LEED as analytical techniques. The ability to cool the surface below room temperature is desirable for elastic scattering studies, and the ability to heat the surface is necessary for cleaning and reactive scattering studies. The incident angle, azimuthal angle, and tilt angle between the incident beam and the surface are easily varied under vacuum by means of a sample manipulator. Differential pumping of the detector, essential for the study of low probability surface reactions, yields a low background signal.

The apparatus shown in Figure 1 is designed to measure mass, angular, velocity, and residence time distributions for reactive or nonreactive processes. Mass and angular distributions are attainable by use of a rotatable quadrupole mass spectrometer detector. The velocity and residence time distributions are measured by a time of flight technique. Modulation of the incident beam yields surface residence time

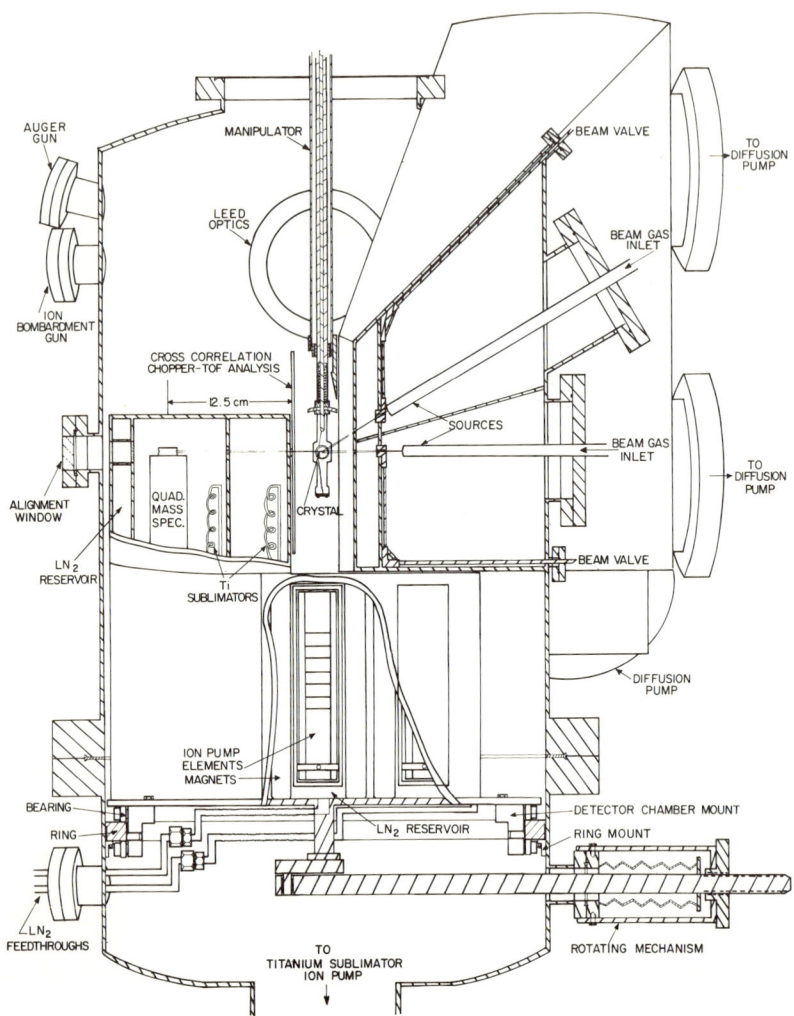

Figure 1 Molecular beam-surface scattering apparatus.

distributions. Modulation of the scattered beam eliminates the possibility of an error in the velocity distribution measurement due to a residence time of the incident species at the surface. Thus, modulation of the scattered beam is essential following a reactive or an inelastic event that has a surface residence time greater than about 5×10^{-6} sec. The ability to measure a velocity distribution as a function of scattering angle is necessary for a clarification of single- vs multiphonon inelastic transitions.

In this review we see how mass, angular, velocity, and surface residence time distributions have produced and will continue to produce an improved understanding of the microscopic details of gas–surface interactions. We discuss the significant advances in experiments dealing with elastic diffractive scattering, energy exchange, and surface reaction mechanisms with particular emphasis on work done in the last five years. Not included are high energy and ion surface scattering studies. Theoretical surface studies are included only as they apply to the interpretation of experimental results or as they predict a feasible experimental investigation. There have been several recent reviews of general experimental (3–6), nonreactive experimental (7–10), reactive experimental (11–15), and theoretical studies (16–19) that may also be of interest to the reader.

ELASTIC SCATTERING–DIFFRACTION STUDIES

There is much interest in the use of elastic scattering of light atoms from surfaces for the extraction of the gas–surface interaction potential by means of the observed structure in the angular distribution measurements. The use of light atom diffractive scattering to deduce surface structure and surface Debye temperature has received less attention, mainly due to the inability of most metal surfaces to exhibit light atom diffraction spectra and to the comparatively easy technique of LEED to obtain similar information.

Helium, Ne, H, and D scattering from the alkali halide single crystals of LiF and NaF have been the only systems intensively studied (2, 20–27) and have shown well-defined diffraction peaks. These experiments are generally of a finer quality than earlier ones, due to the clean, well-characterized surfaces employed (2, 20–23), the use of low surface temperatures, which reduce the inelastic background (2, 23, 24), and the use of well-collimated supersonic (2, 20–24) or velocity selected incident beams (25, 26).

Gas-Surface Interaction Potential

Analysis of these results of well-controlled collisions yields information about the gas–surface interaction potential since it is this interaction potential that determines the collision dynamics. Scattering data that are purely elastic with clear evidence of diffraction ultimately provide the most precise information regarding the interaction. There is however, no unique way to invert the scattering data into the interaction potential. Thus, a reasonable form of the interaction potential must be assumed and then fitted to the scattering data by means of a minimum number of adjustable potential parameters. Consequently, a very important result of the diffraction experiments has been to provide stimulus and guidance for the development of a quantum mechanical theory of gas–surface scattering.

Two types of measurements facilitate the extraction of the gas–surface potential parameters, which are important for the development of theory. First, the scattering distributions need to be measured as a function of the reflected angle so that the relative intensities of the diffraction peaks are known. The relative intensities of the diffraction peaks depend on the strength of the repulsive part of the potential. In addition to rigorous calculations, as shall be seen later, a simple relationship between the strength of the repulsive part of a Lennard-Jones potential and the number of diffraction peaks has been established to allow an approximate calculation of the repulsive strength parameter (28). Secondly, the intensity of a diffraction peak must be monitored as a function of incident or azimuthal angle in order to observe the angular location of the minimum or maximum in intensity that is created when an atom falls into a bound state resonance. The angular location of the bound states is essential to calculation of the energy levels, well depth, and range parameter of the attractive part of the potential.

ALKALI HALIDE SURFACES In this fashion the scattering of He and Ne from LiF(001) was studied (2, 23). The angular locations of the diffraction peaks agree well with the values calculated from a knowledge of the deBroglie wavelength and the surface crystallographic structure. The diffraction probabilities for each channel are extracted from the experimental peak intensities by taking the velocity distribution of the supersonic incident beam into consideration. Satisfactory agreement is obtained between the experimental diffraction probabilities and those calculated by an approximate quantum theory of Levi et al (29–31). The theory employs a simple, corrugated hard wall as the interaction potential but takes into account the accelerating effect of an attractive square well on the incident atom. The incident atom is not allowed to be bound at the surface in this well nor is multiple scattering explicitly allowed. Thus, a comparison of the theoretical values of the diffraction probabilities to the experimental probabilities is only valid at particular incident and azimuthal angles that do not correspond to a resonance condition, and, indeed, the theory does not reproduce the observed variation in diffraction probability as a function of azimuthal angle. Such comparison between experiment and a theory that is not realistic on all aspects of the interaction is, however, valuable to the theorist for an orderly and progressive understanding of the gas–surface interaction. The approximate nature of this quantum mechanical calculation is also easier to carry out compared to a rigorous calculation of the Schrödinger equation of the system. Although not applied to the above set of data, recent approximate quantum theories that have varying degrees of realistic interaction potentials include the semiclassical theory of Doll (32–34) and Masel et al (35, 36), the quasiclassical theory of Ray et al (37), and the CCGM theory of Goodman (38–41). For the most part, these theories await rigorous comparison to good, structured scattering data produced by monoenergetic beam scattering from clean, low temperature surfaces. However, a comparison of these approximate theories to exact quantum results has been presented (42, 43).

Four bound states of He on LiF (001) have been labeled through investigation of the intensity of the specular beam as a function of azimuthal angle (22). These bound state resonances are evidenced by sharp drops in intensity of the specularly scattered beam. A knowledge of the incident and azimuthal angles at which these minima occur, as well as the incident beam energy, enables calculation of the bound state energies. They range in value from 0.1 to 6 meV for He bound to LiF. The four bound state energies of the He/LiF system were found not to fit the energy level spacings of a Morse potential. In fact, a Morse potential predicts the occurrence of only three bound states. Tsuchida (44) previously predicted the energies of the bound states of the He/LiF system by numerically solving the Schrödinger equation that employed a realistic potential obtained from a pairwise summation of a Lennard-Jones 12–6 potential between a He atom and each crystal atom. His calculated values agree well with the observed bound state energies (22). Tsuchida (45) has since taken the experimental values of the four resonance energies and tried to obtain an optimum set of parameters for the pair-wise summed 12–6 potential. He has shown, however, that the bound state energy values alone do not uniquely predict an optimum set of potential parameters. The diffracted intensities were not measured in this study (22).

There has been much conjecture concerning the fate of those atoms that are trapped at the surface in bound state resonances or, in other words, are selectively adsorbed. The observation of minima in the specular intensity leads to the notion that the atom is preferentially inelastically scattered because it has a longer surface residence time and thus is lost from the diffracted beam (20–22, 26, 27, 46). Recently, minima and maxima have been observed to occur simultaneously in different diffracted beams (24, 47), which is indicative of a redistribution in intensities among the diffraction channels once the atom leaves its bound state. It appears that a maximum in the intensity of a diffracted beam occurs when its nearest reciprocal lattice point has participated in a bound state event (47). The ratio of maximum intensity of a diffraction peak created when an atom is selectively adsorbed to the intensity of the diffraction peak not at resonance is temperature independent between 140–420°K, which is the temperature range studied (24). This observation supports the idea that inelastic processes are not involved once the atom is selectively adsorbed, although the minimum intensity of a diffraction peak has not been studied as a function of temperature. Theoretically, the bound state minima and maxima have been produced from a purely elastic scattering process (48) but rigorous comparison of the intensities of these extrema to experiment has not been accomplished. The question of involvement of inelastic processes in the selective adsorption of an atom on a surface needs to be answered experimentally by a more thorough study of the temperature dependence of the scattered background and intensities of the bound state extrema. If inelastic processes are involved, fundamental gas–surface energy transfer information can be obtained, since bound atoms in any resonance condition are trapped at the surface in a known energy level.

Chow & Thompson (49) have calculated the angular positions of the bound state extrema under incident conditions of two degenerate transitions via two different reciprocal lattice vectors to the same bound state energy level for the He/LiF and He/NaF systems. A splitting of the bound state minimum occurs as the result of a perturbation on the bound state energy levels by the periodic part of the potential. The magnitude of the splitting depends on the strength of the Fourier components of the periodic potential, another condition which would have to be met in fitting a potential to the experimental data. The calculated splittings that range from 1–6° in azimuthal angle are well within the possible experimental resolution. These splittings have not been observed, due to the special incident conditions required.

H and D have also been scattered from the alkali halides (26). The three observed bound state energies fit a Morse potential's energy level spacings well. Surprisingly, however, the well depth of H on LiF and H on NaF is identical. The strength of the exponentially repulsive part of the potential is calculated. The results are in fair agreement with the experimental diffracted intensities.

Thus it becomes apparent from comparison to existing data that gas-surface scattering, even for these systems, cannot yet be fully described theoretically. Both a complete quantum mechanical theory and a complete reliable set of data are missing. Progress in theory development, however, is being made with the use of simple corrugated hard wall potentials for optimization of the scattering model (48, 50–52), the extension to more realistic interaction potentials (53–56), and the critical evaluations of the various assumptions in the calculations (37, 40, 41, 57–60). The exact quantum mechanical calculations of Wolken (46) and Tsuchida (44, 45) indicate that a reliable and realistic gas-surface interaction potential can be extracted from scattering data.

METAL SURFACES   The discussion so far has been concerned with the determination of interaction potentials between an atom and an alkali halide crystal. Corresponding studies of single crystal metal surfaces have not been very successful, due to the absence of well-defined diffraction peaks. Beeby has discussed the absence of diffraction from metal surfaces as the result of a smaller value of θ_D, the surface Debye temperature, in contrast to the alkali halides, thus leading to a larger fraction of inelastic transitions (61–63). As first discussed by Weinberg (64), it has become apparent from quantum (65) and semiclassical calculations (36) that the periodic part of the gas-surface interaction potential is too weak to produce strong nonspecular diffraction. This idea is supported by the experimental evidence from the first pure metal single crystal to exhibit diffraction peaks. Helium diffraction was only observed from the W(112) crystal in the direction in which the periodic surface structure is more pronounced (66). A more thorough study of He diffraction has recently been carried out (67). A semiclassical calculation employing a Lennard-Jones 3–9 potential fits the scattering distribution qualitatively. Unfortunately, an effusive thermal He beam was used that makes the theoretical distribution insensitive to variation of the well depth of the potential employed in the calculation.

Two other reports of diffraction from metal surfaces have not been very convincing (68, 69). Very recently however, He and H_2 diffraction has been observed from a Ag(111) surface in high resolution apparatuses (70, 71). In both reports, the surface was cooled to below or near the surface Debye temperature in order to minimize inelastic transitions. The first order diffraction peaks are of the order of 1000 times less intense than the specular peak. Interestingly, the intensity of H_2 diffraction is significantly greater than that for He diffraction. Hopefully, the observation of diffraction from metal surfaces will yield information about the interaction potential of some chemically important systems.

RAINBOW PEAKS   Bilobular scattering or rainbow scattering has been observed from LiF (72, 73) and stepped metal surfaces (74–77). Observation of this structure in the scattering distribution also allows determination of the gas-surface interaction potential. Helium atoms incident along the step edges of a Pt(553) single crystal scatter specularly, while rainbow peaks are evident if the He atoms are incident perpendicular to the step edges (77). Thus in agreement with semiclassical calculations (36), it appears that a strong periodic structure of the surface is as essential as it is for diffraction in order for the rainbow phenomenon to occur. The classical trajectory calculations of McClure (78, 79) and to a lesser extent those of Steele (80) who uses a less realistic interaction potential fit well the experimental scattering distributions of Ne from LiF (72, 73). The potential parameters of McClure (78, 79) are in close agreement with those found by Finzel et al (26). Rainbow peaks have been described as an accumulation of trajectories at two angles that correspond to the maximum and minimum scattering angles introduced by the inflection points of a periodic potential. Jewsbury (81) has recently reinterpreted the bilobate structure as the result of single and double scattering events.

Surface Properties

The study of surface structure as well as the surface Debye temperature by means of atom scattering holds several advantages over LEED. Surfaces unstable under electron bombardment and insulators may be studied more readily by atom scattering. More importantly for the case of surface structure evaluation, the atoms sample only the outermost layer of the surface, unlike electrons, which penetrate into deeper layers and thus complicate the analysis of the diffraction pattern.

STRUCTURE   Atomic diffraction from adsorbed gaseous layers on solid substrates has not been employed very often for the study of surface structure, nor has it been very successful in most cases. Observation of He diffraction from a WC R(3 × 5) surface (82) has indicated a dense carbide structure rather than the dilute one originally proposed from LEED data. An oxygen structure on a WC surface has been tentatively identified (83). The diffraction patterns of ethanol and water adsorbed on a LiF crystal are complicated by surface diffusion, because of the method of introducing the adsorbents (84, 85). Adsorbed layers of CCl_4 (85) and Hg (86) do not show diffraction peaks in a He scattering distribution.

SURFACE DEBYE TEMPERATURE   Measurement of a purely elastic peak intensity versus incident beam energy, angle, or surface temperature allows calculation of the surface Debye temperature. This procedure was carried out on the intensity of atoms diffracted from LiF (2, 87, 88) and NaF (24, 25) surfaces. The surface Debye temperature of LiF (87) was calculated by use of a procedure that accounts for the acceleration of the incoming atom by the attractive potential well as first suggested (61) and since amplified (62) by Beeby. The importance of the inclusion of this effect has been illustrated (25). Goodman (89) has taken the same data (87) and accounted for the velocity spread of the incident Maxwellian beam, since the Debye-Waller factor from which the Debye temperature is calculated assumes a different form after energy averaging (90). A more accurate value of $560°K$ is derived from this treatment.

In many experiments, the surface Debye temperatures of metals as measured by atomic scattering were found to be larger than the bulk values (67, 91). This is unreasonable since surface atoms have an increased amplitude of vibration due to their weaker bond strength. Lapujoulade & Lejay (91) tried to invoke inelastic effects as the reason for the larger Debye temperature. Indeed Lin & Wolken (92) have shown that there is a considerable inelastic flux that contributes to the specular intensity. This portion of the flux must be known and subtracted from the specular intensity in order to obtain the pure elastic flux so that the Debye-Waller formalism can be applied. A metal does have a smaller Debye temperature than an alkali halide, which implies an increased importance of inelastic effects relative to alkali halides. However, the proportion of inelasticity needed to reduce the surface Debye temperature below the bulk value is unreasonably large, based on the velocity measurements of Lapujoulade et al (91). Stoll et al (67) have reasoned that the discrepancy in the surface Debye temperatures measured by atom scattering as compared to LEED is due to the nature of the atomic trajectories. In scattering experiments, atoms move more slowly than electrons and thus remain coupled to the surface for a longer time. In addition, they scatter from the repulsive wall made up of several lattice atoms, and not from the core potentials of individual atoms. An average mean square displacement of the unit cell and not the mean square displacement of individual atoms should thus be used in the Debye-Waller formalism. The surface Debye temperature obtained in this fashion is decreased by a factor of 1.4–2 to a value below the bulk Debye temperature. This treatment gives an improved fit to the data. Comsa offers another explanation (93). If only specular reflection is evident with no higher order diffraction peaks, as is the case for most metals, then the specular beam does not necessarily represent coherent zero order diffraction for which a Debye-Waller formalism applies. Certainly, more careful consideration of the inelastic effects, the nature of the averaged unit cell displacement and an incoherent specular peak are required experimentally and theoretically. It is clear that surface properties can be determined from scattering experiments, but only after the basic interaction mechanism is known.

ENERGY EXCHANGE STUDIES

The energy transfer processes that occur as the result of the initial collision of a gas atom or molecule with the surface can be suitably studied by a molecular beam surface scattering experiment. The ideal experiment would determine the trajectory and the exact quantum state of both the incoming and the outgoing gaseous particle as well as the phonon distribution of the surface before and after interaction with the gaseous particle. This information would paint a detailed picture of the gas-solid energy transfer pathways. At present, such measurements are not available. Trajectories are inferred from the incident angle and final scattering angle of a large number of gaseous particles. The translational velocity of the incident beam is known but usually there is no information on the energy state of the scattered beam or of the surface. As in diffractive elastic scattering experiments however, inelastic scattering data, even in its incomplete state, serve as guides for the development of a theory describing the energy transfer mechanism.

There are three basic interactions of an atom or molecule that can occur with a solid surface. An elastic interaction is an impulsive collision of the scattering atom or molecule with the steep portion of the surface potential. Only momentum exchange occurs during this interaction. The previous section dealt only with elastic interactions. An inelastic interaction involves some degree of energy exchange between the gaseous particle and the surface upon the initial collision with the steep portion of the surface potential. The inelastic interaction may involve more than one collision of the gas particle with the steep repulsive wall, depending on its residence time in the region of the repulsive wall. These elastic and inelastic processes constitute physical interactions. Lastly, a trapping interaction involves a gas particle's loss of enough kinetic energy at the initial collision with the repulsive wall to reside temporarily in the attractive well. The length of time that an atom or molecule resides in this well is commonly called the surface residence time. The incident beam conditions of angle and energy, the interaction potential and the relative energy levels of the gas-surface system determine which of these processes dominates the interaction. Usually all three interactions occur simultaneously, which is a complicating factor in gas-surface energy exchange studies.

Presumably, if a particle is trapped on a surface it can undergo reaction either with another trapped particle, a gas phase particle, or the surface itself. We treat energy exchange studies of reactive encounters as a separate topic.

Nonreactive Studies

ANGULAR DISTRIBUTION MEASUREMENTS

Metal surfaces Much of the present information about gas-surface energy exchange is inferred from angular distribution measurements. The shape of a scattering distribution and its behavior as a function of surface temperature and angle of incidence are used to classify the initial degree of energy exchange (94–96). Merrill et al (94–95) found that the dimensionless well depths of the rare gases scattered from Pt, W, and Ag correlate well with the energy transfer assignments based on the shape and behavior of the scattering distributions. As described by Merrill et al (94–95), an elastic interaction is characterized by an intense but narrow peak centered at the specular angle. The peak intensity attenuates as the surface temperature increases because of thermal roughening of the surface. A broader, less intense scattering distribution, not centered at the specular angle, is the result of an inelastic interaction. The peak intensity attenuates as a function of increased surface temperature and incident angle. Lastly, a trapping dominated interaction is characterized by a very broad scattering distribution with the maximum intensity near the surface normal, presumably due to the large contribution of trapped particles emitted from the surface in a cosine distribution that has a maximum at the surface normal. The maximum intensity increases with surface temperature since fewer particles can be trapped at the surface at higher temperatures.

However, it is not true that particles trapped at the surface are necessarily in thermal equilibrium with the surface (97–99). From the data of the rare gases scattered from Pt, W, and Ag (94, 95), the fraction trapped was estimated as the ratio of the total flux scattered in a cosine distribution (as calculated by fitting a point of the diffuse background to a cosine distribution) to the total scattered flux of the whole angular distribution. Comparison of the fraction trapped with the values calculated from a simple square well semiempirical model (100) shows qualitative agreement, while comparison with the values calculated from the theoretical harmonic soft cube model (101) shows rather uncertain agreement. The discrepancy between the soft cube model and the experimental values is due, for the most part, to the requirement that the fraction trapped accommodate fully with the surface before desorbing (i.e. the angular distribution is assumed to be cosine), while the theoretical values for the fraction trapped include only those particles that have a kinetic energy less than the well depth after the initial collision with the repulsive wall of the surface. That the atom or molecule is trapped at the surface says little about the extent of energy transfer. Modak & Pagni (99) employ the harmonic soft cube model on a variety of gas surface systems to show that for a constant monoenergetic or Maxwell-Boltzmann atomic beam incident on a surface, the trapped fraction's energy distribution differs significantly from complete equilibration with the surface. Thus, the scattering distributions deviate from cosine emission since only those particles in thermal equilibrium with the surface, as in evaporation (102), are emitted in a cosine distribution. It is apparent that a value for the fraction of particles trapped at the surface cannot be derived through separation of a scattering distribution into cosine and noncosine parts.

Other systems whose angular distributions have been investigated in an attempt to extract energy exchange information include He scattering from semiconductors (103) and K scattering from alkali halides (104). Both systems exhibit more diffuse scattering than expected by comparison to rare gas scattering under similar conditions. Surface contamination in both systems, however, cannot be ruled out. It has been shown that atom scattering from a CO-covered surface that is the major contaminant in an ultrahigh vacuum system exhibits a cosine distribution (105). The increased inelasticity as indicated by the angular distributions may be the result of surface contamination. In addition, care must be taken to know the structure of the surface if a more diffuse scattering is said to be the result of increased energy transfer (106). A disordered surface is rougher, which leads to more diffuse scattering even after purely elastic transitions.

Molecules scatter in broad distributions with the maximum intensity at either side of the specular angle (107). These distributions yield no precise information about gas-solid energy transfer since they are the result of a mixture of elastic, inelastic, and trapping interactions including the involvement of internal modes. Thus, the angular distribution is not generally a good indication of the absolute amount of energy transferred.

The most interesting angular distribution measurements, however, have involved a comparison between that of H_2 and D_2 scattered from single crystal surfaces (95, 108). The scattering distributions of D_2 are always less intense and broader than those of H_2, which is indicative of increased energy transfer even though the shape of the interaction potentials of the two isotopes with the same surface is identical. A mean translational velocity measurement corroborates the observed increased inelasticity of D_2 (97). The increased energy transfer has been attributed to the fact that D_2 has rotational modes that can be populated by a one-phonon interaction with the solid (95). The rotational energy level spacings of H_2 are too large to be populated by a single-phonon interaction. The additional energy transfer pathway in D_2 leads to a more diffuse and less intense scattering distribution. Several hard cube models have been used to study diatomic scattering from surfaces (109–111). These calculations yield a shift in the maximum intensity of the scattering distribution away from the specular angle, which is not always observed experimentally. A quantum mechanical one-phonon close-coupling calculation that includes an attractive Morse potential well affords finer agreement with experiment (112, 113). The experiment studied theoretically by Lin & Wolken (113) involved the angular distributions produced by H_2 and D_2 scattering from Ag (111) (95). They found that the coupling of the rotational modes of the diatomic and the surface vibrations was not responsible for the broader and less intense scattering distribution of D_2 but that the mass difference was more important.

Alkali halide surfaces The clearest demonstration of energy exchange during a gas-surface interaction, as seen from an angular scattering distribution, has been observed from diffracting surfaces. Distinct inelastic structures that correspond to phonon emission and adsorption events are evident around all the elastic diffraction peaks in the He and Ne angular scattering data from LiF and NaF (2, 114, 115). From knowledge of the angular positions of the inelastic maxima as a function of out-of-plane scattering angle, Williams was able to construct a surface phonon dispersion curve for LiF (115). Lagos & Birstein (116) present a quantum mechanical calculation that considers one surface phonon exchange only. Their theoretical curve is well fit to the intensity and shape of the He scattering distribution from LiF of Williams (115). This theoretical formalism has been extended to include multiphonon interactions (117), but no comparison with experiment is given. Corresponding velocity measurements (118–120) indicate that bulk phonons predominate in the interaction, although a dispersion curve is not calculated. Benedek (121) discusses the appearance of inelastic transitions in the angular distribution that involve surface phonons at the zone boundary or other special points on the surface Brillouin zone that can easily be distinguished from bulk phonon peaks by observation of their angular location as a function of incident angle. Experimental verification of this argument would aid in resolving the inelastic mechanism.

Inelastic diffraction peaks have been observed in the scattered angular distribution of H_2, D_2 (122–125), and HD (123, 124) from the alkali halide crystals. The peaks correspond to molecules in selected rotational states. In this type of energy exchange, a redistribution of energy occurs between the translational and rotational modes of the molecules without involving phonon interactions. The experimentally observed probabilities for translational-rotational energy exchange are generally comparable to elastic scattering probabilities (124, 125). The occurrence of rotationally inelastic diffraction peaks had previously been predicted (126). Since then several quantum calculations for diatomic scattering from alkali halides have been presented (127–131), but generally imprecise knowledge of the diatomic molecule-surface interaction potential precludes quantitative agreement with experiment. Certainly this type of energy transfer warrants future consideration in the course of delineating the overall view of energy exchange mechanisms.

VELOCITY MEASUREMENTS Mean velocity measurements or velocity distribution measurements provide a more reliable handle with which to decipher the gas-surface energy exchange process. However, few velocity determinations have been made, largely due to the need for an expensive and technologically sophisticated apparatus.

The mean energy of an Ar beam scattered from polycrystalline Pt was measured as a function of scattering angle in the incident plane and also as a function of out-of-plane angle (132). The translational energy of the scattered Ar decreases as a function of increasing scattering angle measured from the surface normal. This trend qualitatively agrees with the predictions of a hard cube model (133, 134). Resolution of the measured velocity into a normal and a tangential component shows that the tangential component is approximately conserved. There is, however, a substantial amount of flux scattered outside of the incident plane that is attributed to surface roughness. In another study of Ar scattering, this time from a W single-crystal surface (135), the shape and intensity of the scattering distributions were found to depend on O_2 coverage, but the mean velocity did not. This observation supports an earlier comment that the shape of a scattering distribution is not a measure of the velocity of the scattered particles.

Measurements from two different laboratories (136, 137) disagree on the velocity of As_4 scattered from a GaAs single crystal. One group has also measured the velocity distribution of As_2 and Ga scattered from the same crystal and found that the As_2 scattered velocity was substantially lower than either the incident beam temperature of $1095°K$ or the surface temperature of $923°K$. The scattered Ga was found to have undergone complete energy exchange with the surface. The lower translational energy of the scattered As_2 molecules is attributed to a channeling of some of the surface vibrational energy into the internal degrees of freedom of the molecule. It is also plausible, however, that the surface temperature was sufficiently larger than the well depth to prevent the molecule from residing in the well (assuming that the molecule is trapped) long enough to come to thermal equilibrium with the surface (138). It would be informative here to know the fraction trapped. A velocity measurement of the scattered beam as a function of incident beam temperature and as a function of surface temperature would yield additional information concerning the energy transfer pathway.

Good examples of the type of energy exchange information that can be obtained by variation of the incident beam and surface temperature are the scattering experiments from pyrolytic graphite surfaces (97) of rare gas atoms and diatomic molecules. The energy of the scattered beam is independent of the energy of the incident beam. The fact that the scattered beam energy does not depend on the incident energy indicates that the fraction trapped is unity for Ar, Xe, Kr, H_2, D_2, and O_2. The scattered beam energy does increase with surface temperature up to a limiting value dependent on the particular gas-surface system. This limiting value then has the characteristics of a well depth for the interaction.

Recent velocity distribution measurements of Ne and Ar scattered from Cu (100) have shown that the tangential velocity is not conserved, especially in the subspecular direction (91). This observation, as long as the interaction is inelastic, indicates the involvement of surface phonons during the interaction. The velocity measurements at any reflected angle of Ne and Ar also indicate that there is a substantial superposition of elastic and inelastic multiphonon processes. It can be seen that velocity measurements, as well as angular distribution measurements, can suffer the problem of multiple interactions occurring simultaneously.

There has been one experiment in which a distinct inelastic scattering peak was observed from a metal surface as well as a narrow, intense specular lobe (139). Helium scattered from an epitaxially grown Ag (111) surface exhibits a distinct peak shifted 20° toward the surface normal from the specular angle. Mean velocity measurements show that the scattered beam is 20% faster than the incoming beam, and the velocity of the scattered beam increases monotonically as the angle is varied from the specular position to the surface normal (140). However, it has been shown (95) that scattering distributions from epitaxially grown crystals are generally broader and less reproducible between different epitaxial crystals than those from different bulk single crystals. Thus, caution is advised in interpreting these distributions until the structure of the epitaxial surfaces is more firmly established, since a distinct inelastic peak was not observed from a single crystal Cu (100) face (91) nor from a single crystal Ag (111) surface (95). The absence of a distinct inelastic peak in the latter case may be due to the broadness of the thermal beam and in the former case due to the relatively high energy of the beam used, in comparison to the energy of the beam employed from the epitaxial Ag surface (139).

Nevertheless, no experiment has ignited more theoretical interest in one- and two-phonon inelastic rare gas scattering from metal surfaces than the observation of a distinct inelastic peak of a He beam scattered from the epitaxial Ag (111) surface. Whether the He atom undergoes a single- or multiphonon scattering process is not clear from comparison of existing perturbation theories with experimental observation. Goodman (141–143) qualitatively reproduces the experimental scattering distributions and speed measurements by a three-dimensional quantum mechanical treatment for one-phonon inelastic scattering of atoms. Beeby (144, 145) claims that the smaller angular resolution, compared to the experimental value, and the monoenergetic incident beam assumed in Goodman's calculations lead to the qualitative agreement with experiment. Beeby's quantum mechanical calculation includes two-phonon scattering of atoms. From his calculated scattering distribution and mean velocity as a function of angle, Beeby concludes that the two-phonon contribution is substantial. Weare (146) points out that Beeby's method is very dependent on the scattering model chosen, although Weare's calculation as well as that of Garibaldi et al (147) also indicate a substantial two-phonon inelastic contribution. In addition Weare shows that the two-phonon contribution to the inelastic lobe can be suppressed by scattering with low beam energies from liquid N_2 cooled surfaces.

Lin & Wolken (92) have recently qualitatively reproduced the results of the scattering experiments from both the epitaxial and single crystal Ag surfaces by use of a one-phonon close coupling quantum mechanical calculation. They treat the surface as an isotropic Debye solid. Another recent classical calculation (148–151) that treats the surface as a large collection of harmonic or anharmonic oscillators agrees with the single crystal results of Lin & Wolken. For both experimental conditions, it is shown (92) that there is indeed a large flux of atoms at the specular angle that undergoes an inelastic transition with the solid. Thus one does not have a precise value for the elastic scattering probability from an angular distribution measurement. It appears that the question of the validity of a one- vs a two-phonon interaction for He will only be solved by a best fit of a model to both the angular and velocity distribution of a monoenergetic beam scattered from a single crystal over a wide temperature range. From these calculations, however, it is apparent that lattice effects are important in the extent of energy transfer (151).

As is evident from the above, simple gas–surface energy exchange information is complicated by the simultaneous availability of three scattering pathways. The extent to which each of these pathways participates in the energy transfer mechanism is not necessarily clear from the macroscopic angular or velocity distribution unless care is taken to eliminate one or more of the possible routes. A well-defined velocity of the incident beam separated in temperature from the surface would perhaps permit experimental distinction of the three energy exchange pathways. This experimental distinction has been obtained with particular systems. Diffracting surfaces more clearly separate the elastic component from the phonon-involved inelastic component and the translational-internal energy exchange component. Some surfaces have been shown to trap completely the impinging gas particles.

Much experimental work has been initiated in an attempt to elucidate the gas–surface energy exchange process. However, very detailed work on any system that includes intensity measurements and velocity distribution measurements as a function of scattering angle in the incident plane, incident beam temperature, surface temperature, incident angle, azimuthal angle between the beam and the surface, and scattering angle out of the incident plane has not been accomplished. The attempt to unravel a particular energy exchange pathway necessitates these measurements.

RESIDENCE TIME MEASUREMENTS   The length of time that a particle stays trapped in the surface well is an important parameter in its degree of energy exchange with the solid (98) and provides complementary information to angular and velocity measurements. Yet few such measurements have been made for nonreactive systems. The residence times of rare gases on metals ($\sim 10^{-13}$ sec) are too small to be experimentally obtainable at present (152). Measurable surface residence times ($>10^{-4}$ sec) include elements other than the rare gases (98) and unreacted molecules (I. E. Wachs & R. J. Madix, unpublished).

Residence times studied as a function of surface temperature yield binding energies and desorption kinetics (153, 154). An interesting measurement of the residence time distribution of K atoms scattered from a polycrystalline W surface has enabled the trapping probability to be calculated (155). Two peaks are evident in the distribution. One peak corresponds to those atoms that have elastically or inelastically scattered from the surface, while a distinct peak at longer times corresponds to those atoms that have been trapped at the surface. Thus the fraction trapped is easily calculated.

Reactive Systems

The only studies of energy exchange associated with reactive encounters between a gas and a solid surface, or with the reaction of two gas species where the surface serves as a catalyst, are those involving the recombination of H or D atoms. Stickney et al (156–158) examined the angular distributions of H_2 desorbed from a variety of surfaces (Fe, Pt, Ni, Cu, Nb, SS304) after atomic recombination. H_2 scattered from the pure, clean surfaces with a cosine distribution. Intentional contamination of the surfaces with S, C, or Si leads to distributions more peaked than cosine but still centered at the surface normal. The pure, clean Cu surface, however, exhibits a highly peaked scattering distribution at the surface normal. Such noncosine emission has been attributed to an energy barrier to adsorption (159). These noncosine scattering distributions are reproduced by a detailed-balance application (138) of the experimental energy and angle dependence of the dissociative adsorption probability of H_2 on Cu (159). A velocity distribution measurement from a similar system shows that D_2 desorbs with a velocity greater than the one that would correspond to the temperature of the surface (160). Since only those D_2 molecules incident on the surface that have enough kinetic energy to surmount the adsorption energy barrier can be trapped, the desorbing molecules will lack the less energetic tail of the Maxwell-Boltzmann distribution in accordance with the quasi-equilibrium and detailed balancing argument (138).

Scoles et al (161, 162) have measured the amount of energy deposited on a low temperature surface by recombining H atoms. The sticking coefficient of the H atom and the amount of energy transferred depend on the H atom coverage; the amount of the energy transferred is a maximum at some value of the coverage. In all coverage cases, however, the accommodation coefficient is nearly one. No angular distribution of the scattered products was measured nor was the incident beam energy varied.

There exists no firm experimental evidence for the fate of the chemical energy released upon atomic recombination. Goodman lends credence to the possibility that the translational energy of the desorbed molecule benefits from the energy released upon recombination with a macroscopic model of an energy conservation scheme (163). One theoretical study predicts that a detectable fraction of recombined As atoms on an As (111) surface are thermally excited (164).

McCreery & Wolken have investigated theoretically the partitioning of a set amount of energy between the translational and internal degrees of freedom of H_2 after recombination on a W (100) surface as well as the angular distributions of the desorbed molecules (165–168). A modified LEPS potential describes the diatomic-surface interaction, while Morse functions approximate the atom-atom and atom-surface interaction (169). They treat two cases: the recombination of two adsorbed atoms and the recombination of an adsorbed atom with a gas phase atom. For the two adatoms case, the angular distribution has its maximum at the surface normal but is more sharply peaked than a cosine distribution. In the latter case, the molecular scattering maximum is shifted from the specular angle by 20–30° toward the surface tangent. This result means that, experimentally, an angular distribution measurement would be able to distinguish the extent of participation of two reaction pathways. That the two systems have the same amount of energy after recombination and desorption is, however, unrealistic and affects the shape of the angular distribution. Almost twice as much energy could conceivably be distributed to translation for the gas atom-adatom interaction as that for the two adatom interaction. The vibration to rotation ratio remains the same in each case with the high rotational energy levels of H_2 desorbed molecules being greatly populated.

The energy exchange experiments do not yet provide any information regarding the internal modes of excitation of the molecules scattered from the surface. Conceivably, laser-induced fluorescence (170) could be employed in probing the internal state distributions of the scattered simple molecules.

REACTION MECHANISM STUDIES

Modulated molecular beams have been used extensively to study chemical catalytic and corrosion reactions of solid surfaces. The modulation technique allows the combined residence time of the reactants, reaction intermediates, and reaction products to be determined. The residence times in turn provide information that can be used to test models describing the elemental chemical reaction steps at the surface. However, the data collection and its subsequent analysis are long and involved tasks. Moreover, for a given set of data it is not possible to develop a unique model of a chemical reaction mechanism, but it is easy to test any proposed alternative model. In order to determine rate constants for a proposed reaction mechanism, the amplitude and phase of the product signal relative to the incident signal must be measured as a function of reactant pressure, reactant temperature, surface temperature, and modulation frequency. Few gas-surface reactive systems have been thoroughly investigated in this manner (171-173). The kinetic mechanism of a surface reaction as indicated by the experimental data basically belongs in one of six categories: first order (12, 174), linear series step (12, 174), linear parallel or branched process (12, 174), surface diffusion limited (12, 174), bulk diffusion controlled (12, 174), or second order and nonlinear reactions (12, 136, 175, 176). The theoretical dependence of the amplitude and phase on the experimental parameters has been derived and reviewed (13, 174).

There are two main areas of investigation in beam-surface reaction studies: studies in which the surface participates as one of the reactants and those in which a surface participates as a catalyst.

Surface Corrosion Studies

Systems studied in this case include the oxidation of Si (177, 178), Ge (177, 178), Mo (179), and graphite (171, 172, 180), the halogenation of Si (180, 181), Ge (180-182), and Ta (183), and the hydrogenation of graphite (173). The kinetics of all of these systems [except the oxidation of Si and Ge (177, 178) where surface diffusion was a fast step] is rate-limited by surface diffusion or dominated by bulk diffusion, as indicated by the sluggish phase response to surface temperature and frequency variation. As exemplified in these experiments, surface diffusion limitation may or may not be an artifact, while bulk diffusion control is an artifact of the low surface-coverage-modulated technique since the bulk lattice or surface active sites do not have a constant equilibrium supply of reactants. Surface diffusion limitation as an artifact depends on the relative rates of the incident particles' arrival, the surface diffusion step, and the reaction. Knowledge of these parameters from the experimental conditions and from the model fitted to the data would allow determination of the surface diffusion step as an artifact of the technique or as an integral part of the reaction.

The oxidation of Mo exemplifies a surface diffusion-limited reaction (179). MoO, MoO_2, and MoO_3 were the only oxides observed in the reaction. Although each product found involved a different number of bond formations, the residence time (phase lag) observed for each product formation was identical. Thus, a rate-limiting step, which is identified from the phase and amplitude behavior as surface diffusion to reactive sites at grain boundaries, precedes the oxide formation. The extraction of rate constants for the individual steps for the oxide formation is very difficult since these rate constants represent small perturbations in the phase and amplitude data. Similarly, the reported results for the halogenation of Si and Ge (180, 182) indicate that the reaction proceeds through a first order series mechanism (180), but the measured rate constant belongs to a surface migration step. The measured low preexponential factor, as compared to a gas phase unimolecular value, results in the assignment of a surface diffusion process as the rate-determining step.

Olander and co-workers fit a reasonable model for a reaction scheme involving diffusion steps to the amplitude and phase response of the products emitted from various graphite (171–173) and Ta surfaces. They varied the magnitudes of the rate constants of the model to obtain the best fit to the data with the least number of parameters. A two-branch mechanism that involves direct reaction of O_2 to form CO at one active site or diffusion of O_{ads} to a second type of active site is invoked to explain the observed hysteresis in CO amplitude with surface temperature during the oxidation of the basal plane of graphite (171). The diffusion step is so labeled, as in the halogenation of Si and Ge (180, 182), because of its calculated low preexponential factor and its small activation energy. Thus, by fitting a model to the data the values for individual rate constants or products of several rate constants for reactive steps in the proposed mechanism are bracketed even in the presence of surface diffusion limitation. Knowledge of an approximate value of the reaction rate and a diffusion rate by virtue of the calculated rate constants would allow comparison with the arrival rate of molecules to the surface in order to determine whether a diffusion process as a rate-limiting or rate-dominating step was introduced by the modulation as discussed above.

The production of CO and CO_2 from a prism plane of graphite (172), the production of methane and acetylene (173) from prism and basal planes of graphite, and the production of TaF_5 from Ta (183) are dominated by a bulk diffusion process. That is, the tendency of the incident molecules to establish equilibrium with the bulk during the on cycle of the beam overrides the probability of reaction. Although bulk diffusion dominates these reactions, the site at which the diffusing species enters the bulk differs. In the oxidation of the prism plane of pyrolytic graphite, a double diffusion model is presumed in which O atoms enter the bulk only at grain boundaries and can then diffuse between the grains. Because of its small size, hydrogen is assumed to enter the bulk at any lattice point of the prism and basal planes of graphite. In the model for the fluorination of Ta, F_2 has no active sites for dissociation. Its atoms can easily diffuse through a supposed TaF_3 "scale" and proceed finally to the Ta bulk-scale interface where they regenerate the scale, which is used up in the rate-determining step for TaF_5 formation. The fitting of a model to the amplitude and phase response does not necessarily demand entrance at a particular site. For example, any two distinct diffusion paths could lead to the double diffusion model as described above. Clearly, supplemental information is required to best depict the mathematical model fit to the data.

In realistic corrosion situations where there is a constant flux impinging on the surface, these surface and bulk diffusional processes, although present, do not necessarily limit or control the surface reaction rates. The modulated beam flux accentuates these transport processes. In this case, careful analysis of the data shows systems in which this technique becomes useful. However, the problem in identifying the diffusing species in reactive situations is paramount, since no model fit to the data uniquely defines it. At present, good chemical intuition is used to name the diffusing species.

The problem of obtaining good values for rate constants for individual reactive steps of a mechanism still remains. Based on the characteristic time for observable reaction and the activation energies for the kinetic process, Madix (12) argues that kinetic phenomena will more likely be observable at temperatures below $1000°K$ for the above systems. For the case of bulk-diffusion-controlled corrosion reactions, the crystal bulk from whose surface the reactant beam scatters could perhaps be saturated by permeation with an isotopic species of the reactant beam from the back side of the surface in order to decrease the solubility of the species in the incident beam, thus allowing the slow reactive steps to make themselves more visible. Clearly, innovation is required here!

Catalytic Studies

Catalytic surface reactions investigated by the modulated molecular beam technique have generally not exhibited the problem of modulation-induced bulk diffusion or rate-limiting surface diffusion that masks reactive kinetic steps. This is perhaps due to the use of single crystals in the catalytic studies, which possess fewer grain boundaries to serve as active sites or entrances to the bulk than the polycrystalline samples used in the corrosion studies. It is also true, however, that the catalytic reactions in general have been less thoroughly investigated and less thoroughly analyzed.

The oxidation of D_2 (184), C_2H_6 (185), C_2H_4 (185), C_2H_2 (185), and CO (186) on a epitaxially grown Pt (111) crystal exemplify experiments of catalytic systems that are amenable to further study of their reaction mechanism with a modulated molecular beam technique. The data for the oxidation of CO imply that at low temperatures and at high CO flux the reaction is limited by surface diffusion of CO across the perimeter of the beam shadow on the surface to some sites on the Pt crystal that have not been poisoned. This interesting phenomenon is discerned only with a modulated beam flux incident on the surface. However, it is a consequence of the spatially limited character of the beam and not an occurrence in a steady state catalysis situation.

The decomposition of ethylene (J. B. Hudson & H. Zuhr, unpublished) and formic acid (187) on Ni (110) and polycrystalline Pt (188), and the decomposition of N_2O on Pt (111) (189), has been studied in various laboratories. The first study reveals that ethylene dissociatively chemisorbs on the surface, and at temperatures above 150°C H_2 is evolved. There is some evidence that the chemisorption of ethylene is followed by equilibrium adsorption of a loosely bound, nondissociatively adsorbed ethylene phase which, the authors point out, is too small to be detected by LEED, AES, or photoemission spectroscopy. A mechanism for formic acid decomposition on Ni (110) is suggested (187), that is mainly based on phase vs temperature data from which rate constants are extracted. The rate constants are then used to calculate the amplitude behavior of the product signal, and this is found to fit well the experimental amplitude vs temperature data for the linear processes. Beam intensity and frequency data, however, are needed to describe the complex process of adsorption of the formic acid and the apparent second order reactions for the formation of H_2 and H_2O. The decomposition probability of formic acid on various overlayers was also investigated (I. E. Wachs & R. J. Madix, unpublished) and substantially depends on the elemental composition of the overlayer, above room temperature. Differences are noted in the desorption kinetics of formic acid on various overlayers between a flash desorption measurement and a modulated beam measurement and are attributed to stronger adsorbate-adsorbate interactions in the higher coverage flash desorption experiments.

H_2–D_2 exchange has been studied on various Pt single crystal surfaces (108, 190). The proposed mechanism suggests that at low temperatures ($<600°K$) the rate-determining step is the diffusion of a D_2 molecule to a step site where HD is formed subsequent to D_2 dissociation. At high temperatures an Eley-Rideal mechanism is proposed. The Pt (111) surface exhibited substantially lower HD production than the stepped surfaces investigated. Recently, the H_2–D_2 exchange data has been reexamined (191). A much improved fit of the amplitude and phase response to surface temperature and frequency variation has been obtained with the simple branched mechanism of two parallel steps for HD formation. No model is, however, specifically proposed. The ratio of reaction probabilities between a stepped crystal and a flat (111) surface is calculated to be 18, in agreement with flash desorption measurements (192). The H_2–D_2 exchange probability has also been observed (193) as a function of step azimuthal orientation to the incident beam. A factor of 2 in increased reactivity is seen when the beam is incident on the open edges of the step as compared to incidence on the back edges or the terraces. Clearly, the H_2–D_2 exchange mechanism is still an open question.

The results of experiments that study chemical interactions of solid surfaces with molecular beams can provide reliable mechanistic data through careful analysis in the absence of long diffusion processes. However, the experimental results have also forced the invocation of detailed diffusion steps to specific reaction sites and opened the question concerning the influence of bulk properties on surface chemical reactions.

ACKNOWLEDGMENTS

Dr. Wigbert J. Siekhaus is sincerely thanked by S.C. for many hours of enthusiastic and valuable discussion. Support from the National Science Foundation and the US Energy Research and Development Administration is also gratefully acknowledged.

Literature Cited

1. Mason, B. F., Williams, B. R. 1972. *Rev. Sci. Instrum.* 43:375–83
2. Boato, G., Cantini, P., Mattera, L. 1976. *Surf. Sci.* 55:141–78
3. Saltsburg, H. 1973. *Ann. Rev. Phys. Chem.* 24:493–514
4. Somorjai, G. A., Brumbach, S. B. 1974. *Crit. Rev. Solid State Sci.* 4:429–54
5. Weinberg, W. H. 1975. *Adv. Colloid Interface Sci.* 4:301–47
6. Bernasek, S. L., Somorjai, G. A. 1975. *Prog. Surf. Sci.* 5:377–440
7. Smith, J. N. 1973. *Surf. Sci.* 34:613–37
8. Toennies, J. P. 1974. *Appl. Phys.* 3:91–114
9. Boato, G., Cantini, P. 1974. *Proc. Int. Sch. Phys. Enrico Fermi, Course 58, Varenna, 1973*, ed. F. O. Goodman, pp. 707–35
10. Nahr, H. 1975. *Surf. Sci., Lect. Int. Conf., Trieste, 1974,* 2:9–62
11. Scoles, G. 1974. See Ref. 9, pp. 588–604
12. Schwarz, J. A., Madix, R. J. 1974. *Surf. Sci.* 46:317–41
13. Madix, R. J. 1977. *Phys. Chem. Fast React.* In press
14. Palmer, R. L., Smith, J. N. 1975. *Catal. Rev.* 12:279–301
15. Olander, D. R. 1977. *J. Colloid Interface Sci.* 58:169–83
16. Nocilla, S. 1973. *Entropie* 49:37–45
17. Celli, V. 1974. See Ref. 9, pp. 331–46
18. Goodman, F. O., Wachman, H. Y. 1974. See Ref. 9, pp. 347–529
19. Beeby, J. L. 1974. See Ref. 9, pp. 736–58
20. Houston, D. E., Frankl, D. R. 1973. *Phys. Rev. Lett.* 31:298–300
21. Meyers, J. A., Houston, D. E., Frankl, D. R. 1974. *Proc. 2nd Int. Conf. Solid Surf., Kyoto, Japan, 1974, J. Appl. Phys.* Suppl. 2, Pt. 2, pp. 563–66
22. Meyers, J. A., Frankl, D. R. 1975. *Surf. Sci.* 51:61–74
23. Boato, G., Cantini, P., Garibaldi, U., Levi, A. C., Mattera, L., Spadacini, R., Tommei, G. E. 1973. *J. Phys. C* 6:L394–98
24. Wood, B., Mason, B. F., Williams, B. R. 1974. *J. Chem. Phys.* 61:1435–54
25. Wilsch, H., Finzel, H. U., Frank, H., Hoinkes, H., Nahr, H. 1974. See Ref. 21, pp. 567–70
26. Finzel, H. U., Frank, H., Hoinkes, H., Luschka, M., Nahr, H., Wilsch, H., Wonka, U. 1975. *Surf. Sci.* 49:577–605
27. Hoinkes, H., Nahr, H., Wilsch, H. 1972. *Surf. Sci.* 30:363–78
28. Doll, J. D. 1974. *Chem. Phys. Lett.* 29:195–98
29. Garibaldi, U., Levi, A. C., Spadacini, R., Tommei, G. E. 1974. See Ref. 21, pp. 549–52
30. Garibaldi, U., Levi, A. C., Spadacini, R., Tommei, G. E. 1975. *Surf. Sci.* 48:649–75
31. Chiroli, C., Levi, A. C. 1976. *Surf. Sci.* 59:325–31
32. Doll, J. D. 1974. *Chem. Phys.* 3:257–64
33. Doll, J. D. 1974. *J. Chem. Phys.* 61:954–57
34. Dion, D. R., Doll, J. D. 1976. *Surf. Sci.* 58:415–28
35. Masel, R. I., Merrill, R. P., Miller, W. H. 1974. *Surf. Sci.* 46:681–88
36. Masel, R. I., Merrill, R. P., Miller, W. H. 1976. *J. Chem. Phys.* 64:45–56
37. Ray, C. J., Bowman, J. M. 1975. *J. Chem. Phys.* 63:5231–34
38. Goodman, F. O., Cabrera, N., Celli, V., Manson, R. 1970. *Surf. Sci.* 19:67–92
39. Goodman, F. O. 1970. *Surf. Sci.* 19:93–108
40. Goodman, F. O., Tan, W. K. 1973. *J. Chem. Phys.* 58:5527–30
41. Goodman, F. O. 1973. *J. Chem. Phys.* 58:5530–33
42. Masel, R. I., Merrill, R. P., Miller, W. H. 1976. *J. Vac. Sci. Technol.* 13:355–59
43. Masel, R. I., Merrill, R. P., Miller, W. H. 1976. *J. Chem. Phys.* 65:2690–99
44. Tsuchida, A. 1974. *Surf. Sci.* 46:611–24
45. Tsuchida, A. 1975. *Surf. Sci.* 52:685–88
46. Wolken, G. 1973. *J. Chem. Phys.* 58:3047–64
47. Liva, M. P., Frankl, D. R. 1976. *Surf. Sci.* 59:643–47

48. Chow, H., Thompson, E. D. 1976. *Surf. Sci.* 54:269–92
49. Chow, H., Thompson, E. D. 1976. *Surf. Sci.* 59:225–51
50. Beeby, J. L. 1973. *J. Phys.* C6:1229–41
51. Beeby, J. L. 1974. See Ref. 21, pp. 541–43
52. Flytzanis, N., Celli, V. 1974. *Surf. Sci.* 42:173–89
53. Cabrera, N., Goodman, F. O. 1972. *J. Chem. Phys.* 56:4899–4902
54. Goodman, F. O., Liu, W. S., Cabrera, N. 1972. *J. Chem. Phys.* 57:2698–2702
55. Davies, R. O., Ullermayer, L. S. 1973. *Surf. Sci.* 39:61–82
56. Goodman, F. O. 1976. *J. Chem. Phys.* 65:1561–64
57. Wolken, G. Jr. 1974. *J. Chem. Phys.* 61:456–60
58. Goodman, F. O. 1976. *J. Chem. Phys.* 64:1051–57
59. Marin, M., Solana, J., Garcia, N. 1976. *J. Chem. Phys.* 64:5311–12
60. Garcia, N. 1976. *J. Chem. Phys.* 65:2824–26
61. Beeby, J. L. 1971. *J. Phys.* C 4:L359–62
62. Beeby, J. L. 1974. See Ref. 21, pp. 537–40
63. Beeby, J. L. 1975. *Comments Solid State Phys.* 7:1–6
64. Weinberg, W. H. 1972. *J. Phys.* C 5:2098–2104
65. Masel, R. I., Merrill, R. P., Miller, W. H. 1975. *Phys. Rev.* B 12:5545–51
66. Tendulkar, D. V., Stickney, R. E. 1971. *Surf. Sci.* 29:516–22
67. Stoll, A. G., Ehrhardt, J. J., Merrill, R. P. 1976. *J. Chem. Phys.* 64:34–40.
68. Chappell, R., Hayward, D. O. 1972. *J. Vac. Sci. Technol.* 9:1052–55
69. Hayward, D. O., Walters, M. R. 1974. See Ref. 21, pp. 587–90
70. Boato, G., Cantini, P., Tatarek, R. 1976. *J. Phys.* F 6:L237–40
71. Horne, J. M., Miller, D. R. 1977. *Surf. Sci.* In press
72. Smith, J. N., O'Keefe, D. R., Palmer, R. L. 1970. *J. Chem. Phys.* 52:315–20
73. O'Keefe, D. R., Palmer, R. L., Smith, J. N. 1971. *J. Chem. Phys.* 55:4572–77
74. Stoll, A. G., Merrill, R. P. 1973. *Surf. Sci.* 40:405–9
75. Stoll, A. G., White, R. E., Ehrhardt, J. J., Masel, R. I., Merrill, R. P. 1975. *J. Vac. Sci. Technol.* 12:192–98
76. White, R. E., Ehrhardt, J. J., Merrill, R. P. 1976. *J. Chem. Phys.* 64:41–44
77. Ceyer, S. T., Gale, R. J., Bernasek, S. L., Somorjai, G. A. 1976. *J. Chem. Phys.* 64:1934–40
78. McClure, J. D. 1972. *J. Chem. Phys.* 57:2810–22
79. McClure, J. D. 1972. *J. Chem. Phys.* 57:2823–29
80. Steele, W. A. 1973. *Surf. Sci.* 38:1–16
81. Jewsbury, P. 1975. *Surf. Sci.* 52:325–39
82. Weinberg, W. H., Merrill, R. P. 1972. *J. Chem. Phys.* 56:2893–2902
83. Horne, J., Miller, D. R. 1975. *J. Vac. Sci. Technol.* 13:351–54
84. Mason, B. F., Williams, B. R. 1972. *J. Chem. Phys.* 56:1895–1904
85. Mason, B. F., Williams, B. R. 1974. *Surf. Sci.* 45:141–62
86. Mason, B. F., Williams, B. R. 1972. *J. Chem. Phys.* 57:872–84
87. Hoinkes, H., Nahr, H., Wilsch, H. 1972. *Surf. Sci.* 33:516–24
88. Hoinkes, H., Nahr, H., Wilsch, H. 1973. *Surf. Sci.* 40:457
89. Goodman, F. O., 1974. *Surf. Sci.* 46:118–28
90. Comsa, G., Comsa, G. H., Fremerey, J. K. 1974. *Z. Naturforsch. Teil A* 29:189–93
91. Lapujoulade, J., Lejay, Y. 1975. *J. Chem. Phys.* 63:1389–1400
92. Lin, Y. W., Wolken, G. Jr. 1976. *J. Chem. Phys.* 65:2634–49
93. Comsa, G. 1973. *J. Phys.* C 6:2648–52
94. Weinberg, W. H., Merrill, R. P. 1972. *J. Chem. Phys.* 56:2881–92
95. Sau, R., Merrill, R. P. 1973. *Surf. Sci.* 34:268–88
96. Weinberg, W. H. 1974. *J. Colloid Interface Sci.* 47:372–74
97. Siekhaus, W. J., Schwarz, J. A., Olander, D. R. 1972. *Surf. Sci.* 33:445–60
98. Weinberg, W. H., Merrill, R. P. 1973. *J. Vac. Sci. Technol.* 10:411–12
99. Modak, A. T., Pagni, P. J. 1976. *J. Chem. Phys.* 65:1327–44
100. Weinberg, W. H., Merrill, R. P. 1971. *J. Vac. Sci. Technol.* 8:718–24
101. Modak, A. T., Pagni, P. J. 1973. *J. Chem. Phys.* 59:2019–31
102. Balooch, M., Dabiri, A. E., Stickney, R. E. 1972. *Surf. Sci.* 30:483–86
103. Houston, D. E., Frankl, D. R. 1974. *J. Chem. Phys.* 60:3268–70
104. Tomoda, S., Kodera, K., Kusunoki, I. 1974. *Surf. Sci.* 45:657–76
105. Bernasek, S. L., Somorjai, G. A. 1974. *J. Chem. Phys.* 60:4552–56
106. West, L. A., Somorjai, G. A. 1971. *J. Chem. Phys.* 54:2864–73
107. West, L. A., Somorjai, G. A. 1972. *J. Chem. Phys.* 57:5143–53
108. Bernasek, S. L., Somorjai, G. A. 1975. *J. Chem. Phys.* 62:3149–61
109. Doll, J. D. 1973. *J. Chem. Phys.* 59:1038–42
110. Nichols, W. L., Weare, J. H. 1975. *J. Chem. Phys.* 62:3754–62

111. Nichols, W. L., Weare, J. H. 1975. *J. Chem. Phys.* 63:379–83
112. Wolken, G. Jr. 1974. *J. Chem. Phys.* 60:2210–19
113. Lin, Y. W., Wolken, G. Jr. 1976. *J. Chem. Phys.* 65:3729–34
114. Williams, B. R. 1971. *J. Chem. Phys.* 55:1315–22
115. Williams, B. R. 1971. *J. Chem. Phys.* 55:3220–25
116. Lagos, M., Birstein, L. 1975. *Surf. Sci.* 51:469–88
117. Lagos, M., Birstein, L. 1975. *Surf. Sci.* 52:391–400
118. Fisher, S. S., Bledsoe, J. R. 1972. *J. Vac. Sci. Technol.* 9:814–18
119. Bledsoe, J. R., Fisher, S. S. 1974. *Surf. Sci.* 46:129–56
120. Mason, B. F., Williams, B. R. 1974. See Ref. 21, pp. 557–62
121. Benedek, G. 1975. *Phys. Rev. Lett.* 35:234–37
122. Boato, G., Cantini, P., Mattera, L. 1974. See Ref. 21, pp. 553–56
123. Rowe, R. G., Ehrlich, G. 1975. *J. Chem. Phys.* 62:735–37
124. Rowe, R. G., Ehrlich, G. 1975. *J. Chem. Phys.* 63:4648–65
125. Boato, G., Cantini, P., Mattera L. 1976. *J. Chem. Phys.* 65:544–49
126. Logan, R. M. 1969. *Mol. Phys.* 17:147–55
127. Wolken, G. Jr. 1973. *Chem. Phys. Lett.* 21:373–79
128. Wolken, G. Jr. 1973. *J. Chem. Phys.* 59:1159–65
129. Wolken, G. Jr. 1975. *J. Chem. Phys.* 62:2730–35
130. Goodman, F. O., Liu, W. S. 1975. *Surf. Sci.* 49:417–32
131. Garibaldi, U., Levi, A. C., Spadacini, R., Tommei, G. E. 1976. *Surf. Sci.* 55:40–60
132. Rudnicki, A. R. Jr., Wachman, H. Y. 1973. *Surf. Sci.* 34:679–92
133. Logan, R. M., Stickney, R. E. 1966. *J. Chem. Phys.* 44:195–201
134. Stickney, R. E., Logan, R. M., Yamamoto, S., Keck, J. C. 1967. In *Fundamentals of Gas-Surface Interactions*, ed. H. Saltsburg, J. N. Smith, Jr., M. Rogers, pp. 422–34. New York: Academic. 557 pp.
135. Stickney, R. E., Tendulkar, D. V., Yamamoto, S. 1972. *J. Vac. Sci. Technol.* 9:819–24
136. Foxon, C. T., Boudry, M. R., Joyce, B. A. 1974. *Surf. Sci.* 44:69–92
137. Arthur, J. R., Brown, T. R. 1975. *J. Vac. Sci. Technol.* 12:200–3
138. Cardillo, M. J., Balooch, M., Stickney, R. E. 1975. *Surf. Sci.* 50:263–78
139. Subbarao, R. B., Miller, D. R. 1969. *J. Chem. Phys.* 51:4679–80
140. Subbarao, R. B., Miller, D. R. 1972. *J. Vac. Sci. Technol.* 9:808–11
141. Goodman, F. O. 1972. *J. Vac. Sci. Technol.* 9:812–13
142. Goodman, F. O. 1972. *Surf. Sci.* 30:1–42
143. Goodman, F. O., Tan, W. K. 1973. *J. Chem. Phys.* 59:1805–31
144. Beeby, J. L. 1972. *J. Phys. C* 5:3438–56
145. Beeby, J. L. 1972. *J. Phys. C* 5:3457–61
146. Weare, J. H. 1974. *J. Chem. Phys.* 61:2900–10
147. Garibaldi, U., Levi, A. C., Spadacini, R., Tommei, G. E. 1973. *Surf. Sci.* 38:269–74
148. Adelman, S. A., Doll, J. D. 1974. *J. Chem. Phys.* 61:4242–45; 62:2518
149. Doll, J. D., Myers, L. E., Adelman, S. A. 1975. *J. Chem. Phys.* 63:4908–14
150. Adelman, S. A., Doll, J. D. 1976. *J. Chem. Phys.* 64:2375–88
151. Adelman, S. A., Garrison, B. J. 1976. *J. Chem. Phys.* 65:3751–61
152. Pagni, P. J. 1973. *J. Chem. Phys.* 58:2940–54
153. Scheer, M. D., Klein, R., McKinley, J. D. 1972. *Surf. Sci.* 30:251–62
154. Müller, R., Wassmuth, H. W. 1973. *Surf. Sci.* 34:249–67
155. Hurkmans, A., Overbosch, E. G., Olander, D. R., Los, J. 1976. *Surf. Sci.* 54:154–68
156. Bradley, T. L., Dabiri, A. E., Stickney, R. E. 1972. *Surf. Sci.* 29:590–602
157. Bradley, T. L., Stickney, R. E. 1973. *Surf. Sci.* 38:313–26
158. Balooch, M., Stickney, R. E. 1971. *Surf. Sci.* 44:310–20
159. Balooch, M., Cardillo, M. J., Miller, D. R., Stickney, R. E. 1974. *Surf. Sci.* 46:358–92
160. Dabiri, A. E., Lee, T. J., Stickney, R. E. 1971. *Surf. Sci.* 26:522–44
161. Marenco, G., Schutte, A., Scoles, G., Tommasini, F. 1972. *J. Vac. Sci. Technol.* 9:824–27
162. Schutte, A., Bassi, D., Tommasini, F., Turelli, A., Scoles, G., Hermans, L. J. F. 1976. *J. Chem. Phys.* 64:4135–42
163. Goodman, F. O. 1972. *Surf. Sci.* 30:525–35
164. Brumbach, S. B., Rosenblatt, G. M. 1972. *Surf. Sci.* 29:555–70
165. McCreery, J. H., Wolken, G. Jr. 1975. *J. Chem. Phys.* 63:4072–73
166. McCreery, J. H., Wolken, G. Jr. 1976. *J. Chem. Phys.* 64:2845–53

167. McCreery, J. H., Wolken, G. Jr. 1976. *J. Chem. Phys.* 65:1310–16
168. Elkowitz, A. B., McCreery, J. H., Wolken, G. Jr. 1976. *Chem. Phys.* 17:423–31
169. McCreery, J. H., Wolken, G. Jr. 1975. *J. Chem. Phys.* 63:2340–49, 65:2922
170. Zare, R. N., Dagdigian, P. J. 1974. *Science* 185:739–47
171. Olander, D. R., Siekhaus, W. J., Jones, R., Schwarz, J. A. 1972. *J. Chem. Phys.* 57:408–20
172. Olander, D. R., Siekhaus, W. J., Jones, R., Schwarz, J. A. 1972. *J. Chem. Phys.* 57:421–33
173. Balooch, M., Olander, D. R. 1975. *J. Chem. Phys.* 63:4772–86
174. Jones R. H., Olander, D. R., Siekhaus, W. J., Schwarz, J. A. 1972. *J. Vac. Sci. Technol.* 9:1429–41
175. Olander, D. R., Ullman, A. 1976. *Int. J. Chem. Kinet.* 8:625–37
176. Schwarz, J. A. 1974. See Ref. 21, pp. 579–82
177. Madix, R. J., Susu, A. A. 1970. *Surf. Sci.* 20:377–400
178. Madix, R. J., Parks, R., Susu, A. A., Schwarz, J. A. 1971. *Surf. Sci.* 24:288–301
179. Madix, R. J., Ullman, A. Z. 1974. *High. Temp. Sci.* 6:342–50
180. Madix, R. J., Schwarz, J. A. 1971. *Surf. Sci.* 24:264–87
181. Madix, R. J., Susu, A. A. 1973. *J. Catal.* 28:316–21
182. Madix, R. J., Susu, A. A. 1972. *J. Vac. Sci. Technol.* 9:915–19
183. Machiels, A., Olander, D. R. 1977. *Surf. Sci.* 65:325–44
184. Smith, J. N. Jr., Palmer, R. L. 1972. *J. Chem. Phys.* 56:13–20
185. Palmer, R. L., Smith, J. N. Jr. 1974. *J. Chem. Phys.* 60:1453–63
186. Palmer, R. L. 1975. *J. Vac. Sci. Technol.* 12:1403–9
187. Wachs, I. E., Madix, R. J. 1977. *Surf. Sci.* 65:287–313
188. Dahlberg, S. C., Fisk, G. A., Rye, R. R. 1975. *J. Catal.* 36:224–34
189. West, L. A., Somorjai, G. A. 1972. *J. Vac. Sci. Technol.* 9:668–72
190. Bernasek, S. L., Siekhaus, W. J., Somorjai, G. A. 1973. *Phys. Rev. Lett.* 30:1202–4
191. Wachs, I. E., Madix, R. J. 1976. *Surf. Sci.* 58:590–96
192. Lu, K. E., Rye, R. R. 1974. *Surf. Sci.* 45:677–95
193. Gale, R. J., Salmeron, M., Somorjai, G. A. 1977. *Phys. Rev. Lett.* 38:1027–29

RESONANCE EFFECTS IN VIBRATIONAL SCATTERING FROM COMPLEX MOLECULES

✣ 2656

Thomas G. Spiro and Paul Stein

Department of Chemistry, Princeton University, Princeton, New Jersey 08540

INTRODUCTION

Resonance Raman spectroscopy has emerged as a new and powerful tool for the study of molecular (1) or solid state (2) vibrations that are coupled to electronic transistions. It is a form of high-resolution vibronic spectroscopy, in which the intensity of vibrational light scattering is examined as a function of proximity of the excitation wavelength to electronic absorption bands of the sample. Much information can be obtained about the potential surfaces of both ground and excited states. Moreover, the technique may be utilized as a probe of local chromophore structure in complex mixtures because of the selectivity conferred by resonance enhancement. As the excitation wavelength is tuned to an electronic transition, the chromophore vibrational modes are lifted above the vibrational background. They can be assigned with reference to the chemical structure of the chromophore, and then used to monitor structural changes without interference from the chemical matrix. Applications to chromophoric groups in biological systems are under active investigation and have recently been reviewed (3, 4). A well-studied chromophore is the prosthetic group of heme proteins. Excitation in the visible absorption bands produces resonance-enhanced porphyrin vibrational modes without interference from the polypeptide chains (5). The frequency shifts of these vibrations have been used to analyze porphyrin structure changes at the protein active sites (6).

Progress in the field has been heavily dependent on laser technology. Because of the weakness of Raman scattering, heroic measures were required, before the advent of lasers, to concentrate enough photons on the sample to produce useful spectra. The sample was surrounded cylindrically by gas discharge lamps, and the scattered photons had to travel the length of the sample tube before entering the spectrometer (7). This geometry required that the sample be highly transparent. While important early investigations of resonance enhancement were carried out (8), absorbing materials generally could not be studied. The laser has changed this situation dramatically by providing an intense directional line source (9). The scattering volume is very small and can be situated at the front surface, thereby minimizing the optical path (10). The sample can be moved rapidly through the laser beam to minimize complications due to heating (11) or photochemistry (12, 13). Finally, available laser wavelengths cover an increasing range of the electromagnetic spectrum, making it possible to examine an increasing variety of chromophores. At the moment most instruments are limited to the visible region, but studies with ultraviolet lasers are increasing.

These technical developments have led to intense experimental activity, and a substantial data base already exists. The parallel development of Raman theory has been equally intense (14–20a,b) and is the subject of a recent review in this series (20b). The objective of the present article is to place selected recent experimental observations within a conceptual framework that may provide guidance in future applications.

RESONANCE SCATTERING

Few matters have exercised spectroscopists as heatedly as the distinction, if any, between resonance Raman scattering and fluorescence, inasmuch as both involve excitation and emission within an absorption band. The debate has been clarified by the emission lifetime measurements of Rousseau & Williams (21) shown in Figure 1, obtained by use of a pulsed laser in exact resonance with an excited state rovibrational level of gaseous iodine, and also by related theoretical treatments (22–27). The decay curve shows a single lifetime, characteristic of the excited state. This is as expected for the process traditionally called resonance fluorescence (28), i.e. sequential absorption and emission associated with a specific excited state. When the laser is detuned slightly from exact resonance, two emission processes are observed, a fast one with the lifetime of the laser pulse, and a slower one with the excited state lifetime. The latter process can be attributed to the Maxwellian distribution of gas phase molecular energies, which ensures that a fraction of the molecules is always in resonance with the laser frequency. Consideration of the lifetime of the off-resonance scattering event is obscured by the finite pulse-width

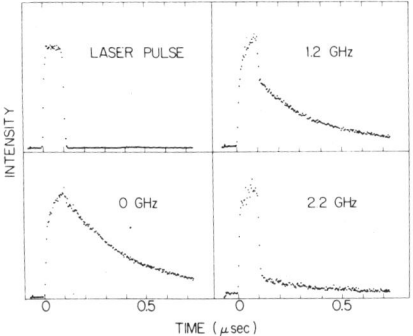

Figure 1 Time decay data for pulsed excitation of molecular I_2 at the Raman shifted S branch of the $43'-0''R(15)$ transition. The detuning interval of the laser is indicated ($3GH_z = 0.1$ cm^1) (From Reference 21a.)

of the laser (or the associated uncertainty in its frequency). Indeed, a principal conclusion of the theoretical treatments (22–27) is that in any real experiment the reemission lifetime is characteristic of either the excited state or of the incident radiation, whichever has the narrower linewidth. Friedman & Hochstrasser (22) suggest that the term resonance fluorescence should be applied to the former situation and resonance Raman scattering to the latter; complex molecules usually have broad absorption bands and fall in the resonance Raman category. It should be emphasized, however, that the distinction is an operational one, reflecting relative linewidths; the physical process is the same in both cases. In particular, the incident and scattered photon retain a constant phase relationship if the resonant excited state is isolated (21).

Dephasing of the photons requires mechanisms for transferring energy within or between molecules. Such mechanisms result in emission that is independent of the absorption, and that is broadened and shifted by virtue of the energy redistribution. The term relaxed fluorescence has been suggested (22) for these processes [hot luminescence is the phrase used by solid state spectroscopists (2, 29)], which are physically distinct from the resonance Raman fluorescence process. The situation is illustrated diagrammatically in Figure 2. An emission spectrum will generally contain contributions from both relaxed and unrelaxed emission, depending on the probability of energy dissipation within the lifetime of the molecular excitation. For gas phase diatoms, collisions provide the main relaxation pathway (21), while for complex molecules the main relaxation pathway is deactivation via the numerous vibrational levels to lower regions of the excited state potential well, or via resonant transfer to lower excited states. Figure 3 shows emission spectra of ferrocytochrome c with excitation at several wavelengths in the region above the first $\pi-\pi^*$ absorption (22). The sharp Raman bands are followed by a broad hump of relaxed emission whose position is independent of the excitation wavelength. The effect on emission

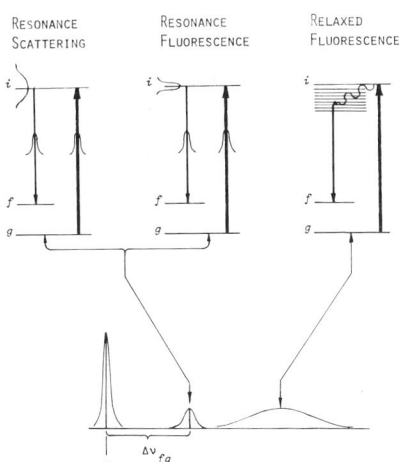

Figure 2 Schematic representation of reemission processes.

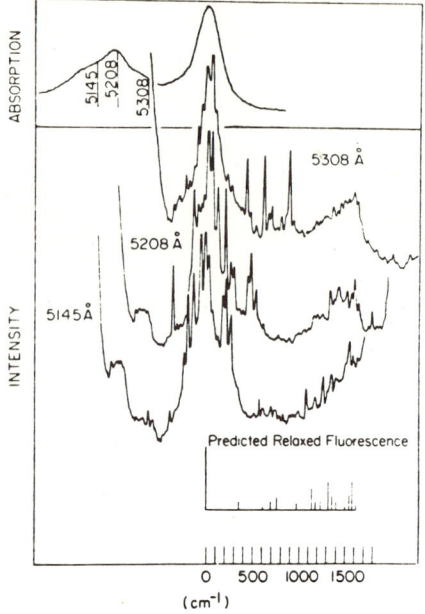

Figure 3 Ferrocytochrome c room temperature Raman spectra, superimposed upon a relaxed fluorescence background, with the indicated excitation wavelengths. The spectra are placed on a common energy scale to emphasize the background, of which the most prominent peak matches the 0–0 absorption band shown above. (From Reference 22.)

processes of progressively shorter electronic lifetimes of heme proteins have been analyzed in detail by Adar et al (30).

In summary, an incident photon may be reemitted from a molecule before or after energy transfer. The former process gives a sharp line spectrum, reflecting energy conservation, the linewidth being characteristic of either the laser or the excited state, whichever is narrower. The latter process usually gives a broadened spectrum, reflecting the energy redistribution, although sharp lines may be observed if the relaxation is constrained to a specific pathway, as in Shpolskii matrices (31). Energy transfer may also result, of course, in radiationless deactivation. Indeed, such processes are essential for the observation of good quality resonance Raman spectra in complex molecules; in their absence the Raman peaks may easily be swamped by the broad, relaxed emission envelope.

SCATTERING MECHANISMS

Adiabatic Treatment

For a Raman transition between states g and f the scattering intensity (14) is

$$I_s = \frac{8\pi v_s^4}{9c^4} I_0 \sum_{\rho\sigma} |(\alpha_{\rho\sigma})_{gf}|^2, \qquad 1.$$

where I_0 is the incident intensity at frequency v_0, v_s is the scattering frequency, c is the velocity of light, and $(\alpha_{\rho\sigma})_{gf}$ is the transition polarizability tensor, with incident and scattered polarizations indicated by ρ and σ. Second order perturbation theory gives the following expression for the polarizability

$$(\alpha_{\rho\sigma})_{gf} = \frac{1}{\hbar}\sum_e \left(\frac{\langle f|\mu_\rho|e\rangle\langle e|\mu_\sigma|g\rangle}{v_{eg} - v_0 + i\Gamma_e} + \frac{\langle f|\mu_\sigma|e\rangle\langle e|\mu_\rho|g\rangle}{v_{ef} + v_0 + i\Gamma_e} \right), \qquad 2.$$

where μ_ρ and μ_σ are dipole moment operators, $g>$ and $f>$ are the initial and final state wave function, $e>$ is the wave function of an excited state, e, of half-bandwidth Γ_e, and v_{eg} and v_{ef} are transition frequencies. When $v_0 \ll v_{eg}$, both terms give comparable contributions, and the polarizability is nearly independent of the incident frequency. As v_0 approaches v_e, the first term becomes dominant and is responsible for resonance effects. In the ensuing discussion the nonresonant term is neglected unless otherwise indicated.

To the extent that the Born-Oppenheimer approximation is valid, the wave functions may be separated into electronic and vibrational parts, giving

$$\langle f|\mu|e\rangle = \langle j|M_e|v\rangle \quad \text{and} \quad \langle e|\mu|g\rangle = \langle v|M_e|i\rangle,$$

where $i>$ and $j>$ are initial and final vibrational wave functions of the ground electronic state and $v>$ is a vibrational wave function of the excited electronic state, e. M_e is the pure electronic transition moment between g and e. It is a weakly varying function of the nuclear coordinates, and may be expanded in a Taylor series about the equilibrium geometry, o:

$$M_e = M_e^0 + (\partial M/\partial Q)^0 Q \ldots, \qquad 3.$$

where Q is a given normal mode of the molecule. The first two terms in the expansion give the following expression for the polarizability (neglecting the nonresonant part and ignoring polarization, for the moment):

$$\alpha = A + B; \qquad 4.$$

$$A = (M_e^0)^2 \frac{1}{\hbar} \sum_v \frac{\langle j|v\rangle\langle v|i\rangle}{v_{vi} - v_0 + i\Gamma_v}, \qquad 5.$$

$$B = M_e^0(\partial M/\partial Q)^0 \frac{1}{\hbar} \sum_v \frac{\langle j|Q|v\rangle\langle v|i\rangle + \langle j|v\rangle\langle v|Q|i\rangle}{v_{vi} - v_0 + i\Gamma_v}. \qquad 6.$$

The transition frequencies v_{vi} are from the ground vibrational level, i, to the vibrational levels, v, of the resonant excited state, e.

A-Term Scattering

Since A is the leading term, it is ordinarily the dominant contributor to the resonance Raman intensity. It can only contribute to scattering from totally symmetric modes, however, since only totally symmetric modes displace the equilibrium nuclear positions, giving nonzero Franck-Condon overlaps. For discrete rovibronic states, as encountered in gaseous I_2, for example, the overlaps vary irregularly for successive levels, and the emission peaks show irregular intensity sequences (21, 28). When the density of states is high, as in the I_2 continuum, the resonance Raman spectrum shows a smooth decrease in the intensity of successive overtones and a regular increase in their bandwidth (21, 28). This is also the pattern commonly shown by complex molecules that have high densities of states within their absorption bands. An example is the resonance Raman (RR) spectrum of the $Mo_2Cl_8^{4-}$ ion (32), shown in Figure 4, excited in an absorption band thought to arise from a δ–δ* transition localized on the quadruple Mo–Mo bond. The dominant feature of the RR spectrum is a band at 346 cm^{-1}, assignable to Mo–Mo stretching, that shows a long overtone progression and even forms a subsidiary progression on another fundamental, v_4 (Mo–Cl stretching). The 5°K absorption spectrum likewise shows a vibrational progression, with a spacing of 337 cm^{-1} (33a,b).

Among the normal modes of a complex molecule, the ones that experience the strongest resonance enhancement via A term scattering are those whose coordinates most closely correspond to the distortions experienced by the molecule in the resonant excited state, as enunciated by Hirakawa & Tsuboi (34). Thus molecular

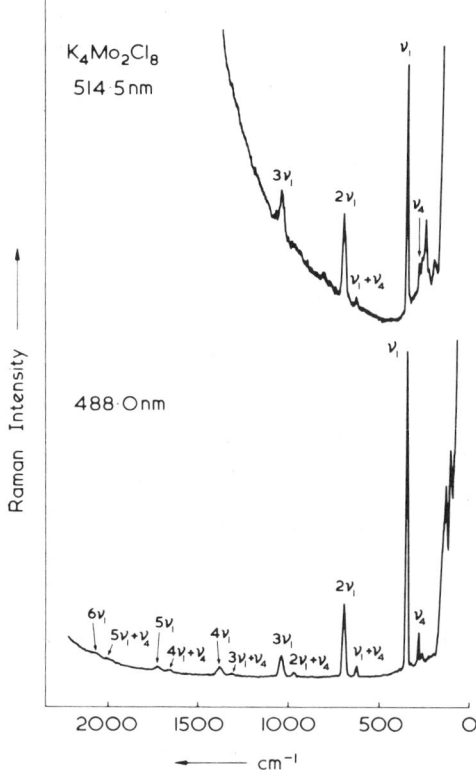

Figure 4 Resonance Raman spectra of $K_4Mo_2Cl_8$, showing progressions in the v_1 (Mo–Mo stretch) mode, and a contribution from relaxed fluorescence. (From Reference 32.)

π–π* transitions provide large enhancements for modes involving the stretching of the π-bonds, consistent with weakening of these bonds in the excited state, but they provide little enhancement for carbon-hydrogen stretching modes, which are not significantly perturbed in the excitation (35). Similarly ligand-metal charge transfer transitions selectively enhance stretching modes of the ligand-metal bonds (36) that are weakened in the excited state, or internal modes of the ligand that may also be affected by the excitation (37). These classes of vibrations can be intuitively seen to have large Franck-Condon overlaps by virtue of the displacement of the excited state potential along the normal coordinates.

A more quantitative approach requires the estimation of the overlap integrals, which, for a complex molecule, requires a normal mode analysis for both the ground and the excited state. Warshel & Karplus (38) have developed a computational scheme for doing this, using a combination of molecular orbital calculations and empirical functions to define the ground and excited state potential surfaces. When applied to the retinals (38) the calculations gave a reasonable account of observed relative intensities of the Raman bands (35). Calculations for porphyrin are currently in progress (39). This approach is particularly attractive in connection with the use of resonance Raman spectroscopy as a structure probe in complex systems. Relative band intensities, which depend upon favorable Franck-Condon overlaps and therefore reflect alterations in excited state structure, are much more sensitive to intermolecular interactions than are changes in vibrational frequency, which reflect distortion of the ground state structure.

B-Term Scattering

Nontotally symmetric modes can gain intensity via the B term, Equation 6, because of the Q-dependent vibrational integrals. In principle, the derivative $(\partial M/\partial Q)^0$ can be directly evaluated (39) if the potential surfaces are accurately known. Alternatively one can expand the excited state wave functions in terms of the normal modes in the Herzberg-Teller approach (14) and obtain

$$B = \frac{(M_e^0)(M_s^0)h_{es}}{(v_s - v_e)} \sum_v \frac{\langle j|Q|v\rangle\langle v|i\rangle + \langle j|v\rangle\langle v|Q|i\rangle}{(v_{vi} - v_0 + i\Gamma_v)}, \qquad 8.$$

where s is a higher-lying excited state, and $h_{es} = \langle s|\partial H/\partial Q|e\rangle$. This integral is nonvanishing for a vibration whose symmetry is contained in the direct product of the two electronic symmetries. (Expansion of the ground-state wave function also leads to a C term (14) that reflects vibronic coupling of ground and excited states. Because of the large energy denominators involved, this term is usually neglected.) B-term scattering may become dominant in resonance with a weakly allowed transition that is vibronically coupled to a nearby strongly allowed transition, in which case $(M_e^0)(M_s^0) \gg (M_e^0)^2$. Striking enhancements are observed for nontotally symmetric modes which are particularly effective in mixing the two transitions. For N-heterocyclic aromatic molecules the lowest electronic transition is n–π* in character and is orbitally forbidden. It mixes with a nearby π–π* via an out-of-plane C–H bending mode, near 900 cm^{-1}, that is strongly enhanced in the Raman spectrum, as shown for pyrazine (which provided the first clear example of B-term scattering) (40a,b, 41), quinoxalane (42), and pyridine (43). [It was recently reported (44), however, that the totally symmetric modes become dominant at exact resonance with the 3360 Å $n-\pi^*$ transition in crystalline pyrazine at 4°K; there appears to be a delicate balance of A and B term contributions in the spectrally narrowed system.] In the resonant state of these molecules the out-of-plane mode is markedly reduced in frequency and is also decidedly anharmonic, again reflecting the strong vibronic coupling (43). Protonation of pyridine renders the mode undetectable in the Raman spectrum, consistent with the elimination of the $n-\pi^*$ state that provides its intensity (43).

Naphthalene and anthracene Raman spectra show strong enhancement of a nontotally symmetric C–C stretching mode that is particularly effective at mixing the long- and short-axis $\pi-\pi^*$ transitions (43). Likewise, porphyrin Raman spectra in resonance with the lowest, weakly allowed $\pi-\pi^*$ transition (α or Q_0) display nontotally symmetric ring modes that are effective in mixing it with the next higher, intense $\pi-\pi^*$ transition (Soret or B) (45). In this case both transitions are degenerate (E_u symmetry in the D_{4h} molecular point group), and the allowed vibrations are $E_u \times E_u = A_{1g} + A_{2g} + B_{1g} + B_{2g}$, but the A_{1g} modes are ineffective for vibronic mixing [although there is some evidence to the contrary (46)]. The A_{2g} modes are of particular interest because their scattering tensors are antisymmetric and the Raman bands display anomalous polarization (see below).

The B term is also expected to be responsible for resonance effects, if any, associated with symmetry-forbidden transitions, such as the $d-d$ bands of transition metal complexes. For centro-symmetric complexes significant effects are shown only by the totally symmetric modes (47), which are presumably coupled to an upper state by a vibronically induced odd-parity component of the resonant $d-d$ transition. The coupling is so weak that it actually shows up as a de-enhancement (antiresonance) due to an interference effect (see below). Noncentro-symmetric complexes can mix p character into the $d-d$ transitions. Considerable intensification of the visible $d-d$ band is seen for tetrahedral Co^{2+} complexes, for which significant RR effects have recently been reported (48, 49). The dominant mode is v_3, the T_{2g} Co-ligand stretch, which presumably mixes the $^4T_1$ resonant state with higher-lying $d-p$ or charge-transfer states of the same symmetry.

Time-Dependent Theory: Off-Resonance Scattering

An alternative to the Herzberg-Teller approach is to use time-dependent perturbation theory as advocated by Peticolas and co-workers (15, 20). Carried out to second order, this method gives the A term, Equation 4, while the third order result (resonant part only) is

$$\alpha' = (M_e^0)(M_s^0)h_{es} \sum_v \sum_u \frac{\langle j|u\rangle\langle u|Q|v\rangle\langle v|i\rangle}{h^2(v_{ui} - v_0 + i\Gamma_u)(v_{vi} - v_0 + i\Gamma_v)}, \qquad 9.$$

where u is a vibrational level of s. This expression looks like the vibronic B term, and it indeed reduces to Equation 8 upon performing the sum over u for the nonresonant state, s. However, there is no restriction that $e \neq s$, and the diagonal term, with $e = s$, provides for single state scattering, i.e. it is a third-order contribution to the

A term:

$$A' = (M_e^0)^2 h_{ee} \sum_v \sum_u \frac{\langle j|v'\rangle \langle v'|Q|v\rangle \langle v|i\rangle}{\hbar^2(v_{v'i} - v_0 + i\Gamma_u)(v_{vi} - v_0 + i\Gamma_v)}, \qquad 10.$$

where v' and v are now different vibrational levels of e, and $h_{ee} = \langle e|\partial H/\partial Q|e\rangle$, which is just the normal coordinate dependence of the excited state energy.

The importance of this contribution is that it does not vanish off-resonance, while the second-order contribution does. If Equation 5 is evaluated off-resonance, where the frequency denominator is essentially independent of v, the summation over v gives

$$A = \frac{(M_e^0)^2 \langle j|i\rangle}{\hbar^2(v_{eg} - v_0)},$$

but $\langle j|i\rangle = 0$ except for Rayleigh scattering ($i=j$). A similar treatment of Equation 10 gives

$$A' = \frac{(M_e^0) h_{ee} \langle j|Q|i\rangle}{\hbar^2(v_{eg} - v_0)^2}, \qquad 11.$$

which remains nonzero because of the Q-dependence. Tang & Albrecht (14) have shown that Equation 11 can be obtained via the vibronic approach if the vibrational coordinate dependence of the energy denominator is explicitly taken into account. In effect A' is that part of the scattering from a single electronic state that depends on $\partial M/\partial Q$ (cf. Equation 6). It can be expected to be a small contribution at resonance, but it gains in importance off-resonance as the Franck-Condon products cancel.

The off-resonance form of the B term is:

$$B = \frac{(M_e^0)(M_s^0) h_{es} \langle j|Q|i\rangle}{\hbar^2(v_{sg} - v_0)(v_{eg} - v_0)}. \qquad 12.$$

The different preresonance frequency dependences predicted by Equations 11 and 12 have been used to distinguish between A- and B-term scattering (50). However, the detuning interval $(v_{eg} - v_0)$ required for Equations 11 and 12 to be good approximations is not well defined. For A-term scattering the rate at which the Franck-Condon factors of Equation 5 cancel, depends in detail on the separation and width of successive excited state vibrational levels (see below).

Nonadiabatic Effects: Jahn-Teller Distortion

In the case of vibronic mixing, the assumption of electron-nuclear separability begins to break down as the mixing states approach one another in energy. The scattering amplitude increases over the value expected from adiabatic theory. The Raman intensity is expected to show extra enhancement in resonance with higher vibrational levels of the excited state manifold, since these are closer in energy to the upper mixing state. Also overtone and combination bands may gain intensity by proximity to the upper state (16). Nonadiabatic corrections to the Herzberg-Teller treatment can be carried out via perturbation theory (16, 17). Johnson et al

(20) have shown that the time-dependent approach gives these corrections automatically if vibrational wave functions appropriate to the excited states are used.

The Born-Oppenheimer approximation breaks down altogether in the excited state Jahn-Teller effect, which can be considered an extreme form of vibronic mixing, in which the coupled states are degenerate. The perturbation approach is inapplicable and the vibronic wave functions must be determined by direct diagonalization of the hamiltonian (51). The normal modes responsible for the distortion can be expected to be especially active in the RR spectrum. An illustration is provided by the squarate and croconate ions (52), whose first allowed $\pi-\pi^*$ transitions show evidence of Jahn-Teller splitting in the absorption spectra. The next allowed transitions are too high in energy for effective vibronic coupling, yet the resonance Raman spectra are dominated by nontotally symmetric in-plane ring deformation modes, which are plausibly the Jahn-Teller active vibrations. Iijima et al (52) noted that these modes have unusual intensity even off resonance. Some time ago Stufkens (53a) had explained unusually strong off-resonance E_g and T_{2g} modes of the octahedral s^2 ions SeX_6^{2-} and TeX_6^{2-} (X = Cl, Br) in terms of a dynamic excited state Jahn-Teller effect, consistent with observed splittings in their ultraviolet absorption bands. Clark & Duarte (53b) suggest, however, that the resonance enhancement of the E_g mode of $TeBr_6^{2-}$ arises from vibronic mixing of $^3T_{1u}$ and $^1T_{1u}$ states.

Intermolecular Effects

Intermolecular interactions can effect changes in either Raman frequencies or intensities. The latter are more probable since only changes in excited state energies or transition moments are required. A well-known example is the weakening of purine and pyrimidine ring mode intensity upon stacking of the bases in nucleic acids. This is connected with the parallel hypochromism of ultraviolet bands with which the Raman modes are in resonance (54, 55). Charge-transfer complexes show RR enhancements for both donor and acceptor groups (56) but only slight frequency shifts with respect to the free molecules. The rhodopsin RR spectrum was reported to contain bands attributable to protein aromatic groups (57) that might be involved in a charge-transfer interaction with the retinal chromophore, although this observation is disputed (13). Excitation within mixed-valence charge transfer bands likewise enhances vibrational modes of both valence partners (36b). In the case of $Cs_2[SbCl_6]$, which contains both $SbCl_6^-$ and $SbCl_6^{3-}$ octahedral anions, the breathing mode, v_1, of $SbCl_6^-$ forms higher-order progressions not only with itself, but also with the breathing mode of $SbCl_6^{3-}$, and even with a lattice mode, deduced to occur at 60 cm^{-1} (58).

POLARIZATION: ANTISYMMETRIC SCATTERING

Rotational Invariants

The polarizability tensor contains nine elements $\alpha_{\rho\sigma}$, each of which, in principle, can be measured independently in an oriented system. For randomly oriented molecules the tensor has three rotational invariants, which can be defined in alternative ways. The set most familiar to Raman spectroscopists consists of the isotropic part, or mean value

$$\bar{\alpha}^2 = \frac{1}{9}\left(\sum_\rho \alpha_{\rho\rho}\right)^2, \qquad 13.$$

the symmetric anisotropy

$$\gamma_s^2 = \frac{1}{2}\sum_{\rho,\sigma}(\alpha_{\rho\rho} - \alpha_{\sigma\sigma})^2 + \frac{3}{4}\sum_{\rho,\sigma}(\alpha_{\rho\sigma} + \alpha_{\sigma\rho})^2, \qquad 14.$$

and the antisymmetric anisotropy

$$\gamma_{as}^2 = \frac{3}{4}\sum_{\rho,\sigma}(\alpha_{\rho\sigma} - \alpha_{\sigma\rho})^2. \qquad 15.$$

These differ only by numerical factors from the trace, quadrupole, and magnetic dipole invariants defined by Placzek (59). Their relative magnitudes are reflected in the polarization properties of the scattered light. If the incident beam is linearly polarized perpendicular to the scattering direction, the intensity of the scattered component that is polarized parallel to the incident polarization is given by

$$I_\| = K(45\alpha^2 + 4\gamma_s^2),$$

where K is a constant of the experiment, and the intensity of the perpendicular component is

$$I_\perp = K(3\gamma_s^2 + 5\gamma_{as}^2). \qquad 16.$$

The depolarization ratio is

$$\rho = I_\perp/I_\| = (3\gamma_s^2 + 5\gamma_{as}^2)/(45\alpha^2 + 4\gamma_s^2). \qquad 17.$$

If, as is generally assumed, the tensor is symmetric, i.e. $\alpha_{\rho\sigma} = \alpha_{\sigma\rho}$, then $\gamma_{as}^2 = 0$ and $\rho \leq \frac{3}{4}$. But if this is not the case, ρ can exceed $\frac{3}{4}$, a situation that has been termed anomalous polarization. The term can also apply to lesser values of ρ that still exceed the value expected for a symmetric tensor, e.g. for totally symmetric modes of a molecule of cubic symmetry ($\alpha_{xx} = \alpha_{yy} = \alpha_{zz}$), the off-diagonal elements are expected to be zero, giving $\rho = 0$.

If the tensor is not symmetric, circular polarization measurements are required to determine all three tensor invariants (59, 60). These should be obtained in forward or backscattering to maintain the circular polarization. The backscattered component rotating in the same sense (corotating) as the incident polarization is given by:

$$I_{co} = K(6\gamma_s^2), \qquad 18.$$

while the contrarotating component is given by

$$I_{contra} = K(45\bar{\alpha}^2 + \gamma_s^2 + 5\gamma_{as}^2). \qquad 19.$$

With a backscattering arrangement in which both the incident and scattered light pass through a single Fresnel rhomb (61) or quarter-wave plate (62) one can obtain $I_\|$, $I_\perp$, I_{co}, and I_{contra} in a single experiment. Any three serve to determine the tensor invariants and the fourth measurement provides a check.

Electronic and Spin-Orbit-Induced Antisymmetry

Antisymmetric scattering is expected to arise for Raman transitions that involve a change in electronic angular momentum, and that are therefore magnetic-dipole-allowed (59). Such transitions produce a rotation of the electric vector. A rich source of examples is provided by the low-lying electronic transitions among states in the spin-orbit manifolds of heavy metal ions (63a), and these formed the basis for testing the prediction of antisymmetric scattering in the electronic Raman effect (63b). The polarized electronic Raman spectra obtained by Porto and co-workers on single crystals of $CeCl_3$ provided the first direct observation of asymmetry in the scattering tensor (64).

If the electronic ground state is degenerate, then it is possible for molecular vibrations to induce antisymmetric scattering, as shown theoretically by Childs & Longuet-Higgins (65a,b). The Raman transition can change the electronic angular momentum via the spin-orbit coupling. This effect was recently observed by Hamaguchi et al (66a), who discovered that all the Raman-active modes of $IrCl_6^{2-}$ (a case actually considered by Childs 65b) show anomalous polarization when excited in the Cl → Ir charge-transfer absorption envelope. Outside this envelope the depolarization ratios are normal for an octahedral complex (0 for the A_{1g} mode and $\frac{3}{4}$ for the E_g and T_{2g} modes), and the electron spin resonance spectrum is iso-

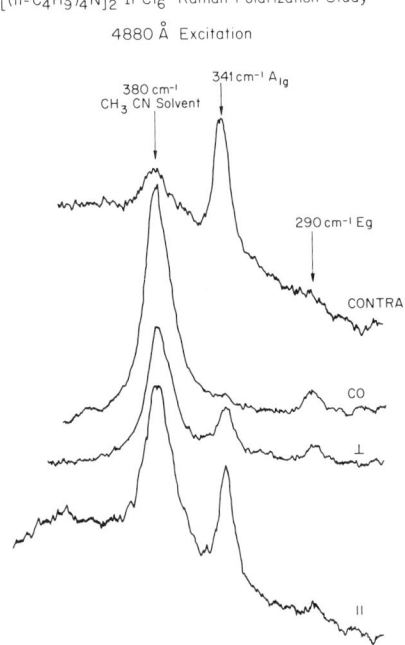

Figure 5 Complete polarization spectra for $IrCl_6^{2-}$ in acetonitrile. (From Reference 68.)

tropic (67). Thus the anomalous polarization cannot be attributed to a ground-state structural distortion. The purely antisymmetric character of the anomalous scattering is demonstrated by circular polarization measurements (68) (Figure 5) that show that all the nonisotropic scattering for the A_{1g} Ir–Cl breathing mode is found in $I_\perp$, and none in I_{co}. Thus $\gamma_s^2 = 0$, while γ_{as}^2 has an appreciable value, and the scattering tensor consists of isotropic diagonal and antisymmetric off-diagonal elements. This is as expected for an effect linked to the electronic degeneracy (the $^2T_{2g}(d^5)$ ground state is split by spin-orbit coupling into a lower energy Kramers doublet and an upper quartet state), since the spin-orbit operator mixes pure rotational (axial vector) symmetry into the orbital motions of the electrons. Hamaguchi & Shimanouchi have also observed anomalous polarization in $IrBr_6^{2-}$ and have calculated depolarization ratios, taking into account the specific nature of the resonant excited spin-orbit states (66b).

A surprising instance of anomalous polarization is that of $FeBr_4^{2-}$; two groups (69, 70) have reported a depolarization ratio near 0.2 for the symmetric stretching mode, whereas a value of 0 is expected for a tetrahedral complex. Structural distortion (70) and spin-orbit effects (69) have both been advanced in explanation. The latter is confirmed by circular polarization measurements that definitely establish an antisymmetric contribution to the scattering (68). In this case, however, the ground state is an orbital singlet (^6S) and not subject to spin-orbit splitting, so that spin-orbit effects in the resonant charge transfer states must be sufficient for antisymmetric scattering.

Vibronic Antisymmetry

Another mechanism for antisymmetric vibrational scattering is vibronic mixing of excited electronic states by vibrational modes that have rotational symmetry themselves, and whose scattering tensor patterns are therefore antisymmetric (60). These modes gain Raman intensity via the B term (Equations 7 and 8). Porphyrins exemplify this effect, and heme proteins provided the first instance of anomalously polarized vibrational Raman bands (45). As mentioned above, antisymmetric vibrations, of A_{2g} symmetry, are effective in mixing the first two π–π^* electronic transitions, which are strongly interacting. Such modes are expected to show an infinite depolarization ratio (inverse polarization), since both $\bar{\alpha}^2$ and γ_s^2, and therefore $I_\parallel$, should be zero. In fact, $I_\parallel$ is within experimental error of zero for these modes in hemoglobin, but appreciable $I_\parallel$ intensities are seen for cytochrome c (45). Circular polarization measurements (61, 62) indicate that these components reflect a significant lowering of the 4-fold porphyrin symmetry.

An alternative viewpoint on A_{2g} scattering has been advanced by Warshel (71), who emphasizes the role of these modes in rotating the two orthogonal transition moments of the degenerate resonant state. He has given a method of evaluating $\partial M/\partial Q$ (Equation 7) and has shown that the parallel component induced into the 1580 cm^{-1} A_{2g} mode of Cu-porphin by attachment of methyl groups at the 1, 3, 5, and 7 ring positions (73a, b) is directly attributable to their kinematic mass effect.

The requirement for degenerate transition moment mixing via a rotational-type vibration is a stringent one, and, to date, antisymmetric vibronic scattering has not been detected in molecules other than porphyrin derivatives. Perrin et al (73a) pointed out some time ago that in cyclic aromatic molecules, with $4n + 2$ π-electrons, the bond stretching modes that alternate phase around the ring, which classify as A_{2g} in porphyrins, are effective in vibronic mixing if the number of atoms is $4n$, rather than $4n + 2$. The reason for this can be seen with reference to the anulene π-electron model of Simpson (73b) in which the first two $\pi-\pi^*$ transitions occur between the degenerate pair of highest filled molecular orbitals, with angular momentum $\mathbf{l} = \pm n$, to the lowest unfilled pair, with $\mathbf{l} = \pm(n + 1)$. The two degenerate transitions have $\Delta \mathbf{l} = 1$ and $\Delta \mathbf{l} = 2n + 1$, and they can be coupled via vibrational modes with $2n$ units of angular momentum, i.e. $2n$ nodes. Bond alternant modes change phase between every pair of atoms, and the number of nodes is therefore half the number of atoms. Consequently, $2n$ nodes require $4n$ atoms in the ring, rather than $4n + 2$. Porphyrin dianions fulfill this requirement since they approximate 16 atom annulenes ($n = 4$) and have 18 electrons. It follows from this rule that neutral aromatics, such as benzene, cannot be expected to show Raman activity for their bond alternant modes, even under resonant conditions.

EXCITATION PROFILES

The dependence on excitation wavelength has held a central role in the interplay between Raman theory and experiment. An excitation profile (74) is a plot of Raman intensity vs laser wavelength. It is similar to a fluorescence excitation spectrum except that individual Raman bands can be separately monitored. Most profiles have been constructed so far from individual spectra obtained with the various Ar^+ and Kr^+ laser lines, and are therefore of low resolution. Some striking effects have nevertheless been observed. The advent of tunable dye lasers makes it possible to increase the resolution to any desired level. Jeanmaire & Van Duyne (75) have shown that a wealth of vibronic detail may be revealed, even for complex molecules with broad absorption bands, by measuring spectra at excitation intervals of ~ 5 cm^{-1}.

A-Term Profiles

For A-term scattering, a peak in the excitation profiles is expected at the origin of the resonant electronic transition and subsidiary peaks at successive excited state vibrational levels, the amplitudes depending on the successive Franck-Condon factors. These can be estimated in the displaced harmonic oscillator approximation as shown by Miyazawa and co-workers (76), who obtained fits to the excitation profiles of three fundamentals, two overtones, and two combination bands of β-carotene, with reasonable values for excited state displacements and damping constants. These parameters were used in turn to obtain a good simulation of the observed $\pi-\pi^*$ absorption band. Similar calculations (77) have been carried out for MnO_4^- (36a) and for cobalamin (18, 78).

It is also possible for an excitation profile to show peaks at positions corresponding to the excited state level of a different normal mode than the one under consideration, as has recently been observed for the 860 and 1400 cm^{-1} modes of azulene (79a) and the 400 cm^{-1} mode of CrTPPCl(79b). This may occur when the displacement in the excited state of the "helping" mode is much larger than the displacement of the normal mode under consideration. Another situation arises when the normal mode vectors change their composition upon electronic excitation, and the transformation between the ground and excited state is no longer diagonal [Dushinsky effect (79c)]; the excitation profile may then become quite complex. An interesting example is revealed by transferrin (37), the iron-binding protein, which has a broad, unstructured visible absorption band, assignable to a charge transfer excitation from phenolate (tyrosine) ligands to the bound Fe^{3+} ion. The RR spectra show enhanced phenolate modes. The four strongest of these give excitation profiles (Figure 6), which show considerable complexity; but *all* of them are dominated by a progression with a 1000 cm^{-1} spacing. This is a reasonable frequency for the excited state C–O stretching mode, lowered from the ground state value of 1281 cm^{-1} by loss of π-conjugation attendant on the transfer of an oxygen π-electron to Fe^{3+} (37). The observation of a common progression for the 1608, 1506, 1281, and 1174 cm^{-1} bands implies that the C–O coordinate is mixed significantly into all four of these modes, but that the excited state C–O stretch is relatively pure.

Thus, changes in normal mode composition between ground and excited state, which arise from changes in force constants and geometry in the excited state or anharmonicity in the normal modes, can add new features to the excitation profile.

B-Term Profiles

For B-term scattering excitation profile maxima are expected at the 0–0 and 0–1 positions for each mixing mode, since the products $\langle j|Q|v\rangle\langle v|i\rangle$ and $\langle j|v\rangle\langle v|Q|i\rangle$

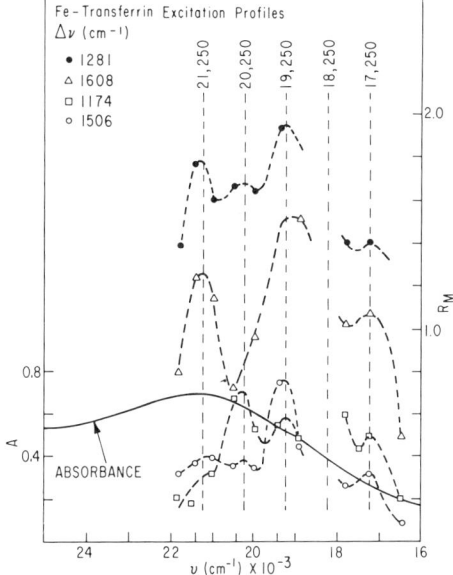

Figure 6 Excitation profiles for four modes (Fe^{3+}-bound phenolate vibrations) showing a common 1000 cm^{-1} progression, attributable to the excited state C–O stretching mode. (From Reference 37.)

are largest for these levels. This was confirmed experimentally for the vibronic modes of porphyrin that show a common first maximum coincident with the α-absorption band (80) and second maxima that are individually shifted from the α-band frequency by the vibrational frequency of the mode (45). Because of the symmetry of Equation 6, the 0–0 and 0–1 enhancements are predicted by vibronic theory to be equal (8). As mentioned above, nonadiabatic effects should increase the 0–1, relative to the 0–0, enhancement, although model calculations suggest that the imbalance is small unless the separation of the electronic states is comparable to the vibrational frequency (19). Such a case is, however, claimed by Shelnutt et al (46) who observed significantly stronger 0–1 than 0–0 maxima for several modes of manganese (III) etioporphyrin I when excited in the region of an intense absorption band (Band V) with both π–π* and porphyrin-metal charge transfer character. The 0–1, 0–0 imbalance was attributed to a nonadiabatic effect arising from another transition (Band Va) only ~ 2100 cm^{-1} away. Asher & Sauer (82), however, suggested that the apparent 0–1 maxima might actually arise from resonance with still another electronic transition that appears as a shoulder on Band V and is resolved in the magnetic circular dichroism (MCD) spectrum.

Interference Effects; Antiresonance

Because scattering from successive resonant states contribute to the Raman tensor, the shape of the excitation profile is influenced by interference effects (83). Raman intensity is proportional to the square of the tensor. If the latter has two contributions, its square contains a cross product, which may be either positive (constructive interference) or negative (destructive interference). The frequency denominators of the intensity equations change signs as the incident frequency passes through resonance with successive scattering levels. Between two levels interference is therefore destructive if the numerators have the same sign and constructive if they have opposite signs. Destructive interference deepens the excitation profile and pushes the adjacent maxima away from one another. Constructive interference fills in the valleys and pulls the adjacent peaks toward one another. The actual shapes depend in detail on the scattering amplitudes and level separations; they also depend on the damping factors (bandwidths) that introduce scattering components that have absorption rather than dispersion lineshapes. Variations in these three parameters can yield a wide assortment of spectral shapes. Higher order interferences, between levels of different vibrational modes, may even lead to suppression of excitation peaks (79a).

For A-term scattering, the successive numerators, i.e. $\langle 1|0\rangle\langle 0|0\rangle$, $\langle 1|1\rangle\langle 1|0\rangle$, $\langle 1|2\rangle\langle 2|0\rangle$, etc, alternate in sign (46), as can be seen from the form of the displaced harmonic oscillator equations (75). Interference between levels is therefore constructive. Off resonance the successive contributions cancel, causing the A term to vanish, except for the A' vibronic contribution, as noted above. For B-term scattering, interference between the 0–0 and 0–1 levels may be destructive or constructive, depending on whether the tensor is symmetric or antisymmetric, since

$$(\alpha_{\rho\sigma} \pm \alpha_{\sigma\rho}) = M_e^0(\partial M/\partial Q)^0 \frac{1}{\hbar}\left(\frac{\langle 1|Q|0\rangle\langle 0|0\rangle}{v_{00} - v_0 + i\Gamma_0} \pm \frac{\langle 1|1\rangle\langle 1|Q|0\rangle}{v_{10} - v_0 + i\Gamma_0}\right).$$

For bands of mixed polarization, the antisymmetric component gains in relative importance between the levels; the depolarization ratio maximizes at $v_0 = (v_{00} + v_{10})/2$ (81), as has been observed for porphyrins (72, 84).

Off resonance, the antisymmetric amplitudes cancel when the detuning interval greatly exceeds the vibrational frequency, so that vibronically induced anomalous polarization vanishes at wavelengths well outside the absorption band (80). This is not true of antisymmetry in electronic Raman scattering, however, since the relevant interferences are between electronic levels that may be far apart (64). In the non-resonant limit, $v_0 = 0$, the tensor must, of course, be symmetric, since the resonant and nonresonant terms in the dispersion equation (Equation 2), which have reversed polarization, are then equal. Antisymmetry cancellation from this source is a slowly varying function of v_0, however, and is far from complete in the visible region (64).

Vibrational Raman scattering can also experience interference from different electronic states. If the scattering amplitude from the resonant state is comparable to that from a stronger off-resonant state, then the cross product, which has a dispersion lineshape, may dominate the excitation profile. This situation is encountered in centrosymmetric transition metal complexes for which the resonance scattering amplitude from the vibronically induced ligand-field states is comparable to the preresonance amplitude from higher energy charge-transfer states (47). The result is a trough in the excitation profile coincident with the ligand-field band envelope (Figure 7). A similar electronic interference has been observed for a cobalt (II) macrocyclic complex with a charge-transfer band near a stronger π–π^* band (85),

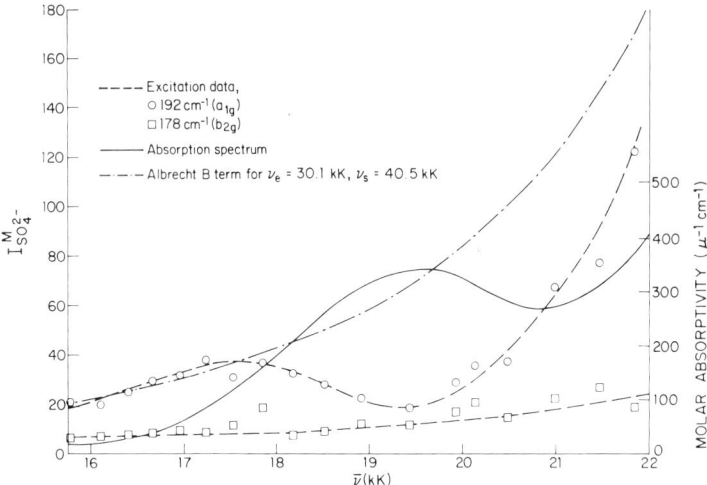

Figure 7 Excitation profiles for $PdBr_4^{2-}$, showing an antiresonance for the A_{1g} mode within the ligand field absorption envelope, attributable to interference from preresonance scattering via the higher charge-transfer states. The dashed curve shows the upper state preresonance frequency dependence without regard to the ligand field transition. (From Reference 47.)

and also between the B and X states of matrix-isolated Br_2 (86). A striking case is that of I_2^- in solution (87). When the solvent is C_6D_6, the excitation shows an antiresonance effect under the profile of the absorption band at ~ 470 nm, due to interference from the higher lying $I_2 - C_6D_6$ charge transfer transition. In $CHCl_3$, where charge-transfer interaction is absent, the excitation peak is coincident with the absorption band.

The sign of the scattering amplitude may be the same or different for adjacent electronic levels. In the latter case the excitation profile may show a minimum at energies below the lower level, an effect recently reported by Ito (43, 88) for a vibronic band of naphthalene.

CONCLUSIONS

The phenomena of vibrational resonance Raman spectroscopy, although complex and novel, are reasonably well comprehended by current theory. In most instances the adiabatic scattering equations provide a satisfactory approximation. The dominant contribution to enhancement by allowed electronic transitions is the A term, which is limited to totally symmetric modes and their overtones. The intensities depend on the oscillator strength and on the Franck-Condon overlaps, the evaluation of which provides the key to understanding relative mode intensities in complex molecules and changes produced by intermolecular interactions. Excitation profiles follow the electronic state vibrational progressions, and can also be approximated by evaluating the Franck-Condon overlaps. In cases where the normal mode composition is appreciably altered in the resonant state, the excitation profile peaks do not necessarily correspond to the vibrational shifts of the ground state mode being monitored. Off resonance, the Franck-Condon factors of the successive vibrational levels cancel, and the remaining intensity is due to the vibronic component, A', which arises from the dependence of the transition frequencies on the normal coordinate.

Nontotally symmetric modes owe their intensity to B-term scattering, which depends on vibronic mixing of excited states. The effect is particularly pronounced when a weakly allowed resonant state mixes with a nearby strongly allowed state. Intensities depend on the vibronic mixing integrals or, equivalently, on the derivatives of the transition moment with respect to the normal coordinates, which can in principle be evaluated directly from the potential functions. Excitation profiles maximize at the 0–0 and 0–1 transition frequencies. The Born-Oppenheimer approximation becomes increasingly inaccurate as the energy gap to the mixing state becomes comparable to the vibronic energy, and it breaks down altogether when the resonant and mixing states are degenerate, i.e. in the Jahn-Teller effect. Jahn-Teller active modes are particularly prominent in the Raman spectrum, even quite far from resonance.

Antisymmetric scattering arises with Raman transitions that involve a change in electronic angular momentum. This can be induced for systems with Kramers degeneracy via the spin-orbit coupling, as in transition metal complexes. This mechanism can produce antisymmetric scattering for both electronic and vibrational transitions. Another mechanism for antisymmetric scattering is vibronic coupling of degenerate excited states by vibrations of rotational symmetry. The conditions for this effect are restrictive, and it has so far only been detected in the porphyrin class of molecules.

While recent experimental and theoretical developments lead to a reasonably coherent picture of resonant Raman scattering, much remains to be learned about the mechanisms involved, and there will no doubt be surprises to come. Applications of the technique to structural problems can be expected to multiply rapidly because of its high selectivity and structural specificity. Proper exploitation of these advantages requires insight into the scattering mechanisms available for the molecules under study.

Literature Cited

1. Bernstein, H. J. 1977. *Mol. Spectrosc.* 5: In press
2. Martin, R. M., Falerov, L. M. 1975. In *Topics in Applied Physics*, ed. M. Cardona, 8:79–145. New York: Springer
3. Spiro, T. G., Loehr, T. M. 1975. In *Advances in Infrared and Raman Spectroscopy*, ed. R. J. H. Clark, R. E. Hester, Vol. 1, Chap. 3, pp. 98–142. London: Heyden
4. Spiro, T. G., Gaber, B. P. 1977. *Ann. Rev. Biochem.* 46:553–72
5. Spiro, T. G. 1975. *Biochim. Biophys. Acta* 416:169–89
6. Spiro, T. G., Burke, J. M. 1976. *J. Am. Chem. Soc.* 98:5482–89
7. Harrison, G. R., Lord, R. C., Lofbourow, J. R. 1948. *Practical Spectroscopy*. New York: Prentice-Hall. pp. 506–30
8. Behringer, J. 1967. In *Raman Spectroscopy*, ed. H. A. Szymanski, 1:168–221. New York: Plenum
9. Porto, S. P. S., Wood, D. L. 1962. *J. Opt. Soc. Am.* 52:251
10. Strekas, T. C., Adams, D. H., Packer, A., Spiro, T. G. 1974. *Appl. Spectrosc.* 28:324; Shriver, D. F., Dunn, J. B. R. 1974. *Appl. Spectrosc.* 28:319
11. Kiefer, W., Bernstein, H. J. 1971. *Appl. Spectrosc.* 25:500; Woodruff, W. H., Spiro, T. G. 1974. *Appl. Spectrosc.* 28:74
12. Mathies, R., Oseroff, A. R., Stryer, L. 1976. *Proc. Natl. Acad. Sci. USA* 73:1–5
13. Callendar, R. H., Doukas, A., Crouch, R., Nakanishi, K. 1976. *Biochemistry* 15:1621–28
14. Tang, J., Albrecht, A. C. 1970. In *Raman Spectroscopy*, ed. H. A. Szymanski, 2:33–67. New York: Plenum
15. Peticolas, W. L., Nafie, L., Stein, P., Fanconi, B. 1970. *J. Chem. Phys.* 52:1576–84
16. Friedman, J. M., Hochstrasser, R. M. 1973. *Chem. Phys.* 1:457–67
17. Mingardi, M., Siebrand, W. 1975. *J. Chem. Phys.* 62:1074–85
18. Garozzo, M., Galluzzi, F. 1976. *J. Chem. Phys.* 64:1720–23
19. Barron, L. D. 1976. *Mol. Phys.* 31:129–45
20a. Johnson, B. B., Nafie, L. A., Peticolas, W. L. 1977. *Chem. Phys.* 19:303–11
20b. Johnson, B. B., Peticolas, W. L. 1976. *Ann. Rev. Phys. Chem.* 27:465–91
21. Rousseau, D. L., Williams, P. F. 1976. *J. Chem. Phys.* 64:3519–37; Williams, P. F., Rousseau, D. L., Divoretsky, S. H. 1974. *Phys. Rev. Lett.* 32:196
22. Friedman, J. M., Hochstrasser, R. M. 1974. *Chem. Phys.* 6:155
23. Berg, J. O., Langhoff, C. A., Robinson, G. W. 1974. *Chem. Phys. Lett.* 29:305
24. Hilborn, R. C. 1975. *Chem. Phys. Lett.* 32:76
25. Mukamel, S., Jortner, J. 1975. *J. Chem. Phys.* 62:3609; Metiu, H., Ross, J., Nitzan, A. 1975. *J. Chem. Phys.* 63:1289
26. Kimble, H. J., Mandel, L. 1975. *Opt. Commun.* 14:167
27. Madden, P. A., Wennerstrom, H. 1976. *Mol. Phys.* 31:1103–15
28. Holzer, W., Murphy, W. F., Bernstein, H. J. 1970. *J. Chem. Phys.* 52:399
29. Shen, Y. R. 1974. *Phys. Rev. B* 9:622
30. Adar, F., Gouterman, M., Aronowitz, S. 1976. *J. Phys. Chem.* 80:2154–91
31. Callis, J. B., Gouterman, M., Jones, Y. M., Henderson, B. H. 1971. *J. Mol. Spectrosc.* 39:410–20
32. Clark, R. J. H., Franks, M. L. 1975. *J. Am. Chem. Soc.* 97:2691

33a. Cowman, C. D., Gray, H. B. 1973. *J. Am. Chem. Soc.* 95:8177
33b. Clark, R. J. H., Stuart, B. To be published
34. Hirakawa, A. Y., Tsuboi, M. 1975. *Science* 188:359
35. Gill, D., Heyde, M. E., Rimai, L. 1971. *J. Am. Chem. Soc.* 93:6289; Heyde, M. E., Gill, D., Kilponen, R. G., Rimai, L. 1971. *J. Am. Chem. Soc.* 93:6776
36a. Kiefer, W., Bernstein, H. J. 1971. *Chem. Phys. Lett.* 8:381–84; 1972 *Mol. Phys.* 23:835
36b. Clark, R. J. H. 1975. See Ref. 3, Chap. 4, pp. 143–72
37. Gaber, B. P., Miskowski, V., Spiro, T. G. 1974. *J. Am. Chem. Soc.* 96:6868
38. Warshel, A., Karplus, M. 1974. *J. Am. Chem. Soc.* 96:5677–89
39. Warshel, A. 1977. *Ann. Rev. Biophys. Bioeng.* 6:273–300
40a. Ito, M., Suzuka, I., Udagawa, Y., Kaya, K., Mikami, N. 1972. *Chem. Phys. Lett.* 16:211
40b. Kamagawa, K., Ito, M. 1976. *J. Mol. Spectrosc.* 60:277–89
41. Kalantar, A. H., Franzosa, E. S., Innes, K. K. 1972. *Chem. Phys. Lett.* 17:335
42. Ohta, N., Ito, M. 1976. *J. Mol. Spectros.* 59:396
43. Ito, M. 1976. *Proc. Int. Conf. Raman Spectrosc., 5th,* pp. 267–76, ed. E. D. Schmid et al. Freiburg: Schulz
44. Hong, H., Jacobsen, C. W. 1976. *J. Chem. Phys.* 65:2470
45. Spiro, T. G., Strekas, T. C. 1972. *Proc. Natl. Acad. Sci. USA* 69:2622
46. Shellnut, J. A., O'Shea, D. C., Yu, N.-T., Cheung, L. D., Felton, R. H. 1976. *J. Chem. Phys.* 64:1156
47. Stein, P., Miskowski, V., Woodruff, W. H., Griffin, J. P., Werner, K. G., Gaber, B. P., Spiro, T. G. 1976. *J. Chem. Phys.* 64:2159–67
48. Chottard, G., Bolard, J. 1975. *Chem. Phys. Lett.* 3:309; 1976. See Ref. 43, pp. 306–7
49. Bosworth, Y. M., Clark, R. J. H., Turtle, P. C. 1975. *J. Chem. Soc. Dalton Trans.,* pp. 2027–31
50. Albrecht, A. C., Hutley, M. C. 1971. *J. Chem. Phys.* 55:4438–43
51. Tsuboi, M., Hirakawa, A. Y. 1975. *J. Mol. Spectrosc.* 56:146–58
52. Iijima, M., Udagawa, Y., Kaya, K., Ito, M. 1975. *Chem. Phys.* 9:229–35
53a. Stufkens, D. J. 1970. *Recl. Trav. Chim. Pays-Bas* 89:1185–1201.
53b. Clark, R. J. H., Duarte, M. L. 1976. *J. Chem. Soc. Dalton Trans.* pp. 2081–87
54. Small, E. W., Peticolas, W. L. 1971. *Biopolymers* 10:69; Pezolet, M., Yu, T.-J., Peticolas, W. L. 1975. *J. Raman Spectrosc.* 3:55; Peticolas, W. L. 1976. *Proc. Int. Conf. Raman Spectrosc., 5th,* p. 163
55. Tsuboi, M., Hirakawa, A. Y., Nishimura, Y. 1974. *J. Raman Spectrosc.* 2:609
56. Jensen, P. W. 1976. *Chem. Phys. Lett.* 39:138
57. Lewis, A., Fager, R. S., Abrahamson, E. W. 1973. *J. Raman Spectrosc.* 1:465–70
58. Clark, R. J. H., Trumble, W. R. 1976. *J. Chem. Soc. Dalton Trans.,* pp. 1145–49
59. Placzek, G. 1934. *Handbuch der Radiologie,* ed. E. Marx, 2:209–372. Leipzig: Akademische Verlagsgesellschaft 6
60. McClain, W. M. 1971. *J. Chem. Phys.* 55:2789–96
61. Nestor, J., Spiro, T. G. 1973. *J. Raman Spectrosc.* 1:539–50
62. Pezolet, M., Nafie, L. A., Peticolas, W. L. 1973. *J. Raman Spectrosc.* 1:455–64
63a. Konigstein, J. A. 1976. *Mol. Spectrosc.* 4:196–224
63b. Mortensen, O. S., Konigstein, J. A. 1968. *Phys. Rev.* 168:75
64. Kiel, A., Damen, T., Porto, S. P. S., Singh, S., Varsanyi, F. 1969. *Phys. Rev.* 178:1518–24
65a. Childs, M. S., Longuet-Higgins, H. C. 1961. *Philos. Trans. R. Soc. London* 254:259
65b. Childs, M. S. 1962. *Philos. Trans. R. Soc. London* 255:31
66a. Hamaguchi, H., Harada, I., Shimanouchi, T. 1975. *Chem. Phys. Lett.* 32:103
66b. Hamaguchi, H., Shimanouchi, T. 1976. *Chem. Phys. Lett.* 38:370
67. Griffiths, G. H. E., Owen, J., Ward, I. M. 1953. *Proc. R. Soc. London Ser. A* 219:526
68. Stein, P., Brown, J., Spiro, T. G. 1977. *Chem. Phys.* In press
69. Clark, R. J. H., Turtle, P. C. 1976. *J. Chem. Soc. Faraday Trans. 2,* pp. 1885–91
70. Forel, M.-T., Mejean, T. 1976. *Proc. Int. Conf. Raman Spectrosc., 5th,* pp. 308–9
71. Warshel, A. 1976. *Chem. Phys. Lett.* 43:273–78
72. Sunder, S., Mendelsohn, R., Bernstein, H. J. 1975. *J. Chem. Phys.* 63:573–80
73a. Perrin, M. H., Gouterman, M., Perrin, C. L. 1969. *J. Chem. Phys.* 50:4137–50
73b. Simpson, W. T. 1949. *J. Chem. Phys.* 17:1218–21

74. Rimai, L., Heyde, M. E., Heller, H. C., Gill, D. 1971. *Chem. Phys. Lett.* 10:207–11
75. Jeanmaire, D. L., Van Duyne, R. P. 1976. *J. Am. Chem. Soc.* 98:4034–39
76. Tasumi, M., Inagaki, F., Miyazawa, T. 1973. *Chem. Phys. Lett.* 22:515–18; Inagaki, F., Tasumi, M., Miyazawa, T. 1974. *J. Mol. Spectrosc.* 50:286–303
77. Mingardi, M., Siebrand, W., Van Labeke, D., Jacon, M. 1975. *Chem. Phys. Lett.* 31:208–11
78. Galluzzi, F., Garozzo, M., Ricci, F. F. 1974. *J. Raman Spectrosc.* 2:351–62
79a. Liang, R., Schnepp, O., Warshel, A. 1976. *Chem. Phys. Lett.* 44:394
79b. Shelnutt, J. A., Cheung, L. D., Chang, R. C. C., Yu, N. T., Felton, R. H. 1977. *J. Chem. Phys.* 66:3387
79c. Dushinsky, F. 1973. *Acta Physicochim. URSS* 1:551
80. Strekas, T. C., Spiro, T. G. 1973. *J. Raman Spectrosc.* 1:387–92
81. Mortensen, O. S. 1975. *Chem. Phys. Lett.* 30:406–9
82. Asher, S., Sauer, K. 1976. *J. Chem. Phys.* 64:4115–23
83. Friedman, J., Hochstrasser, R. M. 1975. *Chem. Phys. Lett.* 32:414–19
84. Collins, D. W., Fitchen, D. B., Lewis, A. 1973. *J. Chem. Phys.* 59:5714
85. Nafie, L. A., Pastor, R. W., Dabrowiak, J. C., Woodruff, W. H. 1976. *J. Am. Chem. Soc.* 98:8007–14
86. Friedman, J. M., Rousseau, D. L., Bondybey, V. E. 1976. *Phys. Rev. Lett.* 37:1610–13
87. Matsuzaki, S., Maeda, S. 1974. *Chem. Phys. Lett.* 28:27
88. Ohta, N., Ito, M. 1976. *J. Chem. Phys.* 65:2907–8

AUTHOR INDEX

A

Abbati, I., 172
Abdel-Gayed, R. G., 83
Abdel-Khalik, S. I., 190, 196, 198, 200
Abe, T., 252
Abgrall, H., 334
Abouaf-Marguin, L., 139, 154
Abragam, A., 48, 51
Abrahamson, E. W., 268, 510
Abramenkov, A. V., 34
Acierno, D., 192, 198, 200
Ackers, G. K., 245
Acrivos, A., 186, 197, 198, 302, 303
Adams, D. H., 502
Adams, D. J., 389
Adamson, A. W., 216
Adar, F., 504
Adelman, S. A., 374, 385, 489
Agostini, G., 58, 61, 65, 69, 70
Ahmad, A., 394
Ahmed, A. I., 235
Ailawadi, N. K., 307
Aiuchi, T., 250
Albano, A. M., 315
Albrecht, A. C., 216, 502, 504, 507, 509
Alder, B. J., 301-5, 309, 373, 379, 389, 391, 393, 400
Aldridge, J. I. III, 133, 144
Aleshin, V. G., 164
Alexander, C. J., 455
Alexander, M. H., 353
Alfano, R. R., 210-12, 220, 222, 227
Alford, N., 177
Alkaitis, S. A., 108
Allen, C. W., 354
Allen, G. G., 304
Allen, L. C., 17, 470
Allen, R. C. Jr., 194
Alley, W. E., 302, 304, 305
Alms, G. R., 211
Alpert, S. S., 236, 237
Altenberger, A. R., 236
Ambartzumian, R. V., 133, 136, 144-51, 155, 156
Ananth, M. S., 373, 375, 376, 381, 383
Ancker-Johnson, B., 241
Andersen, A., 17
Andersen, E. A., 42
Andersen, E. L., 16-18, 34, 39
Andersen, H. C., 303, 373, 374, 381, 385, 393, 399
Anderson, J., 173
Anderson, J. B., 353
Anderson, J. G., 365
Anderson, J. L., 237
Anderson, P. W., 105
Anderson, R. W. Jr., 210, 219
Anderson, V. E., 35
Andreev, E. A., 267
Andres, R. P., 353
Andrews, J. R., 287, 295-97
Andrews, T., 411
Andrich, M., 242
Andrist, A. H., 126
Apatin, V. M., 147
Aragon, S. R., 236, 247
Arbeau, P., 101
Armstrong, R. C., 186, 191, 193, 196, 199, 200
Armstrong, R. L., 303, 373, 375
Arnett, E. M., 455, 457, 458, 461, 462, 470
Arnold, W., 106
Arnoldi, D., 153
Arnon, D. I., 107
Aronowitz, S., 504
Arons, J. D., 423
Arshadi, M., 446, 466, 468-70
Arshadi, M. R., 468
Arthur, J. R., 488
Arthurs, E. G., 222
Arvidson, G., 242
Asai, H., 248
Asher, S., 516
Ashraf El-Bayoumi, M., 225
Ashurst, W. T., 373
Astarita, G., 186
Atabek, O., 336, 337
Auclair, J. M., 136
Aue, D. H., 446, 448, 449, 455, 457
Auer, P. L., 196
Aung, S., 344
Ausloos, P., 446
Aviram, I., 373, 374
Avouris, P., 225, 341
Axelrod, D., 242

B

Baardsen, E. L., 369
Bach, G., 86, 95
Badger, R. M., 152, 335
Baer, M., 336, 337
Baer, Y., 169, 170
Bafos, D. A., 360
Baily, E. D., 249
Baird, D. G., 193
Baker, A. D., 161
Baker, C., 161
Baker, G. A., 416
Balakrishnan, C., 193, 194
Baldwin, J. E., 123, 125, 126
Ballal, B. Y., 194
Balooch, M., 486, 488, 491-93
Balquist, J. M., 117, 123, 124, 127
Balykin, V. I., 148, 152
Bancroft, F. C., 234
Band, Y. B., 334-37
Banks, G., 237
Barajas, L., 306
Baran, G. J., 236, 240
Baranovski, A. P., 148
Barber, M., 167, 179, 180
Barboza, M., 195
Barenholz, Y., 246
Bargeron, C. B., 235
Barisas, B. G., 252
Barker, J. A., 303, 373, 374, 434
Barnard, B. J. S., 193
Barnett, E. F., 166, 297
Barojas, J., 303, 308, 373, 374, 391, 392, 395, 396, 400, 401
Baronavski, A. P., 330
Barron, L. D., 502, 510, 516
Barthes-Biesel, D., 198

523

AUTHOR INDEX

Bartis, J. T., 414, 431, 432, 434, 437
Bartlett, P. D., 111, 128
Bartmess, J. E., 462
Basco, N., 330
Baskin, R. J., 236
Basov, N. G., 139, 154, 155
Bassi, D., 491
Batchelor, G. K., 186, 236
Batt, L., 125
Bauer, D. R., 211, 238, 239, 301-3, 308
Bauer, E., 263, 267, 268, 277, 278
Bauer, R. S., 175
Bauer, S. H., 276
Baxter, J. E., 246
Bayer, L., 179, 180
Bayes, K. D., 329, 345
Bazhin, N. M., 147
Bazhutin, S. A., 148
Beams, J. W., 251
Bearman, R. J., 374, 385, 395
Beasley, G. H., 122
Beauchamp, J. L., 446, 448, 451-55, 457, 463-65, 468, 470
Beavers, G. S., 190
Beck, G., 108
Becker, K. H., 369
Bedeaux, D., 310, 311, 315
Beeby, J. L., 479, 482, 483, 489
Beeman, D., 393
Beens, H., 225
Beggs, D. P., 468
Behrens, R. Jr., 358, 359
Behringer, J., 501, 516
Bekov, G. I., 145
Belenov, E. M., 139, 154, 155
Belkind, A. I., 179
Bell, G. M., 415, 439
Bellemans, A., 373, 374, 388, 406
Benard, D. J., 275
Benbasat, J. A., 234, 239
Benedek, G., 487
Benedek, G. B., 234, 235, 237, 238, 240
Benmair, R. M. J., 149
Bennema, P., 216
Bennett, C. H., 389, 393
Bennett, R. A., 41, 42
Bennett, R. G., 208
Bennett, S. L., 468
Benoit, F., 457
Benoit, H., 240
Ben-Shaul, A., 269, 271,
362
Benson, S. W., 111-17, 123-25, 127, 128, 460
Berendsen, J. C., 379
Berg, J. O., 502, 503
Berglund, C. N., 179
Bergman, A., 219
Bergman, R. G., 111, 118, 123, 124, 127, 128
Bergmann, K., 351, 352, 368
Berker, A. N., 440
Berkowitz, J., 31, 451
Berkowitz, S. A., 234, 236
Berliner, L. J., 248
Bermark, T., 161
Bernasek, S. L., 479, 482, 486, 495
Berne, B. J., 234, 237, 240, 301, 303, 304, 308, 312, 315, 373, 374, 377, 378, 395
Bernengo, J. C., 248, 251
Bernhardt, A., 139, 140
Bernstein, H. J., 501, 502, 505-7, 513, 517
Bernstein, R. B., 265, 269, 271, 336, 337, 351, 360, 362
Berry, M. J., 335, 336, 349
Berry, R. S., 15, 106, 458-60
Bersohn, R., 326, 327, 345
BERSON, J. A., 111-32; 117, 118, 123, 124, 126-28
Bertolotti, M., 234
Bertucci, S. J., 303
Best, P. E., 179
Beswick, J. A., 336, 337, 339, 340
Bevan, M. J., 262
Biedermann, S., 125
Billings, B. H., 152
Biloen, P., 179
Binding, D. M., 194
Biondi, M. A., 278
BIRD, R. B., 185-206; 186, 190-94, 196-201, 233
Birely, J. H., 133, 144
Birnboim, H. C., 244
Birnboim, M. H., 193, 236
Birstein, L., 487
Bishop, M., 304, 305
Bishop, S. G., 173
Bixon, M., 223, 233, 302, 307
Bjerre, A., 267
Black, G., 263, 267, 268
Blackwell, B. A., 364
Blackwell, J., 233
Blair, L. K., 461
Blanks, R. F., 195
Bledsoe, J. R., 487
Blinc, R., 48, 53-56, 61, 62, 64-68, 71
Blint, R. J., 464
Bloembergen, N., 147, 149
Blok, J., 249
BLOOMFIELD, V. A., 233-59; 233-36, 238-40, 249
Blout, E. R., 246
Blum, L., 374, 385
Blumberg, G., 243
Blume, M., 440
Blumenfeld, S. M., 373, 375
Blumstein, C., 116-18
Blustin, P. H., 33, 34
Boato, G., 477, 479, 480, 482, 483, 487
Bock, E., 62-66, 68, 69
Bockris, J. O., 106
Bockris, J. O. M., 106
Boer, Y., 161
Boggs, R. A., 128
Bohme, D. K., 446-49, 451, 452
Bok, J., 179
Bolard, J., 508
Bollinger, L. E., 79, 87
Bondybey, V. E., 518
Bonner, W. D. Jr., 107
Bonnhoeffer, K. F., 324
Bordwell, F. G., 462
Bortner, M. H., 263
Bose, T. K., 286, 288
Bosworth, Y. M., 508
Bottcher, C., 34
Bottreau, A. M., 290
Botts, J., 246
Boublik, T., 374, 386
Boudry, M. R., 488, 492
Bowers, M. T., 446, 448, 449, 455, 457
Bowman, J. M., 480, 482
Boyd, A. W., 222
Boyd, D. B., 17
Bozarth, R. F., 244
Bradley, A. B., 369
Bradley, D., 83
Bradley, D. J., 222
Bradley, J. N., 276
Bradley, T. L., 491
Braicovich, L., 172
Brand, J. C. D., 331
Brandow, B. H., 21
Branscomb, L. M., 15, 23
Brauman, J. I., 17, 41, 42, 128, 155, 211, 239, 301-3,

AUTHOR INDEX 525

308, 460, 461, 463
Braun, W., 175, 269, 276
Braverman, L. W., 271
Brehm, G. A., 235
Breig, E. L., 263, 268
Breit, G., 353
Brennan, M. E., 263, 268
Brenner, H., 186, 197, 302, 303
Brewer, R. G., 211
Brezin, E., 416
Briggs, J. P., 446, 449, 450, 455
Brigham, E. O., 287
Brindley, G., 194
Brinkley, S., 79
Brinkmann, U., 140
Brockelhurst, B., 106
Broida, H. P., 212, 269, 276, 278, 331
Bronskill, M. J., 216
Brophy, J. H., 355, 356, 366, 369
Brot, C., 301, 303, 308, 373, 374, 395, 396, 401
Brouwer, D. M., 463
Brown, G. H., 47-50
Brown, G. R., 414
Brown, J., 512, 513
Brown, J. C., 235, 237
Brown, J. M., 111
Brown, T. R., 488
Brown, W., 242
Brown, W. A. C., 225
Bruce, C., 190, 194
Bruin, C., 305
Bruining, J., 401
Brumbach, S. B., 479, 491
Brundle, C. R., 161, 166
Brunn, P., 200
Buchanan, D. N. E., 168, 172, 176, 178
Bucher, E., 178
Buchwalter, S. L., 125, 126
Buckingham, A. D., 211, 373-75, 384, 397
Bueche, R., 199
Buenker, R. J., 17
Buerger, H., 125
Büldt, G., 238
Bunker, D. L., 304
Burcat, A., 95
Burgar, M., 53-56, 61, 66, 67
Burke, J. M., 501
Burkhardt, T. W., 440
Burley, D. M., 413
Burrow, P. D., 42, 266, 279
Busby, E. T., 194

Busca, G., 222
Busch, G., 169, 170
Busch, G. E., 219, 223, 327-30, 345
Bushaw, B. A., 220
Buslaev, Y. A., 179, 180
Bustin, M., 243
Butcher, R. J., 262, 270
Butler, S., 329, 345
Bykhovskii, K., 263

C

Cabane, B., 47, 48, 53, 61, 62, 69
Cabello, A., 148
Cabrebra, N., 480
Cade, P. E., 16, 23-25, 27, 42
Cadogan, K. D., 216
Calabrese, A., 179, 180
Caldwell, C. D., 354, 368, 369
Callcott, T. A., 165
Callear, A. B., 262, 266, 269
Callen, H. B., 425
Callendar, R. H., 502, 510
Callis, J. B., 242, 504
Callomon, J. H., 331, 339
Calvert, J. G., 331, 332
Calvin, M., 179
Camerini-Otero, R. D., 234, 235, 237
Campagna, M., 178
Campillo, A. J., 222
Canales, E. R., 237
Candau, S., 240
Cann, J. R., 245
Cannell, D. S., 235
Cantini, P., 477, 479, 480, 482, 483, 487
Cantoni, G. L., 233
Cantrell, C. D. III, 133, 144, 149
Capelle, G., 331, 369
Caplan, C. E., 333
Capon, B., 111
Cardillo, M. J., 488, 491
Cardona, M., 164, 172, 175
Carelli, P., 303
Cargill, R. L., 111, 124
Cargle, V. H., 122, 124, 127
Cario, G., 208
Carlson, F. D., 234, 240, 241
Carlson, L. R., 145
Carlson, T. A., 161
Carlsten, J. L., 16, 35, 36

Carovillano, R. L., 353
Carpenter, B. K., 118, 123, 124, 126, 127
Carreau, P. J., 198
Carrington, A., 48, 51
Carroll, T. X., 457
Carter, W., 123, 124, 127
Cartwright, D. C., 133, 144
Carver, T. R., 353, 354
Casassa, E. F., 243
Case, D. A., 369
Casey, C. P., 128
Castleman, A. W. Jr., 304, 447, 466
Catsimpoolas, N., 244
Cavigli, P. R., 127
Cederbaum, L. S., 16-18, 21, 22
Celli, V., 479, 480, 482
Celotta, R. J., 41, 42
Cercignani, C., 316, 317
Cermak, V., 446
CEYER, S. T., 477-99; 482
Chaffey, C. E., 197
Chahine, R., 286, 288
Chambers, M. V., 374, 387
Chan, S. I., 344
Chan Man Fong, C. F., 195
Chance, B., 106, 107
Chandler, D., 301-4, 373, 374, 381, 385, 388, 393, 402-4
Chandra, P., 326, 327, 345
Chandrasekhar, P., 431
Chandrasekhar, S., 47, 48, 50
Chandross, E. A., 216, 217
Chang, C. F., 192, 195
Chang, C. T., 249
Chang, D. B., 241
Chang, K. I., 193
Chang, R. C. C., 508, 514, 516
Chapela, G. A., 373
Chapman, S., 301, 304, 306
Chappell, R., 482
Chapuisat, X., 124
Chebotayev, V. P., 141
Chekalin, N. V., 136, 142, 148, 149, 151, 155
Chen, F. C., 235, 236
Chen, J. M., 168
Chen, T. T., 15-17, 23, 24, 34
Chen, Y. D., 374, 385
Cherepanova, T., 311
Cherry, R. J., 247
Chesick, J. P., 125
Cheung, L. D., 508, 514, 516

AUTHOR INDEX

Cheung, P. S. Y., 373, 374, 392, 395-97, 399-402
Chibrikin, V. M., 106
Child, M. S., 333, 337, 338, 512
Childs, J. D., 244
Chiou, C. S., 193
Chiroli, C., 480
Chistyakov, I. G., 47, 48, 50
Chitumbo, K., 242
Chiu, N. W. K., 118
Chiu, S. C., 317
Chmurny, A. B., 123, 124, 127
Chompff, A. J., 240
Chottard, G., 508
Chow, G. K., 278
Chow, H., 481, 482
Christensen, N. E., 170
Christiansen, R. L., 196, 198
Chrzeszczyk, A., 236
Chu, B., 234-36, 414
Chuang, T. J., 213, 215-17
Chuche, J., 120
Chudgar, A., 236
Chung, S.-Y., 252
Chupka, W. A., 31, 451
Chutjian, A., 339, 344
Ciccotti, G., 379
Cirkel, H.-J., 222
Claasen, T. A. C. M., 288
Claesson, S., 216
Clark, A. H., 287
Clark, J. H., 148
Clark, M. D., 139, 154
Clark, N. A., 238
Clark, R. J. H., 506-8, 510, 513
Clark, R. K., 414, 422, 423, 425
Clark, R. W., 243
Clarke, J., 105
Clarke, J. H. R., 401
Clarke, T. C., 118
Clarkson, T. S., 284, 288
Claverie, J., 245
Clayton, R. K., 106
Clement, M. J. Y., 324
Clementi, E., 16, 17, 29, 32, 373, 374, 468, 469
Closs, G. L., 125, 126
Clouston, J. G., 276
Co, A., 190, 194, 198
Cocks, A. T., 125
Cogoli, A., 247
Cohen, A., 94
Cohen, L., 394
Cohen, L. M., 101

Cohen, M. L., 172
Cohen, R., 245
Cohen, R. J., 235
COLE, R. H., 283-300; 286, 288-90, 294, 297, 298
Coleman, B. D., 190
Coleman, R. V., 105
Coles, H. J., 249
Collins, D. W., 517
Collins, J. G., 446
Collins, R. J., 267
Comsa, G. H., 483, 484
Condon, E. U., 337
Cone, R. A., 250
Connor, J. A., 179, 180
Conti, F., 246
Conti, J. J., 304
Conway, B. E., 106
Conway, D. C., 466, 467, 472
Cooke, R., 233
Coombe, R. D., 262
Coon, J. B., 331
Cooper, F. L., 241
Cooper, M. J., 415
Copeland, D., 216
Cornelius, J. R., 327, 328, 345
Cotter, M. A., 426
Cotterill, R. M. J., 373
Coulson, C. A., 35
Coutanceau, M., 195
Cowdry, R., 17, 42
Cowling, T. G., 301, 304, 306
Cowman, C. D., 506
Cox, P. A., 169
Cox, R. G., 252
Cox, R. J., 216
Cram, D. J., 123, 124, 127
Cramer, W. A., 247
Crawford, O. H., 35
Crawford, R. J., 118-20, 123, 124, 127
Creel, C. L., 329, 345
Criminale, W. O. Jr., 191
Crooker, A. J., 97
Crosignani, B., 234
Crosley, D. R., 365
Crosswhite, H. M., 365
Crothers, D. M., 233
Crouch, R., 501, 510
Croxton, C. A., 373, 375, 376, 380, 384, 386
Cruse, H. W., 350, 356-61, 363, 369
Csanak, G., 17
Csizmadia, I. G., 17
Cukier, R. I., 303, 304
Cumming, J. B., 458-60

Cummings, D. J., 251
Cummins, H. Z., 234
Cundall, R. B., 262
Cuniberti, C., 238
Cunningham, A. J., 447, 466-68, 470
Current, S., 128
Curtiss, C. F., 186, 191, 196-98, 200, 201
Czapski, G., 216
Czworniak, K. J., 303

D

Dabiri, A. E., 486, 491
Dabrowiak, J. C., 517
Dagdigian, P. J., 349-53, 356-63, 369, 492
Dahlberg, S. C., 495
Dahler, J., 301, 306
Dahmen, A., 120
Dalgarno, A., 34, 446
Dalton, L. A., 248
Dalton, L. R., 248
Damen, T., 512, 517
Dandliker, W. B., 246
D'Anna, J. A., 246
Das, G., 16, 17, 33
das Gupta, N. N., 251
Dashevsky, V. G., 373, 374
Datta, S., 152
Datz, S., 262
Davidovits, P., 266, 279
Davidson, E. R., 328, 345
Davidson, W. R., 447, 466, 467, 469, 470
Davies, D. R., 233
Davies, J. M., 194
Davies, R. O., 482
Davis, D. D., 369
Davis, D. R., 107
Davis, H. T., 303, 312, 315
Davis, L. I., 369
Dawkins, J. V., 244
Day, L. A., 234, 236
Deakin, J. J., 262
DeBoer, C. D., 128
Debye, P., 302
deCindio, B., 193
DeCorpo, J. J., 360
de Gennes, P. G., 47-50
de Heij, M. E., 249
DeKeizer, A., 244
Delaage, M. A., 244
DeLaney, D. E., 248
Delbrück, M., 242
DelGreco, F. P., 365
Deloche, B., 69
de Loor, G. P., 290
De Maeyer, L., 234

DeMartini, F., 227
De Michelis, B., 172
Demtröder, W., 351, 352, 368
Denariez-Roberge, M. M., 222
Denizot, F. C., 244
Denn, M. M., 186
Denson, C. D., 194
Depew, R. E., 244
DePristo, A. E., 353
Dervan, P. B., 123, 128
Desai, R. C., 306, 312-14
DeSantis, A., 303
Deutch, J. M., 235, 236, 301, 304, 309, 312, 374, 385
DeVault, D., 106, 107
Devlin, J. L., 458
Devlin, J. L. III, 458
Dewan, R. K., 236
Dewar, M. J. S., 120, 125, 127
Dewey, D. B., 56, 59, 61, 62, 68, 69
Dewey, H. J., 153
Dexter, D. L., 208
Dickens, G., 262, 263
Diehl, P., 48, 49
Dieke, G. H., 365
Diemann, E., 179, 180
Diepen, G. A. M., 423
Diercksen, G. H. F., 468-70
Dietz, R., 235
Dill, K., 252
Dimant, Y., 193
Dimpfl, W. L., 355, 356, 365, 366
Ding, A. M. G., 364
Dion, D. R., 480
Diporto, P., 234
Di Salvo, F. J., 176, 177
Ditchfield, R., 35, 458
Divoretsky, S. H., 502, 503, 505, 506
Dixon, D. A., 32, 358
Doane, J. W., 47-50, 53-56, 58, 61-65, 68-70
Docken, K. K., 39
Dodson, R. W., 106
Doemeny, L. J., 262
Doering, W. v. E., 123, 124, 126-28
Dogonadze, R., 106
Doherty, P., 237
Dolbier, W. R. Jr., 111
Doljikov, V. S., 142, 148, 149, 151, 155
Doll, J. D., 17, 18, 480, 486, 489

Domb, C., 411, 413, 415, 421
Dondes, S., 152
Dong, R. Y., 62-66, 68, 69
Donovan, R. J., 261, 262, 270, 273
Dorfman, J. R., 317, 318
Dorfman, L. M., 216
Dorge, K. J., 83
Dörr, F., 225
Doukas, A., 501, 510
Dourlent, M., 249
Dousma, J., 216
Dreux, H., 245
Drexhage, K. H., 212, 213
Drucker, G. E., 462
Drullinger, R. E., 355
Duarte, M. L., 510
Duben, A. J., 451
Dubin, S. B., 234, 235, 238
Dubost, H., 139, 154
Dubrin, J. W., 144
Ducas, T. W., 145
Ducuing, J., 209, 223, 224, 227
Duerre, D., 139, 140
Dufour, C., 250
Duguay, M. A., 211, 221
Duncan, J. L., 125
Dunkin, F. C., 447, 448
Dunlop, P. J., 304
Dunn, J. B. R., 502
Dunn, O., 152
Duplessix, R., 240
Durden, D. A., 446, 447, 467
Dushinsky, F., 514
Duthler, C. J., 278
Dutton, P. L., 107
Dutuit, Y., 290
Dybowski, C. R., 53, 56, 59, 61, 62
Dymanus, A., 367
Dymond, J. H., 303-5
Dynes, R. C., 105
Dzidic, I., 446, 447, 456, 466, 469, 470
Dzvonik, M. J., 326, 327, 345

E

Eastman, D. E., 162-64, 167, 168, 172-74, 176, 178-80
Easwaran, K. R. K., 62
Ebbesjö, I., 306
Edidin, M., 242
Edse, R., 79, 87
Edward, J. T., 302

Edwards, D. H., 91
Egelstaff, P. A., 373-76, 381, 383, 384, 396, 402-5
Eggarter, T. P., 439
Egger, K. W., 125
Ehrenberg, M., 247
Ehrenson, S., 470
Ehrhardt, J. J., 482-84
Ehrlich, G., 487
Ehrmann, G., 193
Eib, W., 178
Eichinger, B. E., 235
Eisenberg, H., 234, 235, 238
EISENTHAL, K. B., 207-32; 208-10, 212, 213, 215-17, 226, 301, 303
Eland, J. H. D., 161
El-Bayoumi, M. A., 225
Elert, M. L., 343-45
Eley, D. D., 106
Elkowitz, A. B., 491
Elliott, B. J., 296, 299
Elliott, S. P., 127
Eloranta, J., 216
El-Sayed, M. A., 210, 341
Elson, E. L., 241, 242
Emery, V. J., 440
Emsley, J., 69
Emsley, J. W., 48, 49
Engel, C. D., 98
Entine, G., 242
Epstein, A. J., 180
Erbudak, M., 162, 163, 172
Ericksen, J. L., 191
Ermak, D. L., 237
Ermolaev, V. L., 208
Ernst, M. H., 305, 306
Erpenbeck, J. J., 373, 393
Esaki, L., 105
Euler, L., 392
Evans, D. C., 196
Evans, D. J., 373, 377, 390, 392, 393
Evans, E. L., 167, 176
Evans, G. T., 199, 215, 303
Evans, K., 340, 341, 345
Evans, S., 161
Ewing, J. J., 262, 270, 273
EYRING, H., 1-13; 106, 434

F

Fabelinskii, I. L., 211
Fager, R. S., 510
Fagerness, P. E., 70
Fahlman, A., 161
Fairbank, W. M. Jr., 349
Faist, M. B., 351
Falerov, L. M., 501, 503
Falkenstein, W., 223, 224

AUTHOR INDEX

Fanconi, B., 502, 508
Fantina, M. E., 128
Farber, M., 31
Farkas, L., 324
Farquhar, I. E., 374, 385
Farrar, J. M., 350
Faure, J., 222
Feeney, J., 48, 49
Feeney, R. E., 235
Feher, G., 106, 107, 234, 241
Fehsenfeld, F. C., 446-48, 466-68
Feitelson, J., 225
Feke, G. T., 241
Feld, M. E., 68
Feldmann, D., 29, 42
Fellner-Feldegg, H., 166, 286, 288, 297
Felton, R. H., 508, 514, 516
Feng, H. C., 168
Fenn, J. B., 353
Ferapontov, N. B., 139, 154
Ferguson, E. E., 446-48, 466-68
Ferrie, F., 99
Ferry, J. D., 190, 193, 251
Fessenden, R. W., 115, 118
Feuerbacher, B., 170
Few, G. A., 373, 374, 404
Field, F. H., 446, 447, 463, 464, 466-68
Filbey, G. L. Jr., 191
Filson, D. P., 235
Fine, D. H., 460
Fine, J., 262, 263
Finer, E. G., 298
Fink, E. H., 369
Fink, W. H., 17
Finzel, H. U., 479, 481, 483
Fiquet-Fayard, F., 334
Fischer, D. W., 176
Fischer, H., 115
Fischer, M., 216
Fischer, S., 223
Fischkoff, S., 242
Fish, W. W., 244, 248
Fishel, D., 63
Fisher, E. R., 263, 267, 268, 271, 277, 278
Fisher, M. E., 411, 413-16, 421
Fisher, R. A., 211
Fisher, S. S., 487
Fisk, G. A., 495
Fitchen, D. B., 517
Fite, W. L., 277, 278
Fitts, D. D., 374, 387, 404
Fixman, M., 199, 200, 215, 374
Fleischauer, P. D., 216
Fleming, G. R., 214
Fletcher, G. C., 236
Florida, D., 337
Flory, P. J., 195
Flowers, M. C., 127
Floyd, R. A., 107
Flumerfelt, R. W., 194
Flygare, W. H., 234, 237, 239, 303
Flynn, G., 349
FLYNN, G. W., 261-82
Flytzanis, N., 482
Flytzani-Stephanopoulos, M., 374, 387
Fock, W., 434
Foley, R. J., 144
Fong, M. C., 79, 87
Foo, P. D., 262, 270
Forbes, W. F., 62, 63, 68
Ford, N. C., 235, 431
Ford, P. W., 127
Forel, M.-T., 513
Forshey, D. R., 86, 89, 90
Forst, W., 111, 112, 124, 269, 271
Forster, D., 301, 304, 373
Förster, T., 208, 218, 219, 224
Foster, J., 95
Fouassier, J., 222
Fournier, J., 268
Foweraker, A. R., 249
Fox, K., 35
Foxon, C. T., 488, 492
Fradkin, E. H., 439
Franck, J., 208, 215
Franck-Neumann, M., 128
Frank, H., 479, 481, 483
Frankel, N. A., 197
Franken, P. A., 353
Frankl, D. R., 479, 480, 481, 486
Franklin, J. L., 360, 446, 462
Franklin, R. M., 234, 237
Franks, F., 298
Franks, M. L., 506
Franzosa, E. S., 508
Fraser, A. B., 240, 241
Freasier, B. C., 373, 374, 395
Frederick, J. E., 238
Fredericq, E., 248, 249
Fredrickson, A. G., 192
Freed, J. H., 52, 53, 56, 57, 61, 69, 70, 301, 307
Freed, K. F., 16, 17, 223, 334-37, 344
Freedman, A., 358, 359
Freeman, G. R., 125
Freeman, R. R., 145, 353
Freeouf, J. L., 162-64, 167, 172-74, 176, 178
Freifelder, D., 234
Fremerey, J. K., 483
French, M. A., 466-68
Frenkel, J., 105
Freund, S. M., 154
Frey, H. M., 111, 112, 116, 124, 127
Friberg, S., 48
Fricke, J., 277, 278
Friedel, R. A., 225
Friedhoff, L., 240
Friedman, J. M., 502-4, 509, 516, 518
Friedman, L., 446
Frigon, R. P., 245
Friš, P., 196, 199
Fröhlich, H., 197
Fromageot, H. P. M., 235
Fromm, J., 373, 374
Frosch, R. P., 223
Fry, R. S., 95
Fuchs, P., 246
Fujime, S., 248
Fujita, H., 250
Fujita, T., 216, 217
Fulton, A., 108
Fulton, R. W., 251
Fulton, T. A., 105
Fung, B. M., 56, 59, 61, 68, 71
Furtado, P. M., 313, 314
Furuyama, S., 460
Fury, M., 308
Furzikov, N. P., 147, 150, 151, 155
Fushiki, Y., 264, 266
Futrell, J. H., 468

G

Gaber, B. P., 501, 507, 508, 514, 515, 517
Gabler, R., 235
Gadzuk, J. W., 171, 177
Gaedtke, H., 328, 345
Gaily, T. D., 351, 352
Gajewski, J. J., 111
Galanin, M. D., 208
Gale, R. J., 482, 495
Galley, W. C., 247
Gallo, R. J., 194
Galluzzi, F., 502, 514
Gann, R. G., 278
Gans, W. L., 287, 296, 297
Ganz, A., 317

AUTHOR INDEX

Garbuzov, D. Z., 179
Garcia, N., 482
Garcia-Colin, L. S., 306
Garcia de la Torre, J., 249
Garibaldi, U., 479, 480, 487, 489
Garisto, F., 312
Garozzo, M., 502, 514
Garrett, W. R., 35, 41
Garris, C. A., 101
Garrison, B. J., 489
Gass, D. M., 301-3
Gaunt, D. A., 421
Gavezzotti, A., 124
Gavrilina, I. K., 139, 154
Gaydon, A. G., 27, 276
Gear, C. W., 392, 393
Gebelein, H., 325, 337-39
GELBART, W. M., 323-48; 341, 343-45
Gelius, U., 161, 164, 180
Geller, M., 17
Gemmer, R. V., 128
Geny, F., 247
Gerber, B. R., 249
Gerberich, H. R., 127, 128
Gergely, J., 248
Gerischer, H., 106
German, K. H., 369
Germark, T., 161
Gerth, C., 251
Gestblom, B., 287, 288, 291, 295
Giaever, I., 105
Gibbons, R. M., 374, 385, 386
Gibbons, W. A., 216
Gibert, R., 153
Gibson, T. A., 127
Giese, K., 296
Giesekus, H., 190, 192, 196, 197
Gilbert, K. E., 123, 125, 126
Gilbert, R. G., 325, 337-39
Gilbert, T. L., 16, 17
Gilchrist, T. L., 111
Gill, D., 507, 514
Gill, G. B., 111
Gillispie, G. D., 341
Gires, F., 211
Glaeser, R. M., 106
Glass, I. I., 276
Glasser, L., 284
Gleiter, R., 127
Glynn, T. J., 222
Gnadig, K., 216, 217, 234
Gō, N., 200
Gobby, P. L., 173
Gochelashvili, K. S., 155
Godbole, E. W., 446

Goddard, J. D., 187, 197
Goddard, R. D., 460
Goddard, W. A. III, 112, 120, 121, 125
Godfrey, T. S., 225
Gogos, C., 186
Goldanskii, V. I., 107
Golden, D. M., 125, 460
Goldman, A., 164, 175
Goldman, K., 400
Goldman, M., 48, 51
Goldmann, A., 175
Goldschmidt, C. R., 216
Goldstein, B., 243
Goldstein, H., 391
Gole, J. L., 32
Golger, A. L., 141
Golibersuch, D. C., 239
Gommer, R., 216
Gonzalez, M. A., 265, 334
Good, A., 446, 447, 467
Good, W. B., 245
Goodeve, C. F., 330
Goodman, F. O., 479, 480, 482, 483, 487, 489, 491
Goodwin, J. W., 237
Gopal, E. S. R., 431
Gordon, R. G., 303
Gordon, R. J., 193, 194
Goren, S. D., 62, 66
Gorokhov, Yu. A., 136, 147-51, 155
Gortemaker, F. H., 187, 193, 194
Goscinski, O., 16-18
Gosling, E. M., 302
Goswami, D. N., 251
Gotte, L., 251
Gottlieb, M., 197, 200
Goulden, D. D., 194
Gouterman, M., 106, 220, 504, 513
Govan, D. W., 176
Govindarajan, K., 431
Grad, H., 317
Grady, D. L., 194
Graessley, W. W., 186, 199
Graf, V., 54, 61-66, 71
Graham, C., 374, 384
Grant, D. M., 70
Grant, E. H., 250
Gratzel, M., 108
Gravesteyn, H., 290
Gray, C. G., 373-76, 381-84, 387, 395-99, 402-6
Gray, G. W., 47-50
Gray, H. B., 506
Green, A. E., 190
GREEN, J. C., 161-83; 180

Green, M. L. H., 180
Green, M. S., 197, 201, 411, 413, 415, 416
Greer, E., 107
Grellmann, K. H., 225
Greve, J., 249
Grice, R., 350
Griffin, J. P., 508, 517
Griffing, K. M., 16-18, 34
Griffith, A. L., 244
Griffiths, G. H. E., 512
Griffiths, R. B., 411, 413, 415, 418, 426, 435, 439, 440
Grimsrud, E. P., 466, 467
Grobman, W. D., 163, 167, 168, 172, 178-80
Gross, R. W. F., 133, 136
Grossweiner, L. I., 216
Gruler, H., 61, 65
GUBBINS, K. E., 373-410; 301, 373-76, 381-84, 387, 388, 395-99, 403-6
Guermant, C., 245
Guggenheim, E. A., 413
Guirao, C., 86
Gunning, H. E., 152
Gunton, J. D., 411, 416
Gupta, A. K., 250
Gurney, R. W., 105, 108
Gurr, M., 225
Gustafson, B., 243
Gustafson, T. K., 211
Guttman, A. J., 421

H

Haacks, D., 369
Haas, D. D., 239, 240
Haas, Y., 148
Haberland, H., 365
Haga, T., 252
Hagens, O., 142
Haile, J. M., 373, 374, 382, 383, 387, 395-99, 403, 405, 406
Halevi, E. A., 127
Hall, C. K., 414, 415, 431-34, 437
Hall, J. L., 41, 42
Hallett, F. R., 240
Hallsworth, R. S., 148
Halperin, B. I., 415
Halpern, B., 216
Halvorson, H. R., 245
Hamaguchi, H., 512, 513
Hamburger, H., 120
Hammond, D., 166
Hammond, G. S., 128
Hamnett, A., 161, 164

AUTHOR INDEX

Hamrin, K., 161
Han, C. C.-C., 238
Han, C. D., 186
Hand, G. L., 191, 198
Hankey, A., 415
Hanley, H. J. M., 384
Hansch, T. W., 349, 355
Hansen, D. R., 201
Hansen, J. P., 373
Hansen, J. W., 211, 221
Hansen, M. G., 193
Hansen, R. L., 197, 200
Hansma, P. K., 105
Hanss, M., 251
Happel, J., 302, 303
Hara, I., 194
Harada, I., 512
Harada, Y., 179
Harbus, F., 440
Harding, L., 120, 121, 124
Hardy, B., 249
Hariri, A., 269, 271, 273, 274, 279
Harland, P. W., 446
Harley, E. H., 244
Harmon, J., 52, 56, 62
Harp, G. D., 373, 379, 391, 398, 400
Harris, F. E., 16
Harris, J., 414
Harris, J. M., 247
Harrison, A. G., 457
Harrison, G. R., 501
Hart, D., 239
Harteck, P., 152
Hartford, S. L., 234, 239
Hartig, W., 140
Hartnett, J. P., 193
Hase, W. L., 341
Hassager, O., 186, 190, 191, 193, 196-201
Hasselmann, D., 128
Hassler, J. C., 269, 271
Hawley, M. C., 195
Hawley, S. A., 244
Hay, M. H., 86, 89, 90
Hay, P. J., 112, 120, 121, 125
Hayakawa, R., 250
Hayasi, Y., 179
Haydock, R., 166
Hayes, D. M., 331, 343-45
Hayes, E. F., 120
Hayes, R. G., 179, 180
Haynes, R. N., 446, 448, 468
Hayward, D. O., 482
Heaton, M. M., 17, 42
Heden, P. F., 161
Hedges, R. E. M., 269

Hedman, J., 161
Hefter, U., 351, 352, 368
Hehre, W. J., 35, 452-54, 457
Heidrich, F. E., 334
Heine, V., 166
Heinzinger, K., 373, 374
Helfand, E., 426
Helgerson, S. L., 247
Heller, D. F., 340, 341, 343-45
Heller, E. J., 340
Heller, H. C., 514
Heller, P., 411, 413, 414
Hemming, M., 244
Hemsworth, R. S., 446
Henderson, B. H., 504
Henderson, D., 303, 373
Henderson, G. A., 34
Henderson, R. L., 374, 387
Henderson, W. G., 446, 455
Henderson, W. H., 446, 455
Hendrix, J., 234
Hendfling, D., 116-18
Henglein, A., 108
Henkes, W., 142
Henneker, W. H., 17
Henry, B. R., 223, 339
Herbage, D., 248
Herbst, E., 41, 42
Hering, P., 351, 352, 368
Heritage, J. P., 225
Herm, R. R., 358, 359, 361
Herman, P. T., 304, 305
Herman, R., 35
Hermans, L. J. F., 491
Herron, J. T., 269
Herschbach, D. R., 357, 358, 360-62, 369
Hertel, I. V., 266, 279, 351, 352
Herz, S., 240
Herzberg, G., 27, 33, 143, 323-25, 331, 335, 337, 350
Herzfeld, K. F., 337
Hester, R. E., 501
Hewitt, R., 61, 65
Heyde, M. E., 507, 514
Hicks, A. N., 246
Higashitani, K., 192-94
Highsmith, S., 247
Hillborn, R. C., 502, 503
Hildreth, W., 107
Hill, C. T., 192
Hill, T. L., 419, 420, 435
Hillel, Z., 246
Hillier, I. H., 179, 180
Hills, B. P., 312
Hills, L. P., 466, 468

Himpsel, F. J., 170, 171
Hinch, E. J., 197, 198
Hinze, J., 39
Hirakawa, A. Y., 506, 510
Hiraoka, K., 448, 451, 452, 458-61, 466-68, 470-72
Hirata, Y., 225
Hirs, C. H. W., 233
Hirschfelder, J. O., 434
Hitchcock, W. J., 152
Hiyama, T., 107
Ho, P. P., 211, 212
Hochstrasser, R. M., 143, 148, 210, 222, 223, 225, 502-4, 509, 516
Hocker, L. O., 240
Hodgins, M. G., 251
Hodgins, O. C., 251
Hoel, D., 308
Hoffman, G. W., 215
Hoffmann, G., 218, 219
Hoffmann, R., 111, 112, 119, 120, 127
Hofmann, H., 266, 279, 351, 352
Hogan, P., 369
Hogenboom, D. L., 53, 61, 62, 69
Hogeveen, H., 463
Hogg, A. M., 446, 448, 468
Hoggard, P. E., 221
Hogrel, J. F., 249
Hohenberg, P. C., 415
Hoinkes, H., 479, 481, 483
Holbrook, K. A., 111, 112, 115, 124, 341
Holcenberg, J. S., 243
Holder, R., 128
Holdy, K. E., 265, 330, 334
Holneicher, G., 16-18, 22
Holstein, T., 106
Holt, R. A., 351, 352
Holten, D., 220
Holtz, D., 446, 455
Holtzberg, F., 178
Holzer, W., 502, 505, 506
Holzwarth, G., 252
Hong, H., 508
Hooper, G., 91
Hoover, W. G., 305, 373
Hopkins, H. P. Jr., 455
Horiguchi, H., 264, 266, 278
Horiuti, J., 105
Hornbeck, J. A., 106
Horne, J. M., 482, 483
Horne, R. A., 106
Horsley, J. A., 112, 120, 121, 124, 126, 127
Hotop, H., 15, 23, 32

AUTHOR INDEX 531

Houghton, A., 416
Houssier, C., 248, 249
Houston, D. E., 479, 481, 486
Houston, P. L., 148, 331, 333, 342-44
Howard, C. J., 466, 468, 471
Høye, J. S., 374, 385, 387
Hoyt, J. W., 193
Hoytink, G. J., 216
Hsu, C. S., 374, 385, 402, 403
Hsu, D. S. Y., 269-71, 279
Hsu, J. C., 194
Hu, A. S., 242
Hu, C.-M., 302, 307, 308
Huang, W.-N., 238
Hubbard, P., 303
Huber, H., 120
Hudson, B., 239
Hüfner, S., 168, 169, 173
Hui, M. H., 216
Huilgol, R. R., 186, 194
Huisgen, R., 120, 126, 127
Hunt, J. W., 216
Hunt, W. J., 112, 120, 121, 125
Hunten, D. M., 277
Hunter, D. L., 415
Huo, W. M., 23-25, 27
Huppert, D., 216
Hurkmans, A., 490
Hurle, I. R., 276
Husain, D., 261, 262, 267, 270, 273
Hush, N. S., 106
Hutley, M. C., 509
Hutton, J. F., 186
Hwang, L.-P., 52, 56
Hyde, J. W., 248
Hyer, R. C., 222
HYNES, J. T., 301-21; 301, 304, 307-9, 315

I

Ibach, H., 171
Ide, R., 216, 217
Iijima, M., 510
Ikegami, A., 248
Ilver, L., 166
Inagaki, F., 514
Ingham, K. C., 225
Innes, K. K., 148, 331, 508
Inokuchi, H., 179
Ippen, E. P., 209, 212, 219, 220, 222, 224-26
Isakov, V. A., 139, 154, 155
Isbister, D., 374, 385

Isenberg, I., 246
Isenor, N. R., 148
Ishii, T., 173, 175
Iskander, M. F., 289
Ito, M., 508, 510, 518
Itzkan, I., 144
Ivanov, L. N., 144
Iwamoto, K., 193
Izawa, Y., 147

J

Jackels, C. F., 328, 345
Jackson, G., 225
Jackson, W. M., 367, 369
Jacobsen, C. W., 508
Jacobson, J. D., 389, 390
Jacobson, K., 242, 246
Jacon, M., 514
Jacucci, G., 303, 373
Jaffe, H. H., 342, 344
Jaffe, R. L., 331, 343-45
Jakeman, E., 234
James, D. F., 192
Jamieson, A. M., 235
Janak, J. F., 163
Janes, G. S., 144
Janeschitz-Kriel, H., 187, 193
Janik, G. S., 466, 467
Janssone, V. M., 373, 374, 390, 395
Japar, S., 369
Jaraudias, J., 222
Jarry, J. P., 247
Javan, A., 264
Jean, Y., 112, 120, 121, 124, 126
Jeanmaire, D. L., 514, 516
Jedziniak, J. A., 235
Jeffrey, A., 244
Jeffrey, D. J., 186
Jenkin, J. G., 164, 168
Jenkins, J., 123, 128
Jennings, B. R., 249
Jennings, D. A., 269, 276
Jensen, D. E., 31
Jensen, E. J., 373
Jensen, P. W., 510
Jensen, R. J., 148, 149
Jesson, J. P., 115, 118
Jewsbury, P., 483
Joachim, P. J., 180
Job, V. A., 331
Joffe, A., 105
Johansson, G., 161
Johnson, B. B., 502, 508
Johnson, B. R., 362
Johnson, D. L., 53, 61, 62
Johnson, K. H., 176

Johnson, M. W. Jr., 190, 192
Johnson, S. A., 145
Johnson, S. G., 262
Johnson, W. C., 287
Johnston, D. E., 461, 462
Johnston, H. S., 335
Jolly, D., 234, 235, 238, 373, 374, 395
Jonah, C., 326, 327, 345
Jonas, J., 3-8, 301-3
Jones, D. R., 303
Jones, F. M. III, 455, 457, 470
Jones, G. II, 111, 127, 128
Jones, I. T. N., 329, 345
Jones, R. H., 492, 493
Jones, R. P., 216, 224
Jones, V. T., 331
Jones, W. M., 195
Jones, Y. M., 504
Joop, N., 225
Jordan, K. D., 16-18, 22, 34, 36, 39, 40, 42
Jørgensen, C. K., 169
Jørgensen, P., 16, 17, 22
Jortner, J., 325, 334, 336-38, 340, 341, 502, 503
Joschek, H. I., 216
Josefowicz, J., 240
Joseph, D. D., 186, 190, 194
Josephson, B. D., 105, 415
Joussot-Dubien, J., 216
Joyce, B. A., 488, 492
Judge, D. L., 335

K

Kaback, H. R., 246
Kac, M., 411
Kadanoff, L. P., 415, 416
Kahler, C., 329, 345
Kaifu, Y., 225
Kaiser, W., 223, 224, 227, 228
Kalantar, A. H., 508
Kalff, P. J., 277
Kalinin, V. P., 136, 144
Kam, Z., 234
Kamagawa, K., 508
Kanda, H., 250
Kane, E. O., 164
Kaplan, J. H., 239, 240
Kaplan, N. O., 242
Kapral, R., 304, 308, 310, 312-15
Kari, R. E., 17
Karl, G., 263-66, 334
Karl, R. R. Jr., 148
Karlov, N. V., 155

AUTHOR INDEX

Karlsson, S.-E., 161
Karplus, M., 507
Kasha, M., 223
Kassel, T., 265
Kast, W., 49
Kastelyn, P. W., 411
Kataev, D. I., 142
Kato, M., 235
Kato, R., 173
Katz, S., 330
Kaufman, B., 411
Kaufman, F., 278, 365
Kaufman, J. J., 17
Kaufmann, K., 153
Kaufmann, K. J., 227
Kaufmann, R., 240
Kauzmann, W., 233
Kawasaki, M., 326
Kay, K. G., 339, 340
Kaya, K., 508, 510
Kaye, R. L., 128
Kear, K., 268
Kearns, D. R., 179, 262, 268
KEBARLE, P., 445-76; 446-62, 466-72
Keck, J. C., 488
Kegeles, G., 245
Keller, K. H., 237
Keller, R. A., 153
Keller, W., 244
Kelley, P. L., 141, 211
Kellogg, R. E., 208
Kelly, J., 91
Kelly, M. J., 166
Kemeny, P. C., 173
Kenney, J., 16-18, 21, 34
Kenney-Wallace, G., 216
Kerr, K. A., 225
Kestner, N. R., 216
Kevan, L., 107
Keyes, P. H., 415
Keyes, T., 301, 305, 309, 310, 312
Khairutdinov, R. F., 107
Khetrepal, C. L., 48, 49
Khrustov, V. S., 34
Kiefer, W., 502, 507
Kiel, A., 512, 517
Kielich, S., 211
Kierstead, H. A., 420
Kihara, T., 107, 373, 374, 377, 385, 386
Kikuchi, K., 248, 249
Kildal, H., 141
Kilponen, R. G., 507
Kim, K., 304
Kim, K. S., 173
Kimble, H. J., 502, 503
King, D. S., 143, 148, 222

King, G. W., 330, 337
King, K. D., 125, 460
King, T. A., 238, 240
KINSEY, J. L., 349-72; 39, 336, 337, 350, 353, 355, 356, 366
Kirk, A. D., 221
Kirkwood, J. G., 196, 374, 384, 388
Kirsanov, B. P., 211
Kirsch, L. J., 364
Kirschner, L., 227
Kirschner, S., 120, 125, 127
Kistenmacher, H., 373, 374, 468, 469
Kivelson, D., 301-3, 308, 309
Kivelson, M. G., 308
Kiyono, S., 179
Klason, C., 187
Klein, M. L., 373, 374, 379, 391, 393
Klein, R., 490
Kleppner, D., 145, 353
Klotz, L. C., 265, 330, 334
Knaff, D. B., 107
Knapp, J. A., 173
Knox, A., 238, 240
Knox, K., 126
Knystautas, R., 86, 87, 89-91, 95
Kobayashi, S., 249
Kobayasi, T., 175
Koch, A. L., 243
Koch, B. J. D., 35
Koch, E. E., 170, 171
Kodera, K., 486
Koenig, S. H., 234, 235, 237, 246
Kogarko, S. M., 90, 96
Kogovsek, F., 53, 54, 61, 64
Koh, I.-Y., 193
Koide, A., 373, 377
Kokunov, Y. V., 179, 180
Kollman, P. A., 470
Kollman, V. H., 222
Kollmar, H. W., 112, 120
Kolomisky, Yu. R., 142, 149, 151
Kommandeur, J., 216
Kompaneets, A. S., 83
Konigstein, J. A., 512
Kono, S., 173, 175
Koppel, D. E., 234-37, 241, 242
Koppel, I., 449, 452-55, 457, 465
Koren, G., 149

Korn, C., 62, 66
Kotaka, T., 196
Kovac, J., 199
Kovalev, V. V., 179, 180
Kowalczyk, S. P., 164, 167, 168, 172
Kowalsky, A., 106
Kowert, B., 302, 308, 309
Kraemer, W. P., 468-70
Kramer, O., 193
Kraus, M., 325, 339
Krause, H. F., 262, 277, 278
Krause, L., 261
Krause, S., 248
Krauss, M., 17
Krikorian, O., 144
Krinsky, S., 440
Kristensen, W. D., 373
Krusic, P. J., 115, 118
Kubat, J., 187
Kubota, M., 216, 217
Kucherenko, Y. N., 164
Kuhn, H. G., 140
Kuhn, W., 136, 152
Kuizenga, D. J., 214
Kummler, R. H., 263
Kung, A. H., 148
Kuntz, I. D. Jr., 233
Kunwar, A. C., 48, 49
Kunz, A. B., 171
Kuo, A.-L., 62, 68, 69
Kupke, D. W., 251
Kuratani, K., 276
Kürkijasvi, J., 303, 304, 307
Kurokawa, M., 252
Kuščer, I., 316
Kushick, J., 304, 373, 374, 377, 395
Kusunoki, I., 486
Kuzmin, M. G., 107
Kwok, M. A., 136
Kyong, W. H., 95

L

Ladanyi, B. M., 396, 397
Laderman, A. J., 79
Lagos, M., 487
Laidler, K. J., 111, 112, 124, 262
Laiken, N., 251
Lajzerowicz, J., 440
Lakatos-Lindenberg, K., 303
Lalanne, J. R., 212
LaMantia, F. P., 192, 198, 200
Lambert, J. D., 262
Lamotte, M., 153

AUTHOR INDEX 533

Lampe, F. W., 215
Landanyi, B. M., 374, 385
Lang, J. C. Jr., 413
Langelaar, J., 225
Langer, D. W., 172
Langer, R. M., 325
Langhoff, C. A., 502, 503
Langley, K. H., 431
Lantelme, F., 306
Lapeyre, G. J., 173
Lapujoulade, J., 483, 484, 488, 489
Larsen, B., 374, 387
Larsen, D. M., 149
Larson, C. W., 125
Lasker, S. E., 248
Lathan, W. A., 35
Lau, Y. K., 450-55, 458, 459
Laubereau, A., 227, 228
Laufer, J. C., 400
Laun, H. M., 193, 194
Laurence, R. L., 195
Laurent, T. C., 242, 243
Lawetz, V., 339
Lawrence, G. M., 335
Layec-Raphalen, M. N., 193
Leaback, D. H., 244
Leach, B. S., 244
Leach, S. J., 233
Leal, L. G., 197, 198
Lebowitz, J. L., 306, 374, 384, 385
Lee, B. H. K., 89-91
Lee, C. H., 194, 211
Lee, E. K. C., 329, 331, 332, 341, 344, 345
LEE, J. H. S., 75-104; 81, 86-92, 95, 96
Lee, J. K., 373
Lee, L. C., 335
Lee, M., 415
Lee, P. H., 269
Lee, S. J., 326
Lee, S. P., 235
Lee, T. D., 418
Lee, T. J., 491
Lee, T.-Y., 235
Lee, W. I., 235, 237, 238
Lee, Y. T., 350, 351
Lefebvre, R., 336, 337, 339, 340
Legay, F., 139, 154
Le Guillou, J. C., 416
Lehr, R. E., 111
Leighton, S. B., 243
Lejay, Y., 483, 484, 488, 489
Lekey, R. C. G., 168
Lemberg, J., 373, 374, 379, 391, 393
LEMONT, S., 261-82; 274
Lengel, R. K., 365
Lentz, B. R., 246
Leone, S. R., 148, 269, 271, 273, 274, 279
Leonhardt, H., 216
Leonov, Yu. S., 154, 155
Lesclave, R., 216
Lesnow, J. A., 234
Lessing, H. E., 222
LETOKHOV, V. S., 133-59; 133, 136, 141, 142, 144-52, 155, 156
Leung, S. S., 415
Levelt Sengers, J. M. H., 411, 415, 431
Levenson, M. D., 141
Levesque, D., 303, 304, 307, 308, 373, 374, 391, 392, 395-96, 400, 401
Levi, A. C., 479, 480, 487, 489
Levich, V. G., 106
Levin, G., 216
Levin, L., 144
Levin, M., 144
Levin, V. A., 95
Levine, R. D., 265, 269, 271, 334, 336, 337, 362
Levison, S. A., 246
Levitt, D. G., 303, 312
Levstik, I., 53, 54, 61, 64
Levy, D. H., 329, 345, 353
Levy, R., 144
Ley, L., 164, 167, 168, 172
Lewis, A., 510, 517
Lewis, B., 79
Lewis, J. C., 304
Lewis, J. W. E., 373
Lewis, R. S., 332
Li, E. K., 176
Lian, M. S., 447, 466, 470
Liang, R., 514, 516
LIBBY, W. F., 105-10; 105, 107
Lie, G. C., 373, 374
Lie, T-J., 426
Liebes, L., 234, 239
Liebman, P. A., 242
Liebsch, A., 171
Liesegang, J., 168
Lifshitz, A., 95
Light, J. C., 334, 358
Lim, E. C., 341
Lim, T. K., 236, 240
Lin, M. C., 112, 269-72, 279
Lin, S. F., 180
Lin, S. L., 414
Lin, S.-M., 358, 359, 361
Lin, Y. W., 483, 487, 489
Lindau, I., 166
Lindberg, B., 161
Lindblom, G., 242
Linderberg, J., 17, 18
Lindgren, I., 161
Lindman, B., 242
Lindon, J. C., 69
Lineberger, W. C., 15, 16, 22, 23, 35, 36, 41, 42
Ling, J. H., 330
Linnett, J. W., 33, 34, 262, 263
Linschitz, H., 216
Liotta, C. L., 455
Lippert, E., 222
Lipscomb, W. N., 17
Litchfield, E. L., 86, 89, 90
Litman, B. J., 246
Little, D. J., 262
Littman, M. C., 145
Liu, D. D. S., 152
Liu, D. S., 331
Liu, K.-C., 127
Liu, K. S., 373
Liu, T. Y., 235
Liu, W. S., 482, 487
Liuti, G., 152
Liva, M. P., 481
Lloyd, D. R., 166
Lobo, P. F., 193
Lodge, A. S., 186, 187, 190-93, 196-98, 201
Loeb, H. W., 295, 298
Loehr, T. M., 501
Loewenstein, M. A., 236
Lofbourow, J. R., 501
Logan, R. M., 487, 488
Lokhman, V. N., 142, 149, 151, 155
Long, J. W., 462
Longinotti, L. D., 178
Longuet-Higgins, H. C., 512
Lopez-Delgado, R., 339
Lord, R. C., 501
Los, J., 490
Lossing, F. P., 453, 458, 467
Loucks, L. F., 111, 124
Lougnot, D., 222
Lowden, L. J., 374, 385
Lowdermilk, H., 209, 223, 224, 227
Lowe, J. P., 452
Loyalka, S., 317, 318
Lu, K. E., 495

AUTHOR INDEX

Lubensky, T. C., 53, 61, 62
Luckhurst, G., 69
Luken, W., 16, 36, 40
Lukin, L., 311
Lukman, B., 18
Lumley, J. L., 193
Lumry, R. W., 107
Luntz, A. C., 332, 333
Luprinski, J. M., 216
Luschka, M., 479, 481, 483
Lutz, H., 210, 219
Luzar, M., 53-56, 61, 66, 67, 71
Lyerla, J. R. Jr., 251
Lyman, J. L., 148, 149
Lynch, T. R., 123, 124, 127
Lyons, L. E., 108
Lytle, F. E., 247

M

Ma, S. K., 411, 416
Maas, E. T., 167, 168, 172, 179, 180
Maccoll, A., 463, 464
Machiels, A., 492, 493
MacInnes, D. A., 374, 385
Mack, M. E., 221, 222
MacRury, T. B., 373, 378
MacSporran, W. C., 194
Madden, P. A., 502, 503
Madden, W. G., 374, 387, 404
Madge, D., 218-20, 222, 241
Madix, R. J., 479, 492-95
Maeda, S., 518
Mahan, G. D., 164, 165
Mahoney, R. T., 327, 328, 345
Makarov, A. A., 146, 148
Makarov, G. N., 136, 147-51, 155
Maker, P. D., 211
Makino, S., 243
Makkes van der Deijl, G., 216
Malenkov, G. G., 373, 374
Malherbe, R., 123, 128
Mali, M., 53, 55, 56, 61, 66, 67, 71
Malley, M. M., 211, 212, 222, 224
Mal'tzev, A. A., 142
Mandel, L., 502, 503
Mandel, M., 250
Mandell, M. J., 373
Maniatis, T., 244
Manisse, N., 120
Manne, R., 161

Manson, N., 99
Manson, R., 480
Manuccia, Y. J., 139, 154
Marchal, E., 250
Marchand, A. P., 111
Marcus, R. A., 105, 112
Marenco, G., 491
Margitan, J. J., 365
Margolin, Z., 462
Marin, M., 482
Marion, C., 251
Markin, E. P., 139, 154, 155
Markov, V. V., 95
Markowski, L., 120
Marks, S. B., 62, 66
Markstein, G., 78
Marling, J. B., 147
Marrucci, G., 186, 192, 198, 200
Marshall, D. C., 116
Martins, A. F., 62
Martin, C. J., 241
Martin, H., 136, 152
Martin, R. L., 457
Martin, R. M., 262, 501, 503
Martinez, G., 172
Marus, R. J., 106
Masaski, S., 216, 217
Masel, R. I., 480, 482, 483
Mason, B. F., 477, 479, 481, 483, 487
Mason, D. W., 243
Mason, M. G., 175
Mason, S. G., 252
Mataga, N., 216, 217, 221, 225
Matheson, M. S., 216
Mathies, R., 502
Matsui, H., 88-90
Matsui, K., 225
Matsukawa, T., 173
Matsumoto, T., 250
Matsuzaki, S., 518
Mattera, L., 477, 479, 480, 483, 487
Mattheiss, L. F., 169
Matthews, D. F., 106
Matthews, R. C., 67, 69
Matthews, W. S., 462
Mattison, E. M., 353
Mauzerall, D. C., 106, 107
Maxson, V. T., 332, 333
May, C. A., 145
Mayer, G., 211
Mayer, S. W., 136
Mazenko, G. F., 312-14
Mazer, N. A., 237
Mazo, R. M., 196, 247

Mazur, P., 304, 310, 311, 315
Mazzocchi, P., 126
McAdam, J. D. G., 238, 240
McCabe, M., 242, 243
McCammon, J. A., 235
McClain, W. M., 511, 513
McClure, C. F., 317, 318
McClure, J. D., 483
McConnell, H. M., 106
McCool, M., 302
McCormick, J. J., 234, 239
McCreery, J. H., 491
McCullough, D. W., 267
McDaniel, E. V., 446
McDonald, G. G., 242
McDonald, H. J. H., 120
McDonald, I. R., 302, 303, 373, 374, 379, 387, 389, 391, 393, 404, 406
McDonald, J. M., 330, 337
McDonald, J. R., 330
McDonnell, M. E., 235
McElroy, J. D., 106, 107
McEwan, B. C., 434, 436
McFeely, F. R., 164, 167, 168, 172
McGlynn, S. P., 330, 337
McGrath, W. D., 267
McGreer, D. E., 118
McIver, R. T., 446, 448, 449, 452-55, 457, 458, 460, 465
McKean, D. C., 125
McKinley, J. D., 490
McLachlan, A. D., 168
McLaughlin, E., 302
McLaughlin, I., 339, 340
McLean, A. D., 16, 17, 48, 51
McLennan, J. A., 317
McMahon, T. B., 448, 458-62, 464, 466, 467, 471
McNeal, R. J., 263, 268
McQuigg, R. D., 331, 332
McTaque, J. P., 373
Mead, C. A., 105
Mead, R., 236
Meakin, P., 115, 118
Meerts, W. L., 367
Mehaffey, J. R., 306, 312-14
Meiboom, S., 61, 65
Meier, F., 178
Meijering, J. L., 439, 440
Meiron, D. I., 416
Meissner, J., 193, 194
Mejean, T., 513
Mele, A., 335
Meleva, A., 166
Mena, B., 195

AUTHOR INDEX 535

Mendelsohn, R., 513, 517
Mendelson, R. A., 246, 247
Mentall, J. E., 277
Meot-Ner, M., 446, 463, 464, 466, 468
Merkel, P. B., 262, 268
Merchant, V., 148
Mermin, N. D., 415, 426, 427, 429, 431
Merrill, R. P., 480, 482-91
Metiu, H., 307, 502, 503
Metropolis, N., 389
Meyer, H., 414, 420
Meyerhoff, G., 236
Meyers, J. A., 479, 480, 481
Mialocq, J. C., 222
Michaels, I. A., 310
Michejda, J. A., 42
Mickish, D. J., 171
Mies, F. M., 325, 339
Mikami, N., 508
Mikhailov, A. I., 107
Miller, C., 197
Miller, D. R., 482, 483, 489, 491
Miller, J. R., 106, 108
Miller, J. S., 458-60
Miller, R. G., 331, 332, 341, 344
Miller, R. L., 248
Miller, S. J., 246, 249
Miller, W. B., 362
Miller, W. H., 267, 271, 339, 480, 482, 483
Milnes, H. W., 35
Milvy, P., 248
Mims, C. A., 358, 359, 361
Minakata, A., 249
Mingardi, M., 502, 509, 514
Mirmira, S. K. V., 171
Mishimura, Y., 510
Mishin, V. I., 144, 145, 147, 148, 152
Mishra, A., 118, 119
Miskowski, V., 507, 508, 514, 515, 517
Misumi, S., 216, 217
Mitchell, A. C. G., 262
Mitchell, G. W., 247
Mitchell, R. C., 334
Mitchell, R. M., 244, 245
Mitrofanov, V. V., 98
Mitschele, C. J., 219
Mitsky, J., 455
Miyazawa, T., 514
Mizushina, T., 195
Mo, K. C., 374, 388, 406
Modak, A. T., 485, 486, 489
Modena, I., 303

Mohan, R., 240
Molin, Yu. N., 147
Monnerie, L., 247
Monson, P. R., 227
Moody, R., 166
Moore, C. B., 133, 136, 144, 146-48, 156, 331, 333, 341-44, 349
Moore, C. E., 350
Moos, H. W., 136
Morales, M. F., 247
Morawetz, H., 233
Moreau, J., 290
Morgan, J. M., 91
Morokuma, K., 331, 343-45
Morr, C. V., 236
Morris, J. M., 214
Morrison, P. F., 374, 385
Morse, M. D., 336, 337
Morse, R. I., 327, 328, 345
Mortensen, O. S., 512, 517
Morton, T. H., 457
Moruzzi, V. L., 163
Moser, C., 112, 120, 121, 124, 126
Mott, N. F., 105
Mountain, R. D., 301, 307, 314, 373
Mourou, G., 211, 212, 222, 224
Mukamel, D., 440
Mukamel, S., 336, 337, 502, 503
Mulac, W. A., 216
Mulholland, G. W., 415, 431
Müller, A., 179, 180, 221
Muller, B. H., 52, 56, 62
Muller, G., 250
Muller, R., 490
Munch, J. P., 240
Munson, B., 462
Münstedt, M., 194
Murai, N., 250
Murphy, W. F., 502, 505, 506
Murray, R. B., 176
Murrell, J. N., 124
Mustacich, R. V., 239
Myers, L. E., 489

N

Nafie, L. A., 502, 508, 511, 513, 517
Nagakura, I., 173
Nagy, A. F., 263, 268
Nahr, H., 479, 481, 483
Nakajima, H., 200

Nakanishi, K., 501, 510
Nakano, H., 414
Nakashima, N., 216, 221
Nangia, P. S., 127
Narang, H., 374, 387
Narten, A. H., 374, 383
Narvaez, C., 192
Nebenzahl, I., 144
Nee, T. W., 312
Neece, G. A., 414, 422, 423
Neet, K. E., 243
Nefedov, V. I., 179, 180
Neff, J. D., 47-50
Nelson, A. C., 222
Nelson, D. J., 238
Nelson, G. L., 128
Nemoshkalenko, V. V., 164
Nemoto, N., 251
Nesbitt, L. E., 466, 467
Nestler, F. H. M., 193
Nestor, J., 511, 513
Netzel, T. L., 216
Neumann, D., 17
Newman, J., 234, 236, 250
Newton, M. D., 35, 470
Nguyen, A. L., 251
Nichol, L. W., 245
Nicholas, W. L., 482, 486, 487, 489, 491, 494, 495
Nicholls, J. A., 95
Nicholson, D., 389
Nickel, B. G., 416
Nickerson, M. A., 53, 54, 61, 62, 68
Nicolson, A. M., 287, 290, 294-96
Nielsen, P., 180
Niemeijer, T., 416
Nihei, T., 246
Niki, A., 369
Nikitin, E. E., 263, 267
Nilsson, P. O., 166
Nishimura, M., 106, 107
Nishio, K., 194
Nishioka, N., 250
Nitzan, A., 219, 339-41, 502, 503
Noack, F., 54, 60-66, 71
Nocilla, S., 479
Noguchi, N., 147
Noll, W., 190
Noordermeer, J. W. M., 193
Norcross, D. W., 32
Norden, B., 252
Nordheim, L., 105
Nordio, P., 58, 61, 65, 69, 70
Nordling, C., 161
Noreland, E., 287, 288, 291, 295

AUTHOR INDEX

Norrish, R. G. W., 330
Norsdieck, A., 392, 393
Novakov, T., 179
Novros, J. S., 216
Noyes, R. M., 215, 216
Nozaki, Y., 243

O

Oancea, D., 461, 462
O'Brien, E. F., 373, 374
Odell, B. G., 127
O'Dell, J., 303, 315, 373
Ofran, M., 225
Ogan, K., 301
Ogryzlo, E. A., 268
O'Hare, P. A. G., 17
Ohno, K., 179
Ohrn, Y., 16-18, 22
Ohsawa, A., 176
Ohta, N., 508, 518
Okabe, H., 335
Okada, T., 216, 217
Okagawa, A., 252
Okamura, T., 225
O'Keefe, D. R., 482, 483
Okon, M., 149
O'Konski, C. T., 248
Oksuz, I., 16
Okuda, M., 329
Olah, G. A., 463, 464
Olander, D. R., 479, 485, 486, 488, 490, 492, 493
Oldman, R. J., 327, 328, 345
Oldroyd, J. G., 191, 197
O'Leary, T. J., 155
Oliver, C. J., 234
Omura, I., 250
O'Neal, H. E., 111, 112, 114-18, 123-25, 128
Onsager, L., 411
Oosterhoff, L. J., 216
Opella, S. J., 238
Opitz, A. C. L., 373
Oppenheim, A. K., 79, 80, 101
Oppenheim, I., 304, 308-10, 312, 315, 318
Oppenheim, U. P., 149
Oppenheimer, J. R., 105
Oppliger, M., 247
Oraevskii, A. N., 139, 154, 155
Orchard, A. F., 161, 164, 180
Orchin, M., 225
Ore, A., 209
O'Reilly, D. E., 53, 61, 62, 68

Orlandi, G., 339
Orlov, A. N., 155
Orner, G. C., 223
Orwoll, R. D., 56, 59, 61, 68, 71
Osaki, K., 193
Oseroff, A. R., 502
Osgood, R. M. Jr., 264
O'Shea, D. C., 508, 516
Osmers, H. R., 193
Osredkar, R., 53, 55, 56, 61, 66, 67, 71
Osuga, D. T., 235
Ottewill, R. H., 237
Ottinger, C., 32, 33
Ottolenghi, M., 216
Overbeek, J. T. G., 244
Overbosch, E. G., 490
Owen, C. S., 242
Owen, J., 512
Oxtoby, D., 307

P

Pack, J. L., 466, 467
Pack, R. T., 344
Packer, A., 502
Pagam, C. D., 317
Pagni, P. J., 485, 486, 489, 490
Paisner, J. A., 145
Palmer, R. L., 479, 482, 483, 494
Pangritz, D., 81, 83
Papahadjopoulos, D., 246
Parfitt, G. D., 106
Park, H. D., 195
Parker, F. R., 389
Parkes, J. H., 106, 107
Parkin, J. E., 339
Parks, R., 492
Parlee, N. A. D., 303
Parlett, J. L., 127
Parmenter, C. S., 339
Parmon, V. N., 107
Parola, A., 246
Parr, T. P., 358, 359
Parrish, D. D., 358
Parson, J. M., 351
Parsonage, N. G., 389
Parsons, G. H., 95
Partridge, R. B., 353, 354
Parvatiker, R. R., 434, 436
Pasternack, L., 351, 352, 356, 361, 369
Pastor, R. W., 517
Patashinskii, A. Z., 311, 415
Patey, G. N., 373, 374,

387, 389, 395, 403-5
Patterson, T. A., 15, 23, 32, 41, 42
Patumtevapibal, S., 227
Patureau, J. P., 101
Paul, E., 196
Pauly, H., 353
Pawel, D., 79
Payzant, J. D., 446, 447, 451, 466-68, 470
Pearlstein, A. J., 249
Pearson, J. R. A., 186
Pearson, P. K., 17, 41
Pechukas, P., 358, 373, 377
Pecora, R., 211, 234, 236, 238, 239, 247, 301-3, 308, 374, 383
Pedersen, J. B., 301
Pedersen, L. D., 118, 123, 124, 126, 127
Pekeris, C. L., 16
Pendry, J. B., 166
Penzkofer, A., 223-25
Percus, J. K., 374, 384, 385
Perico, A., 238
Perram, J. W., 374, 387, 404
Perrin, C. L., 513
Perrin, M. H., 513
Perry, M. H., 106
Pertel, R., 152
Pertoft, H., 243
Peters, B. A., 243
Peters, D., 339, 340
Petersen, A. B., 269, 271, 273, 274, 279
Petersen, D. C., 250
Peterson, E. M., 53, 61, 62, 68
Peterson, J. R., 16, 35, 36
Peterson, O. G., 144
Peticolas, W. L., 502, 508, 510, 511, 513
Petrie, C. J. S., 186
Petrov, R. P., 155
Petrov, Yu. N., 155
Peyerimhoff, S. D., 16-18, 22
Pezolet, M., 510, 511, 513
Pezzin, G., 251
Pfeiffer, G. V., 17, 42
Phelps, A. V., 466, 467
Phillies, G. D. J., 236, 237, 240, 241
Phillion, D. W., 214
Phillips, M. C., 298
Piaggio, P., 238
Pickup, B. T., 16-18
Picot, C., 240

AUTHOR INDEX 537

Pierce, D. T., 178
Pike, C. T., 144
Pike, E. R., 234
Piña, E., 306
Pincus, P., 53, 61, 62
Pine, A. S., 221
Pintar, M. M., 62, 63, 68
Pipano, A., 17
Pipkin, A. C., 191
Pirs, J., 53-56, 58, 61, 62, 65, 68-70
Pittman, C. U., 463, 464
Pitzer, R. M., 344
Placzek, G., 511, 512
Plock, R. J., 196
Podo, F., 242
Pokrovskii, V. L., 415
Polanyi, J. C., 262-66, 269, 271, 364, 365
Polanyi, M., 105
Pollack, M. A., 330
Pollack, S. K., 458
Pollak, R. A., 167, 168, 172, 178-80
Polnaszek, C. F., 57
Pomeau, Y., 301, 307, 309, 312
Pommelet, J. C., 120
Pong, W., 173
Poole, T., 244
Pope, M., 211
Popkie, H., 373, 374, 468, 469
Popkie, H. E., 17
Pople, J. A., 35, 211, 374, 381, 452-54, 457
Poreh, M., 193
Porter, G., 216, 222, 225
Porter, G. B., 221
Porter, N., 128
Portmann, A. J., 246
Porto, S. P. S., 501, 512, 517
Poshyunaite, N. P., 34
Post, M. F. M., 225
Potts, A. W., 164
Pound, G. M., 373
Powell, C. J., 166
Powell, H. T., 264
Powell, R. L., 193
Powles, J. G., 373, 374, 392, 395-97, 399-402
Pownall, H. J., 242
Prager, S., 196
Preston, B. N., 243
Preston, R. K., 267, 271, 339
Price, W. C., 164
Prins, R., 179, 180
Prins, W., 238, 240, 241

Pritchard, D. E., 353
Pritchard, W. G., 193
Pritt, A. T. Jr., 262
Prokhorov, A. M., 155
Protopapas, P., 303
Provencher, S. W., 246
Pruett, J. G., 356, 363, 364
Pulleyblank, D. E., 244
Puntambekar, P. N., 222
Puretskii, A. A., 136, 144, 145, 147-51, 155
Purvis, G., 16-18, 22
Pusey, P. N., 234, 235, 237
Pyun, C. W., 199

Q

Quentrec, B., 303, 308, 373, 374, 391-93, 395, 396, 400, 401
Quickenden, P. A., 287, 295, 298
Quigley, G. P., 278
Quinn, C. M., 166

R

Rabalais, J. W., 330, 337
Rabani, J., 216
Rabideau, S. W., 147
Rabinovitch, B. S., 112, 125, 127
Raczek, J., 236
Radom, L., 458, 462
Rahman, A., 305, 307, 373, 374, 379, 389, 391, 393
Raimondi, D. L., 16
Raj, T., 237
Ralston, A., 393
Ram, J., 374, 387
Ramachandra, R., 431
Ramamurthi, K., 87-89, 92
Ramme, G., 216
Ramsay, D. A., 324
Rangel, C., 195
Rankin, C., 358
Rapp, D., 265, 334
Rapp, M., 222
Rasaiah, J. C., 374, 387
Raveche, H. J., 373
Ray, C. J., 480, 482
Rayfield, G. W., 216
Redmon, L. T., 18
Reed, C. C., 237
Reed, W. J., 460
Rees, C. W., 111
Reggel, L., 225

Rehm, D., 208
Rehm, R. G., 154
Reichmann, M. E., 234
Reid, E. S., 222
Reilly, J., 148
Reimann, C. W., 458-60
Reinhardt, W. P., 17, 18
Rentzepis, P. M., 216, 219, 223, 224, 339, 340
Resibois, P., 301, 306, 307, 309, 313, 314
Reynolds, J. A., 243
Reynolds, W. L., 107
Ricard, D., 209, 223, 224, 227
Ricci, F. F., 514
Ricci, F. P., 303
Rice, O. K., 325
Rice, S. A., 223, 337, 339, 340, 341, 345
Rich, J. W., 154
Rich, M. A., 234, 239
Richardson, A. W., 330, 337
Richardson, J. H., 17, 41, 42, 460
Richardson, M. C., 148, 211
Richardson, N. V., 166
Richardson, S., 316
Riddle, M. J., 192
Ridge, D. P., 464
Rigatti, G., 58, 61, 65, 69, 70
Rigby, M., 302, 373, 374, 386, 404
Rigler, R., 247
Riley, S. J., 327, 328, 334, 345
Rimai, L., 234, 239, 240, 507, 514
Ringwelski, L., 222
Rink, J., 149
Ritter, J. J., 153, 154
Riveros-Moreno, V., 237
Rivlin, R. S., 190, 194, 197
Rizzo, G., 192, 198, 200
Robb, M. A., 17
Robbins, P. W., 246
Roberts, J., 243
Roberts, J. D., 127
Roberts, W. W., 251
Robertson, J. M., 225
Robieux, J., 136
Robinson, C. P., 149
Robinson, G. W., 214, 223, 227, 502, 503
Robinson, P. J., 111, 112, 115, 124, 341
Roche, M., 342, 344

AUTHOR INDEX

Rockley, M. G., 221
Rockwood, S. D., 142, 147, 149
Rodbard, D., 244
Roddie, A. G., 222
Roetti, C., 29
Rohmer, P., 365
Romanenko, V. I., 139, 154, 155
Rose, M. E., 353, 374, 381, 405
Rosen, B., 350
Rosen, N., 325
Rosenblatt, G. M., 491
Rosenbluth, A. W., 389
Rosenbluth, M. N., 389
Rosner, S. D., 351, 352
Ross, G. F., 290, 295, 296
Ross, I. G., 325, 337-39
Ross, J., 329, 345, 502, 503
Rost, K. A., 266, 279, 351, 352
Rost, K. J., 266, 279
Rouse, P. E. Jr., 199
Rousseau, D. L., 502, 503, 505, 506, 518
Roux, B., 248
Rowe, D. J., 18
Rowe, E. S., 248
Rowe, J. E., 165, 171, 172
Rowe, R. G., 487
Rowland, C., 112
Rowley, L. A., 389
Rowlinson, J. S., 373, 374, 385, 400, 413, 426
Roy, J. K., 210
Rozmarin, M., 53, 54, 61, 64
Rubenstein, I., 243
Rubin, R. J., 304
Rubsamen, R., 315
Ruddock, I. S., 222
Rudnicki, A. R. Jr., 488
Ruelle, D., 426
Ruff, I., 106
Rulis, A. M., 360
Rundle, H. W., 446
Rushbrooke, G. S., 415
Russo, A. L., 276
Rutar, V., 53, 54, 61, 62, 64
Ryabov, E. A., 136, 142, 143, 148, 149, 151
Ryckaert, J. P., 373, 379
Rye, R. R., 495
Rzepecka, M. A., 289

S

Saam, W. F., 415

Sabety-Dzvonik, M. J., 367
Sachdev, K., 123, 124, 126, 127
Sachs, L., 17
Sack, R., 197
Sackett, P. B., 264, 269, 271, 273, 279
Sadovskii, N. A., 107
Sadowski, C. M., 278
Saffman, P. G., 242
Safron, S. A., 358, 362
Sagawa, T., 173, 175
Saile, V., 170, 171
St. John, W. M. III, 120
Sakamoto, M., 250
Sakata, Y., 216, 217
Sakisaka, Y., 173
Sakurai, K., 331
Sala, K., 211
Saleh, B., 234
Salem, L., 112, 120, 121, 124, 126, 127
Sallouta, H., 415
Salmeen, T., 234, 239
Salmeron, M., 495
Saltsburg, H., 479
Samulon, H. A., 294
Samulski, E. T., 56, 59, 61
Sander, R. K., 327, 328, 345, 369
Sandler, S. I., 374, 388, 404
Sandman, D. J., 180
Sarkisov, G. N., 373, 374
Sato, S., 173
Sau, R., 485-87, 489
Sauer, K., 516
Saunders, V. R., 179, 180
Saupe, A., 48, 50
Savadatti, M. I., 216
Savage, C. M., 211
Saville, G., 373
Scandola, M., 251
Scarborough, J., 446, 466, 468, 470
Scesney, P. E., 421, 439
Schadee, A., 354
Schaefer, D. W., 234, 235, 237
Schaefer, H. F. III, 17, 41
Schäfer, F. P., 222, 349
Schafer, I. A., 235
Schafer, T. P., 351
Scharf, G., 317
Schawlow, A. L., 136, 155, 349
Schechter, N. M., 243
Schechtman, B. H., 180
Scheer, M. D., 262, 263,
490
Scheer, W., 120
Scheller, K., 95
Scheps, R., 340, 341, 345
Scheraga, H. A., 200
Scherr, V., 330, 337
Schiff, H. I., 278, 446, 451, 452
Schindler, H., 241
Schlag, E. W., 112, 127, 223
Schlessinger, J., 242
Schleyer, P. v. R., 452-54
Schlossberg, H. R., 141
Schlosser, D., 327, 328, 345
Schlüter, M., 172
Schmeltekopf, A. L., 447, 448, 466, 468, 471
Schmidt, D., 235
Schmidt, K. S., 365
Schmidt, P. P., 106
Schmidt, R. L., 238
Schmidt, W., 356, 359-61
Schmitz, K. S., 235, 238
Schneider, B., 17
Schneider, G., 247
Schneider, G. M., 411, 413
Schneider, S., 225
Schnepp, O., 514, 516
Schofield, P., 301, 304, 306, 393, 415
Schonert, H., 245
Schouten, J. A., 434
Schowalter, W. R., 186, 192, 194, 195, 197
Schrader, U., 249
Schrag, J. L., 193
Schreiber, J. L., 364
Schug, R., 127
Schuldiner, S., 246
Schultz, A., 350, 356-61, 368
Schulz-Hennig, J., 221
Schumaker, V. N., 243
Schurr, J. M., 235, 237, 238
Schutte, A., 491
Schwartz, M., 70
Schwartz, R. N., 337
Schwarz, G., 249
Schwarz, H. A., 216
Schwarz, J. A., 479, 485, 486, 488, 492-94
Schwarz, W. H., 190, 193, 194
Schwentner, N., 170, 171
Scoles, G., 479, 491
Scott, B. A., 167, 168, 172, 179, 180

AUTHOR INDEX 539

Scott, G. W., 210, 219
Searles, S. K., 446, 447
Sears, A. B., 111, 124
Sears, V. F., 374, 383
Segal, G. A., 127, 339
Segalman, D. J., 190, 192
Segre, U., 58, 61, 65, 69, 70
Sehrag, J. L., 251
Seidel, J. C., 248
Seilmeier, A., 227, 228
Seki, K., 179
Seliger, J., 53, 55, 56, 61, 66, 67, 71
Sellen, D. B., 235
Sellier, P., 216
Selser, J. C., 236
Semchishen, V. A., 148, 152
Semeluk, G. P., 453, 458, 467
Semenza, G., 247
Semonov, N. N., 90
Sen, R. K., 106
Sen-Sharma, D. K., 466, 467
Serase, N. G., 400
Sergushin, N. P., 179, 180
Sethuraman, V., 331
Setser, D. W., 127
Seubold, F. H. Jr., 113
Shafer, R. H., 251, 252
Shank, C. V., 209, 212, 219, 220, 222, 224-26
Shannon, C., 294
Shapiro, M., 334
Shapiro, S. L., 210-12, 220, 222, 227
Sharts, C. M., 116-18, 127
Shaw, B. R., 235, 238
Shaw, M. T., 193
Shcerbukhin, V. V., 245
Shchelkin, K. I., 83
Shelnutt, J. A., 508, 514, 516
Sheludchenko, L. M., 164
Shen, M., 201
Shen, Y. R., 212, 503
Shepherd, F. R., 176
Shepherd, I. W., 234
Sheridan, M. E., 107
Shevchik, N. J., 164, 172, 175
Shill, J. P., 243
Shima, T., 171
Shimanouchi, T., 512, 513
Shimizu, F., 211
Shimoda, K., 141, 349
Shinitzky, M., 246
Shirley, D. A., 162, 164, 167, 168, 172, 457

Shirom, M., 216
Shishaev, A. V., 141
Shizuka, H., 225
Shore, H. B., 234
Shore, J. E., 250
Shortridge, R. G., 269, 270, 272, 279
Shriver, D. F., 502
Shukula, K. P., 374, 387
Shure, M., 244
Sichel, M., 91
Siebrand, W., 223, 339, 502, 509, 514
Siegbahn, K., 161, 166
Siegel, A., 356, 359-61
Siegel, S., 209
Siegman, A. E., 214
Siekhaus, W. J., 485, 486, 488, 492, 493, 495
Sigli, D., 195
Silbey, R., 39, 304
Silver, J. A., 355, 356, 366, 369
Silverman, J., 106
Silvers, J. H., 458
Simonetta, M., 124
SIMONS, J., 15-45; 15-18, 20-24, 32, 34, 39, 42
Simons, J. P., 265, 330, 334
Simons, J. W., 112
Simpson, J., 139, 140
Simpson, R. T., 243
Simpson, W. T., 513, 514
Sinanoglu, O., 16
Singer, J. V. L., 373, 374, 395-97, 400, 401
Singer, K., 373, 374, 395-97, 400, 401
Singh, S., 512, 517
Singh, Y., 374, 387
Singwi, K. S., 304
Sinha, M. P., 350, 354, 368, 369
Sirovich, L., 317
Siska, P. E., 351
Siu, A. K. Q., 120
Sivardiere, J., 440
Sizun, M., 334
Skibowski, M., 170
Skinner, G. B., 95
Skinner, H. A., 125
Sköld, K., 306
Skubnevskaya, G. I., 147
Slanger, T. G., 263, 267, 268
Slawsky, Z. I., 337
Sloan, J. J., 364, 365
Sloane, C. S., 341
Small, E. W., 246, 510

Small, H., 244
Small, L. E., 461, 462
Small, R. T., 458
Smalley, R. E., 353
Smerdon, M. J., 246
Smigel, M. D., 248
Smith, B. A., 53, 61, 62, 239
Smith, G. K., 267, 271
Smith, G. P., 356, 357, 359, 361, 362
Smith, I. W. M., 264-66
Smith, J. A., 173
Smith, J. N. Jr., 479, 482-84
Smith, L. C., 242
Smith, N. V., 161-63, 165, 168, 171, 176-78
Smith, S. R., 457
Smith, W. D., 15-18, 20, 22-24, 32, 34
Smith, W. R., 404
Smyth, K. C., 41, 42
Snavely, B. B., 140, 144
Sobel'man, I. I., 140
Soep, B., 369
Solana, J., 482
Solarz, R., 329, 345
Solarz, R. W., 145
Solly, R. K., 460
Solodovnikov, S. P., 106
Solomon, J., 326, 327, 345
Solomon, J. J., 446, 447, 463, 464, 466
Solomon, M., 106
Soloukhin, R. I., 91
SOMORJAI, G. A., 477-99; 479, 482, 486, 495
Sone, Y., 317
Sorensen, J. P., 196
Sorokin, N. I., 147
Soutar, A. K., 242
South, G. P., 250
Sovers, O., 262, 263
Spadacini, R., 479, 480, 487, 489
Spangler, C. W., 111
Spanner, K., 227
Spears, K. G., 223
Speiser, S., 216
Spencer, D. J., 136
Spencer, R. D., 246
Spicer, W. E., 166, 175, 179, 180
SPIRO, T. G., 501-21; 501, 502, 507, 508, 511-14, 516, 517
Srivastava, R. D., 31
Staley, R. H., 449, 452-55, 457, 464, 465, 468, 470

Stanley, H. E., 411, 415, 440
Starr, W. L., 277
Starvanov, V. S., 211
Steele, W. A., 301, 373, 374, 376, 383, 385, 396, 397, 400, 483
Stegeman, G. I. A., 211
Stehrenberger, J., 196
Stein, G., 216
STEIN, P., 501-21; 502, 508, 512, 513, 517
Steinberg, I. Z., 209
Steiner, B., 42, 460
Steiner, G., 127
Steiner, R., 240
Steinfeld, J. I., 261
Steinhardt, J., 248
Steinmann, W., 170
Stell, G., 374, 385, 387, 415
Stepanov, N. F., 34
Stephen, M. J., 47, 48, 50, 237, 240
Stephenson, J. C., 264
Stephenson, L. M., 17, 41, 42, 127, 128, 460
Stern, R. A., 79
Stern, R. C., 140
Stevens, C. G., 331
Stevens, I. D. R., 127
Stevens, R. M., 112, 120, 121, 124, 126
Stevenson, C. D., 273
Stevenson, D. P., 434
Stewart, R. F., 457
Stewart, W. E., 196
Stickney, R. E., 482, 486, 491
Stidham, H. D., 199
Stillinger, F. H., 373, 374, 379, 391-93, 415, 426
Stillinger, F. H. Jr., 426
Stockmayer, W. H., 197
Stogryn, A. P., 373, 376
Stogryn, D. E., 373, 376
Stohrer, M., 54, 60-66, 71
Stoicheff, B. P., 211
Stolarski, R. S., 263, 268
Stoll, A. G., 482-84
Storr, R. C., 111
Strain, R. H., 262, 270
Straley, J. P., 47, 48, 50, 426
Strausz, O. P., 152
Streets, D. G., 164
STREETT, W. B., 373-410; 373, 374, 381, 390, 395-99, 401-3, 405
Strehlow, R. A., 94, 98

Strekas, T. C., 502, 508 513, 516, 517
Stretton, J. L., 337, 338
Struve, W. S., 216
Stryer, L., 502
Stuart, B., 506
Stubbs, G. W., 246
Stuckly, S. S., 289
Stufkens, D. J., 510
Sturges, L., 190
Sturm, J., 249
Subbarao, R. B., 489
Subramanian, G., 303, 312, 315
Subramanyan, S. V., 431
Suchard, S. N., 350
Suess, H., 265
Sugai, S., 250
Suggett, A., 283, 287, 295, 297, 298
Summerhays, K. D., 455, 458
Sunder, S., 513, 517
Sung, C. C., 53, 61, 62
Sung, S., 374, 388, 403, 404
Suppan, P., 225
Sushchinskii, M. M., 118
Susu, A. A., 492
Sutcliffe, L. H., 48, 49
Sutton, J., 222
Suzuka, I., 508
Suzuki, I. H., 277, 278
Swaminathan, S., 127
Sweet, J. R., 373, 376, 396, 397
Swift, P., 167
Swinney, H. L., 250
Sworakowski, J., 179
Sykes, J., 306
Syozi, I., 414
Szu, H.-L., 316
Szwarc, M., 216

T

Taagepera, M., 446, 449, 452-55, 457, 465
Tadmor, Z., 186, 198
Taft, R. W., 446, 449, 452-58, 464, 465
Taguchi, R. T., 264, 265
Takahashi, Y., 179
Tal, D., 149
Tal'rose, V. L., 462
Tamburin, H. J., 126
Tan, W. K., 480, 482, 489
Tanaka, I., 225
Tanaka, T., 235, 240

Tanford, C., 233, 243
Tang, D., 244
Tang, I. N., 447, 466
Tang, J., 502, 504, 507, 509
Tang, K. Y., 332
Tanner, R. I., 191, 192, 195, 196
Tardy, D. C., 125
Tarr, C. E., 53, 54, 61, 62, 68
Tashiro, H., 221, 222
Tasker, P. W., 265, 330, 334
Tasumi, M., 514
Tatarczyk, T., 369
Tatarek, R., 482
Tatum, J. B., 354
Taylor, A., 373, 374, 395-97, 400, 401
Taylor, G. I., 91, 93
Taylor, H. S., 16, 17
Taysum, D. H., 106
Tejeda, J., 164, 172, 175
Telle, H., 140
Teller, A. H., 389
Teller, D. C., 243
Teller, E., 389
Tendulkar, D. V., 482, 488
te Nijenhuis, K., 193
Teramoto, A., 250
Terenin, A. N., 208
Terhune, R. W., 211, 369
Ter Meulen, J. J., 367
Teschke, O., 220, 226
Theodosopulu, M., 301, 306
Thomas, D. D., 248
Thomas, H. T., 216, 217
Thomas, J. M., 167, 176-78
Thomas, T. D., 457
Thompson, E. D., 481, 482
Thompson, M. R., 248
Thompson, S. M., 402
Thompson, T. E., 246
Thornber, J. P., 106
Thrush, B. A., 268
Thulstrup, E. W., 17
Thulstrup, P. W., 17
Thurston, G. B., 249, 251
Tiddy, G. J. T., 48
Tiemann, R., 296
Tiffany, W. B., 136
Tildesley, D. J., 373, 374, 381, 395-99, 401-3
Timasheff, S. N., 233, 245
Tinoco, I. Jr., 233
Tirganov, E. V., 211
Titomanlio, G., 192, 198, 200

AUTHOR INDEX

Tjerneld, F., 252
Tjon, J. A., 304
Tobolsky, A. V., 197, 201
Toennies, J. P., 350, 353, 479
Tomchuk E., 62-66, 68, 69
Tomita, Y., 186
Tommasini, F., 491
Tommei, G. E., 479, 480, 487, 489
Tomoda, S., 486
Tompkins, D. C., 128
Toong, T. Y., 101
Topchian, M. E., 98
Topp, M. R., 223, 224
Torchia, D. A., 251
Torrance, J. B., 178
Torrey, H., 52, 56, 62
Torrie, G. M., 389
Totsuka, T., 248
Townsend, P., 193
Tracey, A. S., 48, 49
Traeger, J. C., 463, 464
Trainor, D. W., 262, 270
Trajmar, S., 344
Trappeniers, N. J., 434
Traum, M. M., 165, 168, 169, 171, 176-78
Treanor, C. E., 154
Tredwell, C. J., 222
Tregear, R. T., 246
Troe, J., 328, 345
Trong Anh, Nguyen, 111
Trotter, J., 225
Truesdell, D., 192
Trumble, W. R., 510
Tsang, W., 453, 463, 464
Tscharnuter, W., 235
Tsirul, Z. Y., 34
Tsuboi, M., 506, 510
Tsuchida, A., 480-82
Tsuchiya, S., 264, 266, 276-78
Tsui, F., 16, 17
Tuccio, S. A., 144
Tully, F. P., 351
Tully, J. C., 267, 268, 339, 358
Tumanov, O. A., 148
Tung, T. T., 195
Turelli, A., 491
Turner, D. W., 161, 180
Turner, J. E., 35
Turq, P., 306
Turtle, P. C., 508, 513
Turtle, R. R., 165
Tuxworth, R. W., 284
Tward, E., 303
Twu, C. H., 374, 384, **387**, 405, 406

U

Udagawa, Y., 508
Ueno, N., 179
Uhlenhopp, E. L., 251
Ukleja, P., 53-56, 58, 61, 62, 65, 68-70
Ullermayer, L. S., 482
Ullman, A., 492
Ullman, A. Z., 492, 493
Umazume, Y., 248
Urch, D. S., 179
Urmston, J. W., 152
Urtiew, P. A., 79, 80
Uselman, W. M., 329, 345
Usui, H., 195
Utting, B. D., 373, 375
Uy, O. M., 31
Uyehara, T., 128
Uzgiris, E. E., 235, 239, 240

V

Valeur, B., 247
Valleau, J. P., 373, 374, 387, 389, 395, 403-5
van Beijeren, H., 306, 317, 318
van der Deijl, G. M., 216
van der Drift, W. P. J. T., 244
Vanderkooi, J. M., 242
van der Touw, F., 250
Van der Waals, J. D., 411, 413
van de Sande, H., 244
Van Duyne, R. P., 514, 516
van Gemert, M. J. C., 284, 288-90, 294, 298
Van Itallie, F. J., 262
Van Itterbeck, A., 400
van Kampen, N. G., 315, 318
Van Kranendonk, J., 373, 375
Van Labeke, D., 514
van Leeuwen, J. M. J., 411, 416
Van Loef, J. J., 302
van Mierlo, G. W. M., 367
Vanta, E., 95
Van Tiggelen, P. J., 79
Van Voorst, J. D., 216
Van Voorst, J. D. W., 225
Varsanyi, F., 512, 517
Vasatko, H., 79
Vasilenko, L. S., 141
Veduta, A. P., 211
Velasco, R., 32, 33
Verbeke, O., 400
Verlet, L., 303, 304, 307, 373, 384, 389, 393, 404
Vestner, H., 315
Vicentini-Missoni, M., 415
Viellard-Baron, J., 373, 374, 390, 395, 404
Vilesov, F. I., 179
Vilfan, M., 53-56, 61, 62, 64-67, 71
Vinje, M. G., 118
Vinograd, J., 244
Visintainer, J. J., 53, 61-66, 68, 69
Voevodskii, V. V., 106
Vogel, P. C., 373, 374
Voitsekhovskii, B. V., 98
von der Linde, D., 227
von Niessin, W., 16-18, 22
Vosberg, P., 244
Vu, B. T., 251

W

Wachman, H. Y., 479, 488
Wachs, I. E., 495
Wada, A., 238
Wada, Y., 200, 250
WADE, C. G., 47-73; 53, 56, 59, 61, 62, 67-69, 71
Wade, L. E., 120
Wagner, H. G., 79, 81, 83
Wagner, P. J., 127
Wahl, A. C., 16, 17, 34
Wahl, P., 247
Wainwright, T. E., 301-3, 305, 309, 389
Waldmann, L., 315
Wales, J. L. S., 187
Walker, D. C., 216
Walker, J. C. G., 263, 268
Wallace, B. Jr., 414, 420
Wallace, D. J., 416
Waller, I., 306
Wallis, R. F., 35
Walmsley, M., 35
Walsh, R., 111, 112, 116, 124
Walter, T. A., 31
Walters, K., 186, 190, 191, 193, 194
Walters, M. R., 482
Walters, W. D., 127, 128
Walther, H., 140, 349
Walton, A. G., 233
Waltz, W. L., 216
Wang, C. C., 369
Wang, C. H., 70, 303
Wang, F. W., 201

AUTHOR INDEX

Wang, J. C., 244
Wang, S. S., 373, 374, 384, 395, 396, 402-5
Ward, I. M., 512
Ware, B. R., 239, 240
Ware, W. R., 216, 225
Warner, H. R. Jr., 196
Warnock, T. T., 362
Warshel, A., 507, 513, 514, 516
Wassmuth, H. W., 490
Watanabe, H., 176
Watanabe, M., 173
Watanabe, N., 248
Waterhouse, W. M., 193
Waters, N. D., 190
Watson, P. J. S., 265, 334
Watt, W. S., 276
Watts, R. O., 373, 374, 377, 384, 390
Waugh, J. S., 48, 49
Weare, J. H., 482, 486, 487, 489, 491, 494, 495
Webb, H. M., 446, 448, 449, 455, 457
Webb, W. W., 241, 242
Weber, G., 208, 246, 247
Weber, J., 312
Weeks, J. D., 373, 393
Weger, M., 53, 61, 62
Wegner, F., 416
Weijland, A., 305
Weil, R., 149, 246
Weill, G., 249
Weiner, B., 235
Weiner, J., 431
Weinberg, M., 304, 308, 310, 315
Weinberg, W. H., 479, 482, 483, 485, 488, 490, 491
Weinstein, N. D., 358
Weinstock, B., 369
Weis, J. J., 373, 374, 395, 400, 401, 404
Weiss, A. W., 16
Weiss, G. H., 244
Weiss, J., 106
Weisshaar, J. C., 148
Weissman, M., 241
Weissman, S. I., 106
Weitz, E., 261, 349
Welch, J., 234
Weller, A., 216, 217, 221, 225
Wendling, L. A., 118
Wennerstrom, H., 242, 502, 503
Werme, L. O., 161
Werner, K. G., 508, 517
Wertheim, G. K., 168, 169,
173, 176, 178
Wertheim, M. S., 374, 385
Wessel, J. E., 223
West, L. A., 486, 495
Westhead, E. W., 235
Westmore, J. B., 460
Wetmur, J. G., 246, 249
Wharton, L., 353
WHEELER, J. C., 411-43; 414, 415, 422, 423, 425, 426, 431, 434, 437, 439
Whipple, B. A., 194
White, D. R., 94
White, L. R., 374, 387, 404
White, R. E., 482
Whittenburg, S. L., 303
Whittingham, T. H., 292
Wiberg, K. B., 112, 125, 127
Widmer, F., 242
Widom, B., 411, 413-16, 420, 422, 423, 426
Wieder, G. M., 112
Wiener, R. S., 239
Wiesenfeld, J. R., 262, 270
Wieting, R. D., 464
Wilburn, B. E., 126
Wilemski, G., 197
Wilf, H., 393
Wilkes, G. L., 251
Wilkinson, F., 208
Wilkinson, R. S., 249, 251
Willcott, M. R. III, 111, 122, 124, 127
Williams, A. R., 163
Williams, B. R., 477, 479, 481, 483, 487
Williams, G., 284, 288
Williams, M. C., 186, 194, 199-201
Williams, M. M. R., 316
Williams, P. F., 502, 503, 505, 506
Williams, P. M., 176
Williams, R. H., 176
Willis, M. R., 106, 111
Wilsch, H., 479, 481, 483
Wilson, A. D., 265
Wilson, A. H., 105
Wilson, K. G., 411, 416
Wilson, K. R., 265, 327-30, 334, 345
Windsor, M. W., 218-22
Winter, H. H., 194
Winzoi, D. J., 245
Wirth, P., 225
Wiszniewska, M., 63
Wittenberg, J. B., 237
Witting, C., 264, 269, 271, 273, 274, 279
Wodarczyk, F. J., 269, 271, 273, 279
Woessner, D., 63
Wolf, J. F., 449, 452-55, 457, 458, 465
Wolfel, W., 54, 60, 61, 71
Wolff, C., 193
Wolff, R. K., 216
Wolfrum, J., 153
Wolga, G. J., 278
Wolk, G. L., 262
Wolken, G., 481, 482
Wolken, G. Jr., 482, 483, 487, 489, 491
Wolynes, P. G., 318
Wong, G. K. L., 212
Wong, K. C. K., 118
Wong, Y. C., 351
Wonka, U., 479, 481, 483
Wood, D. L., 501
Wood, L., 139, 140
Wood, M. H., 179, 180
Wood, P. M., 265
Wood, W. W., 301, 373, 389, 390, 393
Woodruff, W. H., 502, 508, 517
Woodward, R. B., 111
Woolf, L. A., 302
Woolf, M. A., 216
Worden, E. F., 145
Workman, H., 374
Wortis, M., 440
Wright, A. K., 246, 248
Wright, J. S., 112, 120, 121, 124, 126, 127
Wu, C. H., 369
Wu, C. W., 246
Wu, F. Y. H., 246
Wu, Y., 196, 199
Wun, K. L., 238, 240, 241

Y

Yajima, T., 221, 222
Yalabik, M., 416
Yamakawa, H., 199, 233
Yamamoto, H., 176
Yamamoto, M., 186, 191
Yamamoto, S., 488
Yamdagni, R., 446, 449-55, 457, 458, 460, 461, 466-71
Yang, C. N., 411, 418
Yang, J. H., 466, 467
Yang, K., 269
Yang, N. C., 127
Yang, S. C., 326, 327, 345
Yarbrough, L. R., 246
Yardley, J. T., 261
Yaris, R., 17

AUTHOR INDEX 543

Yarlagadda, B. S., 17
Yarovoi, S. S., 34
Yatsiv, S., 262, 270
Yeh, A., 235
Yeh, Y., 235, 236
Yeung, E. S., 136, 147, 331, 341, 344, 345
Yip, S., 313, 314
Yogev, A., 149
Yoo, S. S., 193
Yoshimine, M., 16, 17, 36
Yoshioka, K., 248, 249
Yotis, W. W., 244
Young, A. C., 241
Young, G. M., 295, 298
Youngren, G. K., 302, 303
Yu, H., 197, 238
Yu, N. T., 508, 514, 516
Yu, T.-J., 510
Yu, W., 211, 212

Yuen, M. J., 262, 270
Yum, S. I., 237

Z

Zabramski, J. M., 126
Zagar, V., 53, 54, 61, 64
Zagrubskii, A. A., 179
Zahr, G. E., 267, 271, 339
Zamaraev, K. I., 107
Zandstra, P. J., 216
Zare, R. N., 32, 33, 152, 324, 349, 350, 353-64, 368, 369, 492
Zel'dovich, I., 90
Zel'dovich, Ya. B., 83
Zelikoff, M., 152
Zemansky, M. W., 262
Zemke, W. T., 16, 17, 34
Zener, C., 105

Ziman, J. M., 418
Zimm, B. H., 199, 201, 243, 251
Zimmerman, A. H., 460
Zimmerman, H., 225
Zimmerman, J. H., 460
Zimmerman, J. K., 245
Zimmerman, M. L., 145
Zinn-Justin, J., 416
Zirnitis, U., 118
Zolla, A., 446, 466, 468, 470
Zollweg, J. A., 415, 431
Zughul, M. B., 148
Zumer, S., 56, 62
Zupancic, I., 53, 54, 61, 64
Zwanzig, R., 250, 302, 305, 307, 308, 312
Zwanzig, R. W., 309
Zwolinski, B. J., 106
Zwolle, S., 250

SUBJECT INDEX

A

Acetaldehyde
 photodissociation of, 327
Acetone
 photodissociation of, 327
Acetonitrile
 charge distribution in, 469
Acetylene-oxygen mixtures
 ignition of, 82, 87, 89, 94
Acids, organic
 proton affinities of, 454
Acids, organic, aliphatic
 gas phase acidities of, 461
 halo substituents of, 461
Actin, F-
 complex with meromyosin, 241
Agarose
 macroscopic gel theories and, 241
Albumin, bovine serum
 diffusion coefficient of, 237
Alcohols
 gas phase acidities of, 461
 picosecond studies of, 213
 protein affinities of, 454
Alcohols, aliphatic
 dielectric relaxation of, 297, 298
Aldehydes
 protein affinities of, 454
Alkali halides
 atom scattering and, 487
 structure of, 172, 173
 surfaces of
  atom scattering and, 387-490
  crystallographic structure of, 480
  scattering from, 480-82
Alkali hydride
 rotational-state distributions of, 368
Aluminum
 oxidation of
  electron tunneling and, 105
Aluminum oxide
 formation of
  excitation spectrum of, 357, 358
 laser-induced fluorescence, 356, 357, 359
Amines
 proton affinities of, 457
Amines, alkyl
 basicities of, 455
Ammonia
 central force method applied to, 391
 compounds with proton affinities higher than that of basicities of, 455-58
 nitrogen isotope separation in, 147
 photodissociation of, 324
 proton affinities between those of H_2O and, 452-55
 simulation of structure of, 373
 central force models and, 379
Aniline
 ring-protonation of, 458
Anions
 metastable organic molecular, 42
 neutral parents with large dipole moments, 42
Anisol
 protonation of, 457
Anthracene
 absorption spectra of, 221
 picosecond-laser studies of, 217
 Raman spectrum of, 508
Argon beam
 scattered from polycrystalline Pt
 mean energy of, 488
Argon, solid
 photoelectron energy of, 170
Aryl halides
 photodissociation of, 326, 327
Aurora, Type B red
 excitation of by nitrogen, 277
Azaindole, -7, dimer
 tautomer production in, 225

Azulene
 excitation profile peaks of, 514
 lowest excited singlet state of, 225

B

Bacteriophage, T4D
 attachment of, 239
 electrophoresis of, 244
Barbituric acid
 gas phase acidity of, 462
Barium
 isotopes of
  separation of, 140
 reaction with chlorides
  laser-induced fluorescence, 361, 362
Barium atoms
 ionization potentials of, 360, 361
 reactions of
  laser-induced fluorescence and, 356, 357
Barium fluoride
 formation from B + HF
  laser-induced fluorescence and, 363, 364
Barium halides
 detection and reactions of
  laser-induced fluorescence and, 359, 360
Barium iodide
 excitation spectrum of, 360
Barium oxide
 formation of
  "conventional" molecular beam studies of, 358
  internal excitation of BaO, 358, 359
  laser-induced fluorescence and, 356, 358
BeH⁻ ions
 dissociation energies for, 27
 energies for, 26, 28
 energy vs internuclear separation data for, 25
 HF calculations for, 25
 ionization energy calculations of, 29

544

SUBJECT INDEX 545

photodetachment energies for, 29
potential curves for, 25, 28, 29
spectroscopic parameters for, 27, 28
structure of, 24
thermodynamic ionization energy of, 28, 29
vertical ionization energies for, 25, 28
vibrational energies and, 29
vibrational force constants for, 27
vibrational frequencies of, 27-29
20 STO basis sets for, 24, 25
Benzaldehyde
 protonation of, 457
Benzene
 Gaussian overlap model for, 378
 model for structure of, 308
 pair potential for, 378
 photodissociation of, 324
 predissociation of first singlet state of, 339
 proton affinity of, 465
 simulation of structure of, 373
 site-site model of, 477
Benzenes, 14C-substituted
 mass dependence and, 304
Benzenes, protonated
 proton transfer and, 450
Benzenes, substituted
 proton affinities of, 458
Benzenium
 hydride affinities of, 465
Benzoic acids
 gas phase acidity of, 461, 462
Benzophenone
 intersystem-crossing rate for, 218
 nonexponential decay of, 223
 picosecond experiments with, 210, 212
Benzophenone, 4-(1-naphthyl-methyl)-
 internal conversion in, 219
BeO and BeO$^-$
 potential energy curves of, 38
Biaryls
 excimer formation in, 225
Binaphthyl, 1, 1'-
 spectroscopic and X-ray studies of, 225, 226
Biophysical chemistry
 hydrodynamics in, 233-52
 see also Hydrodynamics in biophysical chemistry
Bis-(4-dimethylaminodithio-benzyl)-Ni(II)
 decay kinetics of, 220
BO$^-$ ions
 charge distributions in, 29
 geometry of, 30
 ground states of, 40
 ground state potential curves for, 30
 ionization potentials of, 31
Boron
 isotope separation of, 147, 154
Boron chloride
 isotope separation and, 142, 149, 154
Bromine
 isotope separation of, 148
 multipole moments of, 376
Bromine atoms
 excited
  reactions of, 272-75
  reactions with CO, 270
Bromine cyanide
 reactions with metals, 361
Bromoacetylene
 vibrational levels of singlet states, 340
Bromobenzene
 picosecond studies on, 212
Butane-n
 simulation of structure of, 373
Butanone, 2-
 diffusion studies of, 238
Butene, iso-
 determining connection with ammonia, 451

C

Calcium
 isotopes of
  separation of, 140, 145
Calcium-binding protein of muscle
 depolarized scattering and, 238
Calcium chloride
 formation from Ca + NaCl, 362
Carbon
 isotope separation of, 148
Carbon dioxide
 dissociation of, 335, 336
 Gaussian overlap model for, 378
 molecular dynamics simulation of experiments, 401
 multipole moments of, 376
Carbon disulfide
 picosecond studies on, 211, 121
Carbonium ions
 α-carbon branching in, 465
 empty binding orbitals and, 471
 heats of formation of, 464, 465
Carbon monoxide
 central force method applied to, 391
 E-V excitation of
  vibrational population distribution, 270, 271
 formation of
  formaldehyde photodissociation, 332, 333
 surface diffusion and, 493, 494
 modeling of, 308
 Monte Carlo simulation of experiment and, 402
 multipole moments of, 376
 vibrational excitation of collisions with excited Na, 271
 sodium excitation and, 278
Carbon tetrachloride
 "cage" lifetime of, 215
 internal excitation of, 361
 isotope separation and, 149
 picosecond studies on, 212
Carbonyls, aliphatic
 photodissociation of, 326, 327
Carbonyl sulfide
 reaction with excited Br, 274
Cerium chloride
 crystals of
  electronic Raman spectra of, 512
Cesium atoms
 donating electron to LiCl, 41
 excited by laser, 269
Cesium chloride
 dipole moments in, 37

SUBJECT INDEX

Chalcogenides
 photoemission studies of, 172
Chlorine
 isotope separation of, 152, 154
Chloroacetylene
 fluorescence lifetime of, 340, 341
Chloroform
 picosecond studies of, 213
Cholesterol esters
 mesophases formed by, 67
Chromatin
 histone composition of, 243
Chromatin multimers
 diffusion studies of, 235
Chromophores
 vibrational modes of, 501
Chymotrypsinogen
 reversible denaturation of, 244
CN^- ions
 basis set optimization for, 30
 basic sets for, 30
 charge distributions in, 29
 EOM ionization potential of, 31
 equilibrium of separation from molecule, 31
 ground state potential curves for, 36
 vertical ionization energy for, 30
CN radicals
 produced by photodissociation
  internal-state distributions and, 367
  ultraviolet photodissociation of, 367, 368
Collagen
 diffusion studies of, 236
 orientation by dipole mechanism, 248
Collagen, reconstituted
 melting behavior of, 251
Copper
 effective Chodorow one-electron potential of, 163
 He - I study of, 166
 photoemission spectrum of, 168
 ultraviolet photoelectron spectra of, 165
 X-ray electron spectra of, 164
Copper-porphin
 attachment of methyl groups to, 513
Coronene
 picosecond-laser studies of, 220
Coumarin 6
 stretching mode of, 228
Cryptocyanine
 fluorescence studies of, 221, 222
 picosecond studies of, 220, 221
Crystals, liquid
 relaxation processes studied in, 212
Crystals, organic
 electric conductivity of electron tunneling and, 106
Crystals, thermotropic
 liquid, NMR relaxation in, 47-71
 carbon-13 relaxation in liquid crystals, 70
 conclusions and future possibilities, 70, 71
 deuteron relaxation in liquid crystals, 68, 69
 theoretical limitations of, 69
 diffusion (SD) mechanism for, 56
 Torrey's theory and, 56
 generalized treatment of relaxation in nematics, 56-59
 general properties of liquid crystals, 49-51
 cholesterics, 49-51
 curvature strains of, 50
 cybotactic groups in, 51
 dynamical properties and, 51
 fluctuations near phase transitions, 51
 nematics and, 49-57, 62, 68-70
 smectics, 50, 51, 66, 67
 static distortions in, 50, 51
 NMR relaxation
 combined mechanisms and, 53
 electron spin resonance in liquid crystals, 57-59
 intermolecular dipolar relaxation and, 52
 mesophases of liquid crystals and, 48
 nature of, 48, 51, 52
 quadrupolar relaxation and, 53
 order director fluctuation mechanism, 54-56
 coordinate system for, 55
 proton relaxation in liquid crystals, 59-68
 chemical formulas of liquid crystals, 59
 cholesterics and, 67, 68
 intermolecular and intramolecular effects, 59-61
 proton relaxation in, MBBA, 62-66
 proton relaxation in PAA and PAA-d6, 60-62
 smectic liquid crystals and, 66, 67
 relaxation in liquid crystals, 53, 54
 historical aspects of, 53, 54
Crystal-violet dye
 internal conversion studies of, 218
 structure of, 218
Cuprous halides
 photoelectron studies of, 175
Cyanine dyes
 picosecond studies of, 220, 221
Cyanomethemoglobin, human
 diffusion studies of, 237
Cyclobutanes
 thermal reorganizations of reaction pathways of, 127-29
Cyclopropane, pi-derived from pyrazoline thermolysis, 119
Cyclopropanes
 thermal reorganizations of reaction pathways of, 111-29
Cytochrome
 bacterium Chromatium oxidation of
 electron tunneling and, 106, 107

D

Dansylgalactoside
 electronic properties of, 246
Dermatan sulfate
 human placental
 glucuronic acid in, 235

SUBJECT INDEX 547

Deuterium
  desorption from surfaces, 491
  scattering from alkali halides, 481, 482, 486, 487
  separation from hydrogen, 147
Dichalcogenides
  structure of, 165
Dichalcogenides, transition metal
  photoemission studies of, 176
  semiconductors among, 176
Dichroism, induced
  soluted molecules studied by, 212
Dideuteriocyclopropane, cis-1,2-
  stereomutation of, 114
Diethylanaline, N, N-
  picosecond-laser studies of, 217
Diethyl-2, 2'-dicarbocyanine iodide
  fluorescence studies of, 221
Diethyloxadicarbocyanine, 3, 3'-
  flash photolysis of, 220
Dimethylanaline
  picosecond studies of, 217
Dimethyl POPOP
  fluorescence emission of, 222
Dimyristoyl-lethecin
  diffusion rate of, 242
Diphenylanthracene
  fluorescence depolarization of, 247
Diphenylpolyenes
  Rayleigh scattering by, 239
DNA
  birefringence measurements on, 257
  conformations of, 252
  dielectric dispersion studies of, 250
  diffuse coefficient of, 237
  molecular weight of, 244
  renaturation of chloroacetaldehyde and, 246
  viscoelastic relaxation studies of, 251
DNA, double stranded
  transition of structure of, 249
Dysprosium
  valence band shape of, 170

E

Elastin
  dynamic studies of, 251
Electrolysis
  electron tunneling through electrodes, 105, 106
Electronic-to-vibrational energy transfer processes, 261-79
  conclusions, 279
  future research directions, 279
  early history of, 262
  models and, 262, 263
  electronic fluorescence quenching, 261, 262
  E-V energy transfer with classical methods, 263-68
  additional E-V studies employing classical methods, 267, 268
  aeronomic implications of, 268
  "bulb" experiments, 267
  cascading to low-lying vibrational states, 264
  crossed beam E-V studies, 266, 267
  excited mercury atoms experiments, 263-66
  flash kinetic spectroscopy, 267
  Hg reactions with NO, 265
  impulsive "half-collision" model for, 265, 266
  intermediate compounds, 265
  mechanisms of, 265
  models of, 265, 266
  overall efficiency of E-V energy transfers, 268
  phase shifts and, 264
  physical vs. chemical E-V transfer, 267
  quenching of excited oxygen, 267
  quenching of Hg atoms by CO, 263
  rate constants for, 264
  relative vibrational state populations, 264
  resonance effects, 266
  Rice-Ramsperger-Kassel-Marcus theory and, 268
  superelastic electron scattering, 266
  theoretical curve-crossing model, 268
  vibrational excitation mechanism, 265
  vibrational fluorescence, 264
laser studies of E-V energy transfer, 268-75
  CO laser probing apparatus, 270
  E-V energy transfer in polyatomic systems, 274
  E-V energy transfer rates, 273
  E-V lasers, 274, 275
  E-V studies using IR laser probes, 270, 271
  laser excitation by photofragmentation with IR fluorescent detection, 271-74
  laser excitation with fluorescence-sensitized probing, 269-71
  lasers forming excited metal oxides, 275
  lasing molecules pumped by E-V energy transfer, 275
  mode preferential excitation by, 275
  resonance role in reactions, 274
  RRK and RRKM theories of unimolecular reactions, 271
  "sensitization" technique for, 274
  surface transitions and, 271
  vibrational population distribution, 270
  visible and infrared fluorescence and, 269
previous reviews of, 262
V-E energy transfer studies, 275-79
  excitation temperatures, 276
  multiphoton absorption and, 279
  resonance importance in, 278
  spectrum-line reversal method, 276
  state-by-state analysis of, 277
  vibrational relaxation measurements, 276
  vibrational state analysis of,

548   SUBJECT INDEX

261-79
Electron tunneling, 105-8
  applications of, 105, 108
  biological applications of, 106, 107
  models for, 108
  recent developments in, 107, 108
  reactions in solids and, 107
  reviews on 107
  theoretical work on, 106, 108
Electroplating
  comparison with condensation processes, 251
Eosin
  fluorescence lifetime of, 222
  orientational relaxation time of, 214
Eosin isothiocyanate
  covalent conjugation of, 247
Erbium
  valence band shape for, 170
Erythrosin
  fluorescence decay time of, 222
  fluorescence lifetime of, 222
Esters
  proton affinities of, 454
Ethanol
  electron transfer in, 108
  intramolecular relaxation studies of, 227
Ethers
  proton affinities of, 454
Ethidium bromide
  reaction with macromolecular DNA, 241
Ethylene
  decomposition of, 495

F

Ferri-ferrocyanide
  electron exchanges between, 105, 106
Ferrocytochrome
  emission spectra of, 503, 504
  Ising model of, 416
Fluids
  simulation methods for, 389
Fluids, decorated lattice-gas models of critical phenomena in, 411-40
  decorated lattice models, 422-38
  Bartis-Hall model, 433, 434
  cells for a one-component gas, 428
  cells of simple lattice-gas model, 429
  closed-loop coexistence curves, 436
  coexistence curve diameter, 426, 429
  coexistence curve for, 432
  critical azeotropy, 423, 438
  critical double points, 431-34
  existence of a phase transition in model, 426
  "gas-gas equilibrium" types, 423
  grand partition function and, 424-30, 435
  law of the rectilinear diameter, 426
  lower critical solution points, 434-37
  "mermaid potential," 434
  Mermin's decorated lattice model, 429-31
  microcanonical formulation and, 423
  mixtures: Mermin's model and hybrids, 431-38
  mixtures: Widom's model, 422-26
  model with a phase transition in both pure components, 437, 438
  orientation-dependent potentials, 434
  penetrable sphere model, 426
  pressure-composition diagram for, 432
  primary and secondary cells in Widom's model, 422
  pure fluid models, 426-31
  shape of secondary cell for Mermin model, 430
  three-component liquid mixture, 423
  upper critical solution temperature, 434
  Widom's model as a pure fluid (Mermin's bar model), 427-29
  extensions and limitations, 438-40
  multiple occupancy of cells, 439
  multiple-phase equilibria, 439, 440
  phase transitions and, 439
  historical resumé of, 411-16
  classical theories of, 411-13
  coexistence curve and, 413
  coexisting phase properties, 414
  critical point behavior of fluids, 411, 412
  decorated lattice-gas models and, 414
  Ising model, 413-15
  magnets and, 415
  multicomponent fluid critical phenomena, 414
  multicomponent van der Waals equation of state, 413
  quasichemical approximation, 413
  renormalization group approach, 416
  universality of critical behavior, 415
  lattice-gas Ising model, 416-21
  magnetism model and, 417
  potential energy density of lattice gas, 420
  reviews on, 421
  square-planar lattices, 417, 418
  "up" vs "down" spins, 418-20
  reviews on, 411
Fluids, dense
  molecular motion in statistical mechanics of, 301-18
  see also Molecular motion in dense fluids, statistical mechanics of
Fluorescein-antifluorescence antibody
  complex formation of, 246
Fluorescein derivatives
  fluorescence lifetimes of, 222
Fluorescence, laser-induced, 349-69

SUBJECT INDEX 549

chemical reactions studied by, 356-67
angular distributions of reactions produced by LIF, 361-63
B + HF→BaF + H, 363, 364
CM cross sections, 362
electron-jump ("harpoon") model for, 360
electron-jump reactions, 359-61
flux densities and, 358
forward-backward symmetry and, 362
fully resolved LIF spectra, 365-67
H + NO_2 → OH + NO, 365-67
M + BrCN reactions, 361
"osculating complex" and, 362
phase-space theory of Pechukas and, 358, 359
potential surface types, 364
reactions producing metal oxides, 357-59
spin-forbidden reactions, 358
state-to-state reaction rates, 363, 364
statistical theories and, 362
transition state theory and, 358
vibrational intensities at scattering angles, 362
conclusions, 369
fluorescent intensities analysis, 353-56
beyond the Breit formula, 355, 356
Breit formula and, 353, 354
"level-crossing" phenomena, 353
polarization dependence of, 353
saturation on optical pumping and, 355
unresolved initial or final states, 354, 355
internal space distributions of photodissociation products, 367, 368
miscellaneous applications of, 369
atmospheric OH detection, 369
radiative lifetimes of vibration-rotation states, 369
molecular beam diagnostics, 368
supersonic nozzle beams and, 368
molecular beams, 350-53
angular distributions in, 352
center-of-mass velocity and, 352
Doppler profiles and, 352
flux density and, 352
internal state distributions, 351
molecular beam geometries, 351
nozzle beam diagnostics, 353
reactive scattering, 351-53
reviews of, 350
two crossed-beam configurations, 352
velocity-angle distributions and, 352
velocity angle maps, 351
velocity distributions in, 352
uses of
conditions for, 350
determination of populations in quantum states, 350
Fluorosulfonic acid
basicity of, 455
Formaldehyde
electronic spectrum of, 331
ground state dissociation threshold, 344
isotope separation and, 147, 148
photochemistry of, 330, 331
photodissociation of, 327, 330-33, 342, 343
photoexcitation of, 341
potential energy surfaces of, 343, 344
radiationless decay of, 341
Formamide
picosecond studies of, 213
Formic acid
decomposition of
scattering experiments and, 495

G

Gadolinium
valence band shapes for, 170
Gaseous detonation initiation, 75-102
conclusions, 101, 102
detonability limits, 96-101
cellular detonation waves and, 96
detonation kernel concept and, 98
multiple discrete reaction centers and, 100
periodic reinitiations and, 97-100
reacting blast "wavelets" and, 96, 97
transverse wave spacing data and, 98, 100
wave motions and, 96, 97
direct initiation of, 83-96
blast wave coordinates, 92-94
critical energy for, 91, 92, 95
critical regime of, 84, 85
"detonation bubble" and, 86
"detonation reestablishment" in, 85
flame kernel and, 92
general model for, 92
igniter energy and, 83-86
induction geometry, 91
induction time of, 90, 91, 94, 95
localized explosion in, 85, 86
quasi-steady period of blast motion, 85
spark energy and, 86-90
spark geometry and, 87-89
subcritical regime of, 84
supercritical regime of, 84
theoretical models for, 90-96
self-initiation, 76-83
amplification factor for, 82
autoignition limits and, 81
criterion for, 81-83
deflagration-detonation transition, 77
flame folding in, 82, 83
general features of, 76
hemispheroidal grid use in, 83
initiation of detonation, 79, 80

parameters determining
  flame travel, 76
predetonation regime, 79
progress of detonation,
  79
shock-flame interaction,
  78, 79
turbulence generation, 76,
  79, 80
turbulent flame speed and,
  82, 83
Gases, dense
  molecular movement in
    statistical mechanics of,
    301-18
Gases, low density
  boundary conditions for, 316
Gas-surface interactions
  see Surface scattering
Germanium
  photoemission spectrum of,
    172
Germanium compounds
  semiconductors
    bond-charge model of, 172
Glasses, solid
  electron tunneling in, 107
Glycoproteins, antifreeze
  diffusion studies of, 235
Gold
  angular-dependence studies
    on, 166
  photoelectron spectroscopic
    calibration, 167
  photoelectron spectroscopy
    of, 162, 168, 169
Graphite, pyrolytic
  scattering from surface of,
    488
Group 1B halides
  photoemission studies of,
    175

H

Halide ions
  transfer of, 465
Halides, polyatomic
  energy disposal by
    laser-induced fluorescence
    and, 361
Halogens
  molecular dynamics simulation of
    experiment and, 401, 402
HCN^- ions
  possible existence of, 34
Helium
  four bound states on LiF,
    480, 481

superfluid, 412, 413
$_3He - _4He$ system, 438
Heme proteins
  electronic lifetimes of,
    504
Hemoglobin solutions
  light-scattering studies of,
    239
Heptyloxyazoxybenzene, 4,
  4-di-n-
  ring deuteron relaxation in,
    68
Hexadecane
  "cage" lifetime of, 215
Hexokinase
  state of aggregation of, 243
Hexyloxybenzoic acid, p, n-
  deuteron relaxation of, 69
$(HF)_2^-$ ions
  possible existence of, 34
  unstable state of, 34
Histones
  conformational changes in,
    246
  effect on internal mode of
    DNA, 238
H_2O^- ions
  possible existence of, 34
Hydrates
  binding energies of, 470
Hydride ions
  transfer equilibria of, 463
Hydrocarbon-oxygen mixture
  autoignition of, 81, 82
Hydrocarbon radicals
  hydride ion equilibria, 463
Hydrocarbons
  solutions in liquid argon
    radiolysis of, 107
Hydrochloric acid
  gas phase acidity of, 462
  multipole moments of, 376
Hydrocyanic acid
  energy transfer from excited Br, 273, 274
  photodissociation of, 31,
    335, 337
Hydrodynamics in biophysical
  chemistry, 233-52
  books and reviews on, 233
  conclusions, 252
  rotational motion and flexibility, 246-52
    birefringence in steady
      flow, 252
    dielectric dispersion,
      250, 251
    differential polarized-
      phase fluorometry, 247
    electric dichroism and,

249
    electrooptics, 248-50
    EPR of spin labels, 247,
      248
    exponential decays in, 246
    fluorescence anisotropy,
      246, 247
    globular proteins and, 248
    high-frequency relaxations,
      250
    instrumentation for, 246
    moment index displacement,
      246
    orientation autocorrelation
      and, 247
    phosphorescence anisotropy and, 247
    polymerization of rodlike
      structures, 252
    reaction dynamics and,
      249
    rotational depolarization
      of fluorescence, 246
    rotational diffusion coefficient, 246
    rotational-relaxation processes, 247
    transient electric birefringence, 249
    viscoelasticity and flow
      orientation, 251, 252
    wavelength dispersion of
      dichroism, 249
  standard relations in, 233,
    234
  translational motion, 234-
    45
    active-enzyme ultracentrifugation, 242
    affinity chromatography,
      244
    anisotropic motion of nonspherical molecules, 238
    books and reviews on, 234
    charge interactions and,
      237
    concentration dependence
      of diffusion, 236
    contribution of internal
      modes to QLS, 238
    depolarized scattering
      and, 238, 239
    electrophoresis and, 244
    electrophoresis with density-gradient stabilized
      bands, 240
    electrophoretic light
      scattering theory, 240
    fluctuation spectroscopy,
      241

SUBJECT INDEX 551

fluorescence correlation spectroscopy, 241
fluorescence methods, 242
fluorescence photobleaching, 241
fluorescence photobleaching recovery, 242
fluorescent particles lateral mobility, 242
free-draining rouse model and, 238
gel permeation chromatography, 243-45
homodyne electrophoresis light-scattering apparatus, 240
immunosedimentation, 243
interpolyion distances, 237
isoelectric focusing and, 244
kinetic applications of QLS, 239
macroscopic gel theories, 240
molecular weight determination and, 235
osmotic compressibility of solutes, 236
polydispersity of polymers, 235-37
polyelectrolyte diffusion theory, 240
polyelectrolyte studies, 237
polymer relaxation times, 238
protein studies, 235-38, 243
pulsed-gradient NMR, 241, 242
QLS combined with electrophoresis, 239, 240
quasi-elastic laser light scattering, 234-41
Rayleigh-Debye particles and, 235, 236
refractive index matching and, 238
Scheraga-Mandelkern approach and, 235
sedimentation in density-stabilized gradients, 243
sedimentation studies of very large molecules, 243
small molecule diffusion measurements, 241
theory for unconnected segments in motion, 238
theory of translational diffusion, 242
translational diffusion coefficient, 234
transport in reacting systems, 245
ultracentrifugation, 242, 243
viscoelastic gels and, 240
Hydrogen
contaminated surfaces scattering from, 491
excited levels of collisions with excited mercury, 269
generation by sunlight, 108
isotope separation of, 147
proton affinities between those of H_2O and, 451, 452
scattering from alkali halides, 481, 482, 486, 487
vibration levels of, 269
Hydrogen atoms
recombination of energy released by, 491
Hydrogen bonds
gas phase studies of, 461, 462
Hydrogen-deuterium exchange atom scattering studies and, 495
Hydrogen fluoride
excitation of molecules of, 278, 279
V-T/R relaxation of, 264, 265
Hydrogen halides
reactions with excited Br, 272-74
Hydrogen-oxygen mixtures
ignition of, 87, 89, 94, 98, 100
Hydrogen selenide
photodissociation of, 324
Hydrogen sulfide
photodissociation of, 324, 325
Hydroxyl
atmospheric detection of, 369
formation from H and NO_2 fully resolved LIF spectra of, 365-67
infrared chemiluminescence and, 365, 366
rotational surprisal plot of, 366
Hydroxyl ions
core orbital of, 24
detachment energy of, 23, 24
Hartree-Fock wave function from, 23
structure of, 22-24

I

Iodine
isotope separation of, 148, 152
molecules of photodissociation of, 215, 216
Iodine atoms
excited reactions with CO, 270
Iodine molecules
time decay for pulsed excitation of, 502
Iodobenzene
photodissociation of, 327
Iodobiphenyl
photodissociation of, 327, 328
Iodocyanide
photodissociation of, 329, 330, 336, 337
Iodomethane
picosecond studies on, 212
Iodonaphthalene, α
photodissociation of, 327, 328
Ion equilibria
solvation from gas phase, 443-73
Ions, negative molecular
see Negative molecular ions, theoretical studies of
Ion thermochemistry, gaseous, 445-73
association and clustering reactions, 465-73
binding energies for N_4^+, O_4^+, and O_4^-, 472
binding energies of compounds, 470
electrostatic bonding, 468, 469
enthalpy and entropy changes for cluster dissociation, 467
enthalpy changes for dissociation of ion hydrates, 466
$(H_3 - H_2)^+$ species energy, 472
Lewis acid (alkali ion) base interactions, 470
methods for study of, 468

552   SUBJECT INDEX

monohydration of NO^+ and NO_2^-, 471
Mulliken electron population analysis, 469
noble gas-type alkali ions, 468
noble gas-type halide ions, 469
positively charged Brönsted acids, 470
reactions of alkyl cations with H_2 and CH_4, 471, 472
reviews on, 468
stepwise hydrations of ions, 469
total single-ion enthalpies of hydration, 469
basicities of compounds with PA (proton affinities) higher than that of ammonia, 455-58
Baker-Nathan effect and, 458
hydrogen affinities and, 457
internal cyclization of, 457
ionization energies and, 457
theoretical quantum mechanical calculations, 457
conclusions, 473
hydride and halide ion affinities, 463, 465
hydride ion transfer equilibria, 463, 465
ion equilibria conditions and apparatus, 446-49
collisional dissociation and, 448
flowing afterglow, 447, 448
mass spectrometry, 447
metallic ions and, 447
pulsed electron beam, 447
pulsed trapped ion ICR spectrometer, 448, 449
ion equilibrium method development, 445, 446
hydride transfer equilibria, 446
ion-molecule equilibria, 446
proton transfer equilibria, 446
reviews of, 446
occurrence-nonoccurrence (bracketing) technique, 462, 463
proton affinities of compounds between H_2 and H_2O, 451, 452
proton affinities of compounds between H_2O and NH_3, 452-55
ladder of proton transfer equilibria, 452
proton transfer equilibria with positive ions: proton affinities, 449-51
entropy changes and, 450
proton affinity types, 450
proton transfer, 449
reviews of, 449
proton transfer to negative ions
acidities in aprotic solvents, 462
acidities of carbon and nitrogen acids, 461, 462
gas phase acidities, 458, 459
intrinsic acidities of AH, 458-62
isodesmic proton transfer, 462
rotational degrees of freedom and, 460
standard acid selected, 458
stabilities of carbonium ions, 464, 465
Iridium
photoemission spectra of, 169
Iridium chloride
Cl to Ir charge transfer, 512, 513
Iron bromide
anomalous polarization in, 513
Isotopes, laser separation of, 133-56
concept and requirements for, 133, 134
conditions for, 134
excitation energy and, 134
conclusions, 155, 156
applications of, 155, 156
future for, 155
new methods proposed, 155
different approaches compared, 137-39
isotopically irreversible methods, 139
particle deflection, 139
photochemical reaction, 137, 138
photopredissociation and photoisomerization, 138, 139
scrambling process and, 138
isotope effect and selective excitation, 139-43
atomic isotope shift vs neutron number, 140
atomic spectra and, 140, 141
Doppler broadening elimination, 140
dynamic cooling and, 142
excitation selectivity increase methods, 140, 141
molecular spectra, 141-43
molecules with Q-branches and, 142
multiple-photon excitation, 142
two-quantum excitation methods, 141
laser separation methods, 135-39
early efforts and, 136
processes causing loss of selectivity, 137
properties required for, 135, 136
reactivity of particles and photon absorption, 136
resonant excitation transfer, 137
scrambling processes, 137
selective photopredissociation and, 136
separation selectivity and, 136, 137
thermal nonselective excitation, 137
two-step photoionization, 136
photochemical methods, 152-55
electronic photochemistry, 152, 153
Raman excitation and, 154
vibrational photochemistry, 153-55
photophysical methods, 143-51
dissociation of SF_6 molecules, 150, 151

SUBJECT INDEX 553

dissociation selectivity and IR laser frequency, 149, 150
ease of ionization and, 145
multiple photon dissociation, 151
multiple-photon dissociation of polyatomics, 148-51
one-step photopredissociation, 147, 148
pilot plan development and, 151
polyatomic molecule dissociation, 149
reviews on, 149
selective multistep photoionization, 143-45
selective two step (IR + Ur) photodissociation, 145-47
types of, 143
uranium isotope separation, 144
vibrational spectrum and, 146

K

Ketones
 proton affinities of, 454
Krypton
 photoelectron energy of, 170, 171

L

Laser-induced fluorescence, 349-69
 see also Fluorescence, laser induced
Lasers
 vibrational scattering and, 501, 502
Laser separation of isotopes, 133-56
 see also Isotopes, laser separation of
Lasers, picosecond
 picosecond spectroscopy and, 207-28
Lattice-gas models
 critical phenomena in fluids and, 411-40
$LiCl^-$ ions
 finite dipole model of, 35
 stable state of, 35
LiF^- ions
 potential energy curves of, 37
LiH

potential energy curves of, 37
 stable anions of, 34
LiH^- ions
 potential energy curves of, 37
Linear molecules, liquids of computer simulation and theory of, 373-406
Liquid crystals
 books and reviews on, 48, 49
 carbon-13 relaxation in, 70
 chemical formulas of, 59
 "clearing point" of, 48
 deuteron relaxation in, 68, 69
 diffusional coefficients of, 56
 general properties of, 49-51
 lyotrophic
  NMR studies on, 48
 mean field theory applied to, 58
 NMR relaxation in, 53, 54, 57
 order director fluctuation mechanisms in, 54, 55
 orientation of, 47, 48
 phases of study of, 48
 proton relaxation in, 59-68
Liquid crystals, thermotropic
 definition of, 47
 see also Crystals, thermotropic liquid, NMR relaxation in
Liquids
 molecules in
 orientational relaxation of, 210-15
 phases of
 isotropic vs nematic, 395
 photodissociation in, 304
Liquids, dense
 molecular motion in statistical mechanics of, 301-18
Liquids of linear molecules computer simulation and theory of, 373-406
Liquids, polymeric, rheology of, 185-202
 conclusions, 201, 202
 areas needing future work, 201, 202
 finite-element methods for complex flow prob-

lems, 195
 flow analysis, 194, 195
 flow problems of industrial interest, 195
 helical flows, 195
 non-Newtonian fluid dynamics, 194
 polymer flow through porous media, 195
 second-order fluid flow, 194
 squeeze flow experiments, 194
 tangential annular flow, 195
 "torsional-balance" rheometer, 194
 viscosity profiles, 194
 kinetic theory of, 185-202
 modern continuum mechanics and, 185
 kinetic theory of concentrated solutions, 200, 201
 Fixman's theory, 201
 stress tensor for, 200
 kinetic theory of concentrated solutions and undiluted polymers, 200, 201
 elastically coupled entanglement model, 201
 network theory, 201
 polymer-polymer interactions in, 201
 kinetic theory of dilute solutions, 195-200
 chainlike models, 199
 Curtiss-Bird-Hassager theory, 200
 dashpot-spring model of polymers, 199
 equilibrium-isolated molecule models, 195
 finitely extendable non-linear elastic dumbbell model, 198
 Hookean dumbbell model, 199
 jointed bead-rod chainlike models, 199, 200
 kernel functions of memory-integral expansions, 195
 Lodge rubberlike liquid molecules, 198
 molecular-model interpretations, 195-97
 molecular orientation and stretching in complex flows, 198
 "random-walk distribution"

of constants, 200
rheological equations of
   state forms, 198
rigid dumbbell studies, 198
slip effect near wall, 200
ultrasimplified models of
   particles, 195
kinetic theory of undiluted
   polymers, 201
measurement of material
   functions, 193, 194
droplet-elongation experiment, 194
elongated stress-growth
   experiments, 194
flow birefringence for, 193
high-shear-rate techniques,
   193
low-density polyethylene
   flow, 194
nonlinear viscoelasticity,
   193
"omegameter," 193
rheometric devices, 193
shear-free flows, 193, 194
"shear thickening" of dilute
   solutions, 193
"stressmeter," 193
time-dependent shear flows,
   193
new books, reviews and journals, 186, 187
rheological equations of
   state, 187-92
   constitutive equation, 187
   continuum mechanics and,
      187, 190
   corotating rate-of-defor-
      mation tensor, 190
   corotational memory-inte-
      gral expansion for simple
      fluid, 188-90
   Criminale-Ericksen-Filbey
      equation, 191
   eight-constant Oldroyd
      model, 191
   equations of change, 187
   friction of entangled chains,
      192
   generalized Newtonian
      fluid, 191
   general linear viscoelastic
      model, 190
   Hand equation, 191, 192
   instrumentation for, 191
   isotropic stress, 187
   Lodge equation, 190
   Lodge's rubberlike-liquid
      model, 192
   "Maxwell orthogonal rhe-
      ometer," 190
   oscillatory-motion experi-
      ment, 190
   polymer fluid dynamics,
      187
   relaxation modules and,
      190
   retarded motion expansion,
      190
   rheological equation of
      state, 187-89
   "rheologically simple
      fluids," 187
   rigid dumbbell solutions,
      191
   "second order fluid," 190
   "specialized" rheological
      equations of state, 190
   steady shear flow, 191
   stress tensor for simple
      fluid, 187
   unsteady-state unidirec-
      tional shear flow, 190,
      191
   "viscometric" flows, 191
   "viscometric" functions,
      191
rheological phenomena, 192,
   193
   "drag reduction" and, 193
   "heat-transfer reduction,"
      193
   polymeric fluid phenom-
      ena, 192, 193
   "vortex inhibition," 193
Lithium
   dissociation energy of Li_2,
      32
   Li_2 formed from, 32
Lithium chloride
   second unoccupied σ orbit-
      al in, 35
Lithium fluoride
   potential energy curves,
      37
Li_2, 32-41
   bond order of, 34
   closed-shell nature of, 32
   ionization of, 34
   molecular properties of,
      33
   optimized 20 STO basis
      set for, 33
   thermodynamic electron
      affinity of, 34
Li_2^+ ions
   dissociation energy of, 34
   spectroscopic parameters
      of, 34
Li_2^- ions
   dissociation energy of, 34
   formation of, 32
   ground state of, 33
   molecular properties of, 37
   potential curves of, 32
   spectroscopic parameters
      of, 34
   thermodynamic electron
      affinity of, 33
   vertical photodetachment
      energy of, 34
   vibrational frequency of, 34
   vibrational wave functions
      of, 34
Lysozyme, hen egg
   reversible unfolding of, 245

M

Magnesium
   XPS spectrum of, 168
Magnesium sulfates
   TDR studies of, 298
Malachite green
   picosecond-laser studies
      of, 219
Malononitrile
   gas phase acidity of, 462
Manganese etioporphyrin I
   modes of, 516
MBBA
   see methoxybenzylidene-n-
      butylanaline
Membranes, biological
   hydrocarbon region of, 246
   lateral diffusion and phase
      transitions in, 242
   rotational and translational
      diffusion
   reviews on, 242
Men, mines, and molecules,
   1-13
Mercury
   excited atoms of
      E-V energy transfer and,
         263-66
      reactions with hydrogen,
         269
   isotope separation of, 152,
      153
Meromysin
   decays of, 248
Metal complexes
   redox reactions between
      electron tunneling and,
         107
Metal oxides
   reactions producing
      laser-induced fluorescence,
         356-59

SUBJECT INDEX 555

Metals
  atom scattering studies of
    angular distribution measurements, 485, 486
    surface DEBYE temperatures of, 483, 484
    surface scattering from, 482
Methane-air mixture
  ignition of, 95, 96
Methoxybenzilidene p-n-butylanaline
  liquid crystal of
    viscosity of, 212
    NMR relaxation of, 54
    parameters obtained by GNS for, 65
    proton relaxation in, 62-66
    ring deuteron relaxation in, 68, 69
    viscosity coefficients of, 62
  Woessner's ellipsoidal model of
    relaxation of, 63
Methylcyanide
  experimental enthalpies of, 469
Methyl iodide
  intramolecular relaxation studies of, 227
  photodissociation of, 327
Molecular ions, negative
  see Negative molecular ions, theoretical studies of
Molecular motion in dense fluids
  boundary conditions, 315-18
    approximate methods for, 317
    Chapman-Enskog solution, 317
    "inner-outer" analysis and, 317
    kinetic or Knudsen layer, 316-18
    kinetic theory approaches, 316, 317
    microscopic details near boundaries, 316
    microscopic kinetic theory, 317, 318
    mode-coupling considerations, 318
    nonequilibrium thermodynamics and hydrodynamics, 315, 316
    phenomenological boundary conditions, 315
    Rayleigh problem, 316
    slip boundary conditions and, 318
    tangential velocity change, 316
    viscous wave vector-dependent eigenmode and, 318
  Enskog and related kinetic descriptions, 306
    contact pair distribution, 306
    dense impulsive systems and, 306
  hydrodynamic and related approaches, 307-9
    angular velocity and orientational correlation times, 308
    correlation functions via hydrodynamics, 307
    friction constants via hydrodynamics, 307, 308
    impulsive interactions, 307
    momentum transfer and, 307
    related approaches, 308, 309
    repulsive ("hard core") anisotropic interactions, 308
    rotational friction, 309
    rough sphere rotational motion, 308
    viscoelasticity and, 307
  introduction, 301-4
    boundary conditions, 302
    Brownian motion limit, 304
    correlated collision effects, 304
    diffusion constant prediction, 303
    Enskog theory, 301
    equations of motion for correlation functions, 303
    experimental results tabulated, 302
    friction constants, 302, 303
    hardsphere molecular dynamics, 301, 303
    hydrodynamic interpretations, 301-3
    impulsive interaction systems, 301
    modeling studies, 304
    theory of equilibrium liquid structure, 303
    translational and rotational diffusion constants, 301
    velocity and angular velocity correlation functions, 301, 303
    viscosity dependence in rotation, 302
  mass dependence, 304-6
    computer studies of, 305
    diffusion constants and, 304, 305
    hard sphere systems, 305
    Lorentz limit, 305
    other systems, 306
    "rattling" and "slipping" motions, 305
    velocity correlation function, 305
  mode coupling theory, 309-12
    diffusion kernel and, 309, 310
    long time tail in VCF, 309
    microscopic mode-coupling theory, 312
    mode coupling structure and applications, 309
    orientational relaxation time, 312
    phenomenological approach to, 310, 311
    rotational motion, 312
    translational diffusion and, 309-12
    transport kernels, 309
  renormalized kinetic theory, 312-15
    cage or back scattering effect, 314
    coupling and longitudinal and transverse velocity modes, 314, 315
    general features of, 313
    recollision contribution to rotational friction, 315
    recollision friction term in, 314
    structure and analysis, 313, 314
    VCF and AVCF applications, 314, 315
    reviews of, 301
    statistical mechanics of, 301-18
Molecules, complex resonance in vibrational scattering, 501-19
  chromophore vibrational modes, 501

SUBJECT INDEX

conclusions, 518, 519
excitation profiles, 514-18
  antisymmetry cancellation, 517
  A-term profiles, 514, 515
  B-term profiles, 515, 516
  destructive vs constructive interferences, 516
  displaced harmonic oscillator, 514
  excitation profiles for four modes, 515
  fluorescence excitation spectrum, 514
  interference effects: antiresonance, 516-18
  off resonance, 516, 517
  vibrational Raman scattering, 517
laser technology and, 501, 502
polarization: antisymmetric scattering, 510-14
  circular polarization measurements, 513
  degenerate transition moment mixing, 513
  electric vector rotation and, 512
  electronic and spin-orbit-induced antisymmetry, 512, 513
  rotational invariants, 510, 511
  spin-orbit coupling and, 512, 513
  vibronic antisymmetry, 513, 514
resonance Raman spectroscopy and, 501
resonance scattering and, 502-4
  emission lifetime measurements, 502
  hot luminescence, 503
  photon dephasing, 503
  reemission processes represented, 503
  relaxed fluorescence, 503, 504
  resonance fluorescence and, 502, 503
  resonance Raman scattering and, 503
  time-decay data for pulsed excitation of I_2, 502
scattering mechanisms, 504-10
  adiabatic treatment of, 504, 505

A-term scattering, 505-7
B-term and resonance effects, 508
B-term scattering, 507, 508
electronic wave functions and, 505
intermolecular effects, 510
nonadiabatic effects: Jahn-Teller distortion, 509, 510
overlap integrals on, 507
quadruple Mo-Mo bond, 506
time-dependent perturbation theory, 508
time-dependent theory: off-resonance scattering, 508, 509
vibrational wave functions, 505
wave function types, 505
Molecules, linear, liquids of theory of, 373-406
  comparison with experiment, 400-2
    nitrogen, 400, 401
    other fluids, 401, 412
  comparison with theory, 402-6
    bulk thermodynamic properties and, 404-6
    correlation function, 402-4
    dipole-dipole forces and, 405
    Gubbins-Gray perturbation expansion and, 403
    Mo-Gubbins theory and, 405, 406
    Monte Carlo simulation and, 404
    other macroscopic properties, 406
    Pople expansion and, 405
    reference interaction site models and, 402, 403
    site-site distribution functions and, 403
  computer simulation of, 373-406
  conclusions, 406
  equilibrium theory of, 373, 374
  intermolecular potential energy models, 374-79
    anisotropic overlap forces, 376
    central force models, 379
    discrete charge models, 376
    Gaussian overlap model, 377, 378
    generalized Stockmayer models, 375, 376
    general Stockmayer potential, 377
    intermolecular pair potential, 374
    intermolecular potential curves for symmetrical molecules, 378
    intermolecular potential curves for unsymmetrical molecules, 379
    intermolecular potentials for rigid molecules, 375-79
    Kihara core model, 377, 378
    Lennard-Jones model, 377
    multipole moments and, 376
    orientation of molecules in space, 375
    separation distances of liquids, 378
    site-site or atom-atom models, 376, 377
    spherical harmonic expansion and, 375, 376
  orientational structure, 394-400
    angular pair correlation function, 395, 396
    atom-atom potentials, 399
    computer calculations of angular pair correlations, 395-97
    conditional pair correlation functions, 396-98
    direct calculation from multidimensional histograms, 395
    generalized Stockmayer potentials, 399
    Lennard-Jones potentials, 399
    orientational ordering, 399
    orientational structure descriptions, 397-400
    partial pain correlation function, 395, 396
    quadrupole-quadrupole term, 399

SUBJECT INDEX 557

spherical harmonic expansions, 396, 397
reviews on, 373
simulation methods, 389-94
"bottleneck effect" and, 390
central force method, 391
Cheung method, 392
computing requirements, 392-94
intramolecular potentials and, 391, 393
intramolecular vibrations and, 391
Lennard-Jones potential, 393, 394
molecular dynamics and, 389-93
Monte Carlo methods for, 389-91
"neighbor list" and, 393
programming devices, 393, 394
random displacements and, 389, 390
rigid particle method, 391, 392
rotational motion problem, 392, 393
Runge-Kutta-Gill method, 393
Saville's method, 394
Stockmayer potentials and, 391
vector differential form, 392
simulation of real vs model liquids, 374
geometric forms of, 374
statistical mechanics and, 379-89
angle-dependent potentials, 387
angular pair correlation functions, 381, 386
approximate theories, 384-89
configurational Helmholtz free energy, 382
correlation functions for molecular centers, 380, 381
electron density distribution, 383
expressions for physical properties, 382-84
Fowler-Kirkwood-Buff model, 382
generalized Stockmayer potentials, 388
harmonic coefficients, 381
integral equation methods, 384, 385
Kirkwood theory, 384
Lennard-Jones potential model, 385, 389
mean spherical approximation, 384, 385
molecular distribution functions, 380-82
Padé approximations for free energy and, 387
pair distribution functions, 380
Percus-Yevick approximation, 384-88
Perram-White expansion, 389
perturbation expansions, 388
Pople reference potentials and, 387, 388
reference interaction site model, 384, 385, 388
repulsive potentials, 389
scaled particle theory, 385, 388
site-site correlation function, 381, 389
system of two fluid phases, 382
Molecules, polyatomic, photodissociation of, 323-46
conclusion, 345, 346
principles of, 323-26
coupled vibrational degrees of freedom of, 323
direct photodissociation, 324
"heterogeneous" predissociations, 324
"metastable" ions and, 325
selection rules for predissociation, 324
spectra of, 324, 325
theoretical work on, 325
types of photodissociation, 324
unimolecular decomposition, 325
theory of collinear triatomics, 333-39
collinear theory of Atabeck, 336
"effective oscillators," 335
ground state potential and, 337
"half-collisions" and, 334
impulsive model and, 334
intra- and interfragment excitation, 333, 336
normal mode change and, 334
optical excitation, 333-37
"quasidiatomic" description, 333
"radiationless transitions" and, 338
thermal excitation and, 337-39
thermal fragmentation of collinear triatomics, 337
vibrational wavefunctions, 338
theory of nonlinear polyatomics, 339-45
consecutive decay and, 340
dynamics of model for, 339
fluorescence rates and, 340
gas-kinetic collisional deactivation, 345
nonsequential photodissociation dynamics, 339
"quasi"-sequential behavior and, 340
theory of photodissociation processes, 339, 346
"bobsledding oscillations" and, 346
time-resolved experimental studies, 326-33
aliphatic carbonyls and aryl halides, 326-28
anisotropy parameter, 326
formaldehyde, 330-33
intrafrequent effects, 329
NO_2, NOCl, and ICN, 328-30
"photolysis mapping" and, 326
repulsive triplets and, 378
theories of collinear triatomic dissociations, 329
Molybdenum
oxidation of
surface diffusion-limited reactions, 493
Molybdenum chloride

Mo-Mo bond properties in, 506
Molybdenum fluoride
  isotope separation and, 149
Molybdenum sulfide
  photoelectric emission of, 177
Monochalcogenides, samarium
  physical properties of, 178
Monochalcogenides, thulium
  interconfiguration fluctuations in, 178
Mucopolysaccharides
  translational diffusion coefficient and, 235
Muscle, skeletal
  proteins of
    coherent light scattered from, 241
Myoblasts
  concavalin A mobility in, 242
Myosin
  cross-bridges in muscle, 246
  fragment S-1 of
    binding to F- actin, 247
Myosin fragments
  dipole behavior, 248

N

NaH and NH^-
  potential energy curves of, 38
Naphthalate ion
  intramolecular proton transfer in, 225
Naphthalene
  moieties of
    relative orientation of, 226
  Raman spectrum of, 508
Negative molecular ions,
  theoretical studies of, 15-41
  anions studied, 16, 17
  charge densities and, 17
  conclusion, 42, 43
  electron correlation effects and, 15-17
  equation-of-motion theory and, 16
  equilibrium geometries and, 17
  Koopman's theorem and, 15
  laser technology for, 16
  orbit relaxation and, 16
  other recent results, 41, 42
  perturbation theory and, 16
  photodetachment and photoelectron spectroscopy and, 16
  review of, 15, 16
  survey of results, 23-42
    BeH^-, 24-29
    BO^- and CN^- ionization potentials, 31
    BO^- and CN^- stability and, 31
    Born-Oppenheimer approximation and, 35, 41
    calculated electron affinities and, 39
    CN^- and BO^-, 29-32
    collision dynamics and, 41
    dipole models and, 40
    dipole moment and electron affinity, 40
    dipole moments of molecules, 41
    double zeta basis sets and, 29
    electron-jump mechanism and, 41
    electropositive role of Li, 40
    equation-of-motion electron affinities and, 36
    fixed finite dipole models and, 39
    ground state anions and, 41
    ground state potential energy curves of, 37
    Kooperman's theorem and, 31, 32, 36
    LiF^-, $LiCL^-$, LiH^-, NaH^+ and BeO^-, 34-41
    Li_2 and, 32-34
    negative-ion spectroscopy, 28
    OH^-, 23, 24
    Padé approximations and, 39
    Slater-type orbitals and, 35
    spectroscopic constants for neutral ions and anions, 39
    STO basis sets for groups, 36
    vertical displacement energies, 42
    theoretical methods for, 17-23
    classes of negative ions discussed, 22
    electron propagator and, 17-22
    equation-of-motion, 17, 18
    hydrides and, 22
    matrix elements in, 20-22
    propagator approximation, 22
    Rayleigh-Schrödinger perturbation theory and, 19-21
    special electron propagator and, 18
    superoperator scalar product and, 18
    third order electron propagator and, 20
    third-order self-energy diagrams, 21, 22
    wave functions and, 22
Neon
  scattering of LiF, 483
Neon, solid
  photoelectron energy of, 170
Nickel
  photoemission spectra of, 168
Nitric oxide
  electronic structure of, 328, 329
  photodissociation of, 328, 329
  reaction with excited Br, 274
  reactions with mercury, 265, 266
Nitrobenzene
  picosecond studies on, 211, 212
Nitrogen
  central force method applied to, 391
  isotope separation of, 148, 154
  models of
    Gaussian overlap model, 378
    simulation of, 374
  molecular dynamic simulations, of, 308
    comparison with experiment, 400, 401
  multipole moments of, 376
  vibrationally excited, 266
Nitrogen atoms
  energy transfer to Na
  aeronomic importance of, 277, 278
Nitrogen bases

SUBJECT INDEX 559

proton transfer equilibria of, 455
Nitrogen vibrational temperature
  earth's upper atmosphere modeling of, 268
Nitromethane
  nitrogen isotope separation and, 142, 149
Nitronaphthalene
  intersystem-crossing rate for, 219
Nitrotoluene, m-
  picosecond studies, 212
Nitrous oxide
  collisional exchange rates for, 337-39
  photodissociation of, 325
  quencher of excited nitrogen, 277
  vibrational levels of, 337
Nitroxide spin labels
  membrane fluidity, 247
NMR relaxation
  see Crystals, thermotropic liquid, NMR relaxation in
NOCl
  photodissociation of, 329, 330
Norbornyl cation
  formation of, 465

O

Octaethylporphinatomin (IV) dichloride
  absorption spectrum of, 220
Octylbiphenyl, 4-4'-d_{17}-n-
  nematic phase of
  deuteron relaxation in, 69
Osmium oxide
  isotope separation and, 142, 147-51
Oxides, linear triatomic
  photodissociation of, 325
Oxyanions
  X-ray emission spectra of, 179, 180
Oxyanions, transition metal
  XPS spectra of, 180
Oxygen
  electronically excited molecules
    reactions of, 268
  isotope separation of, 148
  models of
    simulation of, 374
  molecular dynamics simulation of

comparison with experiment, 401
Oxygen atoms
  excited
    reaction with CO, 270, 271
  vibrationally excited
    flash kinetics studies of, 267
Ozone
  vibrationally excited oxygen atoms of, 267

P

Palladium
  photoemission spectra of, 168, 169
  X-ray electron spectra, 164
Paramyosin
  forms of, 248
Pentanol, 1-
  picosecond studies of, 213
Perylene
  fluorescence emission of, 222, 223
Phenols
  gas phase acidities of, 461
Phenothiazine
  photoionization of
    electron tunneling and, 107
Phenylcyclopentane
  stereomutation of, 123
Phenyl iodide
  photodissociation of, 327
Phosgene
  photon absorption by
    isotope separation and, 136
Photoelectron spectroscopy, 161-81
  chemical studies, 168-80
  alkali halides, 172, 173
  compounds of type A^nB^{8-n}, 172
  enhancement of spectra, 168
  group 1B halides, 175
  metals, 168-70
  model band structures, 169
  molecular solids, 179, 180
  nonmetallic elements, 170-72
  oxyanions and, 179
  rare-earth compounds, 178, 179

rare-earth metals, 169
rare-gas solids, 170, 171
silicon and, 171, 172
transition metal dichalcogenides, 176-78
transition metal oxides and halides, 173, 174
XPS spectra for metals, 169
experimental considerations, 166-68
  calibration, 167
  cesiation, 168
  metal work function, 168
  sample preparation, 167, 168
  synchrotron radiation and, 167
  X-radiation width and, 166, 167
  X-ray photoelectron spectrum, 166
reviews on, 101
summary, 180, 181
three-step model, 161-66
  angular variation of photoemission, 164-66
  Chodorow potentials and, 163
  constant-momentum matrix elements, 163
  density of states, 162, 163
  electron-electron scattering and, 163
  electron escape depths, 166
  energy distribution of joint density of states, 162, 163
  Gilat-Raubenheimer method, 163
  Korringa-Kohn-Rostoker method, 163
  "Mahan cones," 165
  photoemission process, 161, 162
  photoionization cross sections, 164
  relaxation energy and, 164
  ultraviolet photoelectron spectra, 165
  valence-electron states, 164
  variations in momentum matrix elements, 164
  X-ray photoelectron spectrum, 162
Picosecond spectroscopy
  see Spectroscopy, picosecond
Platinum

SUBJECT INDEX

photoemission spectrum of, 168, 169
Polyatomic molecules
　photodissociation dynamics of, 323-46
　see also Molecules, polyatomic photodissociation of
Poly(carbobenzoxy-L-lysine)
　helix-coil transition of, 250
Polyethylene, low-density
　rheology of, 194
Polymeric liquids, rheology of, 185-202
　see also Liquids, polymeric rheology of
Polymer solutions
　rheology of, 195-201
Poly(n-hexylisocyanate)
　depolarized forward scattering of, 238
Polypeptides
　conformational changes in, 237
　molecular weights of, 244
Polypeptides, synthetic
　dispersions of, 250
Polysaccharides
　polydispersity of, 235
Polystyrene
　diffusion studies of, 238
　fluorescence polarization studies of, 247
Polystyrene, random-coil
　gel permeation chromatography, 244
Porphyrins
　absorption spectra of, 220
　Raman spectra of, 508, 513
　structural changes in, 501
　vibrational modes of, 501
　vibronic mixing and, 514
　vibronic modes of, 516
Potassium
　isotope separation of, 145
Potassium chloride
　final-state spectroscopy of, 173
Potassium ion
　temperature dependence of ion equilibria, 469
Propionaldehyde
　photodissociation of, 327
Proteins
　analysis of
　　electrophoresis and, 244
　films of
　　thickness of, 235
　gel permeation chromatography and, 243

model membranes and, 235
molecular weights of, 235
translational diffusion studies of, 235
Proteins, globular
　relaxation times of, 248
Protons
　transfer of
　　ion thermochemistry and, 449
Purine rings
　mode intensity of, 510
Pyrazine
　B-term scattering and, 508
Pyrazolines
　thermal decomposition of, 118
Pyrene excited-state dimer
　rate of formation of, 242
Pyridine
　B-term scattering in, 508
Pyridine, substituted
　free energies of protonation of, 457
Pyrimidine rings
　mode intensity of, 510

Q

Quinoxaline
　B-term scattering in, 508

R

Rearrangements, thermal, 111-29
　cyclobutanes formed, rearranged, and decomposed, 127-29
　alternative pathways involving 1,4-biradicals, 127
　hypothetical biradical intermediate and, 129
　intermediates in, 127-29
　kinetic analysis of alternatives, 128
　cyclopropanes and, 111-29
　activation energy calculations, 112
　A-factor estimates, 115, 116
　barrier to ring closure and, 115-17
　biradical mechanism correlations and predictions, 114-18
　biradical mechanism experimental validation, 114

cyclopropane, π-, 119
　energy surface for, 120
　E_0 estimates, 116-18
　gas-phase reactions of, 112
　hyperconjugative interaction and, 120
　intermediate in reactions, 112, 113
　isomerization of, 1,1,2,2-tetramethylcyclopropane, 116, 117
　mechanism of π-cyclopropane stereomutation, 123
　microscopic reversibility and, 120
　pyrolysis by biradical mechanism, 113, 115
　quantum mechanical calculations, 112
　quantum mechanics predictions and correlations, 118-21
　question of a local minimum, 125, 126
　single and double rotation mechanisms in, 122-24
　spin effects, 126
　stereomutation of, 112
　stereorandom biradical mechanism of stereomutation, 122, 124
　synchronous conrotation and, 120
　theoretical descriptions of mechanisms, 112
　thermal ring openings, 120
　thermochemical-kinetic calculations, 112-18
　trimethyl biradical as intermediate, 112, 113
　"uncoupled" biradical in, 124
　reviews of, 111
Reflectometry, time domain, 283-99
　see also Time domain reflectometry
Rhenium
　photoemission spectra of, 168, 169
Rhodamine B
　vibrational relaxation of, 224
Rhodamine 6G
　fluorescence correlation spectroscopy of, 241
　picosecond experiments with 209, 213, 214
　vibrational relaxation of, 224

SUBJECT INDEX 561

Rhodopsin
  dielectric dispersion studies of, 250
  RR spectrum of, 510
Ribosomes
  characterization of, 236
RNA
  conformations of, 252
RNA, double-stranded
  molecular weight of, 244
RNA polymerase
  gene transcription and, 247
Rose bengal
  orientational relaxation time of, 214
Rubidium
  isotope separation of, 145
Rubidium ion
  hydration of, 470

S

Salicylic ester
  enol-to-keto isomerization, 225
Samarium
  photoemission spectrum of, 169, 170
Silicon
  photoemission spectra of, 171, 172
Silver
  photoemission spectrum of, 168, 169
  X-ray electron spectra of, 164
Silver halides
  photoemission studies of, 175
Sodium
  isotope separation of, 145
  XPS spectrum of, 168
Sodium atoms
  detection by laser-induced fluorescence, 349
  electronically excited, 266, 267
  excited reactions with excited CO_2, 271
  excited by laser, 269
  vibrational excitation of, 276-78
Sodium dimer molecules
  production of
    LiF and, 368
Solids, molecular
  valence electrons of, 179
Spectroscopy
  laser-induced fluorescence and, 349-69
Spectroscopy, photoelectron
  see Photoelectron spectroscopy
Spectroscopy, picosecond, 207-28
  concluding remarks, 228
  conformational changes in excited electronic states, 225, 226
  equilibrium "coplanar" excited state, 226
  excimer formation and, 225
  polarization-dependent kinetics, 226
  electron photoejection and solvation, 216
  excited state charge-transfer complexes, 216-18
    diffusion-controlled reactions and, 216
    exciplex formation and, 217
    molecular fluorescence quenching, 216
    orientational restrictions and, 217
    packing effects and, 217
    picosecond-laser studies, 217
    polar vs nonpolar solvents, 217
    translational motions in liquids, 2
  intermolecular energy transfer, 208-10
    anisotropy decay, 208
    anisotropy development and, 208
    diffusion of reactants, 209
    dipole-dipole interaction, 209
    donor-decay function, 209
    Franck-Condon factors, 210
    ground-state donor absorption, 208
    intersystem crossing, 210
    power density of excitation pulse, 209
    rotational motion and, 210
    singlet-singlet transfer, 208, 209
    transition moments of molecules, 208
    triplet-triplet transfer, 209, 210
    two-component system study, 209
    vibrational relaxation time and, 209
  internal conversion and intersystem crossing, 218-23
    emission measurements of, 221-23
    external heavy-atom effect, 220
    fluorescence phenomena and, 221-23
    mode-locked laser and, 221
    picosecond continuum pulse, 220
    picosecond techniques for, 218
    solvent effects and, 218-20
    two-photo excitation and, 223
  intramolecular proton transfer, 225
    proton-transfer rate constant and, 225
    quantum mechanical tunneling and, 225
  orientational relaxation of molecules in liquids, 210-15
    anisotropic orientational distribution and, 213
    birefringence decay and, 211, 212
    birefringence induced in, 210
    dichroism decay and, 210, 211
    dielectric relaxation measurements and, 212
    fluorescence depolarization method, 214
    fluorescence polarization and, 211
    induced dichroism method, 212-14
    laser-generated ultrafast light gate, 211
    linear scaling and, 213
    mixtures and, 212
    optical Kerr effect and, 210-12
    orientational relaxation and, 211, 212
    orientational relaxation time of excited molecules, 214
    refractive index change and, 211
    relaxation processes in liquid crystals, 212

SUBJECT INDEX

transient grating method, 214, 215
viscosity effects and, 212
photodissociation and the cage effect, 215, 216
geminate and nongeminate processes, 215
kinetic energy of fragments in, 215
rate constant for, 215
theoretical model for, 215
picosecond lasers and, 207-12
vibrational relaxation in excited electronic states, 223-25
rapid deexcitation and, 224
subpicosecond pulse excitation, 224
techniques for, 223, 224
vibrational relaxation in the ground electronic state, 226-28
anti-Stokes spontaneous scattering, 227
dephasing processes, 226, 227
infrared and visible pulse method, 227, 228
linewidth determination, 226, 227
population lifetime in, 226, 227
separation of two relaxation channels, 228
tunable picosecond pulses and, 227
Spectroscopy, vibronic resonance Raman spectroscopy and, 501
Statistical mechanics
molecular motion in dense fluids, 301-18
Strontium ion
hydration of, 470
Sulfhydryl ion
structure of, 22
Sulfur fluoride
isotope separation and, 142, 147-51
Surface scattering, 477-95
elastic scattering diffraction studies, 479-84
acceleration of incoming atom by attractive potential, 483
alkali halide surfaces, 480-82
angular positions of bound state extrema, 481
approximate quantum theory and, 480
bilobar scattering, 482, 483
bound state resonance, 480
degenerate transitions in, 481
diffraction peak intensity and incident angle, 480
four bound states of He on LiF, 480, 481
gas-surface interaction potential, 479-83
inelastic processes in, 481
metal surfaces, 482
Morse potential predictions, 480
rainbow peaks, 482, 483
realistic interaction potentials, 480, 481
redistribution in intensities among diffraction channels, 481
scattered distribution and reflected angle, 479
splitting of bound state minima, 487
surface-bound atoms, 481
surface Debye temperature, 483, 484
surface properties, 483, 484
surface structure, 483
theory and observation in, 480
theory development in, 482
energy exchange studies, 484-92
alkali halide surfaces, 487-90
angular distribution measurements, 485-88
CO-covered surface, 486
complete energy exchange with surface, 488
diatomic molecule-surface interaction potential, 487
elastic interactions, 484, 485
energy released by combinations on surface, 491
hard cube models, 487, 488
harmonic soft cube models, 486
H_2 vs D_2 scattering distributions, 486
inelastic diffraction peaks, 487
inelastic interactions, 485
inelastic scattering peak, 489
interactions of particle with surface, 484, 485
lattice effects, 490
metal surfaces and, 485-87
multiphonon interactions, 487
nonreactive studies, 485-90
one- and two-phonon inelastic rare gas scattering from metal surfaces, 489
quantum mechanical one-phonon close-coupling calculation, 487
reactive systems, 490-92
residence time measurements, 490
scattered beam faster than incoming beam, 489
scattering from pyrolitic graphite surfaces, 488
separation of scattering pathways, 490
square well semiempirical model, 486
surface contamination, 486
surface phonon dispersion curve for LiF, 487
surface residence time, 485
three-dimensional quantum mechanical treatment of, 489
trapping interactions, 485
velocity measurements, 488-90
reaction mechanism studies, 492-95
bulk diffusion controlling reactions, 493, 494
catalytic studies, 494, 495
surface corrosion studies, 492-94
surface diffusion vs rate limiting, 493
types of, 492
technique for, 477, 478

T

TCNQ

SUBJECT INDEX 563

photoemission studies of, 180
Tetracene crystal
triplet energy transfer in, 210
Tetramethyl benzidine, N,N, N,N1,N2-
gas ionization potential of, 108
photoionization of electron tunneling and, 107, 108
Tetramethylcyclopropane 1, 1,2,2,-
structural isomerization of, 116, 117
Tetraphenylethylene dianions sodium salt of
flash-photolysis studies of, 216
Tetrazene, s-
isotope separation and, 143, 148
photochemical dissociation of, 222
Thermal rearrangements
see Rearrangements, thermal
Thermochemistry
ions and, 445-73
Thiatricarbocyanine, 3, 3'-iodide 2,2'-
absorption band of, 220, 221
Time domain reflectometry, 283-99
basic principles of, 284-86
charge and current of a dielectric sample, 285
complex permittivity, 286
electromagnetic equations solutions for, 286
forms of response observed, 285, 286
instrumentation for, 284, 285
basic relations, 287-92
complex reflection coefficient and, 287
dielectric and conductance behavior, 290
"lumped circuit" method, 289
magnetic behavior study, 290
method including all reflections, 287, 288
reflections from open-circuited sample, 289
reflections from sample in a matched line, 287, 288
reflection from short-circuit sample, 290, 291
sample termination method, 289
single-reflection method, 287
"thin-sample" formula, 288
"time referencing," 291, 292
"time windows" for open-circuited method, 290
transmission methods, 291, 292
conclusions, 299
procedures and problems in use, 292-99
alcohols studied by, 297, 298
aliasing errors and, 294
aqueous solutions studied by, 298
conduction currents in sample, 294
data acquisition and processing, 297
"data-acquisition" device, 293
dielectric relaxation of alcohols, 297
distortions of pulse forms, 294, 295
electrode polarization and, 294
Elliot's control for time scanning and, 296, 297
jitter and, 296
Laplace transforms and, 293
noise and drift, 296, 297
pulse forms and their transforms, 293-95
pulse generation and detection equipment, 293
representative results, 297-99
sample cells and, 292, 293
Shannon sampling theorems and, 294
timing errors, 295
"total" complex permittivity, 294
unwanted reflections, 295, 296
reviews on, 283, 284
summary of, 283
attractive features of, 283
dielectric responses, 283
new instrumentation for, 283
TMB
see Tetramethyl benzidine, N,N,N$_1$,N$_2$-
Toluene
picosecond studies on, 212
Toluene-glycerine system
decorated lattice model of, 436
Transcarbamylase, succinyl-ated aspartate
diffusion studies of, 23
Transferrin
excitation profile of, 515
Transhydrogenase, pyridine nucleotide
state of aggregation of, 243
Transition-metal complexes
intersystem crossing in, 221
Transition metal oxides and halides
photoemission studies of, 173, 174
Trimethylene
internal rotation of, 116
Triphenylmethane dyes
model for, 218, 219
Tryptophane
buried in proteins
phosphorescent lifetimes of, 247
Tubulin, calf brain
self-association of, 245
Tungsten
photoemission from, 170
structure of, 165
Tunneling, electron
see Electron tunneling

U

Undecanol, 1-
picosecond studies of, 213
Uranium
isotope separation of, 144

V

Valence bands in solids
see Photoelectron spectroscopy
Vibrational scattering
resonance effects on complex molecules
see molecular complex, resonance in vibrational scattering
Vinylcyclopropane, 1-

stereomutation of, 121
Viruses
  enumeration of, 234
Viruses, bacterial
  electrophoresis studies of, 240
Virus, tobacco mosaic
  anomalous negative birefringence of, 248
Visual pigment
  diffusion in rod outer segment, 242

W

Water
  central force method applied to, 391
  diffusion in animal tissues, 241
  modeling of density maximum in, 415
  photodissociation of, 324
  proton affinities between those of H_2 and, 451, 452
  proton affinities between those of NH_3 and, 452-55
  simulation of structure of, 373, 374
    central force models and, 379
  TDR studies of, 298
Water vs ice
  electron exhanges between electron tunneling and, 106

X

Xenon
  photoelectron energy of, 170, 171

Z

Zinc-blende compounds
  semiconductors, 172

CUMULATIVE INDEXES

CONTRIBUTING AUTHORS VOLUMES 24-28

A

Alder, B. J., 24:325-37
Andersen, H. C., 26:145-66
Anderson, L. L., 26:339-57

B

Bak, T. A., 25:1-10
Barron, L. D., 26:381-96
Bauer, D. R., 27:443-63
Berne, B. J., 25:233-53
Berry, M. J., 26:259-86
Berson, J. A., 28:111-32
Bigeleisen, J., 24:407-40
Bird, R. B., 28:185-206
Bixon, M., 27:65-84
Blaney, B. L., 27:553-86
Bloomfield, V. A., 28:233-59
Bonner, J. C., 27:291-317
Brauman, J. I., 27:443-63
Buckingham, A. D., 26:381-96

C

Carlson, T. A., 26:211-33
Ceyer, S. T., 28:477-99
Chang, K. C., 27:369-85
Cole, R. H., 28:283-300

D

Deutch, J. M., 24:301-23
Dubrin, J., 24:97-120

E

Egelstaff, P. A., 24:159-87
Eisenthal, K. B., 28:207-32
El-Sayed, M. A., 26:235-58
Erpenbeck, J. J., 27:319-48
Ewing, G. E., 27:553-86
Eyring, E. M., 25:255-74
Eyring, H., 25:39-77; 27:45-57; 28:1-13
Eyring, L., 24:189-206

F

Farrar, J. M., 25:357-85
Ferguson, E. E., 26:17-38
Field, R. J., 25:95-119
Finkelstein, R. S., 24:207-34
Flynn, G. W., 25:275-315; 28:261-82
Franklin, J. L., 25:485-526

G

Gasparoux, H., 27:175-201
Gelbart, W. M., 28:323-48
George, T. F., 24:263-300
Gershfeld, N. L., 27:349-68
Goldanskii, V. I., 27:85-126
Gole, J. L., 27:526-49
Gordon, L. G. M., 25:11-38
Green, J. C., 28:161-83
Grunwald, E., 27:369-85
Gubbins, K. E., 28:373-410

H

Harland, P. W., 25:485-526
Haydon, D. A., 25:11-38
Henderson, D., 25:461-83
Hladky, S. B., 25:11-38
Hoffmann, G. W., 26:123-44
Hudson, B., 25:437-60
Hyde, J. S., 25:407-35
Hynes, J. T., 28:301-21

J

Jhon, M. S., 27:45-57
Johnson, B. B., 27:465-91
Johnson, K. H., 26:39-57
Johnston, H. S., 26:315-38
Jonas, J., 26:167-90
Jørgensen, P., 26:359-80

K

Kahlweit, M., 27:59-63
Kaiser, W., 26:83-100

Kebarle, P., 28:445-76
Kellogg, H. H., 27:387-406
King, D. L., 27:407-42
Kinsey, J. L., 28:349-72
Klein, F. S., 26:191-210
Knaap, H. F. P., 26:59-81
Knox, J. H., 24:29-49
Knudtson, J. T., 25:255-74
Kohler, B., 25:437-60
Koningstein, J. A., 24:121-34

L

Lallemand, P., 26:59-81
Laubereau, A., 26:83-100
Lee, J. H. S., 28:75-104
Lee, M. W., 24:407-40
Lee, Y. T., 25:357-85
Leffler, J. E., 27:369-85
Lemont, S., 28:261-82
Letokhov, V. S., 28:133-59
Lewis, J. S., 24:339-51
Libby, W. F., 28:105-10
Lide, D. R. Jr., 24:135-58
Lin, S. H., 25:39-77
Luzzati, V., 25:79-94

M

Mandel, F., 24:407-40
Marcuson, S. W., 27:387-406
Matthews, B. W., 27:493-523
McGlashan, M. L., 24:51-76
Miller, T. A., 27:127-52

N

Nagle, J. F., 27:291-317
Netzel, T. L., 24:473-92
Noyes, R. M., 25:95-119

O

O'Grady, W. E., 26:287-314

565

CONTRIBUTING AUTHORS

P

Palmer, H. B., 24:235-62
Pecora, R., 25:233-53; 27:443-63
Peticolas, W. L., 27:465-91
Porter, R. N., 25:317-55
Prost, J., 27:175-201

R

Rabitz, H., 25:155-77
Rao, Y. K., 27:387-406
Rentzepis, P. M., 24:473-92
Ross, J., 24:1-27; 24:263-300
Rossmassler, S. A., 24:135-58

S

Saltsburg, H., 24:493-14
Saupe, A., 24:441-71
Schaefer, H. F. III, 27:261-90
Secrest, D., 24:379-406
Seery, D. J., 24:235-62
Sen, R. K., 26:287-314
Setser, D. W., 27:407-42
Silbey, R., 27:203-23
Simons, J., 28:15-45
Somorjai, G. A., 28:477-99
Soos, Z. G., 25:121-53
Spiro, T. G., 28:501-21
Stein, P., 28:501-21
Stein, R. S., 24:207-34
Stephens, P. J., 25:201-32
Street, W. B., 28:373-410
Struve, W. S., 24:473-92
Swinton, F. L., 27:153-74

T

Tai, L.-T., 24:189-206
Tardieu, A., 25:79-94
Taylor, R. E., 25:387-405
Toennies, J. P., 27:225-60
Truhlar, D. G., 27:1-43

V

Verkleij, A. J., 26:101-22
Ververgaert, P. H. J. Th., 26:101-22
Von Foerster, H., 24:353-78

W

Wade, C. G., 28:47-73
Weitz, E., 25:275-315
Westenberg, A. A., 24:77-96
Weston, P. E., 24:353-78
Wheeler, J. C., 28:411-43
Wilson, E. B., 24:1-27
Wilson, M. K., 26:1-16
Wiser, W. H., 26:339-57
Wood, W. W., 27:319-48
Wyatt, R. E., 27:1-43

Y

Yamakawa, H., 25:179-200
Yeager, E., 26:287-314

CHAPTER TITLES VOLUMES 24-28

ARTIFICIAL INTELLIGENCE
  Computer Dynamics   B. J. Alder   24:325-37
  Artificial Intelligence and Machines that
    Understand   P. E. Weston, H. Von Foerster   24:353-78
BIOPHYSICAL CHEMISTRY
  Hydrodynamics in Biophysical Chemistry   V. A. Bloomfield   28:233-59
CHROMATOGRAPHY
  High Speed Liquid Chromatography   J. H. Knox   24:29-49
DATA COMPILATION
  Status Report on Critical Compilation of
    Physical Chemical Data   D. R. Lide Jr., S. A. Rossmassler   24:135-58
DIELECTRIC POLARIZATION
  Time Domain Reflectometry   R. H. Cole   28:283-300
ELECTROCHEMISTRY
  Theory of Charge Transfer at Electrochemical Interfaces   R. K. Sen, E. Yeager, W. E. O'Grady   26:287-314
ELECTRON MICROSCOPY
  The Architecture of Biological and Artificial
    Membranes as Visualized by Freeze Etching   A. J. Verkleij, P. H. J. Th. Ververgaert   26:101-22
FUELS
  Transformation of Solids to Liquid Fuels   W. H. Wiser, L. L. Anderson   26:339-57
ION EXCHANGE, IONS
  Gaseous Negative Ions   J. L. Franklin, P. W. Harland   25:485-526
  Ion-Molecule Reactions   E. E. Ferguson   26:17-38
  Ion Thermochemistry and Solvation from
    Gas Phase Ion Equilibria   P. Kebarle   28:445-76
ISOTOPES
  Equilibrium Isotope Effects   J. Bigeleisen, M. W. Lee, F. Mandel   24:407-40

  Lipid Phases: Structure and Structural
    Transitions   V. Luzzati, A. Tardieu   25:79-94
  Oscillatory Chemical Reactions   R. M. Noyes, R. J. Field   25:95-119
  Laser Studies of Vibrational and Rotational
    Relaxation in Small Molecules   E. Weitz, G. Flynn   25:275-315
  Molecular Trajectory Calculations   R. N. Porter   25:317-55
  Chemical Dynamics   J. M. Farrar, Y. T. Lee   25:357-85
  Excitable Membranes   R. E. Taylor   25:387-405
  Laser Separation of Isotopes   V. S. Letokhov   28:133-59
KINETICS - GENERAL
  Reactions of High Kinetic Energy Species   J. Dubrin   24:97-120
  Theory of Rotational and Vibrational Energy
    Transfer in Molecules   D. Secrest   24:379-406
  Dynamical Aspects of Gas-Solid Interactions   H. Saltsburg   24:493-514
  Chemical Reactions at Very Low Temperatures   V. I. Goldanskii   27:85-126
  Phase Transitions—Beyond the Simple Ising
    Model   J. F. Nagle, J. C. Bonner   27:291-317
  Thermal Rearrangements   J. A. Berson   28:111-32
KINETICS - GASES
  Gas Phase Reaction Kinetics   A. A. Westenberg   24:77-96

Theory of Simple Mixtures	D. Henderson	25:461-83
The Calculation and Measurement of Cross Sections for Rotational and Vibrational Excitation	J. P. Toennies	27:225-60
Molecular Dynamics and Monte Carlo Calculations in Statistical Mechanics	W. W. Wood, J. J. Erpenbeck	27:319-48
Reactions of Electronically Excited-State Atoms	D. L. King, D. W. Setser	27:407-42

KINETICS - SOLUTIONS

Picosecond Spectroscopy of Molecular Dynamics in Liquids	A. Laubereau, W. Kaiser	26:83-100
Isotope Effects in Chemical Kinetics	F. S. Klein	26:191-210
Effects of Molecular Mobility on Reaction Rates in Liquid Solutions	E. Grunwald, K. C. Chang, J. E. Leffler	27:369-85
Electron Tunneling in Chemistry and Biology	W. F. Libby	28:105-10

LASERS

Light Scattering by Gases	H. F. P., Knaap, P. Lallemand	26:59-81
Laser Studies of Gas Phase Chemical Reaction Dynamics	M. J. Berry	26:259-86
Vibrational State Analysis of Electronic-to-Vibrational Energy Transfer Processes	S. Lemont, G. W. Flynn	28:261-82
Laser-Induced Fluorescence	J. L. Kinsey	28:349-72

LIQUIDS

The Structure of Simple Liquids	P. A. Egelstaff	24:159-87
Liquid Crystals	A. Saupe	24:441-71
The Structure of Liquids	H. C. Andersen	26:145-66

LUNAR CHEMISTRY AND COSMOCHEMISTRY

Chemistry of the Planets	J. S. Lewis	24:339-51

MAGNETISM

Paramagnetic Relaxation	J. S. Hyde	25:407-35

MICROWAVES

Double Resonance and the Properties of the Lowest Excited Triplet State of Organic Molecules	M. A. El-Sayed	26:235-58

NOMENCLATURE

Internationally Recommended Names and Symbols for Physicochemical Quantities and Units	M. L. McGlashan	24:51-76

NUCLEAR MAGNETIC RESONANCE

Nuclear Magnetic Resonance at High Pressures	J. Jonas	26:167-90
NMR Relaxation in Thermotropic Liquid Crystals	C. G. Wade	28:47-73

OPTICAL ROTATORY POWER

Magnetic Circular Dichroism	P. J. Stephens	25:201-32

ORGANIC REACTIONS

The Stochastic Theory of the Origin of the Genetic Code	G. W. Hoffmann	26:123-44

POLLUTION

Chemistry of Pollutant Formation in Flames	H. B. Palmer, D. J. Seery	24:235-62
Pollution of the Stratosphere	H. S. Johnston	26:315-38

POLYENES

Linear Polyene Electronic Structure and Spectroscopy	B. Hudson, B. Kohler	25:437-60

POLYMERS

Optical Properties of Polymers	R. S. Stein, R. S. Finkelstein	24:207-34
Polymer Statistical Mechanics	H. Yamakawa	25:179-200
Polymer Dynamics in Solution	M. Bixon	27:65-84
Rheology and Kinetic Theory of Polymeric Liquids	R. B. Bird	28:185-206

CHAPTER TITLES

PREFATORY CHAPTERS
Physical Chemistry in Cambridge, Massachusetts — E. B. Wilson, J. Ross — 24:1-27
The History of Physical Chemistry in Denmark — T. A. Bak — 25:1-10
The Top Twenty and the Rest: Big Chemistry and Little Funding — M. K. Wilson — 26:1-16
History of H_3 Kinetics — D. G. Truhlar, R. E. Wyatt — 27:1-43
Men, Mines, and Molecules — H. Eyring — 28:1-13

QUANTUM THEORY
Quantum Dynamical Theory of Molecular Collisions — T. F. George, J. Ross — 24:263-300
Quantum Chemistry — K. H. Johnson — 26:39-57
Molecular and Atomic Applications of Time-Dependent Hartree-Fock Theory — P. Jørgensen — 26:359-80
Molecular Electronic Structure Theory: 1972-1975 — H. F. Schaefer III — 27:261-90
Theoretical Studies of Negative Molecular Ions — J. Simons — 28:15-45

RADIATION CHEMISTRY
Laser Light Scattering from Liquids — B. J. Berne, R. Pecora — 25:233-53
Laser-Induced Chemical Reactions — J. T. Knudtson, E. M. Eyring — 25:255-74

SHOCK WAVES
Initiation of Gaseous Detonation — J. H. S. Lee — 28:75-104

SOLID STATE
Photoelectron Spectroscopy: Study of Valence Bands in Solids — J. C. Green — 28:161-83

SOLUTIONS - NONELECTROLYTES
Solutions of Nonelectrolytes — F. L. Swinton — 27:153-74

SPECTROSCOPY
Raman Spectroscopy Involving Electronic Levels — J. A. Koningstein — 24:121-34
Picosecond Spectroscopy — T. L. Netzel, W. S. Struve, P. M. Rentzepis — 24:473-92
Theory of π - Molecular Charge-Transfer Crystals — Z. G. Soos — 25:121-53
Rotation and Rotation-Vibration Pressure-Broadened Spectral Lineshapes — H. Rabitz — 25:155-77
Photoelectron Spectroscopy — T. A. Carlson — 26:211-33
Rayleigh and Raman Optical Activity — L. D. Barron, A. D. Buckingham — 26:381-96
The Spectroscopy of Simple Free Radicals — T. A. Miller — 27:127-52
Picosecond Spectroscopy — K. B. Eisenthal — 28:207-32
Photodissociation Dynamics of Polyatomic Molecules — W. M. Gelbart — 28:323-48

SPECTROSCOPY - INFRARED
The Resonant Raman Effect — B. B. Johnson, W. L. Peticolas — 27:465-91
Resonance Effects in Vibrational Scattering from Complex Molecules — T. G. Spiro, P. Stein — 28:501-21

STATISTICAL MECHANICS
Statistical Mechanics of Simple Polar Fluids — J. M. Deutch — 24:301-23
Stochastic Processes in Physical Chemistry — S. H. Lin, H. Eyring — 25:39-77
Statistical Mechanics of Molecular Motion in Dense Fluids — J. T. Hynes — 28:301-21
Decorated Lattice-Gas Models of Critical Phenomena in Fluids and Fluid Mixtures — J. C. Wheeler — 28:411-43

STRUCTURE - CRYSTALS
The Structural Chemistry of Extended Defects — L. Eyring, L.-T. Tai — 24:189-206
Liquid Crystals — A. Saupe — 24:441-71
Electronic Energy Transfer in Molecular Crystals — R. Silbey — 27:203-23
X-Ray Crystallographic Studies of Proteins — B. W. Matthews — 27:493-523

STRUCTURE - LIQUIDS

The Structure of Liquids	H. C. Andersen	26:145-66
Liquid Theory and the Structure of Water	M. S. Jhon, H. Eyring	27:45-57
Liquid Crystals	H. Gasparoux, J. Prost	27:175-201
Depolarized Light Scattering from Liquids	D. R. Bauer, J. I. Brauman, R. Pecora	27:443-63
Liquids of Linear Molecules: Computer Simulation and Theory	W. B. Streett, K. E. Gubbins	28:373-410

STRUCTURE - MOLECULES

Molecular Mechanisms of Ion Transport in Liquid Membranes	S. B. Hladky, L. G. M. Gordon, D. A. Haydon	25:11-38
Molecular Trajectory Calculations	R. N. Porter	25:317-55
Van der Waals Molecules	B. L. Blaney, G. E. Ewing	27:553-86

SURFACE CHEMISTRY

Kinetics of Crystallization	M. Kahlweit	27:59-63
Physical Chemistry of Lipid Films at Fluid Interfaces	N. L. Gershfeld	27:349-68
Surface Scattering	S. T. Ceyer, G. S. Somorjai	28:477-99

THERMOCHEMISTRY AND THERMODYNAMICS

Pyrometallurgy	H. H. Kellogg, Y. K. Rao, S. W. Marcuson	27:387-406
High Temperature Chemistry: Modern Research and New Frontiers	J. L. Gole	27:526-49

QD   Annual review of physical
1   chemistry.
.A732, v.28

DATE DUE			

DEC 1 6 1977

W. F. MAAG LIBRARY
YOUNGSTOWN STATE UNIVERSITY
YOUNGSTOWN, OHIO   44555